D1341898

*Playing for Real*

# Playing for Real

## *A Text on Game Theory*

KEN BINMORE

2007

OXFORD
UNIVERSITY PRESS

Oxford University Press, Inc., publishes works that further
Oxford University's objective of excellence
in research, scholarship, and education.

Oxford New York
Auckland Cape Town Dar es Salaam Hong Kong Karachi
Kuala Lumpur Madrid Melbourne Mexico City Nairobi
New Delhi Shanghai Taipei Toronto

With offices in
Argentina Austria Brazil Chile Czech Republic France Greece
Guatemala Hungary Italy Japan Poland Portugal Singapore
South Korea Switzerland Thailand Turkey Ukraine Vietnam



Published by Oxford University Press, Inc.
198 Madison Avenue, New York, New York 10016

www.oup.com

Oxford is a registered trademark of Oxford University Press



Library of Congress Cataloging-in-Publication Data
Binmore, K. G., 1940–
Playing for real : a text on game theory / Ken Binmore.
p. cm.
Includes index.
ISBN 978-0-19-530057-4
1. Game theory. 1. Title.
QA269.B475 2005
519.3—dc22 2005053938

12 11

Printed in the United States of America
on acid-free paper

I dedicate *Playing for Real* to my wife, Josephine

# Preface

There are at least three questions a game theory book might answer:

> What is game theory about?
> How do I apply game theory?
> Why is game theory right?

*Playing for Real* tries to answer all three questions. I think it is the only book that makes a serious attempt to do so without getting heavily mathematical. There are elementary books that offer students the opportunity to admire some game theory concepts. There are cookbooks that run through lots of applied models. There are philosophical works that supposedly address the foundational issues, but none of these address more than two of the questions.

However, answering questions is only part of what this book is about. Just as athletes take pleasure in training their bodies, so there is immense satisfaction to be found in training your mind to think in a way that is simultaneously rational and creative. With all of its puzzles and paradoxes, game theory provides a magnificent mental gymnasium for this purpose. I hope that exercising on the equipment will bring you the same kind of pleasure it has brought me.

*Moving on.* *Playing for Real* isn't my first textbook on game theory. My earlier book, *Fun and Games,* was used quite widely for teaching advanced undergraduate and beginning graduate students. I had originally planned a modestly revised second edition, in which the rather severe introduction would be replaced with a new chapter that would ease students into the subject by running through all the angles on the Prisoners' Dilemma. The remaining chapters were then simply to be broken down into more digestible chunks. But the project ran away with me. I made the improvements I planned to make but somehow ended up with a whole new book.

There are two reasons why. The first is that game theory has moved on since I wrote *Fun and Games*. Some of the decisions on what material to include that seemed a little daring at the time now look totally uncontroversial. So I have tried my luck at guessing which way the subject is going to jump again.

The second reason is that I have moved on as well. In particular, I have done a great deal of consulting work, applying game theory to real-world problems in order to raise money for my research center. The biggest project was the design of a telecom auction that raised $35 *billion*. I always knew that game theory works, but seeing it triumph on such a scale was beyond all expectation! I have also written a book applying game theory to philosophical issues, which taught me a great deal about how and why beginners make mistakes when thinking about strategic issues. Both kinds of experience have contributed to making *Playing for Real* a better book than its predecessor. My flirtation with philosophy even generated a lot of light-hearted exercises that nevertheless make genuinely serious points.

*Material.* As a text on game theory for undergraduates with some mathematical training, *Playing for Real* improves on *Fun and Games* in a number of ways. It continues to be suitable for courses attended by students from a variety of disciplines. (Some of my very best undergraduates at the University of Michigan were from Classics.) It also continues to provide backup sections on the necessary mathematics, so that students whose skills are rusty can keep up with what's going on without too much effort. However, the book as a whole covers *fewer* basic topics in a more relaxed and discursive style, with many more examples and economic applications.

I hope the opening chapter, which uses the Prisoners' Dilemma to provide an undemanding overview of what game theory is all about, will prove to be a particularly attractive feature. Economists will also be pleased to see a whole chapter devoted to the theory of imperfect competition, where I believe I may even have made Bertrand-Edgeworth competition accessible to undergraduates. It is a tragedy that evolutionary game theory had to go, but this important subject has gotten so big that it deserves a whole book to itself.

Although fewer topics are covered, some topics are covered in much more detail than in *Fun and Games*. These include cooperative game theory, Bayesian decision theory, games of incomplete information, mechanism design, and auction theory, each of which now has its own chapter. However, the theory of bargaining has grown more than anything else, partly because I hope to discourage various misunderstandings of the theory that have become commonplace in applied work, and partly because I wanted to illustrate its potential use in ethics and moral philosophy.

→ 1.1

*Teaching.* There is enough material in this book for at least two courses in game theory, even leaving aside the review and other sections that are intended for private reading. I have tried to make things easy for teachers who want to design a course based on a selection of topics from the whole book by including marginal notes to facilitate skipping. For example, the Mad Hatter, who has appeared in the margin, suggests skipping on to the first chapter, on the grounds that there is too much philosophy in this preface.

The exercises are similarly labeled with warnings about their content. Nobody will want to attempt all of the enormous number of exercises, but when I teach, I insist on students trying a small number of carefully chosen exercises every week.

Once they get into the habit, students are often surprised to find that solving problems can be a lot of fun.

By the time the book is published, Jernej Copic will have finished getting his solutions onto a website. Oxford University Press will provide access details to recognized teachers.

*Thanks.* So many people have helped me, with both *Fun and Games* and *Playing for Real,* that I have lost track of them all. I shall therefore mention only the very special debt of gratitude I owe to my long-time coauthor, Larry Samuelson, for both his patience and his encouragement. I also want to thank the California Institute of Technology for giving me the leisure to complete this book as a Gordon Moore Scholar. I should also acknowledge the Victorian artist John Tenniel, whose magnificent illustrations from Lewis Carroll's Alice books I have shamelessly stolen and messed around with.

*Apologies.* Let me aopolgize in advance for the errors that have doubtless found their way into *Playing for Real.* If you find an error, please join the many others who have helped me by letting me know about it at k.binmore@ucl.ac.uk. I will be genuinely grateful.

Finally, I need to apologize not only for my mistakes but also for my attempts at humor. Oscar Wilde reported that a piano in a Western saloon carried a notice saying, "Please don't shoot the pianist. He's doing his best." The same goes for me, too. It isn't easy to write in a light-hearted style when presenting mathematical material, but I did my best.

KEN BINMORE

# Contents

*Playing for Real*

# 1 Getting Locked In

## 1.1 What Is Game Theory?

A game is being played whenever people have anything to do with each other. Romeo and Juliet played a teenage mating game that didn't work out too well for either of them. Adolf Hitler and Josef Stalin played a game that killed off a substantial fraction of the world's population. Kruschev and Kennedy played a game during the Cuban missile crisis that might have wiped us out altogether.

Drivers maneuvering in heavy traffic are playing a game with the drivers of the other cars. Art lovers at an auction are playing a game with the rival bidders for an old master. A firm and a union negotiating next year's wage contract are playing a bargaining game. When the prosecuting and defending attorneys in a murder trial decide what arguments to put before the jury, they are playing a game. A supermarket manager deciding today's price for frozen pizza is playing a game with all the other storekeepers in the neighborhood with pizza for sale.

If all of these scenarios are games, then game theory obviously has the potential to be immensely important. But game theorists don't claim to have answers to all of the world's problems because the orthodox game theory to which this book is devoted is mostly about what happens when people interact in a *rational* manner. So it can't predict the behavior of love-sick teenagers like Romeo or Juliet or madmen like Hitler or Stalin. However, people don't always behave irrationally, and so it isn't a waste of time to study what happens when we are all wearing our thinking caps. Most of us at least try to spend our money sensibly—and we don't do too badly much of the time; otherwise, economic theory wouldn't work at all.

Even when people haven't actively thought things out in advance, it doesn't necessarily follow that they are behaving irrationally. Game theory has had some notable successes in explaining the behavior of insects and plants, neither of which can be said to think at all. They end up behaving rationally because those insects and plants whose genes programmed them to behave irrationally are now extinct. Similarly, companies may not always be run by great intellects, but the market can sometimes be just as ruthless as Nature in eliminating the unfit from the scene.

## 1.2 Toy Games

Rational interaction within groups of people may be worth studying, but why call it game theory? Why trivialize the problems that people face by calling them *games?* Don't we devalue our humanity by reducing our struggle for fulfillment to the status of mere *play* in a game?

Game theorists answer such questions by standing them on their heads. The more deeply we feel about issues, the more we need to strive to avoid being misled by wishful thinking. Game theory makes a virtue out of using the language of parlor games like chess or poker so that we can discuss the logic of strategic interaction *dispassionately*.

Bridge players have admittedly been known to shoot their partners. I have sometimes felt the urge myself. But most of us are able to contemplate the strategic problems that arise in parlor games without getting emotionally involved. It then becomes possible to follow the logic wherever it leads, without throwing our hands up in denial when it takes us somewhere we would rather not go. When game theorists use the language of parlor games in analyzing serious social problems, they aren't therefore revealing themselves to be heartless disciples of Machiavelli. They are simply doing their best to separate those features of a problem that admit an uncontroversial rational analysis from those that don't.

This introductory chapter goes even farther down this path by confining its attention to *toy* games. In studying a toy game, we seek to sweep away all the irrelevant clutter that typifies real-world problems, so that we can focus our attention entirely on the basic strategic issues. To distance the problem even further from the prejudices with which we are all saddled, game theorists usually introduce toy games with silly stories that would be more at home in *Alice in Wonderland* than in a serious work of social science. But although toy games get discussed in a playful spirit, it would be a bad mistake to dismiss them as too frivolous to be worthy of serious attention.

Our untutored intuition is notoriously unreliable in strategic situations. If Adam and Eve are playing a game, then Adam's choice of strategy will depend on what strategy he predicts Eve will choose. But she must simultaneously choose a strategy, using her prediction of Adam's strategy choice. Given that it is necessarily based on such circular reasoning, it isn't surprising that game theory abounds with surprises and paradoxes. We therefore need to sharpen our wits by trying to understand really simple problems before attempting to solve their complicated cousins.

Nobody ever solved a genuinely difficult problem without trying out their ideas on easy problems first. The crucial step in solving a real-life strategic problem nearly always consists of locating a toy game that lies at its heart. Only when this has been solved does it make sense to worry about how its solution needs to be modified to take account of all the bells and whistles that complicate the real world.

## 1.3 The Prisoners' Dilemma

The Prisoners' Dilemma is the most famous of all toy games. People so dislike the conclusion to which game-theoretic reasoning leads in this game that an enormous literature has grown up that attempts to prove that game theory is hopelessly wrong.

There are two reasons for beginning *Playing for Real* with a review of some of the fallacies invented in this critical literature. The first is to reassure readers that the simple arguments game theorists offer must be less trivial than they look. If they were obvious, why would so many clever people have thought it worthwhile to spend so much time trying to prove them wrong? The second reason is to explain why later chapters take such pains to lay the foundations of game theory with excruciating care. We need to be crystal clear about what everything in a game-theoretic model means—otherwise we too will make the kind of mistakes we will be laughing at in this chapter.

### 1.3.1 Chicago Times

The original story for the Prisoners' Dilemma is set in Chicago. The district attorney knows that Adam and Eve are gangsters who are guilty of a major crime but is unable to convict either unless one of them confesses. He orders their arrest and separately offers each the following deal:

> If you confess and your accomplice fails to confess, then you go free. If you fail to confess but your accomplice confesses, then you will be convicted and sentenced to the maximum term in jail. If you both confess, then you will both be convicted, but the maximum sentence will not be imposed. If neither confesses, you will both be framed on a minor tax evasion charge for which a conviction is certain.

In such problems, Adam and Eve are the players in a game. In the toy game called the Prisoners' Dilemma, each player can choose one of two *strategies*, called *hawk* and *dove*. The hawkish strategy is to fink on your accomplice by confessing to the crime. The dovelike strategy is to stick by your accomplice by holding out against a confession.

Game theorists assess what might happen to a player by assigning *payoffs* to each possible outcome of the game. The context in which the Prisoners' Dilemma is posed invites us to assume that neither player wants to spend more time in jail than necessary. We therefore measure how a player feels about each outcome of the game by counting the number of years in jail he or she will have to serve. These penalties aren't given in the statement of the problem, but we can invent some appropriate numbers.

If Adam holds out and Eve confesses, the strategy pair (*dove*, *hawk*) will be played. Adam is found guilty and receives the maximum penalty of 10 years in jail. We record this result by making Adam's payoff for (*dove*, *hawk*) equal to $-10$. If

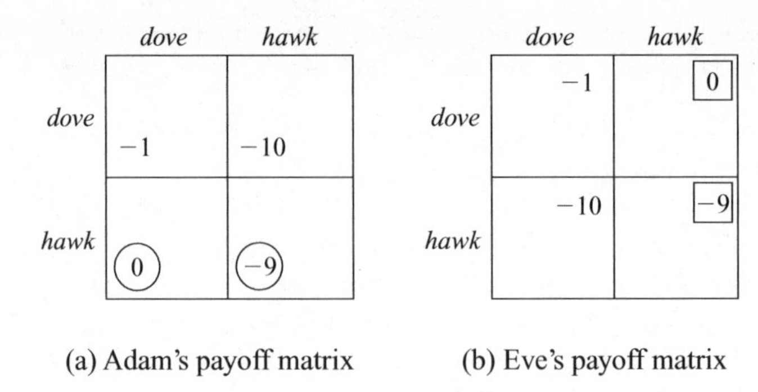


(a) Adam's payoff matrix (b) Eve's payoff matrix

Figure 1.1 Payoff matrices in the Prisoners' Dilemma. Adam's best-reply payoffs are circled. Eve's best replies are enclosed in a square.

Eve holds out and Adam confesses, (*hawk*, *dove*) is played. Adam goes free, and so his payoff for (*hawk*, *dove*) is 0. If Adam and Eve both hold out, the outcome is (*dove*, *dove*). In this case, the district attorney trumps up a tax evasion charge against both players, and they each go to jail for one year. Adam's payoff for (*dove*, *dove*) is therefore −1. If Adam and Eve both confess, the outcome is (*hawk*, *hawk*). Each is found guilty, but since confession is a mitigating circumstance, each receives a penalty of only 9 years. Adam's payoff for (*hawk*, *hawk*) is therefore −9.

The payoffs chosen for Adam in the Prisoners' Dilemma are shown as a *payoff matrix* in Figure 1.1(a). His strategies are represented by the rows of the matrix. Eve's strategies are represented by its columns. Each cell in the matrix represents a possible outcome of the game. For example, the top-right cell corresponds to the outcome (*dove, hawk*), in which Adam plays *dove* and Eve plays *hawk*. Adam goes to jail for 10 years if this outcome occurs, and so −10 is written inside the top-right cell of his payoff matrix.

Eve's payoff matrix is shown in Figure 1.1(b). Although the game is symmetric, her payoff matrix isn't the same as Adam's. To get Eve's matrix, we have to swap the rows and columns in Adam's matrix. In mathematical jargon, her matrix is the transpose of his.

Figure 1.2(a) shows both players' payoff matrices written together. The result is called the payoff table for the Prisoners' Dilemma.[1] Adam's payoff appears in the southwest corner of a cell and Eve's in the northeast corner. For example, −1 is written in the southwest corner of the top-left cell because this is Adam's payoff if both players choose *dove*. Similarly, −9 is written in the north-east corner of the bottom-right cell because this is Eve's payoff if both players choose *hawk*.

The problem for the players in a game is that they usually don't know what strategy their opponent will choose. If they did, they would simply reply by choosing whichever of their own strategies would then maximize their payoff.

[1]Although its entries are vectors rather than scalars, such a table is often called the payoff matrix of the game. Sometimes it is called a bimatrix to indicate that it is really two matrices written together. Most game theorists write the payoffs on one line, so the entry in the cell (*hawk, hawk*) would be (−9, −9). Beginners seem to find my representation less confusing. Thomas Schelling tells me that he has carried out experiments which confirm that payoff tables written in this way reduce the number of mistakes that get made.

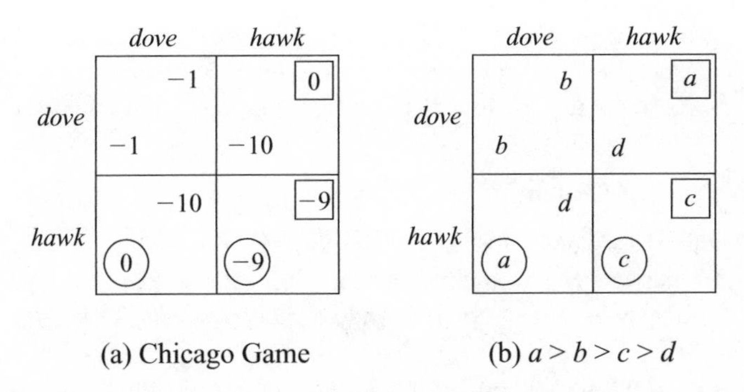


(a) Chicago Game (b) $a > b > c > d$

Figure 1.2 The Prisoners' Dilemma. Adam's payoffs are in the southwest of each cell. Eve's are in the northeast of each cell. Adam's and Eve's best-reply payoffs are enclosed in a circle or a square.

For example, if Adam knew that Eve were sure to choose *dove* in the Prisoners' Dilemma, then he would only need to look at his payoffs in the *first* column of his payoff matrix. These payoffs are −1 and 0. The latter is circled in Figures 1.1(a) and 1.2(a) because it is bigger. The circle therefore indicates that Adam's best reply to Eve's choice of *dove* is to play *hawk*. Similarly, if Adam knew that Eve were sure to choose *hawk*, then he would only need to look at his payoffs in the *second* column of his payoff matrix. These payoffs are −10 and −9. The latter is circled in Figures 1.1(a) and 1.2(a) because it is bigger. Adam's best reply to Eve's choice of *hawk* is therefore to play *hawk*.

In most games, Adam's best reply depends on which strategy he guesses that Eve will choose. The Prisoners' Dilemma is special because Adam's best reply is necessarily the *same* whatever strategy Eve may choose. He therefore doesn't need to know or guess what strategy she will use in order to know what his best reply should be. He should never play *dove* because his best reply is *always* to play *hawk*, whatever Eve may do. Game theorists express this fact by saying that *hawk* strongly dominates *dove* in the Prisoners' Dilemma.

Since Eve is faced by exactly the same dilemma as Adam, her best reply is also *always* to play *hawk*, whatever Adam may do. If both Adam and Eve act to maximize their payoffs in the Prisoners' Dilemma, each will therefore play *hawk*. The result will therefore be that both confess, and hence each will spend nine years in jail—whereas they could have gotten away with only one year each in jail if they had both held out and refused to confess.

People sometimes react to this analysis by complaining that the story of the district attorney and the gangsters is too complicated to be adequately represented by a simple payoff table. However, this complaint misses the point. Nobody cares about the story used to introduce the game. The chief purpose of such stories is to help us remember the relative sizes of the players' payoffs. Moreover, the precise value of the payoffs we write into a table does not usually matter very much. We are interested in the strategic problem embodied in the payoff table rather than the details of some silly story. Any payoff table with the same strategic structure as Figure 1.2(a) would therefore suit us equally well, regardless of the story from which it was derived.

Figure 1.2(b) is the general payoff table for a Prisoners' Dilemma. We need $a > b$ and $c > d$ to ensure that *hawk* strongly dominates *dove*. We need $b > c$ to ensure that both players would get more if they both played *dove* instead of both playing *hawk*.

### 1.3.2 Paradox of Rationality?

Critics of game theory don't like our analysis of the Prisoners' Dilemma because they see that Adam and Eve would both be better off if they came to an agreement to play *dove*. Neither would then confess, and so each would go to jail for only one year.

Naive critics think that this observation is enough to formulate an unassailable argument. They say that there are two theories of rational play to be compared. Their theory recommends that everybody should play *dove* in the Prisoners' Dilemma. Game theory recommends that everybody should play *hawk*. If Alice and Bob play according to the naive theory, each will go to jail for only one year. If Adam and Eve play according to game theory, each will go to jail for nine years. So their theory outperforms ours.

There is admittedly much to be said for asking people who claim to be clever, "If you're so smart, why ain't you rich?" But when you compare how successful two people or two theories are, it is necessary to compare how well each performs under the *same* circumstances. After all, one wouldn't say that Alice was a faster runner than Adam because she won a race in which she was given a head start. Let us therefore compare how well Alice and Adam will do when they play under the same conditions. First imagine what would happen if both were to play against Bob, and then imagine what would happen if both were to play against Eve.

When they play against Bob, Alice goes to jail for one year, and Adam for no years. So game theory wins on this comparison. When they play against Eve, Alice goes to jail for ten years, and Adam for nine years. So game theory wins this on this comparison as well. Game theory therefore wins all around when like is compared with like. Only when unlike is compared with unlike does it seem that the critics' theory wins.

The trap that naive critics fall into is to let their emotions run away with their reason. They don't like the conclusion to which one is led by game theory, and so they propose an alternative theory with nothing more to recommend it than the fact that it leads to a conclusion that they prefer. Game theorists also wish that rational play called for the play of *dove* in the Prisoners' Dilemma. They too would prefer not to spend an extra eight years in jail. But wishing doesn't make it so. As so often in this vale of tears, what we would like to be true is very different from what actually is true.

Of course, most critics are less naive. They continue to deny that game theory is right but recognize that there is a case to be answered by saying that the Prisoners' Dilemma poses a *paradox of rationality* that desperately needs to be resolved. They get all worked up because they somehow convince themselves that the Prisoners' Dilemma embodies the essence of the problem of human cooperation. If this were true, the game-theoretic argument, which denies that cooperation is rational in the Prisoners' Dilemma, would imply that it is never rational for human beings to cooperate. This would certainly be dreadful, but it isn't a conclusion that any game theorist would endorse.

Game theorists think it just plain wrong to claim that the Prisoners' Dilemma embodies the essence of the problem of human cooperation. On the contrary, it represents a situation in which the dice are as loaded against the emergence of cooperation as they could possibly be. If the great game of life played by the human species were the Prisoners' Dilemma, we wouldn't have evolved as social animals! We therefore see no more need to solve some invented paradox of rationality than to explain why strong swimmers drown when thrown in Lake Michigan with their feet encased in concrete. No paradox of rationality exists. Rational players don't cooperate in the Prisoners' Dilemma because the conditions necessary for rational cooperation are absent in this game.

### 1.3.3 The Twins Fallacy

One of the many attempts to resolve the paradox of rationality supposedly posed by the Prisoners' Dilemma tries to exploit the symmetry of the game by treating Adam and Eve as twins. It goes like this:

> Two rational people facing the same problem will come to the same conclusion. Adam should therefore proceed on the assumption that Eve will make the same choice as he. They will therefore either both go to jail for nine years, or they will both go to jail for one year. Since the latter is preferable, Adam should choose *dove*. Since Eve is his twin, she will reason in the same way and choose *dove* as well.

The argument is attractive because there are situations in which it would be correct. For example, it would be correct if Eve were Adam's reflection in a mirror, or if Adam and Eve were genetically identical twins, and we were talking about what genetically determined behavior best promotes biological fitness (Section 1.6.2). However, the reason that the argument would then be correct is that the relevant game would no longer be the Prisoners' Dilemma. It would be a game with essentially only one player.

As is commonplace when looking at fallacies of the Prisoners' Dilemma, we find that we have been offered a correct analysis of some game that isn't the Prisoners' Dilemma. The Prisoners' Dilemma is a two-player game in which Adam and Eve choose their strategies *independently*. Where the twins fallacy goes wrong is in assuming that Eve will make the same choice in the Prisoners' Dilemma as Adam, *whatever* strategy he chooses. This can't be right because one of Adam's two possible choices is irrational. But Eve is an independent rational agent. She will behave rationally whatever Adam may do.

Insofar as it applies to the Prisoners' Dilemma, the twins fallacy is correct only to the extent that rational reasoning will indeed lead Eve to make the same strategy choice as Adam if he chooses rationally. Game theorists argue that this choice will be *hawk* because *hawk* strongly dominates *dove*.

*Myth of the Wasted Vote.* It is worth taking note of the twins fallacy at election time, when we are told that "every vote counts." However, if a wasted vote is one that doesn't affect the outcome of the election, then *all* votes are wasted—unless it turns out that only one vote separates the winner and the runner-up. If they are separated by two or more votes, then a change of vote by a single voter will make no difference at all to who is elected. But an election for a seat in a national assembly is almost never settled by a margin of only one vote. It is therefore almost certain that any particular vote in such an election will be wasted.

Since this is a view that naive people think might lead to the downfall of democracy, reasons have to be given as to why it is "incorrect." We are therefore told that Adam is wrong to count only the impact that his vote alone will have on the outcome of the election; he should instead count the total number of votes cast by all those people who think and feel as he thinks and feels and hence will vote as he votes. If Adam has ten thousand such soulmates or *twins*, his vote would then be far from wasted because the probability that an election will be decided by a margin of ten thousand votes or less is often very high.

This argument is faulty for the same reason that the twins fallacy fails in the Prisoners' Dilemma. There may be large numbers of people who think and feel like you, but their decisions on whether to go out and vote won't change if you stay home and wash your hair.

Critics sometimes accuse game theorists of a lack of public spirit in exposing this fallacy, but they are wrong to think that democracy would fall apart if people were encouraged to think about the realities of the election process. Cheering at a football game is a useful analogy. Only a few cheers would be raised if what people were trying to do by cheering was to increase the general noise level in the stadium. No single voice can make an appreciable difference in how much noise is being made when a large number of people are cheering. But nobody cheers at a football game because they want to increase the general noise level. They shout words of wisdom and advice at their team even when they are at home in front of a television set.

Much the same goes for voting. You are kidding yourself if you vote because your vote may possibly be pivotal. However, it makes perfectly good sense to vote for the same reason that football fans yell advice at their teams. And, just as it is more satisfying to shout good advice rather than bad, so many game theorists think that you get the most out of participating in an election by voting *as though* you were going to be the pivotal voter, even though you know the probability of one vote making a difference is too small to matter (Section 13.2.4). Behaving in this way will sometimes result in your voting strategically for a minor party. The same pundits who tell you that every vote counts will also tell you that such a strategic vote is a wasted vote. But they can't be allowed to have it both ways!

## 1.4 Private Provision of Public Goods

Before looking at more fallacies, it will be useful to tell another story that leads to the Prisoners' Dilemma, so that we can get ourselves into an emotionally receptive state.

Private goods are commodities that people consume themselves. Public goods are commodities that can't be provided without *everybody* being able to consume them. An army that prevents your country being invaded is an example. Streetlights are another. So are radio or television broadcasts. No matter who pays, everybody has access to a public good.

Our taxes pay for most public goods. Advertisers pay for others. But we are interested in the public goods that are paid for by voluntary subscription. Lighthouses were originally funded in this way. Charities still are. Universities depend on endowments from rich benefactors. Public television channels wouldn't survive without the contributions made by their viewers. Young men offered their very lives for what they saw as the public good when volunteering in droves for various armies at the beginning of the First World War.

Utopians sometimes toy with the idea that all public goods should be funded by voluntary subscription. Economists then worry about the *free rider problem.* For example, if people can choose whether or not to buy a ticket when riding on trains, will enough people pay to cover the cost of running the system? Utopians shrug off this problem by arguing that people will see that it makes sense to pay because otherwise the train service will cease to run.

*Free Rider Problem.* The Prisoners' Dilemma can be used to examine the free rider problem in a very simple case. A public good that is worth \$3 each to Adam and Eve may or may not be provided at a cost of \$2 per player. The public good is provided only if one or both of the players volunteer to contribute to the cost. If both volunteer, both pay their share of the cost. If only one player volunteers, he or she must pay *both* shares. Assuming that Adam and Eve care only about how much money they end up with, how will they play this game?

Figure 1.3(a) shows the payoffs in dollars. To play *dove* is to make a contribution. To play *hawk* is to attempt to free ride by contributing nothing. Thus, if Adam and Eve both play *dove*, each will gain $3-2=1$ dollar, since they will then share the cost of providing the public good. If Adam plays *dove* and Eve plays *hawk*, the public good is provided with Adam footing the entire bill. He therefore loses $4-3=1$ dollar. Eve enjoys the benefit of the public good without contributing to the cost at all. She therefore gains \$3.

Since our public goods game has the structure of Figure 1.2(b), it is a version of the Prisoners' Dilemma. As always in the Prisoners' Dilemma, *hawk* strongly dominates *dove*, and so rational players will choose to free ride. The public good will therefore not be provided. As a result, both players will lose the extra dollar they could have made if both had volunteered to contribute.

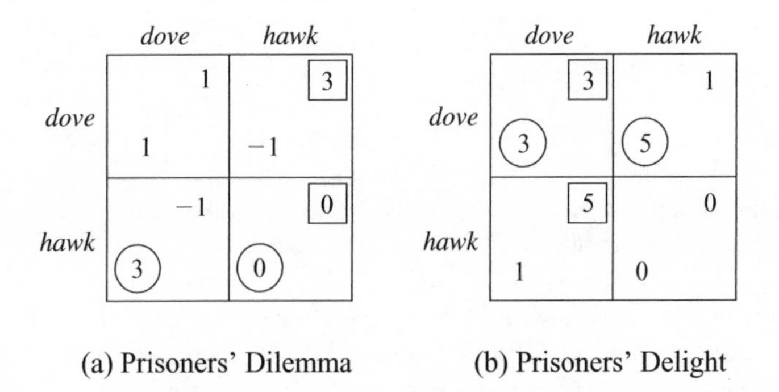


Figure 1.3 The private provision of a public good.

### 1.4.1 Are People Selfish?

Critics get hot under the collar about the preceding analysis. They say that game theorists go wrong in assuming that people care only about money. Real people care about all kinds of other things. In particular, they care about other people and the community within which they live. What is more, only the kind of mean-minded, money-grubbing misfits attracted into the economics profession would imagine otherwise.

But game theory assumes nothing whatever about what people want. It says only what Adam or Eve should do if they want to maximize their payoffs. It doesn't say that a player's payoff is necessarily the money that finds its way into his or her pocket. Game theorists understand perfectly well that money isn't the only thing that motivates people. We too fall in love, and we vote in elections. We even write books that will never bring in enough money to cover the cost of writing them.

Suppose, for example, that Adam and Eve are lovers who care so much about each other that they regard a dollar in the pocket of their lover as being worth twice as much as a dollar in their own pocket. The payoff table of Figure 1.3(a) then no longer applies since this was constructed on the assumption that the players care only about the dollars in their own pockets. However, we can easily adapt the table to the case in which Adam and Eve are lovers. Simply add twice the opponent's payoff to each payoff in the table. We then obtain the payoff table of Figure 1.3(b). The new game might be called the Prisoners' Delight because *dove* now strongly dominates *hawk*. The same principle that says that players should free ride in the Prisoners' Dilemma therefore demands that Adam and Eve should volunteer to contribute in the Prisoners' Delight.

Critics who think that human beings are basically altruistic therefore go astray when they accuse game theorists of using the wrong analysis of the Prisoners' Dilemma. They ought to be accusing us of having correctly analyzed the wrong game. In the case of the private provision of public goods, the evidence would seem to suggest that they would then sometimes be right and sometimes be wrong. This is fine with game theorists, who have no particular attachment to one game over another. You tell us what you think the right game is, and we'll do our best to tell you how it should be played.

*Reason Is the Slave of the Passions.* This is the famous phrase used by David Hume when explaining that rationality is about means rather than ends. As he said, there would be nothing *irrational* about his preferring the destruction of the entire universe to scratching his finger.

Game theory operates on the same premise. It is completely neutral about what motivates people. Just as arithmetic tells you how to add 2 and 3 without asking why you need to know the answer, so game theory tells you how to get what you want without asking why you want it. Making moral judgements—either for or against—is essential in a civilized society, but you have to wear your ethical hat and not your game theory hat when doing it.

So game theory doesn't assume that players are necessarily selfish. Even when Adam and Eve are modeled as money grubbers, who is to say why they want the money? Perhaps they plan to relieve the hardship of the poor and needy. But it is a sad fact that most people are willing to contribute only a tiny share of their income to the private provision of public goods. Numerous experiments confirm that nine out of ten laboratory subjects end up free riding once they have played a game like the Prisoners' Dilemma with large enough dollar payoffs sufficiently often to get the hang of it. Even totally inexperienced subjects free ride half the time.

Governments are therefore wise to think more in terms of the Prisoners' Dilemma than the Prisoners' Delight when legislating tax enforcement measures. Nobody likes this fact about human nature. But we won't change human nature by calling economists mean-minded, money-grubbing misfits when they tell us things we wish weren't true.

### 1.4.2 Revealed Preference

The payoffs in a game needn't correspond to objective yardsticks like money or years spent in jail. They may also reflect a player's subjective states of mind. Chapter 4 is devoted to an account of the modern theory of utility, which justifies the manner in which economists use numerical payoffs for this purpose. This section offers a preview of the basic idea behind the theory.

*Happiness?* In the early nineteenth century, Jeremy Bentham and John Stuart Mill used the word *utility* to signify some notional measure of happiness. Perhaps they thought some kind of metering device might eventually be wired into a brain that would show how many utils of pleasure or pain a person was experiencing. Critics of modern utility theory usually imagine that economists still hold fast to some such primitive belief about the way our minds work, but orthodox economists gave up trying to be psychologists a long time ago. Far from maintaining that our brains are little machines for generating utility, the modern theory of utility makes a virtue of assuming *nothing whatever* about what causes our behavior.

This doesn't mean that economists believe that our thought processes have nothing to do with our behavior. We know perfectly well that human beings are motivated by all kinds of considerations. Some people are clever, and others are stupid. Some care only about money. Others just want to stay out of jail. There are even saintly people who would sell the shirt off their back rather than see a baby cry. We accept that people are infinitely various, but we succeed in accommodating their infinite variety within a single theory by denying ourselves the luxury of speculating about what is going on inside their heads. Instead, we pay attention only to what we see them doing.

The modern theory of utility therefore abandons any attempt to explain *why* Adam or Eve behave as they do. Instead of an explanatory theory, we have to be content with a descriptive theory, which can do no more than say that Adam or Eve will be acting inconsistently if they did such-and-such in the past but now plan to do so-and-so in the future.

*Revealed Preference in the Prisoners' Dilemma.* Analyzing the Prisoners' Dilemma in terms of the modern theory of utility will help to clarify how the theory works. Instead of deriving the payoffs of the game from the assumption that the players are trying to make money or stay out of jail, the data for our problem ultimately comes from the *behavior* of the players.

In game theory, we are usually interested in deducing how rational people will play games by observing their behavior when making decisions in one-person decision problems. In the Prisoners' Dilemma, we therefore begin by asking what decision Adam would make if he knew in advance that Eve had chosen *dove*.

If Adam would choose *hawk*, we would write a larger payoff in the bottom-left cell of his payoff matrix than in the top-left cell. These payoffs may be identified with Adam's utilities for the outcomes (*dove*, *hawk*) and (*dove*, *dove*), but notice that our story makes it nonsense to say that Adam chooses the former because its utility is greater. The reverse is true. We made the utility of (*dove*, *hawk*) greater than the utility of (*dove*, *dove*) because we were told that Adam would choose the former. In opting for (*dove*, *hawk*) when (*dove*, *dove*) is available, we say that Adam reveals a preference for (*dove*, *hawk*), which we indicate by assigning it a larger utility than (*dove*, *dove*).

We next ask what decision Adam would make if he knew in advance that Eve had chosen *hawk*. If Adam again chooses *hawk*, we write a larger payoff in the bottom-right cell of his payoff matrix than in the top-right cell.

On the assumption that we know what choices Adam would make if he knew what Eve were going to do, we have written payoffs for him in Figure 1.2(b) that satisfy $a > b$ and $c > d$. However, the problem in game theory is that Adam usually *doesn't* know what Eve is going to do. To predict what he will do in a game, we need to assume that he is sufficiently rational that the choices he makes in a game are consistent with the choices he makes when solving simple one-person decision problems.

An example will help us here. Professor Selten is a famous game theorist with an even more famous umbrella. He always carries it on rainy days, and he always carries it on sunny days. But will he carry it tomorrow? If his behavior in the future is consistent with his behavior in the past, then obviously he will. The fact that we don't know whether tomorrow will be rainy or sunny is neither here nor there. Our data says that this information is irrelevant to Professor Selten's behavior.

To predict Adam's behavior in the Prisoners' Dilemma, we need to appeal to this Umbrella Principle. Our data says that Adam will choose *hawk* if he learns that Eve is to play *dove* and that he will also choose *hawk* if he learns that she is to play *hawk*. He thereby reveals that his choice doesn't depend on what he knows about Eve's choice. If he is consistent, he will therefore play *hawk* whatever he guesses Eve's choice will be. In other words, a consistent player must choose a strongly dominant strategy.

*Criticism.* Critics respond in two ways to this line of reasoning. The first objection denies the premises of the argument. People say that Adam *wouldn't* choose *hawk* if he knew that Eve were going to choose *dove*. Perhaps he wouldn't—but then we wouldn't be analyzing the Prisoners' Dilemma.

The second objection always puzzles me. The Prisoners' Dilemma is first explained to the critic using some simple story that deduces the players' behavior from the assumption that they are trying to maximize money or to minimize years spent in jail. This allows the mechanism that deduces their payoffs from their behavior in one-person decision problems to be short-circuited. When the critic objects that real people aren't necessarily selfish, he is introduced to the theory of revealed preference and so learns that the logic of the Prisoners' Dilemma applies to everybody, no matter how they are motivated.

Sometimes the attempt to communicate breaks down at this point because the critic can't grasp the idea of revealed preference. Philosophers find the idea particularly troublesome because they have been brought up on a diet of Bentham and Mill.[2] But when critics do follow the argument, a common response is to argue that, if an appeal is to be made to the theory of revealed preference, then nobody need pay attention because the result has been reduced to a tautology. They thereby contrive to reject the argument on the grounds that it is too simple to be wrong!

## 1.5 Imperfect Competition

→ 1.6

The Mad Hatter who has just appeared in the margin is rushing on to Section 1.6 to avoid learning what relevance the Prisoners' Dilemma has for the economics of imperfect competition. However, he will miss out on a lot if he always skips applications of game theory to economics.

It shouldn't be surprising that game theory has found ready application in economics. The dismal science is supposedly about the allocation of scarce resources. If resources are scarce, it is because more people want them than can have them. Such a scenario creates all the necessary ingredients for a game. Moreover, neoclassical economists proceed on the assumption that people will act rationally in this game. Neoclassical economics is therefore essentially a branch of game theory. Economists who don't realize this are like M. Jourdain in Molière's *Le Bourgeois Gentilhomme,* who was astonished to learn that he had been speaking prose all his life without knowing it.

Although economists have always have been closet game theorists, their progress was hampered by the fact that they didn't have access to the tools provided by Von Neumann and Morgenstern when they invented modern game theory in 1944.[3]

As a consequence, they could offer only a satisfactory analysis of imperfect competition in the special case of *monopoly*. A monopoly raises no strategic questions because it can be modeled as a game with only one player. Only with the advent of game theory did it become possible to study other kinds of imperfect competition in a systematic way.

Before looking at how the Prisoners' Dilemma can be used to illustrate a simple problem in imperfect competition, it will he helpful to see how a straightforward monopoly would work under the same circumstances.

### 1.5.1 Monopoly in Wonderland

The hatters of Wonderland make top hats from cardboard. Since the hatters are mad,[4] they give their labor for free, and so the production function therefore only

[2]They can also point to the existence of a modern school of behavioral economists who have revived traditional utility theory in seeking to make sense of psychological experiments. However, such behavioralists don't defend the orthodox analysis of the Prisoners' Dilemma.

[3]Von Neumann was one of the truly great mathematicians of the last century. His contributions to game theory were just a sideline for him. Such a man is surely entitled to call himself whatever he likes, but, in some parts of the German-speaking world, I have been worked over for according him the aristocratic *von* his father purchased from the Hungarian government. So I now write his name as Von Neumann rather than von Neumann.

[4]Lewis Carroll's mad hatter wasn't angry but crazy. The odd behavior for which Victorian hatters were famous is now thought to have been caused by their absorbing strychnine through the skin during the hat-making process.

recognizes cardboard as an input in the hat-making process. It exhibits decreasing returns to scale because hatters are wasteful when hurried. The precise production function to be used is defined by the equation:

$$a = \sqrt{r}.$$

This means that $r$ sheets of cardboard will make $a = \sqrt{r}$ top hats. Only one sheet of cardboard is therefore needed to make one top hat, but four sheets of cardboard are needed to make two top hats.

Alice is a monopolist in the hat business. Cardboard can be bought at one dollar a sheet, and so it costs her one dollar to make one top hat and four dollars to make two top hats. In general, the cost of making $a$ top hats is given by the cost function

$$c(a) = a^2.$$

If Alice can sell top hats at a price of $p$ dollars each, her profit $\pi$ is the revenue $pa$ she derives from selling $a$ hats minus the cost $c(a)$ of making them:

$$\pi = pa - a^2.$$

To know what price maximizes her profit, Alice needs to know the number $a$ of hats that will be bought at each possible price $p$. In Wonderland, this information is given by the demand equation:

$$pa = 30.$$

Since Alice is the only maker of hats, she can meet all the demand at any price. If she makes $a$ hats, she will therefore be able to sell all the hats for $p = 30/a$ dollars each. Writing this value of $p$ into the expression for $\pi$, we find that her profit will be

$$\pi = 30 - a^2.$$

This equation illustrates how monopolists make money. They force the price up by artificially restricting supply. In Wonderland, the effect is extreme. However many hats she sells, Alice's revenue is always $pa = \$30$. So she does best to reduce her cost of $a^2$ by making as few hats as possible. She therefore makes just one hat,[5] which sells for \$30. Since one hat costs only \$1 to make, her profit is then \$29.

### 1.5.2 Duopoly in Wonderland

A classic monopolist is a *price maker*, because she has complete control over the price at which her product is sold. The traders in a perfectly competitive market are *price takers*, because they have no control at all over the market price of the goods they trade. This is usually because all the traders are so small that any action by an individual has a negligible effect on the market as a whole. Most real markets lie

[5]Lewis Carroll would have delighted in pointing out that Alice could do even better by selling no hats at an infinite price, but we assume that the demand equation applies only when $a$ is a positive integer.

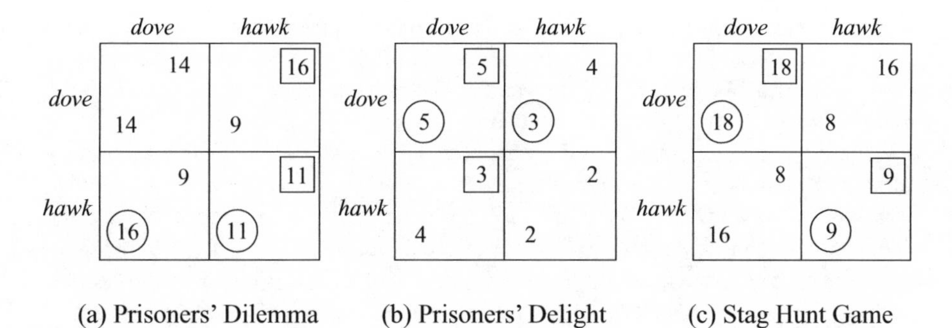


(a) Prisoners' Dilemma (b) Prisoners' Delight (c) Stag Hunt Game

Figure 1.4 Some games that can arise from a duopoly.

between these two extremes. The traders have some partial control over the price at which goods are sold, but their control is limited by competition from their rivals.

A simple example arises when Bob decides to enter the Wonderland hat-making business as a rival to Alice. The market that then arises is called a *duopoly* because it has two competing producers. If Alice produces $a$ hats and Bob produces $b$ hats, each hat will sell for $p = 30/(a+b)$ dollars. If Alice and Bob both care only about maximizing their own profit, how many top hats should each produce?

To keep things simple, assume that Alice and Bob are each restricted to producing either one or two hats. We can then represent their problem as a game in which each player has two strategies called *dove* and *hawk*. The payoff table of the game is shown in Figure 1.4(a). It is yet another example of the Prisoners' Dilemma.

In a duopoly, Alice and Bob can *jointly* make more money by getting together to restrict supply like a monopolist. If they both play *dove* and so supply a total of only two top hats, each will then make a profit of $14.[6]

However, neither player will then be maximizing his or her own *individual* profit. In the Prisoners' Dilemma, *hawk* always strongly dominates *dove*. No matter how many hats Alice is planning to produce, it is therefore always best for Bob to play *hawk* by making two hats on his own. Since the same goes for Alice, both will therefore play *hawk,* and the result will be that each obtains a payoff of only $11.

The outcome illustrates why competition is good for consumers. Bringing in Bob to compete with Alice raises the number of top hats produced from one to four. Simultaneously, the price of a hat goes down from $30 to $7.50. If game theory's critics were right in saying that *dove* is the rational strategy for Alice and Bob in the Prisoners' Dilemma, only two hats would be produced, and they would be sold for $15 each. It is therefore not always such a bad thing that rationality demands the play of *hawk* in the Prisoners' Dilemma!

## 1.6 Nash Equilibrium

Duopolies don't always give rise to the Prisoners' Dilemma. Consider, for example, the effect of decreasing the demand for top hats in Wonderland so that the demand

[6]They make the most money by agreeing to supply only one hat and splitting the profit, but our current model is too crude to take such collusion into account (Section 1.7.1).

equation becomes $p(a+b)=12$. We are then led to the payoff table of Figure 1.4(b). This is another example of the Prisoners' Delight, in which *dove* strongly dominates *hawk*. Rational play will therefore result in the players jointly extracting the maximum amount of money from the consumers.

The Prisoners' Dilemma and the Prisoners' Delight are solved by throwing away strongly dominated strategies, but we can't solve all games this way. To see why, consider the case when Alice's and Bob's production costs are both zero, and the demand equation is $p(a+b)^2=72$. We are then led to the payoff table of Figure 1.4(c). This toy game is called the Stag Hunt Game, after a story told by the philosopher Jean-Jacques Rousseau about how he thought trust works. Like most games, it has no strongly dominant strategy. Adam should play *dove* if he thinks that Eve will play *dove*. He should play *hawk* if he thinks that she will play *hawk*.

What does game theory say about rational play in games with no strongly dominant strategies? This question takes us right back to the origin of the theory of imperfect competition in the work of Augustin Cournot. After formulating the duopoly model we have been studying, he faced the same question. His answer was that we must look for strategies that are in *equilibrium*.

The world wasn't ready for the idea of an equilibrium when David Hume first broached the idea in 1739. It still wasn't ready when Cournot put the idea on a formal footing in 1838. Only after Von Neumann and Morgenstern's *Games and Economic Behavior* appeared in 1944 did the soil became fertile. John Nash's 1951 reinvention of a stripped-down version of Cournot's idea then spread around the world like wildfire.[7] Cournot's contribution is sometimes recognized by calling the idea a Cournot-Nash equilibrium, but the usual practice is simply to speak of a Nash equilibrium.

Like many important ideas, it is almost absurdly simple to explain what a Nash equilibrium is:

> A pair of strategies is a *Nash equilibrium* in a game if and only if each strategy is a best reply to the other.

We have already seen many Nash equilibria. Whenever *both* payoffs in a cell of a payoff table are enclosed in a circle or a square, we are looking at a Nash equilibrium.

For example, (*hawk*, *hawk*) is always a Nash equilibrium in the Prisoners' Dilemma, including the version of Figure 1.4(a) used to model a simple Cournot duopoly. Similarly, (*dove*, *dove*) is a Nash equilibrium in the Prisoners' Delight of Figure 1.4(b). Each of the top-left and the bottom-right cells in the payoff table of the Stag Hunt Game of Figure 1.4(c) have both their payoffs enclosed in a circle or a square. Both (*dove*, *dove*) and (*hawk*, *hawk*) are therefore Nash equilibria in the Stag Hunt Game.

*Why Nash Equilibrium?* Why should anyone care about Nash equilibria? There are at least two reasons. The first is that a game theory book can't authoritatively point to

[7]John Nash was awarded the Nobel Prize for game theory in 1994, along with Reinhard Selten and John Harsanyi. For most of the time between his work on equilibrium theory and the award of the prize, he was incapacitated by a serious schizophrenic illness.

a pair of strategies $(s, t)$ as the solution of a game unless it is a Nash equilibrium. Suppose, for example, that $t$ weren't a best reply to $s$. Eve would then reason that if Adam follows the book's advice and plays $s$, then she would do better not to play $t$. But a book can't be authoritative on what is rational if rational people don't play as it predicts.

Evolution provides a second reason why we should care about Nash equilibria. If the payoffs in a game correspond to how fit the players are, then adjustment processes that favor the more fit at the expense of the less fit will stop working when we get to a Nash equilibrium because all the survivors will then be as fit as it is possible to be in the circumstances.

We therefore don't need our players to be mathematical whizzes for Nash equilibria to be relevant. They often predict the behavior of animals quite well. Nor is the evolutionary significance of Nash equilibria confined to biology. They have a predictive role whenever some adjustment process tends to eliminate players who get low payoffs. For example, stockbrokers who do less well than their competitors go bust. The rules of thumb that stockbrokers use are therefore subject to the same kind of evolutionary pressures as the genes of fish or insects. It therefore makes sense to look at Nash equilibria in the games played by stockbrokers, even though we all know that some stockbrokers wouldn't be able to find their way around a goldfish bowl, let alone a game theory book.

### 1.6.1 Selfish Genes?

phil → 1.7

Because evolution stops working when a Nash equilibrium is reached, biologists say that Nash equilibria are evolutionarily stable.[8] Each relevant locus on a chromosome is then occupied by the gene with maximal fitness. Since a gene is just a molecule, it can't choose to maximize its fitness, but evolution makes it seem as though it had. Game theory therefore allows biologists to get at the final outcomes of an evolutionary process without following each twist and turn that the process might take.

The title of Richard Dawkins's famous *Selfish Gene* expresses the idea in a nutshell. His metaphor is vivid but risky. I particularly enjoyed watching an old lady rebuke him for his effrontery in putting about such evolutionary nonsense, when we can all see that genes are just molecules and thus can't have free will.

### 1.6.2 Blood Is Thicker Than Water

It is a pity that space doesn't allow a proper discussion of the biological applications of game theory, but there is time to consider Bill Hamilton's explanation of why we should expect animals (and people) to get along better with their relatives than with strangers.

To a first approximation, the fitness of a gene is the average number of copies of itself that appear in the next generation. However, a gene in Alice's body would be remiss if its fitness calculation neglected the probability that copies of itself are already present in the bodies of Alice's relatives. After all, if Alice's brother carries

[8]John Maynard Smith defined an evolutionarily stable strategy (ESS) to be a best reply to itself that is a better reply to any alternative best reply than the alternative best reply is to itself. In my experience, biologists seldom worry about the small print involving alternative best replies.

the gene, he will contribute just as many copies of the gene to the next generation on average as Alice herself.

The degree of relatedness $r$ between Alice and Bob is the probability they share any particular gene. If Bob is Alice's full brother, $r = \frac{1}{2}$. If they are full cousins, $r = \frac{1}{8}$. How will $r$ matter if Alice and Bob play a game with each other, like fledglings in a nest?

We only consider the case $r = 1$, so that Alice and Bob are identical twins or clones. If their strategies in the Prisoners' Dilemma are determined by the gene occupying a particular locus, the gene knows that a copy of itself is determining the strategy of its opponent (Exercise 1.13.26). So only one gene is really playing. In this one-player game, the optimal choice is *dove*, and so Alice and Bob cooperate. In brief, the fallacy of the twins ceases to be a fallacy because Alice and Bob really are exact duplicates of each other.

If Alice and Bob are less closely related, a modified version of the lovers' story of Section 1.4.1 applies. The larger $r$ is, the more likely they are to cooperate (Exercise 1.13.29). Hamilton observes that this must be why sociality has evolved separately so many times among the Hymenoptera—ants, bees and wasps. Because of their peculiar sexual arrangements, two sisters in such species have $r = \frac{2}{3}$, rather than $r = \frac{1}{2}$ like us.

## 1.7 Collective Rationality?

Von Neumann and Morgenstern's *Games and Economic Behavior* distinguishes two kinds of game theory. So far we have discussed only *noncooperative* games, in which the players independently choose their strategies to maximize their own payoffs.

Critics of the game-theoretic analysis of the Prisoners' Dilemma sometimes ask why we perversely choose to ignore Von Neumann and Morgenstern's theory of *cooperative* games, in which the players are assumed to negotiate a binding agreement on what strategies to use before play begins. Such critics are usually sold on the idea that rationality resides in groups rather than individuals. They therefore think that rational behavior on the part of an individual player lies merely in agreeing to whatever is rational for the group of players as a whole. Karl Marx is the most famous exponent of this error.[9] The biological version of the mistake is called the group selection fallacy.

*Pareto Efficiency.* A standard assumption in cooperative game theory is that a rational agreement will be *Pareto efficient*. Pareto efficiency comes in a weak form and a strong form. The weak form is easiest to defend. It says that an agreement is Pareto efficient when there is no other feasible agreement that all the players prefer. The argument for assuming that agreements will be weakly Pareto efficient is that rational players won't stop bargaining as long as everybody has something to gain by continuing to negotiate. However, the only one of the four outcomes in the Prisoners' Dilemma that *isn't* Pareto efficient is (*hawk*, *hawk*), which is precisely the outcome that noncooperative game theory says will result from rational play.

[9]Recall that he treated abstractly conceived coalitions like Capital and Labor as though they had the single-minded and enduring aims of individual people.

Philosophers who think that this fact reveals a contradiction between noncooperative and cooperative game theory overlook the importance of the assumption in cooperative game theory that *binding* agreements can be made. It isn't enough that Adam and Eve have promised to honor an agreement. We have all broken our word at one time or another because something else seemed more important at the time. For a truly binding agreement, all the players must know that everybody will have overpowering reasons to keep their word when the time comes. Game theorists say that the players then know that they are all *committed* to honor the agreement.

*Making Commitments Stick.* In real life, our legal system often provides a workable way of enforcing commitments. If Adam and Eve each sign a legally binding contract, then they will be effectively committed to the deal if the penalties for breach of contract outweigh any advantages that either might get from cheating. However, building such opportunities for making commitments into a model inevitably changes the game that is being played and hence removes the contradiction that our critics believe they see.

Suppose, for example, that Adam and Eve have discussed the Prisoners' Dilemma before it is played and agreed that both will play *dove*. We can then relabel their two strategies as play-*dove*-and-keep-your-word and play-*hawk*-and-break-your-word. If the agreement is legally binding, then both players will be liable to a penalty if they break their word. Figure 1.5(a) shows how a penalty of three dollars for breaching the contract changes the Prisoners' Dilemma used to model the private provision of public goods in Figure 1.3(a). The new game is another version of the Prisoners' Delight of Figure 1.3(b), in which *dove* strongly dominates *hawk*. Keeping your word therefore becomes the rational strategy, and so each player's promise to play *dove* is effectively a commitment.

*Modeling Promises.* People who think that game theory is immoral sometimes downplay the need for *external* enforcement by arguing that a player's conscience serves as an *internal* policeman. Game theorists have no difficulty in modeling the fact that most people don't like breaking promises. But how bad does breaking a promise make you feel? I wouldn't feel at all bad about breaking a promise if there

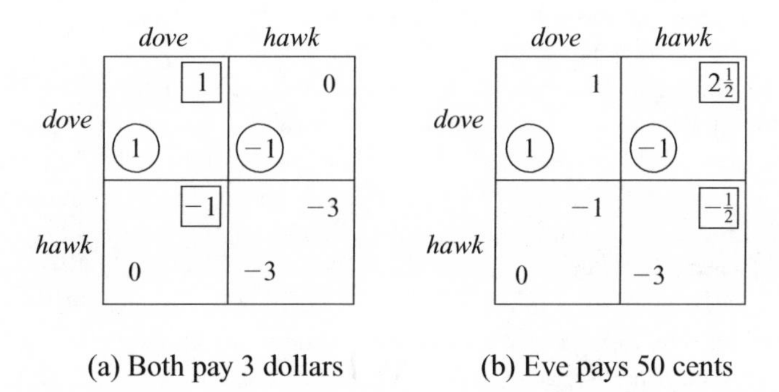


Figure 1.5 Breaking your word. The payoff tables are obtained by subtracting a penalty from a player's payoff when he or she plays *hawk* in the game of Figure 1.3(a), which models the private provision of public goods.

were no other way to get money to feed my starving child. Some people feel the same about all promises—otherwise we wouldn't need to bother with a legal system at all. We therefore need to face up to the fact that the amount that needs to be subtracted from my payoff to capture my distress at breaking a promise may be too small to affect my behavior.

As an example, consider again the Prisoners' Dilemma of Figure 1.3(a) used to model the private provision of public goods. If we only subtract fifty cents from Eve's payoff when she breaks her promise to play *dove* but continue to subtract three dollars from Adam's payoff when he breaks his promise, then we are led to the game of Figure 1.5(b). This is the first asymmetric game we have encountered, but we can still solve it by eliminating strongly dominated strategies. It is rational for Adam to play *dove* and Eve to play *hawk*.

Eve therefore free rides while Adam pays the full cost of providing the public good. But Adam isn't the classic sucker who is never to be given an even break. He predicts that Eve is going to play *hawk* but plays *dove* anyway because he values his peace of mind more than the money he would save by playing *hawk*. If this weren't the case, the theory of revealed preference tells us that three dollars would have been too large a penalty to write into his payoffs.

### 1.7.1 Collusion

People often react badly to the suggestion that it may be rational to cheat and lie. They think that society would collapse if such things were true. Where would we be if we couldn't trust our friends and neighbors? But game theorists don't say that rational people should *never* trust each other. They only say that it is irrational to do something without being able to give a good reason for doing it.

We have good reasons for trusting our friends and neighbors, but we have equally good reasons for distrusting politicians and used-car salesmen. Whether it is sensible to put our trust in other people depends on the circumstances. For example, everybody knows not to trust a stranger who approaches you in a dark alley late at night.

Game theorists argue that it would be unwise for Adam to trust Eve's word if they were about to play the Prisoners' Dilemma. He should get her signature on a legally binding contract before counting on her cooperation. However, if Eve were Adam's wife or sister, they wouldn't be playing the Prisoners' Dilemma. The games we play with those we trust are much more complicated.

An important assumption built into the Prisoners' Dilemma is that the players will never interact again. If Adam and Eve believed they might meet in the future to play again, they would have to take into account the impact that their choice of *dove* or *hawk* in the present might have on the choices their opponent might make in the future. The Prisoners' Dilemma is therefore not capable of modeling long-term relationships in which a player's reputation for honesty can be very valuable—and easily lost. As a dealer in curios put it in the *New York Times* of 29 August 1991 when asked whether he could rely on the honesty of the owner of the antique store that sold his goods on commission: "Sure I trust him. You know the ones to trust in this business. The ones who betray you, bye-bye."

A duopoly is a good setting within which to consider the problem of trust because cooperation among duopolists is commonly illegal. We even use a special word to register our disapproval. When two duopolists agree to cooperate rather than compete, we say that they are *colluding*.

Collusion in a duopoly can't be sustained legally because neither party is going to sue the other for failing to honor a contract that it would be illegal to sign. Nor is it hard to imagine that colluding duopolists will lack moral scruple. After all, it is hardly compatible with an upright nature to enter into a conspiracy whose aim is to screw the consumer. Indeed, in real life, colluding executives seem to relish their shady dealing by choosing to meet in smoke-filled hotel rooms late at night—just like gangsters in the movies.

If Alice and Bob are to collude successfully, they therefore need to have a good reason to trust each other, even though each knows that the other is motivated only by a selfish desire to maximize his or her own profit. A proper explanation of how cooperation can be sustained in an ongoing relationship without internal or external enforcement will have to wait until we study the theory of repeated games (Section 11.3.3). However, it is easy to give the flavor of the explanation while correcting yet another fallacious line of reasoning that has been proposed by philosophers.

*The Transparent Disposition Fallacy.* The transparent disposition fallacy asks us to believe two doubtful propositions. The first is that rational people have the will-power to commit themselves in advance to playing games in a particular way. The second is that other people can read our body language well enough to know when we are telling the truth. If we truthfully claim that we have made a commitment, we will therefore be believed.

If these propositions were correct, our world would certainly be very different! Rationality would be a defense against drug addiction. Poker would be impossible to play. Actors would be out of a job. Politicians would be incorruptible. However, the logic of game theory would still apply.

As an example, consider two possible mental dispositions called CLINT and JOHN. The former is named after the character played by Clint Eastwood in the spaghetti westerns. The latter commemorates a hilarious movie I once saw in which John Wayne played the part of Genghis Khan. To choose the disposition JOHN is to advertise that you have committed yourself to play *hawk* in the Prisoners' Dilemma no matter what. To choose the disposition CLINT is to advertise that you are committed to playing *dove* in the Prisoners' Dilemma if and only if your opponent is advertising the same commitment. Otherwise you will play *hawk*.

If Alice and Bob are allowed to commit themselves transparently to one of these two dispositions before playing the Prisoners' Dilemma of Figure 1.4(a), what should they do? Their problem is a game in which each player has two strategies, CLINT and JOHN. The outcome of this Film Star Game is (*hawk*, *hawk*) unless both players choose CLINT, in which case it is (*dove*, *dove*). The payoff table for their game is therefore given by Figure 1.6(a).

The Film Star Game has no strongly dominant strategies. It is always a best reply for Alice to choose CLINT, but CLINT isn't always her *only* best reply. If Alice predicts that Bob will choose JOHN, then she gets the same payoff whether she chooses CLINT or JOHN. Under such circumstances, we say that CLINT weakly dominates JOHN.

A rational player must play *hawk* in the Prisoners' Dilemma because *hawk* strongly dominates *dove*. We can't say that rational players *must* play CLINT in

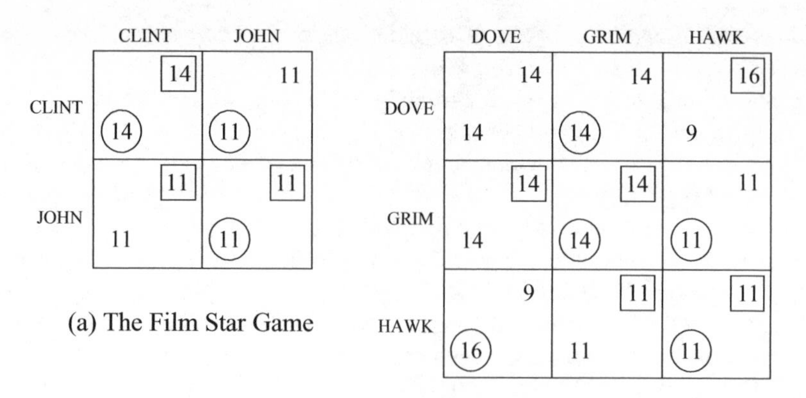


Figure 1.6 Cooperation.

the Film Star Game because it is also a Nash equilibrium for both to play JOHN. However, if Alice or Bob entertains any doubt at all about which strategy the other will choose, he or she does best to play CLINT because CLINT is *sure* to be a best reply, whereas JOHN is only a best reply if the other player also chooses JOHN.

If Alice and Bob can successfully advertise having made a commitment to play like CLINT, then both will play *dove* in the Prisoners' Dilemma. Advocates of the transparent disposition fallacy think that this shows that cooperation is rational in the Prisoners' Dilemma. It would be nice if they were right in thinking that real-life games are really all film star games of some kind—especially if one could choose to be Adam Smith or Charles Darwin rather than John Wayne or Clint Eastwood. But even then they wouldn't have shown that it is rational to cooperate in the Prisoners' Dilemma. Their argument shows only that it is rational to play CLINT in the Film Star Game.

## 1.8 Repeating the Prisoners' Dilemma

If rational cooperation is impossible in the Prisoners' Dilemma, how come duopolists like Alice and Bob often succeed in colluding in real life? The reason is that the real world is more complicated than Wonderland. Real duopolists don't make their decisions once and for all but compete on a day-by-day basis. The Prisoners' Dilemma doesn't capture the essence of such ongoing economic interaction, but we can create a toy game that does by supposing that Alice and Bob must play the Prisoners' Dilemma every day from now until eternity. Their payoffs in this new game are simply their average daily profits.

When we study repeated games seriously, we will find that Alice and Bob have huge numbers of strategies, but we will just look at three: DOVE, HAWK, and GRIM. The first of these is the strategy of always playing *dove*. The second is the strategy of always playing *hawk*. The third is the strategy of playing *dove* as long as your opponent does the same, but switching permanently to *hawk* the day after your opponent first fails to reciprocate.[10]

If our only strategies were DOVE and HAWK, the repeated Prisoners' Dilemma would be the same as the one-shot version, but we also have GRIM to worry about. When GRIM plays DOVE or itself, both players use *dove* every day, and so each gets a daily payoff of fourteen dollars. Things get complicated only when GRIM plays HAWK. The first day will then see one player using *dove* and the other *hawk*. On all subsequent days, both players will use *hawk* because GRIM requires that a failure to reciprocate its play of *dove* on the first day be punished forever. If one player uses GRIM and the other HAWK, each therefore gets an average payoff of 11 because the payoffs Alice and Bob get on the first day are irrelevant when computing averages over an infinite period.

Putting these facts together, we are led to the payoff table of Figure 1.6(b), which is only a tiny part of the true payoff table of the repeated Prisoners' Dilemma, because we have considered only three of the vast number of possible strategies. If we didn't have GRIM in the table, we would be back with the one-shot Prisoners' Dilemma. If we didn't have DOVE, we would be back with the Film Star Game. This perhaps explains why philosophers are so enthusiastic about CLINT. They have seen Clint Eastwood playing a version of the GRIM strategy in the spaghetti westerns, but they didn't notice that he tries to get along with the bad guys before reaching for his gun and that the bad guys totally fail to read the body language with which he conveys his talents as a gunslinger.

Two of the cells of the payoff table of Figure 1.6(b) have both their payoffs enclosed in a circle or a square. These correspond to two Nash equilibria. We are familiar with the equilibrium in which both players use HAWK. But this is now joined by a new equilibrium in which Alice and Bob both use GRIM and hence collude by playing *dove* in each repetition of the Prisoners' Dilemma. They thereby squeeze the maximum possible amount out of the consumer.

The GRIM equilibrium shows how collusion can survive in a duopoly. Alice and Bob need neither a legal system nor a sense of moral obligation to keep them from cheating if they agree to operate a Nash equilibrium. In the case of the GRIM equilibrium, a player who cheats on the agreement will simply provoke the other player into switching to *hawk* on all subsequent days. Neither player therefore has an incentive to cheat.

Sometimes this result is trumpeted as the "solution" to the paradox of rationality raised by the Prisoners' Dilemma. It is certainly important for game theory that we have found a Pareto-efficient Nash equilibrium in the repeated Prisoners' Dilemma. We can thereby explain how cooperation can survive in long-term relationships without the need for external enforcement. But only confusion can result from confounding the *repeated* Prisoners' Dilemma with the Prisoners' Dilemma itself. The only Nash equilibrium in the one-shot Prisoners' Dilemma continues to require that both players use *hawk*.

[10]The GRIM strategy gets its name because it punishes an opponent's transgression relentlessly. Many readers will have heard of the strategy TIT-FOR-TAT. Popular writers are mistaken when they assert that this strategy outperforms all rivals.

## 1.9 Which Equilibrium?

We found two Nash equilibria in both the Stag Hunt Game and the simplified repeated Prisoners' Dilemma of Figure 1.6. The full repeated Prisoners' Dilemma has an infinite number of Nash equilibria. We therefore have to confront what game theorists call the equilibrium selection problem. Which equilibrium should we choose?

No attempt will be made to answer this question here, except to say that nothing says that there must be a "right" equilibrium. After all, nobody thinks there has to be a "right" solution to a quadratic equation. We choose whichever solution fits the problem from which the quadratic equation arose. So why should things be different in game theory?

Advocates of collective rationality don't like this answer. They say that rationality demands the choice of a Pareto-efficient equilibrium in those cases where one exists. But the Stag Hunt Game of Figure 1.4(c) should give them pause. Under the name of the Security Dilemma, experts in international relations use this game to draw attention to the limitations of rational diplomacy.

In the Stag Hunt Game, the Nash equilibrium in which both Alice and Bob play *dove* is Pareto efficient. But suppose their game theory book says that *hawk* should be played. Could rational players persuade each other that the book is recommending the wrong equilibrium? Alice may say that she thinks the book is wrong, but would Bob believe her?

Whatever Alice is planning to play, it is in her interests to persuade Bob to play *dove*. If she succeeds, she will get 18 rather than 8 when playing *dove*, and 16 rather than 9 when playing *hawk*. Rationality alone therefore doesn't allow Bob to deduce anything about her plan of action from what she says because she is going to say the same thing no matter what her real plan may be! Alice may actually think that Bob is unlikely to be persuaded to switch from *hawk* and hence be planning to play *hawk* herself, yet still try to persuade him to play *dove*.

The point of this Machiavellian story is that attributing rationality to the players isn't enough to resolve the equilibrium selection problem—even in a case that seems as transparently straightforward as the Stag Hunt Game. If we see Alice and Bob playing *hawk* in the Stag Hunt Game, we may regret their failure to coordinate on playing *dove*, but we can't accuse them of being irrational because neither player can do any better, given the behavior of their opponent (Section 12.9.1).

## 1.10 Social Dilemmas

Psychologists refer to multiplayer versions of the Prisoners' Dilemma as social dilemmas. You can usually tell that you are in a social dilemma by the fact that your mother would register her disapproval of any hawkish inclination on your part by saying, "Suppose everybody behaved like that?"

Immanuel Kant is sometimes said to be the greatest philosopher of all time, but he too thought that it couldn't be rational to do something if it would be bad if everybody did it. As his famous categorical imperative says:

> Act only on the maxim that you would will to be a universal law.

For example, when waiting at an airport carousel for our bags, we would all be better off if we all stood well back so that we could see our bags coming. The same applies when people stand up at a football match or when they conduct their business in slow motion after reaching the head of a long line.

When large numbers of anonymous folk play such social dilemmas, Kant and your mother are right to predict that things will work out badly if everybody behaves antisocially. But urging people to behave better in such situations is seldom very effective. Why should you lose out by paying heed to your mother when everybody else is ignoring theirs?

### 1.10.1 Tragedy of the Commons

The kind of everyday social dilemma just described can be irritating, but some social dilemmas spell life or death for those who are forced to play them. The standard example is called the Tragedy of the Commons in the political science literature. If you can follow the calculus needed to explain this game properly, you probably know enough mathematics to get started on this book. The Mad Hatter in the margin is there to suggest that readers who find the mathematics challenging would nevertheless be wise not to skip the material.

Ten families herd goats that graze on one square mile of common land. The milk a goat gives per day depends on how much grass it gets to eat. A goat that grazes on a fraction $a$ of the available common land produces

$$b = e^{1-1/10a}$$

buckets of milk a day. This production function has been chosen so that a goat that grazes on one-tenth of the common land gives one bucket of milk. As the fraction of land available for it to graze decreases, the goat's yield progressively declines until a goat without grass to eat gives no milk at all.

A social planner asked to decide the optimal total number $N$ of goats would first note that each goat would occupy a fraction $a = 1/N$ of the common land. Total milk production is then

$$M = Nb = Ne^{1-N/10},$$

which is largest[11] when $N = 10$, making total milk production $M = 10$ buckets a day. If all families are to share equally in the milk produced, the planner would therefore assign the ten families one goat each. Each family would end up with one-tenth of the total milk production, which is one bucket a day per family.

But suppose the planner's edicts can't be enforced. Each family will then make its own decision on the number $g$ of goats to keep. Its own milk production is

$$m = gb = ge^{1-(g+G)/10} = e^{-G/10}ge^{1-g/10},$$

[11]To find where $y = xe^{-x}$ is largest, set its derivative to zero. But $dy/dx = e^{-x} - xe^{-x}$ is zero when $x = 1$. Thus $(N/10)e^{-N/10}$ is largest when $N = 10$. The same is therefore true of $eNe^{-N/10} = Ne^{1-N/10}$.

where $G$ is the total number of goats kept by all the other families. Since $G$ stays constant while our family makes its decision, the solution of its maximization problem is the same as the planner's. It will therefore keep ten goats, regardless of how many goats the other families choose to keep. Since all ten families will do exactly the same, the result will be that one hundred goats are turned loose on the common land, which will therefore be grazed into a desert. When $N = 100$, total milk production is

$$M = 100e^{-9} = 0.012,$$

which is just about enough to wet the bottom of a bucket.

Figure 1.7 makes the connection with the Prisoners' Dilemma in a variety of ways. Figure 1.7(a) substitutes for a player's payoff matrix. It shows a family's milk production as a function of the number $g$ of goats that it keeps and the total number $G$ of goats kept by all the other families. Figure 1.7(b) shows the same data in the

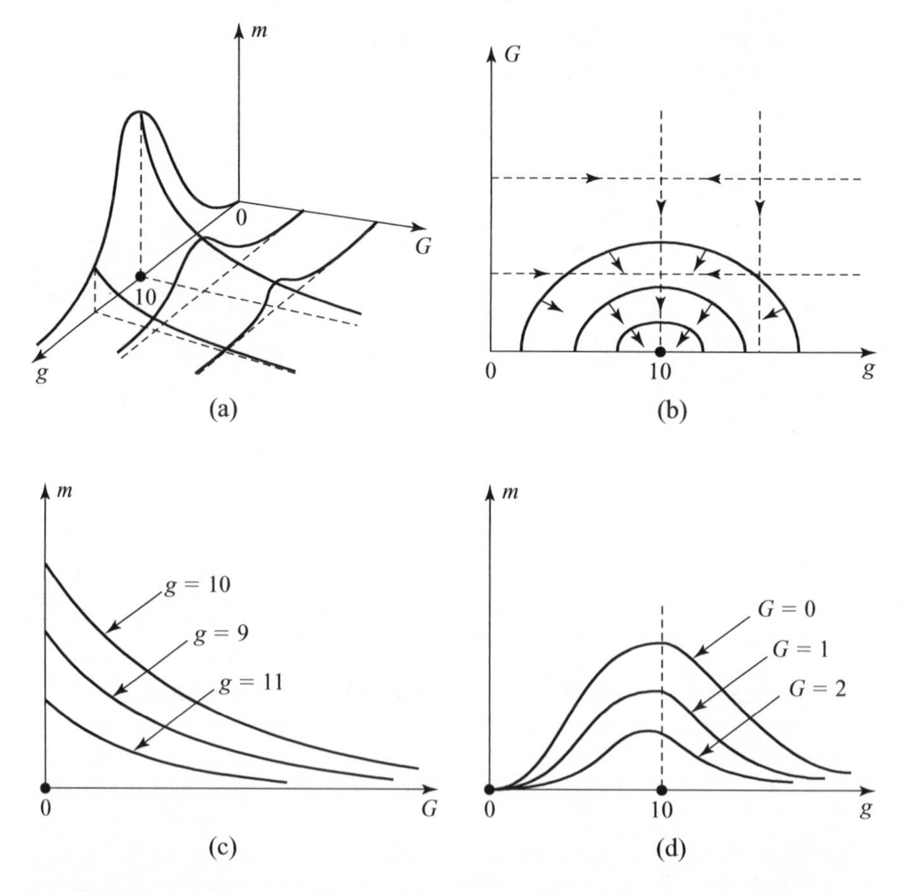


Figure 1.7 Milk production in the Tragedy of the Commons. Figure 1.7(c) shows that it is a strongly dominant strategy to keep ten goats.

form of a contour map. The graphs of Figure 1.7(c) are slices through the milk-production surface of Figure 1.7(a), in which $g$ is held constant. One can think of such slices as representing rows in the payoff matrix. Figure 1.7(d) shows slices through the milk-production surface in which $G$ is held constant. One can think of such slices as columns in the payoff matrix.

A strategy for a family in the Tragedy of the Commons is the number $g$ of goats that it chooses to keep. These strategies are represented as graphs in Figure 1.7(c), or as points on the horizontal axis in Figure 1.7(d). It is easier to see that the hawkish strategy of keeping ten goats is strongly dominant in Figure 1.7(c). One only has to take note of the fact that the graph corresponding to $g = 10$ always lies above each of the graphs corresponding to other strategies. Whatever the value of $G$, a family therefore always gets more milk by keeping ten goats than by keeping any other number of goats. In particular, the hawkish strategy of keeping ten goats strongly dominates the dovelike strategy advocated by the planner of keeping only one goat. Nevertheless, everybody would be far better off if everybody had taken the planner's advice.

The Tragedy of the Commons captures the logic of a whole spectrum of environmental disasters that we have brought upon ourselves. The Sahara Desert is relentlessly expanding southward, partly because the pastoral peoples who live on its borders persistently overgraze its marginal grasslands. But the developed nations play the Tragedy of the Commons no less determinedly. We jam our roads with cars. We poison our rivers and pollute the atmosphere. We fell the rainforests. We have plundered our fishing areas until some fish stocks have reached a level from which they may never recover.

What is to be done about the Tragedy of the Commons? Nobody likes where the logic of the game theory argument leads, but it doesn't help to insist that the logic must therefore be wrong. One might as well complain that arithmetic must be wrong because seven loaves and two fishes won't feed a multitude. Nor does there seem much point in arguing that we can rely on people caring for each other to get us out of such messes. If we could, the mess wouldn't have arisen in the first place.

Game theorists prefer a more positive approach. When they are convinced that they have gotten the game right but don't like the answer to which its analysis leads, they ask whether it may be possible to *change the game*.

### 1.10.2 Mechanism Design

The rules of a game are sometimes called a mechanism. Mechanism design is therefore the branch of game theory in which one asks whether games can be invented that rational people will play in socially beneficial ways.

It is realistic to think of changing the game only if a government or some other powerful planning agency is able to monitor and enforce the new rules, but central planners are notorious for knowing less about what needs to be done than the people they order around. In a good design, the planner therefore doesn't tell everybody what to do. The decisions are left to the people who have the necessary knowledge and expertise. The role left for the planner is to guide their decisions in a socially desirable direction by enforcing a carefully designed system of incentives and constraints. We can then get the logic of game theory to work for us instead of against us.

It will come as no surprise that working out the best system of incentives and constraints can often be difficult, but we can use the Tragedy of the Commons to get the general idea. We have seen that a planner who knew as much about keeping goats as a goat herder would issue each family a license to keep one goat. However, a real planner would be unlikely to know that ten licenses is the socially optimal number.

Suppose, for example, that the planner knows only that each goat's milk production function is of the form

$$b = e^{1-1/Aa},$$

but that you need to have herded goats all your life to be aware that $A = 10$. The planner can work out that the socially optimal number of goats is $A$, but you can't issue $A$ licenses if you don't know what $A$ is. A stupid planner might guess at the value of $A$ and issue that many licenses, but a clever planner will exploit the goat herders' knowledge and experience and let them make the decision on how many goats to keep themselves.

We know that the goat herders will choose in a disastrous way unless the planner intervenes somehow. There are various ways the planner might manipulate their choice. If it is possible for the planner to confiscate the entire milk production and then divide it equally among the ten families, the outcome is particularly benign because each family's aims then become the same. They no longer have an incentive to put one over on their neighbors by sneaking an extra goat onto the common. Their common goal is now to maximize the total amount of milk produced.

To be pedantic, each of the ten families forced to play the planner's confiscation game will now choose $g$ to maximize

$$m = \left(\frac{g+G}{10}\right) e^{1-(g+G)/A},$$

which is largest when $g + G = A$. If each family makes a best reply to the strategies chosen by their opponents—so that a Nash equilibrium is played—the total number $g + G$ of goats that graze the common land will then be socially optimal. However, the planner will find out that the socially optimal number is ten only after counting the number of goats that get turned loose on the common after the new rules are introduced.

### 1.10.3 Second Best

It shouldn't be thought that it is always possible for a social planner to find a way to get to the socially optimal outcome. For example, the mechanism we have just considered won't work if the planner can't monitor how much milk each goat produces since the goat herders have an incentive to keep back some of the milk for their own private use.

Economists express the fact that the best workable mechanism may fail to match up with what an omniscient and omnipotent planner would be able to achieve by saying that, when the *first-best* outcome isn't available, we have to be satisfied with the *second-best* outcome.

People who insist that it must be rational to cooperate in the Prisoners' Dilemma also reject second-best outcomes. When they insist on nothing less than the first-best, economists believe that they are denying the most elementary principle of decision theory—one must first decide what is *feasible* before thinking about which of the feasible alternatives is *optimal.*

The feasible solutions to a problem are those that will work. For example, feasible solutions to reaching a high shelf would be to stand on a chair or to use a broom to lengthen your reach. An infeasible solution would be to swallow the contents of a bottle called Drink-Me in the hope that it will make you grow taller. The optimal solution to the problem is the feasible alternative that costs you least in time and trouble. Standing on a chair is therefore probably optimal, even though putting the chair in the right place and climbing up on it will be a nuisance. However, if you emulate Alice by trying to find a bottle labeled Drink-Me, you will never reach the high shelf at all. In rejecting the second-best outcome in favor of an illusory first-best outcome, you condemn yourself to a third-best or worse outcome.

Planners are particularly likely to make this kind of error when reforming human organizations. They fail to see that people will change their behavior in response to the new incentives created by the reform.

The U.S. Congress made precisely such a mistake in 1990 when it passed an act intended to ensure that Medicare wouldn't pay substantially more for its drugs than private health providers. The basic provision of the act said that a drug must be sold to Medicare at no more than 88% of the average selling price. The problem was created by an extra provision that said that Medicare must also be offered at least as good a price as any retailer. This provision would work as its framers intended only if drug manufacturers could be relied upon to ignore the new incentives created for them by the act. But why would drug manufacturers ever sell a drug to a retailer at less than 88% of the current average price if the consequence is that they must then sell the drug at the same price to a huge customer like Medicare? However, if no drugs are sold at less than 88% of the current average, then the average price will be forced up!

Mechanism design corrects this kind of error by using game theory to predict *how* people's behavior will adapt after a reform has been implemented. Only then can we know what outcomes are genuinely feasible and so make a reasoned choice of what is optimal.

## 1.11 Roundup

Each chapter in this book ends with a summary of the material it covers. Usually, the vital definitions and results are reviewed to give a sense of what is of primary importance. This introductory chapter is exceptional in that the concepts it introduces are dealt with again more carefully in later chapters. The lessons that need to be learned from this chapter are philosophical.

Don't despise toy games. Even a game as simple as the Prisoners' Dilemma is the object of an ongoing controversy. The fact that rational players won't cooperate in the Prisoners' Dilemma isn't a paradox of rationality. People who think this usually make the mistake of imagining that the Prisoners' Dilemma captures the essentials of what matters about human interaction in general, but the one-shot Prisoners'

Dilemma is actually a game whose structure is exceptionally hostile to the emergence of cooperation. In games that better capture the circumstances under which people cooperate in real life, rational players won't necessarily double-cross each other. For example, in the game created by repeating the Prisoners' Dilemma infinitely often, we identified a Nash equilibrium in which the players always cooperate.

When critics offer rival analyses of the Prisoners' Dilemma, they usually fail to notice that they are substituting some other game for the Prisoners' Dilemma. They often mistakenly believe that game theory requires that people care only about how much money they have in their own pockets. They seem never to understand that the payoffs in game theory are derived in principle from the theory of *revealed preference*. This assumes nothing whatever about what motivates people but simply asks that people make decisions consistently. Game theory is neutral on moral and psychological issues.

The basic concept of game theory is called a *Nash equilibrium*. It arises when all players choose a strategy that is a best reply to the strategies chosen by the other players. It is important for two reasons. The first is that a great book of game theory that listed the "rational solutions" of all games would never list a strategy profile that isn't a Nash equilibrium. If it did, at least one player would have an incentive to deviate from the book's advice, and so its advice wouldn't be authoritative. The second reason is evolutionary. An evolutionary process—economic, social, or biological—that acts to maximize the fitness of the players will cease to operate when it reaches a Nash equilibrium. Part of the success of game theory lies in the possibility of switching back and forth between the two interpretations. In particular, we can use the language of rational optimization when talking about the end product of trial-and-error processes of evolutionary adaptation.

Although human interactions that can effectively be modeled using variants of the Prisoners' Dilemma are rare, the results can be disastrous when they do arise. The Tragedy of the Commons is a particularly sad case. In such situations, game theorists don't bury their heads in the sand by pretending that some more amenable game is being played—they ask whether it is actually possible to change the rules to create a more amenable game.

The science of designing new games that rational people will play in a desirable way is called *mechanism design*. Perhaps it will one day become a routine instrument of good government. In the meantime, game theorists advocate its use wherever we understand what is going on well enough to be able to predict how people will respond to the novel incentives created by a newly designed game.

## 1.12 Further Reading

*Thinking Strategically*, by Barry Nalebuff and Avinash Dixit: Norton, New York, 1991. This best-selling book is written for a popular audience. It contains many examples of game theory in action, both in business and in everyday life.

*Playing Fair: Game Theory and the Social Contract I*, by Ken Binmore: MIT Press, Cambridge, MA, 1995. Chapter 3 discusses many fallacies of the Prisoners' Dilemma that circulate in the philosophical literature.

*A Beautiful Mind*, by Sylvia Nasar: Simon and Schuster, New York, 1998. Few of us will experience the highs and lows that are described in this biography of John Nash. There is now a movie with the same title.

*John Von Neumann and Norbert Wiener*, by Steve Heine: MIT Press, Cambridge, MA, 1982. People who knew Von Neumann say he was so clever that it was like talking to someone from another planet.

*Evolution and the Theory of Games*, by John Maynard Smith: Cambridge University Press, Cambridge, UK, 1982. This beautiful book introduced game theory to biology.

*Behavioral Game Theory*, by Colin Camerer: Princeton University Press, Princeton, NJ, 2003. Some bits of game theory work well in the laboratory, and some don't. This book surveys the evidence and looks at possible psychological explanations of deviations from the theory.

## 1.13 Exercises

1. The simplest strategic story that yields the Prisoners' Dilemma arises when Adam and Eve both have access to a pot of money. Both are independently allowed either to give their opponent \$2 from the pot, or to put \$1 from the pot in their own pocket. Write down the payoff table of the game on the assumption that the players care only about how many dollars they make. Which strategy is strongly dominant?
2. A feasible outcome is (weakly) Pareto efficient if there is no other feasible outcome that all the players prefer. Explain why only the outcome (*hawk*, *hawk*) isn't Pareto efficient in the Prisoners' Dilemma. What are the Pareto-efficient outcomes in the Stag Hunt Game?
3. A sealed-bid auction is to be used to sell a collection of ten old coins to the highest bidder at the price he or she bids. The only bidders are Alice and Bob, who both value each coin at \$10. If both make the same bid, each pays half their bid for half the coins. Assuming they are restricted to bidding only \$97 or \$98, show that they are playing a Prisoners' Dilemma in which the strongly dominant strategy is to bid high. Show that the same is true if the only possible bids are \$99.97 and \$99.98.
4. Tenants who sweep the hallways in apartment buildings without a janitor provide a public good. Formulate a version of the Prisoners' Dilemma based on this story.
5. The classic toy game called Chicken derives from the James Dean movie *Rebel without a Cause,* in which two teenage boys drive cars toward a cliff edge to see who chickens out first. The same game is played by middle-aged drivers who approach each other in streets too narrow for them to pass without someone slowing down.

   Explain why the payoff table of Figure 1.8(a) fits both stories. Enclose the payoffs that correspond to best replies in a circle or a square. Explain why neither player has a dominant strategy. Why are (*slow*, *speed*) and (*speed*, *slow*) Nash equilibria? What are the Pareto-efficient outcomes in this game?
6. A couple on their honeymoon in New York are separated in the crowds without having agreed on where they should go in the evening. At breakfast, they had discussed either a visit to the ballet or a boxing match.

   Explain why the Battle of the Sexes of Figure 1.8(b) might be used to model their dilemma.[12] Enclose the payoffs that correspond to best replies in a circle

[12]The sexist assumption that the row player is the husband is usually made, but my wife and I are at least one couple that the stereotype doesn't fit.

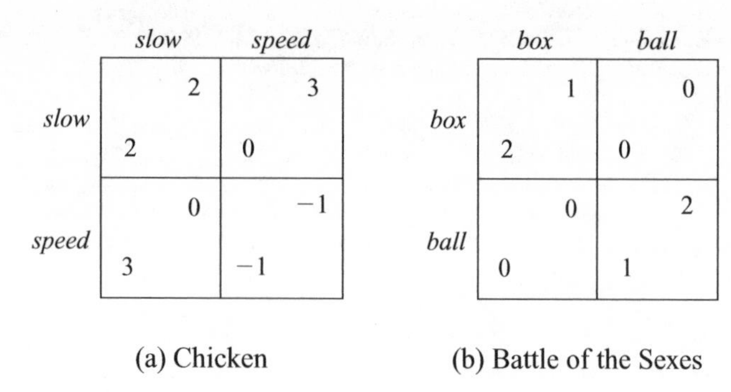


(a) Chicken (b) Battle of the Sexes

Figure 1.8 Two famous toy games.

or a square. Explain why neither player has a dominant strategy. Why are (*box*, *box*) and (*ball*, *ball*) Nash equilibria? What are the Pareto-efficient outcomes in this game?

7. The favorite toy game of evolutionary biologists is called the Hawk-Dove Game. Two birds of the same species are competing for a scarce resource. Each can behave aggressively or passively. Payoffs are measured in terms of a bird's fitness—the extra number of offspring the bird will have on average as a result of the way the game was played. If one bird is aggressive and the other is passive, the aggressive bird takes the entire resource. The aggressive bird then gets a payoff of $V > 0$, and the passive bird gets 0. If both birds are passive, the resource is shared, and each bird gets a payoff of $\frac{1}{2}V$. If both birds are aggressive, there is a fight, and both birds receive a payoff of $W$.

   If $0 < W < \frac{1}{2}V$, show that the Hawk-Dove Game is an example of the Prisoners' Dilemma. If the damage a bird is likely to receive in a fight is sufficiently large, then $W < 0$. Show that the Hawk-Dove Game then reduces to a version of the game Chicken, introduced in Exercise 1.13.5.
8. Adapt Exercise 1.13.1 to obtain an asymmetric version of the Prisoners' Dilemma. Confirm that *hawk* is a strongly dominant strategy but that the outcome (*hawk*, *hawk*) is Pareto inefficient.
9. In Section 1.4.1, the Prisoners' Dilemma of Figure 1.3(a) was converted to the Prisoners' Delight of Figure 1.3(b) by changing the assumption that Adam and Eve care only about themselves to the assumption that they care twice as much about their partner as they do about themselves. What happens if Adam and Eve both care $r$ times as much about their partner as they care about themselves? Show that:
   a. They are still playing the Prisoners' Dilemma when $0 \leq r < \frac{1}{3}$.
   b. They are playing the Prisoners' Delight when $r > 1$.
   c. They are playing a version of Chicken when $\frac{1}{3} < r < 1$.
10. Explain why neither *hawk* nor *dove* is strongly dominant when $\frac{1}{3} \leq r \leq 1$ in the previous problem. For what values of $r$ does the game have a weakly dominant strategy?

econ

11. Section 1.5.1 describes Alice operating a monopoly in Wonderland. Instead of a single Alice acting as a price maker, assume that there are fifteen hat manu-

facturers acting as price takers. Analyze this example of perfect competition,[13] and show that each manufacturer makes one hat, which sells for \$2. What is the total profit of the manufacturers? How does this compare with Alice's profit?

12. In Section 1.5.2, the sum of the profits of the duopolists who make one hat each is \$28. A monopolist who made two hats would obtain a profit of only \$26. Trace this apparent anomaly to the fact that the production function has decreasing returns to scale. **econ**
13. Discuss monopoly and duopoly in the example of Section 1.5 when the production function is $a = r^2$, which has increasing returns to scale. Why is it problematic to attempt an analysis of perfect competition along the lines of Exercise 1.13.11? **econ**
14. Section 1.5.2 derives the Prisoners' Dilemma from a problem in which Alice and Bob compete in a market with demand equation $p(a+b) = X$. Show that the Prisoners' Dilemma arises when $X > 18$, and the Prisoners' Delight when $X < 18$. What happens when $X = 18$? **econ**
15. Why can the following situations be thought of as social dilemmas?
    a. Everybody talking louder and louder in a restaurant until nobody can hear what anybody is saying.
    b. Watering your garden in a drought.
    c. Sneaking excess hand baggage onto a crowded airplane.Think of at least one more everyday example.
16. Suppose that the milk production function in the Tragedy of the Commons takes the form given in Section 1.10.2. Verify that the socially optimal number of goats is $A$.
17. Each of $n$ farmers can costlessly produce as much wheat as he or she chooses. If the total amount of wheat produced is $W$, the price at which wheat sells is determined by the demand equation $p = e^{-W}$. **econ**
    a. Show that the strategy of producing one unit of wheat strongly dominates all of a profit-maximizing farmer's other strategies. Verify that the use of this strategy yields a profit of $e^{-n}$ for a farmer.
    b. Explain why the best agreement that treats each farmer equally requires each to produce only $1/n$ units of wheat. Verify that a farmer's profit is then $1/en$. Why would such an agreement need to be binding for it to be honored by profit-maximizing farmers?
    c. Confirm that $xe^{-x}$ is largest when $x = 1$. Deduce that all the farmers would make a larger profit if they all honored the agreement rather than each producing one unit and so flooding the market.

    This problem has the same structure as the Tragedy of the Commons of Section 1.10.1, but the consumers are unlikely to regard it as tragic if the farmers are unable to agree to restrict their production to $1/n$ units of wheat. What term will the consumers use to describe the farmers' agreement if they succeed in making it stick?

[13]Maximize a manufacturer's profit for a given $p$ by differentiating $\pi = pa - a^2$, keeping $p$ constant. Total output $A$ at price $p$ is fifteen times the amount each manufacturer produces when maximizing profit at this price. The demand equation $pA = 30$ then allows the market-clearing price to be determined.

18. Political scientists regard the following "wasted vote" problem as a relative of the Tragedy of the Commons. Of 100 people who live in a village, 51 support the conservative candidate, and 49 support the liberal candidate. Villagers get a payoff of $+10$ if their candidate gets elected and a payoff of $-10$ if the opposition candidate gets elected. But voting is a nuisance that results in a unit being subtracted from the payoff that a voter would otherwise receive. Those who stay at home and don't vote evade this cost but are rewarded or punished just the same as those who shoulder the cost of voting.
 a. Why is it not a Nash equilibrium for everybody to vote?
 b. Why is it not a Nash equilibrium for nobody to vote?
19. As a primitive exercise in mechanism design, imagine you are a planner who would like Adam and Eve to cooperate when playing the Prisoners' Dilemma. Since you can change the game by imposing fines on one or both of the players, it would be easy to achieve your objective if you were fully informed of everything that matters. You could simply impose a heavy fine on any player who chooses *hawk*. Your problem is that you never get to see the payoff table, and the labeling of the strategies has gotten jumbled up, with the result that you don't know whether the cooperative strategy is *hawk* or *dove*.

 Can you think of a way of creating a game in which it is a Nash equilibrium for Adam and Eve to cooperate, without the need for you to know which strategy is which? The fallacy of the twins may provide some inspiration.
20. As in the previous problem, you are a planner who doesn't know which strategy is which in the Prisoners' Dilemma of Figure 1.3(a). You have probably figured out that you can make it rational for the players to choose the same strategy by fining them both if they choose different strategies. What will the payoff table of the resulting game look like to the players if you make the fine equal to (a) fifty cents; (b) four dollars. In which of the two games is it a Nash equilibrium to cooperate? Find another Nash equilibrium of this game. Which equilibrium is better for both players than the other?
21. Continuing the previous problem, find a fine that makes the new game into a version of the Stag Hunt Game.
22. You are a planner in the Tragedy of the Commons who is unable to redistribute the milk produced and doesn't know the milk production function. Use the idea introduced in the preceding problems to find a way that might lead rational players to use the common land efficiently.
23. Robert Nozick, a Harvard philosopher, believed that Newcomb's paradox shows that maximizing your payoff can be consistent with using a strongly dominated strategy. If true, this would be a disaster for game theory.[14] Newcomb's paradox involves two boxes that possibly have money inside. Adam is free to take either the first box or both boxes. If he cares only for money, which choice should he make? This seems an easy problem. If *dove* represents taking

phil

[14]This exercise draws attention to one of the flaws in Nozick's analysis without addressing the more fundamental issues. My book *Playing Fair* explains why it makes as much sense to pose Newcomb's paradox as to ask who shaves the barber who shaves every man in a town who doesn't shave himself. As Bertrand Russell observed, we are led to a contradiction both if we assume that he shaves himself and if we don't. No such barber therefore exists. Nor can there be an Eve who is sure to predict in advance choices that Adam freely makes.

| | *dd* | *dh* | *hd* | *hh* |
|---|---|---|---|---|
| *dove* | 2 | 2 | 0 | 0 |
| *hawk* | 3 | 1 | 3 | 1 |

Figure 1.9 Adam's payoff matrix in the Newcomb paradox: Does *hawk* dominate *dove*?

only the first box and *hawk* represents taking both boxes, then Adam should choose *hawk* because this choice always results in his getting at least as much money as *dove*. Nozick says that *hawk* therefore "dominates" *dove*.

However, there is a catch. It is certain that there is one dollar bill in the second box. The first box may contain nothing, or it may contain two dollar bills. The decision about whether there should be money in the first box is made by Eve, who knows Adam so well that she is always able to make a perfect prediction of what he will do. Like Adam, she has two choices, *dove* and *hawk*. Her dovelike choice is to put two dollar bills in the first box. Her hawkish choice is to put nothing in the first box. Her motivation is to catch Adam out. She therefore plays *dove* if and only if she predicts that Adam will choose *dove*. She plays *hawk* if and only if she predicts that Adam will choose *hawk*.

Adam's choice of *hawk* now doesn't look so good. If he chooses *hawk*, Eve predicts his choice and puts nothing in the first box, so that Adam gets only the single dollar in the second box. If Adam chooses *dove*, Eve predicts his choice and puts two dollars in the first box for Adam to pick up. But how can it be right for Adam to choose *dove* when this choice is supposedly strongly dominated by *hawk*?

Explain the payoffs in Adam's payoff matrix of Figure 1.9. Notice that Eve has four strategies: *dd*, *dh*, *hd*, and *hh*. For example, the strategy *hd* means that she plays *hawk* if Adam plays *dove* and *dove* if he plays *hawk*. We are told that she will actually choose *dh*, which means that she plays *dove* if Adam plays *dove* and *hawk* if he plays *hawk*. However, for *hawk* to dominate *dove*, it must be at least as good as *dove* for *all* of Eve's strategies. Is this true?

24. The late David Lewis, a Princeton philosopher, believed that Adam's payoff matrix in Newcomb's paradox should be assumed to be the same as his payoff matrix in the Prisoners' Dilemma of Exercise 1.13.1. Why doesn't such a model take account of the fact that Eve always predicts Adam's choice correctly, whatever it may be? **phil**
25. Relate the model of Newcomb's paradox illustrated in Figure 1.9 to the Transparent Disposition fallacy. If Lewis's model of Newcomb's paradox from the previous problem is combined with the assumption that Eve always mirrors his choice, why are we back with the twins fallacy? **phil**
26. Section 1.6.2 talks about a gene *knowing* something. How would you explain what this means to an old lady who objects that this evolutionary talk is nonsense because genes are just molecules and thus can't know anything at all? **phil**

phil

27. Evolutionary games between relatives are considered in Section 1.6.2. Why is $r = \frac{1}{8}$ the degree of relationship between full cousins?

phil

28. Why did the biologist J. B. S. Haldane joke that he would jump in a river at the risk of his own life to save two brothers or eight cousins?
29. Alice's and Bob's payoffs in an evolutionary game are their biological fitnesses. If Alice and Bob were unrelated, the game would be the Prisoners' Dilemma of Figure 1.3(a). If their degree of relationship is $r = \frac{2}{3}$, show that their payoff table is a version of the Stag Hunt Game.[15]

fun

30. Douglas Hofstadter used the column he once wrote for *Scientific American* to argue for a version of the twins fallacy (Section 1.3.3). The magazine followed up by proposing a Million Dollar Game. The rules of the game specify that if $n$ readers enter the competition, then a prize of $1/n$ million dollars is awarded to a randomly chosen entrant.

    If entry is costless, what is a strictly dominant strategy for a reader? The selfless strategy is for a reader not to enter, but why can the categorical imperative not recommend this strategy? (Section 1.10) Why will readers all have to enter with the same positive probability in order to follow the categorical imperative? What considerations may be relevant in determining what this probability should be?[16]

[15]But the evolutionarily stable outcomes aren't simply the Nash equilibria of this payoff table because a selfish gene will know that the other player is a copy of itself two-thirds of the time (Section 1.6.2).

[16]In the event, many readers entered, but the game was wrecked because the magazine got cold feet and allowed readers to submit multiple entries. Inevitably, some joker entered a googolplex number of times.

# 2
# *Backing Up*

## 2.1 Where Next?

Popular accounts of game theory seldom go beyond the simple payoff tables of the previous chapter, leaving all kinds of problems hanging in the air. How do the players of a game figure out what their strategies are? For a game like chess, this is a task of immense complexity. How do the players know what payoffs they will receive after each has chosen a strategy? What do the payoffs mean? As our discussion of the Prisoners' Dilemma in the previous chapter shows, we need to think of the payoffs as being measured in utils rather than dollars. But what precisely is a unit of utility?

This chapter is the first of three in which these questions are answered systematically. Much of the fascination of game theory lies in learning how to handle the problems of timing, risk, and information that need to be solved in coming up with the answers.

The current chapter concentrates on *timing*. How do we cope with games like chess, whose outcome is decided only after long sequences of moves? The next chapter concentrates on *risk*. How do we handle games like poker, in which the outcome is partly determined by chance? No matter how well you play your cards, you are not going to win if your opponents keep getting dealt better hands. The subject of *information* is too important to be hurried, and so we get by with saying as little as possible until it can be discussed with the attention it deserves in Chapter 12. The equally important subject of *utility* is more urgent, and so we study it in Chapter 4 immediately after discussing risk in Chapter 3. In the meantime, all talk of payoffs is avoided.

Some backing up on the previous chapter is therefore necessary. We need to reformulate ideas introduced in Chapter 1 without making premature appeals to the theory of utility. The expedient I employ is to express the ideas directly in terms of the players' preferences over the outcomes of a game. To simplify this task, it is necessary to restrict attention temporarily to *strictly competitive* games. These are two-player games in which Adam's and Eve's interests are diametrically opposed. A major advantage of this restriction is that the principle of *backward induction* can then be introduced in a context in which its role in analyzing games is least problematic.

## 2.2 Win-or-Lose Games

The simplest kind of strictly competitive game allows only winning or losing. In such games, Adam and Eve distinguish only two outcomes, $\mathscr{W}$ and $\mathscr{L}$. The symbol $\mathscr{W}$ denotes a win for Adam and a loss for Eve. Similarly, $\mathscr{L}$ denotes a loss for Adam and a win for Eve. I can remember desperately trying to lose when playing board games with my young children, but Adam and Eve are assumed to be more simply motivated. Whenever offered a choice between winning and losing, each player chooses to win. Economists summarize this behavior by saying that it *reveals* a preference for winning over losing.

The assumptions over Adam's and Eve's preferences that we are making in win-or-lose games can be expressed in formal terms by writing:

$$\mathscr{L} \prec_A \mathscr{W} \quad \text{and} \quad \mathscr{W} \prec_E \mathscr{L}.$$

To write $\mathscr{L} \prec_A \mathscr{W}$ is to say that Adam strictly prefers winning to losing. In operational terms, he never chooses to lose when it is possible for him to win. Remember that writing $\mathscr{W} \prec_E \mathscr{L}$ also means that Eve strictly prefers winning to losing because, for her, $\mathscr{W}$ counts as a loss and $\mathscr{L}$ as a win.

### 2.2.1 The Inspection Game

The Inspection Game is an example of a win-or-lose game that matters in real life. It is used here as a vehicle for introducing the basic ideas to be explored in this chapter in an informal way. The rest of the chapter then ties the ideas down more carefully.

An unscrupulous firm has committed itself to discharging effluent into a river either today or tomorrow. It knows that the local environmental agency will be aware that it has made such a decision, but it isn't too worried because it can be convicted only if caught red handed by an inspector on the spot. However, the agency's resources are so overstretched that it can afford to dispatch an inspector on only one of the two days. The problem for the agency is whether to send its inspector today or tomorrow.

Matching Pennies is a playground game that poses an identical strategic problem. Adam covers a penny with his hand. Eve guesses whether he is hiding a head or a tail. She wins the penny if she guesses right. He wins the penny if she guesses wrong.

The timing structure of the Inspection Game is illustrated in Figure 2.1(a). The firm's opening move is represented by the node at the foot of the diagram. The two lines leading away from the node are labeled $t$ for today and $T$ for tomorrow. They

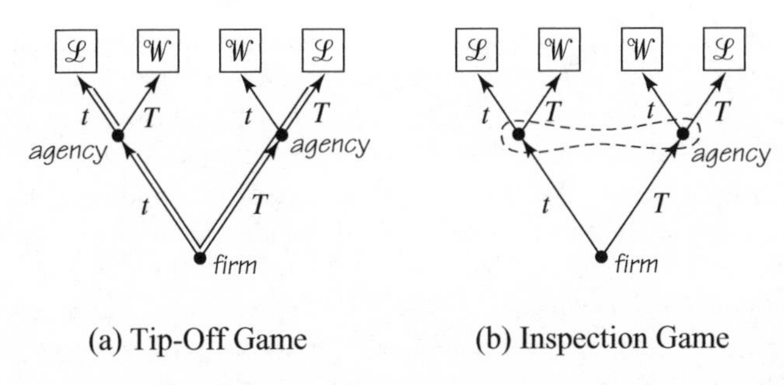


Figure 2.1 Inspection Game. Figure 2.1(a) shows what the structure of the game would be if the agency were sure to be warned in advance of the firm's decision. In the Inspection Game, there is no tip-off. It is therefore necessary to introduce an explicit information set, as in Figure 2.1(b).

represent the firm's two choices of action: to pollute the river today or to pollute it tomorrow. Either of these decisions leads to a node representing a move for the environmental agency. In each case, the agency can decide whether to inspect today or tomorrow. The game ends after each player has moved. Each outcome of the game is labeled with $\mathscr{W}$ or $\mathscr{L}$ to represent a win or a loss for the firm.

The same figure will do equally well to describe the timing structure of Matching Pennies. Simply replace the firm and the agency by Adam and Eve. The symbol $t$ will then have to stand for *heads*, and $T$ for *tails*.

Something very important is missing from Figure 2.1(a). To represent the problem faced by the environmental agency properly, we need to indicate what the agency *knows* when it makes its decision. Game theorists use *information sets* for this purpose.

An appropriate information set for the Inspection Game has been drawn in Figure 2.1(b). This information set includes both of the agency's decision nodes. Including both nodes in one information set means that, when the agency makes its decision at one of these nodes, it doesn't know which of these two nodes the game has reached. That is to say, when the agency decides whether to inspect today or tomorrow, it doesn't know in advance whether the firm has decided to pollute the river today or tomorrow.

When no information set has been drawn around a particular decision node, the assumption is that the player deciding at that node will know for sure that the game has reached that node when making a decision. In this case, one should properly draw a singleton information set that contains only that node, but life is usually too short for such niceties. As drawn, Figure 2.1(a) therefore represents the game in which some whistleblower can be counted on to call the agency before it decides on which day to inspect, with a reliable tip-off about the day on which the firm is going to pollute the river.

The equivalent situation in Matching Pennies would occur if Adam failed to hide his coin successfully, so that Eve could see what it was. Adam would be foolish to be so careless, but no more foolish than the folks who regularly play poker without ever learning to hold their cards close to their chests! If such infringements of the informational rules occur, it is important to recognize that we are not playing Matching Pennies or poker any more. We are playing some other game, which needs a new name—like Peeking Pennies or Suckers' Poker. Our name for the new game created by changing the rules of the Inspection Game to allow a tip-off is the Tip-Off Game.

It isn't hard to figure out what the agency should do in the Tip-Off Game. If the tip-off is that the firm has played $t$, then the agency should play $t$. If the tip-off is that the firm has played $T$, then the agency should play $T$. Whatever choice the firm makes, the agency will then win. The winning actions for the agency are indicated in Figure 2.1(a) by doubling the lines that represent them. Assuming that the firm knows that the agency will be tipped off, it will predict that the agency will choose the doubled line at whichever decision node it finds itself. If the firm plays $t$, it will therefore anticipate that the agency will also play $t$, with the result that the firm will lose. If the firm chooses $T$, it will anticipate that the agency will play $T$, with the result that the firm loses again. Either way, the firm loses. Since both of its choices lead to the same outcome, the firm will be indifferent between them. Both lines at its decision node have therefore been doubled in Figure 2.1(a).

The process of working backward through a game from the outcomes to the initial move, doubling the lines representing the best moves at each decision node, is called *backward induction* or *dynamic programming*. We don't need such heavy machinery to solve the Tip-Off Game, but games don't need to get much more complicated before it becomes useful to apply the principle of backward induction systematically.

However, we can't solve all games by using backward induction. In particular, we can't use it to solve the Inspection Game because the information set in Figure 2.2(b) prevents the agency from knowing which decision node the game has reached when it makes its decision. When deciding what action to take, it therefore doesn't know which of $t$ and $T$ will generate the better outcome.

The information set that distinguishes Figures 2.1(a) and 2.1(b) therefore makes a big difference. The difference is reflected in the strategies available to the players in the different games obtained by assuming that there is or is not a tip-off. In both cases, the firm simply chooses $t$ for today or $T$ for tomorrow. In the Inspection Game, the agency also has only two strategies, $t$ and $T$. Its outcome table therefore takes the simple form shown in Figure 2.2(b).

Drawing an outcome table for the Tip-Off Game isn't so simple because the agency's choice of action will depend on the whistleblower's information about the firm's choice. As a consequence, it is necessary to distinguish four strategies for the agency: $tt$, $tT$, $Tt$, and $TT$. The first letter in each pair says what action the agency plans to take if tipped off that the firm has chosen $t$. The second letter says what action the agency plans to take if tipped off that the firm has chosen $T$. We are then led to the outcome table of Figure 2.2(a).

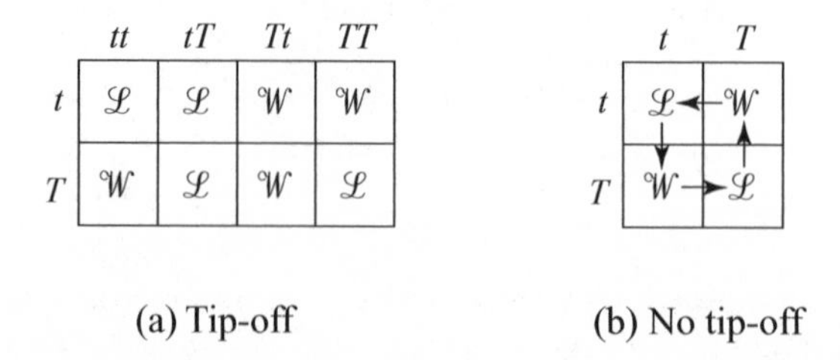

| | $tt$ | $tT$ | $Tt$ | $TT$ |
|---|---|---|---|---|
| $t$ | $\mathcal{L}$ | $\mathcal{L}$ | $\mathcal{W}$ | $\mathcal{W}$ |
| $T$ | $\mathcal{W}$ | $\mathcal{L}$ | $\mathcal{W}$ | $\mathcal{L}$ |

(a) Tip-off

| | $t$ | $T$ |
|---|---|---|
| $t$ | $\mathcal{L}$ | $\mathcal{W}$ |
| $T$ | $\mathcal{W}$ | $\mathcal{L}$ |

(b) No tip-off

Figure 2.2 Outcome tables for the Tip-Off Game and the Inspection Game. The vertical arrows in Figure 2.2(b) show the firm's preferences. The horizontal arrows show the agency's preferences.

We have already seen that the solution of the Tip-Off Game is for the agency to play the strategy *tT*, which calls for the agency to inspect on whatever day the tip-off says that the firm will pollute the river. It then doesn't matter what the firm does because the agency will always win. In the outcome table of Figure 2.2(a), the column corresponding to the strategy *tT* correspondingly contains only the symbol $\mathscr{L}$. In the language of the previous chapter, *tT* is a weakly dominant strategy for the agency.

However, the agency doesn't get a tip-off in the Inspection Game. So what does game theory then recommend? To answer this question, we need to introduce mixed strategies.

### 2.2.2 Mixed Strategies

When Sherlock Holmes was puzzling about which station to leave the train when pursued by the evil Professor Moriarty, they were playing a version of the Inspection Game. But literature offers a more thoughtful analysis in Edgar Allan Poe's *Purloined Letter*. The villain has stolen a letter, and the problem is where to look for it. Poe identifies the essence of the problem by first analyzing a playground game akin to Matching Pennies.

Poe imagines a boy who is such a good natural psychologist that he successfully predicts the thought processes of his opponents most of the time. He knows that a dull-witted opponent who chose *heads* last time will have just enough ingenuity to play *tails* when the game is played now but that a more subtle opponent will reason that such a switching strategy will be too easy to predict and so will stay with *heads*. A yet more subtle opponent will predict that the boy expects him to play *heads* for this reason and hence will play *tails*. An even more subtle opponent will play *heads*. And so on. Poe's boy is therefore successful because he can extend chains of reasoning of the form

She thinks that I think that she thinks that I think . . .

one step further than his opponents.

When games are played in real life, this psychological element is paramount. Winning big in poker is about little else. For example, the poker column of the *Independent* newspaper of 20 May 1999 has this to say about whether Furlong should have called a half-million-dollar raise by Seed in the world poker championship: "Furlong knew that Seed knew that he was punting on all sorts of hands, and that Seed was primed to go over the top and blast him out. Seed probably knew that Furlong knew this. But what he did not know was that Furlong is the sort of man who virtually never folds an ace, no matter what."

But how can one *rational* player outthink another? If Eve is rational, then she reasons optimally, and so Adam has only to figure out his opponent's optimal line of reasoning to know precisely what she will be thinking. If he has trouble in doing so, he can look the answer up in a game theory book. Psychological questions therefore have no place in a discussion of the rational play of games. If everybody played poker rationally, there wouldn't be a world poker championship because the winners and losers would be entirely determined by what cards the players were lucky enough to be dealt.

After the psychological escape route has been closed, the Inspection Game seems to leave game theory with a seemingly insoluble problem. If each player can predict how the other will reason, what prevents their thoughts revolving forever around the vicious circle shown in Figure 2.2(b)? The vertical arrows show the firm's preferences, and the horizontal arrows show the agency's preferences. None of the four cells of the outcome table can correspond to a solution of the game because each cell has an arrow leading away from it.

For example, if a game theory book were to recommend the strategy pair $(t, T)$ as the solution of the Inspection Game, the agency wouldn't follow its recommendation to play $T$ because it would do better to play $t$ if it thought that the firm were likely to follow the book's recommendation by playing $t$. Similarly, $(T, T)$ can't be the solution because the firm would not play $T$ if it thought that the agency were going to play $T$. In the language of Section 1.6, none of the four strategy pairs of Figure 2.2(b) can count as a solution to the Inspection Game because none of them are a Nash equilibrium. At a Nash equilibrium, each player's strategy choice must be a best reply to the strategy choices of the other players.

Does it follow that the Inspection Game has *no* solution? This wouldn't be particularly paradoxical. After all, there is no real number $x$ that solves the quadratic equation $x^2 + 1 = 0$. However just as mathematicians extended the set of real numbers to the set of complex numbers to ensure that all quadratic equations have roots, so game theorists extend the set of *pure* strategies to the set of *mixed* strategies to ensure that all finite games have Nash equilibria.

A player uses a mixed strategy when his or her choice of pure strategy is made at random. For example, Adam might choose *heads* in Matching Pennies with probability $\frac{1}{3}$ and *tails* with probability $\frac{2}{3}$. But how can it ever be rational to choose at random?

In Matching Pennies, the answer is easy. The whole point of the game is to make your choice unpredictable. But if you want to be unpredictable, you can't do better than to delegate your choice to a randomizing device like a roulette wheel or a pack of cards.[1] Your only problem is to decide the probabilities with which each of your pure strategies is to be chosen.

In Matching Pennies, every child knows that the answer is to choose *heads* and *tails* with equal probability. Indeed, on the playground, Adam often makes a show of tossing his coin to make it clear to Eve that *heads* and *tails* are equally likely. Whatever strategy Eve chooses, she will then end up guessing right half the time. Since all of her strategies produce exactly the same result, they are *all* best replies to Adam's choice of the mixed strategy in which he hides *heads* and *tails* with equal probability. In particular, it is a best reply for Eve to choose the mixed strategy in which she too guesses *heads* and *tails* with equal probability. But then Adam's strategy is a best reply to Eve's strategy for the same reason that her strategy is a best reply to his. We are therefore looking at a Nash equilibrium of Matching Pennies in mixed strategies.

The same unremarkable pair of mixed strategies solves the Inspection Game. The firm tosses a coin to decide whether to pollute the river today or tomorrow. The agency tosses another coin to decide whether to inspect today or tomorrow. Each

[1]People are spectacularly bad at coming up with random sequences in their heads. Quite simple computer programs suffice to detect patterns in the sequences they compose.

player's choice guarantees that they can't do worse than win half the time. Nor can either player do better, given the mixed strategy choice of the other.

The use of mixed strategies therefore short-circuits the vicious circle that arises when following up chains of best replies in the Inspection Game. No matter how clever the players may be at duplicating the reasoning of their opponents, it won't do them any good if all they are able to figure out is that their opponent is going to decide what to do by tossing a coin!

Using mixed strategies is easy in the Inspection Game, but randomizing in an optimal way usually requires a lot more than just tossing a fair coin. The probabilities that a mixed strategy assigns to each of a player's pure strategies usually have to be calculated very carefully. We will therefore leave the subject on a back burner until Chapter 6, by which time we will have met the techniques necessary to handle mixed strategies efficiently. In the meantime, we still have a great deal to learn about games that have Nash equilibria in pure strategies.

## 2.3 The Rules of the Game

This section starts to introduce the mathematics used when modeling the rules of a game. A natural reaction is to ask whether we really need such heavy machinery. The following cautionary story demonstrates the value of proceeding systematically when analyzing a new game. The Mad Hatter in the margin invites you to skip forward to Section 2.3.2 if you don't need any convincing.

### 2.3.1 The Surprise Test

→ 2.3.2

In an airwaves auction I helped design, the telecom companies bid all the way up to a total of $35 billion for the licenses offered. Everybody was surprised at this enormous amount—except for the media experts, who got the figure roughly right in the end by predicting a bigger number whenever the bidding in the auction falsified their previous prediction.

Everybody can see the fraud perpetrated by the media experts on the public in this story, but the fraud isn't so easily detected when it appears in one of the many versions of the surprise test paradox, through which most people first learn of backward induction.

Eve is a teacher who tells her class that they are going to be given a test one day next week, but the day on which the test is given will come as a surprise. Adam is a pupil who has read Section 2.2.1 and so knows all about backward induction. He therefore works backward through the days of the coming school week. If Eve hasn't given the test by the time school is over on Thursday, Adam figures that Eve will then have no choice but to give the test on Friday—this being the last day of the school week. If the test were given on Friday, Adam would therefore not be surprised. So Adam deduces that Eve can't plan to give the test on Friday. But this means that the test must be given on Monday, Tuesday, Wednesday or Thursday. Having reached this conclusion, Adam now applies the backward induction argument again to eliminate Thursday as a possible day for the test. Once Thursday has been eliminated, he is then in a position to eliminate Wednesday. Once he has eliminated all the days of the school week by this method, he sighs with relief and makes no attempt to study over the weekend. But then Eve takes him by surprise by giving the test first thing on Monday morning!

This isn't really a paradox because Adam shouldn't have been so quick to sigh with relief. If the backward induction argument is correct, then the two statements made by Eve are inconsistent, and so at least one of them must be wrong. But why should Adam assume that the wrong statement is that a test will be given and not that the test will come as a surprise? This observation is usually brushed aside because what people really want to hear about is whether the backward induction argument is right. But what they should be asking is whether backward induction has been applied to the right game.

In the game that people imagine is being analyzed, Eve chooses one of five days on which to give the test, and Adam predicts which of the five days she will choose. If his prediction is wrong, then he will be taken by surprise. The solution of this five-day version of the Inspection Game is that Adam and Eve both choose each day with equal probability. The result is that Adam is surprised four times out of five. But this isn't the conclusion we reached using backward induction! Why not?

The reason is that the surprise test paradox applies backward induction to a game in which Adam is always allowed to predict that the test will be today, even though he may have wrongly predicted that it was going to take place yesterday.[2] In this bizarre game, Adam's optimal strategy is therefore to predict Monday on Monday, Tuesday on Tuesday, Wednesday on Wednesday, Thursday on Thursday, and Friday on Friday. No wonder Adam is never surprised by having the test occur on a day he didn't predict!

The surprise test paradox has circulated ever since I can remember. Occasionally it gets a new airing in newspapers and magazines. It has even been the subject of learned articles in philosophical journals. The confusion persists because people fail to ask the right questions. One of the major virtues of adopting a systematic formalism in game theory is that asking the correct questions becomes automatic. You then don't need to be a genius like Von Neumann to stay on the right track. Von Neumann's formalism does the thinking for you.

### 2.3.2 Perfect Information

The rest of this chapter is confined to games of perfect information without chance moves. This restriction allows us to delay saying any more about probability until the next chapter.

In a game of *perfect information*, the players know everything they might wish to know about what has happened in the game so far when they make a move. Each information set therefore reduces to a singleton containing only one decision node. As in the Tip-off Game of Section 2.2.1, we usually therefore don't bother drawing them at all.

The Tip-off Game is a game of perfect information without chance moves, but the Inspection Game isn't. It has no chance moves, but it has an information set containing two decision nodes, and so it is a game of imperfect information. When the

[2]The first step in the backward induction argument shows that Adam should predict that the test will take place on Friday, if Friday is reached without the test already having been given. The next step shows that he should predict that the test will take place on Thursday, if Thursday is reached without the test having been given. But if his prediction that the test will take place on Thursday proves wrong, we have already seen that his strategy requires that he now predict that the test will be given on Friday. Exercise 2.12.23 looks at the details of the argument.

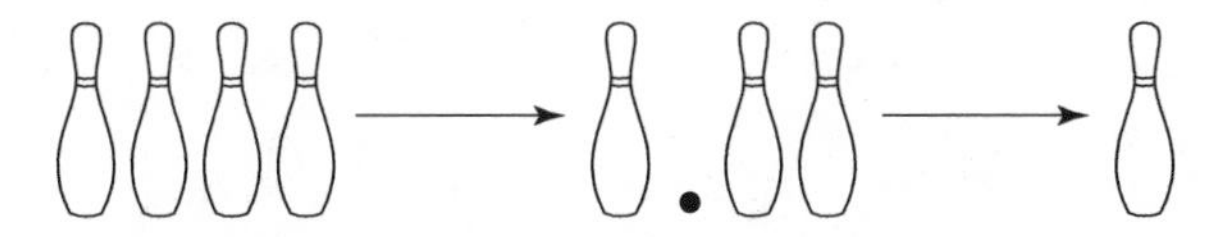

Figure 2.3 A possible play of Kayles with four adjacent skittles. Player I opens the game by taking the second skittle. Player II responds by taking the third and fourth skittles. Player I then loses, since he is forced to take the one skittle that remains.

agency decides whether to inspect today or tomorrow, it doesn't know whether the firm has committed to polluting the river today or tomorrow.

Chess is the most famous game of perfect information without chance moves. Backgammon, Monopoly, and Parcheesi are all games of perfect information, but a chance move takes place whenever the dice are rolled. Poker is a game that has both chance moves and imperfect information.

Chess is too complicated to use as our standard example of a game of perfect information without chance moves. So we will use instead a variant of a game that mathematicians call Kayles.

In our version of Kayles, the players alternate in removing skittles from a row of skittles that may have some gaps. When it is your turn, you must take either one or two *adjacent* skittles. The loser is the player who takes the last skittle. Figure 2.3 shows a possible play in the case when the game begins with four adjacent skittles.

### 2.3.3 Game Trees

The rules of a game need to tell us *who* can do *what*, and *when* they can do it. They must also say who gets *how much* when the game is over. The structure used to convey such information in game theory is called a *tree*.

Combinatorial mathematicians say that a tree is a special case of a graph. Such a graph is simply a set of nodes (or vertices), some of which are linked by edges. As illustrated in Figure 2.4(c), a tree is a connected graph with no cycles, in which a particular node has been singled out to be its *root*.

I pursue the botanical analogy by saying that the edges are *branches* of the tree. A terminal node of a finite tree is reached by starting at the root and moving along branches until one reaches a node from which no further progress is possible without retracing one's steps. Such terminal nodes are sometimes called *leaves*.

*When?* The leaves of the tree correspond to the possible outcomes of the game. A *play* of a finite game is a connected chain of branches that starts at the root and ends at a leaf. A tree for a version $G$ of Kayles is shown in Figure 2.5. The play shown in Figure 2.3 is indicated by thickening appropriate branches. Figure 2.6 shows a streamlined version of Kayles that suppresses forced moves and makes no reference to skittles.

*What?* Nodes in the tree other than leaves are called *decision nodes.* They represent the possible moves in the game. The root of the tree represents the first move of the game. The root of Kayles in Figure 2.6 is labeled $a$.

The branches leading away from a node represent the choices or actions available at that move. There are four choices available at the first move in the game $G$ of

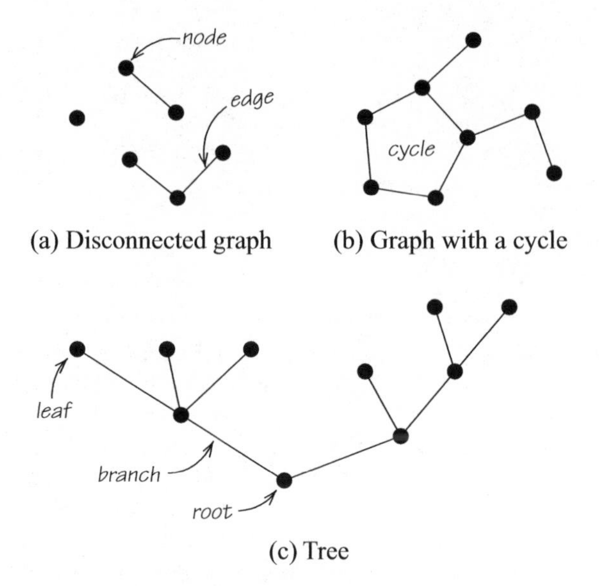

node
edge
cycle
(a) Disconnected graph
(b) Graph with a cycle
leaf
branch
root
(c) Tree

Figure 2.4 Some graphs.

Figure 2.6. These have been labeled *l*, *m*, *n*, and *r*. For example, *n* corresponds to the action in which player I opens the game *G* by taking one of the middle skittles.

*Who?* Each decision node is assigned a player's name or number, so that we know who makes the choice at that move. In the game tree of Figure 2.6, player I chooses at the first move. If he chooses action *n*, then player II makes the next move. She has three choices labeled *L*, *M*, and *R*. If she chooses action *R*, then the game ends with a victory for her.

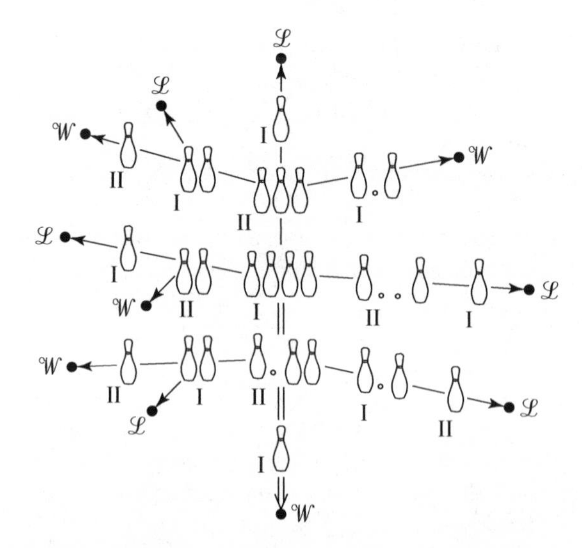

Figure 2.5 Kayles. The game shown is a simplification of Kayles in which moves that lead to the same configuration of skittles are identified.

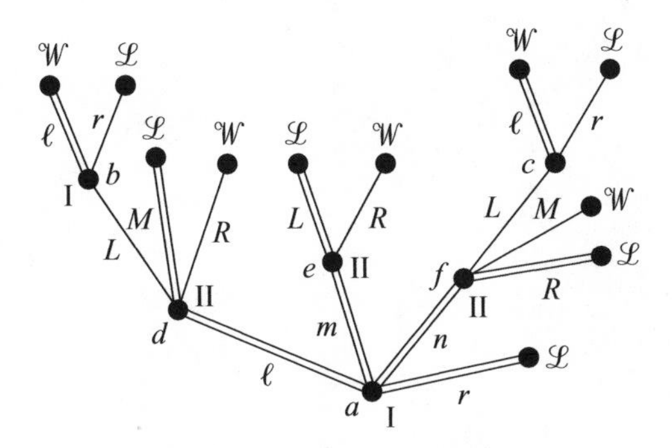


Figure 2.6 Streamlined Kayles. The game $G$ shown further simplifies the version of Kayles of Figure 2.5 by omitting forced moves. The doubled lines indicate the result of applying backward induction.

*How Much?* Each leaf must be labeled with the consequences for each player if the game ends in the outcome to which it corresponds. The game $G$ is a win-or-lose game, and so its leaves are labeled with the symbols $\mathscr{W}$ and $\mathscr{L}$.

### 2.3.4 Two Examples

Kayles is a modern game invented by combinatorial mathematicians as a showcase for their talents. However, archeology reveals that games of perfect information are as old as civilization. Tic-Tac-Toe and Nim are examples of games of perfect information without chance moves that still get played.

*Tic-Tac-Toe.* Everybody knows the rules of Tic-Tac-Toe (or Noughts and Crosses). Its game tree is very large in spite of the simplicity of its rules. Figure 2.7 therefore shows only part of the tree. The labels $\mathscr{W}$, $\mathscr{L}$, and $\mathscr{D}$ indicate a win, loss, and a draw respectively for player I.

*Nim.* Unlike Tic-Tac-Toe, Nim is a win-or-lose game. It begins with several piles of matchsticks. Two players alternate in moving. When it is your turn to move, you must select one of the piles and remove at least one matchstick from that pile. In contrast to our version of Kayles, the last player to take a matchstick is the winner.

A dull art movie called *Last Year in Marienbad* consists largely of the characters playing Nim very badly. Perhaps their ineptitude is intended as a comment on the human condition. However, the only time I have seen Nim played for money, the guy in the bar who proposed playing seemed to know the optimal strategy given in Section 2.6 perfectly well!

## 2.4 Pure Strategies

We have already had a lot to say about strategies. When studying the Inspection Game, we even looked at mixed strategies in a game of imperfect information. But the time has now come to study pure strategies seriously.

A pure strategy for Alice in a game specifies an action at *each* of the information sets at which it would be her duty to make a decision if that information set were

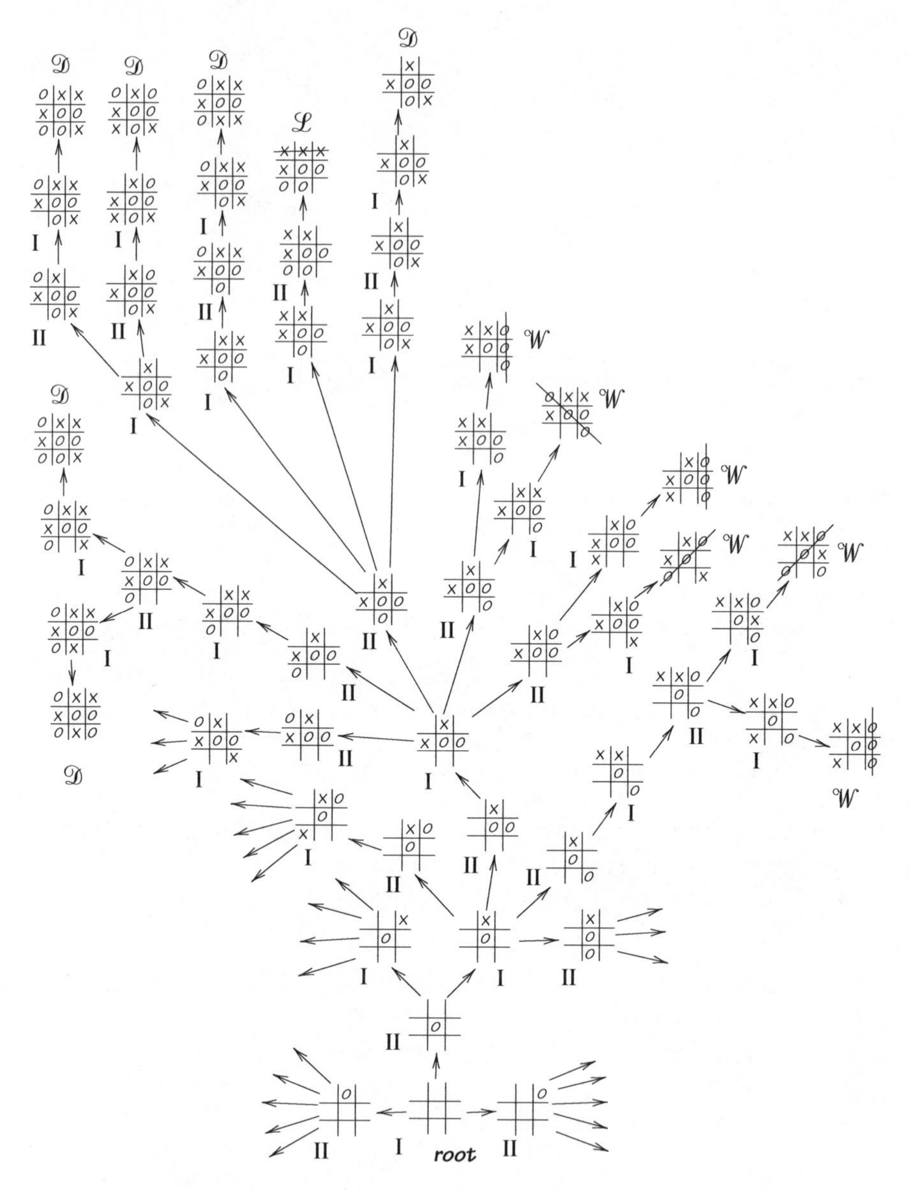


Figure 2.7 Tic-Tac-Toe. Only part of the tree is drawn. At most of the nodes shown, some of the choices have been omitted.

actually reached. If all the players in a game select a pure strategy and stick with it, then their decisions totally determine how a game without chance moves will be played.

In what remains of this chapter, we are considering only games of perfect information. In such a game, everybody knows exactly what point the game has reached whenever they make a decision. It is then relatively easy to draw the extensive form because we don't need to bother with information sets at all. But Section 2.2.1 teaches us that games of imperfect information are easier in at least one respect—they have fewer pure strategies. This is because there can't be more information sets than decision nodes. For example, the firm has two pure strategies in the Inspection Game of Figure 2.1(b). But when we delete the firm's information set to obtain the Tip-Off Game of Figure 2.1(a), the firm's number of pure strategies increases to four.

To determine a pure strategy in a game of perfect information, we must specify a plan of action at each and every node at which the player would have to make a decision if that node were reached. The version of Kayles shown as the game $G$ in Figure 2.6 will serve as an example.

The nodes at which it would be up to player I to make a decision are labeled $a$, $b$, and $c$. A pure strategy for player I must therefore specify actions for him at each of these three nodes. Since there are 4 actions for player I at node $a$, 2 actions at node $b$, and 2 actions at node $c$, player I has a total of $4 \times 2 \times 2 = 16$ pure strategies. These 16 pure strategies can be labeled:

$$lll, \quad llr, \quad lrl, \quad lrr, \quad mll, \quad mlr, \quad mrl, \quad mrr,$$
$$nll, \quad nlr, \quad nrl, \quad nrr, \quad rll, \quad rlr, \quad rrl, \quad rrr.$$

For example, the pure strategy labeled $mlr$ means that action $m$ is to be used if node $a$ is reached, action $l$ is to be used if node $b$ is reached, and action $r$ is to be used if node $c$ is reached.

If player I uses pure strategy $rrr$, then it is *impossible* that nodes $b$ or $c$ will be reached, whatever player II may do. However, the formal definition of a strategy still requires the specification of an action at nodes $b$ and $c$, even though the actions specified at these nodes will never have any affect on how the game gets played.

The nodes at which it would be up to player II to make a decision are labeled $d$, $e$, and $f$ for the game $G$ of Figure 2.6. A pure strategy for player II must therefore specify actions for player II at each of these three nodes. Since there are 3 available actions for player II at node $d$, 2 actions at node $e$, and 3 actions at node $f$, player II has a total of $3 \times 2 \times 3 = 18$ pure strategies. These 18 pure strategies can be labeled:

$$LLL, \quad LLM, \quad LLR, \quad LRL, \quad LRM, \quad LRR,$$
$$MLL, \quad MLM, \quad MLR, \quad MRL, \quad MRM, \quad MRR,$$
$$RLL, \quad RLM, \quad RLR, \quad RRL, \quad RRM, \quad RRR.$$

The pure strategy labeled $MLR$ means that action $M$ is to be used if node $d$ is reached, action $L$ is to be used if node $e$ is reached, and action $R$ is to be used if node $f$ is reached.

The play of Kayles shown in Figure 2.5 begins at the root $a$ of the game $G$ of Figure 2.6 with player I choosing action $n$. This leads to node $f$, at which player II chooses action $R$, which brings the game to an end at a leaf labeled with $\mathscr{W}$ to indicate a win for player I. Such a play of the game will be denoted by the sequence $[nR]$ of actions that generates it.[3]

[3] The square brackets emphasize that a play isn't the same thing as a strategy.

| | *LLL* | *LLM* | *LLR* | *LRL* | *LRM* | *LRR* | *MLL* | *MLM* | *MLR* | *MRL* | *MRM* | *MRR* | *RLL* | *RLM* | *RLR* | *RRL* | *RRM* | *RRR* |
|---|---|---|---|---|---|---|---|---|---|---|---|---|---|---|---|---|---|---|
| *ℓℓℓ* | $\mathscr{W}$ | $\mathscr{W}$ | $\mathscr{W}$ | $\mathscr{W}$ | $\mathscr{W}$ | $\mathscr{W}$ | $\mathscr{L}$ | $\mathscr{L}$ | $\mathscr{L}$ | $\mathscr{L}$ | $\mathscr{L}$ | $\mathscr{L}$ | $\mathscr{W}$ | $\mathscr{W}$ | $\mathscr{W}$ | $\mathscr{W}$ | $\mathscr{W}$ | $\mathscr{W}$ |
| *ℓℓr* | $\mathscr{W}$ | $\mathscr{W}$ | $\mathscr{W}$ | $\mathscr{W}$ | $\mathscr{W}$ | $\mathscr{W}$ | $\mathscr{L}$ | $\mathscr{L}$ | $\mathscr{L}$ | $\mathscr{L}$ | $\mathscr{L}$ | $\mathscr{L}$ | $\mathscr{W}$ | $\mathscr{W}$ | $\mathscr{W}$ | $\mathscr{W}$ | $\mathscr{W}$ | $\mathscr{W}$ |
| *ℓrℓ* | $\mathscr{L}$ | $\mathscr{L}$ | $\mathscr{L}$ | $\mathscr{L}$ | $\mathscr{L}$ | $\mathscr{L}$ | $\mathscr{L}$ | $\mathscr{L}$ | $\mathscr{L}$ | $\mathscr{L}$ | $\mathscr{L}$ | $\mathscr{L}$ | $\mathscr{W}$ | $\mathscr{W}$ | $\mathscr{W}$ | $\mathscr{W}$ | $\mathscr{W}$ | $\mathscr{W}$ |
| *ℓrr* | $\mathscr{L}$ | $\mathscr{L}$ | $\mathscr{L}$ | $\mathscr{L}$ | $\mathscr{L}$ | $\mathscr{L}$ | $\mathscr{L}$ | $\mathscr{L}$ | $\mathscr{L}$ | $\mathscr{L}$ | $\mathscr{L}$ | $\mathscr{L}$ | $\mathscr{W}$ | $\mathscr{W}$ | $\mathscr{W}$ | $\mathscr{W}$ | $\mathscr{W}$ | $\mathscr{W}$ |
| *mℓℓ* | $\mathscr{L}$ | $\mathscr{L}$ | $\mathscr{L}$ | $\mathscr{W}$ | $\mathscr{W}$ | $\mathscr{W}$ | $\mathscr{L}$ | $\mathscr{L}$ | $\mathscr{L}$ | $\mathscr{W}$ | $\mathscr{W}$ | $\mathscr{W}$ | $\mathscr{L}$ | $\mathscr{L}$ | $\mathscr{L}$ | $\mathscr{W}$ | $\mathscr{W}$ | $\mathscr{W}$ |
| *mℓr* | $\mathscr{L}$ | $\mathscr{L}$ | $\mathscr{L}$ | $\mathscr{W}$ | $\mathscr{W}$ | $\mathscr{W}$ | $\mathscr{L}$ | $\mathscr{L}$ | $\mathscr{L}$ | $\mathscr{W}$ | $\mathscr{W}$ | $\mathscr{W}$ | $\mathscr{L}$ | $\mathscr{L}$ | $\mathscr{L}$ | $\mathscr{W}$ | $\mathscr{W}$ | $\mathscr{W}$ |
| *mrℓ* | $\mathscr{L}$ | $\mathscr{L}$ | $\mathscr{L}$ | $\mathscr{W}$ | $\mathscr{W}$ | $\mathscr{W}$ | $\mathscr{L}$ | $\mathscr{L}$ | $\mathscr{L}$ | $\mathscr{W}$ | $\mathscr{W}$ | $\mathscr{W}$ | $\mathscr{L}$ | $\mathscr{L}$ | $\mathscr{L}$ | $\mathscr{W}$ | $\mathscr{W}$ | $\mathscr{W}$ |
| *mrr* | $\mathscr{L}$ | $\mathscr{L}$ | $\mathscr{L}$ | $\mathscr{W}$ | $\mathscr{W}$ | $\mathscr{W}$ | $\mathscr{L}$ | $\mathscr{L}$ | $\mathscr{L}$ | $\mathscr{W}$ | $\mathscr{W}$ | $\mathscr{W}$ | $\mathscr{L}$ | $\mathscr{L}$ | $\mathscr{L}$ | $\mathscr{W}$ | $\mathscr{W}$ | $\mathscr{W}$ |
| *nℓℓ* | $\mathscr{W}$ | $\mathscr{W}$ | $\mathscr{L}$ | $\mathscr{W}$ | $\mathscr{W}$ | $\mathscr{L}$ | $\mathscr{W}$ | $\mathscr{W}$ | $\mathscr{L}$ | $\mathscr{W}$ | $\mathscr{W}$ | $\mathscr{L}$ | $\mathscr{W}$ | $\mathscr{W}$ | $\mathscr{L}$ | $\mathscr{W}$ | $\mathscr{W}$ | $\mathscr{L}$ |
| *nℓr* | $\mathscr{L}$ | $\mathscr{W}$ | $\mathscr{L}$ | $\mathscr{L}$ | $\mathscr{W}$ | $\mathscr{L}$ | $\mathscr{L}$ | $\mathscr{W}$ | $\mathscr{L}$ | $\mathscr{L}$ | $\mathscr{W}$ | $\mathscr{L}$ | $\mathscr{L}$ | $\mathscr{W}$ | $\mathscr{L}$ | $\mathscr{L}$ | $\mathscr{W}$ | $\mathscr{L}$ |
| *nrℓ* | $\mathscr{W}$ | $\mathscr{W}$ | $\mathscr{L}$ | $\mathscr{W}$ | $\mathscr{W}$ | $\mathscr{L}$ | $\mathscr{W}$ | $\mathscr{W}$ | $\mathscr{L}$ | $\mathscr{W}$ | $\mathscr{W}$ | $\mathscr{L}$ | $\mathscr{W}$ | $\mathscr{W}$ | $\mathscr{L}$ | $\mathscr{W}$ | $\mathscr{W}$ | $\mathscr{L}$ |
| *nrr* | $\mathscr{L}$ | $\mathscr{W}$ | $\mathscr{L}$ | $\mathscr{L}$ | $\mathscr{W}$ | $\mathscr{L}$ | $\mathscr{L}$ | $\mathscr{W}$ | $\mathscr{L}$ | $\mathscr{L}$ | $\mathscr{W}$ | $\mathscr{L}$ | $\mathscr{L}$ | $\mathscr{W}$ | $\mathscr{L}$ | $\mathscr{L}$ | $\mathscr{W}$ | $\mathscr{L}$ |
| *rℓℓ* | $\mathscr{L}$ | $\mathscr{L}$ | $\mathscr{L}$ | $\mathscr{L}$ | $\mathscr{L}$ | $\mathscr{L}$ | $\mathscr{L}$ | $\mathscr{L}$ | $\mathscr{L}$ | $\mathscr{L}$ | $\mathscr{L}$ | $\mathscr{L}$ | $\mathscr{L}$ | $\mathscr{L}$ | $\mathscr{L}$ | $\mathscr{L}$ | $\mathscr{L}$ | $\mathscr{L}$ |
| *rℓr* | $\mathscr{L}$ | $\mathscr{L}$ | $\mathscr{L}$ | $\mathscr{L}$ | $\mathscr{L}$ | $\mathscr{L}$ | $\mathscr{L}$ | $\mathscr{L}$ | $\mathscr{L}$ | $\mathscr{L}$ | $\mathscr{L}$ | $\mathscr{L}$ | $\mathscr{L}$ | $\mathscr{L}$ | $\mathscr{L}$ | $\mathscr{L}$ | $\mathscr{L}$ | $\mathscr{L}$ |
| *rrℓ* | $\mathscr{L}$ | $\mathscr{L}$ | $\mathscr{L}$ | $\mathscr{L}$ | $\mathscr{L}$ | $\mathscr{L}$ | $\mathscr{L}$ | $\mathscr{L}$ | $\mathscr{L}$ | $\mathscr{L}$ | $\mathscr{L}$ | $\mathscr{L}$ | $\mathscr{L}$ | $\mathscr{L}$ | $\mathscr{L}$ | $\mathscr{L}$ | $\mathscr{L}$ | $\mathscr{L}$ |
| *rrr* | $\mathscr{L}$ | $\mathscr{L}$ | $\mathscr{L}$ | $\mathscr{L}$ | $\mathscr{L}$ | $\mathscr{L}$ | $\mathscr{L}$ | $\mathscr{L}$ | $\mathscr{L}$ | $\mathscr{L}$ | $\mathscr{L}$ | $\mathscr{L}$ | $\mathscr{L}$ | $\mathscr{L}$ | $\mathscr{L}$ | $\mathscr{L}$ | $\mathscr{L}$ | $\mathscr{L}$ |

Figure 2.8 The strategic form of the game $G$. Player II can guarantee winning by playing *MLR* no matter what pure strategy player I may choose, because every entry in the column corresponding to the pure strategy *MLR* is $\mathscr{L}$.

What are the strategies that result in the play [*nR*] of $G$? The pair of strategies chosen by the players must be of the form (*nxy*, *XYR*), where *nxy* stands for any strategy for player I in which $n$ is chosen at node $a$. There are 4 such strategies, namely *nll*, *nlr*, *nrl*, and *nrr*. Similarly, *XYR* stands for any strategy for player II at which $R$ is chosen at node $f$. There are 6 such strategies, namely *LLR*, *LRR*, *MLR*, *MRR*, *RLR*, and *RRR*. So the total number of strategy pairs that result in the play [*nR*] is $4 \times 6 = 24$.

Figure 2.8 shows the *strategic form* of our variant of Kayles. The representation of $G$ in Figure 2.6 as a game tree is called its *extensive form*. For each pair of strategies, the strategic form indicates what the outcome of the game will be if that pair of strategies is used. The rows of the matrix represent player I's pure strategies, and the columns represent player II's pure strategies. Thus, the cell in row *nll* and column *LLR* contains the letter $\mathscr{L}$. This indicates that player I will lose the game if he uses pure strategy *nll* and player II uses pure strategy *LLR*. This fact was checked out in the previous paragraph by tracing the play [*nR*] that results from the use of strategy pairs of the form (*nxy*, *XYR*).

Von Neumann and Morgenstern called the strategic form of a game its *normal form* because they thought that the "normal" procedure in analyzing a game should be to discard its extensive form in favor of its strategic form. However, the sheer size of the strategic form of Figure 2.8 provides at least one reason why modern game theorists don't always take their advice.

## 2.5 Backward Induction

In the strategic form of Figure 2.8, all the entries in the column corresponding to player II's pure strategy *MLR* are $\mathscr{L}$. So if player II chooses *MLR* in our variant of Kayles, player I is doomed to lose, no matter what strategy he plays.

It turns out that one of the players in a win-or-lose game of perfect information without chance moves *always* has a pure strategy that guarantees victory no matter what the other player may do, but it isn't by any means obvious that the strategic form of such a game must have either a column whose entries are all $\mathscr{L}$ or else a row whose entries are all $\mathscr{W}$. This fact becomes obvious only when we apply backward induction to the extensive form of the game.

We used backward induction to solve the Tip-Off Game in Section 2.2.1. It requires starting from the end of the game and then working backward to its beginning. In this section, we offer an analysis of our variant of Kayles that shows how the same method may always be used to show that one or the other of the two players can guarantee victory in any win-or-lose game of perfect information without chance moves.

### 2.5.1 Subgames

In a game of perfect information, each node $x$ other than a leaf determines a *subgame*.[4] The subgame consists of the node $x$ together with all of the game tree that follows $x$. Figure 2.9 shows the six subgames of the game $G$ of Figure 2.6. (Notice that the definition makes $G$ a subgame of itself.)

### 2.5.2 Values

The value $v(H)$ of a subgame $H$ of $G$ is $\mathscr{W}$ if player I has a strategy for $H$ that wins the game $H$ for him whatever strategy player II may use. Similarly, the value $v(H)$ of the subgame $H$ is $\mathscr{L}$ if player II has a strategy that wins the game $H$ for her whatever strategy player I may use.

When we get to Von Neumann's minimax theorem in Chapter 7, we will learn how to assign values to any two-player game in which the players have diametrically opposed preferences. The minimax theorem applies to all such strictly competitive games, including those with imperfect information and chance moves. But it is very unusual for a game that isn't strictly competitive to have a value at all.

### 2.5.3 Analyzing the Game $G$

Consider first the one-player subgames $G_2$, $G_4$, and $G_5$ of Figure 2.9. Player II wins $G_4$ by choosing action $L$, and so $v(G_2) = \mathscr{L}$. (Recall that an outcome is labeled with $\mathscr{L}$ when player II *wins*.) Player I wins $G_4$ or $G_5$ by choosing action $l$, and so $v(G_4) = v(G_5) = \mathscr{W}$.

Next consider the game $G'$ shown in Figure 2.10. This game is obtained from $G$ by replacing the subgames $G_2$, $G_4$, and $G_5$ with leaves labeled with their values. If $G'$ has a value, then $G$ has a value as well, and $v(G') = v(G)$.

[4]It isn't true that each node of a game of *imperfect* information determines a subgame. Each subgame must have a single node to serve as its root, but we can't separate one node from its fellows in an information set for this purpose.

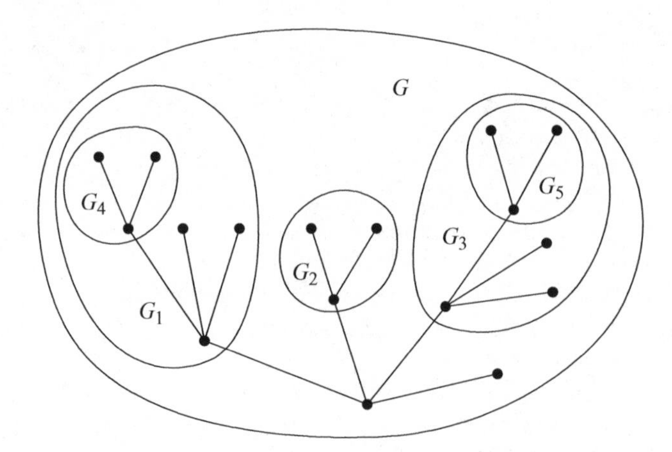


Figure 2.9 The subgames of $G$.

To prove this in the case when player I is the winner, we need to show that, if player I has a strategy $s'$ that always wins in game $G'$, then he necessarily has a strategy $s$ that always wins in $G$. Why is this? Whatever strategy player II uses, player I's choice of $s'$ in $G'$ results in a play of $G'$ that leads to a leaf $x$ of $G'$ labeled with $\mathscr{W}$. Such a leaf $x$ may correspond to a subgame $G_x$ of $G$. If so, then $v(G_x) = \mathscr{W}$. Hence player I has a winning strategy $s_x$ in $G_x$. It follows that player I has a winning strategy $s$ in $G$, which consists of playing according to $s'$ until one of the subgames $G_x$ is reached and then playing according to $s_x$.

Next consider the game $G''$ shown at the foot of Figure 2.10. This game is obtained from $G'$ by replacing the one-player subgames $G'_1$ and $G'_3$ by leaves labeled with their values. By the reasoning used before, if $G''$ has a value, then so does $G'$, and $v(G'') = v(G')$.

All of player I's actions in the one-player game $G''$ lead to a leaf at which he loses. So the value of $G''$ is $\mathscr{L}$. It follows that $G$ also has a value, and

$$v(G) = v(G') = v(G'') = \mathscr{L}.$$

That is to say, player II has a strategy that wins the game $G$, no matter what strategy is used by player I.

### 2.5.4 Finding a Winning Strategy

One way of finding a winning strategy for player I in $G$ is to read it off from the strategic form given in Figure 2.8. However, except in very simple cases, this isn't a sensible way of locating a winning strategy because the heavy labor involved in constructing the strategic form makes the method impractical.

A better way of finding a winning strategy is to mimic the method by means of which it was proved that a winning strategy exists for $G$. Begin by looking at the smallest subgames of $G$ (those with no subgames of their own). In each such subgame, double the branches that correspond to optimal choices in the subgame. Next pretend that the *undoubled* branches in these subgames don't exist. This creates a

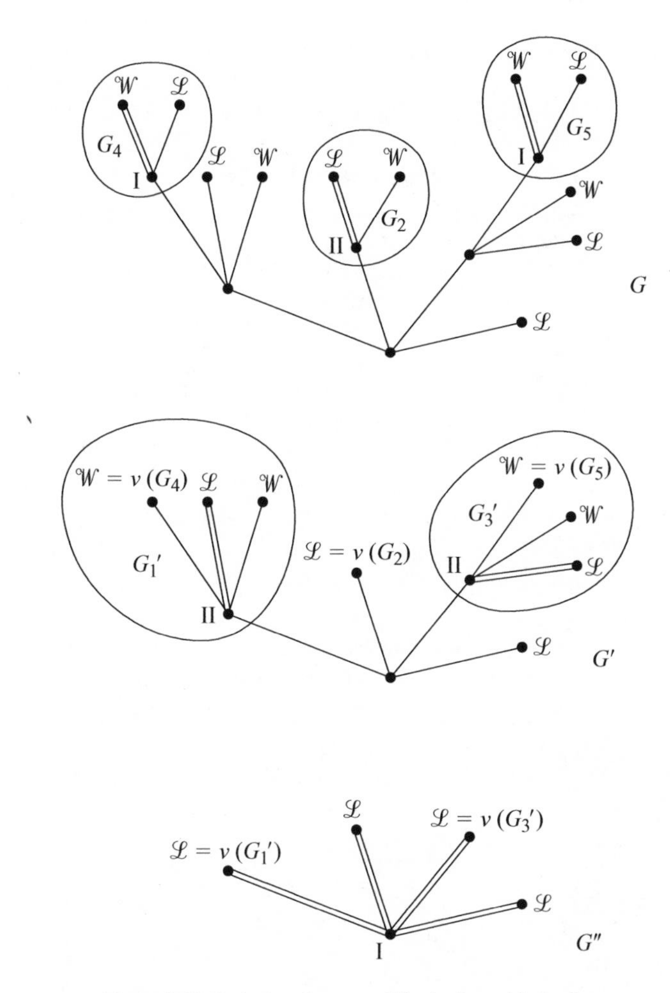


Figure 2.10 Reducing the game $G$ by backward induction.

new game $G^*$. Now repeat the procedure with $G^*$ and continue in this way until there is nothing left to do. At the end of the procedure there will be at least one play of $G$ whose branches have all been doubled. These are the only plays that can be followed if it is common knowledge between the players that each will always try to win under all circumstances.

This procedure has been carried through for the game $G$ in Figure 2.6. Four plays of the game have all their branches doubled, and each leads to a win for player II, thus confirming that she has a winning strategy.

A winning pure strategy can be read off directly from the diagram by choosing one of the doubled branches at each of player II's decision nodes. In the case of $G$, the $M$ branch is doubled at node $d$, the $L$ branch at node $e$, and the $R$ branch at node $f$. Player II therefore has only one winning pure strategy, namely $MLR$. If more than one branch were doubled at some of her decision nodes, player II would have multiple winning strategies.

## 2.6 Solving Nim

The procedure just described could also be carried out for Nim. However, as with Tic-Tac-Toe, it is hard work even to write down its game tree.

In the case of Nim, there is an elegant way of proceeding that avoids the necessity of constructing a game tree. This is illustrated using the version of Nim given in Figure 2.11. In this figure, the numbers of matchsticks in each pile have first been converted into decimal notation and then into binary notation.[5]

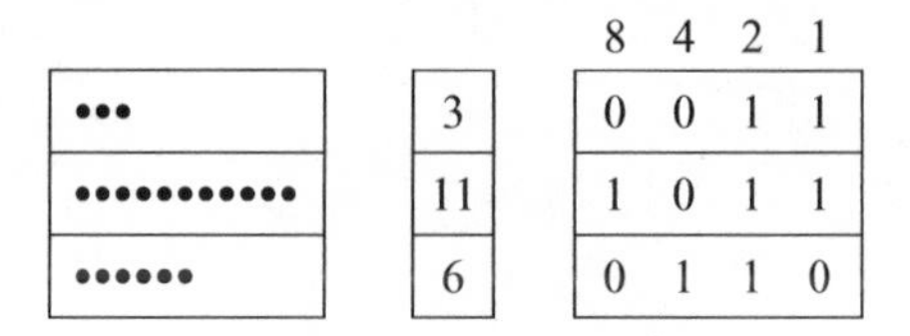


Figure 2.11 Nim with three piles of matchsticks.

Call a game of Nim *balanced* if each column of the binary representation has an even number of 1s and *unbalanced* otherwise. The example of Figure 2.11 is unbalanced because the eights column has an odd number of 1s (as do the fours column and the twos column). It is easy to verify that *any* admissible move in Nim converts a balanced game into an unbalanced game.[6]

The player who moves first in a balanced game can't win immediately because a balanced game must have matchsticks in at least two piles. The player moving

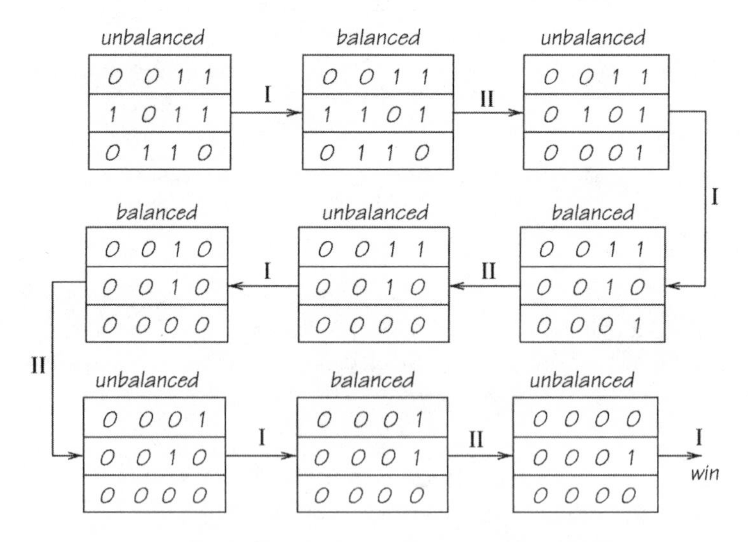


Figure 2.12 Player I uses a winning strategy in Nim.

[5]For example, the number whose decimal representation is 11 is the sum of 1 eight, 0 fours, 1 two, and 1 one. So its representation in binary form is 1011.

[6]At least one 1 in the binary representation of the pile from which matchticks are taken will necessarily be changed to a 0. If the column in which this occurs had $2n$ ones, it will have $2n - 1$ ones afterward.

therefore can't pick up the last matchstick right away because he or she is allowed to take matchsticks from only one pile at a time.

One of the players therefore has a winning strategy, which consists of always converting an unbalanced configuration into a balanced configuration. Using such a strategy guarantees that my opponent can't win on the next move. Since this is true at every stage in the game, my opponent can't win at all. But someone must pick up the last matchstick. If it isn't my opponent, it must be me. So I must be using a winning strategy.

Since most games of Nim start out unbalanced, it is usually the first player to move who has a winning strategy. But if the original configuration of matchsticks is balanced, then the second player has a winning strategy.

Figure 2.12 shows a possible play of the version of Nim given in Figure 2.11. Player I is using a winning strategy. It is worth noticing that, once player I is faced with only two piles of matchsticks with equal numbers of matchsticks in each, then he can win by "strategy stealing." All he need do is to take as many matchsticks from one pile as player II just took from the other.

## 2.7 Hex

→ 2.8

The game of Hex was invented by Piet Hein in 1942. The same John Nash who formulated the idea of a Nash equilibrium came up with an identical set of rules in 1948. Nash is said to have been inspired by the hexagonal tiling in the men's room of the Princeton mathematics department, but he thinks this story is apochryphal.

Hex is a game played between Circle and Cross on a board made up of $n^2$ hexagons arranged in a parallelogram, as illustrated in Figure 2.13(a). At the beginning of the game, each player's territory consists of two opposite sides of the board. The players take turns in moving, with Circle going first. A move consists of taking possession of a vacant hexagon on the board by labeling it with your emblem.

The winner is the first to link their two sides of the board with a continuous chain of hexagons labeled with their emblem. In the game that has just concluded in Figure 2.13(b), Cross was the winner.

Aside from its association with Nash, Hex is interesting for two reasons. The first point of interest is that Hex is a win-or-lose game, although it seems possible at first sight that it might end in a draw. Since all win-or-lose games of perfect information without chance moves have a value, we know that one of the players has a pure strategy for Hex that guarantees victory whatever the other player may do. It isn't known what the winning strategy is when $n$ is reasonably large, but the second interesting feature of Hex is that we can nevertheless show that the player with the winning strategy is Circle.

### 2.7.1 Why Hex Can't End in a Draw

Think of Circle's hexagons as water and Cross's hexagons as land. When all the hexagons have been labeled, either water will then flow between the two lakes originally belonging to Circle, or else the channel between them will be dammed. Circle wins in the first case, and Cross in the second.

This simple argument is intuitively compelling, but it turns out not to be so easy to back it up with a rigorous proof. So why do mathematicians bother? The answer is that the history of mathematics is awash with propositions that seemed obviously

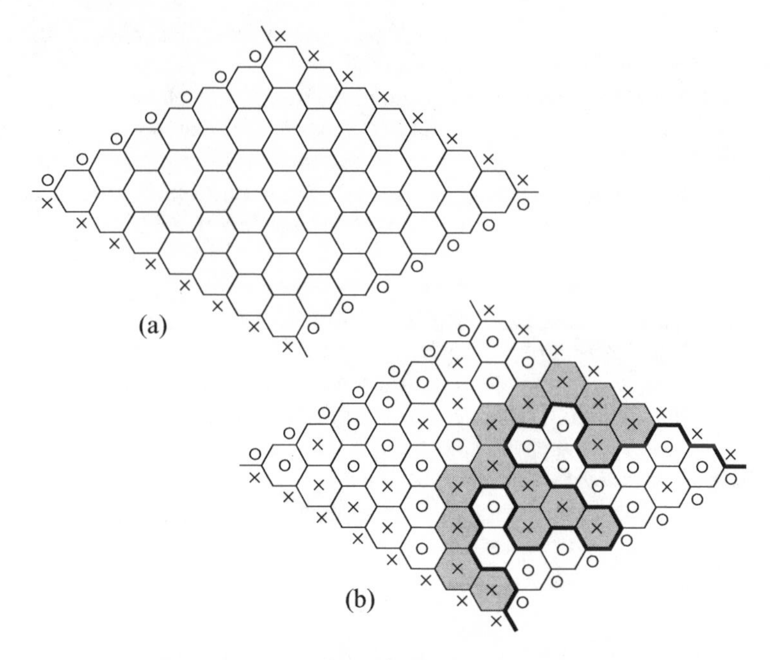


Figure 2.13 Hex.

→ 2.7.2

true but eventually turned out to be false. However, the Mad Hatter in the margin invites you to skip forward to Section 2.7.2 if you aren't interested in the following sketch of David Gale's proof that Hex can't end in a draw.

Gale uses an algorithm that requires starting from a point off the corner of the board, as shown in Figure 2.14(a). You must then trace out a path so that the next segment of the path always has a circled hexagon on one side and a crossed hexagon on the other. You could do this by immediately going back the way you just came, but retracing your steps in this way isn't allowed.

We need to show that such a path can neither terminate on the board, nor return to a point it has visited before. Since the Hex board is finite, the path must then terminate at one of the points off the corners of the board other than that from which it started. It follows, as illustrated in Figure 2.13(b), that one of the two opposite sides of the board must be linked. So Hex can't end in a draw.

Figure 2.14(a) shows a path that has reached a point $p$ in the interior of the board. We need to show that the path can be continued. To reach $p$, the path must have just

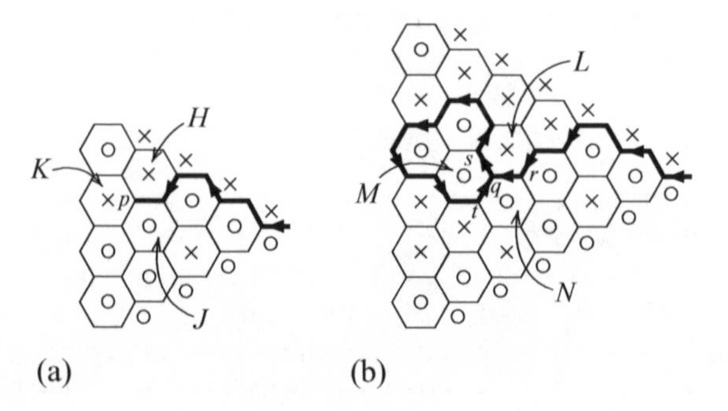


Figure 2.14 Gale's algorithm for Hex.

passed between a crossed hexagon $H$ and a circled hexagon $J$. Since $p$ is in the interior of the board, there has to be a third hexagon $K$ for which $p$ is a vertex. If $K$ is crossed, as in Figure 2.14(a), the path can be continued by passing between $J$ and $K$. If $K$ is circled, the path can be continued by passing between $H$ and $K$.

If $p$ is on the edge of the board, the argument has to be modified slightly, but it still works. The argument fails only if $p$ is one of the four points off the corners of the board. So these are the only points where the path can terminate.

Figure 2.14(b) shows a path returning to an interior point $q$ that it has visited before. To do this, the path violates the rule that it must keep a crossed hexagon on one side and a circled hexagon on the other. To prove by contradiction that a path can never loop back on itself without violating this rule, let $q$ be the first point that gets revisited.

For $q$ to be visited at all, the three hexagons $L$, $M$, and $N$ with a common vertex at $q$ can't all have the same label. Suppose that $L$ is crossed, and the other two hexagons are circled, as in Figure 2.14(b). The path must then have passed between $L$ and $M$, and between $L$ and $N$ on its first visit. Since $q$ is the *first* revisited point on the path, the path can't have gotten back to $q$ via the point $r$ or the point $s$. It can have gotten back to $q$ only via $t$. But $M$ and $N$ are both circled, and so this is impossible. As before, the argument has to be adapted slightly if $q$ is on the edge of the board, but it still works.

### 2.7.2 Why Circle Has a Winning Strategy

→ 2.8

Nash gave a "strategy-stealing" argument that shows that if Cross has a winning strategy, then so does Circle. Since it's impossible for *both* players to win, it therefore can't be true that Cross has a winning strategy. But *someone* has a winning strategy. Since it isn't Cross, it must be Circle.

If Cross has a winning strategy, how would Circle steal it? Nash argued that Circle could follow the following instructions:

1. At the first move, circle a hexagon at random.
2. At later moves, pretend that the last hexagon you circled is unlabeled. Next pretend that the remaining circled hexagons are all crossed and the crossed hexagons are all circled. You have now imagined yourself into a position to which Cross's winning strategy applies. Circle the hexagon that Cross would choose in this position if she were to use her winning strategy. The only possible snag is that this hexagon may be the hexagon you are only pretending is unlabeled. If so, then you don't need to steal Cross's winning move for the position because you have *already* stolen it. Just circle a free hexagon at random instead.

This strategy wins for Circle because he is simply doing what supposedly guarantees Cross a win—but one move earlier. The presence on the board of an extra hexagon labeled with a Circle may result in his winning sooner than Cross would have, but we won't hear him complaining if this should happen!

## 2.8 Chess

Computers can beat anybody at checkers, but world-class players can still beat computers at chess most of the time. However, when computer programs are eventually developed that beat even the best human players, it won't be because game theorists have worked out the optimal way to play. Chess is so complicated that its solution will probably never be known for certain—and this is just as well for people who play for fun. What would be the point of playing at all if you could always look up the optimal next move in a book?

However, game theory isn't entirely helpless. Nobody can find Bigfoot or the Loch Ness Monster because they don't exist, but this isn't the reason that game theorists can't find the solution to chess. We can at least prove that chess actually does have a value.

*Strictly Competitive Games.* The games studied so far in this chapter have nearly all been win-or-lose games. The exception was Tic-Tac-Toe, which can end in a draw. Chess also has three possible outcomes: $\mathscr{W}$, $\mathscr{L}$, and $\mathscr{D}$. We take player I to be White and player II to be Black, and so $\mathscr{W}$ denotes a win for White and a loss for Black.

To write $a \preceq_i b$ means that player $i$ likes $b$ at least as much as $a$. To write $a \prec_i b$ means that player $i$ strictly prefers $b$ to $a$. That is to say, he or she never chooses $a$ when $b$ is on the table. To write $a \sim_i b$ means that player $i$ is indifferent between $a$ and $b$. To say that $a \preceq_i b$ is therefore the same as saying that either $a \prec_i b$ or else $a \sim_i b$.

In a *strictly competitive* game, the players' aims are diametrically opposed. Whatever is good for one is bad for the other. In mathematical terms,[7] this means that for each outcome $a$ and $b$,

$$a \prec_1 b \quad \Leftrightarrow \quad b \prec_2 a.$$

Chess is therefore a strictly competitive game, as the players' preferences are:

$$\mathscr{L} \prec_1 \mathscr{D} \prec_1 \mathscr{W},$$
$$\mathscr{L} \succ_2 \mathscr{D} \succ_2 \mathscr{W}.$$

→ 2.8.1

The fact that chess has a value will be deduced from a more general theorem that tidies up the account of backward induction given in Section 2.5. When the theorem says that player $i$ can *force* an outcome in a set $S$, it means that player $i$ has a strategy that guarantees that the outcome will be in the set $S$, whatever the other player does. The notation $\sim S$ is used for the complement of a set $S$.[8] In the theorem, $\sim T$ therefore consists of all outcomes of the game that aren't in the set $T$.

[7]The notation $P \Rightarrow Q$ means that $P$ implies $Q$, so that the truth of $Q$ can be deduced from the truth of $P$. The notation $P \Leftrightarrow Q$ means that both $P \Rightarrow Q$ and $Q \Rightarrow P$ are true, so that $P$ is true if and only if $Q$ is true. When people say that "$P$ is a *sufficient* condition for $Q$," they simply mean $P \Rightarrow Q$. Similarly, "$P$ is a *necessary* condition for $Q$" means that $Q \Rightarrow P$. To say that "$P$ is a necessary and sufficient condition for $Q$" is therefore just a long-winded way of saying $P \Leftrightarrow Q$.

[8]The notation $x \in S$ means that $x$ is an element (or a member) of the set $S$. The notation $x \notin S$ means that $x$ isn't an element of $S$. The complement $\sim S$ of a set $S$ can therefore be defined symbolically as $\sim S = \{x : x \notin S\}$. For the definition to be meaningful, it is necessary to know the range of the variable $x$ in advance. In the text, the range is understood to be the set $U$ of all outcomes under study.

THEOREM 2.1 *Let T be any set of outcomes in a finite*[9] *two-player game of perfect information without chance moves. Then, either player I can force an outcome in T, or player II can force an outcome in* $\sim T$.

*Proof* Forget all about the players' preferences in the game. We are then free to relabel all the outcomes in $T$ with $\mathcal{W}$, and all the outcomes in $\sim T$ with $\mathcal{L}$. The theorem then reduces to showing that any finite, win-or-lose game has a value. The argument of Section 2.5.3 can be recycled for this purpose, but since we are now proving a formal theorem, we ought to be more careful about the mathematical details.

**Step 1.** The rank of a game is the number of branches in its longest possible play. So a game of rank 1 consists of just a root and some leaves. If player I chooses at the root, then he can win immediately if one of the leaves is labeled with $\mathcal{W}$. Otherwise, all the leaves of a win-or-lose game are labeled with $\mathcal{L}$, and so player II can force a win without doing anything at all (as in the game $G''$ of Figure 2.10). Either way the game has value. Since similar reasoning applies if player II chooses at the root, it follows that any win-or-lose game $H$ of rank 1 has a value $v(H)$ (Section 2.5.2).

**Step 2.** Now suppose that, for some value of $n$, all win-or-lose games of rank $n$ have a value. We will show that any win-or-lose game $H$ of rank $n+1$ must then have a value as well.

Locate the last decision node $x$ on each play of length $n+1$ in $H$. Now throw away anything that follows such a node. The nodes $x$ then become leaves of a new game $H'$ when we label each $x$ with the value $v(H_x)$ of the subgame $H_x$ of $H$ rooted at $x$. Such subgames are of rank 1 and hence must have a value by Step 1.

The game $H'$ is of rank $n$, and so it has a value. Suppose it is player I who has a strategy $s'$ that wins $H'$ whatever player II may do. The use of $s'$ then guarantees that $H'$ will end at a leaf of $H'$ labeled with $\mathcal{W}$. If this leaf corresponds to a subgame $H_x$ of $H$, then $v(H_x) = \mathcal{W}$, and so player I has a winning strategy $s_x$ in $H_x$. So player I can force a win in $H$ by playing $s'$ in $H'$ and $s_x$ in each subgame $H_x$ for which he has a winning strategy. The same reasoning applies if it is player II who has a winning strategy in $H'$. Thus one of the players can force a win in $H$, and so $H$ has a value.

**Step 3.** The final step is to apply the Principle of Induction.[10] Step 1 says that all win-or-lose games of rank 1 have a value. Step 2 then implies that all win-or-lose games of rank 2 also have a value. Step 2 can then be applied again to show that all win-or-lose games of rank 3 have a value. And so on.

All finite win-or-lose games of perfect information without chance moves therefore have a value, and so the theorem is proved.

## 2.8.1 Values of Strictly Competitive Games

A Mad Hatter in the margin is usually running away to another section, and beginners would be advised to follow him. Here he isn't running away, although he

[9]This just means that the game tree has a finite number of nodes.

[10]If $P(n)$ is a proposition defined for each positive integer $n$, and

1. $P(1)$ is true
2. For each $n$, $P(n) \Rightarrow P(n+1)$ is true then $P(n)$ is true for all values of $n$.

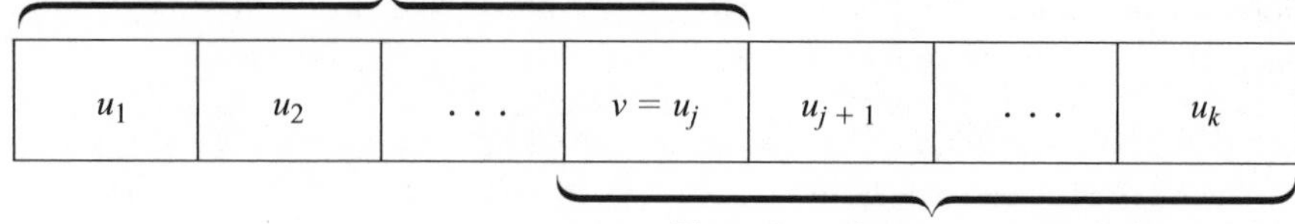


Figure 2.15 The value $v$ of a strictly competitive game in which $u_1 \prec_1 u_2 \prec_1 \prec \cdots \prec_1 u_k$.

looks as though he would like to. This means that something tougher than usual is coming up, but that the urge to rush on by should be resisted.

An outcome $v$ is said to be a value of a two-player game $G$ if and only if player I can force an outcome in the set $W_v = \{u : u \succeq_1 v\}$, and player II can simultaneously force an outcome in the set $L_v = \{u : u \succeq_2 v\}$.

For example, if White has a strategy that can force a draw or better for him and Black has a strategy that can force a draw or better for her, then the value of chess is $\mathscr{D}$. In this case, $W_v = \{\mathscr{D}, \mathscr{W}\}$ and $L_v = \{\mathscr{L}, \mathscr{D}\}$. If it turns out that the value of chess is $\mathscr{W}$, then $W_v = \{\mathscr{W}\}$ and $L_v = \{\mathscr{L}, \mathscr{D}, \mathscr{W}\}$.

Without loss of generality, it will be assumed that player I isn't indifferent between any pair of outcomes of $G$. Thus the outcomes in the set $U = \{u_1, u_2, \ldots, u_k\}$ of all possible outcomes of $G$ can be labeled so that

$$u_1 \prec_1 u_2 \prec_1 \cdots \prec_1 u_k.$$

Player II's preferences then satisfy $u_1 \succ_2 u_2 \succ_2 \cdots \succ_2 u_k$. Figure 2.15 illustrates what it means for such a game to have a value $v$.

Corollary 2.1 *Any finite, strictly competitive game of perfect information without chance moves has a value.*

*Proof* Let $W_v$ be the smallest set into which player I can force the outcome.[11] If $v = u_j$, player I can't force the outcome to be in $W_{u_{j+1}}$ because this is a smaller set than $W_v$. So player II must be able to force an outcome in $\sim W_{u_{j+1}} = L_v$, by Theorem 2.1.

Corollary 2.2 *Chess has a value.*

*Proof* Chess is a finite, strictly competitive game of perfect information without chance moves.

### 2.8.2 Saddle Points

A strategy pair $(s, t)$ is a saddle point of the strategic form of a strictly competitive game if the outcome that results from the use of $(s, t)$ is no worse for player I than any

[11]Mathematicians want to be sure that there is at least one set with this property before talking about the smallest such set. But player I can certainly force the outcome to lie in the set $W_{u_1}$, because this contains all outcomes of the game.

outcome in the column corresponding to $t$ and no better for him than any outcome in the row corresponding to $s$.

Corollary 2.3 *The strategic form of a finite, strictly competitive game of perfect information without chance moves always has a saddle point* $(s, t)$.

*Proof* Let $s$ be a strategy that guarantees player I an outcome no worse than the value $v$ of the game. Then each entry in row $s$ of the strategic form must be no worse than $v$ for player I. Let $t$ similarly guarantee player II an outcome no worse than $v$. Then each entry in column $t$ must be no worse than $v$ for player II. Because the game is strictly competitive, each entry in column $t$ is therefore no better than $v$ for player I. The actual outcome that results from the play of $(s, t)$ must therefore be no worse and no better for player I than $v$. Since players are assumed not to be indifferent between outcomes in this section, the result of playing $(s, t)$ must therefore be exactly $v$.

Theorem 2.2 *If the strategic form of a strictly competitive game G has a saddle point* $(s, t)$ *for which the corresponding outcome is* $v$*, then the value of G is* $v$.

*Proof* Since $v$ is the worst outcome in its row for player I, he can force an outcome at least as good as $v$ by playing $s$. Since $v$ is the best outcome in its column for player I, it is the worst in its column for player II, so she can force an outcome at least as good for her as $v$ by playing $t$.

I find that serious chess players are curiously uninterested in game theory, but when they can be persuaded to offer an opinion, they always guess that the value of chess is $\mathscr{D}$, which would mean that both players have strategies that can force a draw or better.

Figure 2.16 is a notional strategic form for chess drawn on the assumption that the experts are right. In this figure, the strategy $s$ is a pure strategy that forces a draw or better for player I, and $t$ is a pure strategy that forces a draw or better for player II. By Corollary 2.3, the pair $(s, t)$ is then a saddle point of the strategic form of chess.

## 2.9 Rational Play?

What advice should a game theory book give to two people about to play a strictly competitive game $G$ of perfect information without chance moves?

| | | | | $t$ | | |
|---|---|---|---|---|---|---|
| | | | | $\mathscr{L}$ | | |
| | | | | $\mathscr{D}$ | | |
| | | | | $\mathscr{L}$ | | |
| | | | | ⋮ | | |
| $s$ | $\mathscr{D}$ | $\mathscr{W}$ | ⋯ | $\mathscr{D}$ | ⋯ | $\mathscr{W}$ |
| | | | | ⋮ | | |
| | | | | $\mathscr{L}$ | | |

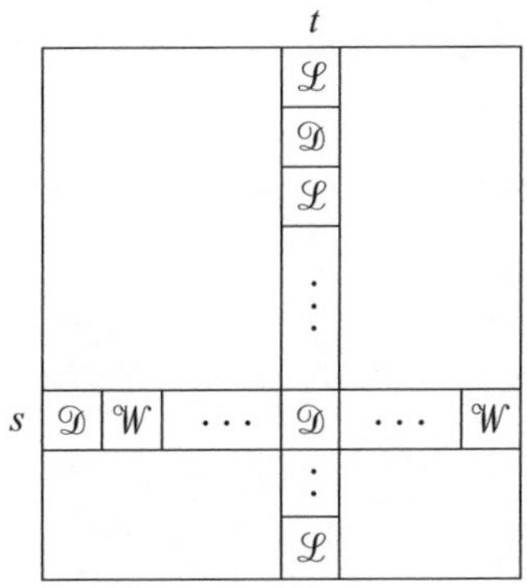

Figure 2.16 A possible strategic form for Chess.

If the game has value $v$, the answer may seem easy. Surely both players should simply choose pure strategies that guarantee each an outcome no worse than $v$. If such a pair $(s,t)$ of pure strategies is used, then the game will end in some outcome that both players regard as being equivalent to $v$.[12] But things are seldom so easy in game theory!

### 2.9.1 Nash Equilibrium

The pair $(s,t)$ certainly meets one of the criteria that must be satisfied if it is to be proposed by a game theory book for general adoption as the rational solution of a game. The criterion is that $(s,t)$ should be a *Nash equilibrium*. This means that each of the pure strategies in the pair $(s,t)$ must be a best reply to the other (Section 1.6).

In a strictly competitive game, a pair $(s,t)$ is a Nash equilibrium if and only if it is a saddle point of the strategic form of the game. The fact that $v$ is best in its column makes $s$ a best reply to $t$ for player I. Since the two players have opposing preferences, the fact that $v$ is worst in its row for player I makes it best in its row for player II. Thus $t$ is a best reply to $s$ for player II.

For example, in the strategic form of Figure 2.8, all pure strategy pairs in which player II uses *MLR* are Nash equilibria. That is to say, *every* outcome in the ninth column of the strategic form corresponds to a saddle point.

It would be self-defeating for a game theorist to publish a recommendation for each player that wasn't a Nash equilibrium. If the advice were generally adopted, then it would be common knowledge how the game would be played. However, if player I knows that player II is sufficiently rational to carry out the book's advice by playing $t$, then he would be stupid to follow the book's advice to play $s$ unless $s$ is a best reply to the strategy $t$ that he knows player II is going to choose. Similarly, if player II knows that player I is sufficiently rational to carry out the book's advice by playing $s$, then she would be stupid to follow the book's advice to play $s$ unless $s$ is a best reply to $t$.

Critics sometimes complain that the idea of a Nash equilibrium gets used even when there isn't any reason to suppose that the players will behave as though they were rational. I think that such attempts to apply game theory in situations to which it isn't applicable deserve all the criticism they get. In particular, rational players who know that their opponents are irrational won't necessarily be content to play so as to guarantee themselves the value of a strictly competitive game. They will want to exploit the folly of their opponent in an attempt to get more than its value.

### 2.9.2 When Are People Rational?

→ 2.9.3

Traditional economics is somewhat shakily founded on the assumption that rationality commonly reigns in the commercial and business world, but modern economists are much less ready than their predecessors to assume that economic agents will always behave rationally.

Perhaps the fact that real people often behave irrationally is just as well for those games that are played mostly for fun. Watching two people play poker optimally

[12]We now admit the possibility that players may be indifferent between some outcomes.

would be about as interesting as watching paint dry—and nobody would play chess at all if it were known how to play it optimally.

However, if we can't count on the players in a game behaving rationally, then we have seen that orthodox game theory won't help us predict how they will play. So when is it reasonable to assume that the players in a game will behave as though it were common knowledge that they are all rational?

Other game theorists are sometimes more optimistic, but my own view is that it is very risky to use game theory for predictive purposes when none of the following criteria are satisfied:

- The game is simple.
- The incentives for playing well are adequate.
- The players have played the game many times before,[13] and hence have had much opportunity for trial-and-error learning.

In laboratory experiments with human subjects, Nash equilibrium normally predicts human behavior quite well when all three criteria are satisfied. The explanation usually offered is that nothing then obstructs the convergence of trial-and-error adjustment processes like those mentioned in Section 1.6. After the process has converged on a Nash equilibrium, the players are seldom able to explain *why* their final choice of strategy is optimal, but it is enough that they are behaving as though they had made a rational choice.

Outside the laboratory, it isn't so easy to tie down the environment within which a game is played. However, the second and third criteria are satisfied, for example, when poker is played by experts at the world poker championships. Moreover, while poker isn't as simple as Tic-Tac-Toe or Nim, it is simple when compared to chess. That is to say, all its many variants, like Texas Hold'em or Seven Card Stud, can be analyzed successfully in principle. The first criterion is therefore also satisfied to some degree. So it is reassuring that play at these championships is much closer to what game theory predicts for rational players than in nickel-and-dime neighborhood games. For example, game theory recommends much bluffing on very bad hands (Section 15.2). Champions know this, but nickel-and-dime players tend to bluff only on middle-range hands that might win anyway.

In biological games, neither the first nor the second criterion commonly holds. Sometimes the advantage that accrues to the fitter of two strategies is so slight as to be imperceptible when a game is played just once. But the third criterion applies with a vengeance since evolution may have had millions of years to learn the optimal strategy by trial and error. Evolutionary biology is therefore an important area of application for the idea of a Nash equilibrium.

In telecom auctions, licenses to broadcast on specified chunks of the radio spectrum have sometimes been sold for several billion dollars. In this context, it is the second criterion that applies with a vengeance, and the third criterion doesn't apply at all. However, the telecom companies use the idea of a Nash equilibrium in deciding how to bid because they don't expect anyone to bid stupidly when such large amounts of money are on the table.

[13]Against different opponents each time. If you play repeatedly against the same opponent, the repeated situation must be modeled as a single "supergame."

### 2.9.3 Subgame-Perfect Equilibrium

The strategy pair (*mlr*, *MLR*) is a Nash equilibrium in the strategic form of Kyles given in Figure 2.8, but you won't come up with this strategy pair by applying backward induction in the extensive form of the game given in Figure 2.6. The strategy pairs selected by backward induction are those that correspond to branches that are doubled in this figure. Backward induction therefore always selects *MLR* for player II but leaves player I free to choose between any strategy of the form *xll*. However, *mlr* doesn't take this form.

Backward induction doesn't select *mlr* because it requires player I to plan to make an *irrational* choice at node $c$. Choosing $r$ at node $c$ is irrational because player I can win at node $c$ by playing $l$ rather than losing by playing $r$. The fact that such an irrational plan is built into *mlr* doesn't prevent the strategy being part of a Nash equilibrium because, if player II uses her Nash equilibrium strategy *MLR*, then node $c$ won't be reached. So player I will never actually be called upon to make the irrational choice that he *would* make if node $c$ *were* reached.

The lesson is that Nash equilibria only ensure that players will behave rationally at nodes on the equilibrium path—the play of the game followed when the players use their equilibrium strategies. Off the equilibrium path, Nash equilibria allow the players to plan to behave in all kinds of crazy ways.

For example, if the value of chess is $\mathscr{D}$, then White has a pure strategy $s$ that guarantees him a draw or better, but he can't do any better than a draw if Black uses the pure strategy $t$ that guarantees her a draw or better. However, real people sometimes make mistakes. What if Black makes a momentary error that results in a subgame being reached that wouldn't have been reached if she hadn't deviated from $t$? The use of strategy $s$ still guarantees a draw or better for White because $s$ guarantees a draw whether Black plays well or badly, but it may be that White can now do better than forcing a draw. Perhaps he has a winning strategy in the subgame $H$ reached as a result of Black's blunder. Why should he then stick with $s$? If another strategy $s'$ guarantees a victory for White in $H$, he does better by switching from $s$ to $s'$.

A game theory book would therefore fail in its duty if it were content to recommend *any* Nash equilibrium of Chess as its solution. The book should offer more refined advice. The conservative candidates for such a refinement are the strategy pairs $(s, t)$ selected by backward induction. Such a strategy pair isn't only a Nash equilibrium in the whole game, it also induces Nash equilibrium play in every subgame $H$—whether or not $H$ is reached in equilbrium.

Following Reinhard Selten, a pair of strategies with this property is called a *subgame-perfect equilibrium*. A Nash equilibrium can fail to be subgame perfect only if it is certain that some subgame won't be reached when the equilibrium strategies are used, but this often happens.

→ 2.10

### 2.9.4 Exploiting Bad Play?

We will use subgame-perfect equilibria a great deal, and so it is important to ask when it is safe to recommend a subgame-perfect equilibrium as the solution of a game. Section 2.9.1 reminds us that orthodox game theory assumes that we begin

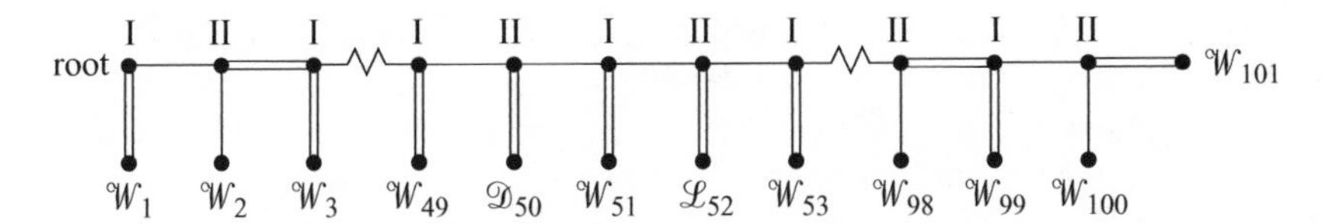


Figure 2.17 A Chesslike game.

playing a game with strong evidence that all the players are rational. But what if one of the players contradicts this evidence by playing badly?

Consider the example of Figure 2.17, which is like chess to the extent that players I and II move alternately, and the labels $\mathcal{W}$, $\mathcal{L}$, or $\mathcal{D}$ refer to a win, draw, or loss for player I. However, unlike chess, the players are assumed to care about how long the game lasts. Player I's preferences are given by

$$\mathcal{W}_1 \succ_1 \mathcal{W}_2 \succ_1 \cdots \succ_1 \mathcal{W}_{101} \succ_1 \mathcal{D}_{50} \succ_1 \mathcal{L}_{52}.$$

Player II is assumed to hold opposing preferences. This makes the game strictly competitive. The doubled branches in Figure 2.17 show the result of applying backward induction.

Since only one branch is doubled at each node, there is only one subgame-perfect equilibrium. This calls on player II to play *down* at node 50. Is this good advice? The answer depends on what she knows about player I. The advice is sound if she is so sure that he is rational that no evidence to the contrary will change her mind. A rational player I would certainly play *down* if he found himself at node 51 because this results in an immediate victory for him. Hence player II had better not let node 51 be reached. She should settle instead for a draw by playing *down* at node 50.

However, node 50 wouldn't have been reached if player I hadn't played *across* on twenty-five consecutive occasions when it was rational to play *down*. This fact isn't consistent with player II's original belief that player I is rational. However, she may reason that even Nobel prize winners sometimes make mistakes. If so, then she can attribute player I's behavior in always playing *across* to twenty-five independent random errors.

At each move, she can argue, player I *intended* to play *down*, but fate intervened by distracting his attention or jogging his elbow, so that he ended up playing *across*. She will assign only a small probability $p$ to his making each such blunder, and so the probability $p^{25}$ of his making twenty-five independent mistakes will be almost infinitesimal.[14] But it remains logically coherent for her to put her faith in this extremely unlikely eventuality, rather than give up believing that her opponent is highly likely to play rationally in the future.

Of course, in real life, nobody seeking to explain the behavior of an opponent in chess who has just made twenty-five consecutive bad moves would think it plausible that he really meant to make a good move each time but somehow always contrived to moved the wrong piece by mistake. The natural conclusion to draw from

[14]With less than one chance in ten of making one mistake, there is less than one chance in one billion billion billion of making twenty-five such mistakes.

observing bad play is that the opponent is a weak player. The question then arises as to how to take advantage of his weakness.[15]

In the game of Figure 2.17, player I's weakness seems to be a fixation on always playing *across*. If player II thinks this explanation of his behavior is likely on finding herself at node 50, she may care to chance playing *across* herself. The risk is that player I may deviate from his previous pattern of behavior by playing *down* at node 51. If so, then player II has passed up the chance for a draw to no avail. However, if player I continues to play *across* at node 51, then she can win at node 52 by playing *down*.

The moral is that subgame-perfect equilibria are fully defensible only in certain games. In short games, there won't be enough time for sufficient evidence to accumulate to reverse the players' initial belief that everyone is rational. In games with enough chance moves and information sets, the leading explanation for play having reached unanticipated subgames will usually be the vagaries of chance, rather than stupid play by other players.

However, even in long games of perfect information, subgame-perfect equilibria may still be useful. Section 14.4 explains how such games can be modified by introducing chance moves and information sets into the rules of the game, so as to model the systematic irrationalities of their opponents that the players would otherwise use to explain arriving at unanticipated subgames. We thereby *construct* a game in which it is sensible to study subgame-perfect equilibria.

When critics attack the idea of a subgame-perfect equilibrium, the appropriate response for a game theorist is therefore similar to what was said in Section 1.4.1 when responding to the criticism that game theorists assume that people are selfish. Such critics would usually do better to stop attacking the methodology of game theory and start criticizing the relevance of the particular game being studied to the real-world problem that it supposedly models.

## 2.10 Roundup

This chapter has looked at strictly competitive games of perfect information with no chance moves. These games have been studied without appealing to utility theory by expressing the players' preferences directly in terms of the possible outcomes of the game. Chess and Tic-Tac-Toe are examples.

A strictly competitive game has two players whose preferences over the possible outcomes of the game are diametrically opposed. The simplest kind of strictly competitive game is a win-or-lose game. In such games, there must be a winner and a loser, and both players prefer winning to losing. Examples of win-or-lose games about which we had something to say are Nim and Hex.

To write down the rules of a game in a precise form, it is necessary to begin by asking the questions *who*, *what*, *when*, and *how much?* The answers are recorded with the help of a game tree. Chance moves arise when the answer to the question *who* is that the relevant decision is made by rolling dice or using some other randomizing device. Shuffling and dealing in poker is a good example of chance move.

[15]It may sometimes be risky to do so because your opponent could be a hustler setting you up for a sting. But no possible advantage can accrue to player I here from playing *across* twenty-five times in a row when he can win immediately on each occasion just by playing *down*.

Once a game tree has been constructed, further vital questions need to be asked. We need to be told what the players *know* and when they know it. Information sets are used to record the answers. A game tree with its associated information sets is called the extensive form of a game. It tells us everything available about the rules of the game.

To include a number of decision nodes in the same information set is to specify that a player doesn't know which of the nodes within that information set the game has reached when he or she decides what action to take next. The game of Matching Pennies provides an example. When Eve guesses *heads* or *tails,* she doesn't know whether Adam previously hid a *head* or a *tail*. Her two decision nodes therefore belong in the same information set.

Matching Pennies is an example of a game of imperfect information because it has an information set that contains more than one decision node. In such games, a player isn't informed about some aspects of the past history of the game that might be useful when making a move. In games of perfect information like chess, all the past history of the game is always an open book. Every information set is therefore a singleton, containing exactly one decision node. When a decision node in a game tree isn't enclosed in an information set, the implication is that the information set hasn't been drawn because it is a singleton. Game trees drawn with no information sets at all should therefore be assumed to be games of perfect information.

A pure strategy specifies an action at each of a player's information sets in the extensive form of a game. Once the players have chosen their pure strategies, the outcome of a game without chance moves is then completely determined. The strategic form of a game is a table that records the outcome corresponding to each possible profile of pure strategies the players might choose. A Nash equilibrium is a strategy profile in which each player's choice of strategy is a best reply to the strategies chosen by the other players. In order to qualify as a candidate for the solution of a game, a strategy profile must be a Nash equilibrium.

In a game of imperfect information like Matching Pennies or the Inspection Game, it sometimes makes sense to delegate your choice of action to a randomizing device. A player who does so is said to be using a mixed strategy. A player who makes a deterministic choice is then said to be using a pure strategy. This chapter avoids saying much about probability by not allowing chance moves and restricting attention to games of perfect information for which mixed strategies are not needed.

Strictly competitive games of perfect information can be solved by backward induction. You take subgames whose solution is known and replace them in the game tree by new leaves labeled with the solution outcome of the subgame. Starting with the smallest subgames and reducing larger and larger subgames, you eventually end up with a game that has only one node, which is labeled with the solution outcome of the game with which you started.

A subgame-perfect equilibrium is a strategy profile that isn't only a Nash equilibrium in the whole game but also calls for a Nash equilibrium to be played in every subgame—whether or not the subgame is reached when everybody plays their equilibrium strategies. Not all Nash equilibria are subgame perfect. Nash equilibria that aren't subgame perfect involve at least one strategy that calls for suboptimal play in a subgame that lies off the equilibrium path. The strategy therefore passes the best-reply test in the game as a whole but fails the best-reply test in some unreached subgame. Backward induction necessarily generates subgame-perfect equilibria.

Backward induction is unproblematic in win-or-lose games. The only time it fails to find a winning strategy for you is when you have no possibility of winning at all against a rational opponent. In strictly competitive games like chess that have more than two possible outcomes, backward induction will find the value of the game, together with a pure strategy whose play guarantees that the outcome will be no worse for you than the game's value. The guarantee applies whether or not your opponent plays rationally. If your opponent is rational, then you can get no more than the value of the game because backward induction will also find a pure strategy that guarantees an outcome for her that is no worse than the game's value. You will then both be playing a subgame-perfect equilibrium that generates the value of the game.

However, opponents are not always rational. Sometimes they can be very stupid indeed. It is therefore not necessarily a good idea to use your backward induction strategy because it sacrifices any chance you might have of exploiting any systematic mistakes you might observe your opponent making. But remember that it is risky to deviate from the backward induction strategy because the world is full of hustlers who pretend to be stupid precisely in order to make money off of those who try to exploit them.

## 2.11 Further Reading

*Lectures on Game Theory*, by Robert Aumann: Westview Press (Underground Classics in Economics), Boulder, CO, 1989. These are the classroom notes of one of the great game theorists.

*Winning Ways for your Mathematical Plays*, by Elwyn Berlekamp, John Conway, and Richard Guy: Academic Press, New York, 1982. This is a witty and incredibly inventive book, which is largely about solving complicated games by backward induction.

*Mathematical Diversions* and *Hexaflexagons*, by Martin Gardner: University of Chicago Press, Chicago, 1966 and 1988. The books gather together many delightful games and brainteasers from the author's long-standing column in *Scientific American*.

*The Game of Hex and the Brouwer Fixed-Point Theorem*, by David Gale: *American Mathematical Monthly* 86 (1979), 818–827. Who would have thought that the fact that Hex can't end in a draw is equivalent to the Brouwer fixed-point theorem?

## 2.12 Exercises

1. Figure 2.18 shows the tree of a strictly competitive game $G$ of perfect information without chance moves.
   a. How many pure strategies does each player have?
   b. List each player's pure strategies using the notation of Section 2.5.
   c. What play results from the use of the pure strategy pair $(rll, LM)$?
   d. Find all pure strategy pairs that result in the play $[rRl]$.
   e. Write down the strategic form of $G$.
   f. Find all the saddle points.
2. Two players alternate in placing dominoes on an $m \times n$ chess board so as to cover two squares exactly. The first to be unable to place a domino is the loser. Draw the game tree for the case $m = 2$ and $n = 3$.
3. Figure 2.19 is a skeleton for the tree of a game called Blackball. A committee of three club members (I, II, and III) has to select one from a list of four candidates (A, B, C, and D) as a new member of the club. Each committee

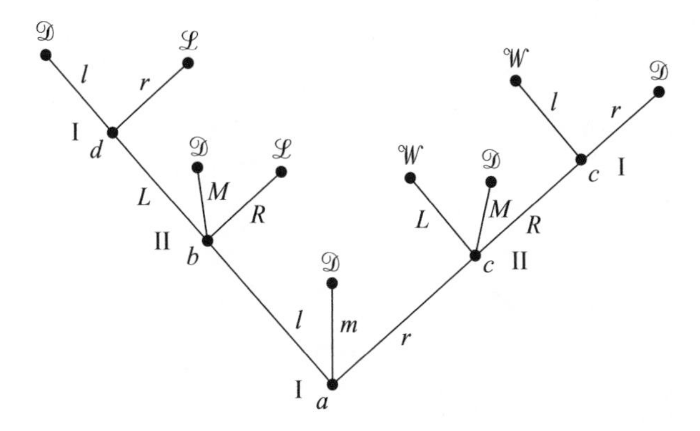


Figure 2.18 The game for Exercise 2.12.1.

member is allowed to blackball (veto) one candidate. This right is exercised in rotation, beginning with player I and ending with player III. Why is Blackball not a strictly competitive game?

Label each decision node on a copy of Figure 2.19 with the numeral of the player who decides at that node. The branches representing choices at the node should be labeled with the candidates who have yet to be blackballed. Each leaf should be labeled with the letter of the candidate elected to the club if the game ends there. How many pure strategies does each player have? What information hasn't been supplied that is necessary to analyze the game?

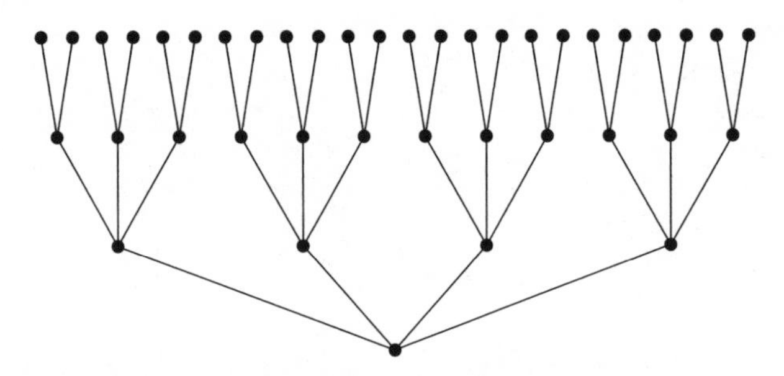

Figure 2.19 A skeleton for the tree of Blackball.

4. Begin to draw the game tree for chess. Include at least one complete play of the game in your diagram.
5. Two players alternate in choosing either 0 or 1 forever. A play of this *infinite* game can therefore be identified with a sequence of 0s and 1s. For example, the play 101000 ... began with player I choosing 1. Then player II chose 0, after which player I chose 1 again. Thereafter both players always chose 0. A sequence of 0s and 1s can be interpreted as the binary expansion of a real number $x$ satisfying $0 \leq x \leq 1$.[16] For a given set of $E$ of real numbers, player I wins if $x \in E$ but loses if $x \in \sim E$. Begin to draw the game tree.

math

[16]For example, $\frac{5}{8} = .101000\ldots$ because $\frac{5}{8} = 1(\frac{1}{2})+0(\frac{1}{2})^2+1(\frac{1}{2})^3+\cdots$.

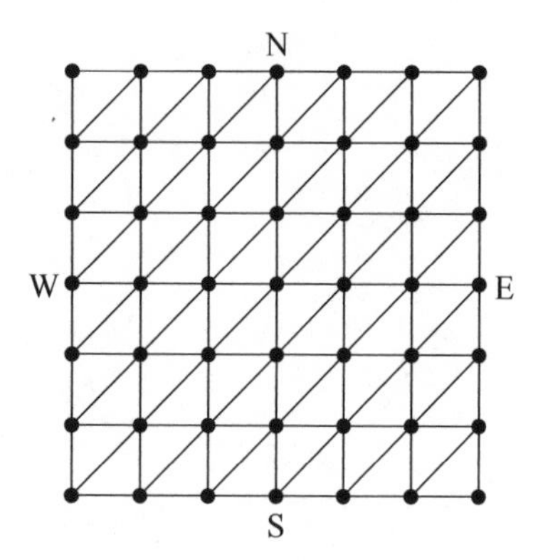


Figure 2.20 A city street plan.

6. Apply backward induction to the game $G$ of Exercise 2.12.1. What is the value of $G$? What is the value of the subgame starting at node $b$? What is the value of the subgame starting at node $c$? Show that the pure strategy $rrr$ guarantees that player I gets the value of $G$ or better. Why is this pure strategy not selected by backward induction?

fun

7. Apply backward induction to the $2 \times 3$ version of the domino-placing game of Exercise 2.12.2. Find the value of the game, and determine a winning strategy for one of the players.

fun

8. Who would win a game of Nim with $n \geq 2$ piles of matchsticks of which the $k$th pile contains $2^{k-1}$ matchsticks?[17] Describe a play of the game in which $n = 3$, and the winner plays optimally while the loser always takes one matchstick from a pile with the median number of matchsticks. (The median pile is the middle-sized pile.) Do the same for $2^n - 1$ piles, of which the $k$th pile contains $k$ matchsticks.

fun

9. Who wins in the domino-placing game of Exercise 2.12.2 when (a) $m$ and $n$ are even; (b) $m$ is even and $n$ is odd; (c) $m = n = 3$?

fun

10. What are the winning opening moves in $3 \times 3$, $4 \times 4$, and $5 \times 5$ Hex?

fun

11. If the first player has to link the more distant sides of an $n \times (n+1)$ Hex board, show that the second player has a winning strategy.[18]

phil

12. Explain why the strategy-stealing argument of Section 2.7.2 doesn't imply that the first player can win after playing anywhere at his first move. Beck's Hex is the same as ordinary Hex, except that it begins with a circle in an acute corner of the board, and Cross moves first. Confirm that Cross has a winning strategy

math

13. The game board of Figure 2.20 represents the downtown street plan of a city. Players I and II represent groups of gangsters. Player I controls the areas to the

[17]Try this with particular values of $n$ to begin with. For example, $n = 3$.

[18]Mathematicians at Princeton apparently used to amuse themselves by inviting visitors to play this game as Circle with a computer playing Cross. The board was shown on the screen in perspective to disguise its asymmetry, and so the visitors thought they were playing regular Hex, but to their frustration and dismay, somehow the computer always won!

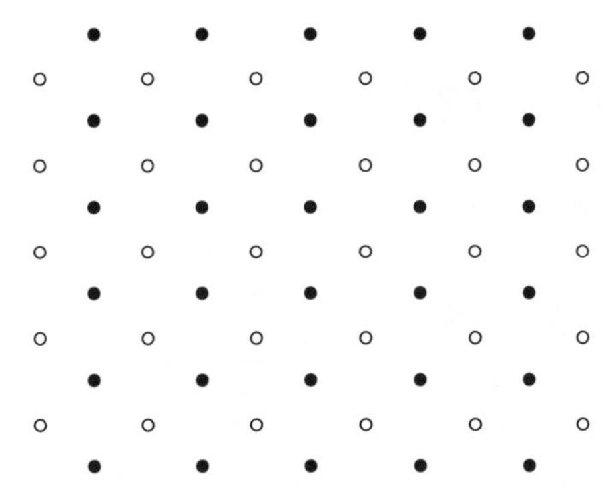

Figure 2.21 The board for Bridgit.

north and south of the city. Player II controls the areas to the east and west. The nodes in the street plan represent street intersections. The players take turns labeling nodes that haven't already been labeled. Player I uses a circle as his label. Player II uses a cross. A player who manages to label both ends of a street controls the street. Player I wins if he links the north and south with a route that he controls. Player II wins if she links the east and west. Why is this game entirely equivalent to Hex?

14. The game of Bridgit was invented by David Gale. It is played on a board like that shown in Figure 2.21. Black tries to link top and bottom by joining neighboring black nodes horizontally or vertically. White tries to link left and right by joining neighboring white nodes horizontally or vertically. Neither player is allowed to cross a linkage made by the other.

math

   a. Find an argument like that used for Hex which shows that the game can't end in a draw.
   b. Why does it follow that someone can force a win?
   c. Why is it the first player who has a winning strategy?
   d. What is a winning strategy?

15. Two players alternately remove nodes from a connected graph $\mathscr{G}$. Except in the case of the first move, a player may remove a node only if it is joined by an edge to the node removed by the previous player. The player left with no legitimate vertex to remove loses. Explain why the second player has a winning strategy if there exists a set $E$ of edges with no endpoint in common such that each node is the endpoint of an edge in the set $E$. Show that no such set $E$ exists for the graph of Figure 2.22. Find a winning strategy for the first player.

math

16. A strategy-stealing argument shows that if the second player to move in Tic-Tac-Toe has a winning strategy, then so does the first player. Why does it follow that the second player can't have a winning strategy? In Hex, one can deduce that the first player has a winning strategy, but the second player can guarantee a draw in Tic-Tac-Toe. How does she guarantee a draw after the first player occupies the middle square? What is the value of Tic-Tac-Toe?
17. The value of chess is unknown. It may be $\mathscr{W}$, $\mathscr{D}$, or $\mathscr{L}$. Explain why a simple strategy-stealing argument can't be used to eliminate the possibility that the value of chess is $\mathscr{L}$.

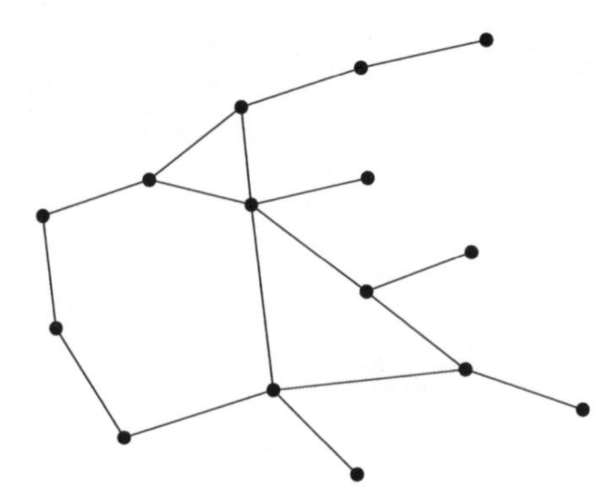

Figure 2.22 A graph $\mathcal{G}$ for Exercise 2.12.15.

math

18. Explain why player I has a winning strategy in the number construction game of Exercise 2.12.5 when $E = \{x : x > \frac{1}{2}\}$. What is player I's winning strategy when $E = \{x : x \geq \frac{2}{3}\}$? What is player II's winning strategy when $E = \{x : x > \frac{2}{3}\}$? Explain why player II has a winning strategy when $E$ is the set of all rational numbers.[19] (A rational number is the same thing as a fraction.)
19. Let $(s,t)$ and $(s',t')$ be two different saddle points for a strictly competitive game. Prove that $(s,t')$ and $(s',t)$ are also saddle points.
20. Find all Nash equilibria in the game $G$ of Exercise 2.12.1. Which of these are subgame perfect?
21. Find the subgame-perfect equilibria for Blackball of Exercise 2.12.3 in the case when the players' preferences satisfy $A \succ_1 B \succ_1 C \succ_1 D$; $B \succ_2 C \succ_2 D \succ_2 A$; $C \succ_3 D \succ_3 A \succ_3 B$. Who gets elected to the club if a subgame-perfect equilibrium is used? Find at least one Nash equilibrium that isn't subgame perfect.
22. In the Inspection Game of Section 2.2.1, each player can choose today or tomorrow on which to act. Write down an outcome table for a five-day version of the Inspection Game in which each player can act on Monday, Tuesday, Wednesday, Thursday, or Friday. If the firm uses the mixed strategy in which each of its five pure strategies is used with equal probability, then it will win four times out of five, no matter what strategy the agency chooses. If the agency uses the same mixed strategy, show that it will win one time out of five, no matter what strategy the firm may use. Why is this pair of mixed strategies a Nash equilibrium?

phil

23. Nothing in the surprise test paradox of Section 2.3.1 hinges on the school week having five days, and so we simplify the story by supposing that only today and tomorrow are available. As in Section 2.2, *today* is denoted by $t$ and *tomorrow* by $T$. Explain why Figure 2.23 models the resulting situation as a game between Adam and Eve. (Pay close attention to the role of the information sets.) Solve the game by using backward induction. In doing so, assume that Eve will

[19] One may ask whether this infinite game always has a value whatever the set $E$ may be. The answer is abstruse. If one assumes a set-theoretic principle called the Axiom of Choice, then there are sets $E$ for which the game has no value. However, but some mathematicians have proposed replacing the Axiom of Choice with an axiom that would imply that the game has a value for every set $E$.

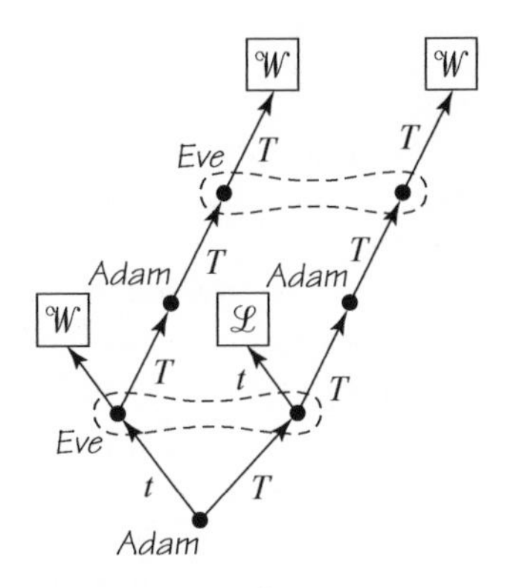


Figure 2.23 The two-day surprise test.

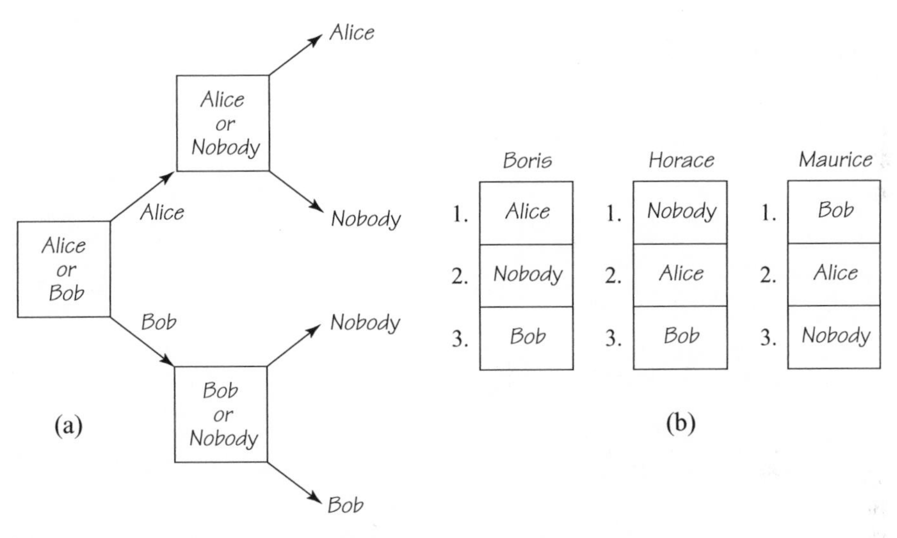


Figure 2.24 Strategic voting.

choose whatever action leaves open the possibility that she might win at her lower information set.[20]

Observe that backward induction selects a pure strategy for Adam in which he will predict that the test will be tomorrow when tomorrow comes, even though he might already have wrongly predicted that the test will be today.

24. Find the strategic form of the game of Figure 2.23. What result is obtained by deleting weakly dominated strategies?

phil

25. In 1961, the philosopher Quine pointed out one of the logical tricks of the surprise test paradox by considering the one-day case. What was the trick he thereby exposed? Make up a similar paradox in which the evil Dr. X promises your worst possible outcome unless you act irrationally.

phil

[20]When doubling branches, remember that Eve has no choice but to select the *same* action at each decision node in the same information set because she can't tell the difference between such decision nodes.

fun

26. The rhyming triplets, Boris, Horace, and Maurice, are the membership committee of the very exclusive Dead Poets Society. The final item on their agenda one morning is a proposal that Alice should be admitted as a new member. No mention is made of another possible candidate called Bob, so an amendment to the final item is proposed. The amendment says that Alice's name should be replaced by Bob's. The rules for voting in committees call for amendments to be voted on in the reverse order to which they are proposed. The committee therefore begins by voting on whether Bob should replace Alice. If Alice wins, they then vote on whether Alice or Nobody should be made a new member. If Bob wins, they then vote on whether Bob or Nobody should be made a new member. Figure 2.24(a) is a diagrammatic representation of the order in which the voting takes place. Figure 2.24(b) shows how the three committee members rank the three possible outcomes.

    Who will win the vote if everybody just votes according to their rankings? Why should Horace switch to voting for the candidate he likes least at the first vote? What happens if everybody votes strategically?

# 3 Taking Chances

## 3.1 Chance Moves

This chapter introduces chance moves into our scheme for writing down the rules of a game. This is no big deal in itself. We simply invent a mythical player called Chance, who randomizes among the actions at her decision nodes. The difficulty lies in modeling the response of rational players to the risks they face in games with chance moves. This problem is postponed until the next chapter by confining attention to win-or-lose games, in which a rational player simply maximizes the *probability* of winning.

### 3.1.1 Monty Hall Problem

This example derives from an old quiz show run by Monty Hall. His role is taken over here by the Mad Hatter to remind us that we are only looking at a toy version of the problem. He asks Alice to choose among three boxes. Two are empty, and the other contains a prize. Alice doesn't know which contains the prize, but the Mad Hatter does.

Alice chooses *Box 2*. To generate some excitement, the Mad Hatter then opens one of the other boxes. When this box turns out to be empty, he invites Alice to change her mind about her choice of box. What should she do?

People usually say it doesn't matter whether Alice changes her mind. The probability of getting the prize was one-third when she chose *Box 2* because there was then an equal chance of the prize being in any of the three boxes. After one of the other boxes is shown to be empty, the probability that *Box 2* contains the prize

Figure 3.1 Which box? Alice chooses *Box 2.* The Mad Hatter then reveals that *Box 3* is empty. Should Alice now switch to *Box 1?*

goes up to one-half because there is now an equal chance that the prize is in one of the two unopened boxes. If she switches boxes, her probability of winning will therefore still be one-half. So why bother changing?

This popular argument is wrong. It would be correct if the Mad Hatter opened boxes at random and just happened not to open a box containing the prize. But he *deliberately* opened an empty box. This strategic behavior conveys information to Alice. If she makes proper use of the information, she will always switch boxes. To see why, it is a good idea to represent Alice's problem of whether to switch boxes as a game tree with a chance move. In Figure 3.2, she is player I.

The root of the game tree is a chance move, represented by a square rather than a circle. The three branches leading away from the root represent the three choices Chance can make. At this opening move, Chance can choose to put the prize in *Box 1, Box 2,* or *Box 3.* Each possibility occurs with probability $\frac{1}{3}$. If the Mad Hatter didn't intervene, Alice's choice of *Box 2* would therefore win the prize with probability $\frac{1}{3}$.

The Mad Hatter is player II. He isn't allowed to open *Box 2.* Nor is he allowed to open one of the other boxes if it contains the prize. He therefore has room for maneuver only if the prize is in *Box 2.*

Alice moves next as player I. She knows which box has been opened but not which of the remaining boxes contains the prize. Her knowledge at this stage is represented by two information sets, one in which she knows that *Box 1* is empty, and one in which she knows that *Box 3* is empty.

The doubled lines in Figure 3.2 show the actions Alice takes at each of her decision nodes if she always switches boxes. To find her overall probability of winning with this strategy, return to the original chance move. The play of the game that starts with Chance putting the prize in *Box 1* ends with the outcome $\mathscr{W}$. So does the play that starts with Chance putting the prize in *Box 3.* So the switching strategy ensures that Alice wins the prize two-thirds of the time. The other third of the time she loses because both plays that start with Chance putting the prize in *Box 2* end with the outcome $\mathscr{L}$. On the other hand, if she sticks with *Box 2,* she will win only one-third of the time.

A cleverer way to see that Alice wins with probability $\frac{2}{3}$ by switching is to note that this is the probability that Alice would *lose* if the Mad Hatter didn't intervene at all. It is therefore also the probability she will *win* if she switches after learning which of the other boxes is empty. But you don't need to be clever if you let Von Neuman's formalism do most of the thinking for you.

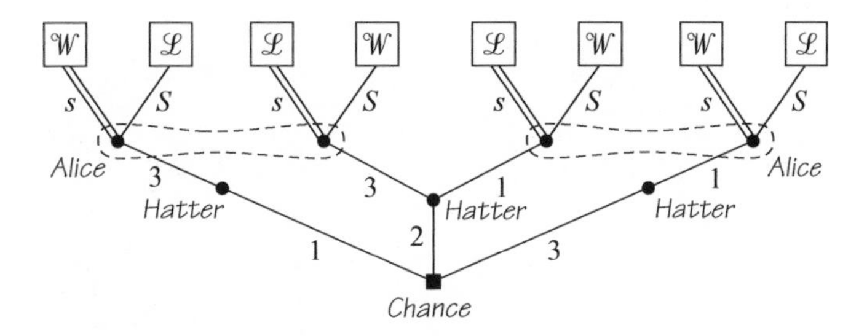


Figure 3.2 The Monty Hall Game. The chance move is shown as a square. Alice's switching choice is denoted by $s$, and her staying choice by $S$. Her optimal choice of switching is indicated by doubling the appropriate branches.

## 3.2 Probability

→ 3.3

When dice are rolled, statisticians say that the set

$$\Omega = \{1, 2, 3, 4, 5, 6\}$$

of all possible outcomes is a *sample space*. Decision theorists call $\Omega$ the *world* within which their decision problems arise. The numbers 1, 2, 3, 4, 5, or 6 are then said to be the possible *states* of the world. The *events* that can result from rolling the dice are identified with the subsets of $\Omega$. Thus the event that the dice shows an even number is the set $E = \{2, 4, 6\}$.

A *probability measure* is a function defined on the set $S$ of all possible events.[1] The number prob($E$) is said to be the probability of the event $E$.

To qualify as a probability measure, the function $\text{prob}: S \to [0, 1]$ must satisfy three properties. The first property is that prob $(\emptyset) = 0$. Since $\emptyset$ is the set with no elements, this means that the probability of the impossible event that nothing at all will happen is zero. The second property is that prob $(\Omega) = 1$, which means that the probability of the certain event that something will happen is 1.

The third property says that the probability that one or the other of two events will occur is equal to the sum of their separate probabilities—provided that the two events can't both occur simultaneously. The set $E \cap F$ represents the event that both events $E$ and $F$ occur at the same time. So $E \cap F = \emptyset$ means that $E$ and $F$ can't occur simultaneously, as in Figure 3.3(b). The set $E \cup F$ represents the event that at least one of $E$ or $F$ occurs. So the third property can be expressed formally by writing

$$E \cap F = \emptyset \quad \Rightarrow \quad \text{prob}(E \cup F) = \text{prob}(E) + \text{prob}(F).$$

A fair die is equally likely to show any of its faces when rolled, and so $\text{prob}(1) = \text{prob}(2) = \cdots = \text{prob}(6) = \frac{1}{6}$. The probability of the event $E = \{2, 4, 6\}$ that an even number will appear is therefore

[1]A function $f: A \to B$ is a rule that assigns a unique $b \in B$ to each $a \in A$. The object $b$ assigned to $a$ is denoted by $f(a)$. It is said to be the value of the function at the point $a$. The notation $[a, b]$ represents the set $\{x : a \leq x \leq b\}$ of real numbers. The function $\text{prob}: S \to [0, 1]$ therefore assigns a unique real number $x = \text{prob}(E)$ satisfying $0 \leq x \leq 1$ to each event $E \in S$.

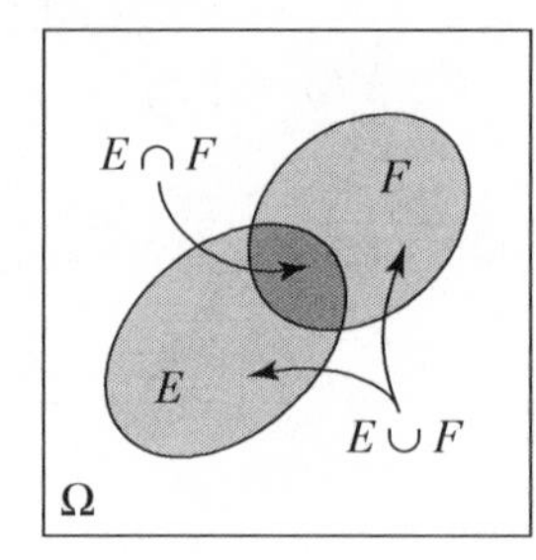


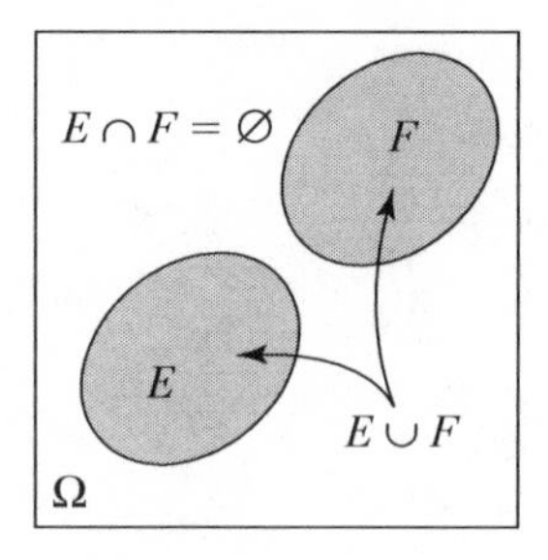


Figure 3.3 Venn diagrams of $E \cup F$.

$$\text{prob}(E) = \text{prob}(2) + \text{prob}(4) + \text{prob}(6) = \tfrac{1}{6} + \tfrac{1}{6} + \tfrac{1}{6} = \tfrac{1}{2}.$$

The proper interpretation of probabilities is a subject endlessly debated by philosophers. For the purposes of game theory, it is usually enough to say that a statement like $\text{prob}(\{4\}) = \frac{1}{6}$ means that there is one chance in six of 4 being rolled.

Gamblers express the fact that $\text{prob}(\{4\}) = \frac{1}{6}$ by saying that the odds are $5:1$ against rolling a 4. If the odds against an event occurring are $a:b$, then the probability that the event will occur is $b/(a+b)$.

For each dollar that you bet on a horse at odds of $5:1$ against its winning, you get back five dollars if the horse wins (plus the dollar you bet). Of course, bookies wouldn't cover their costs in the long run if they quoted the *true* odds against horses winning. They therefore shade the odds in their favor. You might find a bookie who offers odds of $4:1$ against rolling a 4 with a fair die, but hell will freeze over before you are offered odds of $6:1$!

### 3.2.1 Independent Events

If $A$ and $B$ are sets, then $A \times B$ is the set of all pairs $(a, b)$ with $a \in A$ and $b \in B$.[2] Figure 3.4(a) shows the sample space $\Omega^2 = \Omega \times \Omega$ obtained when two *independent* rolls of the dice are observed. In this diagram, (6, 1) represents the event that 6 is rolled with the first dice, and 1 with the second. This isn't the same event as $(1,6)$, which means that 1 is rolled with the first dice, and 6 with the second. The event $E \times F$ has been shaded. It is the event that 3 or more is thrown with the first dice, and 3 or less with the second dice.

There are $36 = 6 \times 6$ possible outcomes in the square representing $\Omega \times \Omega$. If the two dice are rolled independently, each outcome is equally likely. The probability of each is therefore $\frac{1}{36}$. So the probability of $E \times F$ must be

$$\text{prob}(E \times F) = \tfrac{12}{36} = \tfrac{1}{3}.$$

Notice that $\text{prob}(E) = \frac{2}{3}$ and $\text{prob}(F) = \frac{1}{2}$. Thus,

$$\text{prob}(E \times F) = \text{prob}(E) \times \text{prob}(F).$$

[2]In this context, the notation $(a, b)$ means the pair of real numbers $a$ and $b$, with $a$ taken first. If the order of the numbers were irrelevant, one would simply use the notation $\{a, b\}$ for the set containing $a$ and $b$.

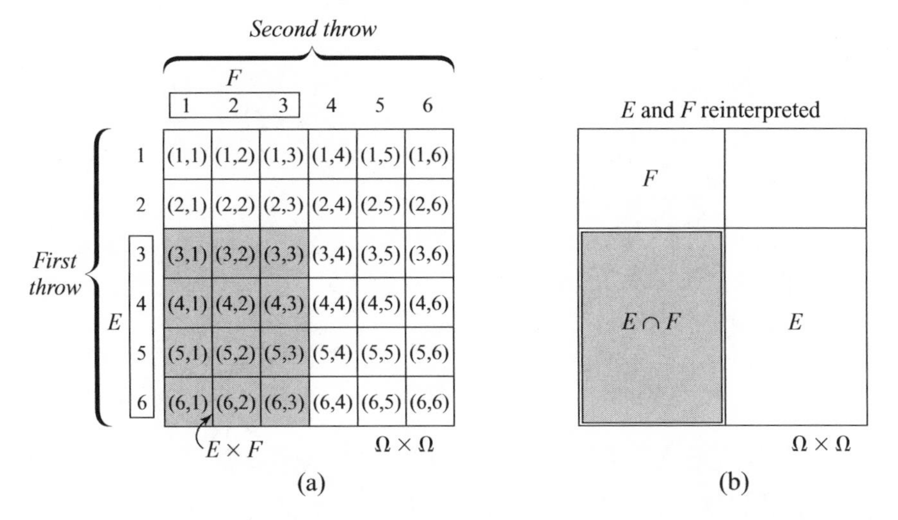


Figure 3.4 The sample space $\Omega \times \Omega$ for two independent rolls of a die.

This equation holds whenever $E$ and $F$ are independent events. The conclusion is usually expressed as

$$\text{prob}(E \cap F) = \text{prob}(E)\,\text{prob}(F),$$

which says that the probability that two independent events will both occur is the *product* of their separate probabilities.

Strictly speaking, writing prob $(E \cap F) =$ prob $(E)$ prob$(F)$ requires reinterpreting $E$ and $F$ as events in $\Omega \times \Omega$ as indicated in Figure 3.4(b). In this diagram, $E$ is no longer the subset of $\Omega$ that represents the event that the first die will show 3, 4, 5, or 6. It is instead the subset of $\Omega \times \Omega$ corresponding to the event in which the first dice shows 3, 4, 5, or 6, and the second die shows anything whatever. Similarly $F$ becomes the subset of $\Omega \times \Omega$ corresponding to the event that the first die shows anything whatever, and the second die shows 1, 2, or 3.

### 3.2.2 Paying Off a Loan Shark

To avoid getting his legs broken, Bob needs to come up with \$1,000 tomorrow to pay off a loan shark. With the \$2 remaining in his wallet, he therefore buys two lottery tickets for \$1 each in two independent lotteries. The winner in each lottery gets a prize of \$1,000 (and there are no second prizes). If the probability of winning in each lottery is $q = 0.0001$, what is the probability that Bob will still be walking around next week?

Let $\mathscr{W}_1$ and $\mathscr{L}_1$ be the events that Bob wins or loses the first lottery. Let $\mathscr{W}_2$ and $\mathscr{L}_2$ be the events that he wins or loses the second lottery. Then prob$(\mathscr{W}_1) =$ prob$(\mathscr{W}_2) = q$, and prob$(\mathscr{L}_1) =$ prob$(\mathscr{L}_2) = 1 - q$.

We need prob$(\mathscr{W}_1 \cup \mathscr{W}_2)$. This isn't prob$(\mathscr{W}_1)+$prob$(\mathscr{W}_2)$ because $\mathscr{W}_1$ and $\mathscr{W}_2$ can occur simultaneously. However, none of the events $\mathscr{W}_1 \cap \mathscr{W}_2$, $\mathscr{W}_1 \cap \mathscr{L}_2$, or $\mathscr{L}_1 \cap \mathscr{W}_2$ can occur simultaneously, and so

$$\text{prob}(\mathscr{W}_1 \cup \mathscr{W}_2) = \text{prob}(\mathscr{W}_1 \cap \mathscr{W}_2) + \text{prob}(\mathscr{W}_1 \cap \mathscr{L}_2) + \text{prob}(\mathscr{L}_1 \cap \mathscr{W}_2).$$

Multiplying the probabilities of the independent events on the right, we find that $\text{prob}(\mathscr{W}_1 \cup \mathscr{W}_2) = q^2 + q(1-q) + (1-q)q = 0.00019998$. So Bob's ambulatory prospects aren't very good. He has less than two chances in ten thousand of coming up with the money.

It is often easier in such problems to work out the probability that the event in question *won't* happen. This is the event $\mathscr{L}_1 \cap \mathscr{L}_2$ that Bob loses both lotteries. We then get the same answer more simply as

$$1 - \text{prob}(\mathscr{L}_1 \cap \mathscr{L}_2) = 1 - (1-q)^2 = 0.00019998.$$

## 3.3 Conditional Probability

After an investigation into a major plane crash proved inconclusive, the *New York Times* carried a sequence of letters about the chances of a meteor strike. The first argued that the probability of a meteor striking an aircraft may be small, but it isn't negligible.[3] The second made fun of the first, arguing that what matters is the incredibly smaller probability that a meteor would strike at the particular time and place of the crash. The third pointed out that the previous letters should have estimated *conditional* probabilities. What really matters is the probability of a meteor strike at the time and place of the crash—conditional on the crash having taken place without any other identifiable cause.

After you observe that an event $F$ has happened, your knowledge base changes. The only states of the world that are now possible lie in the set $F$. You must therefore replace $\Omega$ by $F$, which is the new world in which your future decision problems will be set. The new probability $\text{prob}(E \mid F)$ you assign to an event $E$ after learning that $F$ has occurred is called the *conditional probability* of $E$ given $F$.

For example, we know that $\text{prob}(4) = \frac{1}{6}$ when a fair die is rolled. If we learn that the outcome was even, this probability must be adjusted. The event $F = \{2, 4, 6\}$ that the outcome is even contains three equally likely states. The probability of rolling a 4, given that $F$ has occurred, is therefore $\frac{1}{3}$. Thus,

$$\text{prob}(4 \mid F) = \tfrac{1}{3}.$$

The principle on which this calculation is based is embodied in the formula

$$\text{prob}(E \mid F) = \text{prob}(E \cap F)/\text{prob}(F).$$

### 3.3.1 Peeking in Poker

While playing poker with Bob, Alice hears a bystander whisper that he has a red queen in his hand. Would it make any difference to her estimate of the chances of his

[3] The letter included estimates of the rate at which meteors reach the ground and the proportion of the Earth's surface area taken up by aircraft in flight.

holding a second queen if the bystander had identified the red queen as the queen of hearts? To answer this question, we need to compare prob $(E \mid F)$ and prob $(E \mid G)$, where $E$ is the event that Bob holds two queens, $F$ is the event that he holds the queen of hearts, and $G$ is the event that he holds a red queen.

To simplify the problem, suppose that Alice and Bob are playing poker with a six-card deck, two of which are dealt to each player. The cards that aren't dealt to Alice are ♠A, ♡Q, ♢Q, and ♣8. Alice begins by conditioning on this event and deduces that Bob is equally likely to be holding any of the hands shown in Figure 3.5.

There are six hands in which Bob is holding ♡Q. In two of these, Bob is holding two queens. So prob$(E|F) = \frac{1}{3}$. Similarly, prob$(E|G) = \frac{1}{5}$, because there are two chances in ten that $E$ will occur, given that Bob is only known to be holding a red queen.

As in the Monty Hall problem, even mathematically sophisticated people often get this wrong. They don't see why it should matter whether the red queen is the queen of hearts or not. The lesson is that big brains aren't always an asset. Instead of thinking clever thoughts, it is sometimes better simply to enumerate all the possibilities. If it is a work of great labor to do so, one can always begin with a toy version of the problem, as we did here.

→ 3.4

### 3.3.2 Knowledge and Belief

If you are playing a game, your decision-theoretic world is the set of all possible plays of the game. As the game proceeds, you will usually learn more and more about which play of the game will actually be realized. Von Neumann ingeniously modeled this learning process using information sets. On reaching an information set $F$, you now know that the realized play of the game must pass through one of the decision nodes in $F$.

Game theorists distinguish what you *know* as a result of reaching an information set $F$ from what you *believe* after reaching $F$. Your knowledge is determined by the rules of the game. Your beliefs are determined by your attempts to quantify the uncertainty created by the gaps in your knowledge.

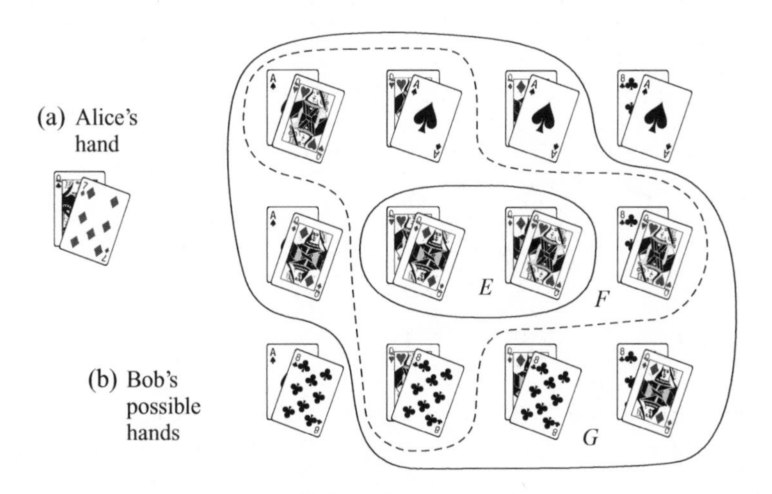


Figure 3.5 Peeking in Poker.

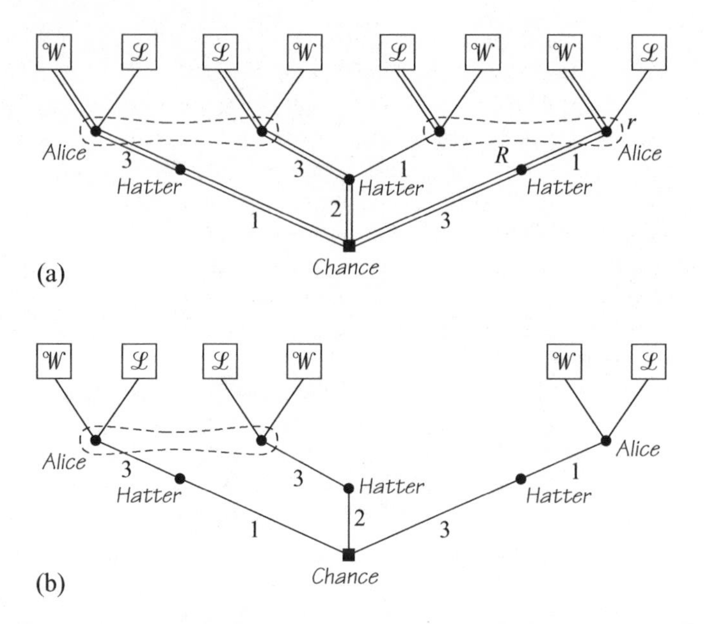


Figure 3.6 The Monty Hall Game again. Figure 3.6(a) shows the three equally likely plays of the game that Alice thinks are possible, if she *believes* that the Mad Hatter never opens *Box 3* when the prize is in *Box 2*. Figure 3.6(b) shows how the rules of the game would need to be altered if Alice *knew* this fact.

The Monty Hall Game, which is shown again in Figure 3.6(a), will serve as an example. Suppose that Alice believes that the Mad Hatter will never open *Box 3* when the prize is in *Box 2*. If she always switches boxes, Alice therefore thinks that only the plays of the game shown with doubled branches in Figure 3.6(a) are possible before the game begins. Since each play is equally likely, she starts by attaching probability $\text{prob}(l) = \frac{1}{3}$ to the event that the realized play will pass through the left decision node $l$ in her left information set $L$.

If the Mad Hatter opens *Box 3*, Alice now *knows* that one of the two plays of the game passing through a decision node in her left information set $L$ has occurred. She therefore replaces the probability $\text{prob}\,(l) = \frac{1}{3}$ by $\text{prob}\,(l \mid L) = 1$ because she now believes that the other play that passes through $L$ is impossible.

Figure 3.6(b) shows a game whose rules say that Alice *knows* that the Mad Hatter never chooses *Box 3* when the prize is in *Box 2*. This game obviously won't do as a vehicle for analyzing the Monty Hall problem because we wouldn't need to write a game down at all if we were so sure beforehand of what Alice believes about the Mad Hatter that we could reclassify her beliefs as knowledge.

### 3.3.3 Updating in the Monty Hall Game

If Alice believes that the Mad Hatter never opens *Box 3* when the prize is in *Box 2*, then she updates her probability of being at $l$ in Figure 3.6(a) to $\text{prob}\,(l \mid L) = 1$ after finding herself at the information set $L$. But what is the value of $\text{prob}\,(l \mid L)$ if the Mad Hatter uses a mixed strategy in which he opens *Box 1* with probability $1 - p$ and *Box 3* with probability $p$?

We need to find prob($E \mid F$) = prob($E \cap F$)/prob($F$) when $E = \{l\}$ and $F = L = \{l, r\}$. Things simplify in this case because $\{l\}$ is a subset of $L$, and so $E \cap F = E$. Thus,

$$\text{prob}(l \mid L) = \frac{\text{prob}(l)}{\text{prob}(l) + \text{prob}(r)} = \frac{\frac{1}{3}}{\frac{1}{3} + \frac{1}{3}p} = \frac{1}{1+p}.$$

To see that prob $(r) = p \times \frac{1}{3}$, we appeal again to the formula prob($E \cap F$) = prob ($E \mid F$)prob($F$), but now $F$ is the event that the prize is in *Box 2,* and $E$ is the event that the Mad Hatter opens *Box 3.*

Notice that it isn't true that Alice will win with probability $\frac{2}{3}$ in Figure 3.1 by switching boxes. This is her probability of winning *before* the Mad Hatter opens a box. Without any information about the Mad Hatter's strategy, all we can say about her probability of winning *after* the Mad Hatter opens a box is that it lies somewhere between $\frac{1}{2}$ and 1.

## 3.4 Lotteries

I never buy lottery tickets because I prefer to not to gamble when the odds are heavily stacked against me. But everybody understands how lotteries work. It therefore makes sense to use the analogy of a lottery when talking about what you might win or lose as a result of a chance move.

For example, a bookie may offer you odds of 3 : 4 against an even number being rolled with a fair die. If you take the bet, you win \$3 if an even number appears and lose \$4 if an odd number appears. Accepting this bet is equivalent to choosing the lottery **L** shown in Figure 3.7(a). The top row shows the possible final outcomes or *prizes*, and the bottom row shows the respective probabilities with which each prize is awarded.

The lottery **M** of Figure 3.7(b) has three prizes. You have five chances in every twelve of winning the big prize of \$24.

### 3.4.1 Random Variables

Mathematicians talk about random variables rather than lotteries. I remember being mystified by random variables when I first studied statistics, but a kindly mathematics professor finally put me straight by explaining that a random variable is simply a function $X : \Omega \to \mathbb{R}$.[4]

For example, the lottery of Figure 3.7(a) is equivalent to the random variable $X : \Omega \to \mathbb{R}$ defined by

$$X(\omega) = \begin{cases} 3, \text{ if } \omega = 2, 4, \text{ or } 6 \\ -4, \text{ if } \omega = 1, 3, \text{ or } 5. \end{cases}$$

In this case, the relevant sample space is $\Omega = \{1, 2, 3, 4, 5, 6\}$.

[4]The set of real numbers is denoted by $\mathbb{R}$, so $X(\omega)$ is a real number.

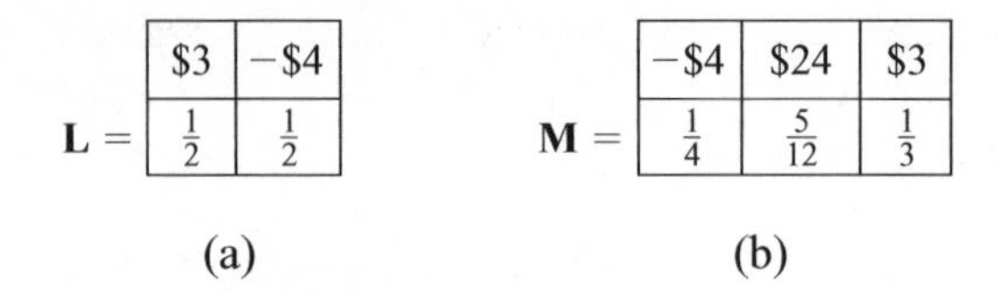


Figure 3.7 Two lotteries.

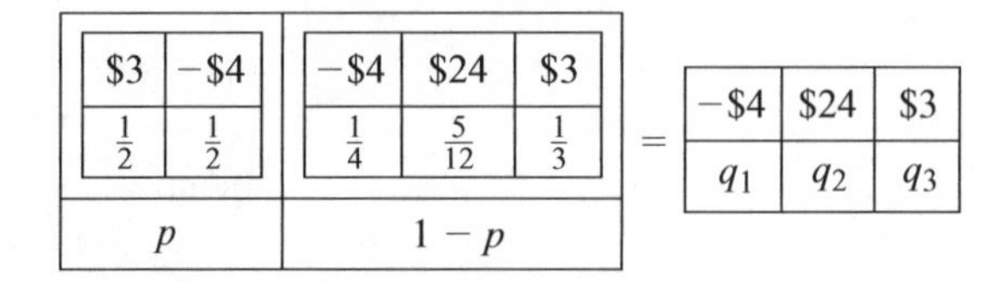


Figure 3.8 The compound lottery $p\mathbf{L}+(1-p)\mathbf{M}$.

If you take the bet represented by the random variable $X$, your probability of winning \$3 is $\text{prob}(X=3)=\text{prob}(\{2,4,6\})=\frac{1}{2}$. Your probability of losing \$4 is $\text{prob}(X=-4)=\text{prob}(\{1,3,5\})=\frac{1}{2}$.

### 3.4.2 Compound Lotteries

One of the prizes in a raffle at an Irish county fair is sometimes a ticket for the Irish National Sweepstake. If you buy a raffle ticket, you are then participating in a *compound* lottery, in which the prizes may themselves be lotteries. It is important to remember that we always assume that all the lotteries involved in a compound lottery are *independent* of each other.

Figure 3.8 illustrates the compound lottery $p\mathbf{L}+(1-\text{p})\mathbf{M}$. The notation means that you get the lottery **L** with probability $p$ and the lottery **M** with probability $1-p$.

A compound lottery can always be reduced to a simple lottery by computing the total probability with which you get each prize. In the case of Figure 3.8:

$$q_1 = p\times\tfrac{1}{2}+(1-p)\times\tfrac{1}{4}=\tfrac{1}{4}-\tfrac{1}{4}p;$$
$$q_2 = (1-p)\times\tfrac{5}{12}=\tfrac{5}{12}-\tfrac{5}{12}p;$$
$$q_3 = p\times\tfrac{1}{2}+(1-p)\times\tfrac{1}{3}=\tfrac{1}{3}+\tfrac{1}{6}p.$$

To find $q_3$, begin by noting that the probability of winning the prize **L** in the compound lottery is $p$. The probability of winning \$3 in the lottery **L** is $\frac{1}{2}$. These events are independent, and so the probability of the event $E$ that they both occur is $p\times\frac{1}{2}$. Similarly, the event $F$ that **M** is won in the compound lottery and that \$3 is won in the lottery **M** has probability $(1-p)\times\frac{1}{3}$. Since $E$ and $F$ can't both happen, the event $E\cup F$ that you win \$3 has probability $q_3=\text{prob}(E)+\text{prob}(F)=p\times\frac{1}{2}+(1-p)\times\frac{1}{3}$.

→ 3.6

## 3.5 Expectation

The expectation or expected value $\mathscr{E}X$ of a random variable $X$ is defined by

$$\mathscr{E}X=\sum k\,\text{prob}(X=k),$$

where the summation extends over all values of $k$ for which prob$(X = k)$ isn't zero. If many independent observations of the value of $X$ are taken, the law of large numbers[5] says that the probability that their long-run average will differ significantly from $\mathcal{E}X$ is small.

Your expected dollar winnings in the lottery **L** of Figure 3.7 are

$$\begin{aligned}\mathcal{E}\mathbf{L} &= \sum k \operatorname{prob}(X = k) \\ &= 3 \times \tfrac{1}{2} + (-4) \times \tfrac{1}{2} = -\tfrac{1}{2}.\end{aligned}$$

If you bet over and over again on the roll of a fair die, winning \$3 when the outcome is even and losing \$4 when the outcome is odd, you are therefore likely to lose an average of about 50¢ per bet in the long run. The expected dollar value of the lottery **M** of Figure 3.7 is

$$\mathcal{E}\mathbf{M} = (-4) \times \tfrac{1}{4} + 24 \times \tfrac{5}{12} + 3 \times \tfrac{1}{3} = 10.$$

If you repeatedly paid \$3 for a ticket in this lottery, you would be likely to win an average of about \$7 per trial in the long run.

### 3.5.1 The Monte Carlo Fallacy

The relation between the expected value of a random variable and its long-run average is frequently misunderstood. Figure 3.9 illustrates the relationship for the case of a fair coin. The expected number of heads in a single throw is $\frac{1}{2}$. If we tossed the coin independently many times, we would be surprised if we didn't see heads appear approximately half the time.

Figure 3.9 shows the $2^7 = 128$ equally likely outcomes that can result when the coin is tossed seven times. The event $F$ consists of all outcomes in which 2, 3, 4, or 5 heads are thrown. Since we are concerned with the average number of heads thrown, observe that $F$ is the event in which this average differs from $\frac{1}{2}$ by less than $\frac{7}{32}$.

There are 112 outcomes in $F$, and so prob$(F) = 112/128 = \frac{7}{8}$, confirming that the average number of heads approximates its expected value of $\frac{1}{2}$ with high probability. Many more throws would be necessary to get a probability of 0.9 that the average is within 0.1 of $\frac{1}{2}$. Even more throws would be needed to get a probability of 0.99 that the average is within 0.01 of $\frac{1}{2}$.

Gamblers in Monte Carlo or Las Vegas commonly attribute the law of large numbers to some mystical influence that acts to keep the average close to $\frac{1}{2}$. When they notice that a large number of heads have been thrown, they fallaciously reason that it is more likely that a tail will be thrown next time.

It is easy to pinpoint the mistake in the Monte Carlo fallacy. Suppose that six heads are thrown with a fair coin. This is the event $E$ in Figure 3.9. What is the probability that the next coin will be a tail? Since each toss of the coin is independent

[5] This is the weak law of large numbers. The strong law says that the *limit* of the average number of heads as the total number of observations becomes infinite is equal to the expected value with probability one.

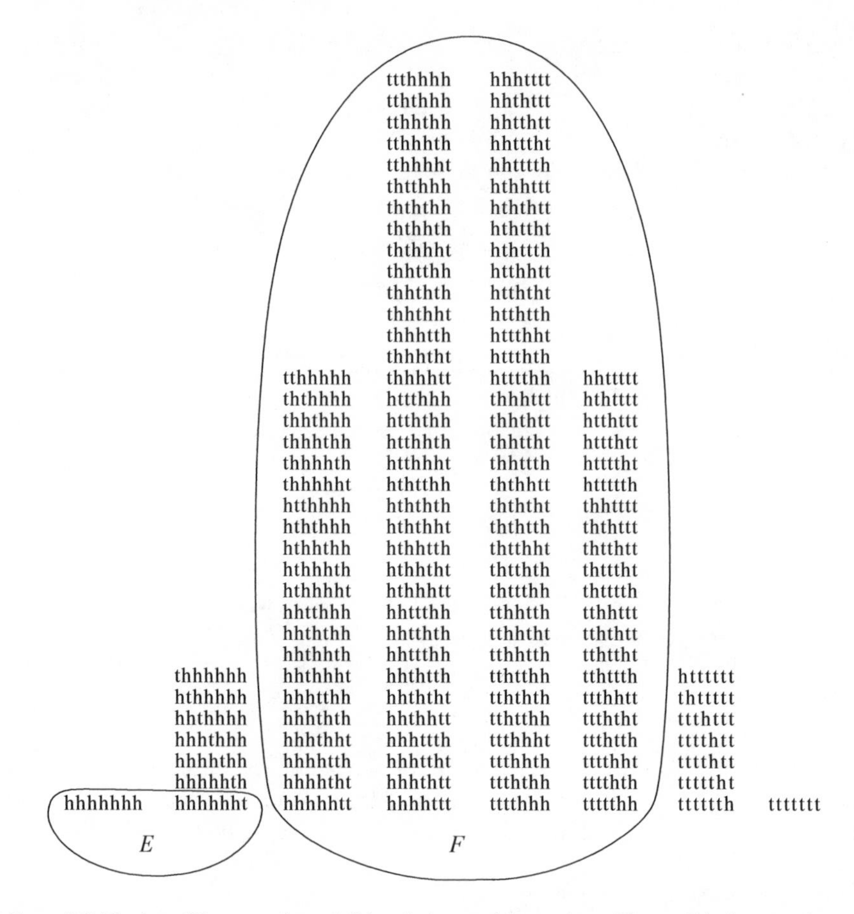


Figure 3.9 The law of large numbers. A fair coin is tossed seven times. The set $F$ is the event in which the average number of heads thrown differs from $\frac{1}{2}$ by less than $\frac{7}{32}$. The set $E$ is the event that the first six tosses are heads.

of the others, we know in advance that the answer must be $\frac{1}{2}$, no matter how many heads may have already been thrown.

Alternatively, we can use Figure 3.9 to verify that prob($hhhhhht\,|E$) $= \frac{1}{2}$. It then becomes obvious that the law of large numbers has nothing to do with the question because $E$ lies outside the set $F$, within which the average number of heads is close to $\frac{1}{2}$.

### 3.5.2 Martingales

→ 3.6

A martingale was originally the betting system in which you double your stake after every loss. When a novice who had fallen for his charms entrusted her family diamonds to his care, Casanova thought he was going to make himself rich by playing this system in a Venetian gambling den. Like many others through the centuries, he underestimated the chances of hitting a long streak of bad luck. If Casanova had been trained in modern mathematics rather than the amatory arts, he would have known that

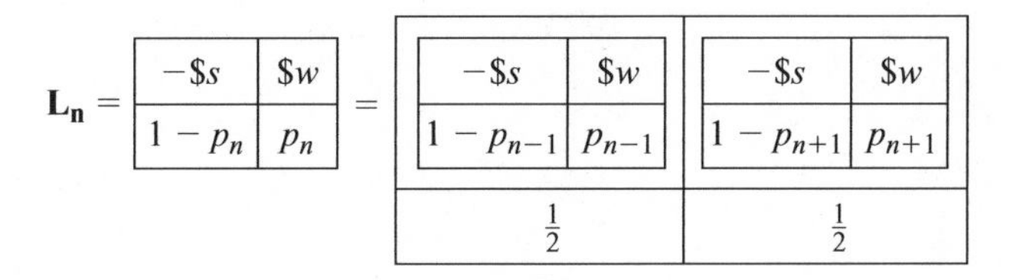


Figure 3.10 A betting system. A gambler repeatedly bets \$1 on a fair coin until he wins \$$w$ or loses his original stake of \$$s$. If he reaches a stage when his current holdings are \$$n$, then he is facing the lottery $\mathbf{L_n}$.

no betting system can beat a casino's odds. Nowadays, we use the word *martingale* in a way that illustrates this sad fact.

Suppose, for example, that Bob uses a system when betting repeatedly on the fall of a fair coin. His wealth then varies over time according to how the coin falls. In mathematical terms, it is a sequence of random variables. Whatever Bob's system may be, this sequence is a martingale in the modern sense because, no matter what he may have won or lost up to now, his expected loss or gain on the next toss of the coin is always a big round zero.

When the idle rich return from Las Vegas boasting about paying for their vacation by using a clever roulette system, they are just fooling themselves. Even if roulette were fair, all they would have done is to trade a high probability of winning a small amount for a low probability of losing a large amount.

To see how this works, we study the most popular betting system of all. You enter a casino with a stake of \$$s$ and plan to bet \$1 repeatedly that heads will be thrown with a fair coin until you have either won \$$w$ or lost your stake of \$$s$. What is your probability of success?

If you currently have \$$n$ at some time, you are facing a lottery $\mathbf{L_n}$ in which your probability of eventually being successful and winning \$$w$ is $p_n$ and your probability of eventually failing and losing \$$s$ is $1-p_n$. To find $p_n$, first notice that $\mathbf{L_n}$ is the compound lottery of Figure 3.10. Because you have half a chance of winning or losing a dollar at the next toss of the coin,

$$p_n = \tfrac{1}{2}p_{n-1} + \tfrac{1}{2}p_{n+1}.$$

Solutions to this difference equation have the form $p_n = An + B$, where $A$ and $B$ are constants.[6] To determine $A$ and $B$, use the fact that you will fail for sure when your stake is lost and succeed for sure if you hit your target amount. Thus $p_0 = 0$ and $p_{s+w} = 1$. It follows that $A = 1/(s+w)$ and $B = 0$. Your probability of success when your stake is \$$s$ is therefore

$$p_s = \frac{s}{s+w}.$$

If the stake you are willing to risk is large compared with your target winnings, you have a high probability of being successful. However, you don't thereby beat the

[6]Substitute $p_n = An + B$ into the difference equation and see whether it works. Or try starting with $p_0$ and $p_1$ and seeing what $p_2$, $p_3$, and so on have to be.

odds. To see this, it is only necessary to compute your expected winnings when you start with a stake of \$$s$:

$$\mathscr{E}\mathbf{L_s} = -s\left(\frac{w}{s+w}\right) + w\left(\frac{s}{s+w}\right) = 0.$$

Whatever betting system we used, this result would have been the same. It follows that casinos wouldn't make any money on average if their games were fair. Most of their games are therefore unfair. For example, you get odds of 35 : 1 against any particular number coming up at roulette, but there are 37 equally likely numbers (including zero). Blackjack used to be an exception, provided you were willing to delay playing until most of the cards remaining in the dealing shoe were favorable. But the management regarded such strategic play as cheating and would throw you out of the casino or worse if they caught you at it! Nowadays shuffling machines have put paid to even this small opportunity to beat the dealer.

Like Bob in Section 3.2.2, you sometimes have no alternative but to bet when the odds are unfair. The law of large numbers is then your enemy. Fooling around with betting systems does you no good at all. Instead of dividing your stake among different bets, you do best to go for the sudden-death option of betting your entire stake on a single trial.

## 3.6 Values of Games with Chance Moves

Every strictly competitive game of perfect information without chance moves has a value $v$ (Corollary 2.1). That is, player I has a pure strategy $s$ that guarantees him an outcome that is at least as good for him as $v$, while player II has a pure strategy $t$ that guarantees her an outcome that is at least as good for her as $v$.

For games with chance moves, neither player will usually be able to guarantee doing at least as well as some pure outcome $v$ every time that the game is played. If you are unlucky, you may lose no matter how cleverly you play. Even the best poker players reckon to lose one session in three.

We therefore have to cease thinking about what can be achieved for certain. A pure strategy pair only determines a *lottery* over the pure outcomes. Instead of asking what pure outcomes can be achieved for certain, we need to ask what lotteries can be achieved for certain. The value of a strictly competitive game with chance moves will therefore normally be a lottery.

Matters are simplified in the current chapter by confining our attention to win-or-lose games. A lottery then takes the form

$$\mathbf{p} = \begin{array}{|c|c|} \hline \mathscr{W} & \mathscr{L} \\ \hline p & 1-p \\ \hline \end{array}$$

A useful trick is to use the boldface notation $\mathbf{p}$ for the lottery in which $\mathscr{W}$ occurs with probability $p$ and $\mathscr{L}$ occurs with probability $1-p$. For example, Figure 3.11 illustrates the fact that the compound lottery $p\,\mathbf{q} + (1-p)\mathbf{r}$ is equivalent to the simple lottery $\mathbf{pq} + (\mathbf{1} - \mathbf{p})\,\mathbf{r}$.

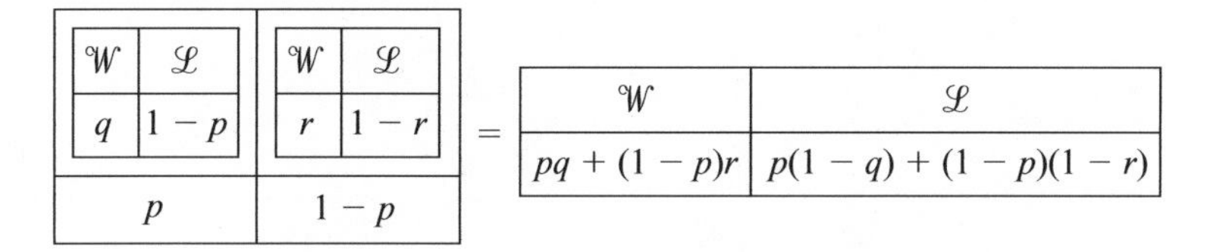

Figure 3.11 The identity $p\mathbf{q} + (1-p)\mathbf{r} = \mathbf{pq} + (\mathbf{1}-\mathbf{p})\mathbf{r}$.

In win-or-lose games, a rational player will seek to maximize the probability of winning. Player I's preferences can then be described by saying that he likes the lottery $\mathbf{p}$ at least as much as the lottery $\mathbf{q}$ if and only if $p \geq q$. The lottery $\mathbf{p}$ assigns player II a probability of $1-p$ of winning. She therefore likes the lottery $\mathbf{p}$ at least as much as the lottery $\mathbf{q}$ if and only if $p \leq q$. A win-or-lose game is therefore necessarily strictly competitive even if it has chance moves. That is to say,

$$\mathbf{p} \preceq_1 \mathbf{q} \quad \Leftrightarrow \quad \mathbf{p} \succeq_2 \mathbf{q}.$$

The argument of Theorem 2.1 can now be recycled to show that we don't need to exclude chance moves when claiming that all win-or-lose games of perfect information have a value. When we have to write down the value of a subgame $H$ whose root is a chance move, we first identify all the smaller subgames that Chance might choose at the root. The value of $H$ is then simply the lottery that yields the values of these smaller subgames with the probabilities with which Chance chooses them.

### 3.6.1 Monty Hall's Value

The Monty Hall problem provides an example in which it is easy to work out the value of a win-or-lose game with a chance move.

The Mad Hatter didn't get equal billing with Alice in Section 3.1.1, but he is a player, too. In accordance with the instructions from the studio that prevent his opening *Box 2* or a box containing the prize, we assume that his aim is to minimize Alice's probability of winning.

We use $s$ to mean that Alice switches from *Box 2* and $S$ to mean that she stays with *Box 2*. Alice has two information sets in Figure 3.2. At her left information set she knows that *Box 3* is empty. At her right information set, she knows that *Box 1* is empty. At each information set she must choose between the actions $s$ and $S$. (Remember that she can't choose different actions at different decision nodes in the same information set because she doesn't know which decision node in the information set has been reached when she chooses an action.)

Alice's four pure strategies are denoted by $ss$, $sS$, $Ss$, and $SS$. For example, $sS$ means that Alice switches to *Box 1* if she is shown that *Box 3* is empty and stays with *Box 2* if she is shown that *Box 1* is empty. The Mad Hatter has only two pure strategies, which we label 1 and 3. Strategy 1 is to open *Box 1* if the prize is in *Box 2*. Strategy 3 is to open *Box 3* if the prize is in *Box 2*. If the prize is in *Box 1* or *Box 3*, he isn't free to choose at all.

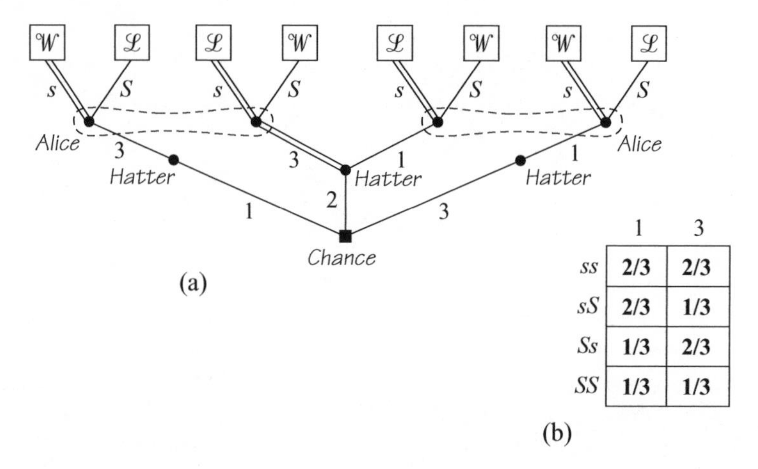


| | 1 | 3 |
|---|---|---|
| *ss* | **2/3** | **2/3** |
| *sS* | **2/3** | **1/3** |
| *Ss* | **1/3** | **2/3** |
| *SS* | **1/3** | **1/3** |

Figure 3.12 The strategic form of the Monty Hall Game is shown in Figure 3.12(b). Both of the cells in the top row correspond to saddle points. The value of the game is therefore **2/3**. Figure 3.12(a) is drawn as an aid in calculating the outcome **1/3**, which occurs when the strategy pair $(sS, 3)$ is used.

Figure 3.12(b) shows the strategic form of the Monty Hall Game. The argument given in Section 3.1.1 shows that the entries in the first and fourth rows of the outcome table must be the lotteries **2/3** and **1/3** respectively. The same mode of reasoning also allows us to fill in the other entries in the table. For example, the pure strategy pair $(sS, 3)$ is indicated in Figure 3.12(a) by doubling appropriate branches. To see that the outcome that results from the use of this strategy pair is **1/3**, one needs only to follow the play that will result from each of the three choices Chance can make at the opening move. Two of these lead to $\mathscr{L}$ and the other to $\mathscr{W}$. When $(sS, 3)$ is played, Alice therefore wins the prize with probability $\frac{1}{3}$.

Recall from Section 2.8.2 that a Nash equilibrium of a strictly competitive game occurs at a saddle point of the outcome table. To find the pure-strategy Nash equilibria of a strictly competitive game, one therefore looks for the entries in the outcome table that are best in their column and worst in their row (from player I's point of view). At a saddle point in a strictly competitive game, each player will then be making a best reply to the other.

Figure 3.12(b) shows that the Monty Hall Game has two saddle points, $(ss, 1)$ and $(ss, 3)$. The entry in the outcome table at each saddle point is **2/3**, and so this is the value of the game. If Alice and the Mad Hatter play optimally, Alice therefore wins the prize with probability $\frac{2}{3}$.

Alice's optimal strategy *ss* requires that she always switch from *Box 2* to whichever box hasn't been opened. As both his pure strategies are optimal, the Mad Hatter has a less exacting task. In fact, he needn't do any thinking at all since all of his mixed strategies are optimal as well.[7]

[7]In Section 3.3.3, we let the Mad Hatter play pure strategy 3 with probability $p$. This mixed strategy is optimal for him because he still gets the outcome **2/3** when Alice plays *ss*.

## 3.7 Waiting Games

The contestants in bicycle races sometimes behave very strategically. They start by maneuvering very slowly for position until someone suddenly breaks away in an attempt to create a decisive advantage. The waiting games of this section have a similar character. There is a waiting phase, followed by a sudden all-or-nothing winning bid by one of the players.

### 3.7.1 Product Races

→ 3.7.2

Two firms sometimes race to be the first to get their product on the market. How long should a firm develop its product before going for broke and seeing whether its current product is good enough to grab the market? Races in which two firms try to be the first to get a new idea into a patentable form have a similar structure.

Here is a toy model of a product race between Alice and Bob. If Alice gets her product on the market first, it will be successful with probability $p_1$. If so, she will then have such a hold on the market that Bob's product won't be able to get off the ground at all when marketed later. On the other hand, if Alice's product fails when first marketed, nobody will want to buy her later attempts to improve the product. Bob can therefore take as long as he needs to come up with a product that is sure to be successful. So Bob wins with probability $1 - p_1$ when Alice gets her product on the market first.

If Bob gets his product on the market first, he wins with probability $p_2$, and Alice wins with probability $1 - p_2$. We don't need to assume much about what happens if both players market their products simultaneously, except that one will then win and the other lose.

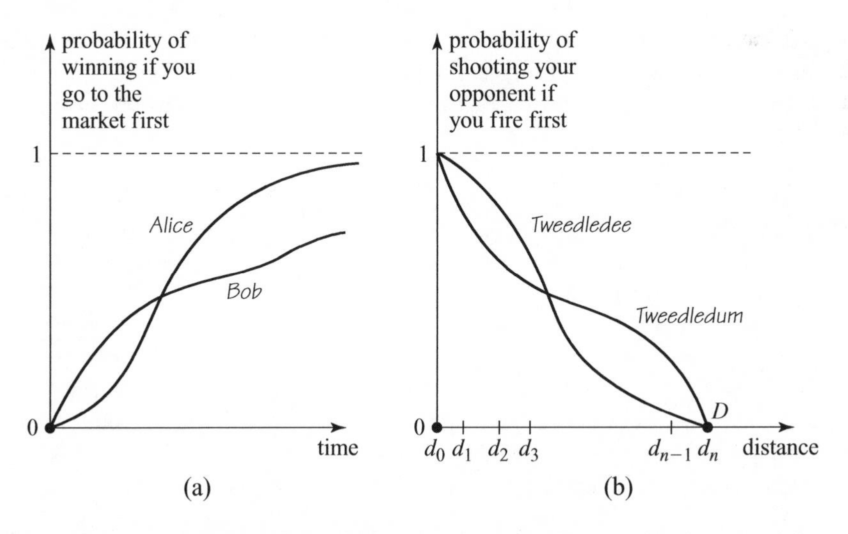


Figure 3.13 Success probabilities: Figure 3.13(a) shows the probability of a player's product being successful if it is first on the market at time $t$. Figure 3.13(b) shows the probability that a player in Duel will hit the other if he fires first when the players are $d$ apart.

A player's probability of winning when first on the market goes up with time. We require that $p_1$ and $p_2$ be continuous and strictly increasing functions of time.[8] As shown in Figure 3.13(a), we also require that both functions start out at zero and eventually approach one.

We assume that Alice and Bob have already sunk the costs of developing their products and that whoever wins the market will be able to exploit it for such a long time that any losses caused by a delay in winning the market are negligible. Alice and Bob are then playing a win-or-lose game in which each seeks to maximize the probability of winning. How should they play?

If the players can monitor each other's progress, so that we are talking about a game of perfect information with many chance moves, the solution isn't hard to find. Rational play requires that Alice and Bob put their products on the market simultaneously as soon as

$$p_1 + p_2 = 1.$$

Several steps are needed to explain why:

**Step 1.** The solution can't say that one player should move before the other. Alice wouldn't follow any advice to move in advance of Bob, because she can always risklessly raise her probability of winning by cutting her lead time by a little. So both players must put their products on the market simultaneously.

**Step 2.** If Alice and Bob put their products on the market simultaneously when their probabilities of winning would be $p_1$ and $p_2$ if they moved first, then Alice will win with some probability $q_1$. We can't have $p_1 > q_1$ since Alice's probability of winning by going first would decrease but still be larger than $q_1$ if she moved a tiny bit sooner than Bob. Thus $p_1 \leq q_1$. Since $p_2 \leq q_2$ for similar reasons, we have that $p_1 + p_2 \leq q_1 + q_2 = 1$.

**Step 3.** We also can't have $1 - p_2 > q_1$ because Alice's probability of winning by going second would remain $1 - p_2$ if she moved later than Bob. Thus $1 - p_2 \leq q_1$. Similarly, $1 - p_1 \leq q_2$, and so $2 - p_1 - p_2 \leq q_1 + q_2 = 1$. It follows that $p_1 + p_2 \geq 1$.

**Step 4.** Since $p_1 + p_2 \leq 1$ and $p_1 + p_2 \geq 1$, it follows that $p_1 + p_2 = 1$.

This argument isn't a proof because it takes too much for granted. But it is solid enough to explain what is going on in the more careful arguments possible in particular cases like the game of Duel, which follows.

### 3.7.2 Duel

Tweedledum and Tweedledee have agreed to fight a duel. Armed with dueling pistols loaded with just one bullet, they walk toward each other. The probability of either hitting the other increases the nearer the two approach. How close should

[8]A real-valued function $f$ is continuous on an interval if its graph can be drawn without lifting the pen from the paper. Actually $p_1$ and $p_2$ can be the realizations of a stochastic process, provided they are continuous and strictly increasing with probability one. Exercise 3.11.24 looks at a case in which $p_1$ and $p_2$ increase in discrete jumps at random times.

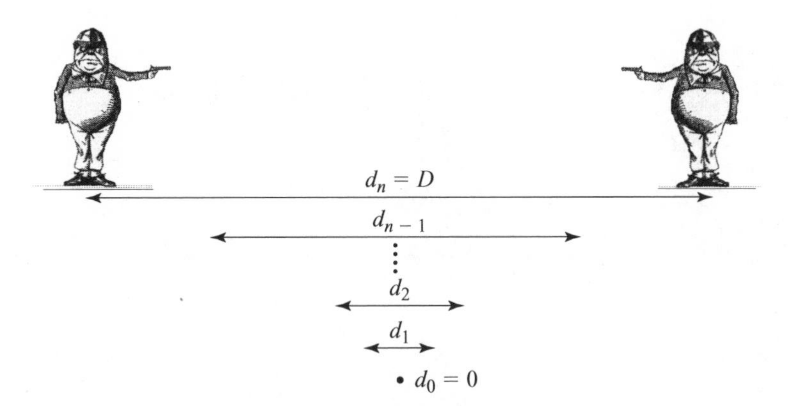


Figure 3.14 Dueling with pistols.

Tweedledum get to Tweedledee before firing? This is literally a question of life and death because, if he fires and misses, Tweedledee will be able to advance to point-blank range with fatal consequences for Tweedledum.

One way of modeling the problem is shown in Figure 3.14. The initial distance between the players is $D$. Points $d_0, d_1, \ldots, d_n$ have then been chosen with $0 = d_0 < d_1 < \cdots < d_n = D$ to serve as decision nodes in the finite game of Figure 3.15(a). We assume that the distance between each pair of neighboring points is very small with a view to taking the limit as $n \to \infty$ at the end of the analysis.

In Figure 3.15(a), Tweedledum is player I and Tweedledee is player II. Thus $\mathscr{W}$ means that Tweedledum lives and Tweedledee dies. Similarly, $\mathscr{L}$ means that Tweedledee lives and Tweedledum dies.

The square nodes are chance moves. At these nodes, Chance determines whether a player will hit or miss his opponent after firing his pistol. Figure 3.13(b) shows the probability $p_i(d)$ that player $i$ will hit his target when he fires from distance $d$. We assume that $p_i$ is continuous and strictly decreasing on $[0, D]$, with $p_i(0) = 1$ and $p_i(D) = 0$.[9] Differences in the hitting probabilities between the two players reflect their differing skills with a dueling pistol.

*Solving the game.* All finite win-or-lose games of perfect information have a value **v**. Since **v** is a lottery in this case, player I has a strategy $s$ that guarantees his survival with probability $v$ or more. Player II has a strategy $t$ that guarantees his survival with probability $1 - v$ or more. We use backward induction to determine these optimal strategies.

**Step 1.** First look at the smallest subgames in Figure 3.15(a). These are all no-player games rooted at a chance move reached after someone fires his pistol. If player I survives in such a subgame with probability $p$, then the value of the subgame is simply the lottery **p**. Each subgame may therefore be replaced with a leaf labeled with the symbol **p**. This first step in the backward induction process has been carried through in reduced game of Figure 3.15(b).

[9]The function is decreasing rather than increasing as in Section 3.7.1 because it is now a function of distance rather than time.

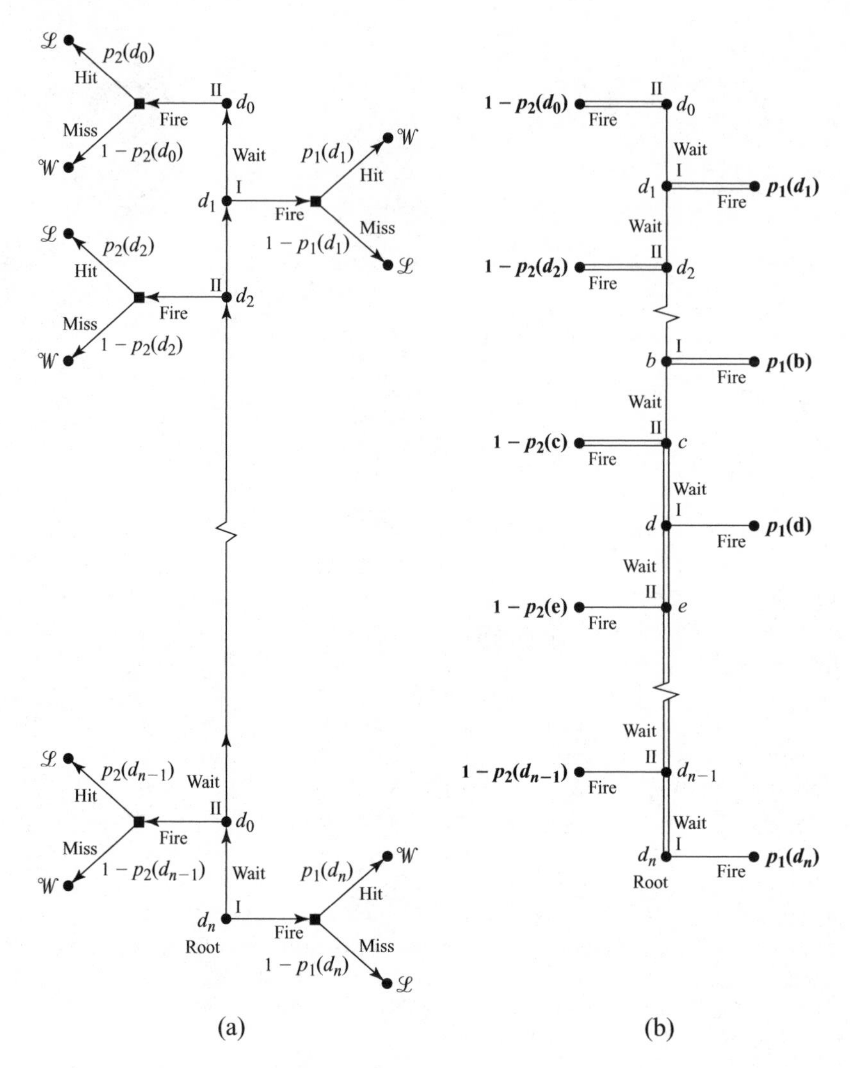


Figure 3.15 Extensive forms for Duel.

**Step 2.** If we ignore the subgame rooted at $d_0$, where player II's only choice is to fire, the smallest subgame in Figure 3.15(b) is rooted at $d_1$. Player I has a choice between firing and waiting at this node. Firing leads to the lottery $\mathbf{p_1}$ $(\mathbf{d_1})$. Waiting leads to the lottery $\mathbf{1} - \mathbf{p_2}(\mathbf{d_0})$. He therefore fires if

$$p_1(d_1) > 1 - p_2(d_0),$$
$$p_1(d_1) + p_2(d_0) > 1.$$

This inequality holds because our assumptions make $p_1(d_1) + p_2(d_0)$ nearly equal to 2. So player I will fire at node $d_1$. The branch that represents this choice has therefore been doubled in Figure 3.15(b).

**Step 3.** It is optimal for player II to fire at node $d_2$ if

$$1 - p_2(d_2) < p_1(d_1)$$
$$p_1(d_1) + p_2(d_2) > 1.$$

This inequality holds because $p_1(d_1) + p_2(d_2)$ is only slightly less than $p_1(d_1) + p_2(d_0)$. So player II will fires at node $d_2$. The branch that represents this choice has therefore been doubled in Figure 3.15(b).

**Step 4.** All the firing branches get doubled in this way until the *first* time that neighboring nodes $c$ and $d$ are reached for which

$$p_1(d) + p_2(c) \leq 1.$$

This must happen eventually because $p_1(d_n) + p_2(d_{n-1})$ is nearly 0.

**Step 5.** From now on, only the case when $c < d$ and $p_1(d) + p_2(c) < 1$ illustrated in Figure 3.15(b) will be considered in detail. In this case, the waiting branch at node $d$ must be doubled because

$$1 - p_2(c) > p_1(d),$$

and so it is optimal for player I to wait at node $d$.

**Step 6.** The waiting branch has also been doubled at the smallest node $e$ larger than $d$. It is optimal for player II to wait at node $e$ because firing leads to the lottery $\mathbf{1} - \mathbf{p_2(e)}$, in which he survives with probability $p_2(e)$, whereas waiting leads to the lottery $\mathbf{1} - \mathbf{p_2(c)}$, in which he survives with probability $p_2(c)$. He prefers the latter because $p_2(c) > p_2(e)$.

**Step 7.** All the waiting branches get doubled in this way whenever the players are more than $d$ apart. If they play optimally, both players will therefore plan to wait until they are distance $d$ apart and to fire thereafter at the earliest opportunity.

**Step 8.** Since $c$ and $d$ are the first pair of neighboring nodes for which $p_1(d) + p_2(c) \leq 1$, it must be true that $p_1(b) + p_2(c) > 1$. But the functions $p_1$ and $p_2$ are continuous, and we have assumed that the points $b$, $c$, and $d$ are all close to each other. It follows that all three points must also be close to the point $\delta$ at which

$$p_1(\delta) + p_2(\delta) = 1.$$

*Conclusion.* Backward induction selects a pure strategy for each player that consists of waiting until the opponent is approximately $\delta$ away and then planning to fire at all subsequent opportunities. The value of the game is approximately $\mathbf{v}$, where $v = p_1(\delta) = 1 - p_2(\delta)$. If the players use their optimal strategies, Tweedledum will therefore survive with probability about $v$, and Tweedledee will survive with probability about $1 - v$.

The closer together we place the decision nodes, the better the approximations become in this analysis. In the limiting case as $n \to \infty$, we recover the conclusion of our product race example.

In the case when $p_1(d) = 1 - d/D$ and $p_2(d) = 1 - (d/D)^2$, the players should wait until they are $d$ apart, where

$$d/D + (d/D)^2 = 1.$$

The positive root of this quadratic equation is $d/D = \frac{1}{2}(\sqrt{5} - 1)$. So nothing will happen until Tweedledum and Tweedledee are about 61% of their original distance apart, when each will fire simultaneously. Tweedledee will be more likely to survive because the probability of his hitting Tweedledum at a given distance is always greater than the probability of Tweedledum hitting him.

## 3.8 Parcheesi

→ 3.9

When visiting India, I was taken to a palace of the Grand Mogul to see the giant marble board on which Akbar the Great played Parcheesi using beautiful maidens as pieces.[10] Parcheesi (or Ludo) is still popular, ranking third after Monopoly and Scrabble on the best-seller list of board games, but the box you buy at the mall contains no beautiful maidens. All you get is a folding board like that in Figure 3.16(a), sixteen counters, and two dice. The toy version to be studied here is even less exotic. It is played on the simplified board of Figure 3.16(b) with just two counters and a fair coin.

Parcheesi is an infinite game in that the rules allow it to continue forever. However, such an eventuality occurs with zero probability and so is irrelevant to an analysis of the game.[11] In any case, this and other technical issues will be ignored. We will simply take for granted that our toy version of Parcheesi and all its subgames have values and focus on determining what these values are.

### 3.8.1 Simplified Parcheesi

Simplified Parcheesi is played between White and Black on the board shown in Figure 3.16(b). The winner is the first to reach the shaded square following the routes indicated. The players take turns, starting with White. The active player either moves his or her counter or leaves it where it is.[12]

If the counter is moved, it must be moved *one* square if tails is thrown with a toss of a fair coin. If heads is thrown, the counter must be moved *two* squares. The last rule has an exception: if the winning square can be reached in one move, the winning move is allowed even when heads has been thrown.

What makes Parcheesi fun to play is the final rule. If a player's counter lands on top of the opponent's counter, then the opponent's counter is sent back to its starting place.

[10]Instead of dice, he threw six cowrie shells. If all six shells landed with their open parts upward, one could move a piece twenty-five squares—hence *parcheesi,* which is derived from the Hindi word for twenty-five.

[11]A zero probability event needn't be impossible. If a fair coin is tossed an infinite number of times, it is possible that the result might always be tails, but this event has zero probability.

[12]If both players choose never to move their counters from some point on, the game is a standoff. The winner is then determined simply by tossing the coin.

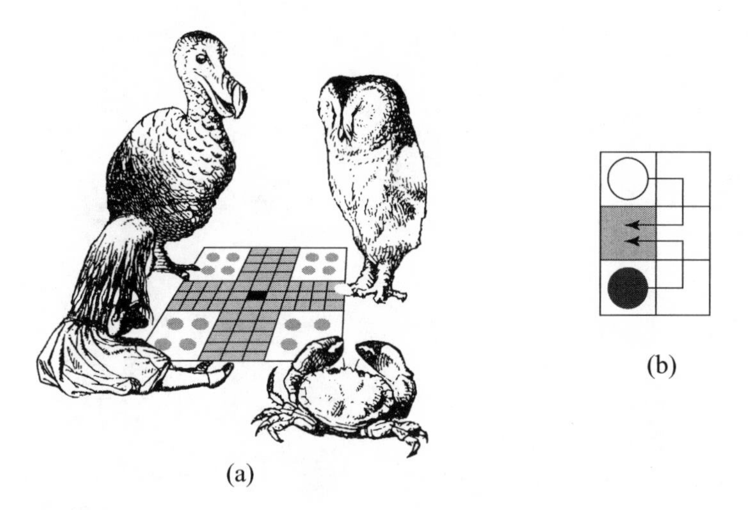


Figure 3.16 Boards for Parcheesi.

### 3.8.2 Possible Positions in Simplified Parcheesi

The eight possible positions that White might face when it is his turn to move are listed in Figure 3.17. The value corresponding to each position is written beneath it. Positions 1 and 2 therefore have the lottery **1** written beneath them because White can win for certain if these positions are reached when it is his turn to move.

The eight positions that Black might face when it is her turn to move are listed in Figure 3.18. Their values can be determined from Figure 3.17. For example, position

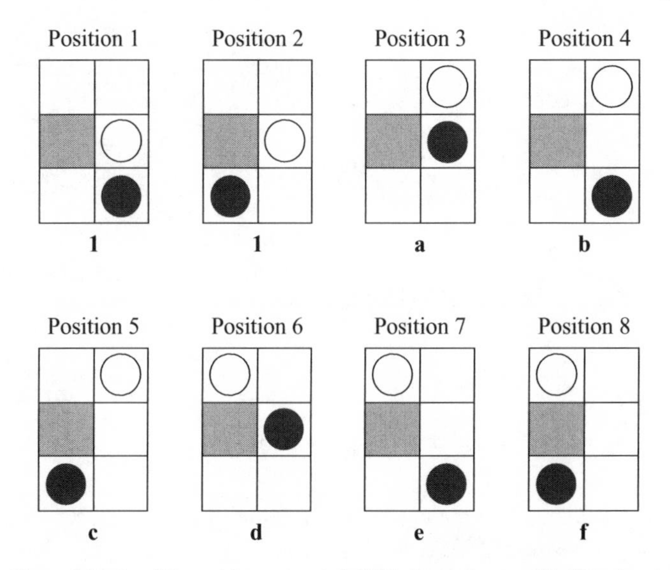


Figure 3.17 Possible positions when it is White's turn in simplified Parcheesi.

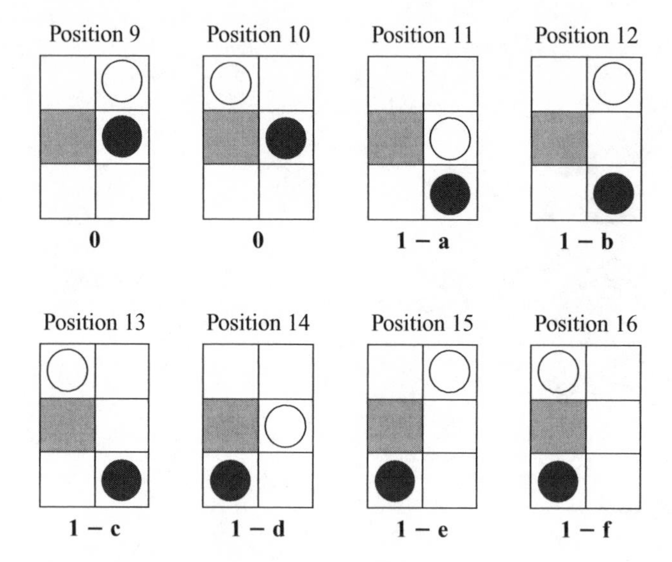


Figure 3.18 Possible positions when it is Black's turn in simplified Parcheesi.

11 looks the same to Black as position 3 looks to White. Since position 3 has value **a**, the value for position 11 must therefore be $\mathbf{1} - \mathbf{a}$.

The value for simplified Parcheesi is **f** since the game starts in this position with White to move. But we can't work out $f$ by backward induction without also determining the values of $a$ through $e$ along the way.

### 3.8.3 Solving Simplified Parcheesi

We will again use backward induction to solve the game, but this time we have to work harder than usual.

**Step 1.** The subgame rooted at position 3 in Figure 3.19 shows the optimal actions for White after the coin is tossed. Thus $\mathbf{a} = \frac{1}{2}\mathbf{1} + \frac{1}{2}(\mathbf{1} - \mathbf{d})$, and so

$$a = \tfrac{1}{2}(1) + \tfrac{1}{2}(1 - d)$$

$$a + \frac{1}{2}d = 1. \tag{3.1}$$

**Step 2.** Position 6 in Figure 3.19 can be treated in the same way. Thus,

$$d = \tfrac{1}{2}(1 - d) + \tfrac{1}{2}(0)$$

$$d = \tfrac{1}{3}$$

$$a = \tfrac{5}{6} \qquad \text{(by equation 3.1)}$$

**Step 3.** It isn't immediately obvious whether White should move his counter after throwing a tail in position 4 of Figure 3.19. If $1 - b \leq \frac{1}{6}$ (and so $b \geq \frac{5}{6}$), it would be optimal for White to move. But then

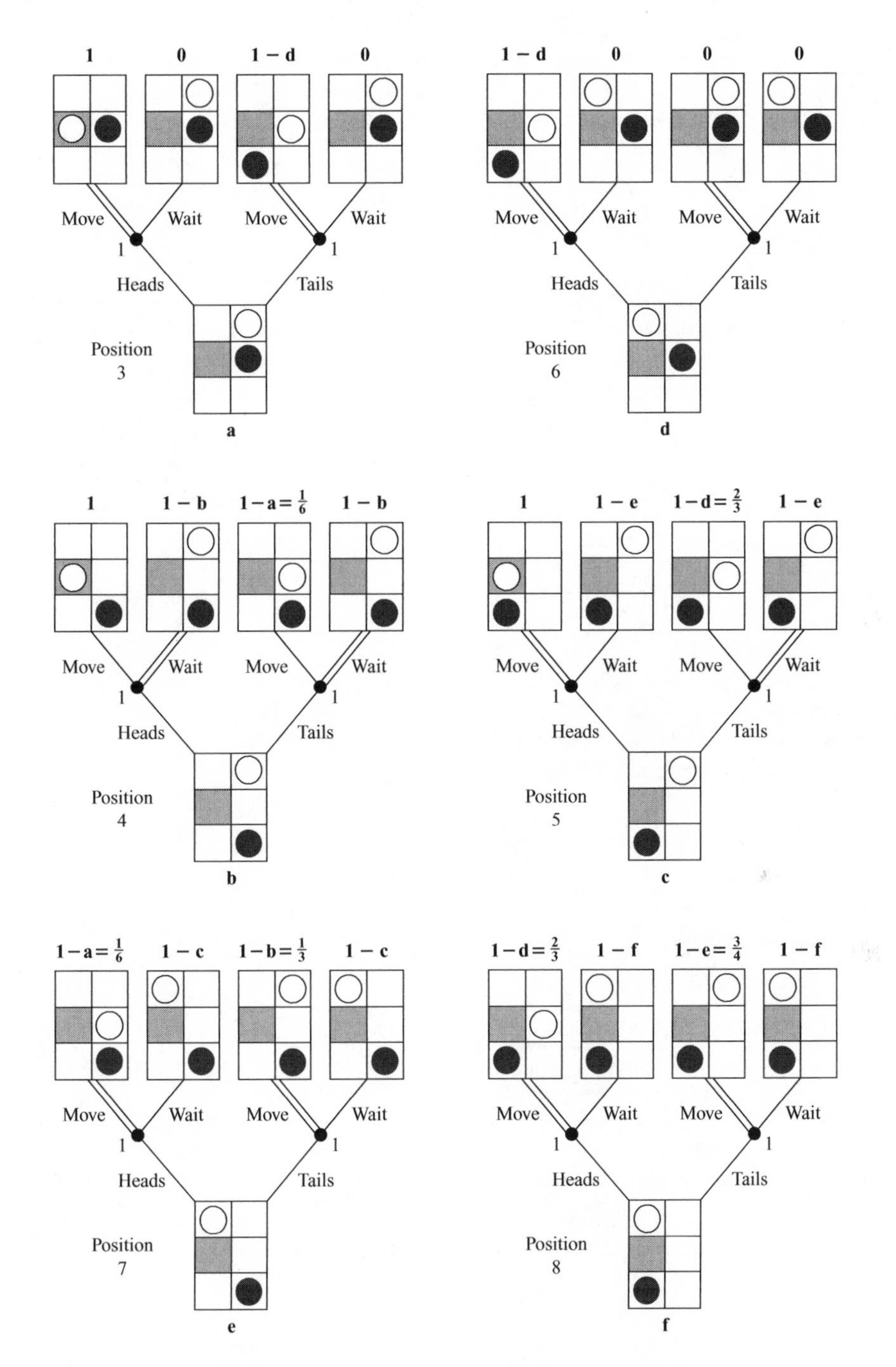


Figure 3.19 Reaching one Parcheesi position from another.

$$\begin{aligned} b &= \tfrac{1}{2}(1) + \tfrac{1}{2}(1-a) \\ &= \tfrac{1}{2}(1) + \tfrac{1}{2}(\tfrac{1}{6}) \\ b &= \tfrac{7}{12}, \end{aligned}$$

which is a contradiction. So it is optimal *not* to move, and

$$\begin{aligned} b &= \tfrac{1}{2}(1) + \tfrac{1}{2}(1-b) \\ b &= \tfrac{2}{3}. \end{aligned}$$

**Step 4.** We take positions 5 and 7 in Figure 3.19 together. If $1 - e \geq \frac{2}{3}$ (and so $e \leq \frac{1}{3}$), an examination of position 5 shows that

$$\begin{aligned} c &= \tfrac{1}{2}(1) + \tfrac{1}{2}(1-e) \\ c + \tfrac{1}{2}e &= 1. \end{aligned} \tag{3.2}$$

But then $1 - c = \frac{1}{2}e \leq \frac{1}{6}$, and so, from position 7,

$$\begin{aligned} e &= \tfrac{1}{2}(1-a) + \tfrac{1}{2}(1-b) \\ &= \tfrac{1}{2}(\tfrac{1}{6}) + \tfrac{1}{2}(\tfrac{1}{3}) \\ e &= \tfrac{1}{4} && (3.3) \\ c &= \tfrac{7}{8} \qquad \text{(by equation 3.2)} && (3.4) \end{aligned}$$

Equations (3.3) and (3.4) were obtained on the assumption that $e \leq \frac{1}{3}$. But it may be that $e > \frac{1}{3}$. If so, position 5 tells us that

$$\begin{aligned} c &= \tfrac{1}{2}(1) + \tfrac{1}{2}(1-d) \\ &= \tfrac{1}{2}(1) + \tfrac{1}{2}(\tfrac{2}{3}) = \tfrac{5}{6}, \end{aligned}$$

and so, from position 7,

$$e = \tfrac{1}{2}(\tfrac{1}{6}) + \tfrac{1}{2}(\tfrac{1}{3}) = \tfrac{1}{4},$$

which contradicts the hypothesis that $e > \frac{1}{3}$. So equations (3.3) and (3.4) do in fact hold.

**Step 5.** If $f < \frac{1}{2}$, White would steal Black's optimal strategy by refusing to move at his first turn, whatever the coin toss showed. It follows that $f \geq \frac{1}{2}$, and so $1 - f \leq \frac{1}{2}$. We can therefore deduce from position 8 that

$$\begin{aligned} f &= \tfrac{1}{2}(1-d) + \tfrac{1}{2}(1-e) \\ &= \tfrac{1}{2}(\tfrac{2}{3}) + \tfrac{1}{2}(\tfrac{3}{4}) \\ f &= \tfrac{17}{24}. \end{aligned}$$

*Conclusion.* White can guarantee winning simplified Parcheesi with a probability of at least $\frac{17}{24}$. He should always move his counter unless a tail is thrown in positions 4,

5, or 6. In positions 4 and 5 he shouldn't move his counter if a tail is thrown. In position 6, his decision doesn't matter. Black's optimal strategy is a mirror image of White's. With this strategy, she guarantees winning with a probability of at least $\frac{7}{24}$. The value of the game is the lottery **17/24**.

## 3.9 Roundup

This chapter is about chance moves, at which a mythical player called Chance makes choices according to a predetermined probability measure. The Monty Hall problem shows that paradoxes can easily be avoided by adopting a systematic modeling methodology.

A probability measure assigns a real number prob($E$) between 0 and 1 to each event $E$. The probability that one of two events $E$ and $F$ will occur when both can't occur simultaneously is prob($E$) + prob($F$). The probability that both of two independent events $E$ and $F$ will occur is prob($E$) $\times$ prob($F$). We need conditional probabilities when $E$ and $F$ aren't independent. A conditional probability prob($E \mid F$) gives the probability that $E$ will occur, given that $F$ has already occurred.

A random variable can be thought of as a lottery ticket. The prizes in some lotteries are tickets for other lotteries. Any such compound lottery can be reduced to a simple lottery using the laws for combining probabilities. When the prizes are given in numerical terms, one can compute the expected value $\mathcal{E}\mathbf{L}$ of a lottery **L**. It is equal to the sum of the values of each prize weighted by the probability of winning the prize. If you repeatedly participate in the lottery, your average winnings will be close to $\mathcal{E}\mathbf{L}$ with high probability in the long run.

Win-or-lose games are necessarily strictly competitive even if they have chance moves. The value **p** of such a game is a lottery in which player I wins with probability $p$ and player II wins with probability $1-p$.

The classical waiting game is called Duel. Economic games in which the players race to be the first to patent an idea or to get a product on the market have the same basic structure. A backward induction analysis shows that both players act when their probabilities of winning sum to one. The intuition is that you should act *immediately* before your opponent unless you are more likely to win by letting him shoot first.

## 3.10 Further Reading

*How to Gamble If You Must*, by Lester Dubbins and Leonard Savage: McGraw-Hill, New York, 1965. This is a mathematical classic.

*Theory of Gambling and Statistical Logic*, by Richard Epstein: Academic Press, New York, 1967. This book is more fun than the book by Dubbins and Savage and fits better into a game theory context, but it still requires some mathematical sophistication.

*Introduction to Probability Theory*, by William Feller: Wiley, New York, 1968. The first volume is a wonderful general introduction to probability theory, but you still need to know some mathematics.

*New Games Treasury*, by Merilyn Mohr: Houghton Mifflin, New York, 1997. How to play an enormous number of games for fun.

*Beat the Dealer*, by Edward Thorp: Blaisdell, New York, 1962. A statistician explains how he beat the dealer at blackjack.

## 3.11 Exercises

fun

1. Marilyn Vos Savant used to write a column in *Parade* magazine based on her reputation of having the highest IQ ever recorded. Various mathematical gurus laughed her to scorn when she answered a question about the Monty Hall problem by saying that switching is always optimal. In reply, she observed that switching would obviously be right if 98 boxes out of 100 were opened. Why is the answer obvious in this case?

fun

2. Martin Gardner used his column in *Scientific American* to get in on the Monty Hall act. He observed that Monty Hall might choose to open a box only when the contestant would lose by switching. Without getting formal, replace the game of Section 3.1.1 by another game in which the Mad Hatter has the option of not opening a box at all. Why is always switching no longer an equilibrium strategy for Alice?

review

3. Explain why the number of distinct hands in straight poker is

$$\binom{52}{5} = \frac{52!}{5!47!} = \frac{52 \times 51 \times 50 \times 49 \times 48}{5 \times 4 \times 3 \times 2 \times 1}.$$

(A deck of cards contains 52 cards. A straight poker hand contains 5 cards. You are therefore asked how many ways there are of selecting 5 cards from 52 cards when the order in which they are selected is irrelevant.)

What is the probability of being dealt a royal flush in straight poker? (A royal flush consists of the A, K, Q, J, and 10 of the same suit.)

review

4. You are dealt ♡A K Q 10 and ♣2. In draw poker, you get to change some of your cards after the first round of betting. If you discard the ♣2, hoping to draw the ♡J, what is the probability that you will be successful? What is the probability of drawing a straight?[13] (Any J will suffice for this purpose.)

review

5. Bob is prepared to make a bet that Punter's Folly will win the first race when the odds are 2:1 against. He is prepared to make a bet that Gambler's Ruin will win the second race when the odds are 3:1 against. He isn't prepared to bet that both horses will win when the odds for this event offered are 15:1 against. If the two races are independent, is Bob consistent in his betting behavior?
6. Find the expected value in dollars of the compound lottery:

<table>
<tr><td>$3</td><td>−$2</td><td>−$2</td><td>$12</td><td>$3</td></tr>
<tr><td>$\frac{1}{2}$</td><td>$\frac{1}{2}$</td><td>$\frac{1}{2}$</td><td>$\frac{1}{6}$</td><td>$\frac{1}{3}$</td></tr>
<tr><td colspan="2">$\frac{1}{3}$</td><td colspan="3">$\frac{2}{3}$</td></tr>
</table>

7. The game of Figure 3.20 has only chance moves that represent independent tosses of a fair coin. Express the situation as a simple lottery. How does your

[13] Drawing to an inside straight is the classic act of folly—but it isn't foolish if the other players don't force you to pay to make the attempt.

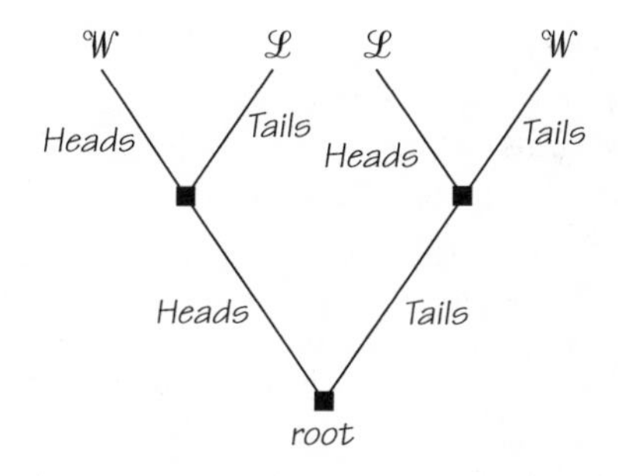


Figure 3.20 A game with only chance moves.

representation change when the chance moves are *not* independent but all refer to a single toss of the same coin?

8. The following table shows the probabilities of the four pairs $(a,c)$, $(a,d)$, $(b,c)$, and $(b,d)$: **review**

| | $c$ | $d$ |
|---|---|---|
| $a$ | 0.01 | 0.09 |
| $b$ | 0 | 0.9 |

The random variable $x$ can take either of the values $a$ or $b$. The random variable $y$ can take either of the values $c$ or $d$. Find:
a. prob $(x=a)$
b. prob $(y=c)$
c. prob $(x=a$ and $y=c)$
d. prob $(x=a$ or $y=c)$

9. In a faraway land long ago, boys were valued more than girls. So couples kept having babies until they had a boy. The frequency of boys and girls in the population as a whole remained equal, but what was the expected frequency of girls per family?[14] (Assume that each sex is equally likely.) **math**

10. Alice learns that the *first* card dealt to Bob is a red queen in the problem of Section 3.3.1. What is her probability that Bob is holding a pair of queens? How would this probability change if she had seen that his first card was the queen of hearts? **review**

11. Alice is dealt ♠A and ♢7 from the deck of Figure 3.4. What is her probability that Bob has a pair of queens if she learns that he has a red queen in his hand? How would this probability change if she had learned that the red queen was the queen of hearts? **review**

[14]It may help to observe that for $0 \le x < 1$,

$$\sum_{n=0}^{\infty} \frac{1}{n+1} x^n = \int_0^x \sum_{n=0}^{\infty} y^n \, dy = \int_0^x \frac{dy}{1-y} = -\ln(1-x).$$

review

12. Bob is the proud father of two children, one of whom is a girl. What is the probability that the other child is a girl? What would the probability have been if you knew that his *older* child were a girl?

math

13. Suppose that Casanova bets one Venetian sequin on the fall of a fair coin and keeps doubling up his stake until he wins. If he wins for the first time on the $n$th toss of the coin, show that he will win precisely one sequin overall. How many sequins will he need to have started with to carry out this strategy when $n = 20$?
14. As long as Casanova has any money in his pocket, he always bets \$1 on the fall of a fair coin until he runs out of money or succeeds in winning a total of \$1. When he loses, he doubles his previous stake. If he begins with \$31 and always bets on heads to win, explain why he will succeed in his aim with any of the sequences that begin *H*, *TH*, *TTH*, *TTTH*, or *TTTTH* but fail with any sequence that begins *TTTTT*. What lottery does he face? Why is its expected dollar value zero?

math

15. The coin tossed in Section 3.5.2 is no longer fair. It lands heads with probability $q$, and the odds are now $m$: 1 against a head. Show that

$$p_{n+1} = q p_{n+m+1} + (1-q) p_n \, .$$

If $r = (1-q)/q$, deduce that the probability of success is

$$p_s = \frac{1 - r^s}{1 - r^{s+w}} \, .$$

16. Player I can choose $l$ or $r$ at the first move in a game $G$. If he chooses $l$, a chance move selects $L$ with probability $p$ or $R$ with probability $1-p$. If $L$ is chosen, the game ends in the outcome $\mathscr{L}$. If $R$ is chosen, a subgame identical in structure to $G$ is played. If player I chooses $r$, then a chance move selects $L$ with probability $q$ or $R$ with probability $1-q$. If $L$ is chosen, the game ends in the outcome $\mathscr{W}$. If $R$ is chosen, a subgame is played that is identical to $G$ *except* that the outcomes $\mathscr{W}$ and $\mathscr{L}$ are interchanged together with the roles of players I and II
    a. Begin the game tree.
    b. Why is this an infinite game?
    c. With what probability will the game continue forever if player I always chooses $l$?
    d. If the value of $G$ is $\mathbf{v}$, show that $v = q + (1-q)(1-v)$ and work out the probability $v$ that player I will win if both players use optimal strategies.
    e. What is $v$ when $q = \frac{1}{2}$?
17. Analyze Nim when the players don't alternate in moving but always toss a fair coin to decide who moves next.

math

18. In the product race of Section 3.7.1, the probability that a player will win if he or she puts their product on the market after $t$ days is

$$p(t) = 1 - e^{-t/100} \, .$$

Show that both will market their products after 69.3 days.

19. In the product race of Section 3.7.1, why is there a unique time at which $p_1 + p_2 = 1$? What implicit assumption about the probabilities that Alice and Bob will win at this time is made in the text in order to ensure the existence of a solution? math
20. How close to the opponent before firing should one get in Duel when $p_1(d) = p_2(d) = 1 - (d/D)^2$?
21. The analysis of Duel of Section 3.7.2 looks in detail only at the case when $c < d$ and $p_1(d) + p_2(c) < 1$. How do things change if $p_1(c) + p_2(d) < 1$? What happens when $c < d$ and $p_1(d) + p_2(c) = 1$? math
22. How does the analysis of Duel change if $p_1(D) + p_2(D) > 1$? What if $p_1(0) + p_2(0) < 1$? What if $p_1(d) + p_2(d) = 1$ for all $d$ satisfying $\frac{1}{3}D \leq d \leq \frac{2}{3}D$? math
23. How does the analysis of Duel change if extra nodes are introduced between $d_k$ and $d_{k+1}$, all of which are assigned to the player who decides at node $d_k$? math
24. What does optimal play look like in Duel if the player who gets to fire at any node is decided by a chance move that assigns equal probabilities to both players? math
25. We return to the product race game of Section 3.7.1 to consider a version in which the probabilities $p_1$ and $p_2$ progress in a sequence of discrete jumps determined by Chance.

    At random times, Chance picks either Alice or Bob with equal probability and increments his or her current value of $p_i$ by $\frac{1}{3}$ until $p_1 = 1$, $p_2 = 1$, or a player has stopped the game by putting their product on the market. Begin to draw a game tree in which chance moves represent some player getting an increment. After such a chance move, assume that the player who gets an increment moves first and the other player moves second. Forget about the random times at which these chance moves occur. Draw enough of the game tree to allow a backward induction analysis.[15] Show that it is always optimal for either Alice or Bob to go to the market when $p_1 + p_2 = 1$.
26. What is the probability that the simplified Parcheesi of Section 3.8.1 will continue for five moves or more if both players always move their counters the maximum number of squares consistent with the rules? fun
27. What is the strategy-stealing argument appealed to at Step 5 in Section 3.8.3 during the analysis of simplified Parcheesi? What strategy-stealing argument shortens the argument at Step 3? fun
28. No mention is made in Section 3.8.3 of the possibility that neither player may choose to move at all on consecutive turns. Why does this possibility not affect the analysis? fun
29. Analyze the simplified Parcheesi game of Section 3.8.1 with the modification that, when a head is thrown, a player may move 0, 1, or 2 squares at his or her discretion. Assume that the other rules remain unchanged. fun
30. Analyze the simplified Parcheesi game of Section 3.8.1 with the modification that, when a counter is exactly one square from the winning square, then only fun

[15] The whole game tree is large, but you don't need to draw it all because some subgames are repeated many times over, and Alice and Bob are in symmetric situations.

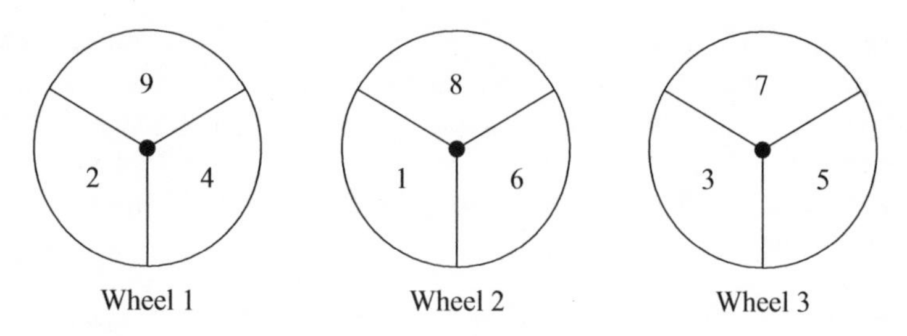


Figure 3.21 Gale's Roulette wheels.

the throw of a tail permits it to be advanced.[16] Assume that the other rules remain unchanged.

31. When a "roulette wheel" from Figure 3.21 is spun, each number on it is equally likely to result. In Gale's Roulette, player I begins by choosing a wheel and spinning it. While player I's wheel is still spinning, player II chooses one of the remaining wheels and spins it. The player whose wheel stops on the larger number wins, and the other player loses.
    a. If player I chooses wheel 1 and player II chooses wheel 2, the result is a lottery **p**. What is the value of *p?* (Assume that the wheels are independent.)
    b. Draw an extensive form for Gale's Roulette.
    c. Reduce the game tree to one without chance moves, as was done for Duel in Section 3.7.2.
    d. Show that the value of the game is **4/9**, so that player II wins more often than player I when both play optimally.
    e. A superficial analysis of Gale's Roulette would suggest that player I should choose the best wheel. Player II will then have to be content with the second-best wheel. But this can't be right because player I would then win more often than player II. What is the fallacy in the argument?[17]
32. Let $\Omega = \{1,2,3,\ldots,9\}$. If player I chooses wheel 2 in Gale's Roulette of the previous exercise, he is selecting a lottery $\mathbf{L_2}$ with prizes in $\Omega$. Express this lottery as a table of the type given in Figure 3.6. Show that

$$\mathscr{E}\mathbf{L_1} = \mathscr{E}\mathbf{L_2} = \mathscr{E}\mathbf{L_3} = 5\,.$$

Let $\mathbf{L_1} - \mathbf{L_2}$ denote the lottery in which the winning prize is $\omega_1 - \omega_2$ if the outcome of lottery $\mathbf{L_1}$ is $\omega_1$ and the outcome of lottery $\mathbf{L_2}$ is $\omega_2$. What is the probability of the prize $-2 = 4 - 6$ in the lottery $\mathbf{L_1} - \mathbf{L_2}$? Why is it true that $\mathscr{E}(\mathbf{L_1} - \mathbf{L_2}) = \mathscr{E}\mathbf{L_1} - \mathscr{E}\mathbf{L_2}$? Deduce that

$$\mathscr{E}(\mathbf{L_1} - \mathbf{L_2}) = \mathscr{E}(\mathbf{L_2} - \mathbf{L_3}) = \mathscr{E}(\mathbf{L_1} - \mathbf{L_3}) = 0\,.$$

[16]This modification makes the game more like real Parcheesi. The new version can be solved by the same method as the original version, but the algebra is a little harder. In particular, positions 1 and 2 of Figure 3.15 no longer have value **1**. If their values are taken to be **g** and **h** respectively, you will be able to show that a contradiction follows unless $d < g < h$.

[17]This exercise provides an advance example of an *intransitive* relation (Section 4.2.2).

Figure 3.22 Which finesses?

33. In an alternative version of Gale's Roulette, each of the three roulette wheels is labeled with *four* equally likely numbers. The numbers on the first wheel are 2, 4, 6, and 9; those on the second wheel are 1, 5, 6, and 8; and those on the third wheel are 3, 4, 5, and 7. If the two wheels chosen by the players stop on the same number, the wheels are spun again and again until someone is a clear winner.
    a. If player I chooses the first wheel and player II chooses the second wheel, show that the probability $p$ that player I will win satisfies $p = \frac{1}{2} + \frac{1}{16}p$.
    b. What is the probability that player I will win the whole game if both players choose optimally?
34. This exercise is for bridge fiends. West is declarer in three no trumps for the deal of Figure 3.22. To keep things simple, assume that she somehow knows that the diamond suit is equally split between her opponents. After a spade lead, West sees that she can win for sure if she can make at least one trick from two finesses in hearts and diamonds. Experts advise taking both finesses in diamonds.

fun

    a. By examining all combinations of cards that North and South might hold, show that the probability that the first diamond finesse succeeds is $\frac{1}{5}$. The probability that either North or South holds ◇ K is $\frac{1}{2}$. The same goes for ◇ Q. So why isn't the answer $\frac{1}{4} = \frac{1}{2} \times \frac{1}{2}$? Why would the answer be nearly $\frac{1}{4}$ if there were a hundred cards per suit?
    b. Show that West's probability of winning at least one trick from two diamond finesses is $\frac{4}{5}$. Show that West's probability of winning at least one trick from one diamond finesse and one heart finesse is $\frac{3}{5}$.
    c. Show that the probability of winning a second diamond finesse after losing the first is $\frac{3}{4}$. Show that the probability of winning a heart finesse after losing a diamond finesse is $\frac{1}{2}$.
    d. Experts appeal to the preceding fact when justifying their advice to take both finesses in diamonds, but they usually say that the probability of winning a second diamond finesse after losing the first is $\frac{2}{3}$. Why would they be about right if there were a hundred cards per suit?
    e. In actual play, the relevant probability after losing the first diamond finesse needs to be conditioned on whether the finesse loses to ◇ K or ◇ Q. Show that this probability can vary between $\frac{3}{5}$ and 1, depending on the probabilities with which South plays ◇ K or ◇ Q when holding ◇ K Q.
    f. In the subgame that follows West's losing the first diamond finesse, explain why it is a strongly dominated strategy for West to take the heart finesse.
35. If all the players in a game become better informed, they may suffer. Confirm this observation by studying a game in which Adam and Eve each choose *dove* or *hawk* without observing the roll of a fair die. Unless a six is rolled, a player who chose *dove* receives a payoff of 1, and a player who chose *hawk* receives a payoff of 0. If a six is rolled, the payoffs are determined by the payoff table for the Prisoners' Dilemma given in Figure 1.3(a). Show that the players get a

phil

smaller expected payoff if the roll of the dice becomes common knowledge before they choose.

fun

36. Lyle Stuart was a big-time gambler who wrote a book on how to win at baccarat and craps. For example, always go to Las Vegas by yourself—you aren't there for fun and games! This exercise is sacred to the memory of Mannie Kemmel, who would apparently wait patiently at the dice table until a number didn't show up for 40 rolls or so and then begin to bet that number every roll. If it failed to come up in another 30 rolls, he would increase his bet. We are told that Mannie rarely failed to walk away with a profit. The story could well be true. If so, does it imply that Mannie found a way around the martingale theorem? (Section 3.5.2)

fun

37. Another of Lyle Stuart's stories concerns a gambler whose son became a mathematician. When the son explains that there is no way to beat the dealer, his father asks where he thinks the money came from to pay for his college education. How should the son reply?

# 4 Accounting for Tastes

## 4.1 Payoffs

In explaining how risk and time enter into the rules of a game, the previous two chapters made no appeal to the theory of utility. But the time has now come to provide a proper account of the way that game theorists use payoffs to model how the players of a game choose between the alternatives available to them.

Chapter 1 explains why it is important to be careful when introducing payoffs. Popular accounts of game theory often try to short-circuit the necessary explanations by simply saying that payoffs are sums of money. This creates no problem if the players are actually trying to make as much money for themselves on average as they can. But game theorists don't restrict themselves to saying what is rational for money grubbers. Our results apply to all rational players, however they are motivated. It follows that payoffs can't be measured just in dollars. In the general case, they are measured in units of utility called *utils.*

To speak of utility is to raise the ghost of a dead theory. Victorian economists thought of utility as measuring how much pleasure or pain a person feels. Nobody doubts that our feelings influence the decisions we make, but the time has long gone when anybody thought that a simple model of a mental utility generator is capable of capturing the complex mental process that swings into action when a human being makes a choice. The modern theory of utility has therefore abandoned the idea that a util can be interpreted as one unit more or less of pleasure or pain.

One of these days, psychologists will doubtless come up with a workable theory of what goes on in our brains when we decide something. In the interim, economists get by with *no theory at all* of why people choose one thing rather than another. The modern theory of utility makes no attempt to *explain* choice behavior. It assumes that we already know what people choose in some situations and uses this data to deduce what they will choose in others—on the assumption that their behavior is consistent.

In game theory, we take as our data the choices that the players would make when solving one-person decision problems by themselves and seek to deduce the choices that they will make when they play games together.

## 4.2 Revealed Preference

Students of economics usually first meet utility theory when modeling the behavior of consumers. Pandora buys a bundle of goods on each of her weekly visits to the supermarket. Since her household budget and the supermarket prices vary from week to week, the bundle she purchases isn't always the same. However, after observing her shopping behavior for some time, it becomes possible to make an educated guess about what she will buy next week, once one knows what the prices will be and how much she will have to spend.

In making such inferences, two assumptions are implicitly understood. The first is that Pandora's choice behavior is *stable.* We obviously won't be able to predict what she will buy next week if something happens today that makes our data irrelevant. If Pandora loses her heart to a football star, who knows how this might affect her shopping behavior? Perhaps she will buy no pizza at all and instead fill her basket with deodorant.

Pandora's choice behavior must also be *consistent.* We certainly won't be able to predict what she will do next if she just picks items off the shelf at random, whether or not they are good value, or satisfy her needs. But what are the criteria that determine whether her behavior is consistent or not? This chapter is largely devoted to the manner in which this question is answered by modern utility theory.

### 4.2.1 Money Pumps

The following example illustrates the kind of way in which economists justify the consistency assumptions they attribute to rational players.

Adam has an apple. Eve offers to exchange his apple for a fig plus a penny. Adam agrees, and now he has a fig. Eve next offers to exchange his fig for a lemon plus a penny. Adam agrees, and now he has a lemon. Eve now offers to exchange his lemon for an apple plus a penny. Adam agrees, and so he ends up with the apple with which he started—minus three pennies that are now in Eve's purse.

If Adam's choice behavior is stable, Eve can now repeat the cycle over and over again until she has extracted every cent he has. A rational player obviously wouldn't fall victim to such a money pump. What do we have to assume about Adam's choice behavior to eliminate the possibility that he might?

Economists say that the choices that Adam makes *reveal* his preferences. If he trades an apple for a fig plus a penny, he reveals a strict preference for a fig over an apple. As in Section 2.2, we then write apple $\prec$ fig. This notation allows us to summarize his revealed choice behavior as:

$$\text{apple} \prec \text{fig} \prec \text{lemon} \prec \text{apple}.$$

It is then evident that Adam fell victim to Eve's money pump because his revealed preferences go around in a circle. Eliminating such cycling from a rational player's choice behavior is therefore our first priority.

### 4.2.2 Full and Consistent Preferences

The crudest way to specify the preferences revealed by a player's choices is to use a preference relation $\preceq$. We assume that a rational player will reveal preferences that satisfy the following criteria:

$$
\begin{array}{llll}
a \preceq b & \text{or} & b \preceq a & \text{(totality)} \\
a \preceq b & \text{and} & b \preceq c \Rightarrow a \preceq c & \text{(transitivity)}
\end{array}
$$

for all $a$, $b$, and $c$ in the set $\Omega$ of all possible outcomes.

The transitivity that prevents cycling is the only genuine consistency requirement. Totality merely says that the player is always able to express a preference between any two outcomes.[1]

A preference relation $\preceq$ shouldn't be confused with the relation $\leq$ used to indicate which of two numbers is larger. The latter satisfies an extra condition:

$$a \leq b \text{ and } b \leq a \quad \Leftrightarrow \quad a = b,$$

which we certainly don't want all preference relations to satisfy. Instead of making this assumption, we define the indifference relation $\sim$ by:

$$a \preceq b \text{ and } b \preceq a \quad \Leftrightarrow \quad a \sim b.$$

The strict preference relation $\prec$ is defined by:

$$a \preceq b \text{ and not}\,(a \sim b) \quad \Leftrightarrow \quad a \prec b.$$

## 4.3 Utility Functions

In making a rational decision, Pandora faces two tasks. The first is to identify the *feasible* set—the subset $S$ of $\Omega$ consisting of those outcomes that are currently available. The second task is to find an *optimal* outcome in $S$. This is an outcome in $S$ that she likes at least as much as any other outcome in $S$.

The problem of finding an optimal $\omega$ looks easy when stated in this abstract way, but it can be hard to solve in practice if $\Omega$ is a complicated set, and so Pandora's preference relation $\preceq$ is difficult to describe.

[1]In mathematics, a relation satisfying totality and transitivity is a pre-ordering. If totality is replaced by $a \preceq a$ (reflexivity), then $\preceq$ becomes a *partial* pre-ordering.

Utility functions are a mathematical device introduced to simplify the optimization problem. A preference relation $\preceq$ is represented by such a utility function $u:\Omega \to \mathbb{R}$ if and only if

$$u(a) \leq u(b) \quad \Leftrightarrow \quad a \preceq b.$$

Finding an optimal $\omega$ then reduces to solving the maximization problem:

$$u(\omega) = \max_{s \in S} u(s),$$

for which many mathematical techniques are available. A maximizing $\omega$ may not exist if $S$ is an infinite set, but we won't need to worry much about such technical difficulties. Nor is there any need to get hung up about the fact that there may sometimes be more than one maximizing $\omega$.

→ 4.3.2

## 4.3.1 Optimizing Consumption

Pandora likes to drink martinis before dinner. It isn't good for her health, but in spite of the title of this chapter, there is no accounting for tastes. Philosophers sometimes say that one consistent set of preferences can be more rational than another, but Section 1.4.1 explains why economists don't join them in telling people what they ought to like. For us, Pandora's preference relation $\preceq$ is part of what makes her a person, like the length of her nose or the color of her hair.

Pandora regards gin and vodka as perfect substitutes for making martinis. This means that she is always willing to exchange one for the other at a fixed rate. In this example, she is always willing to trade at a rate of three bottles of gin for four bottles of vodka.

Let $\Omega$ be the set of all commodity bundles $(g, v)$ consisting of $g$ bottles of gin and $v$ bottles of vodka. The choices Pandora makes when deciding between bundles in $\Omega$ can be expressed in terms of a revealed preference relation $\preceq$, whose structure is indicated in Figure 4.1 by drawing its indifference curves, together with little arrows that show which indifference curves she prefers.[2]

The simplest utility function $U : \Omega \to \mathbb{R}$ that represents Pandora's preference relation is given by

$$U(g, v) = 4g + 3v.$$

For example, the fact that she is indifferent between the commodity bundles $(3, 0)$ and $(0, 4)$ is reflected in the fact that $U(3, 0) = U(0, 4) = 12$.

Pandora can buy vodka at \$10 a bottle and gin at \$15 a bottle. If she has \$60 to spend on feeding her martini habit, how will she split the money between gin and vodka?

If we ignore the fact that liquor stores usually sell their merchandise only in whole numbers of bottles, Pandora's feasible set $S$ consists of all bundles (g, v) with $g \geq 0$ and $v \geq 0$ that lie on or below her budget line: $10g + 15v = 60$. We need to

[2]An indifference set for $\preceq$ consists of all $s \in \Omega$ that satisfy $s \sim \omega$ for some given $\omega$. Such a set is usually a curve in economics examples.

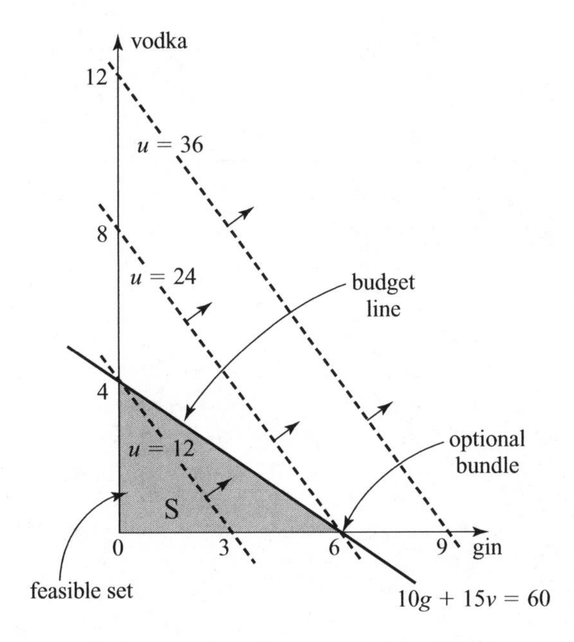


Figure 4.1 What kind of martini is optimal?

find her optimal bundle in this feasible set. This is a very simple example of a linear programming problem, in which a linear function must be maximized subject to a set of linear inequalities (Section 7.6).

Assuming that any money she doesn't spend is wasted, her optimal bundle $\omega = (g, v)$ lies on her budget line. Her utility at this bundle is therefore

$$U(g, 4 - \tfrac{2}{3}g) = 4g + 3(4 - \tfrac{2}{3}g) = 12 + 2g,$$

which is largest when $g$ is biggest. She therefore buys no vodka at all. Since her \$60 will buy six bottles of gin, her optimal bundle is $\omega = (6, 0)$.

Figure 4.1 illustrates the solution. Pandora's indifference curves correspond to contours of her utility function. Just as the height of a hill is constant along a contour on a map, so Pandora's utility is constant along a contour like $U = 12$. Contours like $U = 36$ that don't have a point in common with the feasible set $S$ correspond to unattainable utility levels. The contour with the highest utility that intersects with $S$ is $U = 24$. Its unique point of intersection with $S$ is $\omega = (6, 0)$, which is Pandora's optimal bundle.

### 4.3.2 Constructing Utility Functions

Pandora's choice behavior reveals that she has consistent preferences over the six commodity bundles $a$, $b$, $c$, $d$, $e$, and $f$. Her preferences are

$$a \prec b \sim c \prec d \prec e \sim f.$$

Thus, if Pandora's feasible set is $\{a, b, c\}$, she won't choose $a$, but she might choose either $b$ or $c$. If her feasible set is $\{b, c, d\}$, then only $d$ is optimal.

| $x$ | $a$ | $b$ | $c$ | $d$ | $e$ | $f$ |
|---|---|---|---|---|---|---|
| $U(x)$ | 0 | $\frac{1}{2}$ | $\frac{1}{2}$ | $\frac{3}{4}$ | 1 | 1 |
| $V(x)$ | −123 | 18 | 18 | 19 | 2,947 | 2,947 |

Figure 4.2 Constructing utility functions. The method always works for a consistent preference relation defined over a finite set of outcomes, because there is always another real number between any pair of real numbers.

It is easy to find a utility function $U:\{a,b,c,d,e,f\} \to \mathbb{R}$ that represents Pandora's preferences. She regards the bundles $a$ and $f$ as the worst and the best available. We therefore set $U(a) = 0$ and $U(f) = 1$. Since she is indifferent between $e$ and $f$, we must also set $U(e) = 1$. Next pick any bundle intermediate between the worst bundle and the best bundle, and take its utility to be $\frac{1}{2}$. In Pandora's case, $b$ is a bundle intermediate between $a$ and $f$, and so we set $U(b) = \frac{1}{2}$. Since $b \sim c$, we must also set $U(c) = \frac{1}{2}$. Only the bundle $d$ remains. This is intermediate between $c$ and $e$, and so we set $U(d) = \frac{3}{4}$ because $\frac{3}{4}$ is intermediate between $U(c) = \frac{1}{2}$ and $U(e) = 1$.

The utilities assigned to bundles in Figure 4.2 are ranked in the same way as the bundles themselves. In making choices, Pandora therefore behaves *as though* she were maximizing the value of $U$. But she also behaves as though she were maximizing the value of the alternative utility function $V$ given in Figure 4.2. This observation signals the fact that there are *many* ways in which we could have assigned utilities to the bundles in a manner consistent with Pandora's preferences. The only criterion that is relevant when picking one of the infinity of utility functions that represent a given preference relation is that of mathematical convenience.

→ 4.4

### 4.3.3 Rational Choice Theory?

Outside economics, the use of utility theory is controversial. In political science, the debate over "rational choice theory" often gets quite heated.

However, both sides in such debates commonly subscribe to the causal utility fallacy, which says that decision makers choose $a$ over $b$ because the utility of $a$ exceeds that of $b$. But modern economists don't argue that a person's choice of $a$ over $b$ is *caused* by the utility of $a$ exceeding that of $b$. On the contrary, it is because the preference $a \succ b$ has been revealed that we choose a utility function satisfying $u(a) > u(b)$.

For people to behave *as though* their aim were to maximize a utility function, it is only necessary that their choice behavior be consistent. To challenge the theory, you therefore need to argue that people behave *inconsistently,* rather than that they don't really have utility generators inside their heads. As for the critics who claim that economists believe that people have little cash registers in their heads that respond only to dollars, they haven't bothered to study the theory they are criticizing at all.

## 4.4 Dicing with Death

The game of Russian Roulette will allow us to review some of the ideas that we met in Chapters 2 and 3 while focusing our attention on the inadequacy of what has been said so far about utility functions.

Boris and Vladimir are officers in the service of the czar who have both fallen in love with a beautiful Muscovite maiden called Olga. They agree that it doesn't make sense for both to press their claims simultaneously but disagree on who should back down. Eventually they decide to settle the matter with a game of Russian Roulette, with Boris as player I and Vladimir as player II.

In Russian Roulette, a bullet is loaded at random into one of the chambers of a six-shooter, as illustrated in Figure 4.3(a). The players then take turns pointing the revolver at their heads. When it is your turn, you can either pull the trigger or chicken out. Chickening out and death disqualify you from chasing after Olga any more. One might think that only crazy people would play such a game, but the superlatively creative French mathematician Evariste Galois died at the age of twenty while playing something very similar. Perhaps this is why Russians call the game French Roulette.

Neither Boris nor Valdimir cares about the welfare of the other, so each player distinguishes only three outcomes, $\mathscr{L}$, $\mathscr{D}$, or $\mathscr{W}$, which we can think of as death, disgrace, or triumph. Player $i$'s preferences over these outcomes satisfy

$$\mathscr{L} \prec_i \mathscr{D} \prec_i \mathscr{W}.$$

The outcome $\mathscr{L}$ corresponds to a player shooting himself. The outcome $\mathscr{W}$ corresponds to his being left to woo Olga undisturbed. The outcome $\mathscr{D}$ corresponds to a player chickening out. He will then be forced to sit alone, morosely drinking vodka in the officer's club, while his rival trifles with Olga's affections.

### 4.4.1 Version 1 of Russian Roulette

A natural way of drawing the game tree for Russian roulette is shown in Figure 4.4. The act of loading the single bulllet into the gun is represented by a single chance move that opens the game. Each of the six chambers of the revolver corresponds to one of the six choices available to Chance at this node. The chambers are labeled 1 through 6, according to the order in which they will be reached as the trigger is pulled. Each chamber is equally likely to be chosen, and so the probability that the bullet is in any particular chamber is $\frac{1}{6}$.

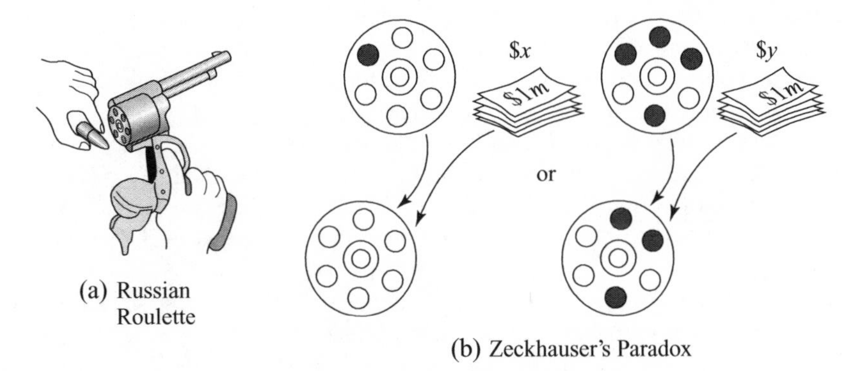


Figure 4.3 Where are the bullets?

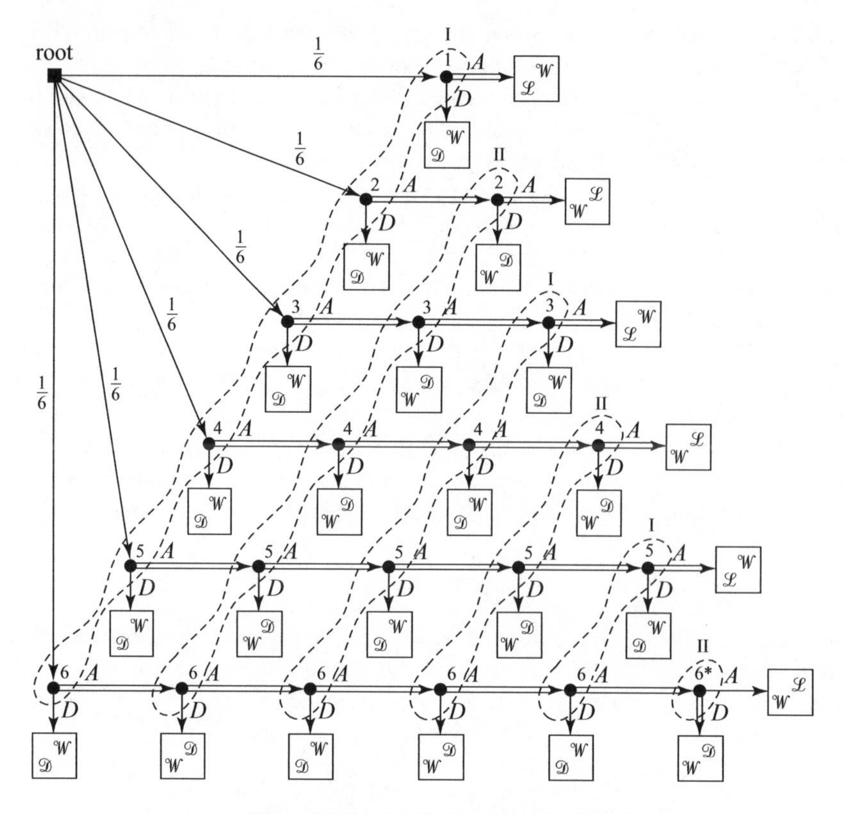


Figure 4.4 Russian Roulette—version 1.

The branches at decision nodes are labeled *A* (for across) and *D* (for down). Playing *down* corresponds to chickening out. Playing *across* corresponds to a player pulling the trigger.

The nodes at which a player chooses between *A* or *D* are labeled with the number of the chamber that contains the bullet. The information sets in Figure 4.4 indicate the fact that the players don't know this information when they decide whether or not to pull the trigger.

Since all but one of the information sets contain more than one decision node, this version of Russian Roulette is a game of *imperfect information*. A pure strategy in a game of imperfect information specifies an action only at each of a player's information sets—not at each of his decision nodes.

The pure strategy pair $(AAA, AAD)$ is indicated in Figure 4.4 by doubling appropriate branches. All six *across* branches have therefore been doubled at player I's first information set. He can't plan to play differently at different nodes in the same information set because he won't be able to distinguish between them when he makes his decision.

Once Boris and Vladimir have chosen their pure strategies, the course of the game is entirely determined, except for the initial decision made by Chance. If Chance puts the bullet in chamber 6, the resulting play of the game starts at the root and proceeds vertically downward to the first node labeled with a 6, where it is

Boris's turn to move. His choice of pure strategy *AAA* requires that he take action *A* at his first move. Accordingly, he pulls the trigger but survives because the bullet isn't in chamber 1. We therefore move on to the second node labeled with a 6, where it is Vladimir's turn to move. His choice of pure strategy *AAD* requires that he take action *A* at his first move. So he pulls the trigger but survives because the bullet isn't in chamber 2.

The play continues horizontally in this way until it reaches the node labeled with 6* at the bottom right of Figure 4.4, where it is Vladimir's move.

Vladimir now knows that the bullet is in chamber 6, and so he is sure to shoot himself if he pulls the trigger. Fortunately, his choice of the pure strategy *AAD* requires that he chicken out by taking action *D* at his third move. This action concludes the play that started with Chance putting the bullet in chamber 6 by taking it downward to a payoff box in which Boris gets the outcome $\mathscr{W}$ and Vladimir gets the outcome $\mathscr{D}$.

While following this play, *we* always knew where the bullet was, but the *players* were in suspense until node 6* was reached. For example, Vladimir didn't know he was about to pull the trigger on an empty chamber at his second move. We knew the game had reached node 6, but Vladimir thought that nodes 4 and 5 in his second information set were just as likely. When he pulled the trigger, he therefore thought he would shoot himself with probability $\frac{1}{3}$ since this is the conditional probability of being at node 4, given that Vladimir's second information set has been reached.

### 4.4.2 Version 2 of Russian Roulette

Figure 4.5 shows an alternative game tree for Russian Roulette. No information sets appear because the new version is a game of perfect information. The price paid for this simplification is that we have to include six chance moves: one for each chamber of the six-shooter.

On the other hand, the new game has lots of subgames that we will exploit when using backward induction to solve the game in Section 4.7. By contrast, version 1 of Russian roulette has only two subgames: the whole game and the one-player subgame rooted at node 6*. No decision node with companions in its information set can serve as the root of a subgame because we can't distentangle such a node from its companions without making nonsense of the informational assumptions of the game.

The strategy pair (*AAA*, *AAD*) has been indicated by doubling branches in Figure 4.5. Its use results in the various leaves being reached with the probabilities written beneath them. Boris ends up with the outcome $\mathscr{W}$ half the time and with $\mathscr{L}$ the rest of the time. If the strategy pair (*DDD*, *AAD*) were used instead, Boris would get $\mathscr{D}$ for certain.

If Boris knows or guesses that Vladimir will choose *AAD*, which of *AAA* or *DDD* is better for him? It is important to recognize that we can't answer this question without knowing more about Boris's preferences.

All we have been told so far is that $\mathscr{L} \prec_1 \mathscr{D} \prec_1 \mathscr{W}$, but this information doesn't help us decide whether Boris prefers $\mathscr{D}$ for certain to the lottery in which he is equally likely to get $\mathscr{W}$ or $\mathscr{L}$. If Boris were young and romantic like Evariste Galois, he might be willing to risk death rather than abandon his beloved, but disillusioned old gentlemen like me won't see the potential reward as being worth much of a risk.

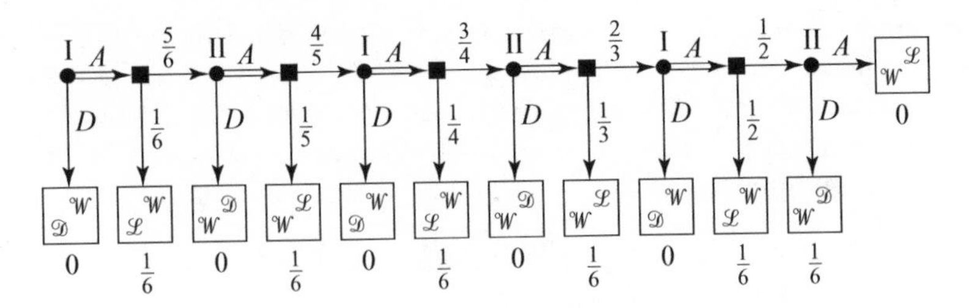

Figure 4.5 Russian Roulette—version 2.

However, both of us will agree that $\mathscr{D}$ is an outcome intermediate between $\mathscr{W}$ and $\mathscr{L}$.

## 4.5 Making Risky Choices

How do we describe a player's preferences over lotteries that involve more than two prizes? A naive approach would be to replace all the prizes in the lotteries by their worth to the player in money. Wouldn't a rational person then simply prefer whichever of two lotteries has the larger dollar expectation?

The story coming up next explains why such an approach won't work. Like Russian Roulette, it is set in the last days of the czars.

### 4.5.1 The St. Petersburg Paradox

Nicholas Bernouilli proposed the following paradox about a casino in St. Petersburg that was supposedly willing to run any lottery whatever, provided that the management could set the price of a ticket to participate.[3]

In the lottery of Figure 4.6, a fair coin is tossed until it shows heads for the first time. If the first head appears on the $k$th trial, you win $\$2^k$. How much should you be willing to pay in order to participate in this lottery?

Since each toss of the coin is independent, the probability of winning $\$2^k$ is calculated as shown below for the case $k = 4$:

$$\text{prob}(TTTH) = \text{prob}(T)\times\text{prob}(T)\times\text{prob}(T)\times\text{prob}(H) = \left(\tfrac{1}{2}\right)^4 = \tfrac{1}{16}.$$

The expectation in dollars of the St. Petersburg lottery **L** is therefore

$$\begin{aligned}\mathscr{E}(\mathbf{L}) &= 2\,\text{prob}(H) + 4\,\text{prob}(TH) + 8\,\text{prob}(TTH) + \cdots \\ &= 2\times\tfrac{1}{2} + 4\times\tfrac{1}{4} + 8\times\tfrac{1}{8} + \cdots \\ &= 1+1+1+1+\cdots,\end{aligned}$$

[3]However, the paradox probably got its name for the more prosaic reason that his brother Daniel published it in the proceedings of the St. Petersburg Academy of 1738.

which implies that its expected dollar value is "infinite." Should Olga therefore be willing to sell off all she owns and borrow as much as she can in order to buy a lottery ticket? Since the probability is $\frac{7}{8}$ that she will end up with no more than \$8, she is unlikely to find the odds attractive.

The moral isn't that the policy of always choosing the lottery with the largest expectation in dollars is necessarily irrational. The St. Petersburg story merely casts doubt on the claim that no other policy can be rational.

The same goes for any theory that claims that there is only one rational way to respond to risk. An adequate theory needs to recognize that the extent to which Olga is willing to bear risk is as much a part of her preference profile as her relative liking for the songs that Boris and Vladimir sing when they play their balalaikas late at night beneath her bedroom window.

### 4.5.2 Von Neumann and Morgenstern Utility

Rationality doesn't require that Olga try to maximize her expected *dollar* value when choosing between lotteries. However, Von Neumann and Morgenstern gave a list of consistency postulates about preferences in risky situations that imply that Olga will behave as though maximizing the expected value of *something* when acting rationally. We call this something the Von Neumann and Morgenstern utility of a lottery.

The first postulate repeats the rationality assumption of Chapter 3:

POSTULATE 1 *A rational player prefers whichever of two win-or-lose lotteries offers the larger probability of winning.*

Postulate 1 is about win-or-lose lotteries, in which the only prizes are drawn from the set $\Omega = \{\mathscr{L}, \mathscr{W}\}$. A utility function $u : \Omega \to \mathbb{R}$ that represents the preference $\mathscr{W} \succ \mathscr{L}$ must have $a = u(\mathscr{L}) < u(\mathscr{W}) = b$.

The set of lotteries with prizes drawn from the set $\Omega$ will be denoted by lott($\Omega$). The win-or-lose lottery **p** in which Olga wins with probability $p$ therefore belongs to lott $(\{\mathscr{W}, \mathscr{L}\})$. The expected utility of **p** is

$$\mathscr{E}u(\mathbf{p}) = p\,u(\mathscr{W}) + (1-p)\,u(\mathscr{L}) = a + p(b-a). \tag{4.1}$$

Since $b - a > 0$, $\mathscr{E}u(\mathbf{p})$ is largest when the probability $p$ of winning is largest.

Equation (4.1) tells us that $\mathscr{E}u$ is a utility function for Olga's preferences over lott($\Omega$) when $\Omega = \{\mathscr{W}, \mathscr{L}\}$. Postulate 1 therefore implies that Olga necessarily acts as though maximizing expected utility when making decisions involving only lotteries whose prizes are $\mathscr{L}$ or $\mathscr{W}$.

| prize | \$2 | \$4 | \$8 | \$16 | ... | $\$2^k$ | ... |
|---|---|---|---|---|---|---|---|
| coin sequence | *H* | *TH* | *TTH* | *TTTH* | ... | *TT...TH* | ... |
| probability | $\frac{1}{2}$ | $\frac{1}{4}$ | $\frac{1}{8}$ | $\frac{1}{16}$ | ... | $\left(\frac{1}{2}\right)^k$ | ... |

Figure 4.6 The St. Petersburg lottery.

Matters become more complicated when there are prizes intermediate between $\mathscr{W}$ and $\mathscr{L}$. It then ceases to be true that $\mathscr{E}u$ is a utility function for Olga's preferences over lotteries whenever $u$ is a utility function for her preferences over prizes. If $u:\Omega \to \mathbb{R}$ is to be a Von Neumann and Morgenstern utility function—so that $\mathscr{E}u$ represents Olga's preferences over *lotteries*—we need to select $u$ very carefully from the large class of utility functions that represent Olga's preferences over *prizes*.

POSTULATE 2 *Each prize $\omega$ between the best prize $\mathscr{W}$ and the worst prize $\mathscr{L}$ is equivalent to some lottery involving only $\mathscr{W}$ and $\mathscr{L}$.*

The postulate says that, for each prize $\omega$ in $\Omega$, there is a probability $q$ for which

$$w \sim \mathrm{q} = \begin{array}{|c|c|} \hline \mathscr{W} & \mathscr{L} \\ \hline q & 1-q \\ \hline \end{array} \qquad (4.2)$$

The second postulate makes it possible to construct a Von Neumann and Morgenstern utility function $u:\Omega \to \mathbb{R}$. The function $u$ is defined so that the value of $u(\mathscr{W})$ is the probability $q$ in (4.2). That is to say, $q = u(\mathscr{W})$ is defined to make Olga indifferent between getting $\omega$ for certain and getting the lottery that yields $\mathscr{W}$ with probability $u(\mathscr{W})$ and $\mathscr{L}$ with probability $1 - u(\mathscr{W})$.

For example, we might begin an experiment to elicit Olga's preferences over risky prospects by asking her whether she will pay \$20 for a ticket for the lottery $\mathbf{q}$ of (4.2) in the case when the best possible prize is $\mathscr{W} = \$100$ and the worst possible prize is $\mathscr{L} = \$0$. If she stops saying *no* and starts saying *yes* when $q$ passes through the value 0.4, then $u(20) = 0.4$.

As we increase the price \$X of a ticket from \$0 to \$100, $u(X)$ will increase from $u(0) = 0$ to $u(100) = 1$. As we will see, the shape of the graph of $u$ will tell us everything we need to know about Olga's attitude to taking risks.

To confirm that $u:\Omega \to \mathbb{R}$ is a Von Neumann and Morgenstern utility function, we need to verify that $\mathscr{E}u:\mathrm{lott}(\Omega) \to \mathbb{R}$ is a utility function for Olga's preferences over lotteries. Figure 4.7 illustrates the two steps in the argument that justifies this conclusion. Each step requires a further postulate.

POSTULATE 3 *Rational players don't care if a prize in a lottery is replaced by another prize that they regard as equivalent to the prize it replaces.*[4]

The prizes available in the arbitrary lottery $\mathbf{L}$ of Figure 4.7 are $\omega_1, \omega_2, \ldots, \omega_n$. By Postulate 2, Olga regards each such prize $\omega_k$ as the equivalent of some win-or-lose lottery $\mathbf{q_k}$. Postulate 3 is then used to justify replacing each prize $\omega_k$ by the corresponding $\mathbf{q_k}$. We then need a final assumption to reduce the resulting compound lottery to a simple lottery.

[4]Critics often forget that, if one of the prizes is itself a lottery, then it is implicitly assumed that this lottery is independent of all other lotteries involved. Without such an independence assumption, the postulate wouldn't make much sense.

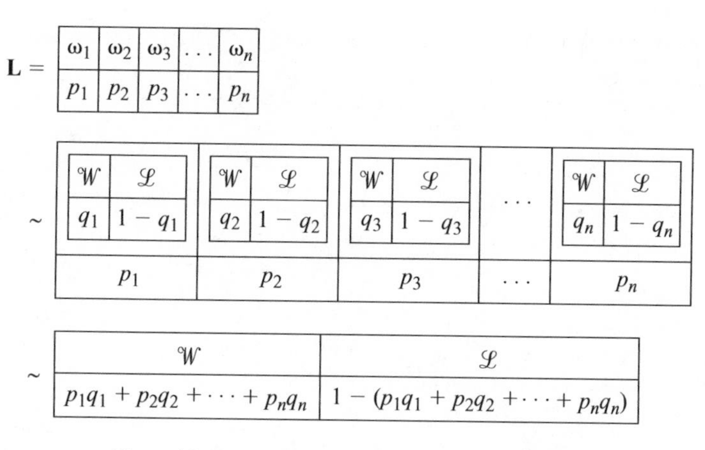

Figure 4.7 Von Neumann and Morgenstern's argument.

POSTULATE 4 *Rational players care only about the total probability with which they get each prize in a compound lottery.*

The total probability of $\mathscr{W}$ in Figure 4.7 is $r = p_1q_1 + p_2q_2 + \cdots + p_nq_n$. Postulate 4 then says that we can replace the compound lottery by the simple lottery **r**, thereby justifying the second of the two steps the figure illustrates.

By Postulate 1, Olga prefers whichever of two lotteries like **L** in Figure 4.7 has the larger value of $r = p_1q_1 + p_2q_2 + \cdots + p_nq_n$. She therefore acts as though seeking to maximize

$$\begin{aligned} r &= p_1q_1 + p_2q_2 + \cdots + p_nq_n \\ &= p_1u(\omega_1) + p_2u(\omega_2) + \cdots + p_nu(\omega_n). \\ &= \mathscr{E}u(\mathbf{L}). \end{aligned}$$

Thus $\mathscr{E}u : \text{lott}(\Omega) \to \mathbb{R}$ is a utility function that represents Olga's preferences in lotteries. But this is what it means to say that $u : \Omega \to \mathbb{R}$ is a Von Neumann and Morgenstern utility function for her preferences over prizes.

### 4.5.3 Attitudes to Risk

How does Von Neumann and Morgenstern's theory deal with the St. Petersburg paradox? Suppose that Olga's utility for money is given by the Von Neumann and Morgenstern utility function[5] $u : \mathbb{R}_+ \to \mathbb{R}$ defined by

$$u(x) = 4\sqrt{x}. \tag{4.3}$$

[5]The set $\mathbb{R}_+ = \{x : x \geq 0\}$ consists of all nonnegative real numbers. Note also that:
1. $\sqrt{a^n} = (a^n)^{1/2} = a^{n/2} = (\sqrt{a})^n$;
2. $\sqrt{b}/b = 1/\sqrt{b}$;
3. If $|r| < 1$, the geometric series $1 + r + r^2 + \ldots$ adds up to something finite. Its sum $s$ satisfies $s = 1 + r + r^2 + \ldots = 1 + r(1 + r + \ldots) = 1 + rs$. Hence, $s = 1/(1 - r)$.

Her expected utility for the St. Petersburg lottery **L** of Figure 4.6 is then

$$\begin{aligned}\mathscr{E}u(\mathbf{L}) &= \tfrac{1}{2}u(2)+(\tfrac{1}{2})^2u(2^2)+(\tfrac{1}{2})^3u(2^3)+\cdots\\ &= 4\{\tfrac{1}{2}\sqrt{2}+(\tfrac{1}{2})^2\sqrt{2^2}+(\tfrac{1}{2})^3\sqrt{2^3}+\cdots\}\\ &= \tfrac{4}{\sqrt{2}}\{1+\left(\tfrac{1}{\sqrt{2}}\right)+\left(\tfrac{1}{\sqrt{2}}\right)^2+\cdots\}\\ &= \frac{4}{\sqrt{2}-1} \approx 4\times 2.42.\end{aligned}$$

Olga is indifferent between the lottery **L** and $X if and only if their utilities are the same. So $X is the dollar equivalent of the lottery **L** if and only if

$$\begin{aligned}u(X) &= \mathscr{E}u(\mathbf{L})\\ 4\sqrt{X} &\approx 4\times 2.42\\ X &\approx (2.42)^2 = 5.86\end{aligned}$$

Thus Olga won't pay more than $5.86 to participate in the St. Petersburg lottery—which is a lot less than the infinite amount she would pay if her Von Neumann and Morgenstern utility function were $u(x)=x$. We will see that the reason we get such a different result is that Olga's new Von Neumann and Morgenstern utility function makes her risk averse instead of risk neutral.

→ 4.5.1

*Paradox of the Infinite?* Is the St. Petersburg paradox really resolved? If $u(x)\to\infty$ as $x\to\infty$, we can revive the paradox simply by choosing a different lottery **L** for which $\mathscr{E}u(L)$ is infinite.[6]

Mathematicians control such problems of the infinite by imposing extra postulates that ensure that a Von Neumann and Morgenstern utility function is bounded when the number of prizes is allowed to be infinite. For example, we could insist that rational players are never caught out by the Box Swapping paradox of Exercise 4.11.27.

However, nothing prevents our working with unbounded utility functions, provided we do only those things that are sanctioned by Von Neumann and Morgenstern's postulates. In particular, we must stick to lotteries that lie between some worst outcome $\mathscr{L}$ and some best outcome $\mathscr{W}$, although there is no harm in allowing lotteries with an infinite number of prizes when this constraint is observed. We can even allow $\mathscr{L}$ and $\mathscr{W}$ themselves to be such infinite lotteries since the Von Neumann and Morgenstern methodology will necessarily assign them both a finite expected utility. What this means in practice is that you don't need to worry that a Von Neumann and Morgenstern utility function is unbounded if you only plan to consider lotteries whose expected utility is finite. This is why the standard resolution of the St. Petersburg paradox with $u(x)=4\sqrt{x}$ is legitimate.

It doesn't help to try to make $\mathscr{W}$ and $\mathscr{L}$ the limits of infinite lotteries whose probabilities are progressively shifted outward toward dollar prizes that are

[6]Choose the prizes $\omega_n$ in **L** so large that $u(\omega_n)\geq 2^n$ $(n=1,2,\ldots)$. Then make the probability with which $\omega_n$ is chosen equal to $2^{-n}$.

increasingly positive or negative. The limiting value of the probability assigned to any particular prize would then be zero, but $\mathscr{W}$ and $\mathscr{L}$ can't have zero probabilities assigned to *all* their prizes.[7] (Exercise 4.11.28)

### 4.5.4 Risk Aversion

The *dollar* expectation of the lottery **M** in Figure 4.8 is

$$\mathscr{E}\mathbf{M} = \tfrac{3}{4} \times 1 + \tfrac{1}{4} \times 9 = 3\,.$$

If Olga's Von Neumann and Morgenstern utility for \$$x$ continues to be $u(x) = 4\sqrt{x}$, as in equation (4.3), her expected *utility* for **M** is

$$\mathscr{E}u(\mathbf{M}) = \tfrac{3}{4}u(1) + \tfrac{1}{4}u(9) = \tfrac{3}{4} \times 4\sqrt{1} + \tfrac{1}{4} \times 4\sqrt{9} = 6\,.$$

It follows that

$$u(\mathscr{E}\mathbf{M}) = u(3) = 4\sqrt{3} \approx 6.93 > 6 = \mathscr{E}u(\mathbf{M})\,,$$

and so Olga would rather not participate in the lottery if she can have its expected dollar value for certain instead.

If Olga would always sell a ticket for a lottery with money prizes for an amount equal to its expected dollar value, she is *risk averse* over money. If she would always buy a ticket for a lottery for an amount equal to its expected dollar value, then she is *risk loving*. If she is always indifferent between buying and selling, she is *risk neutral*.

The graphs of utility functions that represent risk-averse, risk-neutral and risk-loving preferences are shown in Figure 4.9. As we saw in Figure 4.8, chords drawn to the graph of the utility function of a risk-averse person lie on or below the graph. Mathematicians say that such functions are *concave*.[8] A function whose chords lie on or above its graph is *convex*. A person with a convex Von Neumann and Morgenstern utility function is risk loving.

A function with a straight-line graph is commonly said to be "linear," but the proper mathematical term is *affine*. If Olga has an affine Von Neumann and Morgenstern utility function, she is always indifferent between buying or selling a lottery for an amount equal to its expected value in dollars and so is simultaneously risk loving and risk averse.

The fallacy that makes the St. Petersburg story seem paradoxical is that rational people are *necessarily* risk neutral. If Olga were risk neutral (or risk loving), she

[7]The only way to escape pesky restrictions is to allow $\mathscr{W}$ and $\mathscr{L}$ to be something like tickets to heaven or hell, so that *all* lotteries with an infinite number of prizes can be squeezed between them. Infinite expected utilities can't then arise.

[8]A differentiable function $u$ is concave on an interval $I$ if and only if its derivative $u'$ is decreasing inside $I$. Economists refer to $u'(x)$ as a *marginal utility*. A risk-averse player therefore has decreasing marginal utility for money. Each extra dollar is worth less than its predecessor to such a player.

A differentiable function is decreasing on $I$ if and only if $u'(x) \leq 0$ for $x$ inside $I$. Thus, if $u$ can be differentiated twice, it is concave on $I$ if and only if $u''(x) \leq 0$ for $x$ inside $I$. A function $u$ is convex on $I$ if and only if $-u$ is concave on $I$. Thus a criterion for a function $u$ to be convex on $I$ is that $u''(x) \geq 0$ for $x$ inside $I$.

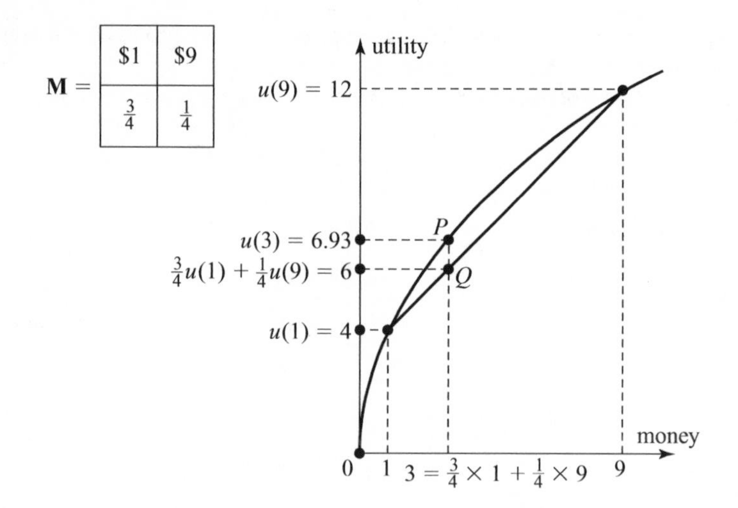


Figure 4.8 A lottery whose dollar expectation is \$3. Olga prefers to have $\$\varepsilon\mathbf{M} = \$3$ for certain to participating in the lottery M. The fact that $u(\mathscr{E}\mathrm{M}) > \varepsilon u(\mathrm{M})$ is equivalent to $P$ lying above $Q$ in the figure.

would indeed be prepared to liquidate all her assets to buy a ticket for the St. Petersburg lottery. But most people are risk averse when faced with similar choices. As we have seen, if Olga has the square-root utility function of equation (4.3), then she will pay no more than \$5.86 for a ticket.

→ 4.6

### 4.5.5 Taste for Gambling?

The shape of Olga's Von Neumann and Morgenstern utility function $u$ determines her attitude toward taking risks. Critics sometimes imagine that this turn of phrase means that $u$ measures the thrill that Olga derives from the act of gambling. They then ask why $u(a) > u(b)$ should be thought to have any relevance to how Olga chooses between $a$ and $b$ in riskless situations.

However, Von Neumann and Morgenstern's fourth postulate takes for granted that Olga is entirely *neutral* about the actual act of gambling. She doesn't bet because she enjoys betting—she bets only when she judges that the odds are in her favor. If she liked or disliked the act of gambling itself, we would have no reason to

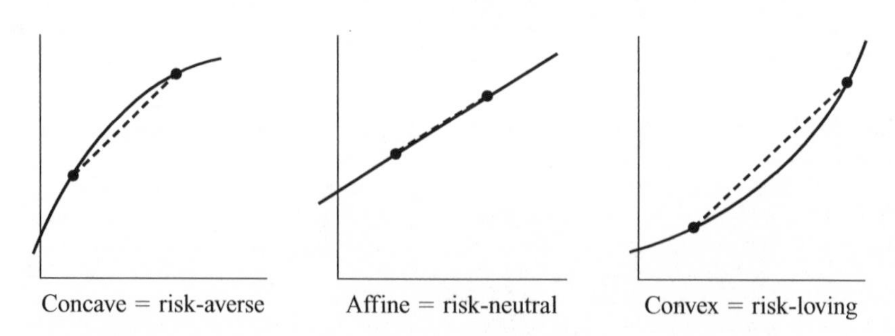


Figure 4.9 The shape of Olga's utility function reveals her attitude to risk.

assume that she is indifferent between a compound lottery and a simple lottery in which the prizes are available with the same probabilities.

To be rational in the sense of Von Neumann and Morgenstern, one needs to be as unemotional about gambling as the proverbial Cool Hand Luke. Alice may bet at the racetrack because she enjoys the excitement of the race. Bob may refuse to bet at all because he believes gambling is wicked. Neither satisfy the Von Neumann and Morgenstern postulates because they each like or dislike gambling for its own sake.

### 4.5.6 Does the End Justify the Means?

In game theory, $\Omega$ can usually be identified with the set of all outcomes of whatever game is being played. For example, when we used the theory of revealed preference in Section 1.4.2 to interpret the payoffs in the Prisoners' Dilemma, the outcomes were the four cells of the payoff table.

More generally, if Alice is a player in a game, we find her payoffs by asking her what she would do if she were free to choose between various pairs of lotteries whose prizes are outcomes in the game. This approach sometimes troubles purists, who feel that the theory of revealed preference should be applied in game theory only when all the players are choosing at once. But they then forget that the avowed purpose of orthodox game theory is to deduce what rational players will do in multiplayer games from the way they solve decision problems in which they are the only player.

Since the outcomes of a game can be identified with the terminal nodes (or leaves) of its extensive form, some philosophical critics complain that game theorists immorally proceed as though the end justifies the means. But this criticism overlooks the fact that each leaf is determined by the play that leads to it. So Von Neumann's formalism doesn't allow us to distinguish an outcome from the sequence of events that brought it about. Far from arguing that the end justifies the means, game theorists therefore take for granted that means and ends are inseparable.

## 4.6 Utility Scales

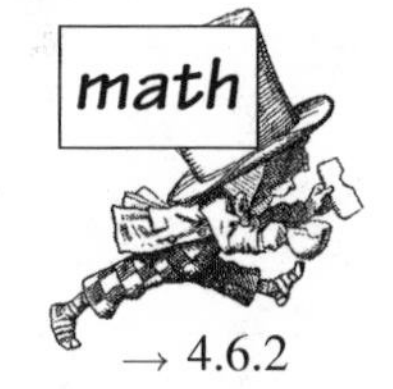

→ 4.6.2

For $u$ to be a utility function that represents the preference relation $\preceq$, we need that $a \preceq b \Leftrightarrow u(a) \leq u(b)$. But $u$ is never the only utility function that represents $\preceq$. There is always an infinite number of possible utility functions for any consistent preference relation.

For example, if we define $v$ and $w$ by $v(s) = \{(u(s)\}^3$ and $w(s) = 3u(s) + 7$, we obtain two alternative utility functions that represent $\preceq$ because

$$u(a) \leq u(b) \quad \Leftrightarrow \quad \{(u(a)\}^3 \leq \{(u(b)\}^3 \quad \Leftrightarrow \quad 3u(a)+7 \leq 3u(b)+7.$$

The same freedom of choice isn't available with a Von Neumann and Morgenstern utility function $u : \Omega \to \mathbb{R}$. It is true that $(\mathcal{E}u)^3$ and $3(\mathcal{E}u)+7$ represent Olga's preferences over lotteries just as well as $\mathcal{E}u$. It is also true that $u^3$ represents Olga's preferences over *prizes* just as well as $u$. But you will be very lucky if $u^3$ turns out to be a Von Neumann and Morgenstern utility function. That is to say, it isn't usually true that $\mathcal{E}(u^3)$ represents Olga's preferences over *lotteries*.

On the other hand, for any constants $A > 0$ and $B$,

$$\mathscr{E}(Au + B) = A\mathscr{E}u + B,$$

and so maximizing $\mathscr{E}u$ is the same as maximizing $\mathscr{E}(Au+B)$. Thus, $3u+7$ is necessarily a Von Neumann and Morgenstern utility function whenever the same is true of $u$.

### 4.6.1 Affine Transformations

If $A > 0$, the function $Au + B$ is a strictly increasing, affine transformation of $u$. The next theorem implies that we get *all* Von Neumann and Morgenstern utility functions that represent a given preference relation by taking strictly increasing, affine transformations of one such representation.

THEOREM 4.1 *If* $u_1 : \Omega \to \mathbb{R}$ *and* $u_2 : \Omega \to \mathbb{R}$ *are alternative Von Neumann and Morgenstern utility functions for a preference relation* $\preceq$ *defined on lott* $(\Omega)$, *then we can find constants* $A > 0$ *and* $B$ *such that*

$$u_2 = Au_1 + B.$$

*Proof* Pick $A_i > 0$ and $B_i$ to make the Von Neumann and Morgenstern utility function $U_i = A_i u_i + B_i$ satisfy $U_i(\mathscr{L}) = 0$ and $U_i(\mathscr{W}) = 1$. For any prize $\omega$ in $\Omega$, there is a probability $q$ for which $\omega \sim \mathbf{q}$ by Postulate 2. Thus,

$$U_i(\omega) = \mathscr{E}U_i(\mathbf{q}) = qU_i(\mathscr{W}) + (1-q)U_i(\mathscr{L}) = q.$$

Thus $A_1u_1(\omega) + B_1 = U_1(\omega) = U_2(\omega) = A_2u_2(\omega) + B_2$. The conclusion of the theorem follows on solving this equation for $u_2(\omega)$.

### 4.6.2 Utils

It follows from Theorem 4.1 that the origin and unit of a Von Neumann and Morgenstern utility scale can be chosen in any way you like, but you have then exhausted your room for maneuvering. Von Neumann and Morgenstern pointed out that things are much the same when measuring temperature.

The Centigrade or Celsius scale assigns 0°C to the freezing point of water and 100°C to its boiling point (at a stated atmospheric pressure). The Centigrade value for all other temperatures is fully determined by these choices. The Fahrenheit scale assigns 32°F to the freezing point of water and 212°F to its boiling point. Once these choices have been made, the Fahrenheit value for all other temperatures is fully determined. As with alternative utility scales, the Fahrenheit temperature $f$ is a strictly increasing affine function of the Centigrade temperature $c$. (In fact, $f = \frac{9}{5}c + 32$).

We can similarly set up an alternative Von Neumann and Morgenstern utility scale by recalibrating the scale determined by the original Von Neumann and Morgenstern utility function $u : \Omega \to \mathbb{R}$ as follows. First pick an outcome $\omega_0$ in $\Omega$ to correspond to the origin of the new utility scale. Then pick another outcome $\omega_1$ in $\Omega$ with $\omega_1 \prec \omega_0$ to determine the unit of the new scale.

It remains to choose a new von Neumann and Morgenstern utility function $U : \Omega \to \mathbb{R}$ with $U(\omega_0) = 0$ and $U(\omega_1) = 1$. Since $U = Au + B$ by Theorem 4.1, this step requires only that we choose $A$ and $B$ so that

$$0 = Au(\omega_0) + B;$$
$$1 = Au(\omega_1) + B.$$

We needn't worry about what values of $A$ and $B$ solve this pair of linear equations. All that matters is that they have a solution, and so we can always set up a new Von Neumann and Morgenstern utility scale with whatever origin and unit we find convenient.[9]

Just as the unit on a temperature scale is called a degree, so the unit on a Von Neumann and Morgenstern utility scale is called a *util.*

For example, we usually choose the utility scale of a risk-neutral player so that her preferences over money are represented by the simple utility function $u : \mathbb{R}_+ \to \mathbb{R}$ defined by $u(x) = x$. A util on the corresponding utility scale is then the same as a dollar. But we aren't able to get away with this simplifying trick when a player is risk averse because each extra util then corresponds to more dollars than the last, no matter what origin and unit we choose.

### 4.6.3 Interpersonal Comparison of Utility

We need to be careful in talking about units of utility called utils because the usage risks our falling prey to various fallacies, of which the most important is that which assumes Adam's utils can automatically be compared with Eve's.

For example, you would be making an unwarranted assumption if you blithely rated each of Adam's utils as being worth exactly the same as each of Eve's utils, without knowing anything about how the choice of origin and unit was made on Adam's and Eve's utility scales. You might as well claim that two rooms are equally warm because the Celsius thermometer in one room is showing the same temperature as the Fahrenheit thermometer in the other.

This observation is sometimes taught to economics students as the dogma that interpersonal comparisons of utility are *intrinsically* meaningless. It is true that we don't know how Adam's pleasure or pain can be compared with Eve's, but the utils of modern utility theory aren't units of pleasure and pain. It is also true that Von Neumann and Morgenstern's postulates provide no basis for making interpersonal comparisons of utility. However, as we will see in Chapter 19, nothing prevents our

[9] A property of a function $u : \Omega \to \mathbb{R}$ that is invariant under strictly increasing transformations is said to be *ordinal*. That is, for any strictly increasing $f : \mathbb{R} \to \mathbb{R}$, the composite function $f \circ u : \Omega \to \mathbb{R}$ defined by $f \circ u(s) = f(u(s))$ must retain the same property. A *cardinal* property is only invariant under strictly increasing, *affine* transformations. That is, for any $A > 0$ and any $B$, the function $Au + B$ must retain the same property. So the property of defining a temperature scale is cardinal, as is that of being a Von Neumann and Morgenstern utility function. The property of being any utility function at all is ordinal.

making further assumptions that correspond to requiring that the thermometers in different rooms all employ the same temperature scale when we use them to compare how warm the rooms are.

## 4.7 Dicing with Death Again

Section 4.4.2 explains that we need information about Boris's and Vladimir's attitudes to taking risks to solve the game of Russian Roulette. How do Von Neumann and Morgenstern utility functions take care of this problem?

The set of outcomes for each player in Russian Roulette is $\Omega = \{\mathscr{L}, \mathscr{D}, \mathscr{W}\}$. Their attitudes to taking risks are built into their Von Neumann and Morgenstern utility functions: $u_1 : \Omega \to \mathbb{R}$ and $u_2 : \Omega \to \mathbb{R}$. It is usually convenient to calibrate the utility scales so that the utility of the worst outcome is zero and the utility of the best outcome is one. We therefore suppose that

$$\begin{aligned} u_1(\mathscr{L}) &= 0, & u_1(\mathscr{D}) &= a, & u_1(\mathscr{W}) &= 1, \\ u_2(\mathscr{L}) &= 0, & u_2(\mathscr{D}) &= b, & u_2(\mathscr{W}) &= 1. \end{aligned}$$

Recall that $u_i(\mathscr{D}) = q$ means that player $i$ will swap $\mathscr{D}$ for the lottery $\mathbf{q}$ in which he gets $\mathscr{L}$ with probability $1 - q$ and $\mathscr{W}$ with probability $q$. Players who are more ready to take a risk therefore have smaller values of $u_i(\mathscr{D})$. So if $a > b$, then Boris is more cautious then Vladimir.

If you feel that the awfulness of being dead is undervalued by setting the utility of $\mathscr{L}$ to zero, think again! It wouldn't make any difference to the analysis if we set the utility of $\mathscr{L}$ equal to $-1{,}000{,}000$ instead. We would merely be recalibrating the utility scales, as explained in Section 4.6.2. It would be totally unrealistic to take $u_i(\mathscr{L}) = -\infty$, even if this were allowed by the Von Neumann and Morgenstern theory. Such a choice would imply that a player would never dare cross a road—even if offered a billion dollars to do so.[10]

After Chapter 3, it is child's play to solve version 2 of Russian Roulette using backward induction. Figure 4.10 shows the solution for three different pairs of values of the parameters $a$ and $b$. The boxes above each node show what the players' expected payoffs would be if the node were reached. They are filled in from right to left as the backward induction proceeds.

Begin by filling in the rightmost box that lies above the last decision node in Figure 4.10(a). The branch labeled $D$ is first doubled because a payoff of 0.55 is better for player II than 0. Thus, if the last decision node is reached, player II will play $D$, and so the outcome will be (1,0.55). This payoff pair is therefore written into the box above the last decision node. The preceding decision node is a chance move. If it is reached, player I's expected payoff is $0.5 \times 0 + 0.5 \times 1 = 0.5$, and player II's expected payoff is $0.5 \times 1 + 0.5 \times 0.55 = 0.775$. Rounding to two decimal places, we therefore write the payoff pair $(0.5, 0.78)$ into the box above the penultimate decision node of the game. At the preceding node, the branch labeled $A$ is now doubled because a payoff of 0.5 is better for player I than a payoff of 0.25.

[10]No matter how much care he took, there would still remain some small but positive probability of his being run over. The player's expected utility from taking up the offer would therefore remain $-\infty$.

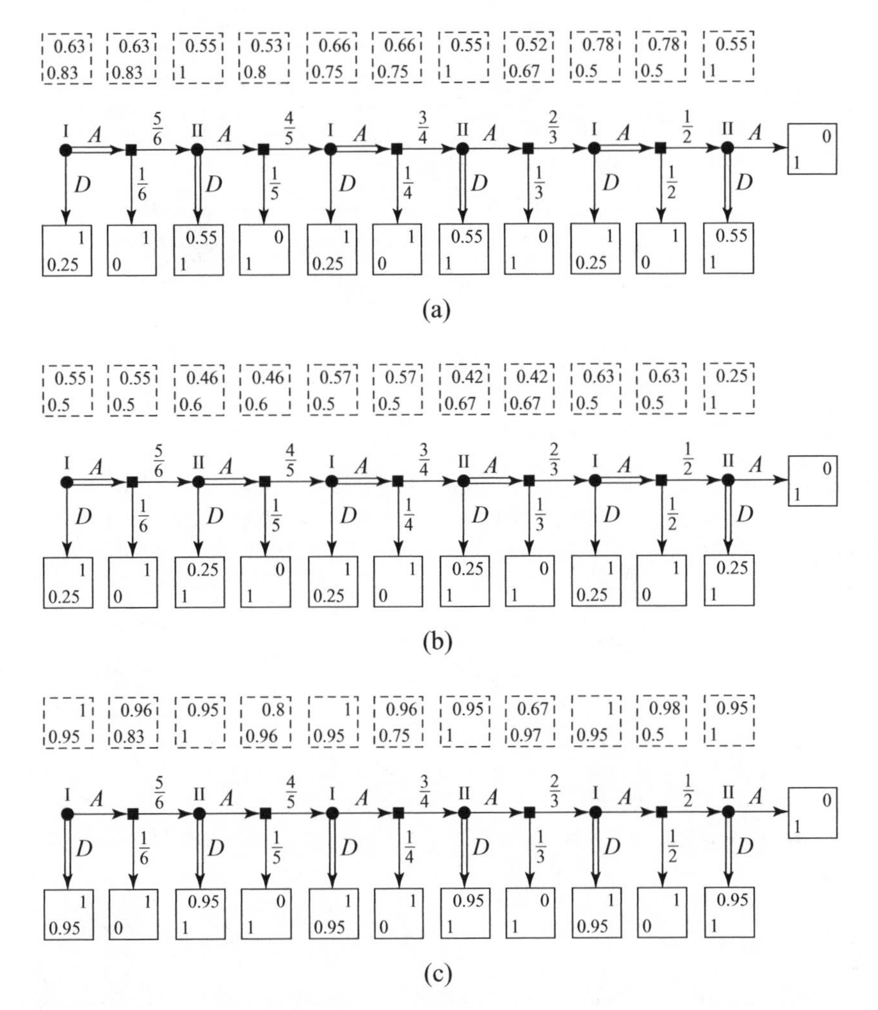


Figure 4.10 Backward induction in Russian Roulette. In Figure 4.10(a), $a = 0.25$ and $b = 0.55$, which makes Boris reckless and Vladimir mildly cautious. In Figure 4.10(b), $a = b = 0.25$, so that both players are reckless. In Figure 4.10(c), $a = b = 0.95$, so that both players are very cautious.

Continuing in this way, we find that player I will use the pure strategy *AAA,* and player II will use the pure strategy *DDD.* The payoffs they then expect to get appear in the leftmost box, above the first decision node of Figure 4.10(a).

*Conclusions.* The players' attitudes to taking risks make a big difference in the way the game is played. As Figure 4.11 indicates, cautious players chicken out a lot. Reckless players keep on pulling the trigger.

Is it better to be reckless or cautious? This is a question the model can't answer. Without building in some extra apparatus, it doesn't make any sense to compare different players' utils (Section 4.6.3).

For example, both players get a payoff of about 1 in case 3, while both players get a payoff of only about $\frac{1}{2}$ in case 2. But we aren't entitled to conclude that Boris and

| | parameter values | | player I | player II |
|---|---|---|---|---|
| I reckless, II cautious | $a = 0.25$ | $b = 0.55$ | $AAA$ | $DDD$ |
| both reckless | $a = 0.25$ | $b = 0.25$ | $AAA$ | $AAD$ |
| both cautious | $a = 0.95$ | $b = 0.95$ | $DDD$ | $DDD$ |

Figure 4.11 Comparing behavior in the three cases studied.

Vladimir would be better off playing Russian Roulette when they are old. For how sweet is an old man's triumph? Not nearly as sweet perhaps as half a chance of victory may seem to a hot-blooded youth—even if the downside is half a chance of getting shot.

→ 4.9

## 4.8 When Are People Consistent?

Von Neumann and Morgenstern's theory of decision making under risk has been much criticized. Some critics attack their consistency postulates. Others draw attention to the data from psychological laboratories, which show that real people often behave inconsistently. Both types of critic make free use of examples in which our gut feelings are at variance with the theory.

### 4.8.1 Allais' Paradox

Leonard Savage developed Von Neumann and Morgenstern's ideas into what is now called Bayesian decision theory (Chapter 13). When Savage was visiting Paris, Maurice Allais asked him to compare lotteries like those of Figure 4.12. When Savage made inconsistent replies, Allais triumphantly deduced that not even Savage believed his own theory!

Like Savage, most people express the preference $\mathbf{J} \succ \mathbf{K}$ because **J** guarantees \$1 million for sure, whereas **K** carries the risk of getting nothing at all. Again like Savage, most people express the preference $\mathbf{M} \succ \mathbf{L}$. Here the risk of ending up with nothing at all can't be avoided. On the contrary, the risk of this final outcome is high in both cases. But if the probability .89 in **L** is rounded up to .90 and .11 is rounded down to .10, then someone who understands what is going on will prefer **M** to the new **L**. If the new **L** is thought to be essentially the same as the old **L**, one then has a reason for preferring **M** to the old **L**.

The preferences $\mathbf{J} \succ \mathbf{K}$ and $\mathbf{M} \succ \mathbf{L}$ violate the Von Neumann and Morgenstern postulates. Otherwise they could be described with a Von Neumann and Morgenstern utility function $u : \Omega \to \mathbb{R}$. But the following argument shows that this is impossible.

Two points on a utility scale can be fixed in an arbitrary manner. In this case, it is convenient to fix $u(0)=0$ and $u(5)=1$. What can then be said about Savage's value for $x=u(1)$? Observe that

$$\begin{aligned}\varepsilon u(\mathbf{J}) &= u(0)\times 0.0 + u(1)\times 1.0 + u(5)\times 0.0 = x\\ \varepsilon u(\mathbf{K}) &= u(0)\times .01 + u(1)\times .89 + u(5)\times .10 = .89x + .10\\ \varepsilon u(\mathbf{L}) &= u(0)\times .89 + u(1)\times .11 + u(5)\times 0.0 = .11x\\ \varepsilon u(\mathbf{M}) &= u(0)\times .90 + u(1)\times 0.0 + u(5)\times .10 = .10.\end{aligned}$$

Since $\mathbf{J}\succ\mathbf{K}$, we have that $x>.89x+.10$, and so $x>\frac{10}{11}$. Since $\mathbf{L}\prec\mathbf{M}$, we also have that $.11x<.10$, and so $x<\frac{10}{11}$. But it can't be simultaneously true that $x<\frac{10}{11}$ and $x<\frac{10}{11}$. So the preferences that Savage expressed can't be described with a Von Neumann and Morgenstern utility function.

### 4.8.2 Zeckhauser's Paradox

I wasn't caught out by Allais' Paradox when it was first put to me, but everyone goes wrong when faced with the following problem, which is particularly apt in a chapter featuring Russian Roulette.

Some bullets are loaded into a revolver with six chambers, as illustrated in Figure 4.3(b). The cylinder is then spun and the gun pointed at your head. Would you be prepared to pay more to get one bullet removed when only one bullet was loaded or when four bullets were loaded? People usually say they would pay more in the first case because they would then be buying their lives for certain. But the Von Neumann and Morgenstern theory says that you should pay more in the second case, provided that you prefer life to death and more money to less.

To see why, suppose that you are just willing to pay \$$X$ to get one bullet removed from a gun containing one bullet and \$$Y$ to get one bullet removed from a gun containing four bullets. Let $\mathscr{L}$ mean death and $\mathscr{W}$ mean being alive after paying nothing. Let $\mathscr{C}$ mean being alive after paying \$$X$ and $\mathscr{D}$ mean being alive after paying \$$Y$.

You are indifferent between $\mathscr{C}$ and the lottery in which you get $\mathscr{L}$ with probability $\frac{1}{6}$ and $\mathscr{W}$ with probability $\frac{5}{6}$. Thus,

$$u(\mathscr{C}) = \tfrac{1}{6}u(\mathscr{L}) + \tfrac{5}{6}\,u(\mathscr{W})\,.$$

**J** =

| \$0m | \$1m | \$5m |
|---|---|---|
| 0 | 1 | 0 |

**K** =

| \$0m | \$1m | \$5m |
|---|---|---|
| .01 | .89 | .10 |

**L** =

| \$0m | \$1m | \$5m |
|---|---|---|
| .89 | .11 | 0 |

**M** =

| \$0m | \$1m | \$5m |
|---|---|---|
| .9 | 0 | .1 |

Figure 4.12 Lotteries for Allais's Paradox. The prizes are given in millions of dollars to dramatize the situation.

Similarly, you are indifferent between the lottery in which you get $\mathscr{L}$ and $\mathscr{D}$, each with probability $\frac{1}{2}$, and the lottery in which you get $\mathscr{L}$ with probability $\frac{2}{3}$ and $\mathscr{W}$ with probability $\frac{1}{3}$. Thus,

$$\tfrac{1}{2}u(\mathscr{L})+\tfrac{1}{2}u(\mathscr{D}) = \tfrac{2}{3}u(\mathscr{L})+\tfrac{1}{3}u(\mathscr{W}).$$

Simplify by taking $u(\mathscr{L}) = 0$ and $u(\mathscr{W}) = 1$. Then $u(\mathscr{C}) = \frac{5}{6}$ and $u(\mathscr{D}) = \frac{2}{3}$. Thus $\mathscr{D} \prec \mathscr{C}$, and thus $X < Y$.

After seeing the calculation, the result begins to seem more plausible. Would I be willing to pay more to get a bullet removed from a six-shooter containing *one* bullet than to get a bullet removed from a six-shooter containing *six* bullets? Definitely not! But getting a bullet removed when there are six bullets isn't so different from getting a bullet removed when there are five bullets, which isn't so different from getting a bullet removed when there are four bullets. *How* different is the difference between each of these cases? Appealing to our gut feelings doesn't get us very far when such questions are asked. We need to calculate.

### 4.8.3 Conclusions?

What conclusion should be drawn from such conflicts between our gut feelings and the Von Neumann and Morgenstern theory? Few people want to admit that their gut feelings are irrational and should therefore be amended. They prefer to deny that the Von Neumann and Morgenstern postulates characterize rational behavior. But consider the following informal experiment.

Would you prefer $96 \times 69$ dollars or $87 \times 78$ dollars? Most people say the former. But $96 \times 69 = 6{,}624$ and $87 \times 78 = 6{,}786$. How should we react to this anomaly? Surely not by altering the laws of arithmetic to make $96 \times 69 > 87 \times 78$! So why should we contemplate altering the Von Neumann and Morgenstern postulates after observing experiments that show they don't correspond with the gut feelings of the man in the street? But if real people don't honor the Von Neumann and Morgenstern assumptions when making risky decisions, how are we to predict their behavior in games?

The answer is similar to that given when we asked why anyone should care about Nash equilibria (Section 1.6). Orthodox game theory can't predict irrational behavior. It works only when players act rationally for some reason. For example, it wouldn't be very surprising to find a large insurance company systematically seeking to maximize its long-term average profit. Such companies employ teams of mathematicians to make sure that everything gets thought out properly. Nor should we be surprised to find animals that have been shaped by evolution over eons acting as though they were seeking to maximize their long-term average fitness.

However, what about games played by people like you and me? Although we are neither genetic robots nor mathematical wizards, we aren't stupid or incapable of adjusting our behavior to new circumstances. If the three criteria of Section 2.9.2 are satisfied, one might therefore hope that our play would evolve toward rationality in at least some games. However, it is necessary to face up to the fact that the laboratory evidence suggests that trial-and-error learning is especially difficult when the feedback from our choices is confused by chance moves.

Fortunately, we don't just learn by trial and error. We also learn from books. Just as it is easier to predict how educated kids will do arithmetic, so the spread of game theory into our universities and business schools will eventually make it easier to predict how decisions get made in economic life. If Pandora knows that $96 \times 69 = 6{,}624$ and $87 \times 78 = 6{,}786$, she won't make the mistake of choosing $96 \times 69$ dollars over $87 \times 78$ dollars—unless she sometimes likes to throw her money away. Once Allais had taught Savage that his choice behavior was inconsistent, Savage changed his mind about how to choose in Allais' Paradox. Similarly, I learned from Zeckhauser that I don't really want to pay more to get a bullet removed from a gun with one bullet than from a gun with four bullets.

In brief, economic theory in general and game theory in particular are useful predictive tools only when the conditions are favorable. Enthusiasts somehow manage to convince themselves that the theory always applies to everything, but such enthusiasm succeeds only in providing ammunition for skeptics looking for an excuse to junk the theory altogether. The unwelcome truth in the case of theories of human behavior under risk is that they have so far *all* performed badly in laboratory experiments. The best that can be said for expected utility theory is that it doesn't perform as badly overall as any of the behavioral theories that have been proposed as alternatives.

## 4.9 Roundup

The modern theory of utility takes choice behavior as basic. From the choices players make in one set of situations, we deduce the choices they will make in others, on the assumption that their behavior is stable and consistent. In the absence of risk, consistency is expressed in terms of the preference relation a player reveals. Rational preference relations are transitive and total. They need to be transitive to immunize players against money pumps.

A rational preference relation $\preceq$ can be described using a utility function $u$. This means that

$$u(a) \leq u(b) \quad \Leftrightarrow \quad a \preceq b\,.$$

Many utility functions describe the same preference relation.

Modern utility theory is commonly confused with a Victorian theory that sought to identify a util with a unit of pleasure or pain. Such a theory would explain our motivations when making choices. But the modern theory eschews all explanatory pretensions. It is a fallacy to say that Alice is motivated to choose $a$ over $b$ because $u(a) > u(b)$. We make $u(a) > u(b)$ because we already know that Alice always chooses $a$ when $b$ is available.

The game of Russian Roulette shows that one usually needs to know the players' attitudes to taking risks to predict how they will play a game. The St. Petersburg paradox shows that it isn't adequate to assume that players will simply maximize their expected gain in dollars. If they are consistent in the sense of Von Neumann and Morgenstern, they will maximize the expected value of a Von Neumann and Morgenstern utility function. The consistency assumptions are four in number:

1. In win-or-lose problems, players maximize their probability of winning.
2. For each outcome, there is a win-or-lose lottery such that a player is indifferent between the outcome and the lottery.
3. Players who are indifferent between two lotteries are willing to substitute one for the other when they appear as prizes in a compound lottery.
4. Players honor the laws of probability when evaluating compound lotteries.

Given a lottery with prizes expressed in dollars, risk-averse players prefer to replace the lottery with its expected value in dollars. Such players have concave Von Neumann and Morgenstern utility functions. Risk-loving players prefer the lottery to its expected value in dollars. They have convex Von Neumann and Morgenstern utility functions. Risk-neutral players are indifferent between the lottery and its expected value in dollars. Such players behave as though maximizing their expected dollar gain.

A Von Neumann and Morgenstern utility function is unique up to a strictly increasing affine transformation. This means that utility scales are related to each other in the same way as temperature scales. One can choose the zero and the unit arbitrarily, but then a utility scale is fixed. Because we may be measuring different people's utility on different scales, it isn't meaningful to compare different people's utils without adding something to the Von Neumann and Morgenstern theory.

The Von Neumann and Morgenstern theory describes rational behavior under risk, but the Allais and Zeckhauser paradoxes show that our gut feelings aren't always rational. Caution is therefore wise in evaluating economic work that takes for granted that ordinary people are maximizers of expected utility.

## 4.10 Further Reading

*Games and Decisions*, by Duncan Luce and Howard Raiffa: Wiley, New York, 1982. This is an old book, but its treatment of the Von Neumann and Morgenstern theory of risk has never been surpassed.

*Notes on the Theory of Choice*, by David Kreps: Westview Underground Classics in Economics, Boulder, CO, 1988. A great deal is explained without getting tangled up in more mathematics than necessary.

*Analytics of Uncertainty and Information*, by Jack Hirshleifer and John Riley: Cambridge University Press, New York, 1992. This is a book for the working economist that avoids technicalities when possible.

*Games and Economic Behavior*, by John Von Neumann and Oskar Morgenstern: Princeton University Press, Princeton, NJ, 1944. At a time when economists held that cardinal utility functions were meaningless, Von Neumann spent an afternoon at Morgenstern's behest inventing the consistency postulates of Section 4.5.2 that overturned the current orthodoxy. Their appendix on the subject is still relevant.

## 4.11 Exercises

phil

1. If Pandora is rational, she first determines which alternatives are feasible and then chooses an optimal alternative from her feasible set. Explain why Pandora can never be made worse off by adding new alternatives to her feasible set if this leaves the old alternatives unchanged. The following example of Amartya

Sen points out the importance of the final proviso. A respectable lady is inclined to accept an invitation to tea until she is told that she will also have an opportunity to snort cocaine. Her feasible set has expanded, but she now declines the invitation. How has her view of the original alternative changed?[11]

2. Rational players stay on the equilibrium play in a game because of what they predict would happen if they were to deviate. One might therefore stretch a point by arguing that the means that prevent a deviation determine the end reached in equilibrium (Section 4.5.5). Show how one can accommodate a critic who doesn't want the end to justify the means (even in this abstruse sense) by changing the payoffs in the strategic form of the game (Section 2.4). **phil**
3. Show that one and only one of **math**

$$a \prec b, \quad a \sim b, \quad a \succ b$$

holds when $\preceq$ is a rational preference relation (Section 4.2.2).

4. Show that any consistent preference relation $\preceq$ is reflexive. That is, for any $a$, $a \preceq a$. **math**
5. If $\preceq$ is a rational preference relation and $\sim$ is the associated indifference relation, show that $\sim$ satisfies reflexivity and transitivity. Show that the associated strict preference relation $\prec$ satisfies only transitivity. **math**
6. If $\preceq$ is a rational preference relation, show that **math**

$$a \prec b \text{ and } b \preceq c \quad \Rightarrow \quad a \prec c\,.$$

7. This exercise describes Condorcet's Voting Paradox (Sections 18.3.2 and 19.3.1). Horace, Boris, and Maurice vote honestly on who should be admitted to their club: Alice, Bob, or Nobody.[12] Their preferences are

$$\begin{array}{c} A \prec_1 B \prec_1 N \\ B \prec_2 N \prec_2 A \\ N \prec_3 A \prec_3 B. \end{array}$$

Who wins a vote between Alice and Bob? Who wins between Bob and Nobody? Who wins between Nobody and Alice?

If we think of the voting as determining a social preference $\preceq$, show that this preference is intransitive, and so democratic societies are collectively irrational in some situations.

8. Solve Pandora's optimization problem of Section 4.3.1 in the case when $U : \Omega \to \mathbb{R}$ is defined by **econ**

$$(a)\ U(g,v) = gv \qquad (b)\ U(g,v) = g^2 + v^2\,.$$

[11] One can always eliminate such apparent paradoxes by carefully separating a player's action, belief, and consequence spaces when writing a model (Section 13.4).

[12] The rhyming triplets voted strategically in Exercise 2.12.26.

econ

9. Construct two different utility functions that represent the preferences

$$a \sim b \prec c \prec d \prec e \sim f\,.$$

econ

10. Pandora can buy gin and vodka in only one of the four following packages: $A=(1,2)$, $B=(8,4)$, $C=(2,16)$, or $D=(4,8)$. When purchasing, she always has precisely \$24 to spend. If gin and vodka are both sold at \$2 a bottle, she sometimes buys $B$ and sometimes $D$. If gin sells for \$4 a bottle and vodka for \$1 a bottle, then she always buys $C$. Find a utility function $U:\{A,B,C,D\} \to \mathbb{R}$ that is consistent with this behavior.
11. Pandora's preferences satisfy $\mathscr{L} \prec \mathscr{D}_1 \prec \mathscr{D}_2 \prec \mathscr{W}$. She regards $\mathscr{D}_1$ and $\mathscr{D}_2$ as being equivalent to certain lotteries whose only prizes are $\mathscr{W}$ or $\mathscr{L}$. The appropriate lotteries are given in Figure 4.13. Find a Von Neumann and Morgenstern utility function that represents these preferences. Use this to determine Pandora's preference in the lotteries **L** and **M** of Figure 4.13 on the assumption that she is rational.
12. Alice's preferences over money are represented by a Von Neumann and Morgenstern utility function $u:\mathbb{R}_+ \to \mathbb{R}$ defined by $u(x)=x^a$. What would be implied about her preferences if $a<0$? What if $a=0$? Explain why Alice is risk averse if $0 \le a \le 1$ and risk loving if $a \ge 1$.

   If $a=2$, explain why Alice would pay \$1 million for the opportunity to participate in the lottery **K** of Figure 4.12. What is her dollar equivalent for the lottery **K**?

phil

13. In what sense is each extra dollar worth more to a risk-loving player than the previous dollar?
14. Pandora's Von Neumann and Morgenstern utility function is chosen so that her utility for dollars satisfies $u(0)=0$ and $u(10)=1$.
    a. If Pandora is risk averse, explain why $u(1) \ge 0.1$ and $u(9) \ge 0.9$.
    b. In one lottery **L**, the prizes \$0, \$1, \$9, and \$10 are available with respective probabilities 0.4, 0.3, 0.2, and 0.1. In a second lottery **M**, the same prizes are available with respective probabilities 0.5, 0.2, 0.1, and 0.2. Explain why a risk-averse Pandora would violate the Von Neumann and Morgenstern rationality assumptions if she expressed the preference $\mathbf{L} \prec \mathbf{M}$.
15. Bob's kindly but dissolute uncle offers him a choice for his birthday present. Two independent lotteries are taking place today and tomorrow. In each lottery, there is a single prize of \$1,000. Bob can have either one ticket in both

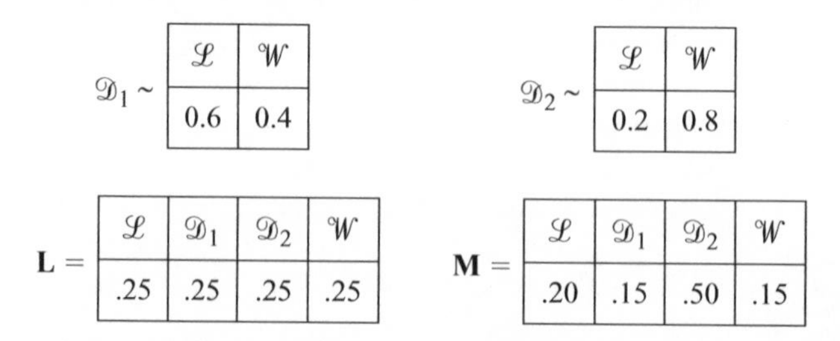


Figure 4.13 Lotteries for Exercise 4.11.1.

lotteries or two tickets in one lottery. If he is risk averse, show that he will prefer the latter option. Although most people are risk averse when it comes to taking out insurance policies, they nevertheless seem to prefer the former option. Offer a possible explanation based on Section 4.5.4.

16. In the previous problem, Bob desperately needs $1,000 to pay off a loan shark. He therefore regards all amounts in excess of $1,000 as being equivalent. Show that he will necessarily prefer the second option. Relate the answer to the advice offered at the end of Section 3.5.2.
17. If applying backward induction to the version of Russian Roulette shown in Figure 4.4 yields that player I uses strategy *AAD* and player II uses strategy *DDD*, what can be said about the values of $a$ and $b$?
18. Version 1 of Russian Roulette has only one chance move located at the beginning of the game. All games with chance moves can be expressed as an extensive form with this structure, provided that care is taken in specifying where the information sets go. Draw an extensive form of Gale's Roulette of Exercise 3.11.31 in which Chance moves only once at the beginning of the game. To simplify the task, assume that the casino has rigged the wheels so that the numbers on which they stop always sum to 15.
19. The rules of Gale's Roulette of Exercise 3.11.29 are changed so that the loser must pay the winner an amount in dollars equal to the difference in their scores. If both players are risk neutral over money, explain why they won't care which choices they make in the game (Exercise 3.11.32).
20. In the version of Gale's Roulette of Exercise 4.11.19, player I's preferences are altered so that his utility for money is described by the Von Neumann and Morgenstern utility function $\phi_1 : \mathbb{R} \to \mathbb{R}$ given by $\phi_1(x) = 3^x$. Denote the event that player I chooses wheel $i$ and player II chooses wheel $j$ by $(\mathbf{L}_i, \mathbf{L}_j)$. List the six possible events of this type. For each such event, find player I's dollar expectation and the utility that he assigns to getting a dollar amount equal to this expectation. Also find player I's expected utility for each of the six events. Is player I risk averse? Is player II risk averse if her Von Neumann and Morgenstern utility function $\phi_2 : \mathbb{R} \to \mathbb{R}$ is given by $\phi_2(x) = -3^{-x}$?
21. A charity is to sponsor a garden party to raise money, but the organizer is worried about the possibility of rain, which will occur on the day chosen for the event with probability $p$. She therefore considers insuring against rain. Her Von Neumann and Morgenstern utility for money $u : \mathbb{R} \to \mathbb{R}$ satisfies $u'(x) > 0$ and $u''(x) < 0$ for all $x$. Why does she like more money rather than less? Why is she strictly risk averse? Why is the function $u'$ strictly decreasing?

econ

If it is sunny on the day of the event, the charity will make \$$y$. If it rains, the charity will make only \$$z$. The insurance company offers full insurance against the potential loss of \$$(y - z)$ from rain at a premium of \$$M$, but the organizer may decide against full coverage by paying only a fraction $f$ of the full premium. This means that she pays \$$Mf$ before the event, and the insurance company repays \$0 if it is sunny and \$$(y - z)f$ if it rains. (Keep things simple by *not* making the realistic assumption that $f$ is restricted to the range $0 \leq f \leq 1$.)

a. What is the insurance company's dollar expectation if she buys full insurance? Why does it make sense to call the insurance contract fair if $M = p(y - z)$?

b. Why does the organizer choose $f$ to maximize $(1-p)u(y-Mf)+pu(z+(y-z)f-Mf)$? What do you get when this expression is differentiated with respect to $f$?
c. Show that the organizer buys full insurance ($f=1$) if the insurance contract is fair.
d. Show that the insurance contract is fair if the organizer buys full insurance.
e. If the insurance contract is unfair, with $M > p(y-z)$, show that the organizer definitely buys less than full insurance ($f < 1$).
f. How would the organizer feel about taking out a fair insurance contract if she were risk neutral?

econ

22. Reverse the prizes $0 million and $5 million in the lotteries of Figure 4.12. Are Savage's original preferences still inconsistent?
23. The cylinder of a six-shooter containing *two* bullets is spun, and the barrel is then pointed at a rich man's head (Section 4.8.2). He is now offered the opportunity of paying money to have the two bullets removed before the trigger is pulled. It turns out that the payment can be made as high as $10 million before he becomes indifferent between paying and taking the risk of getting shot.
   a. Why would the rich man also be indifferent between having the trigger pulled when the revolver contains *four* bullets and paying $10 million to have *one* of the bullets removed before the trigger is pulled? (Assume that he is rational in the sense of Von Neumann and Morgenstern.)
   b. Why wouldn't the rich man be willing to pay as much as $10 million to have *one* bullet removed from a revolver containing only one bullet?

fun

24. A misanthropic billionaire enjoys seeing people make mistakes. Claiming to be a philanthropist, he shows Pandora two closed boxes containing money. Pandora is to keep the money in whichever box she chooses to open. The billionaire explains that, however much she finds in the box she opens, the probability that the other box will contain twice as much is $\frac{1}{2}$. Since the boxes are identical in appearance, Pandora opens one at random. It contains $\$n$. Being risk neutral, she now calculates the expected dollar value of the other box as $\frac{1}{2}(\frac{1}{2}n)+\frac{1}{2}(2n) = 5n/4$. When she laments at having chosen wrongly, the misanthropic billionaire departs chuckling with glee.
   a. Could Pandora have chosen better?
   b. What is paradoxical about this story?
   c. Did Pandora calculate the expected dollar value of the other box correctly?
   d. Suppose that the billionaire actually chose the boxes so that the probability of one containing $\$2^k$ and the other containing $\$2^{k+1}$ is $p_k$ ($k=0,\pm 1, \pm 2,\ldots$). If Pandora knew this and opened a box containing $\$n=2^k$, explain why her conditional probability that the other box contains $\$2n$ would be $p_k/(p_k+p_{k-1})$. What would be her conditional probability that the other box contains $\$\frac{1}{2}n$?
   e. Continuing (d), which law of probability would the probabilities $p_k$ fail to satisfy if what the billionaire said to Pandora were correct?
25. The billionaire of the previous exercise is displeased at being exposed as a liar, and so he proposes another choice problem for Pandora. He chooses a natural number $k$ with probability $p_k > 0$ ($k=1,2,\ldots$) and then puts $\$M_k$ in one box

and $\$M_{k+1}$ in the other. Pandora again selects a box at random. If the billionaire arranges matters so that $M_2 > M_1$ and

$$M_{k+1}p_k + M_{k-1}p_{k-1} > M_k p_k + M_k p_{k-1} \quad (k = 1, 2, \ldots),$$

explain why Pandora will always regret not having chosen the other box. Verify that the choices $M_k = 3^k$ and $p_k = (\frac{1}{2})^k$ suffice to make the billionaire's plan work.

26. Suppose that Pandora is no longer risk neutral as in the previous exercise. Instead, $M_k$ now represents her Von Neumann and Morgenstern utility for whatever the billionaire puts in a box. Explain why her expected utility *before* she looks in a box is given by

math

$$\tfrac{1}{2}p_1 M_1 + \textstyle\sum_{k=2}^{\infty} \tfrac{1}{2}(p_k + p_{k-1})M_k.$$

If this expected utility is *finite*, show how summing the displayed inequality of the previous exercise between appropriate limits leads to the conclusion that $M_{k-1} > M_k$ $(k = 2, 3, \ldots)$.

Explain why it follows that the billionaire can't play his trick on Pandora unless her initial expected utility is infinite. Relate this conclusion to the St. Petersburg paradox of Section 4.5.1.

27. Explain why Pandora will be immune to the billionaire's trick in the Box Swapping paradox of the previous exercise only if her Von Neumann and Morgenstern utility for money is *bounded*. If she is immune, why does it follow that she can't always be risk loving when choosing among lotteries whose prizes are monetary amounts?

math

28. Pandora finds herself in Hell, but the Devil offers her a way out. She gets one chance to participate in a lottery in which the prizes are an eternity in either Heaven or Hell. If she says *yes* to the lottery on her $n$th day in Hell, she gets Heaven with probability $(n-1)/n$ and Hell with probability $1/n$. The philosophical paradox is that if she always waits one more day to improve her chances of Heaven, she will spend eternity in Hell anyway.

fun

Explain why the paradox neglects the disutility of spending an extra day in Hell. Demolish the objection that this disutility must be negligible compared with an eternity in Hell because eternity consists of an infinite number of days. The moral is that if it doesn't matter *when* you get something, then it doesn't matter *if* you get it.

29. Pascal's Wager represents a more serious attempt to use probabilistic arguments in theology than the previous exercise. Pandora can choose to follow the straight and narrow path of rectitude (good) or she can indulge her passions (bad). If there is an afterlife, the ultimate reward for living a good life and the punishment for living a bad life will be infinitely more important than anything that might happen on this earth. Pascal's argument is therefore that Pandora ought to be good, even if she believes that the probability of an afterlife is very small.

phil

Explain why its use of infinite magnitudes means that Pascal's Wager can't be accommodated within the Von Neumann and Morgenstern theory. Omitting the word *infinitely* from Pascal's assumptions, formulate a version of the wager that shows it is rational for Pandora to be good if the probability of an afterlife isn't too small.

Of course, Pandora may doubt Pascal's implicit assumption that only his religion is viable. Analyze a version of the wager in which two religions offer diametrically opposed views on what counts as good or bad.

# 5 Planning Ahead

## 5.1 Strategic Forms

A game defined in terms of a tree is said to be given in extensive form. A pure strategy in the extensive form of a game specifies an action at each of a player's information sets. A pure strategy *profile* specifies a pure strategy for each player. If the players stick with these pure strategies, the resulting play of the game is entirely determined in a game without chance moves.

In a game with chance moves, a pure strategy profile determines a *lottery* over the possible plays of the game. We assess such lotteries using Von Neumann and Morgenstern utilities that we call payoffs. Rational players then act as though attempting to maximize their expected payoff in the game.

The strategic form of a game tells us what payoff a player will get for each strategy profile that might be played. In a two-player game, we usually specify a strategic form with a table. We have already seen many outcome tables, but we stopped giving the outcomes in terms of payoffs after Chapter 1. However, now that we understand what game theorists mean by a payoff, we can can proudly point to the Prisoners' Dilemma as the most famous example of the strategic form of a game.

Von Neumann and Morgenstern invented both the extensive and the strategic form of a game. They called the latter a *normal form* in the belief that one would normally use the extensive form only as a transitional stage in constructing the strategic form. Such an approach amounts to arguing that one can always assume that the players begin a game by making a firm preplay commitment to a particular strategy. But things have moved on since the time of Von Neumann and Morgenstern. Game theorists learned from Thomas Schelling that one needs to be much more careful when modeling credible commitments. When the basics of working with strategic forms have been nailed down, the chapter looks at some examples in which credibility and commitment are important.

## 5.2 Payoff Functions

If player I chooses pure strategy $s$ and player II chooses pure strategy $t$, then the course of a two-player game is entirely determined, except for the game's chance moves. The pair $(s,t)$ therefore determines a *lottery* $\mathbf{L}$ over the set $\Omega$ of pure outcomes of the game. The payoff $\pi_i(s,t)$ that player $i$ gets when the pair $(s,t)$ is used is the expected utility of the lottery $\mathbf{L}$. That is to say,

$$\pi_i(s,\ t) = \mathscr{E}u_i(\mathbf{L}).$$

If $S$ is the set of all player I's pure strategies and $T$ is the set of all player II's pure strategies, then $\pi_i : S \times T \to \mathbb{R}$ is player $i$'s *payoff function*.

A profile of payoff functions is an algebraic way of representing the strategic form or payoff table of a game. If $S = \{s_1, s_2\}$ and $T = \{t_1, t_2, t_3\}$, the payoff table has two rows and three columns. If the payoff functions are given by

$$\pi_1(s_i,\ t_j) = ij\,,$$
$$\pi_2(s_i,\ t_j) = (i-2)(j-2)\,,$$

then the entries in the payoff table are as shown in Figure 5.1. Player I's payoff $\pi_1(s,t)$ goes in the southwest corner of the cell in row $s$ and column $t$. Player II's payoff $\pi_2(s,t)$ goes in the northeast corner.

A strategic form is sometimes called a *bimatrix game* because it is determined by two *payoff matrices*. In Figure 5.1, player I's payoff matrix is A, and player II's payoff matrix is B, where

$$A = \begin{bmatrix} 1 & 2 & 3 \\ 2 & 4 & 6 \end{bmatrix}; \qquad B = \begin{bmatrix} 1 & 0 & -1 \\ 0 & 0 & 0 \end{bmatrix}.$$

In a game with more than two players, a player's payoff function can't be represented as a two-dimensional array like a matrix. With $n$ players, we need an $n$-dimensional array. Figure 5.2(a) shows a three-dimensional payoff array for player I in a game with two pure strategies for each of three players. We usually think of such an array as a stack of matrices. The whole strategic form can then be

| | $t_1$ | $t_2$ | $t_3$ |
|---|---|---|---|
| $s_1$ | 1, 1 | 2, 0 | 3, −1 |
| $s_2$ | 2, 0 | 4, 0 | 6, 0 |

Figure 5.1 A bimatrix game.

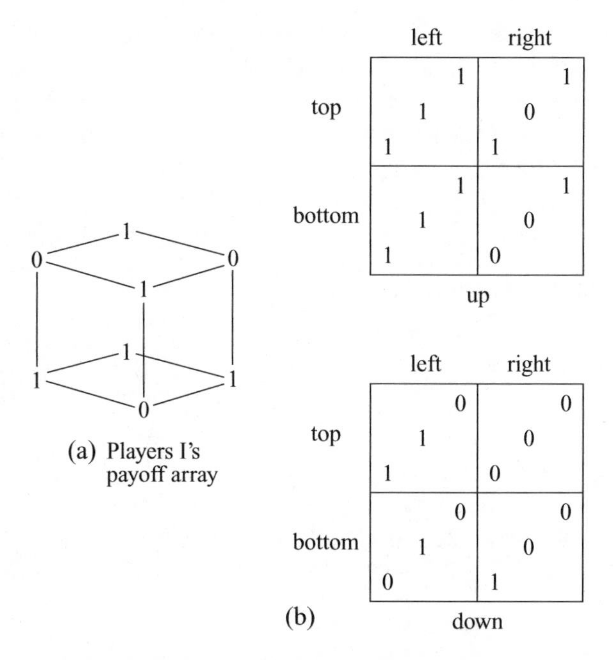


Figure 5.2 The strategic form of a game with three players. Player I chooses a row. His payoffs are at the bottom left of each cell. Player II chooses a column. Her payoffs are in the middle of each cell. Player III chooses a 'matrix.' His payoffs are in the top right of each cell.

represented as in Figure 5.2(b). Player I chooses the row. Player II chooses the column. Player III is usually said to choose the "matrix."[1]

Payoff matrices appeared for the first time in Section 1.3.1 when the Prisoners' Dilemma was introduced, so nothing is new here except for the notation. However, it isn't always easy to compute a player's payoff function when a complicated game is given in extensive form. Some examples may help to show how one goes about this task.

### 5.2.1 A Strategic Form for Duel

Recall that Tweedledum is player I and Tweedledee is player II in the game Duel of Section 3.7.2. The outcome $\mathscr{W}$ is the event that player II gets shot. The outcome $\mathscr{L}$ is the event that player I gets shot. The lottery in which $\mathscr{W}$ occurs with probability $q$ and $\mathscr{L}$ with probability $1-q$ is denoted by $\mathbf{q}$.

*Payoff Functions.* Calibrate the players Von Neumann and Morgenstern utility functions $u_i{:}\{\mathscr{L},\mathscr{W}\} \to \mathbb{R}$ so that $u_1(\mathscr{L}) = u_2(\mathscr{W}) = 0$, and $u_1(\mathscr{W}) = u_2(\mathscr{L}) = 1$. We then have $\mathscr{E}u_1(\mathbf{q}) = q$ and $\mathscr{E}u_2(\mathbf{q}) = 1-q$, which is just a fancy way of saying that both players want to maximize the probability of surviving. Notice that the players' payoffs always sum to one.

[1]When people talk about the payoff matrix of a game without saying whose payoff matrix it is, they usually mean the payoff table of the game.

What matters in Duel is how close you get to your opponent before pulling the trigger. A pure strategy that calls for a player to plan to open fire at node $d$ will be denoted by $d$. There are many such strategies that differ in what they specify at later nodes, but they would be indistinguishable from each other if we included them all in the strategic form of Duel (Section 2.4).

If player I uses pure strategy $d$ and player II uses pure strategy $e$, then the outcome of the game depends on who fires first. If $d > e$, so that player I fires first, the result is the lottery $\mathbf{p_1(d)}$. If $d < e$, so that player II fires first, the result is the lottery $1 - \mathbf{p_2(e)}$. Player I's payoff function is therefore given by

$$\pi_1(d,\, e) = \begin{cases} p_1(d)\,, & \text{if} \quad d > e, \\ 1 - p_2(e)\,, & \text{if} \quad d < e. \end{cases} \tag{5.1}$$

Player II's payoff function is given by $\pi_2(d, e) = 1 - \pi_1(d, e)$.

*Payoff Table.* To obtain a payoff table with numerical entries, we have to assign values to the parameters of the game. We begin by setting $D = 1$ and

$$d_k = 0.1\,k \quad (k = 0,\, 1,\, 2 \ldots 10)\,.$$

The probabilities $p_1(d)$ and $p_2(d)$ are taken to be the same as in the final paragraph of Section 3.7.2. That is to say, $p_1(d) = 1 - d$ and $p_2(d) = 1 - d^2$. The payoffs that go in row $d_2$ and column $d_5$ of Figure 5.3 are therefore

| | $d_9 = 0.9$ | $d_7 = 0.7$ | $d_5 = 0.5$ | $d_3 = 0.3$ | $d_1 = 0.1$ |
|---|---|---|---|---|---|
| $d_{10} = 1.0$ | 1.00 / 0.00 | 1.00 / 0.00 | 1.00 / 0.00 | 1.00 / 0.00 | 1.00 / 0.00 |
| $d_8 = 0.8$ | 0.19 / 0.81 | 0.80 / 0.20 | 0.80 / 0.20 | 0.80 / 0.20 | 0.80 / 0.20 |
| $d_6 = 0.6$ | 0.19 / 0.81 | 0.51 / 0.49 | 0.60 / 0.40 | 0.60 / 0.40 | 0.60 / 0.40 |
| $d_4 = 0.4$ | 0.19 / 0.81 | 0.51 / 0.49 | 0.75 / 0.25 | 0.40 / 0.60 | 0.40 / 0.60 |
| $d_2 = 0.2$ | 0.19 / 0.81 | 0.51 / 0.49 | 0.75 / 0.25 | 0.91 / 0.09 | 0.20 / 0.80 |
| $d_0 = 0.0$ | 0.19 / 0.81 | 0.51 / 0.49 | 0.75 / 0.25 | 0.91 / 0.09 | 0.99 / 0.01 |

Figure 5.3 A strategic form for Duel. The payoff table is strictly a *reduced* strategic form, as we have identified all the pure strategies that call on a player to fire at distance $d$. Note the unique Nash equilibrium $(d_6, d_5)$.

$$\pi_1(d_2, d_5) = 1 - p_2(d_5) = 1 - (1 - d_5^2) = d_5^2 = (0.5)^2 = 0.25,$$
$$\pi_2(d_2, d_5) = 1 - \pi_1(d_2, d_5) = 0.75\,.$$

*Nash Equilibria.* A pair $(\sigma, \tau)$ of strategies is a Nash equilibrium in a two-player game if $\sigma$ is a best reply to $\tau$ and $\tau$ is simultaneously a best reply to $\sigma$ (Section 1.6). This is the same as requiring that the inequalities

$$\left.\begin{array}{l}\pi_1(\sigma,\tau) \geq \pi_1(s,\tau) \\ \pi_2(\sigma,\tau) \geq \pi_2(\sigma,t)\end{array}\right\} \tag{5.2}$$

hold for all pure strategies $s$ and $t$. The first inequality says that player I can't improve on $\sigma$ if player II doesn't deviate from $\tau$. The second inequality says that player II can't improve on $\tau$ if player I doesn't deviate from $\sigma$.

Circles and squares have been used to show best-reply payoffs in Figure 5.3 (Section 1.3.1). For example, 0.80 is enclosed in a square four times in row $d_8$ to indicate that $d_7$, $d_5$, $d_3$, and $d_1$ are all best replies for player II to the choice of $d_8$ by player I.

The only cell with both payoffs enclosed in a circle or a square lies in row $d_6$ and column $d_5$. So $(d_6, d_5)$ is the only Nash equilibrium in pure strategies.[2]

*Conclusion.* How does this result compare with our previous analysis of Duel? Section 3.7.2 used backward induction to determine a subgame-perfect equilibrium for the game. The method used here is less refined in that it finds all Nash equilibria in pure strategies. Recall that any subgame-perfect equilibrium is also a Nash equilibrium, but some Nash equilibria aren't subgame perfect (Section 2.9.3). However, we have only one Nash equilibrium in this case, and so it must coincide with the subgame-perfect equilibrium that an application of backward induction would uncover.

Section 3.7.2 observes that rational players open fire when they are about distance $\delta = (\sqrt{5} - 1)/2 = 0.62$ apart, provided the nodes $d_0$, $d_1$, ..., $d_n$ are closely spaced. In the version of Duel studied here, the distance between nodes is 0.1, so the spacing isn't particularly close. Nevertheless, player I opens fire at distance $d_6 = 0.60$, which isn't too far from $\delta$.

### 5.2.2 A Strategic Form for Russian Roulette

It is necessary to work a little harder to compute the payoff functions in the Russian Roulette game of Section 4.7.

Figure 5.4(a) repeats version 2 of the extensive form of Russian Roulette from Section 4.4.2. Figure 5.4(b) is a reduced strategic form in which only four of each player's eight pure strategies have been included. Russian Roulette is a waiting game like Duel. All that really matters is how long a player is prepared to wait before chickening out. As in Duel, we therefore really need only one pure strategy for each possible waiting time.

[2]The pair $(d_6, d_5)$ is a saddle point of player I's payoff matrix, but only in strictly competitive games like Duel do saddle points always correspond to Nash equilibria (Section 2.8.2).

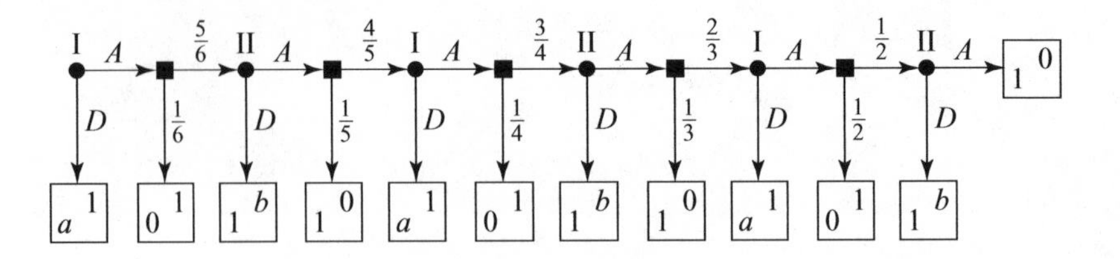


(a) Extensive form

| | $DDD$ | $ADD$ | $AAD$ | $AAA$ |
|---|---|---|---|---|
| $DDD$ | $a$, $1$ | $a$, $1$ | $a$, $1$ | $a$, $1$ |
| $ADD$ | $\frac{5}{6}$, $\frac{1}{6}+\frac{5}{6}b$ | $\frac{1}{6}+\frac{2}{3}a$, $\frac{5}{6}$ | $\frac{1}{6}+\frac{2}{3}a$, $\frac{5}{6}$ | $\frac{1}{6}+\frac{2}{3}a$, $\frac{5}{6}$ |
| $AAD$ | $\frac{5}{6}$, $\frac{1}{6}+\frac{5}{6}b$ | $\frac{2}{3}$, $\frac{1}{3}+\frac{1}{2}b$ | $\frac{1}{3}+\frac{1}{3}a$, $\frac{2}{3}$ | $\frac{1}{3}+\frac{1}{3}a$, $\frac{2}{3}$ |
| $AAA$ | $\frac{5}{6}$, $\frac{1}{6}+\frac{5}{6}b$ | $\frac{2}{3}$, $\frac{1}{3}+\frac{1}{2}b$ | $\frac{1}{2}$, $\frac{1}{2}+\frac{1}{6}b$ | $\frac{1}{2}$, $\frac{1}{2}$ |

(b) Reduced strategic form

| plays | $[Ad]$ | $[AaAd]$ | $[AaAaAd]$ | $[AaAaAaD]$ |
|---|---|---|---|---|
| payoffs | $0$, $1$ | $1$, $0$ | $0$, $1$ | $1$, $b$ |
| probabilities | $\frac{1}{6}$ | $\frac{5}{6}\times\frac{1}{5}=\frac{1}{6}$ | $\frac{5}{6}\times\frac{4}{5}\times\frac{1}{4}=\frac{1}{6}$ | $\frac{5}{6}\times\frac{4}{5}\times\frac{3}{4}=\frac{1}{2}$ |

(c) The lottery corresponding to (*AAD*, *ADD*)

Figure 5.4 A reduced strategic form for Russian Roulette.

Figure 5.4(c) illustrates a method for finding the entries in the strategic form for the pure strategy pair (*AAD*, *ADD*). When this pure strategy pair is used, the possible plays of the game that might result depend on the choices made by Chance. Her choices are denoted by $a$ for *across* and $d$ for *down*.

The play [*AaAaAd*] occurs if Chance plays $a$ at the first and second chance moves and then $d$ at the third chance move. The probability of this play is prob(*aad*) $= \frac{5}{6}\times\frac{4}{5}\times\frac{1}{4}=\frac{1}{6}$, which is the probability that the bullet is in the third chamber of the revolver.

The expected utility of the lottery resulting from the use of $(AAD, ADD)$ is obtained by multiplying each of a player's payoffs by the probability with which it occurs and then summing the resulting products. Thus,

$$\pi_1(AAD,\ ADD) = 0\times\tfrac{1}{6}+1\times\tfrac{1}{6}+0\times\tfrac{1}{6}+1\times\tfrac{1}{2}=\tfrac{2}{3},$$
$$\pi_2(AAD,\ ADD) = 1\times\tfrac{1}{6}+0\times\tfrac{1}{6}+1\times\tfrac{1}{6}+b\times\tfrac{1}{2}=\tfrac{1}{3}+\tfrac{1}{2}b\,.$$

## 5.3 Matrices and Vectors

→ 5.4

We don't need to know much about matrices to study bimatrix games. Even the material surveyed here is more than is really essential.

### 5.3.1 Matrices

An *$m \times n$ matrix* is a rectangular array of numbers with $m$ rows and $n$ columns. In the following examples, $A$ is a $2 \times 3$ matrix and B is a $3 \times 2$ matrix:

$$A = \begin{bmatrix} 3 & 0 & 1 \\ 1 & 0 & -2 \end{bmatrix}; \qquad B = \begin{bmatrix} 2 & 3 \\ 1 & 0 \\ 0 & -3 \end{bmatrix}.$$

The standard notation sometimes invites confusion between a matrix and a number. In particular, the *zero matrix,* whose entries are all zero, is always denoted by 0, whatever its dimensions may be. You have to deduce from the context whether 0 is the zero number or a zero matrix. However, it is always important to be quite clear about what a number is and what a matrix is.

The difference between numbers and matrices is sometimes emphasized by referring to numbers as *scalars*. Our scalars are always real numbers, but they are often complex numbers in other contexts.[3]

*Transposition.* To obtain the *transpose* $M^\top$ or M$'$ of a matrix $M$, you swap its rows and columns. For example,

$$A^\top = \begin{bmatrix} 3 & 1 \\ 0 & 0 \\ 1 & -2 \end{bmatrix}; \qquad B^\top = \begin{bmatrix} 2 & 1 & 0 \\ 3 & 0 & -3 \end{bmatrix}.$$

If $M$ is a $1 \times 1$ matrix, then $M = M^\top$. It is always true that $(M^\top)^\top = M$.

If $M$ is an $m \times n$ matrix, then $M = M^\top$ can hold only if $m = n$, so that $M$ is a square matrix. A square matrix $M$ for which $M = M^\top$ is said to be symmetric. Some examples are

[3]However, scalars must belong to some algebraic field. It follows that a payoff table isn't properly a matrix because a multidimensional vector space isn't a field.

$$I = \begin{bmatrix} 1 & 0 \\ 0 & 1 \end{bmatrix}; \qquad J = \begin{bmatrix} 1 & 2 & 3 \\ 2 & 1 & 3 \\ 3 & 3 & 1 \end{bmatrix}.$$

*Symmetric Games.* A symmetric game is one that looks the same to all the players. In a two-player game, the rows of player I's payoff matrix $A$ must therefore be the same as the columns of player II's payoff matrix $B$. Thus $B$ must be the *transpose* of $A$, so that $B = A^\top$ (and $A = B^\top$).

Although the payoff matrices in a symmetric game must be square, they usually aren't themselves symmetric. For example, the Prisoners' Dilemma is a symmetric game whose payoff matrices aren't symmetric.

### 5.3.2 Vectors

An $n$-dimensional vector is a list of $n$ real numbers $x_1, x_2, \ldots, x_n$ that are called its coordinates. The set of all $n$-dimensional vectors with real coordinates is denoted by

$$\mathbb{R}^n = \mathbb{R} \times \mathbb{R} \times \cdots \times \mathbb{R}.$$

We are accustomed to writing $x = (x_1, x_2, \ldots, x_n)$, but when using matrix algebra, it should always be assumed that $x$ is an $n \times 1$ matrix called a column vector. The corresponding $n \times 1$ row vector is then $x^\top$, so that:

$$x = \begin{bmatrix} x_1 \\ x_2 \\ \vdots \\ x_n \end{bmatrix}; \qquad x^\top = [\, x_1 \quad x_2 \quad \cdots \quad x_n \,].$$

As in Figure 5.5(a), a vector $x = (x_1, x_2)$ in $\mathbb{R}^2$ can be identified with a point in a plane referred to as Cartesian axes. The zero vector $0 = (0, 0)$ then lies at the origin of the pair of axes.

We can also regard $x$ as the displacement that moves everything $x_1$ units to the right and $x_2$ units up. As in Figure 5.5(b), the displacement can be represented as an arrow with its blunt end at the origin and its sharp end at the location $x$. However, any arrow with the same length and direction represents exactly the same displacement, and so we are free to put arrows anywhere convenient when drawing diagrams.

*Ordering Vectors.* If $x_1 \le y_1$, $x_2 \le y_2, \ldots, x_n \le y_n$, then we write $x \le y$. For example,

$$\begin{bmatrix} 3 \\ 0 \\ -1 \end{bmatrix} \le \begin{bmatrix} 3 \\ 2 \\ 0 \end{bmatrix} \qquad (5.3)$$

The set of all $x$ in $\mathbb{R}^2$ with $x \le y$ is shown in Figure 5.6(a). The set of all $x$ in $\mathbb{R}^2$ with $x \ge y$ is shown in Figure 5.6(b). These two sets don't make up the whole of $\mathbb{R}^2$,

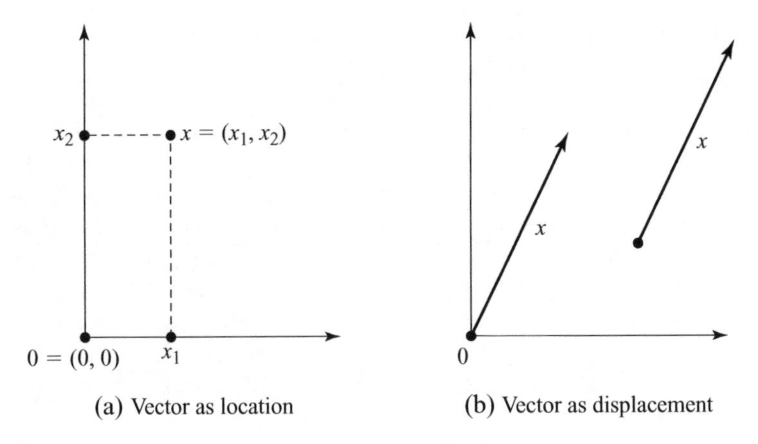


(a) Vector as location (b) Vector as displacement

Figure 5.5 Vectors as locations or displacements.

because the relation $\leq$ is only a *partial* ordering since it doesn't satisfy the totality requirement of Section 4.2.2. For example, neither of the inequalities $(1, 2) \leq (2, 1)$ or $(2, 1) \geq (1, 2)$ is true.

The notation $x < y$ is sometimes used to mean that $x_1 < y_1$, $x_2 < y_2, \ldots, x_n < y_n$, but this book uses the notation $x \ll y$ for this purpose. We use the notation $x < y$ to mean that $x \leq y$ but $x \neq y$. We can therefore replace $\leq$ in (5.3) by $<$ but not by $\ll$.

## 5.4 Domination

Alice doesn't care whether the companies in which she invests actually make money or not. She is only interested in whether their shares go up in value. Whether they go up in value depends on what other people believe about the shares. Investors like Alice are therefore really investing on the basis of their beliefs about other people's beliefs. If Bob plans to exploit investors like Alice, he will need to take account of his beliefs about what she believes about what other people believe. If we want to

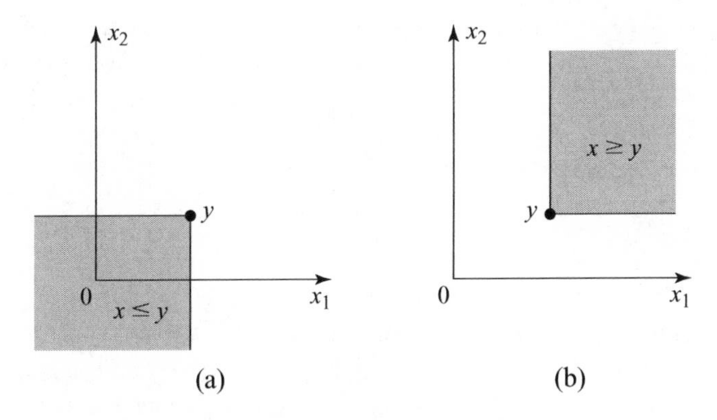


(a) (b)

Figure 5.6 Ordering vectors in $\mathbb{R}^2$.

exploit Bob, we will need to ask what we believe about what Bob believes about what Alice believes about what other people believe.

John Maynard Keynes famously used the beauty contests run by newspapers of his time to illustrate how these chains of beliefs about beliefs get longer and longer the more one thinks about the problem. The aim in these newspaper contests was to choose the girl chosen by most other people. Game theorists prefer to illustrate the problem with a game in which the winners are the players who choose a number that is closest to two-thirds of the average of all the numbers chosen by the players.

If the players are restricted to whole numbers between 1 and 10 inclusive, only a foolish player will choose a number above 7 because the average can be at most 10, and $\frac{2}{3}\times 10 = 6\frac{2}{3}$. You therefore improve your chances of winning by playing 7 instead of 8, 9, or 10. In the language of Section 1.7.1, strategies 8, 9, and 10 are weakly dominated by strategy 7.

However, if nobody thinks that anyone is stupid enough to play 8, 9, or 10, then everybody believes that the average will be at most 7, and $\frac{2}{3}\times 7 = 4\frac{2}{3}$. It would therefore be foolish to play more than 5. But if nobody thinks that anyone is stupid enough to play above 5, then the average will be at most 5, and $\frac{2}{3}\times 5 = 3\frac{1}{3}$. It would then be unwise to play more than 3. Continuing in this way, we find that everybody will choose 1—provided that everybody believes that everybody is clever enough to work through all the necessary steps.

This method of solving a game is called the successive or iterated deletion of dominated strategies.

### 5.4.1 Strong and Weak Domination

We met strongly dominant strategies in Section 1.3.1 when studying the Prisoners' Dilemma. Weakly dominant strategies appeared in the Film Star Game of Section 1.7.1. We now need to put these ideas on firmer ground.

Player I has two pure strategies in the game of Figure 5.1. Pure strategy $s_2$ strongly dominates pure strategy $s_1$. The former is therefore better than the latter for player I *whatever* player II may do. In algebra:

$$[2 \quad 4 \quad 6] \gg [1 \quad 2 \quad 3].$$

None of player II's pure strategies in the game of Figure 5.1 are strongly dominated, but pure strategy $t_1$ weakly dominates pure strategy $t_2$. The former is therefore never worse than the latter, and there is at least one strategy that player II could choose that would make it strictly better. Similarly, $t_1$ weakly dominates $t_3$, and $t_2$ weakly dominates $t_3$. In algebra:

$$\begin{bmatrix}1\\0\end{bmatrix} > \begin{bmatrix}0\\0\end{bmatrix}; \quad \begin{bmatrix}1\\0\end{bmatrix} > \begin{bmatrix}-1\\0\end{bmatrix}; \quad \begin{bmatrix}0\\0\end{bmatrix} > \begin{bmatrix}-1\\0\end{bmatrix}.$$

If we had included all the pure strategies for Duel in the strategic form of Figure 5.3 (instead of picking one representative pure strategy for each decision node $d$), then the payoff table would have had many identical rows and columns. But neither of the two strategies that correspond to such identical rows or columns is said to weakly dominate the other.

Nor is it true that saying that $s$ weakly dominates $t$ excludes the possibility that $s$ strongly dominates $t$—any more than saying that Pandora is somewhere in the house excludes the possibility that she is in the kitchen. Since this small point is a perennial source of confusion, it is fortunate that everybody understands that to say *$s$ dominates $t$* covers both the case in which the domination is strong and the case in which the domination is weak but not strong.

### 5.4.2 Deleting Dominated Strategies

A rational player will never use a strongly dominated strategy. Critics who argue to the contrary for games like the Prisoners' Dilemma usually don't understand how a payoff in a game is defined (Section 1.4.2).

In seeking the Nash equilibria of a game, it therefore makes sense to begin by deleting all the rows and columns corresponding to strongly dominated strategies. For example, row $s_1$ may be deleted in the game of Figure 5.1. We are then left with the simple $1 \times 3$ bimatrix game of Figure 5.7.

In the $1 \times 3$ bimatrix game of Figure 5.7, none of player II's pure strategies are dominated, not even in the weak sense. No further reductions are therefore possible using domination arguments. The remaining strategy pairs $(s_2, t_1)$, $(s_2, t_2)$, and $(s_3, t_3)$ are all Nash equilibria of the game of Figure 5.1, but it certainly isn't always true that only Nash equilibria are left after all dominated strategies have been deleted.

*Duel.* Figure 5.8 demonstrates the use of the same technique with the $6 \times 5$ bimatrix game of Figure 5.3. Domination considerations are used to reduce the game to the single cell $(d_6, d_5)$ that Section 5.2.1 identified as the unique Nash equilibrium of this version of Duel. The steps in the reduction are:

**Step 1.** Delete row $d_{10}$ because it is *strongly* dominated by row $d_8$.

**Step 2.** In the $5 \times 5$ bimatrix game that remains, delete column $d_9$ because it is *strongly* dominated by column $d_7$.

**Step 3.** In the $5 \times 4$ bimatrix game that remains, delete row $d_8$ because it is *strongly* dominated by row $d_6$.

**Step 4.** In the $4 \times 4$ bimatrix game that remains, delete column $d_7$ because it is *strongly* dominated by column $d_5$.

**Step 5.** In the $4 \times 3$ bimatrix game that remains, delete row $d_0$ because it is *strongly* dominated by row $d_6$.

We now have a $3 \times 3$ bimatrix game with no strongly dominated pure strategies. To make further progress, strategies that are only *weakly* dominated must be deleted, but some caution is necessary when you go down this road.

| | $t_1$ | $t_2$ | $t_3$ |
|---|---|---|---|
| $s_2$ | 2, 0 | 4, 0 | 6, 0 |

Figure 5.7 A simplified version of Figure 5.1

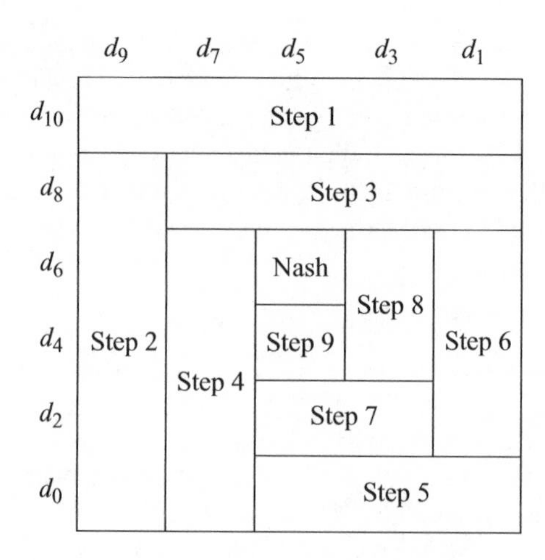


Figure 5.8 Successively deleting dominated strategies in Duel.

It never hurts Pandora to throw away her weakly dominated strategies, but it doesn't follow that it is necessarily irrational for her to choose a weakly dominated strategy. Games often have Nash equilibria that require the play of weakly dominated strategies. Such Nash equilibria are lost if we always delete *any* dominated strategy. However, the simplified game that remains after the process of deleting all dominated strategies is over always retains at least one Nash equilibrium of the original game.

**Step 6.** In the $3 \times 3$ bimatrix game remaining after Step 5, delete column $d_1$ because it is *weakly* dominated by column $d_3$.

**Step 7.** In the $3 \times 2$ bimatrix game that remains, delete row $d_2$ because it is *strongly* dominated by row $d_6$.

**Step 8.** In the $2 \times 2$ bimatrix game that remains, delete column $d_3$ because it is *weakly* dominated by column $d_5$.

**Step 9.** In the $2 \times 1$ bimatrix game that remains, delete row $d_4$ because it is *strongly* dominated by row $d_6$.

This long sequence of deletions leaves the $1 \times 1$ bimatrix game consisting of the single cell of the original game that lies in row $d_6$ and column $d_5$. Since the final game must retain at least one Nash equilibrium of the original game, we have therefore shown yet again that $(d_6, d_5)$ is a Nash equilibrium of Duel.

### 5.4.3 Knowledge and Dominated Strategies

Tweedledum doesn't need to know anything about Tweedledee to decide that it isn't a good idea to use a strongly dominated strategy in Duel. The two brothers famously have a low opinion of each other, but it is irrational to use a strongly dominated strategy even if your opponent is a chimpanzee.

However, to justify deleting column $d_9$ at Step 2 in Section 5.4.2, Tweedledee has to know that Tweedldum is sufficiently rational that he can be relied upon not to use the strongly dominated strategy $d_{10}$. To justify deleting row $d_8$ at Step 3, Tweedledum has to know that Tweedledee will delete column $d_9$ at Step 2. Thus Tweededum has to know that Tweedledee knows that Tweedledum isn't so irrational as to play a strongly dominated strategy. To justify the deletion of column $d_7$ at Step 4, Tweedledee has to know that Tweedledum knows that Tweedledee knows that Tweedledum isn't so irrational as to play a strongly dominated strategy.

To justify an arbitrary number of deletions, we need to assume it to be *common knowledge* that no player is sufficiently irrational as to play a strongly dominated strategy. This isn't the first time that common knowledge has been mentioned. Nor will it be the last, but we will do no more at this stage than to note the technical sense in which game theorists use the term.

Something is common knowledge if everybody knows it; everybody knows that everybody knows it; everybody knows that everybody knows that everybody knows it; and so on.

It isn't always necessary, but game theorists usually take for granted that the rules of a game and the preferences of the players are common knowledge. In analyzing games, they often also need to assume it to be common knowledge that all the players subscribe to appropriate rationality principles—although they seldom say so explicitly. The weakest of all such rationality principles is that which counsels against the use of strongly dominated strategies.

### 5.4.4 Backward Induction and Dominated Strategies

Backward induction has been our most powerful technique for solving games up to now, but it depends heavily on having access to an extensive form. So what happens when we move on to the strategic form of a game? Must we then throw backward induction out of the window? The answer is *no.* We can always mimic the backward induction process by deleting dominated strategies in the appropriate order.

The Tip-Off Game of Section 2.2.1 provides a simple example. Figure 5.9 repeats Figures 2.1(a) and 2.2(a), except that payoffs are now assigned to the outcomes. The firm gets 1 for the outcome $\mathscr{W}$ and 0 for the outcome $\mathscr{L}$. The agency gets 0 for $\mathscr{W}$ and 1 for $\mathscr{L}$.

To solve the Tip-Off Game by backward induction, begin by doubling the agency's action $T$ at the decision node in the extensive form reached after the firm plays $T$. This procedure is equivalent to deleting the pure strategies $tt$ and $Tt$ from the strategic form because these are all the pure strategies in which the agency plays $t$ after the firm plays $T$. The next step is to double the agency's action $t$ at the decision node in the extensive form reached after the firm plays $t$. This procedure is equivalent to deleting the pure strategies $Tt$ and $TT$ from the strategic form because these are all the pure strategies in which the agency plays $T$ after the firm plays $t$.

We are then left with a $2 \times 1$ game that can't be reduced any further. Both of the two cells in this reduced game correspond to subgame-perfect equilibria of the original game because, if the agency plays pure strategy $tT$, then the firm gets a payoff of 0 whatever it does.

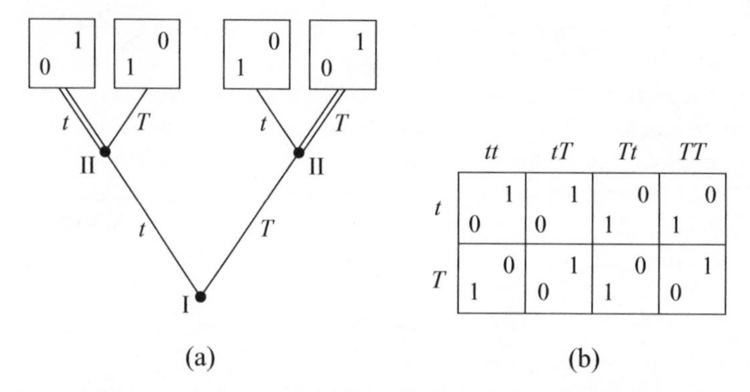


Figure 5.9 Extensive and strategic forms for the Tip-Off Game. Outcomes are given in terms of payoffs to the firm and the agency. Doubling the action $T$ at the agency's right node in Figure 5.9(a) corresponds to deleting the strategies $tt$ and $Tt$ in Figure 5.9(b). Doubling the action $t$ at the agency's left node corresponds to deleting the strategies $Tt$ and $TT$.

### 5.4.5 Problems with Domination

At one time, game theorists were more enthusiastic about the successive deletion of dominated strategies. Even today, the method is still sometimes recommended without reservation for "solving" games in which its use leads to a unique strategy profile. Such authors treat the fact that it isn't necessarily irrational to use a *weakly* dominated strategy as the minor irritant it would be if all players were forced to use each of their pure strategies with some tiny minimal probability. However, both experimental work and evolutionary theory confirm that caution is necessary when weakly dominated strategies are deleted, lest something that matters is thrown away. Nobody doubts the value of the technique as a computational device, but it needs to be used with discretion.

Figure 5.10(a) provides an example of a Nash equilibrium that is eliminated when weakly dominated strategies are deleted. Usually the equilibria that get eliminated deserve no better fate because no rational player would ever think of using them, but one can't count on this being the case. For example, the Nash equilibrium eliminated in Figure 5.10(a) is the one in which the players get a payoff of 100 each. Subgame-

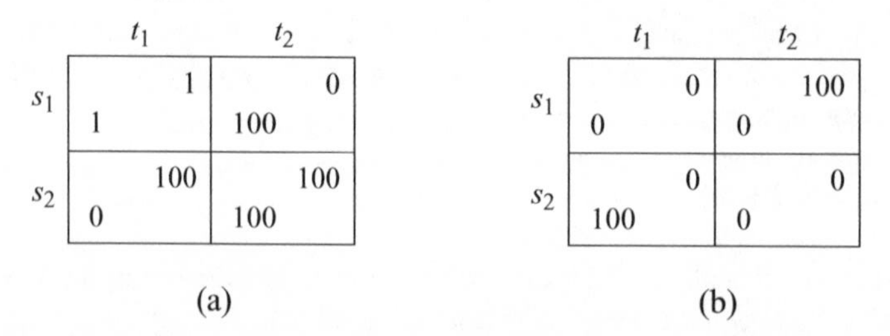


Figure 5.10 Deleting weakly dominated strategies. The Pareto-efficient Nash equilibrium is eliminated in Figure 5.10(a). The order of deletion matters in Figure 5.10(b).

perfect equilibria can also get eliminated if one isn't careful about the order in which strategies are deleted.[4]

It doesn't matter in which order we delete strongly dominated strategies, but Figure 5.10(b) shows that the same isn't true for weakly dominated strategies. Depending on whether we first eliminate player I's first pure strategy or player II's first pure strategy, we are led to different reduced games with different properties.

## 5.5 Credibility and Commitment

→ 5.6

So far, we have mostly applied backward induction and the successive deletion of dominated strategies to strictly competitive games, where their use is relatively uncontroversial. However, their application becomes debatable when more general games are considered.

We already met one of the lines of criticism in Section 1.7.1 when considering the transparent disposition fallacy. We begin by reviewing this fallacy in the context of the Wonderland hat market of Section 1.5.2.

### 5.5.1 Follow the Leader

As in Section 1.5.2, Alice and Bob are hat producers. Alice can only produce either $a=4$ or $a=6$ hats. Bob can only produce $b=3$ or $b=4$ hats. Both players are interested only in maximizing their profit in dollars.

We simplify the cost assumptions of Section 1.5.2 by making Alice's and Bob's cost functions linear. Each faces a constant unit cost of $3, so it costs each player $3h$ dollars to make $h$ hats. The demand equation is also simplified to $p+h=15$, where $p$ is the price at which each hat sells when the total number of hats produced is $h=a+b$.

*Cournot's Model.* Cournot studied the case in which Alice and Bob are both already in the market and independently decide how many hats to produce without knowing the production decision of the other (Section 1.5.2). We then say that they are playing a *simultaneous-move* game—although their decisions may not be made at literally the same moment.

Our experience with the Inspection Game in Section 2.2.1 makes it easy to draw both extensive and strategic forms for the simultaneous-move game. Figures 5.11(a) and 5.11(b) are equivalent extensive forms for the game that differ in the player to whom the root of the game is assigned. It doesn't matter who nominally moves first at the root because the second player moves without knowing anything about the first player's decision. They therefore might as well be moving simultaneously.

The cell that arises when Alice and Bob each produce four hats has both payoffs enclosed in a circle or a square in Figure 5.11(c). It follows that the strategy profile $(4,4)$ is a Nash equilibrium of the game. We could also have found the Nash equilibrium by successively deleting strongly dominated strategies. (First delete

[4]To ensure that subgame-perfect equilibria aren't lost, delete weakly dominated strategies in the same order as they would be deleted when applying backward induction.

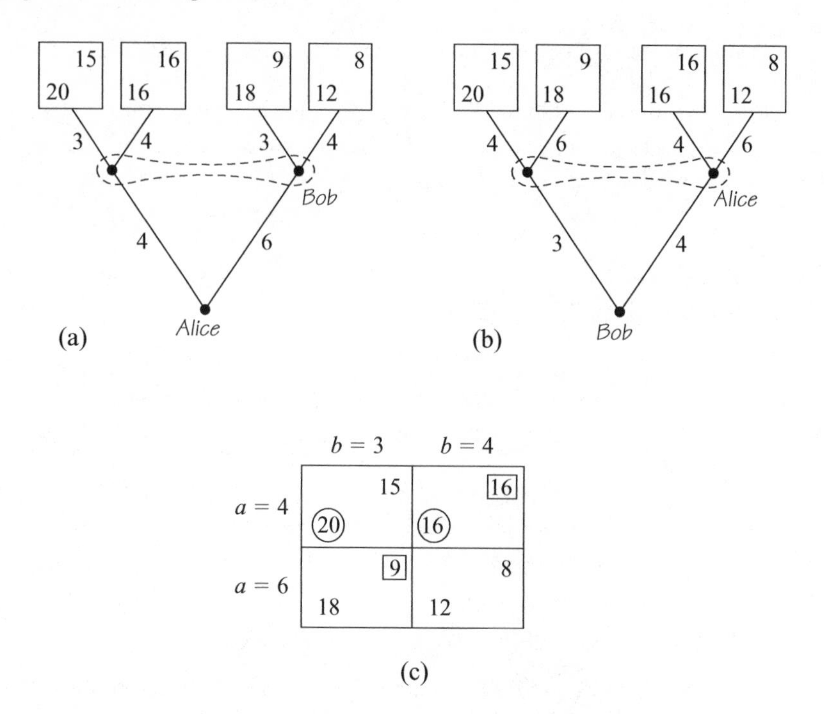


Figure 5.11 The Cournot model as a simultaneous-move game.

Alice's second pure strategy because it is strongly dominated by her first pure strategy. Then delete Bob's first pure strategy in the reduced game that results because it is strongly dominated by his second pure strategy.)

*Stackelberg's Model.* Von Stackelberg pioneered the study of *entry* in imperfectly competitive markets. We can capture his idea by ceasing to assume that Alice and Bob are already in the market when the game begins.

In the Stackelberg setup, Alice is the leader. Although she begins by entering a market that hasn't been previously exploited, she can't act as a monopolist (as we implicitly assumed in Section 3.7.1) because she knows that Bob will follow her into the market to contest her profits.

We assume that the cost functions and the demand equation are unchanged from the Cournot case. All the numbers needed to analyze Stackelberg's leader-follower model are therefore summarized in the payoff table of Figure 5.11(c). Economists commonly argue that Alice first chooses a row in this table. Bob observes her choice and then chooses the column that is his best reply.

If Alice produces 4 hats, Bob's best reply is to produce 4 hats. Alice's payoff is then \$16. If Alice produces 6 hats, Bob's best reply is to produce 3 hats. Alice's payoff is then \$18. She therefore chooses to produce 6 hats, and Bob responds by producing 3 hats. Economists call the strategy profile $(6, 3)$ a Stackelberg equilibrium of the leader-follower model. Notice that the Stackelberg profile $(6, 3)$ is quite different from the Nash equilibrium $(4, 4)$ of the simultaneous-move game.

Although the analysis is very simple, the standard way that economists talk about leader-follower models risks creating confusion. The basic problem is that Figure

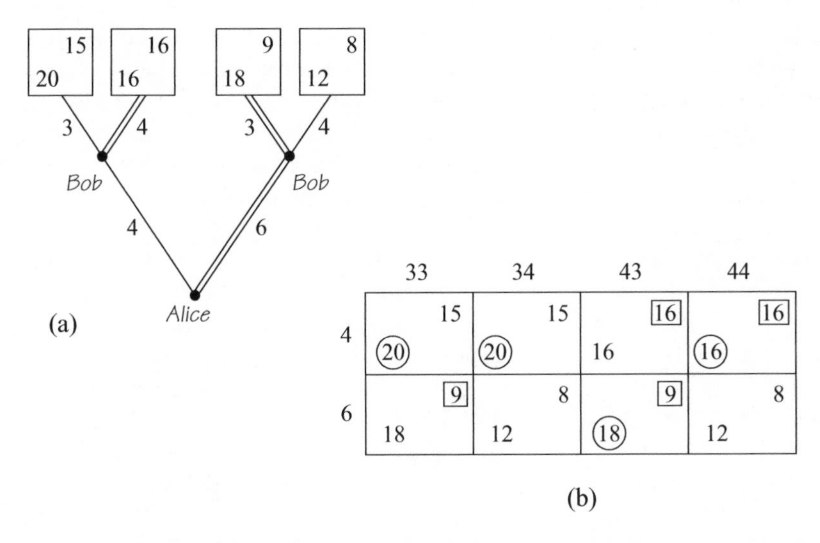


Figure 5.12 The Stackelberg model as a leader-follower game.

5.11(c) isn't the strategic form of the leader-follower game that Alice and Bob are playing.

Our study of the Tip-Off Game in Section 2.2.1 makes it easy to work out the correct strategic form from the extensive form of the leader-follower game shown in Figure 5.12(a). Once we have the strategic form, we can enclose the payoffs that correspond to best replies in circles or squares. The cells in which both payoffs get enclosed then correspond to the game's Nash equilibria in pure strategies. Our leader-follower game has two Nash equilibria: $(6, 43)$ and $(4, 44)$. We therefore have two candidates for the solution of the game.

Applying backward induction in the extensive form of the leader-follower game, we find that $(6, 43)$ is the unique subgame-perfect equilibrium. To mimic backward induction in the strategic form of Figure 5.12(b), first delete the dominated strategies 33, 43, and 44. Then delete the dominated strategy 4 in the reduced game that results. Along the way, the Nash equilibrium (4, 44) is eliminated, and economists therefore usually neglect the possibility that it might be used in practice.

The analysis makes it clear that it is a misnomer to call $(6, 3)$ a Stackelberg equilibrium. It isn't even a strategy profile. It should be written as $[6, 3]$ and identified as the *play* that results when the subgame-perfect *equilibrium* $(6, 43)$ is used in the leader-follower game.

In brief, von Stackelberg adds nothing to the equilibrium ideas that we have been studying. What he contributes is the idea that it is interesting to study duopoly games in which one player moves before the other. Rather than talking about Stackelberg equilibria, we will therefore use Stackelberg's name to refer to the class of leader-follower games whose study he initiated.

### 5.5.2 Incredible Threats

Section 1.7.1 warns against trusting strangers who approach you in dark alleys. In this section, the stranger is carrying a bomb. He threatens to blow you both up if you don't give him your wallet. The threat is worrying, but your wallet contains $100. Do you hand it over? If you have reason to believe that the stranger is rational and wants to live, then his threat is incredible. If you don't hand over your wallet, he won't blow you both to smithereens because he doesn't want to die.

We can run the same argument through our Stackelberg game when evaluating the following attempt to legitimize the Nash equilibrium $(4, 44)$ we eliminated when successively deleting dominated strategies in Figure 5.12(b).

Bob doesn't like the low payoff of $9 that he gets with the subgame-perfect equilibrium $(6, 43)$. Before Alice decides how many hats to produce, Bob therefore threatens that if she produces 6 hats, he will respond by producing 4 hats—even though he would thereby reduce his profit to $8 by not playing his best reply. If Alice believes him, she won't produce 6 hats because her profit will then only be $12. Instead, she will do the equivalent of handing over her wallet by reducing her production to 4 hats. Bob will then reply by producing 4 hats as well. Each will then make a profit of $16—a loss of $2 for Alice when compared with the subgame-perfect equilibrium, but a gain of $7 for Bob.

Game theorists argue that Alice shouldn't believe Bob. His threat is incredible because, if she did produce 6 hats, he would have a choice between $9 and $8 in the subgame that follows. If he is someone who always chooses more money rather than less, then he will necessarily choose $9—whatever he may have told Alice he would do if she were to ignore his threat. He will therefore play according to the subgame-perfect equilibrium $(6, 43)$ and produce 3 hats. One can respond that Bob may be the commercial equivalent of a suicide bomber, but he would then be either irrational or motivated by something other than profit.

The transparent disposition fallacy claims that this defense of subgame-perfect equilibrium is wrong (Section 1.7.1). It says that Bob should make it clear to Alice that he is *committed* to carrying out his threat. But can people really precommit themselves to actions they won't want to take if the occasion arises? And even if they can, how do they convince other people that they have made such a commitment?

Game theorists don't pretend to know the answers to such psychological questions. Our attitude has already been outlined in Section 1.4.1. You tell us what you think the right game is, and we'll do our best to tell you how it should be played. If you think that the players can make precommitments, then let us rewrite the rules of the game to include commitment *moves*. If you think that the players can read each other's body language so well that they will know when a commitment has been made,[5] then we can leave certain information sets out of the new game.

Those who have lost their shirts playing poker or been betrayed by an unfaithful lover may have reservations about the realism of the game you want analyzed. A mathematician will have similar reservations if you ask him to work out the orbit of a planet on the assumption that gravity satisfies an inverse cube law, but he will

[5]Charles Darwin's *Expression of the Emotions* is sometimes cited in support of the contention that our involuntary facial muscles make it impossible to conceal our emotional state from those who know what to look for—although he actually held the opposite view, and all but one of the photographs in his book are of Victorian actors convincingly simulating various emotional states.

come up with an answer. It won't accord with what you see when you look through a telescope,[6] and you may try to persuade your tame mathematician to alter the theory of differential equations because you would prefer an answer that fits the facts better. But his attitude will be that you should formulate your problem properly, rather than trying to squeeze out the right answer by trying to persuade him to analyze the wrong problem wrongly.

Game theorists feel much the same about the way they analyze games. We are impervious to criticism that depends on the assumption that rational players can read each other's minds or convert themselves into irrational robots by exerting enough willpower. It is fine with us if you want to write transparent commitments into the rules of a game. We will do our best to solve your game no matter how unrealistic we think your assumptions are. But you won't persuade us to mess up the way we analyze games by pretending that rationality somehow endows people with superhuman powers.

*Stackelberg Games with Transparent Commitment.* It is easy to modify the Stackelberg game of Figure 5.12(a) to allow Bob to choose whether or not to make a precommitment to retaliate by producing 4 hats if Alice produces 6 hats. We only need to add an extra move at the beginning of the game, as in Figure 5.13(a). If Alice didn't know whether Bob had made the commitment when it is her turn to move, it would be necessary to enclose her two decision nodes in an information set. Omitting such an information set corresponds to assuming that she can read Bob's body language.

A backward induction analysis of our new game produces the unsurprising result that Bob will commit to his threat, and Alice will submit. Nobody need therefore get het up about game theory being wedded to mistaken psychological ideas. You write the psychology that you think appropriate into the rules of a game, and ordinary game-theoretic reasoning will generate the answers that make sense for your psychological assumptions.

*Economic and Legal Commitments.* Economists argue that objective enforcement mechanisms matter more in economic contexts than the subjective commitment mechanisms we have been considering so far.

We think that people who hand over large sums of money to scam artists without getting a legal contract in return are stupid. If Bob doesn't honor a contract he has signed, then Alice can sue him for noncompliance. When using game theory to study law, one may wish to model the whole legal process—with appropriate chance moves to capture the uncertainty involved when legal precedents are scarce—but when the penalty is large and the probability of the guilty party losing the case is high, cheating on the deal becomes a strongly dominated strategy for Bob (Section 1.7). In humdrum economic applications, it therefore often makes more sense to short-circuit the legal hassle by modeling the act of signing a contract as a simple commitment move.

Even without formal commitment moves, the players in an economic game may be able achieve the same effect by irretrievably sinking costs. For example, Alice

[6] With an inverse cube law instead of Newton's inverse square law, Cotes showed that the planets would spiral down into the sun.

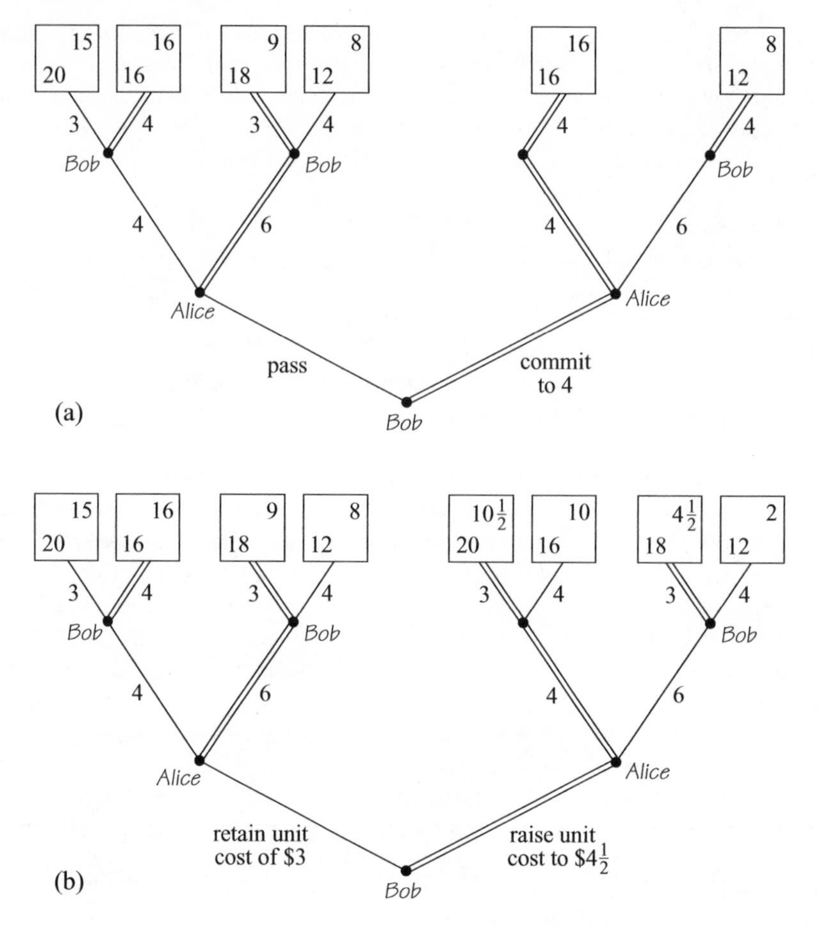


Figure 5.13 Stackelberg games with commitment.

might strategically invest money to improve the production efficiency of her factory. Such a lowering of her costs effectively commits her to producing more hats when playing a Stackelberg game with Bob. In cases like the Chain Store Game of Exercise 5.9.17, Bob may then be deterred from entering the market at all.

A less obvious stratagem is for Bob to *increase* his costs by firing some of his skilled workers or wrecking some machinery. This may seem crazy, but consider the game of Figure 5.13(b), in which Bob has the choice of sticking with a unit cost of $3 or raising his unit cost to $\$4\frac{1}{2}$.

After Bob raises his costs, the question is no longer whether Alice will believe Bob's *threat* to retaliate by overproducing if she chooses a high production schedule but whether she will believe his *promise* to keep his production down if she does the same. As a backward induction analysis of the game shows, such a promise is credible if Bob's unit cost is $\$4\frac{1}{2}$, but not if it is $3.

By increasing his unit cost to $\$4\frac{1}{2}$, Bob moves play to a subgame whose subgame-perfect equilibrium yields him a profit of $\$10\frac{1}{2}$, which is better than the $9 that results when a subgame-perfect equilibrium is played in the subgame in which Bob's unit cost is $3. After she learns that Bob has increased his costs, Alice produces only

4 hats, and Bob then keeps his promise by producing only 3 hats.[7] Alice also does better in the subgame in which Bob has higher costs. Her profit is \$20 instead of \$18. The victim is the consumer. After Bob raises his costs, 7 hats are produced instead of 9, and their price rises from \$6 to \$8.

As we saw in Section 1.5.1, a monopolist makes money by restricting supply to force up the price. Her problem when competitors appear is that they may not cooperate in keeping supply low. By raising his costs, Bob convinces Alice that he won't simply mop up any demand that she leaves unsatisfied. He too will restrict his supply. Alice and Bob therefore succeed in jointly screwing their customers without overtly colluding at all.

## 5.6 Living in an Imperfect World

→ 5.7

Talking about credible threats is just another way of explaining why we focus on the subgame-perfect equilibria studied in Section 2.9.3.

The Nash equilibrium (4, 44) isn't a subgame-perfect equilibrium in the Stackelberg game of Figure 5.12. It doesn't induce equilibrium play in the one-player subgame that would be reached if Alice were to produce six hats. Bob's strategy of 44 requires that he play 4 in this bad subgame, but his optimal action is 3. Although the strategy profile (4, 44) doesn't induce a Nash equilibrium in this bad subgame, it is nevertheless a Nash equilibrium in the whole game because the bad subgame isn't reached when (4, 44) is played. Alice produces four hats, which sends play to the good subgame, where Bob does optimize.

If Alice went to the good subgame because she thinks that Bob wouldn't optimize in the bad subgame, then she believes something that contradicts our standing assumption that the players are rational. In other words, she has given credence to an incredible threat. If the players always reject such incredible threats, then they will necessarily play a subgame-perfect equilibrium

This defense of subgame-perfect equilibrium depends on everyone's believing that all the players will *always* behave rationally, both now and in the future. We certainly want the players to start by believing this, but does it make sense for them to persist in this belief after reaching a subgame that wouldn't have been reached without someone who will move in the subgame having played irrationally in the past? The chesslike game of Section 2.9.4 presses this point by drawing our attention to subgames that can be reached only if one player systematically makes the same mistake over and over again. Shouldn't we then try to exploit the irrationality that such bad play reveals?

Purists say that we should forget about past irrationalities when analyzing what will happen in a subgame. Our initial evidence against anyone's being irrational should be taken to be so strong that any bad play we observe should be attributed to some extraneous cause that needn't be specified. Although this approach is theoretically watertight, it limits the arena for practical applications of game theory to cases like the Stackelberg games of the preceding section, which aren't long enough to allow evidence of systematic irrationality to accumulate. If we want to apply

[7]The smallest unit cost for Bob that makes the argument work is \$4. He is then indifferent between producing 3 or 4 hats after Alice produces 4 hats.

game theory more widely, we therefore have no choice but to find some way of dealing with human error.

### 5.6.1 Bounded Rationality

It has been a long time since Herbert Simon pioneered the investigation of economic theories of bounded rationality by introducing the notion of *satisficing,* but advances in this area remain notoriously elusive.

*Satisficing.* In satisficing models, the players don't optimize down to the last penny. Rather than spending time and energy looking for something better, they declare themselves satisfied when they come across a strategy that is only approximately optimal.

We capture the satisficing idea in game theory by introducing a constant $\varepsilon > 0$ that measures how good an approximation must be before the players are satisfied. The criterion (5.3) for a Nash equilibrium can then be modified to say that a pair $(\sigma, \tau)$ of strategies is an approximate Nash equilibrium when

$$\pi_1(\sigma, \tau) \geq \pi_1(s, \tau) - \varepsilon$$
$$\pi_2(\sigma, \tau) \geq \pi_2(\sigma, t) - \varepsilon$$

for all pure strategies $s$ and $t$. Moving to a satisficing framework therefore potentially increases the number of strategy profiles that count as equilibria.

The idea of an approximate equilibria is admittedly crude, but it will serve to show that the purist attitude to subgame-perfect equilibria sometimes leads to predictions about how games will be played that aren't very realistic.

### 5.6.2 The Holdup Problem

As a small child, I remember wondering why store clerks hand over the merchandise after being paid. Why don't they just pocket the money? This is a simple version of the *holdup problem* that arises in the theory of incomplete contracts.

For example, Alice is considering investing in Bob's firm on the condition that he work harder. But after he has secured her money, what ensures that he will keep his promise? Exercise 5.9.18 models this situation as a simple leader-follower game, like those of the previous section. Unless Bob has reason to fear some penalty if he doesn't deliver on his end of the deal,[8] a subgame-perfect analysis shows that Alice would be unwise to cooperate with Bob at all. The opportunity for the pair to cooperate in creating an economic surplus will therefore be lost. But if this kind of holdup argument always works, how did evolution manage to make us into social animals?

[8]Sanctions that might apply are the risk of losing his commercial reputation or provoking an action for breach of contract. But how does Alice convince the world at large that her money was lost through Bob's neglect rather than a commercial mishap? Only Bob knows for sure how hard he worked. In the language of incomplete contract theory, one can write a contract only on the basis of events that can be *publicly verified.*

Biology offers us an exotic example of sex among the hermaphroditic sea bass as one of many ways the trick might be managed. When sea bass mate, they take turns in laying their own eggs and fertilizing their partner's eggs. However, eggs are expensive to produce, and sperm is cheap. If a sea bass trustingly laid all its eggs at the outset of a romantic encounter, it could be held up by an exclusively male mutant that fertilized the eggs and then swam off to fertilize the eggs of other sea bass without making an equivalent investment in the future of their joint children. When two sea bass mate, each therefore alternates in laying *small* batches of eggs for the other to fertilize, so that neither needs to trust the other very much.

Essentially the same story can be told of two criminals who have agreed to exchange a quantity of heroin for a sum of money. Adam is to end up with Eve's heroin, and Eve with Adam's money. How is this transition to be engineered if both are free to walk away at any time, carrying off whatever is currently in their possession? In real life, matters would be complicated by the threat of physical violence, but we will assume that no sanctions at all for noncompliance are available.

We have seen that there is no point in Adam's handing over the agreed price and waiting for the goods. Like sea bass, our criminals have to arrange a flow between them, so that the money and the drug change hands *gradually*. Such a transaction can be modeled using a version of Rosenthal's Centipede Game.

*The Centipede Game.* Adam's and Eve's payoffs for the commodity bundle $(d,h)$ consisting of $d$ dollars and $h$ grains of heroin are respectively $\pi_1(d,h) = 0.01d + h$ and $\pi_2(d,h) = d + 0.01h$. Thus Adam wants to exchange dollars for heroin, and Eve wants to exchange heroin for dollars. Adam starts with 100 dollars and Eve with 100 grains of heroin. Since neither trusts the other very much, they agree to alternate in handing over single dollars and single grains of heroin until the transaction is complete.

The Centipede Game gets its name because the extensive form of Figure 5.14(a) has a hundred pairs of legs. To play *across* is to honor the deal. To play *down* is to cheat by leaving with what one currently has.

The Centipede Game has only one subgame-perfect equilibrium, which requires that both players always plan to cheat. No trade then takes place. To see this, consider what is optimal in the subgame that arises if the rightmost decision node is reached. Eve must then choose between 100.01 and 100 and thus cheats by choosing the former. In the subgame that arises if the penultimate decision node is reached, Adam predicts that Eve will cheat on the next move, and so his choice is between 99.01 and 99. He therefore cheats by choosing the former. Since the same backward induction argument works at every decision node, the result of a subgame-perfect analysis is that both players plan always to cheat. They therefore both end up with a payoff of 1, rather than the payoff of 100 that each would have obtained if both had honored their agreement.

Figure 5.14(b) shows a reduced strategic form in which the players' pure strategies specify how many times they plan to honor the deal before cheating. Successively deleting weakly dominated strategies in this payoff table mimics the backward induction process. We begin by deleting Eve's first column. Then we delete Adam's first row from the payoff table that remains. Next we delete Eve's second column and then Adam's second row. This process continues until we are left only with each player's last pure strategy, which requires cheating immediately.

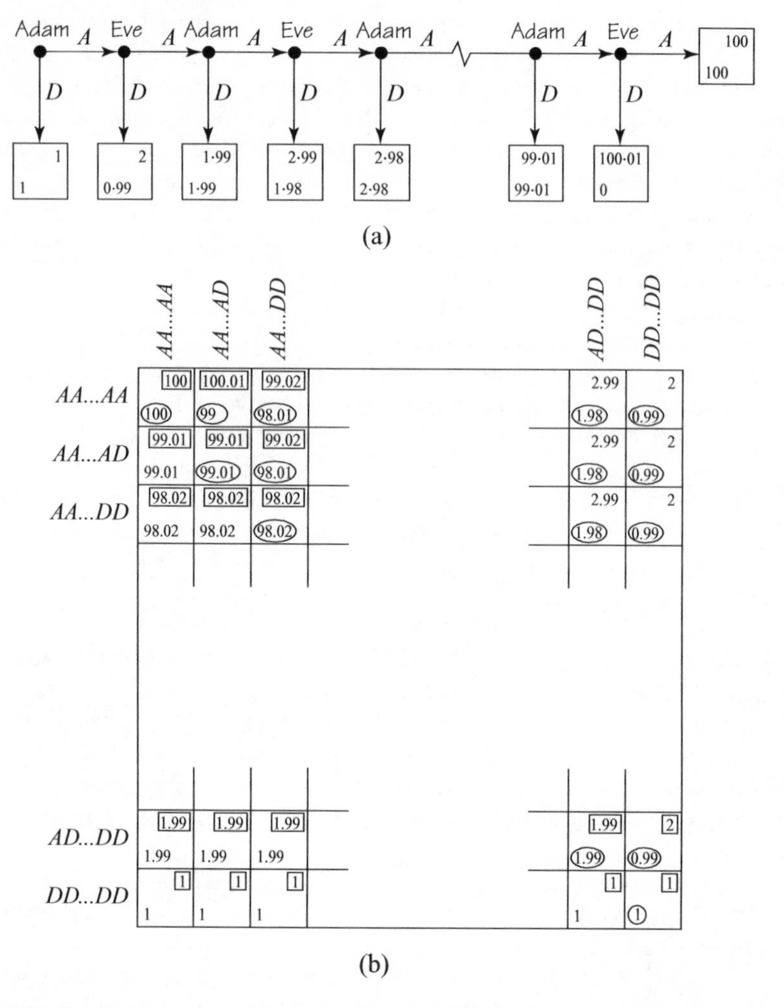


Figure 5.14 The Centipede Game. It is used here to model a trustless exchange of money for heroin between two criminals. The circled and squared payoffs in Figure 5.14(b) indicate *approximate* best replies when $0.01 < \varepsilon < 0.02$. There are many approximate Nash equilibria, including one in which both players always plan to play *across*.

The conclusion that rational players will cheat in the Centipede Game reminds philosophers of the fact that rational players can't cooperate in the Prisoners' Dilemma—but there is a big difference. In the Centipede Game, the result isn't robust to the introduction of tiny imperfections into our specification of the problem.

The real world is imperfect in many ways. The Centipede Game takes account of the imperfection that real money isn't infinitely divisible. But real people are even more imperfect than real money. In particular, they aren't infinitely discriminating. What is one cent more or less to anybody?

Introducing satisficing into the Centipede Game has a dramatic effect when $0.01 < \varepsilon < 0.02$. As shown in Figure 5.14(b) by enclosing approximate best replies, large numbers of equilibria suddenly appear, including an approximate equilibrium in which both players honor their deal and hence secure a payoff of 100 each.

The same result is obtained whenever the trading units are smaller than the threshold that makes a satisficing player sit up and pay attention. However, Adam and Eve will have chosen their trading units with this fact in mind. If dollars and grains are too large, they can deal in cents and hundredths of a grain.[9]

If we want an idealized model from which all imperfections have been eliminated, we are free to allow both the size $\delta > 0$ of the trading units and the perception threshold $\varepsilon > 0$ to tend to zero. Cooperation will then survive as an equilibrium in the limit, provided that we keep $\delta < \varepsilon$ as we take the limit. If one wants to insist that the players always optimize up to the hilt, then $\varepsilon$ must tend to zero *first,* in which case only the cheating equilibrium survives. But this purist approach risks leading us astray since we end up analyzing a model that ignores the players' psychological limitations.

## 5.7 Roundup

The chapter began by legitimizing the strategic form of a game introduced in Chapter 1 when studying the Prisoners' Dilemma. Once the players have chosen their pure strategies, the course of the game is determined except for the game's chance moves. A pure strategy profile therefore assigns an expected Von Neumann and Morgenstern utility to each player. A payoff function tells us what this expected utility is for all pure strategy profiles of the game.

A strategic form for a two-player game is determined by two payoff matrices. The entry in the $i$th row and $j$th column of player $k$'s payoff matrix is given by the value $\pi_k(i,j)$ of player $k$'s payoff function.

A Nash equilibrium $(\sigma, \tau)$ is characterized in terms of payoff functions by the requirement that the inequalities

$$\pi_1(\sigma, \tau) \geq \pi_1(s, \tau)$$
$$\pi_2(\sigma, \tau) \geq \pi_2(\sigma, t)$$

hold for all pure strategies $s$ and $t$.

Dominance relations are also easily expressed in terms of payoff functions. For example, player I's pure strategy $s_1$ is strongly dominated by his pure strategy $s_2$ if

$$\pi_1(s_2, t) > \pi_1(s_1, t)$$

for all player II's pure strategies $t$. Player II's pure strategy $t_2$ is weakly dominated by her pure strategy $t_1$ if

$$\pi_2(s, t_1) \geq \pi_2(s, t_2)$$

[9]Perhaps this is one of the reasons that the smallest unit of currency is always small enough that nobody cares about one unit more or less.

for each value of player I's pure strategy $s$, with *strict* inequality for at least one value of $s$.

The successive deletion of strongly dominated strategies is a powerful method of simplifying games. Its use draws attention to our standing assumption that the players' rationality is common knowledge at the outset of the game. The deletion of weakly dominant strategies is more problematic since the order in which they are deleted can matter, and Nash equilibria may disappear along the way.

Stackelberg games have the same payoff structure as Cournot games, but one of the players moves first. The object that economists call a Stackelberg equilibrium is actually the play that will be followed if the players use a subgame-perfect equilibrium in a Stackelberg game.

Backward induction and the successive deletion of weakly dominated strategies fail to be plausible tools of analysis if the players can make credible threats or promises outside the structure of the game. The answer isn't to scrap our methods of analysis but to change the rules of the game so that credible threats or promises are modeled as formal commitment moves within the game.

Economists are skeptical about the extent to which transparent commitments can be made by willpower alone, but they recognize that one can often achieve the same effect by signing a contract or sinking an investment. Cheating on a commitment may then become too expensive to make it worth bothering to model the possibility in a game.

A major criticism of backward induction is that its validity depends on the players always believing that their opponents will play rationally in the future, even though they may have been observed to play irrationally in the past. As with the commitment problem, this difficulty can sometimes be tackled by incorporating any irrational quirks that afflict the players into the rules of the game. As in the case of the Centipede Game, introducing only a little irrationality can sometimes change the outcome of a game dramatically.

## 5.8 Further Reading

*Game Theory and Economic Modelling*, by David Kreps: Oxford University Press, New York, 1990. Listen to what daddy says on economic modeling, and you won't go far wrong.

*Game Theory for the Social Sciences*, by Hervé Moulin: New York University Press, New York, 1986. This book contains many thought-provoking examples. It is particularly useful on dominated strategies.

*The Strategy of Conflict*, by Thomas Schelling: Harvard University Press, Cambridge, MA, 1960. This classic makes it clear that the power to make commitments is very valuable but not easy to acquire.

*Passions within Reason*, by Bob Frank: Norton, New York, 1988. An economist makes a case for the transparent disposition fallacy.

## 5.9 Exercises

1. Construct a simplified strategic form for Duel just as in Section 5.2.1 but taking $p_1(d) = p_2(d) = 1 - d^2$. (This case was studied in Exercise 3.11.20, but here $D = 1$.) Circle the best payoff for player I in each column. Enclose the best payoff to player II in each row in a square. Hence locate a Nash

equilibrium. How close will the players be when someone fires? Who will fire first?

2. Use the method of successively deleting dominated strategies in the simplified strategic form obtained in the previous exercise. Why is the result a subgame-perfect equilibrium?
3. In this version of the Inspection Game, Jerry can hide in the bedroom, the den, or the kitchen. Tom can search in one and only one of these locations. If he searches where Jerry is hiding, he catches Jerry for certain. Otherwise Jerry escapes.
   a. Assign appropriate Von Neumann and Morgenstern utilities to the possible outcomes.
   b. Draw the game tree for the case in which Tom can see where Jerry is hiding before he starts searching. Find the $3 \times 27$ bimatrix game that is the corresponding strategic form. (Jerry is player I)
   c. Draw the game tree for the case in which Jerry can see where Tom is searching before he hides. Find the $27 \times 3$ bimatrix game that is the corresponding strategic form.
   d. Draw two game trees that both correspond to the case in which Tom and Jerry each make their decisions in ignorance of the other's choice. Find the $3 \times 3$ bimatrix game that is the corresponding strategic form.
   e. In each case, find all pure strategy pairs that are Nash equilibriuma.
4. Write down the transposes of the following matrices:

review

$$A = \begin{bmatrix} 2 & 1 & 3 \\ -1 & 4 & 0 \end{bmatrix}, \quad B = \begin{bmatrix} 1 & 2 \\ 0 & -1 \\ 3 & 0 \end{bmatrix}, \quad C = \begin{bmatrix} 0 & 1 \\ -1 & 2 \\ 0 & 4 \end{bmatrix}.$$

5. Write down the payoff matrices for the two players in the bimatrix games of Figure 5.15. Which of the four payoff matrices are symmetric? Which of the two bimatrix games are symmetric?

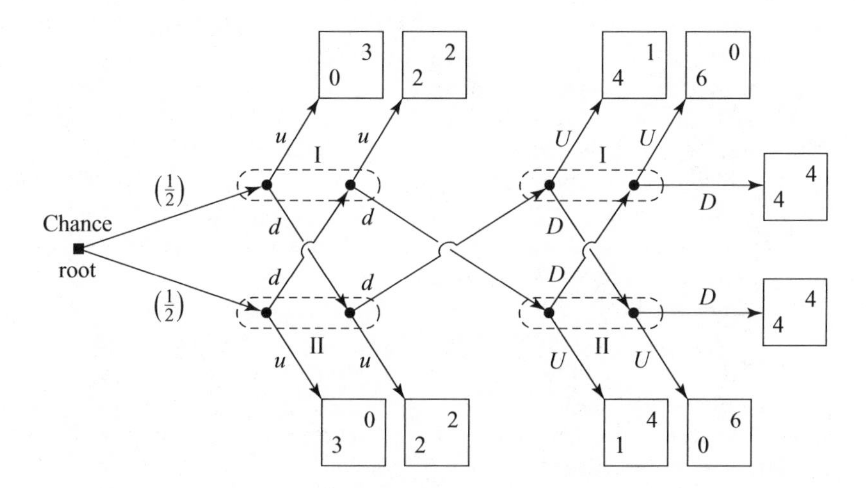


Figure 5.15 The extensive form for Exercise 5.9.10.

review

6. For each $1 \times 2$ vector $y$, the sets

$$A = \{x : x \geq y\} \qquad B = \{x : x > y\} \qquad C = \{x : x \gg y\}$$

represent regions in $\mathbb{R}^2$. Sketch these regions in the case $y = (1, 2)$. For each of the following $1 \times 2$ vectors $z$, decide whether $z$ is a member of $A$, $B$, or C:

(a) $z = (2, 3)$ (b) $z = (2, 2)$ (c) $z = (1, 2)$ (d) $z = (2, 1)$

econ

7. If the pure strategy pair $(d_6, d_5)$ were to be defended as the solution of the bimatrix game of Figure 5.3 on the basis of statements like:

Everybody knows that everybody knows that ... everybody knows that nobody ever uses a weakly dominated strategy,

what is the smallest number of times that the phrase "everybody knows" would need to appear? Bear in mind that several strategies can often be eliminated *simultaneously* during the deletion process.
8. Construct a finite game of perfect information in which a subgame-perfect equilibrium is lost if weakly dominated strategies are deleted from the strategic form in a suitable order. (Your game tree need not be very complicated.)
9. In version 2 of Russian roulette as studied in Section 5.2.2, explain why

$$\pi_1(ADD, AAD) = \tfrac{1}{6} + \tfrac{2}{3}a$$
$$\pi_2(ADD, AAD) = \tfrac{5}{6}.$$

10. Obtain the $4 \times 4$ strategic form of the game whose extensive form is given in Figure 5.15. By deleting dominated strategies, show that $(dU, dU)$ is a Nash equilibrium. Are there other Nash equilibria?
11. Colonel Blotto can send each of his five companies to one of ten locations whose importance is valued at 1, 2, 3, ..., 10, respectively. No more than one company can be sent to any one location. His opponent, Count Baloney, must simultaneously do the same with his four companies. A commander who attacks an undefended location captures it. If both commanders attack the same location, the result is a standoff at that location. A commander's payoff is the sum of the values of the locations he captures minus the sum of the values of the locations captured by the enemy. What would Colonel Blotto do in the unlikely event that he knew what a dominated strategy was?

econ

12. How does the analysis of the Stackelberg model of Section 5.5.1 change if Bob becomes the leader and Alice the follower?

econ

13. The Cournot and Stackelberg models of Figures 5.11 and 5.12 are changed to allow transparent precommitment by the players. In both cases, show that:
    a. If Alice precommits before Bob, the model reduces to a Stackelberg game with Alice as the leader.
    b. If Bob precommits before Alice, the model reduces to a Stackelberg game with Bob as the leader.

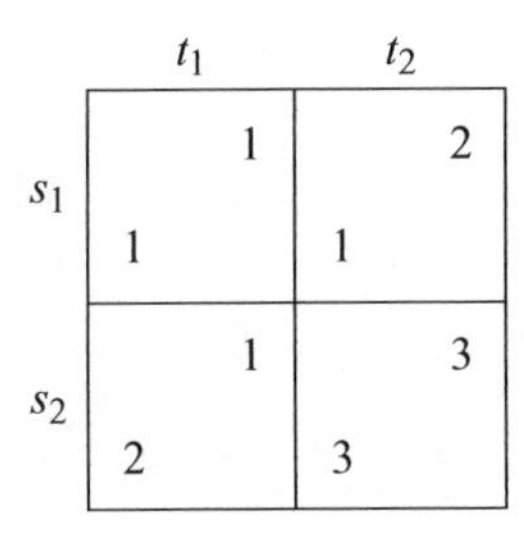

| | $t_1$ | $t_2$ |
|---|---|---|
| $s_1$ | 1, 1 | 2, 1 |
| $s_2$ | 2, 1 | 3, 3 |

Figure 5.16 The bimatrix games for Exercise 5.9.12.

c. If both players precommit simultaneously, the model reduces to a Cournot game.

14. Elaborate the Stackelberg model of Figure 5.12 with Alice as leader so as to allow Alice and Bob a simultaneous preplay opportunity to make a transparent precommitment to one of their strategies—if they so choose. Explain why this change creates a game with the strategic form of Figure 5.17 where □ means that the player chooses not to make a precommitment. The game has three Nash equilibria, which correspond respectively to the Cournot case and the Stackelberg cases with Alice and Bob as leaders. Show that the equilibrium that survives the successive deletion of weakly dominated strategies corresponds to the case in which Bob is the leader rather than Alice.

econ

15. Selten's Chain Store Game is often used to illustrate the logic of entry deterrence in imperfectly competitive markets. Alice and Bob are industrialists who care only about maximizing their expected dollar profit. Alice is an incumbent monopolist, who makes \$5 million if left to enjoy her privileged position undisturbed. Bob is a firm that could enter the industry but earns \$1 million if he chooses not to enter. If Bob decides to enter, then Alice can do one of two things: she can fight by flooding the market with her product so as to force down the price, or she can acquiesce and split the market with Bob. A fight is damaging to both players. They then each make only \$0 million. If they split the market, each will make \$2 million.

econ

a. Why does the Chain Store Game have the extensive form shown in Figure 5.18(a)? Show that the only subgame-perfect equilibrium is (*in*, *acquiesce*).

| | 3 | 4 | □ |
|---|---|---|---|
| 4 | 20, 15 | 16, 16 | 16, 16 |
| 6 | 18, 9 | 12, 8 | 18, 9 |
| □ | 20, 15 | 16, 16 | 18, 9 |

Figure 5.17 Transparent precommitment in a Stackleberg game.

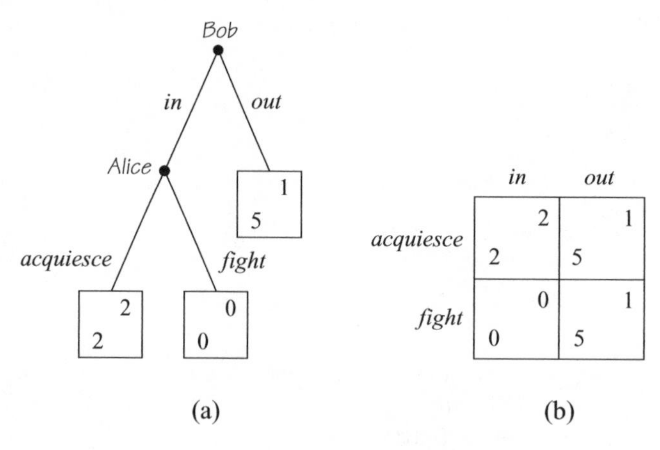


Figure 5.18 The Chain Store Game.

b. Why does the Chain Store Game have the strategic form shown in Figure 5.18(b)? Show that there are two Nash equilibria in pure strategies. Which of these is lost after the successive deletion of weakly dominated strategies?

c. Alice will threaten to fight Bob if he disregards her warning to keep out of the industry. Why will he not find her threat credible? What is the implication for the two Nash equilibria of the game?

econ

16. How would matters change in the Chain Store Game of the previous exercise if the incumbent monopolist could prove to the potential entrant that she had made an irrevocable commitment to fight if he enters?

a. Write down a new game tree in which play of the Chain Store Game is preceded by a commitment move at which Alice decides whether or not to make a commitment to fight if Bob enters.

b. Find a subgame-perfect equilibrium of the new game.

c. Can you think of ways in which Alice could make an irrevocable commitment to fighting? If so, how would she convince Bob that she was committed?

econ

17. The point of the last item in the previous exercise is that it is very hard in real life to commit yourself to a plan of action for the future that won't be in your interests should the occasion arise to carry it out. Just *saying* that you are committed won't convince anyone who believes that you are rational. However, sometimes it is possible to find irreversible actions that have the same effect as making a commitment. As in the story that follows, such actions usually need to be costly, so that the other players can see that you are putting your money where your mouth is. Suppose that the incumbent monopolist can decide, before anything else happens, to make an irreversible investment in extra capacity. This will involve a dead loss of \$2 million if she makes no use of the capacity—and the only time that the extra capacity would get used is if she decides to fight the entrant. Alice will then make \$1 million (inclusive of the cost of the extra capacity) instead of \$0 million, because her extra capacity will make it cheaper for her to flood the market. Bob's payoffs remain unchanged.

a. Draw a new game tree illustrating the changed situation. This will have five decision nodes, of which the first represents Alice's investment decision. If she invests, the payoffs resulting from later actions in the game will need to be modified to take into account the costs and benefits of the extra capacity.
b. Determine the unique subgame-perfect equilibrium.
c. Someone who knows no game theory might say that it is necessarily irrational to invest in extra capacity that you don't believe you will ever use. Why is this wrong?

18. In a simple version of the Holdup Problem, Alice has \$3 million, which she is thinking of investing in Bob's company. If she makes the investment, Bob can either work or slack. If he slacks, he consumes Alice's investment, and she gets nothing. If he works, Alice's doubles her investment, and Bob nets \$2 million. Explain why Alice won't make the investment unless there is some way that she can commit Bob to working.

econ

19. Reinhard Selten, who invented subgame-perfect equilibria, is far from being a purist. He proposed the Chain Store paradox to show that it would be a mistake always to use subgame-perfect equilibria when trying to predict how real players will perform in a game. In the paradox, Alice is an incumbent monopolist who owns the only store in 100 hick towns. Bob, Chris, and ninety-eight other players are potential entrants in the 100 towns. If Bob sets up a rival store in the first town, Alice must play the Chain Store Game with Bob. If Chris later sets up a rival store in the second town, Alice must play the Chain Store Game with Chris. And so on.
    a. Draw an extensive form for the game in which the only potential entrants are Bob and Chris. Show that the unique subgame-perfect equilibrium requires that Alice always acquiesce.
    b. Why will the conclusion be the same with 100 potential entrants?
    c. Why would it make more sense in real life for Alice to fight Bob and Chris in the game with 100 potential entrants? In what respect does real life fail to satisfy the assumptions necessary to justify using backward induction in the Chain Store paradox?

econ

20. An eccentric philanthropist is prepared to endow a university with up to a billion dollars. He invites the presidents of Yalebridge and Harford to a hotel room where he has the billion dollars in a suitcase. He explains to his guests that he would like the two presidents to play a version of the Centipede Game in order to decide whose university gets endowed. The first move consists of an offer of \$1 by the philanthropist to player I (Yalebridge), who can accept or refuse. If he refuses, the philanthropist offers \$10 to player II (Harford). If she refuses, \$100 is then offered to player I, and so on. After each refusal, an amount ten times larger is offered to the other player. If there are nine refusals, player II will be offered the whole billion dollars. If she refuses, the philanthropist takes his money back to the bank.
    a. Analyze this game using backward induction and hence find the unique subgame-perfect equilibrium. What would be the result of successively deleting weakly dominated strategies in the game?
    b. Is it likely that the presidents of Yalebridge and Harford are so sure of each other's rationality that one should expect to see the subgame-perfect equilibrium actually played? What do you predict the president of Yalebridge

would do when offered $100,000 if both presidents had refused all smaller offers?

c. How would you play this game?

21. In Basu's Travelers' Dilemma, an airline loses Adam's and Eve's luggage. Adam and Eve were each carrying home one of a pair of identical jewels. The airline suspects that Adam and Eve may be tempted to inflate the value of the jewels when making a claim for compensation. Having read Section 1.10.2 on mechanism design, the airline tells them that it will pay compensation without any legal hassle, provided that they agree to abide by the following rules. Each must separately name a whole number of dollars between $1,000 and $1,000,000 as the value of their lost jewel. The airline will then pay the *minimum* of the two claims to each player. If one player claims less than the other, the player who made the smaller claim will receive a bonus of $2 that is taken from the player who made the higher claim.
    a. Show that a version of the Prisoners' Dilemma is obtained by allowing only claims of either $999,999 or $1,000,000.
    b. Show that successively deleting weakly dominated strategies in the strategic form of the full simultaneous-move game leaves a Nash equilibrium in which both players claim only $1,000.
    c. If the players are unwilling pay attention to $1 more or less, show that there is an approximate Nash equilibrium in which each player claims $1,000,000.
    d. Is the airline's attempt at mechanism design likely to pay off?

22. The Prisoners' Dilemma of Figure 1.3(a) is repeated $n$ times. The payoffs of the repeated games are the *average* of the payoffs in the stage games. If $n$ is sufficiently large, show that a pair of GRIM strategies (Section 1.8) is an *approximate* Nash equilibrium for the repeated game in which the players cooperate at every stage. How large does $n$ need to be as a function of $\varepsilon$? (Section 5.6.1)

fun

23. Robert Louis Stevenson's *Imp in the Bottle* features a fabulous bottle whose owner will be granted any wish. The snag is that someone who buys the bottle must then sell it to someone else *at a lower price* or else suffer all the pains of hell.
    a. Assuming that the smallest possible unit of currency is a cent, propose a game that represents the sale of the bottle to successive owners. Analyze the game using backward induction.
    b. Would you buy the bottle if it were offered to you for $1,000? If your answer isn't consistent with the backward induction analysis, explain your reasoning.

phil

24. Is it always a good idea to be better informed? Pandora's information sets in a game partition her set of decision nodes. A refinement of this partition is obtained by breaking down one or more of the sets of which it is formed into disjoint subsets. If we make Pandora better informed by refining her information partition, show that she will then have more strategies. Why will Pandora be no worse off if she is the only player, or if the other players are unaware of the possibility that she may have become better informed? Why might Pandora suffer from becoming better informed if the other players learn that she has become better informed?

25. Use the Cournot game of Figure 5.11(c) as an example of a situation in which it isn't desirable to be better informed (Exercise 5.9.24). If Bob learns Alice's strategy before choosing himself, then he will be no better off if she is unaware of his industrial espionage. However, if Bob's espionage becomes common knowledge, the game becomes a leader-follower game in which his equilibrium payoff is reduced from 16 to 9.

phil

# 6 *Mixing Things Up*

## 6.1 Mixed Strategies

To solve a game, we need to close the chains of reasoning that begin:

"Adam thinks that Eve thinks that Adam thinks that Eve thinks..."

After following such a chain for two or three steps, most people begin to mutter darkly about infinite regressions and vicious circles. Perhaps the most important achievement of the early game theorists was to recognize that we needn't get into this kind of tizzy. Focusing on Nash equilibria cuts through the difficulties. Any other strategy profile will be destabilized as soon as the players start thinking about what the other players are thinking.

But what happens when there are no pure equilibria? We answered this question when studying Matching Pennies (Section 2.2.2). Adam makes himself unpredictable by using a *mixed* strategy, in which he randomizes between *heads* and *tails*, choosing each with equal probability. If Eve does the same, the players will be using a Nash equilibrium. Both players then win half the time, which is the best they can do, given the strategy choice of the other.

This chapter introduces the apparatus needed to study mixed strategies in a systematic way. But first we need to look at some less trivial examples than Matching Pennies to make it clear that the effort is worthwhile.

### 6.1.1 A Sealed-Bid Auction

econon

→ 6.2

Pandora is committed to selling her house to the highest bidder in a conventional sealed-bid auction. It is common knowledge that there are two risk-neutral bidders, Alice and Bob, who both value the house at \$1 million. What bids will they seal in their envelopes?

Unless they collude, Alice and Bob are screwed. Counting bids in fractions of a million dollars, they must both bid 1 in equilibrium. If Alice gets the house as a result of winning the resulting coin toss, she then pays Pandora \$1 million and makes a profit of zero. But it can't be in equilibrium for Alice to bid $x < 1$ because Bob would then bid some fractionally larger $y$.

Things change if we model the costs of entering the auction. Such costs include having the house surveyed or arranging the necessary financing. Pandora may even charge a fee to enter her auction. It matters whether Alice and Bob know whether the other has entered the auction when they seal a bid into their envelopes. We assume that they don't.

If Alice and Bob both enter for sure, then they must both bid 1 for the same reason as before. But the winner will now make an overall loss of $c$ and thus would have done better not to to enter at all. On the other hand, if Alice stays out of the auction for sure, then Bob's best reply is to enter with a bid of 0 (negative bids aren't allowed). But if Alice uses this strategy, then Bob's best reply is to enter as well with a bid of fractionally more than 0.

All the pure strategy possibilities are therefore ruled out as possible Nash equilibria in the game between Alice and Bob. But there is a Nash equilibrium in which both players use the same *mixed* strategy. In this equilibrium, Alice and Bob keep each other guessing about whether they are going to enter. Each player stays out of the auction with probability $p$.

If her randomizing device tells Alice to enter the auction, what should she bid? A bid of more than $1-c$ always makes a loss whatever happens, and so she would have done better to stay out in the first place. A bid of exactly $1-c$ is no good either because her payoff will then be 0, but she can get more by bidding 0 and picking up a profit on those occasions when Bob doesn't enter. Nor can a bid of $x < 1-c$ be right. If it were, Bob could do even better by bidding a fractionally larger $y$. So Alice and Bob have more mixing to do.

Consider what happens if Bob stays out with probability $p = c$ and then chooses a bid $y \leq 1-c$ so that

$$\text{prob } (y \leq x) = \frac{cx}{(1-c)(1-x)}.$$

What is Alice's best reply? If she enters and bids $x \leq 1-c$, she expects

$$-c + p(1-x) + (1-p)(1-x) \text{ prob } (y \leq x) = -c + c(1-x) + cx = 0.$$

It follows that Alice gets a payoff of 0 whether she stays out or enters with a bid of $x \leq 1-c$. These pure strategies are all *best* replies to Bob's mixed strategy because her other pure strategies always make a loss.

If Alice makes 0 with all her best replies, then she will also make 0 if she chooses randomly among them. Any mixed strategy that assigns a positive probability only to these best replies is therefore also a best reply. In particular, if Alice plays the same mixed strategy as Bob, she will be making a best reply to his choice of strategy. But since Bob is in exactly the same position as Alice, he will simultaneously be making a best reply to her choice of strategy. We have therefore found a Nash equilibrium in mixed strategies for the game.

Alice and Bob therefore have to work a lot harder when there are entry costs, but their fate is the same. Pandora gets all the available surplus, and they are left with nothing.[1]

*Computing Mixed-Strategy Equilibria.* How did we know what mixed strategy to assign to Bob in the preceding example? The answer is the key to working out mixed-strategy equilibria in general.

We are looking for a symmetric mixed-strategy equilibrium in which Alice and Bob randomize between staying out and bidding anything between 0 and $1-c$. To find the probability $p$ with which Bob stays out and the probability $Q(x)$ that he bids below $x$ after entering, we use the fact that the unknowns need to be chosen to make Alice *indifferent* between staying out and entering with any bid $x \leq 1-c$.

Since Alice gets nothing if she stays out, her indifference is expressed by the equation

$$0 = -c + p(1-x) + (1-p)Q(x)(1-x). \tag{6.1}$$

But $Q(0) = 0$,[2] and so $p = c$. Replacing $p$ by $c$ in (6.1), we then have an equation that can be solved for $Q(x)$.

Why must Alice be indifferent between staying out and entering with any bid $x \leq 1-c$? The reason is simple. If she prefers one of her pure strategies to another, it can't be optimal for her to mix between them. Rather then playing each of two pure strategies *some* of the time, she would do better to play her preferred pure strategy *all* of the time.

## 6.2 Reaction Curves

It is often useful to think about Nash equilibria in terms of what economists call reaction curves. In this section, we first illustrate their use with pure strategies and then with mixed strategies.

### 6.2.1 Reaction Curves with Pure Strategies

Whenever we circled some of player I's payoffs in the strategic form of a game to indicate his best replies, we were constructing his reaction curve in pure strategies. Player II's reaction curve was indicated by enclosing her best reply payoffs in

[1] More twists on this problem appear in Exercises 6.9.4 through 6.9.7.

[2] We have assumed throughout that Bob's probability distribution assigns zero probability to any particular bid $y$. If it didn't, we would say that the distribution has an *atom* at $y$. A symmetric equilibrium can't admit an atom at $y < 1$ in our game because the other player would do better to shift the atom to some fractionally larger bid $z$ than keep it at $y$. In particular, there is no atom at $y = 0$, and so $Q(0) = 0$.

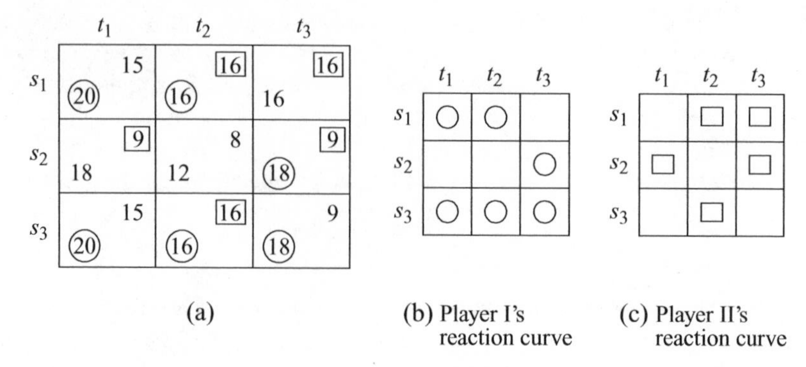


Figure 6.1 Reaction curves.

squares. Since a Nash equilibrium occurs when a cell has both payoffs circled or squared, it follows that the pure Nash equilibria of a two-player game occur where the players' pure reaction curves cross. In Section 6.2.2, we will extend this observation to mixed strategies.

Figure 6.1(a) shows a game we came across in Exercise 5.9.14 whose pure reaction curves are more complicated than usual.

The reaction curves shown separately in Figures 6.1(b) and 6.1(c) are more properly called best-reply correspondences. If we restrict ourselves to pure strategies, player I has the best-reply correspondence $R_1 : T \to S$, and player II has the best reply correspondence $R_2{:}S \to T$ defined by[3]

$$\begin{aligned} R_1(t_1) &= \{s_1, s_3\}, & R_2(s_1) &= \{t_2, t_3\}, \\ R_1(t_2) &= \{s_1, s_3\}, & R_2(s_2) &= \{t_1, t_3\}, \\ R_1(t_3) &= \{s_2, s_3\}, & R_2(s_3) &= \{t_2\}. \end{aligned}$$

For example, $R_1(t_1) = \{s_1, s_3\}$ is the set of best replies by player I to the choice of $t_1$ by player II. Similarly, $R_2(s_3) = \{t_2\}$ is the set of best replies by player II to the choice of $s_3$ by player I.[4]

A pair $(s, t)$ of strategies is a Nash equilibrium if and only if $s$ is in the set $R_1(t)$ of all best replies to $t$, and $t$ is in the set $R_2(s)$ of all best replies to $s$. But to say that $s \in R_1(t)$ and $t \in R_2(s)$ just means that $(s, t)$ is one of the places where the reaction curves cross. The game of Figure 6.1(a) therefore has precisely three Nash equilibria in pure strategies because its pure reaction curves cross precisely three times.

### 6.2.2 Reaction Curves with Mixed Strategies

Figure 6.2(a) shows a strategic form of the Inspection Game of Section 2.2, in which payoffs have been assigned to the outcomes. The reaction curves in pure strategies

[3]We don't call $R_1$ a function because $R_1(s)$ isn't an *element* of $T$ but a *subset* of $T$.

[4]Although we mostly ignore such mathematical niceties, the singleton set $\{t_2\}$ isn't the same thing as its single element $t_2$.

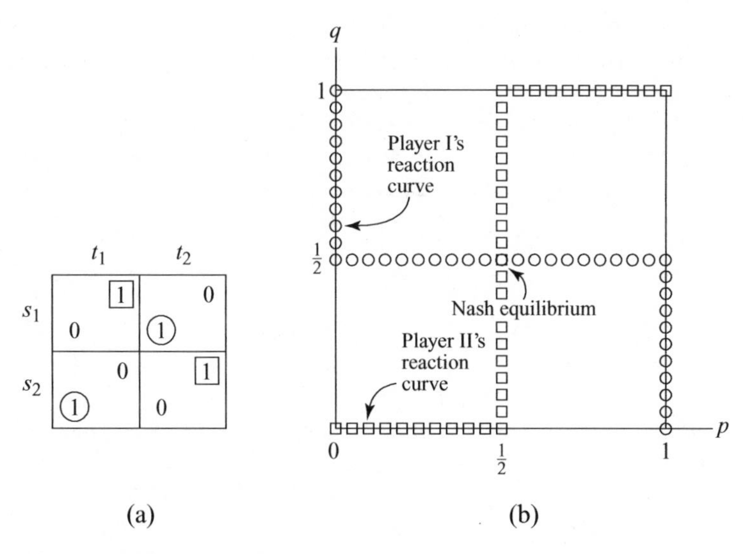


Figure 6.2 Reaction curves with mixed strategies. It is unfortunate that the two reaction curves look like a swastika, but there isn't much that can be done about it.

don't cross at all. Since the game is identical to Matching Pennies, it is no surprise that it has only mixed Nash equilibria. To study these, we look at the game's reaction curves in mixed strategies, which are fortunately easy to draw in the $2 \times 2$ case.

A mixed strategy for player I is a vector $(1-p, p)$, in which $1-p$ is the probability with which he plays $s_1$ and $p$ is the probability with which he plays $s_2$. Each of his mixed strategies therefore corresponds to a real number $p$ in the interval $[0, 1]$. Each mixed strategy for player II similarly corresponds to a real number $q$ in the interval $[0, 1]$. A pair of mixed strategies therefore corresponds to a point $(p, q)$ in the square of Figure 6.2(b).

We need to find player I's best replies to player II's choice of the mixed strategy corresponding to $q$. There is always at least one best reply in pure strategies, and so we look first at his expected payoff $E_i(q)$ when he uses his $i$th pure strategy:

$$E_1(q) = 0(1-q) + q = q,$$
$$E_2(q) = (1-q) + 0q = 1-q.$$

Player I's first pure strategy is therefore better if $q > \frac{1}{2}$. His second pure strategy is better if $q < \frac{1}{2}$.

What if $q = \frac{1}{2}$? Both of player I's pure strategies are then best replies, and so any mixture of them is also a best reply. We met the general principle in Section 6.1.1:

> A mixed strategy is a best reply to something if and only if each of the pure strategies to which it assigns positive probability is also a best reply to the same thing. A player who optimizes by using a mixed strategy will therefore necessarily be *indifferent* between all the pure strategies to which the mixed strategy assigns positive probability.

If there were another strategy $t$ that was definitely a better reply than $s$, nobody would ever want to make a reply that used $s$ with positive probability. Whenever you were called upon to play $s$, you would do better to play $t$ instead.

In summary, player I's best reply when $q < \frac{1}{2}$ is his second pure strategy, which corresponds to $p = 1$. His best reply when $q > \frac{1}{2}$ is his first pure strategy, which corresponds to $p = 0$. Any mixed strategy is a best reply when $q = \frac{1}{2}$. So his best-reply correspondence $R_I : [0, 1] \to [0, 1]$ is given by

$$R_1(q) = \begin{cases} \{1\}, & \text{if } 0 \le q < \frac{1}{2}, \\ [0,1], & \text{if } q = \frac{1}{2}, \\ \{0\}, & \text{if } \frac{1}{2} < q \le 1. \end{cases}$$

The reaction curve representing this correspondence is shown with small circles in Figure 6.2(b). For example, player I's best replies to $q = \frac{1}{4}$ are the values of $p$ at which the horizontal line $q = \frac{1}{4}$ cuts player I's reaction curve. Only $p = 0$ has this property, and so $p = 0$ is the only best reply to $q = \frac{1}{4}$.

Player II's reaction curve is shown with small squares in Figure 6.2(b). For example, player II's best replies to $p = \frac{3}{4}$ are the values of $q$ at which the vertical line $p = \frac{3}{4}$ cuts player II's reaction curve. Only $q = 1$ has this property, and so $q = 1$ is the only best reply to $p = \frac{3}{4}$.

To verify that Player II's reaction curve is correctly drawn, we first look at her expected payoff $F_i(p)$ when she uses her $i$th pure strategy and player I uses the mixed strategy corresponding to $p$:

$$F_1(p) = (1 - p) + 0p = 1 - p,$$
$$F_2(p) = 0(1 - p) + p = p.$$

Player II's second pure strategy is therefore best when $p > \frac{1}{2}$. Her first pure strategy is best when $p < \frac{1}{2}$. If $p = \frac{1}{2}$, any of her mixed strategies is a best reply. So her best-reply correspondence $R_2 : [0, 1] \to [0, 1]$ is given by

$$R_2(p) = \begin{cases} \{0\}, & \text{if } 0 \le q < \frac{1}{2}, \\ [0,1], & \text{if } p = \frac{1}{2}, \\ \{1\}, & \text{if } \frac{1}{2} < p \le 1. \end{cases}$$

Figure 6.2(b) shows that the two reaction curves cross only at $(\tilde{p}, \tilde{q}) = (\frac{1}{2}, \frac{1}{2})$, so this is the only Nash equilibrium of the game. As we saw in Section 2.2.1, each player then keeps the other guessing by acting today or tomorrow with equal probability.

### 6.2.3 Hawk or Dove?

The Hawk-Dove Game of Figure 6.3(a) will give us a chance to practice our skills at computing Nash equilibria in mixed strategies.

Two birds of the same species are competing for a scarce resource whose possession will add $V > 0$ to the evolutionary fitness of its owner. The birds play a

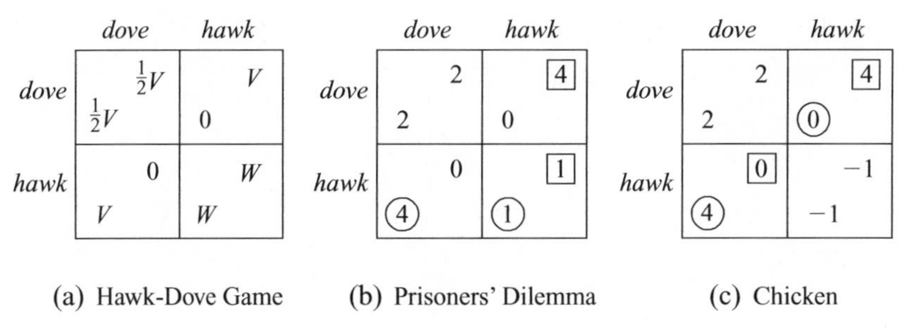


(a) Hawk-Dove Game (b) Prisoners' Dilemma (c) Chicken

Figure 6.3 Hawk-Dove Games.

simultaneous-move game in which each player can adopt a hawkish or a dovelike strategy. If both behave like doves, they split the resource equally. If one behaves like a dove and the other like a hawk, the hawk wins the resource. If both behave like hawks, there is a fight. Each bird is equally likely to win the fight and hence gain the resource, but a fight is a costly enterprise because of the risk of injury. The evolutionary fitness of a bird that has to fight is therefore $W = \frac{1}{2}V - C$, where $C > 0$ is the cost of fighting.

Recall that Chicken is a toy game played by drivers who approach each other in streets that are too narrow for them to pass without someone slowing down. As explained in Exercise 1.13.7, the Hawk-Dove Game reduces to the Prisoners' Dilemma when $W > 0$ and to Chicken when $W < 0$. The versions of the Prisoners' Dilemma and Chicken that appear in Figures 6.3(b) and 6.3(c) are obtained by taking $V = 4$ and $W = 1$ or $W = -1$. Pure reaction curves for the games are shown with circles and squares.

It is nothing new that (*hawk, hawk*) is a Nash equilibrium for the Prisoners' Dilemma. Chicken has two Nash equilibria in pure strategies: (*hawk, dove*) and (*dove, hawk*), but perhaps further Nash equilibria will emerge when mixed strategies are considered. In fact, since games typically have an *odd* number of Nash equilibria, we ought to look especially closely at the mixed strategies for Chicken. No further Nash equilibria will be found for the Prisoners' Dilemma because *dove* is strongly dominated by *hawk*, and hence no rational player will ever choose to play *dove* with positive probability.

Figure 6.4 shows reaction curves for the Prisoners' Dilemma and Chicken when we allow mixed strategies. In the Prisoners' Dilemma, the reaction curves cross only where $(\tilde{p}, \tilde{q}) = (1,1)$, which confirms that the unique Nash equilibrium is for both players to play *hawk*. In Chicken, the reaction curves cross in three places: where $(\tilde{p}, \tilde{q}) = (0,1)$, $(\tilde{p}, \tilde{q}) = (1,0)$, and $(\tilde{p}, \tilde{q}) = (\frac{2}{3}, \frac{2}{3})$. The first and second of these alternatives are the pure equilibria that we know about already. The third alternative is a mixed-strategy Nash equilibrium in which both players use *dove* with probability $\frac{1}{3}$ and *hawk* with probability $\frac{2}{3}$.

Player I's reaction curve for Chicken is vertical when player II uses $\tilde{q} = \frac{2}{3}$. Player II's reaction curve is horizontal when player I uses $\tilde{p} = \frac{2}{3}$. The players are therefore *indifferent* between all the pure strategies that they should play with positive probability when using the mixed equilibrium.

To find the mixed Nash equilibrium in Chicken without drawing the reaction curves, look for the $\tilde{p}$ that makes player I indifferent between *dove* and *hawk* and the

$\tilde{q}$ that makes player II indifferent between *dove* and *hawk*. These requirements generate the equations:

$$2(1-\tilde{p})+0\tilde{p}=4(1-\tilde{p})+(-1)\tilde{p},$$
$$2(1-\tilde{q})+0\tilde{q}=4(1-\tilde{q})+(-1)\tilde{q},$$

which have the unique solution $\tilde{p}=\tilde{q}=\frac{2}{3}$.

*Polymorphic Equilibria.* Chicken has two Nash equilibria in pure strategies, so why should we care about its mixed equilibrium? Biologists care because it is the only *symmetric* equilibrium of the game.

The pure equilibrium (*dove*, *hawk*) isn't symmetric because the row player doesn't use the same strategy as the column player. But how would animals know who is choosing a row and who is choosing a column? Sometimes Nature supplies the means—as when player I is already occupying a territory and player II is an intruder making a takeover bid. But only symmetric equilibria are relevant when Nature simply matches up pairs of animals at random because symmetric equilibria are the only equilibria that can be played without anyone needing to know who is player I and who is player II.

Animals can't roll dice or shuffle cards, so how can they use mixed strategies? The answer is that no animal has to randomize at all for a mixed strategy to be biologically meaningful.

Suppose that two genotypes are present in a population of animals, one of which plays *dove* and the other *hawk*. If there are twice as many hawks as doves, then a randomly chosen opponent will play *dove* with probability $\frac{1}{3}$ and *hawk* with probability $\frac{2}{3}$. Such an opponent is indistinguishable from a player who uses the mixed strategy $(\frac{1}{3},\frac{2}{3})$. Any strategy in Chicken is optimal against this mixed strategy, and

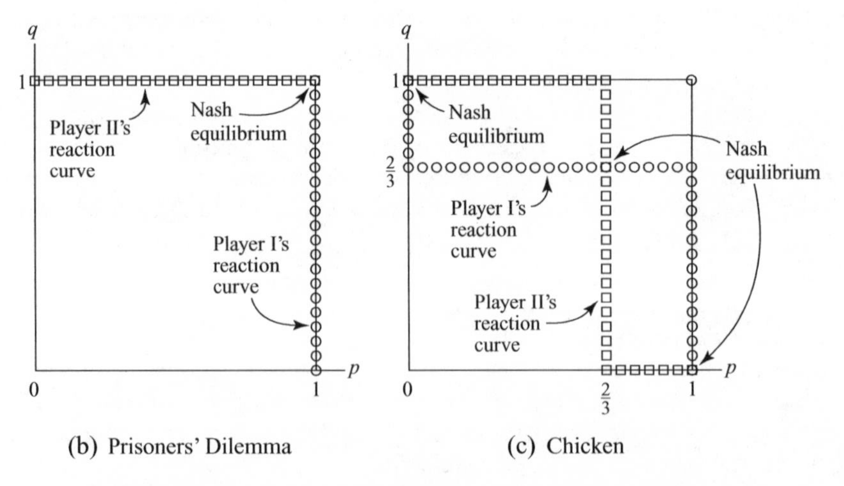


Figure 6.4 Reaction curves for the Prisoners' Dilemma and Chicken.

hence there is no evolutionary pressure against either *dove* or *hawk*. Our mixture of genotypes can therefore survive.

In a biological context, it is sometimes a good idea to focus on the big game being played by the whole population of animals. This game has as many players as there are animals. Each player chooses either *hawk* or *dove*. A chance move then selects two of the players at random to play Chicken. Players who aren't selected get nothing.

Our analysis shows that the population game has a Nash equilibrium in *pure* strategies. Any strategy profile in which $\frac{1}{3}$ of the players choose *dove* and the other $\frac{2}{3}$ choose *hawk* suffices for this purpose. Such equilibria are common in nature. Biologists call them *polymorphic equilibria* because two or more types of behavior coexist together. Each such polymorphic equilibrium of the population game corresponds to a symmetric mixed equilibrium of Chicken.

## 6.3 Interpreting Mixed Strategies

Mixed strategies were introduced in Section 2.2.2 as a way of making yourself unpredictable when playing an opponent who is good at detecting patterns in your behavior. Critics respond that someone who makes serious decisions at random must be crazy. In war, for example, a good commander must keep the enemy guessing, but if things work out badly and a court martial ensues, an officer who wants to stay out of a mental hospital would be wise to deny having based his decision of whether or not to attack on the toss of a coin.

→ 6.4

However, although people are commonly opposed to deciding important matters by rolling dice, they don't slavishly follow some fixed rule that would make their behavior in a game easy to predict. As argued in Section 1.6, evolutionary forces—both social and biological—would tend to eliminate such stupid behavior. The result is that people end up playing mixed equilibria without being aware that they are doing so. This can happen because it doesn't matter whether you really choose at random, provided your choice is unpredictable.

Suppose, for example, that we deny Eve access to a randomizing device when she plays Matching Pennies with Adam. Is she now doomed to lose? Not if she knows her Shakespeare well! She can then make each choice of *head* or *tail* contingent on whether there is an odd or even number of speeches in the successive scenes of *Titus Andronicus*. Of course, Adam might in principle guess that this is what she is doing—but how likely is this? He would have to know her initial state of mind with a quite absurd precision in order to settle on such a hypothesis. Indeed, I don't know myself why I chose *Titus Andronicus* from all Shakespeare's plays to make this point. Why not *Love's Labour's Lost* or *The Taming of the Shrew?* To outguess me in such a matter, Adam would need to know my own mind better than I know it myself.

With this story, a mixed equilibrium need involve no explicit randomization at all. Chance chooses from many different types of people when selecting player I. Some types use *Titus Andronicus* when deciding between *heads* or *tails*. Less literary folk may prefer the incidence of muggings in Milwaukee last September or the number of raindrops they can see on the windowpane.

Whatever their reasons, some fraction of the population from which player I is chosen will play *heads,* and the rest will play *tails.* If the fractions are equal in both this population and the population from which player II is drawn, then we are looking at a *polymorphic* equilibrium in a population game whose players are everybody that Chance might call upon to play Matching Pennies. Although all persons in both populations may make up their minds about whether to choose *heads* or *tails* in an entirely deterministic manner, it will seem to anyone watching Matching Pennies being played that a mixed equilibrium is in use.

Game theorists say that the mixed equilibrium of Matching Pennies has been *purified* when it is interpreted in terms of a polymorphic equilibrium in pure strategies of a larger population game (Section 15.6). The strategies in the mixed equilibrium then cease to say what a rational player will *do* when playing Matching Pennies. They now tell us only what the players *believe* about the distribution of types in the two populations. A purified equilibrium is therefore an equilibrium in beliefs rather than an equilibrium in actions.

→ 6.5

## 6.4 Payoffs and Mixed Strategies

So far, we have managed to get by without much mathematics in this chapter, but we need to be more systematic if the use of mixed strategies is to find a regular place in our toolkit.

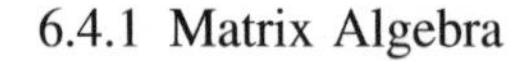

### 6.4.1 Matrix Algebra

→ 6.4.2

Matrices were introduced in Section 5.3 when studying strategic forms. We now need to learn how they are added and multiplied.

*Matrix Addition.* To add two matrices with the same dimensions, just add the corresponding entries. With the examples $A$ and $B$ of Section 5.3.1:

$$A+B^{\top} = \begin{bmatrix} 3 & 0 & 1 \\ 1 & 0 & -2 \end{bmatrix} + \begin{bmatrix} 2 & 1 & 0 \\ 3 & 0 & -3 \end{bmatrix} = \begin{bmatrix} 5 & 1 & 1 \\ 4 & 0 & -5 \end{bmatrix};$$

$$B+0 = \begin{bmatrix} 2 & 3 \\ 1 & 0 \\ 0 & -3 \end{bmatrix} + \begin{bmatrix} 0 & 0 \\ 0 & 0 \\ 0 & 0 \end{bmatrix} = \begin{bmatrix} 2 & 3 \\ 1 & 0 \\ 0 & -3 \end{bmatrix}.$$

We made sense of the expression $B+0$ by interpreting 0 as the $3 \times 2$ zero matrix, but it is never meaningful to try to add matrices that don't have the same dimensions. For example, it doesn't make any sense to write

$$A+B = \begin{bmatrix} 3 & 0 & 1 \\ 1 & 0 & -2 \end{bmatrix} + \begin{bmatrix} 2 & 3 \\ 1 & 0 \\ 0 & -3 \end{bmatrix}.$$

*Scalar Multiplication.* To multiply a matrix by a scalar, just multiply each matrix entry by the scalar. For example,

$$3A = 3\begin{bmatrix} 3 & 0 & 1 \\ 1 & 0 & -2 \end{bmatrix} = \begin{bmatrix} 9 & 0 & 3 \\ 3 & 0 & -6 \end{bmatrix}$$

$$B - A^{\top} = \begin{bmatrix} 2 & 3 \\ 1 & 0 \\ 0 & -3 \end{bmatrix} + (-1)\begin{bmatrix} 3 & 1 \\ 0 & 0 \\ 1 & -2 \end{bmatrix} = \begin{bmatrix} -1 & 2 \\ 1 & 0 \\ -1 & -1 \end{bmatrix}.$$

*Matrix Multiplication.* In order for the matrix product $CD$ to make sense, it is essential that $C$ have the same number of columns as $D$ has rows. If $C$ is an $m \times n$ matrix and $D$ is an $n \times p$ matrix, then $CD$ is an $m \times p$ matrix.

In the examples we are using, $A$ is a $2 \times 3$ matrix and $B$ is a $3 \times 2$ matrix, and so $AB$ is a $2 \times 2$ matrix and $BA$ is a $3 \times 3$ matrix. To find the entry of $AB$ that lies in its second row and first column of $AB$, we first identify the second row of $A$ and the first column of $B$, as shown in Figure 6.5. The answer 2 is then obtained by summing the products of corresponding entries in this row and column to obtain

$$1 \times 2 + 1 \times 0 - 2 \times 0 = 2.$$

Four such calculations need to be made for the matrix $AB$ and nine for the matrix $BA$:

$$AB = \begin{bmatrix} 6 & 6 \\ 2 & 9 \end{bmatrix}; \qquad BA = \begin{bmatrix} 9 & 0 & -4 \\ 3 & 0 & 1 \\ -3 & 0 & 6 \end{bmatrix}.$$

Some care is needed when multiplying matrices. It isn't even guaranteed that the product of two matrices is a meaningful object. For example, one can't multiply a $2 \times 3$ matrix by another $2 \times 3$ matrix, and so it doesn't make sense to write $AB^{\top}$. Even when all the matrix products involved are meaningful, only some of the usual laws of multiplication are valid. It is always true that $(LM)N = L(MN)$ when all the products are meaningful, but you will be lucky if $LM = ML$, even when both sides make sense. The two matrices $AB$ and $BA$ don't even have the same dimensions.

*Vector Arithmetic.* Vectors can be represented as matrices, and so we can add them together and multiply them by scalars.

In particular, if $\alpha$ and $\beta$ are scalars, we can talk about a *linear combination* $\alpha x + \beta y$ of two vectors $x$ and $y$ that have the same dimension. For example, if $x$ and $y$ are vectors in $\mathbb{R}^2$, then

$$\alpha x + \beta y = \alpha(x_1, x_2) + \beta(y_1, y_2) = (\alpha x_1 + \beta y_1, \alpha x_2 + \beta y_2).$$

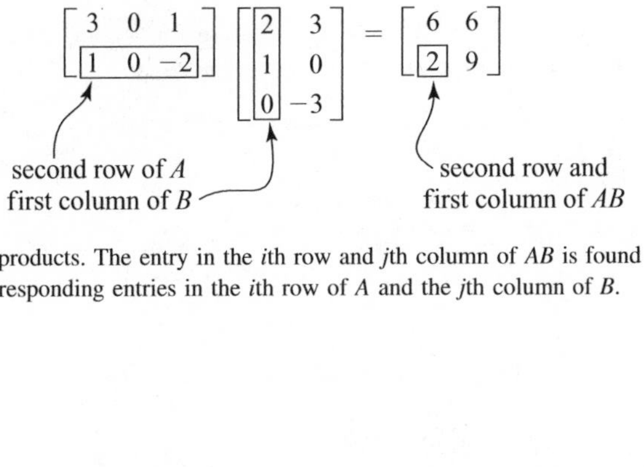

$$\begin{bmatrix} 3 & 0 & 1 \\ \boxed{1 \quad 0 \quad -2} \end{bmatrix} \begin{bmatrix} \boxed{2} & 3 \\ \boxed{1} & 0 \\ \boxed{0} & -3 \end{bmatrix} = \begin{bmatrix} 6 & 6 \\ \boxed{2} & 9 \end{bmatrix}$$

second row of $A$
first column of $B$

second row and
first column of $AB$

Figure 6.5 Matrix products. The entry in the $i$th row and $j$th column of $AB$ is found by summing the products of the corresponding entries in the $i$th row of $A$ and the $j$th column of $B$.

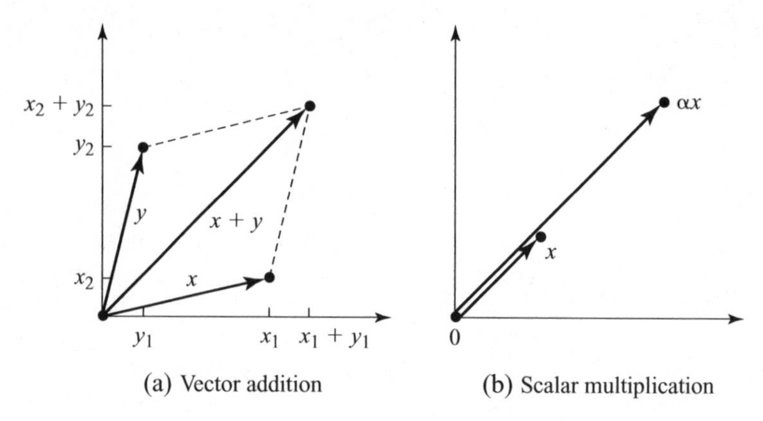


Figure 6.6 Vector addition and scalar multiplication.

Note that $x+y$ can be interpreted as the displacement that results from first using the displacement $x$ and then using the displacement $y$. Figure 6.6(a) illustrates the idea. It also makes it obvious why the rule for adding two vectors is called the *parallelogram law*.

*Orthogonal Vectors.* We can't simply multiply two $n$-dimensional column vectors $x$ and $y$ because the product of two $n \times 1$ matrices is meaningful only when $n=1$. However, it makes sense to multiply the $1 \times n$ matrix $x^\top$ by the $n \times 1$ matrix $y$ to obtain the $1 \times 1$ matrix $x^\top y$. This scalar is given by

$$x^\top y = \begin{bmatrix} x_1 & x_2 & \cdots & x_n \end{bmatrix} \begin{bmatrix} y_1 \\ y_2 \\ \vdots \\ y_n \end{bmatrix} = x_1 y_1 + x_2 y_2 + \cdots + x_n y_n.$$

Mathematicians say that $x^\top y$ is the *inner product* or the *scalar product* of the vectors $x$ and $y$.[5]

The geometric interpretation of inner products is important. A necessary and sufficient condition for two vectors $x$ and $y$ to be *orthogonal* (or perpendicular, or at right angles) is that their inner product $x^\top y$ is zero.

$$\|x\|^2 = x^\top x = x_1^2 + x_2^2 + \cdots + x_n^2.$$

The case $n=2$ is illustrated in Figure 6.7(a). Pythagoras's theorem then tells us that $\|x\|$ is simply the length of the arrow that represents $x$ when this is thought of as a displacement.

[5]The notation $(x,y) = x^\top y$ is frequently used in spite of the risk of confusion with other uses of $(x, y)$. Sometimes $x^\top y$ is written as $x \cdot y$ and called a *dot product*.

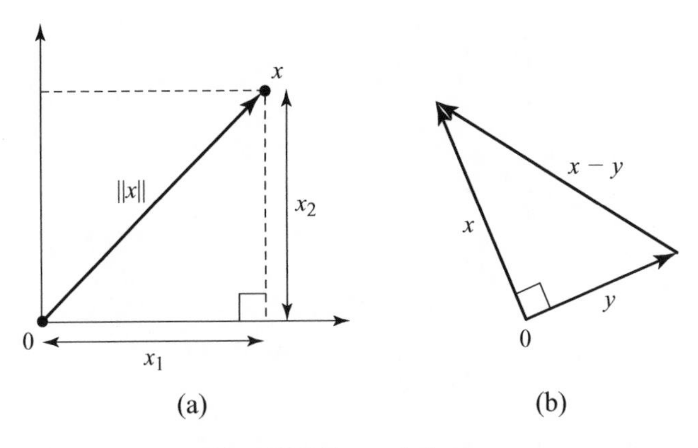


Figure 6.7 Pythagoras's theorem.

We can now apply Pythagoras's theorem to the right-angled triangle of Figure 6.7(b) to verify that the inner product of the orthogonal vectors $x$ and $y$ is zero:

$$\begin{aligned} \|x-y\|^2 &= \|x\|^2+\|y\|^2 \\ (x-y)^\top(x-y) &= x^\top x+y^\top y \\ x^\top x - y^\top x - x^\top y + y^\top y &= x^\top x+y^\top y \\ x^\top y &= 0. \end{aligned}$$

Note that $y^\top x = x^\top y$ because both sides of the equation are equal to $x_1y_1 + x_2y_2 + \cdots + x_ny_n$. More elegantly, we can use the fact that $(CD)^\top = D^\top C^\top$ always holds when the product $CD$ makes sense. Moreover, $y^\top x$ is a scalar and thus equal to its own transpose. Thus, $y^\top x = (y^\top x)^\top = x^\top (y^\top)^\top = x^\top y$.

### 6.4.2 The Algebra of Mixed Strategies

In algebraic terms, a mixed strategy for player I in an $m \times n$ bimatrix game is an $m \times 1$ column vector $p$ with nonnegative coordinates that sum to one. The coordinate $p_j$ is to be understood as the probability with which player I's pure strategy $s_j$ is used. Similarly, a mixed strategy for player II is an $n \times 1$ column vector $q$. The coordinate $q_k$ is the probability with which player II's pure strategy $t_k$ is used. The set of all player I's mixed strategies will be denoted by $P$, and the set of all player II's mixed strategies by $Q$.

Consider the $2 \times 3$ bimatrix game of Figure 6.8(a). The $2 \times 1$ column vector $p = (\frac{3}{4}, \frac{1}{4})^\top$ is an example of a mixed strategy for Adam in this game. To implement this choice of mixed strategy, Adam might draw a card from a well-shuffled deck of cards and use his second pure strategy $s_2$ if he draws a heart and his first pure strategy $s_1$ otherwise. An example of a mixed strategy for Eve is the $3 \times 1$ column vector $q = (\frac{1}{2}, \frac{1}{2}, 0)^\top$. She may implement this mixed strategy by tossing a fair coin and using her first pure strategy $t_1$ if heads appears and her second pure strategy $t_2$ if tails appears.

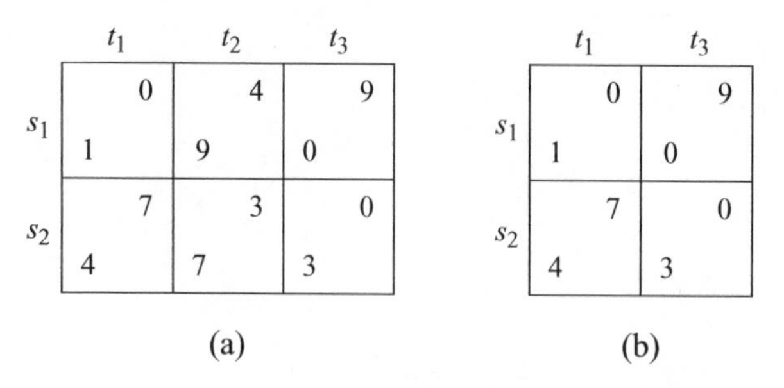


Figure 6.8 Domination by a mixed strategy.

*Domination and Mixed Strategies.* As an example of the use of mixed strategies, we now look at a game that has a pure strategy that is dominated by a mixed strategy but not by any pure strategy.

None of Eve's pure strategies dominates any other in the bimatrix game of Figure 6.8(a). However, Eve's pure strategy $t_2$ is strongly dominated by her *mixed* strategy $q = (\frac{1}{2}, 0, \frac{1}{2})$, which attaches probability $\frac{1}{2}$ to $t_1$ and probability $\frac{1}{2}$ to $t_3$. To see this requires some calculation.

If Eve uses $q$ and Adam uses $s_1$, each of the outcomes $(s_1, t_1)$ and $(s_1, t_3)$ will occur with probability $\frac{1}{2}$. Thus Eve's expected payoff is $0 \times \frac{1}{2} + 9 \times \frac{1}{2} = 4\frac{1}{2}$. Since $4\frac{1}{2} > 4$, Eve does better with $q$ than with $t_2$ when Adam uses $s_1$. Eve also does better with $q$ than with $t_2$ when Adam uses his other pure strategy $s_2$ because $7 \times \frac{1}{2} + 0 \times \frac{1}{2} = 3\frac{1}{2} > 3$. Thus $q$ is better for Eve than $t_2$ whatever Adam does. This means that $q$ strongly dominates $t_2$.

The game that is left after column $t_2$ has been eliminated is shown in Figure 6.8(b). In this reduced game, $s_2$ strongly dominates $s_1$. After row $s_1$ has been deleted, $t_1$ strongly dominates $t_3$. The method of successive deletion of dominated strategies therefore leads to the pure strategy pair $(s_2, t_1)$. Since only strongly dominated strategies were deleted along the way, $(s_2, t_1)$ is the unique Nash equilibrium of the game.

### 6.4.3 Payoff Functions for Mixed Strategies

When working with mixed strategies, we need to replace the payoff function $\pi_i : S \times T \to \mathbb{R}$ introduced in Section 5.2 by a more complicated payoff function: $\Pi_i : P \times Q \to \mathbb{R}$. Just as $\pi_i(s, t)$ is player $i$'s expected payoff when player I uses pure strategy $s$ and player II uses pure strategy $t$, so $\Pi_i(p, q)$ is player $i$'s expected payoff when player I uses mixed strategy $p$ and player II uses mixed strategy $q$.

The first step toward finding a formula for $\Pi_i(p, q)$ is to note that we are usually interested in the case in which Adam and Eve choose their strategies *independently*. So any random devices the players use to implement their mixed strategies must be statistically independent in the sense of Section 3.2.1.

If Adam's mixed strategy is the $m \times 1$ column vector $p$, his second pure strategy $s_2$ gets played with probability $p_2$. If Eve's mixed strategy is the $n \times 1$ column vector $q$, her first pure strategy $t_1$ gets played with probability $q_1$. The pure strategy pair $(s_2, t_1)$ will therefore get played with probability $p_2 \times q_1$.

For example, if $p = (\frac{1}{3}, \frac{2}{3})^\top$ and $q = (\frac{2}{3}, 0, \frac{1}{3})^\top$ in the game of Figure 6.8(a), the probability that $(s_2, t_1)$ gets played is $p_2 q_1 = \frac{2}{3} \times \frac{2}{3} = \frac{4}{9}$. Adam's payoff when this happens is $\pi_1(s_2, t_1) = 4$, and Eve's payoff is $\pi_2(s_2, t_1) = 7$.

We can work out the probability of each of Adam's and Eve's payoffs in the same way, and so it is easy to write down a formula for their expected payoffs when using mixed strategies in terms of the entries in their payoff matrices:

$$\Pi_1(p,q) = p^\top Aq; \qquad \Pi_2(p,q) = p^\top Bq.$$

When $p = (\frac{1}{3}, \frac{2}{3})^\top$ and $q = (\frac{2}{3}, 0, \frac{1}{3})^\top$ in the bimatrix game of Figure 6.8(a), the expected payoffs to Adam and Eve are

$$\Pi_1(p,q) = p^\top Aq = \begin{bmatrix} \frac{1}{3} & \frac{2}{3} \end{bmatrix} \begin{bmatrix} 1 & 9 & 0 \\ 4 & 7 & 3 \end{bmatrix} \begin{bmatrix} \frac{2}{3} \\ 0 \\ \frac{1}{3} \end{bmatrix} = 4.$$

$$\Pi_2(p,q) = p^\top Bq = \begin{bmatrix} \frac{1}{3} & \frac{2}{3} \end{bmatrix} \begin{bmatrix} 0 & 4 & 9 \\ 7 & 3 & 0 \end{bmatrix} \begin{bmatrix} \frac{2}{3} \\ 0 \\ \frac{1}{3} \end{bmatrix} = 12\tfrac{1}{3}.$$

These formulas are correct because each payoff $\pi_i(s_j, t_k)$ gets multiplied by the right probability, namely $p_j q_k$. For example, when $p^\top Bq$ is expanded, $\pi_2(s_2, t_1) = 7$ gets multiplied by $p_2 q_1 = \frac{4}{9}$.

### 6.4.4 Representing Pure Strategies

It is often necessary to talk about pure strategies while using the notation introduced for mixed strategies. For this purpose, we need the column vectors $e_i$ that have a one in their $i$th row and zeros elsewhere. The column vector $e$ with a one in every row is also sometimes helpful.

As with the zero vector, the dimensions assigned to $e_i$ or $e$ depend on the context. When they stand for $3 \times 1$ vectors:

$$e_1 = \begin{bmatrix} 1 \\ 0 \\ 0 \end{bmatrix}; \quad e_2 = \begin{bmatrix} 0 \\ 1 \\ 0 \end{bmatrix}; \quad e_3 = \begin{bmatrix} 0 \\ 0 \\ 1 \end{bmatrix}; \quad e = \begin{bmatrix} 1 \\ 1 \\ 1 \end{bmatrix}.$$

If the $m \times n$ matrix $A$ is Adam's payoff matrix in a game, then the $m \times 1$ column vector $e_i$ represents the mixed strategy in which he plays his $i$th pure strategy with probability one. Playing $e_i$ is therefore the same as playing your $i$th pure strategy. Similarly, the $n \times 1$ column vector $e_j$ represents Eve's $j$th pure strategy.

If Adam and Eve choose $e_i$ and $e_j$, Eve's payoff is the entry $b_{ij}$ in the $i$th row and $j$th column of her payoff matrix $B$. In the example of Section 6.4.3,

$$\Pi_1(e_2, e_1) = e_2^\top Ae_1 = \begin{bmatrix} 0 & 1 \end{bmatrix} \begin{bmatrix} 0 & 4 & 9 \\ 7 & 3 & 0 \end{bmatrix} \begin{bmatrix} 1 \\ 0 \\ 0 \end{bmatrix} = 7.$$

The $i$th entry in the vector $p^\top A$ is $p^\top Ae_i$, which is Adam's payoff when he uses the mixed strategy $p$ and Eve uses her $i$th pure strategy. So $p^\top A$ lists the payoffs that

Adam can get when Eve replies to his choice of $p$ with a pure strategy. Similarly, $Aq$ lists the payoffs that Adam can get by playing a pure strategy when Eve uses the mixed strategy $q$. The vectors $Bq$ and $p^\top B$ have similar interpretations in terms of Eve's payoffs.

For example, we can express the fact that Adam can't get less than $\alpha$ when he plays $p$ by writing

$$p^\top A \geq \alpha e^\top. \tag{6.2}$$

This inequality implies that $p^\top Aq \geq \alpha$ for all mixed strategies $q$ because $e^\top q = q_1 + q_2 + \cdots + q_n = 1$. Similarly, Eve always gets the same payoff of $\beta$ by playing $q$ when

$$Bq = \beta e \tag{6.3}$$

because then we have that $p^\top Bq = \beta p^\top e = \beta$ for all mixed strategies $p$.

### 6.4.5 O'Neill's Card Game

Barry O'Neill used this game in some experimental work because it is the simplest asymmetric, win-or-lose game without dominated strategies.

Alice and Bob each have the A, K, Q, and J from one of the suits in a deck of playing cards. They simultaneously show a card. Alice wins if both show an ace or if there is a mismatch of picture cards. Bob wins if both show the same picture card or if one shows an ace and the other doesn't. If we assign each player a payoff of 1 when they win and 0 when they lose, the players' payoff matrices are:

$$A = \begin{bmatrix} 1 & 0 & 0 & 0 \\ 0 & 0 & 1 & 1 \\ 0 & 1 & 0 & 1 \\ 0 & 1 & 1 & 0 \end{bmatrix}; \qquad B = \begin{bmatrix} 0 & 1 & 1 & 1 \\ 1 & 1 & 0 & 0 \\ 1 & 0 & 1 & 0 \\ 1 & 0 & 0 & 1 \end{bmatrix}.$$

We seek an equilibrium $(p, q)$ in which Alice's and Bob's mixed strategies $p$ and $q$ assign a positive probability to each of their pure strategies. Both players will then be indifferent between all their pure strategies.

We know from Section 6.4.4 that $Aq$ lists the payoffs that Alice gets from playing each of her pure strategies when Bob plays $q$. When each of these payoffs is the same, there is an $\alpha$ for which

$$Aq = \alpha e.$$

With the equation $e^\top q = 1$ (which says that the coordinates of $q$ sum to one), we then have five linear equations for the five unknowns $q_1$, $q_2$, $q_3$, $q_4$, and $\alpha$.

The crudest way of solving these equations is to use a computer to calculate the inverse matrix $A^{-1}$. Then,

$$q = \alpha A^{-1} e = \alpha \begin{bmatrix} 1 & 0 & 0 & 0 \\ 0 & -\frac{1}{2} & \frac{1}{2} & \frac{1}{2} \\ 0 & \frac{1}{2} & -\frac{1}{2} & \frac{1}{2} \\ 0 & \frac{1}{2} & \frac{1}{2} & -\frac{1}{2} \end{bmatrix} \begin{bmatrix} 1 \\ 1 \\ 1 \\ 1 \end{bmatrix} = \alpha \begin{bmatrix} 1 \\ \frac{1}{2} \\ \frac{1}{2} \\ \frac{1}{2} \end{bmatrix}.$$

The coordinates of $q$ sum to one, and so $\alpha = \frac{2}{5}$. It follows that Bob's mixed strategy in the equilibrium is

$$q = (\tfrac{2}{5}, \tfrac{1}{5}, \tfrac{1}{5}, \tfrac{1}{5})^\top.$$

However, nobody ever inverts a matrix if they can help it. In this case, it is a lot easier to notice that $q_2$, $q_3$, and $q_4$ appear in a symmetric way, so there must be a solution with $q_2 = q_3 = q_4$. The vector equation $Aq = \alpha e$ then reduces to the equations $q_1 = \alpha$ and $2q_2 = \alpha$, which solve themselves.

We leave it as an exercise to check that Bob is similarly indifferent between all his pure strategies when Alice plays the mixed strategy

$$p = (\tfrac{2}{5}, \tfrac{1}{5}, \tfrac{1}{5}, \tfrac{1}{5})^\top.$$

## 6.5 Convexity

To see how mixed strategies can be handled using geometric methods, we need to resume the study of vectors that began in Section 6.4.1.

### 6.5.1 Convex Combinations

The linear combination $w = \alpha x + \beta y$ of $x$ and $y$ becomes an *affine* combination when $\alpha + \beta = 1$. Thus

→ 6.5.3

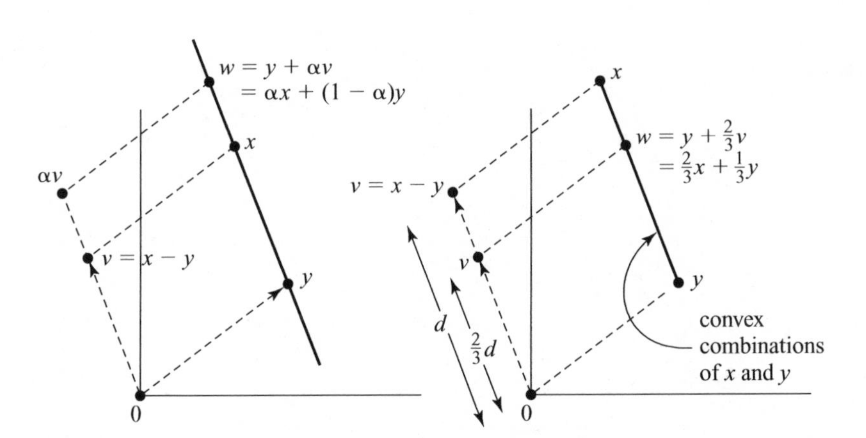


Figure 6.9 Affine and convex combinations.

$$w = \alpha x + (1-\alpha)y = y + \alpha(x-y)$$

is an affine combination of $x$ and $y$. Figure 6.9(a) shows that the set of all affine combinations of $x$ and $y$ is the straight line through the points located at $x$ and $y$. This is the same as the straight line through $y$ in the direction of the vector $v = x - y$.

A *convex* combination of $x$ and $y$ is a linear combination $w = \alpha x + \beta y$ in which $\alpha + \beta = 1$ and also $\alpha \geq 0$ and $\beta \geq 0$. Figure 6.9(b) shows that the set of all convex combinations of $x$ and $y$ is the straight-line segment joining $x$ and $y$.

If the length of the vector $v = x - y$ in Figure 6.9(b) is $\|v\| = d$, then the length of the vector $\frac{2}{3}v$ is $\frac{2}{3}d$. It follows that

$$w = \tfrac{2}{3}x + \tfrac{1}{3}y$$

lies at the point on the line segment joining $x$ and $y$ whose distances from $x$ and $y$ are $\frac{1}{3}$ and $\frac{2}{3}d$ respectively. It therefore lies one-third of the way down the line segment from $x$.

If we think of the line segment as a weightless piece of rigid wire with a mass $\frac{2}{3}$ at $x$ and a mass $\frac{1}{3}$ at $y$, then the point $w$ lies at its center of gravity. As shown in Figure 6.10(a), the wire will balance if supported at $w$.

In the general case, the linear combination

$$w = \alpha_1 x_1 + \alpha_2 x_2 + \cdots + \alpha_k x_k$$

is an affine combination of $x_1, x_2, \ldots, x_k$ when $\alpha_1 + \alpha_2 + \cdots + \alpha_k = 1$. It is a convex combination when we also have $\alpha_1 \geq 0, \alpha_2 \geq 0, \ldots, \alpha_k \geq 0$. In the latter case, $w$ lies at the center of gravity of a system with masses $\alpha_i$ located at the points $x_i$, as shown in Figure 6.10(b).

*Commodity Bundles.* Economists use vectors to describe commodity bundles (Section 4.3.1). If $(1, 3)$ is the bundle in which Pandora gets 1 bottle of gin and 3 bottles of vodka and $(5, 3)$ is the bundle in which she gets 5 bottles of gin and 3 bottles of vodka, then the convex combination

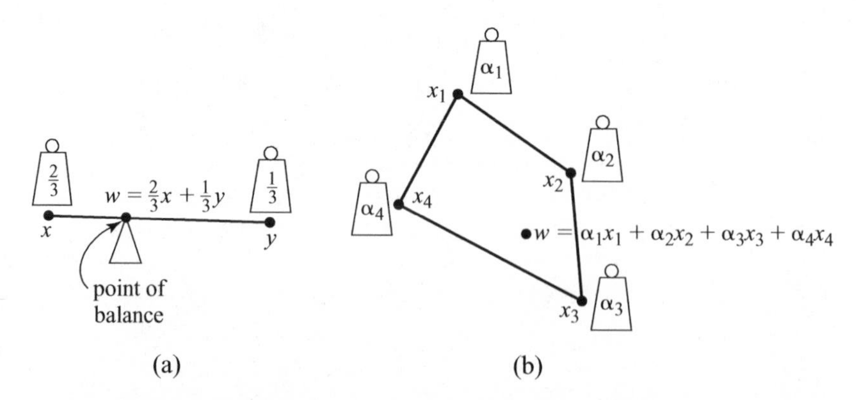


Figure 6.10 Centers of gravity. The center of a gravity of a system is the point where it would balance if supported there.

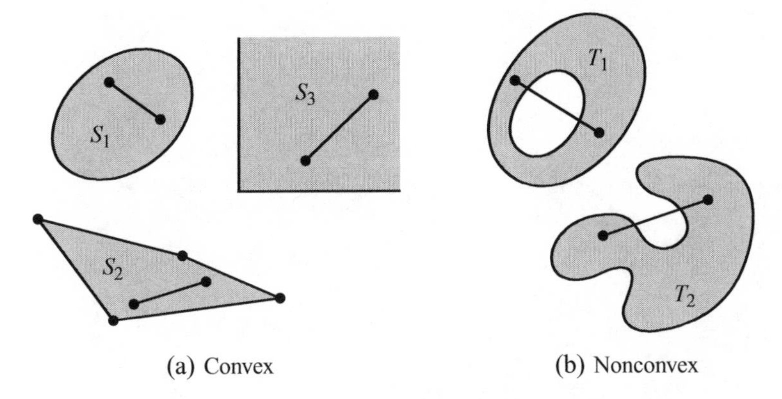


Figure 6.11 Convex and nonconvex sets.

$$\tfrac{3}{4}(1,3)+\tfrac{1}{4}(5,3)=(2,3)$$

is the physical mixture of the two bundles obtained by taking $\frac{3}{4}$ of each commodity from the first bundle and $\frac{1}{4}$ of each commodity from the second.

### 6.5.2 Convex Sets

A set $C$ is *convex* if it contains the line segment joining $x$ and $y$ whenever it contains $x$ and $y$. Figure 6.11 shows some examples of sets that are convex and sets that aren't.

→ 6.6

If $x$ and $y$ lie in a convex set $C$, then so does any convex combination $\alpha x+\beta y$ of $x$ and $y$. In fact, a convex set contains all of the convex combinations of any number of its elements.

The *convex hull* conv($S$) of a set $S$ is the set of all convex combinations of points in $S$. It is therefore the smallest convex set containing $S$. Some examples are shown in Figure 6.12.

### 6.5.3 Representing Mixed Strategies Geometrically

In an $m\times n$ bimatrix game, take $m$ points $s_1, s_2, \ldots s_m$ in some convenient space to represent Alice's $m$ pure strategies. The set $P$ of Alice's mixed strategies can then be identified with the convex hull of $s_1, s_2, \ldots s_m$.

→ 6.5.4

In a space of dimension $m-1$ or more, we will be unlucky if we have made $s_1$, $s_2, \ldots s_m$ affinely dependent.[6] If not, each point $p$ in the convex hull of the points representing Alice's pure strategies can be expressed in just one way as a convex combination $p=p_1s_1+p_2s_2+\ldots p_ms_m$ of $s_1, s_2, \ldots s_m$. We then regard the point $p$ as representing the mixed strategy $(p_1, p_2, \ldots, p_m)$.

When $m=2$, the convex hull $P$ of Alice's two pure strategies is the line segment joining $s_1$ and $s_2$, as shown in Figure 6.13(a). If $\pi$ represents the mixed strategy

[6]This means that one of the points can be expressed as an affine combination of the others. Three points in $\mathbb{R}^2$ are affinely dependent if they all lie on the same straight line. Four points in $\mathbb{R}^3$ are affinely dependent if they all lie in the same plane.

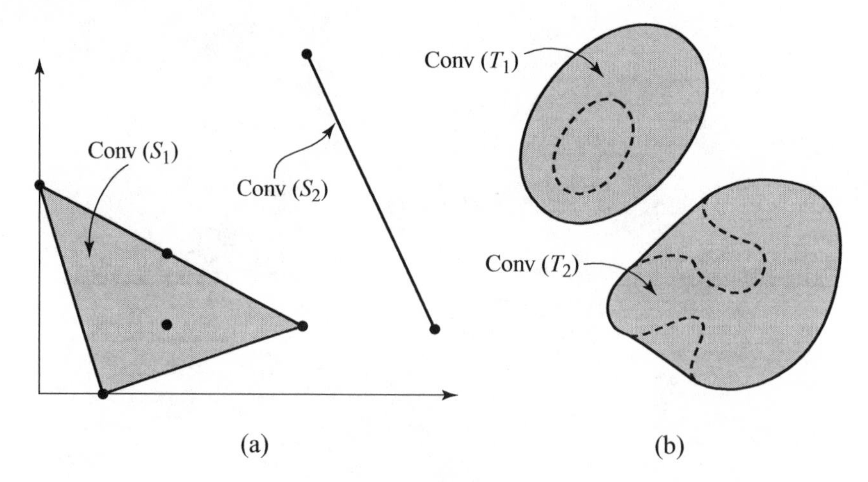


Figure 6.12 Convex hulls. Figure 6.12(a) shows the convex hulls of the sets $S_1 = \{(1,0),(0,3), (2,1), (2,2), (4,1)\}$ and $S_2 = \{(4,5),(6,1)\}$. Figure 6.12(b) shows the convex hulls of the sets $T_1$ and $T_2$ of Figure 6.11(b).

$(\pi_1, \pi_2)$, recall that the distance from $\pi$ to $s_2$ is simply $\pi_1$ of the whole distance from $s_1$ to $s_2$.

Figure 6.13(b) illustrates the case when $m = 3$. The convex hull of Alice's three pure strategies is then a triangle. When making an orthogonal journey from the line $p_3 = 0$ to the line $p_3 = 1$, one encounters the line $p_3 = \pi_3$ after traveling $\pi_3$ of the distance.[7] When $m = 4$, Figure 6.13(c) shows that the convex hull of Alice's four pure strategies is a tetrahedron. Because three-dimensional diagrams are a pain, one often unfolds such tetrahedrons and lays them flat on the page, as in Figure 6.13(d).

We choose the points that represent Alice's pure strategies in any way that is convenient. An unimaginative choice in the case $m = 3$ begins by labeling the three axes of $\mathbb{R}^3$ as $p_1$, $p_2$, and $p_3$. Alice's three pure strategies $s_1$, $s_2$, and $s_3$ then correspond to the points (1, 0, 0), (0, 1, 0), and (0, 0, 1) (Section 6.4.4). As shown in Figure 6.13(e), their convex hull $P$ lies in the plane $p_1 + p_2 + p_3 = 1$. With this special representation, we get the barycentric coordinates of a point $\pi$ in $P$ for free since these are the same as the Cartesian coordinates of $\pi$. But who wants to fuss with a three-dimensional diagram when one can do the same job with a two-dimensional diagram? Instead of drawing Figure 6.13(e), we therefore usually throw away everything but the triangle $P$ and lay this flat on the page, as in Figure 6.13(b).

What happens when we want to represent both players' mixed strategies simultaneously? We did this for a $2 \times 2$ bimatrix game in Figure 6.2. Player I's set $P$ of mixed strategies is represented by the line segment joining $(0,0)$ and $(1,0)$ in $\mathbb{R}^2$. Player II's set $Q$ of mixed strategies is represented by the line segment joining $(0,0)$ and $(0,1)$. The set of all pairs of mixed strategies can then be represented by the square $P \times Q$, illustrated in Figure 6.14(a).

[7]Mathematicians say that $(p_1, p_2, p_3)$ are the barycentric coordinates of the point it represents in the triangle. Three coordinates are then used to locate a point in a two-dimensional space, but remember that $p_1 + p_2 + p_3 = 1$.

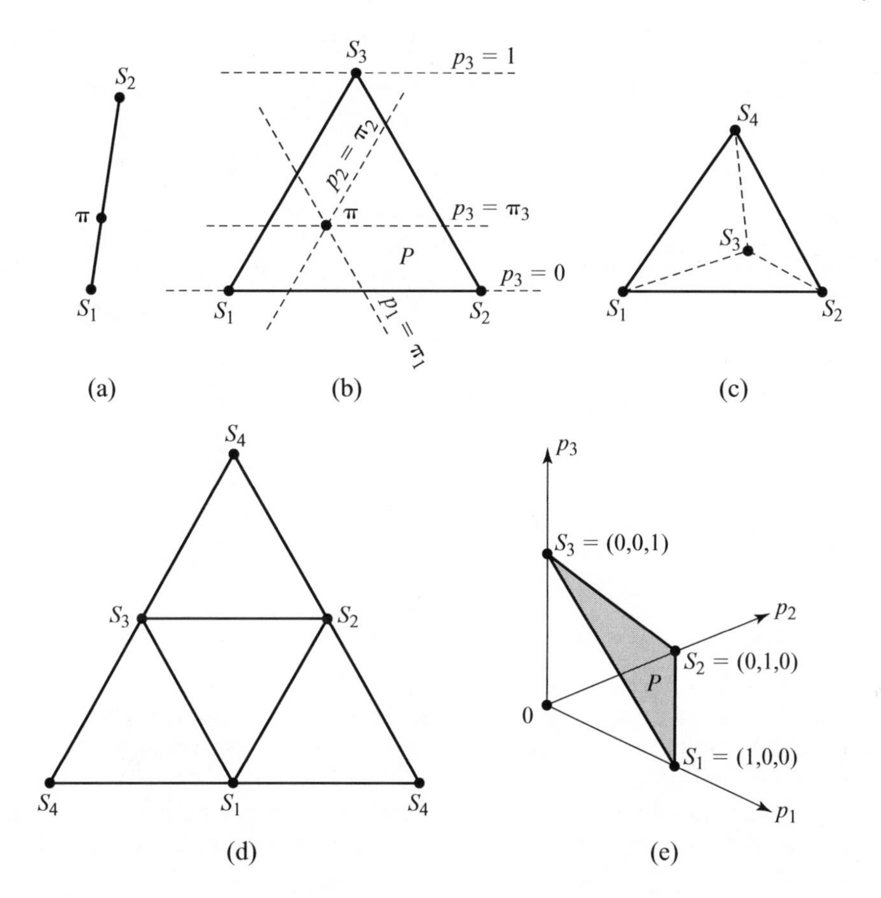


Figure 6.13 Spaces of mixed strategies. A contour labeled $p_i = \pi_i$ in Figure 6.13(b) consists of all points $p = p_1 s_1 + p_2 s_2 + p_3 s_3$ with $p_i = \pi_i$ and $p_1 + p_2 + p_3 = 1$. These contours are straight lines (Exercise 6.9.25). The faces of the tetrahedron of Figure 6.13(c) that meet at the vertex $s_4$ have been peeled away and the whole laid flat on the page to produce Figure 6.13(d). The point $s_4$ therefore appears three different times in the latter figure. One can similarly think of Figure 6.13(b) as the triangle $P$ of Figure 6.13(e) laid flat on the page.

In the case of a $2 \times 3$ bimatrix game, player I's set $P$ of mixed strategies can be represented by a straight-line segment. Player II's set $Q$ of mixed strategies can be represented by a triangle. Figure 6.14(b) shows that the set $P \times Q$ of all pairs of mixed strategies is then a prism.

### 6.5.4 Concave, Convex, and Affine Functions

When we first met concave functions in Section 4.5.3 while studying risk aversion, we noted that chords to their graphs lie on or below the graph. We could equally well have said that the set of points on or below the graph of a concave function is convex.

This geometry translates into an algebraic criterion for a function $f : C \to \mathbb{R}$ to be concave on a convex set $C$. The criterion is that, for each $x$ and $y$ in $C$,

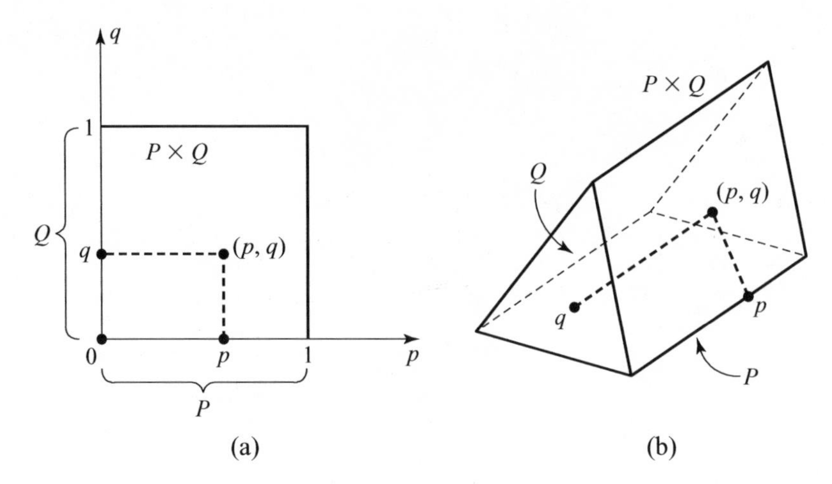


Figure 6.14 Representing mixed-strategy profiles.

$$f(\alpha x+\beta y) \geq \alpha f(x)+\beta f(y) \tag{6.4}$$

whenever $\alpha+\beta=1$, $\alpha \geq 0$, and $\beta \geq 0$.

The concave function $u : \mathbb{R}_+ \to \mathbb{R}$ defined by $u(x) = 4\sqrt{x}$ that we last saw when trying to resolve the St. Petersburg paradox will serve as an example (Section 4.5.3). In Figure 4.7, the chord joining the points $(1, u(1))$ and $(9, u(9))$ lies on or below the graph of the function. Points on this chord are convex combinations of $(1, u(1)) = (1, 9)$ and $(9, u(9)) = (9, 12)$. The point $Q$ of Figure 4.7 is the convex combination

$$\tfrac{3}{4}(1\,, u(1)) + \tfrac{1}{4}(9\,, u(9)) = (3,\, \tfrac{3}{4}u(1) + \tfrac{1}{4}u(9)).$$

Since $Q$ lies below the point $P$ on the graph,

$$u(3) = u(\tfrac{3}{4} \times 1 + \tfrac{1}{4} \times 9) \geq \tfrac{3}{4}u(1) + \tfrac{1}{4}u(9),$$

which is a particular case of the inequality (6.4).

The criterion for a convex function is that, for each $x$ and $y$ in $C$,

$$f(\alpha x+\beta y) \leq \alpha f(x)+\beta f(y),$$

whenever $\alpha+\beta=1$, $\alpha \geq 0$, and $\beta \geq 0$. This criterion is equivalent to saying that the set of points on or above the graph of the function is convex.

For an affine function, we need that, for each $x$ and $y$ in $C$,

$$f(\alpha x+\beta y) = \alpha f(x)+\beta f(y),$$

whenever $\alpha+\beta=1$, $\alpha \geq 0$, and $\beta \geq 0$.[8]

[8]If $C = \mathbb{R}^n$, we don't need to require that $\alpha \geq 0$ and $\beta \geq 0$. Without the requirement that $\alpha+\beta=1$, the condition $f(\alpha x+\beta y) = \alpha f(x) + \beta f(y)$ characterizes a linear function.

Affine functions are therefore characterized by the fact that they preserve convex combinations. If $w$ is a convex combination of $x$ and $y$, this means that $f(w)$ is the same convex combination of $f(x)$ and $f(y)$. That is to say, $w = \alpha x + \beta y \Rightarrow f(w) = \alpha f(x) + \beta f(y)$.

## 6.6 Payoff Regions

A payoff region is the set of all payoff profiles that can occur in a game under various hypotheses about what the players are allowed to do. Figure 6.15 shows versions of Chicken and the Battle of the Sexes from Exercises 1.13.5 and 1.13.6 that will provide instructive examples.

### 6.6.1 Preplay Randomization

The players of a game will frequently find it to their advantage to get together before playing the game to consider whether they might advantageously coordinate their strategy choices. Whole books are devoted to various conventions that bridge players agree to use in such preplay discussions. Our concern here is with how preplay randomizing might arise.

*Cooperative Payoff Regions.* While at breakfast in their honeymoon suite, Adam and Eve realize that they might get separated later in the day. Adam suggests that they should then meet at this evening's big boxing match. Eve suggests meeting instead at a performance of *Swan Lake.* Rather than spoil their honeymoon with an argument, they settle the issue by tossing a coin. What is this agreement worth to each player?

In terms of the Battle of Sexes, the agreement is to play each of (*box*, *box*) and (*ball*, *ball*) with probability $\frac{1}{2}$. Adam gets a payoff of 2 when the coin lands *heads* and a payoff of 1 when it lands *tails.* His expected payoff is therefore $1\frac{1}{2} = \frac{1}{2} \times 2 + \frac{1}{2} \times 1$. Eve gets a payoff of 1 when the coin lands *heads* and a payoff of

(a) Chicken (payoffs listed as Adam's payoff, Eve's payoff)

| | *slow* | *speed* |
|---|---|---|
| *slow* | 2, 2 | 0, 3 |
| *speed* | 3, 0 | −1, −1 |

(b) Battle of the Sexes (payoffs listed as Adam's payoff, Eve's payoff)

| | *box* | *ball* |
|---|---|---|
| *box* | 2, 1 | 0, 0 |
| *ball* | 0, 0 | 1, 2 |

Figure 6.15 Two toy games. Chicken is a game played by two drivers who approach each other on a street that is too narrow for them to pass without someone slowing down. The Battle of the Sexes is a coordination game played by two separated honeymooners trying to get back together.

2 when it lands *tails*. Her expected payoff is therefore $1\frac{1}{2} = \frac{1}{2} \times 1 + \frac{1}{2} \times 2$. It follows that the payoff pair that corresponds to their agreement is the convex combination

$$(1\tfrac{1}{2}, 1\tfrac{1}{2}) = \tfrac{1}{2}(2,1) + \tfrac{1}{2}(1,2)$$

of the payoff pair $(2,1)$ they get when the coin lands *heads* and the payoff pair $(1,2)$ they get when it lands *tails*.

Adam and Eve could also have used other random devices to generate other compromises between the pure outcomes of the Battle of the Sexes. Each such randomization generates a convex combination of the payoff pairs in the game's payoff table. The set of all such convex combinations is the *cooperative payoff* region $C$ of the game.

Since the set $C$ is just the convex hull of the payoff pairs in a game's payoff table, it is easy to draw. Figure 6.16 shows the cooperative payoff regions for both the Battle of the Sexes and the version of Chicken given in Figure 6.15(a).

*Noncooperative Payoff Regions.* When Adam and Eve toss a coin to decide whether to meet at the boxing match or the ballet, they aren't choosing their strategies independently. Far from implementing their mixed strategies using independent random devices as assumed in Section 6.4.3, they cooperate in using the *same* random device.

When finding the noncooperative payoff region $N$ of a game, we rule out all such cooperative activity and allow Adam and Eve to use only *independent* mixed strategies. Thus $N$ is the set of all payoff pairs

$$(x,y) = (p^\top Aq\,, p^\top Bq),$$

when $p$ and $q$ vary over all mixed strategies in $P$ and $Q$ respectively.

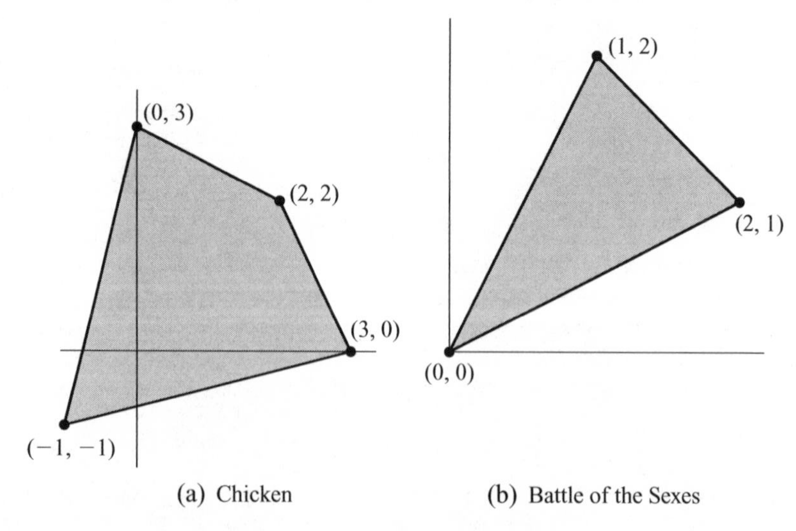


Figure 6.16 Cooperative payoff regions.

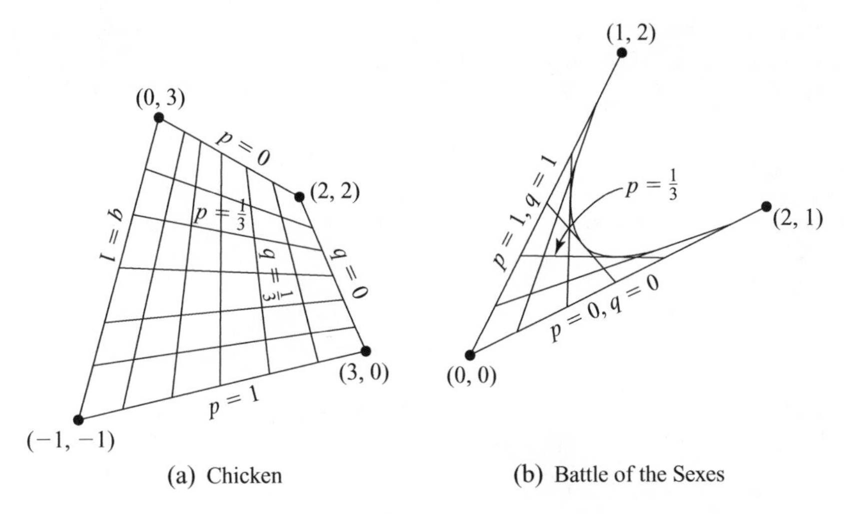


Figure 6.17 Noncooperative payoff regions.

It is instructive to build up the set $N$ one strategy at a time. A mixed strategy $(1-p,p)^{\top}$ for Adam in the Battle of the Sexes traces a line segment in payoff space. To find the line segment when $p = \frac{1}{3}$, begin by locating its endpoints. They occur where Eve uses one of her two pure strategies.

If Eve plays her first pure strategy, Adam's use of $p = \frac{1}{3}$ generates the payoff pair $\frac{2}{3}(2,1) + \frac{1}{3}(0,0)$, which is located one-third of the way down the line segment joining $(2,1)$ and $(0,0)$. If Eve plays her second pure strategy, Adam's use of $p = \frac{1}{3}$ generates the payoff pair $\frac{2}{3}(0,0) + \frac{1}{3}(1,2)$, which is located one-third of the way down the line segment joining $(0,0)$ and $(1,2)$. Mark these two points on the diagram, and then join them with a line segment. This line segment is the set of all payoff pairs that are possible when Adam uses the mixed strategy corresponding to $p = \frac{1}{3}$.

Figure 6.17(b) shows the line segments that correspond to all of Adam's and Eve's mixed strategies when $p$ or $q$ is a multiple of $\frac{1}{6}$. Enough of these line segments are drawn to make it clear that $N$ is very far from convex. The curved part of its boundary is actually a parabola, which is tangent to the straight parts of the boundary.[9]

The payoff pair that results from the play of the mixed strategy profile $(p,q)$ is the point at which the line segments corresponding to $p$ and $q$ cross. (Where both line segments are the same, the payoff pair lies at the point of tangency with the bounding parabola.)

The Nash equilibria of the game can be located by looking hard at the diagram. Two pure equilibria occur where $(p,q) = (0,0)$ and $(p,q) = (1,1)$. A mixed equilibrium occurs where $(p,q) = (\frac{1}{3}, \frac{2}{3})$. The line segment that corresponds to Adam's playing $p = \frac{1}{3}$ is horizontal, and so Eve gets the same payoff whatever she does. Similarly, the line segment that corresponds to Eve's playing $q = \frac{2}{3}$ is vertical, and so Adam gets the same payoff whatever he does.

[9]The parabola is the envelope of all the line segments that correspond to either Adam's or Eve's mixed strategies. This means that it touches each of these segments.

Figure 6.17 shows the noncooperative payoff regions for both the Battle of the Sexes and the version of Chicken given in Figure 6.15(a). The latter is much simpler to draw.

### 6.6.2 Self-Policing Agreements

Honeymooners are unlikely to cheat on any agreement they make on how to play the Battle of the Sexes. But what if we replace Adam and Eve by two suspicious strangers, Alice and Bob?

*Cheap Talk.* The only viable agreements between players who don't trust each other are those in which they agree to coordinate on an equilibrium (Section 1.7.1). Neither player then has an incentive to cheat. One might therefore think that Alice and Bob must agree on one of the three Nash equilibria of the Battle of the Sexes, but the fact that Alice and Bob are able to talk to each other before playing the Battle of the Sexes changes their game.

The messages that Alice and Bob exchange during a preplay negotiation are called *cheap talk* because it doesn't cost Alice or Bob anything to lie. Cheap talk can nevertheless be useful. For example, it allows Alice and Bob to toss a coin together. They can then emulate Adam and Eve by agreeing to play (*box, box*) if the coin lands *heads* and (*ball, ball*) if it lands *tails*. Neither has an incentive to cheat on the deal after the coin has fallen because the agreement always specifies that a Nash equilibrium be played.

We can model the situation by creating a new game $G$ that begins with a chance move. Each choice that Chance can make leads to a subgame of $G$ that is a copy of the Battle of the Sexes. A subgame-perfect equilibrium of $G$ requires that a Nash equilibrium be played in each of these subgames—but it needn't be the *same* Nash equilibrium in every subgame.

We have looked at a case in which Alice and Bob use the Nash equilibrium (*box, box*) in some subgames and the Nash equilibrium (*ball, ball*) in others. When the subgames in which each of these equilibria are to be played are reached with probability $\frac{1}{2}$, Alice and Bob achieve the payoff pair $(1\frac{1}{2}, 1\frac{1}{2})$ in the game as a whole. But the Battle of the Sexes has *three* Nash equilibria. Alice and Bob could agree to play any of these three equilibria in subgames reached with any probabilities they like.

So Alice and Bob don't need to trust each other to achieve any payoff pair in the convex hull of the payoff pairs $(2,1)$, $(1,2)$, and $(\frac{2}{3}, \frac{2}{3})$, which are the payoff pairs corresponding to the three Nash equilibria of the Battle of the Sexes. All they need to do to achieve any payoff pair in this set is to make their choice of a Nash equilibrium in the Battle of the Sexes contingent on a suitable random event that they can observe together.

Figure 6.18 shows the convex hull $H$ of the set of Nash equilibria of both the Battle of the Sexes and the version of Chicken given in Figure 6.15(a). The latter is more interesting because Alice and Bob would like to agree on the payoff pair $(2,2)$, but it isn't in the set $H$. Is there anything that Alice and Bob can do about this?

*Correlated Equilibria.* When Alice and Bob don't trust each other, the first-best payoff pair $(2,2)$ is beyond their reach in Chicken. But the payoff pair $(1\frac{1}{2}, 1\frac{1}{2})$ isn't

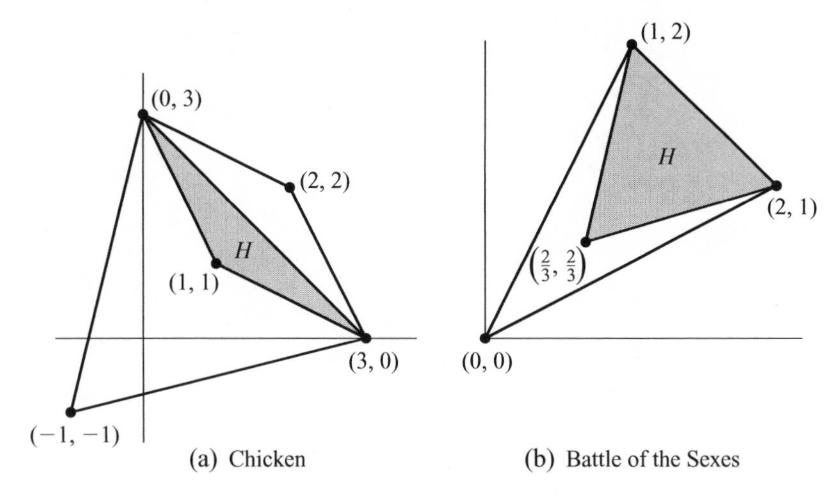


Figure 6.18 The convex hull $H$ of the Nash equilibrium outcomes for Chicken and the Battle of the Sexes. By using a jointly observed random device to coordinate their choice of a Nash equilibrium, Alice and Bob can achieve any payoff pair in $H$ without needing to trust each other. In Chicken, the players would like to agree on $(2, 2)$, but it isn't in the set $H$.

their second-best alternative. With the help of a reliable referee, they have an incentive-compatible means of achieving the pair $(1\frac{2}{3}, 1\frac{2}{3})$.

The referee is needed to operate the opening chance move in a game $G$ that Alice and Bob agree to play in a preplay cheap-talk session. Each choice made by Chance at the opening move of $G$ leads to a copy of Chicken. Since Alice and Bob only care about whether the outcome of the chance move requires them to play *slow* or *speed,* we need only distinguish the four events: $e = (slow, slow)$, $f = (slow, speed)$, $g = (speed, slow)$, and $h = (speed, speed)$.

The chance move wouldn't help matters if Alice and Bob were to see its outcome, but the referee is instructed to tell Alice and Bob only what they need to know: namely, the strategy that Chance has chosen for them to play in Chicken. As shown in Figure 6.19(b), Alice therefore knows only that either the event $A$ in which she is told to play *slow* has occurred or else the event $B$ in which she is told to play *speed.* Bob knows only that either the event $C$ in which he is told to play *slow* has occurred or else the event $D$ in which he is told to play *speed.*

Why should Alice and Bob do what the referee tells them? Their agreement to do so was just cheap talk. Nobody expects them to honor the deal if they can get a higher payoff by doing something else. For the deal to stick, it must therefore always require behavior that is compatible with their incentives.

For Alice and Bob to have an incentive-compatible deal, the probabilities with which Chance chooses the four events $e$, $f$, $g$, and $h$ need to be determined very carefully (Exercise 6.9.30). We will check only that it is enough to make

$$\text{prob}\,(e) = \text{prob}\,(f) = \text{prob}\,(g) = \tfrac{1}{3},$$
$$\text{prob}\,(h) = 0$$

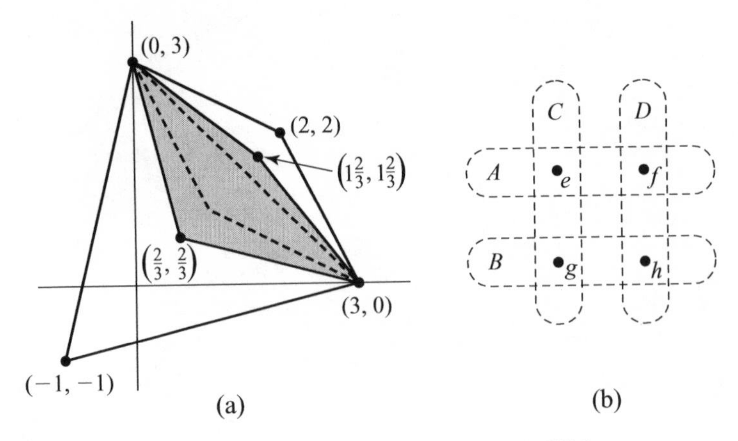


Figure 6.19 Correlated equilibrium outcomes in Chicken.

in Figure 6.19(b). The conditional probabilities introduced in Section 3.3 will be important for the proof. For example, Bob's probability for $A$ after learning that $C$ has occurred is

$$\text{prob}\,(A \mid C) = \frac{\text{prob}\,(A \cap C)}{\text{prob}\,(C)} = \frac{\text{prob}\,(e)}{\text{prob}\,(e) + \text{prob}\,(g)} = \frac{\frac{1}{3}}{\frac{1}{3} + \frac{1}{3}} = \tfrac{1}{2}.$$

For this choice of probabilities to yield an incentive-compatible agreement, we need that neither Alice nor Bob can ever gain anything by cheating on the agreement. We verify this only for Bob since Alice is in an entirely symmetric situation. Two steps are necessary. We must confirm that Bob will honor the deal both when told to play *slow* and when told to play *speed.*

**Step 1.** If the referee tells Bob to play *slow,* he calculates

$$\text{prob (Alice hears } \textit{slow} \mid \text{Bob hears } \textit{slow}) = \frac{\frac{1}{3}}{\frac{1}{3} + \frac{1}{3}} = \tfrac{1}{2},$$

$$\text{prob (Alice hears } \textit{speed} \mid \text{Bob hears } \textit{slow}) = \frac{\frac{1}{3}}{\frac{1}{3} + \frac{1}{3}} = \tfrac{1}{2}.$$

His expected payoff from honoring his agreement to play *slow* when told to do so is therefore $\frac{1}{2} \times 2 + \frac{1}{2} \times 0 = 1$. His expected payoff from cheating on the agreement and playing *speed* when told to play *slow* is $\frac{1}{2} \times 3 + \frac{1}{2} \times (-1) = 1$. He therefore loses nothing by honoring the deal when told to play *slow*.

**Step 2.** If the referee tells Bob to play *speed,* he calculates

$$\text{prob (Alice hears } \textit{slow} \mid \text{Bob hears } \textit{speed}) = \frac{\frac{1}{3}}{\frac{1}{3} + 0} = 1,$$

$$\text{prob (Alice hears } \textit{speed} \mid \text{Bob hears } \textit{speed}) = \frac{0}{\frac{1}{3} + 0} = 0.$$

It is again optimal for him to honor the deal by playing *speed* because

$$1 \times 3 + 0 \times (-1) = 3 > 2 = 1 \times 2 + 0 \times 0.$$

What payoff does Bob get in the self-policing agreement we have found? Returning to Chicken's payoff table, we find that Bob's expected payoff is

$$2 \times \text{prob}\,(e) + 0 \times \text{prob}\,(f) + 3 \times \text{prob}\,(g) = 2 \times \tfrac{1}{3} + 0 \times \tfrac{1}{3} + 3 \times \tfrac{1}{3} = 1\tfrac{2}{3}.$$

Since Alice's expected payoff is the same, we have shown how the players can achieve the payoff pair $(1\frac{2}{3}, 1\frac{2}{3})$.

The set $P$ of all payoff pairs that can be achieved with a self-policing agreement is shown in Figure 6.19(a). The fact that this set is larger than the set $H$ of Figure 6.18(a) was discovered by Robert Aumann. He refers to the Nash equilibrium of the game $G$ as a *correlated equilibrium* of Chicken.

→ 6.6

*Mental Poker.* A problem in implementing correlated equilibria is that it may not be easy to find an incorruptible referee. Philosophers complain about the cynicism they think such remarks imply, but we must remember that Alice and Bob might represent the two firms of Section 1.7.1 seeking to collude on an illegal price-fixing deal.

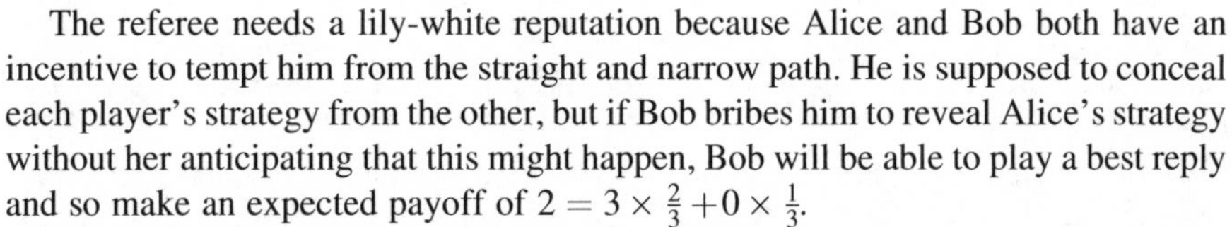

The referee needs a lily-white reputation because Alice and Bob both have an incentive to tempt him from the straight and narrow path. He is supposed to conceal each player's strategy from the other, but if Bob bribes him to reveal Alice's strategy without her anticipating that this might happen, Bob will be able to play a best reply and so make an expected payoff of $2 = 3 \times \frac{2}{3} + 0 \times \frac{1}{3}$.

Is there some way that Alice and Bob can dispense with a human referee? The wonders of modern technology make it possible to answer *yes* to this question, but one has to suspend disbelief when listening to the reason because the same technology makes it possible to play poker over the telephone. How can this be possible? Surely the players would always report that they just happened to have been dealt a royal flush!

As an example, consider the case of Adam and Eve playing the Battle of the Sexes. They would like to toss a coin to decide whether to meet at the boxing match or the ballet, but they can communicate only by telephone. Eve tosses a coin and reports that it has fallen *tails,* and so they should meet at the ballet, but Adam is distrustful. Eve therefore asks him whether he will agree to meet at the boxing match if he can solve a mathematical problem she will give him and at the ballet otherwise. Since he is the world's greatest mathematician, he agrees. Eve then uses her computer to multiply the big prime numbers

$$a = 56123699566021020558766279166381074847903158831451;$$
$$b = 57654165390541998801236990031588314500065809801648 9.$$

The number $c = a \times b$ has ninety-nine digits. The problem Eve gives Adam is to say whether the remainder left after dividing the largest of $c$'s prime factors by four is odd or not.

Adam can use all the computer wizardry he likes, but he will still be unable to factor Eve's number because the necessary computation will take longer than his lifetime. He can therefore do no better than guess at the answer. She then tells him whether he is right or wrong. If he doesn't believe her, she sends him her two prime numbers so that he can verify her claim for himself.

This solution to the coordination problem uses the trick on which modern cryptography is based. Eve's problem has a one-way trapdoor. It is computationally feasible to check that her two numbers are prime and to compute their product, but it isn't computationally feasible to reverse the process.

## 6.7 Roundup

When mixed equilibria are used, a player is indifferent between each pure strategy that is assigned positive probability. This observation often provides the answer to computing mixed equilibria. It can be successful even in complicated cases like the sealed-bid auction of Section 6.1.1.

A reaction curve plots a player's best reply to each of the opponent's strategies. Nash equilibria occur where the reaction curves cross, as each player is then making a best reply to the other.

The Hawk-Dove Game is a toy game used by biologists. Its mixed equilibrium is of interest when regarded as representing a polymorphic equilibrium of a large population game. In such a game, each member of the population chooses a pure strategy, and a chance move then selects a pair from the population to play the Hawk-Dove Game.

If Bob is chosen at random from a population in which a fraction $1-p$ have chosen pure strategy $s$ and a fraction $p$ have chosen $t$, then Alice might as well be playing an opponent using the mixed strategy in which $s$ and $t$ are chosen with probabilities $p$ and $1-p$. A mixed equilibrium can therefore always be interpreted as a polymorphic equilibrium of a large population game. Purifying a mixed equilibrium consists of proposing a population game within which such an interpretation makes sense.

In mathematical terms, a mixed strategy for player I in an $m \times n$ bimatrix game is an $m \times 1$ column vector $p$ with nonnegative coordinates that sum to one. A mixed strategy for player II is an $n \times 1$ column vector $q$. The players' payoff functions are given by

$$\Pi_1(p,q) = p^\top Aq,$$
$$\Pi_2(p,q) = p^\top Bq,$$

where $A$ and $B$ are player I's and player II's $m \times n$ payoff matrices.

The vector $e_i$ has 1 as its $i$th entry and 0s elsewhere. It stands for the mixed strategy in which players use their $i$th pure strategy for certain. The vector whose entries are all 1 is denoted by $e$. One can express the fact that the probabilities listed in the mixed strategy $p$ sum to one by writing $p^\top e = 1$. The vector $Aq$ lists the payoffs that player I will get from playing each of his pure strategies when player II uses the mixed strategy $q$. Similarly, $p^\top A$ lists the payoffs that player I can get when player II responds to his choice of the mixed strategy $p$ by playing a pure strategy.

Preplay randomization may consist of more than the players independently rolling dice or spinning roulette wheels. The set of payoff profiles achievable when the players can condition their choice of strategy on any jointly observed random event is called the *cooperative* payoff region. The set of payoff profiles achievable without the opportunity to condition on a jointly observed random event is called the *noncooperative* payoff region.

When the players lack the apparatus to make binding preplay agreements, anything they say to each other before the game is just cheap talk. Such talk may be cheap, but it can nevertheless be valuable when it allows the players to coordinate on a self-policing agreement that may involve the use of a carefully chosen random event that is at least partially observed by all the players.

The set of payoff profiles that become available when both players fully observe the random event is the convex hull of the game's equilibrium outcomes. Tossing a coin to decide who gets the more favorable equilibrium in the Battle of the Sexes is the simplest example. A larger set sometimes becomes available when a referee can be found who doles out information in a carefully restricted way. The behavior induced in a game when this trick is used is called a *correlated equilibrium.*

## 6.8 Further Reading

*Tracking the Automatic Ant*, by David Gale: Springer, New York, 1998. Along with many mathematical puzzles and games, this book discusses the mechanics of playing mental poker.

## 6.9 Exercises

1. Suppose that player I has a $4 \times 3$ payoff matrix. What vector represents the mixed strategy in which he never uses his second pure strategy and uses each of his other pure strategies with equal probabilities? What random device could player I use to implement this mixed strategy?
2. The $n$ players in the Good Samaritan Game all want an injured man to be helped. They each get a payoff of 1 if someone helps him and a payoff of 0 if nobody helps him. The snag is that anyone who offers help must subtract $c$ from their payoff $(0 < c < 1)$.

   If $n = 1$, the injured man will be helped for sure. If the players walk past the injured man one by one, he will also be helped for sure (by the *last* player to go by). But if $n \geq 2$ and offers of help are made simultaneously, each player will hope that someone else will do the helping. In a symmetric Nash equilibrium, show that each player will refuse to help with probability $c^{1/(n-1)} \to 1$ as $n \to \infty$. Show that the probability the man is helped at all is $1 - c^{n/(n-1)}$, which decreases to $1 - c$ as $n \to \infty$. Where would you rather find yourself in need of help: a big city or a small village?
3. In national lotteries, the jackpot is usually shared equally among all the holders of the winning combination of numbers. If you buy a ticket, you therefore want to avoid popular combinations. In Canada, where a punter chooses six different numbers between 1 and 49, the frequency with which each number was chosen in previous lotteries is published. The least chosen numbers in decreasing order of popularity are often 45, 20, 41, 48, 39, and 40. People who notice this fact

therefore sometimes choose the combination (45, 20, 41, 48, 39, 40), which paradoxically makes it one of the most popular combinations!

In a simple model of a national lottery, there are only three equally likely combinations, *a, b,* and *c*. Six punters each choose one of these combinations in the hope of winning a share of the jackpot. Two punters are known always to choose *a,* and one is known always to choose *b.* The other three punters act like players in a game and therefore don't automatically choose *c*. Instead, they seek to maximize their expected winnings, taking the behavior of the first three punters as given.

It is easy to find a pure Nash equilibrium of the game played by the three strategic punters. One punter chooses *b,* and the others choose *c*. But how do the players know which of the three should choose *b*?

A symmetric Nash equilibrium exists in which each strategic punter uses the same mixed strategy, choosing *a*, *b*, and *c* with probabilities 0, $p$, and $1-p$. In this equilibrium, each strategic punter will be indifferent between *b* and *c,* provided that the other wise punters stick to their equilibrium strategies. Show that $3p^2+8p-2=0$, and hence $p$ is approximately 0.23. Confirm that each strategic punter strictly prefers choosing *b* or *c* to *a* if the other strategic punters stick to their equilibrium strategies.

econ

4. Sketch the pure-strategy reaction curves for the sealed-bid auction game with entry costs given in Section 6.1.1 and so show that they don't cross. (Assume bids are always made in whole numbers of dollars.) Why does it follow that there is no Nash equilibrium in pure strategies?

econ

5. In the sealed-bid auction game with entry costs given in Section 6.1.1, explain why entering and bidding more than $1-c$ is a strongly dominated strategy.

econ

6. In the sealed-bid auction game with entry costs given in Section 6.1.1, explain why it can't be in equilibrium for a player to make any particular bid with positive probability after entering the auction.

econ

7. The rules of the sealed-bid auction game with entry costs given in Section 6.1.1 are changed so that Alice and Bob now know whether the other has entered the auction *before* sealing a bid in their envelopes. Analyze the game that results.
8. Show that the reaction curves in a bimatrix game remain unchanged if a constant is added to each of player I's payoffs in some column. Show that the same is true if a constant is added to each of player II's payoffs in some row.
9. Draw mixed-strategy reaction curves for the versions of the Battle of the Sexes and Chicken given in Figure 6.15. Hence find all Nash equilibria of both games.
10. The version of Chicken given in Figure 6.3(c) has a mixed equilibrium in which each player uses *hawk* with probability $\frac{2}{3}$. This mixed equilibrium can be interpreted in terms of the polymorphic equilibria of a population game. If the population is of finite size $N$, why will it only be an approximate equilibrium for one-third of the population to play *dove* and the other two-thirds to play *hawk?* How many of these approximate equilibria exist when $N=6$?

review

11. Given

$$A=\begin{bmatrix} 2 & 1 & 3 \\ -1 & 4 & 0 \end{bmatrix}, \quad B=\begin{bmatrix} 1 & 2 \\ 0 & -1 \\ 3 & 0 \end{bmatrix}, \quad C=\begin{bmatrix} 0 & 1 \\ -1 & 2 \\ 0 & 4 \end{bmatrix}$$

decide which of the following expressions are meaningful. Where they are meaningful, find the matrix they represent.

(a) $A+B$ (b) $B+C$ (c) $A+B$
(d) $3A$ (e) $3B-2C$ (f) $A-(B+C)^T$

12. Answer the following questions for the matrices

$$A=\begin{bmatrix}0 & 2\\ 4 & 1\\ 0 & 3\end{bmatrix},\quad B=\begin{bmatrix}0 & 1\\ 2 & 0\end{bmatrix},\quad C=\begin{bmatrix}1 & 2\\ 2 & 1\end{bmatrix}.$$

review

a. Why is $AB$ meaningful but not $BA$? Calculate $AB$.
b. Why are both $BC$ and $CB$ meaningful? Is it true that $BC=CB$?
c. Work out $(AB)C$ and $A(BC)$, and show that these are equal.
d. Verify that $(BC)^\top = C^\top B^\top$.

13. Show that the system of "linear equations"

review

$$\left.\begin{aligned}2x_1 - x_2 &= 4\\ x_1 - 2x_2 &= 3\end{aligned}\right\}$$

can be expressed in the form $Ax=b$, with

$$A=\begin{bmatrix}2 & -1\\ 1 & -2\end{bmatrix},\quad x=\begin{bmatrix}x_1\\ x_2\end{bmatrix},\quad \text{and}\quad b=\begin{bmatrix}4\\ 3\end{bmatrix}.$$

14. Given the $2\times 1$ column vectors

review

$$x=\begin{bmatrix}2\\ 1\end{bmatrix},\quad y=\begin{bmatrix}4\\ -3\end{bmatrix},\quad z=\begin{bmatrix}0\\ 2\end{bmatrix},$$

find

(a) $x+y$ (b) $3y$ (c) $-2z$ (d) $-z$ (e) $2x+y$

Illustrate each result geometrically.

15. If $x$ and $y$ are $n\times 1$ column vectors, explain why $x^\top y$ and $xy^\top$ are always both defined, but $x^\top y \neq xy^\top$ unless $n=1$. Why is it true that $x^\top y = y^\top x$ for all $n$?

review

16. Given the $3\times 1$ column vectors

review

$$x=\begin{bmatrix}3\\ 2\\ 1\end{bmatrix},\quad y=\begin{bmatrix}-3\\ 1\\ -2\end{bmatrix},\quad z=\begin{bmatrix}1\\ -1\\ -2\end{bmatrix},$$

find

$$(a)\ x^\top x \quad (b)\ x^\top y \quad (c)\ x^\top z \quad (d)\ y^\top z \quad (e)\ \|x\| \quad (f)\ \|x-y\|$$

Verify that $x^\top(3y+2z) = 3x^\top y + 2x^\top z$.

review

17. Use the results of Exercise 6.9.16 to determine each of the following:
    a. the distance from 0 to $x$
    b. the distance from $x$ to $y$
    c. which two of the vectors $x$, $y$, and $z$ are orthogonal
18. In four different games, Player II has the following payoff matrices:

$$A = \begin{bmatrix} 1 & 2 \\ 3 & 4 \end{bmatrix}; \qquad B = \begin{bmatrix} 1 & 3 \\ 4 & 2 \end{bmatrix};$$
$$C = \begin{bmatrix} 2 & 4 & 6 & 3 \\ 6 & 2 & 4 & 3 \\ 4 & 6 & 2 & 3 \end{bmatrix}; \quad D = \begin{bmatrix} 3 & 2 & 1 & 1 \\ 2 & 3 & 1 & 1 \\ 2 & 2 & 3 & 1 \end{bmatrix}.$$

In which of the games does player II have a pure strategy that is strongly dominated by a mixed strategy but not by any pure strategy? What is the dominated pure strategy? What is the dominating mixed strategy?

math

19. Write down a vector inequality that says that Eve can't get a payoff of more than $\beta$ by playing the mixed strategy $q$. Write down a vector equation that says that Adam's choice of the mixed strategy $p$ makes Eve indifferent between all her pure strategies.
20. Find a mixed strategy $p$ for Alice in O'Neill's Card Game that makes Bob indifferent between all his pure strategies.

math

21. Player I has payoff matrix $A$ in a finite, two-player game. Explain why his mixed strategy $\tilde{p}$ is a best reply to some mixed strategy for player II if and only if

$$\exists q \in Q\ \forall p \in P\ (\tilde{p}^\top Aq \geq p^\top Aq),$$

where $P$ is player I's set of mixed strategies and $Q$ is player II's set of mixed strategies.[10] Explain why $\tilde{p}$ is strongly dominated (possibly by a mixed strategy) if and only if

$$\exists p \in P\ \forall q \in Q\ (p^\top Aq > \tilde{p}^\top Aq).$$

Deduce that $\tilde{p}$ is *not* strongly dominated if and only if

$$\forall p \in P\ \exists q \in Q\ (p^\top Aq \leq \tilde{p}^\top Aq).$$

review

22. Explain why the vector $w = (3-2\alpha, 2, 1+2\alpha)$ is the location of a point on the straight line through the points $x = (1,2,3)$ and $y = (3,2,1)$. For what value of

[10] The notation "$\exists q \in Q$" means, "there exists a $q$ in the set $Q$ such that." The notation "$\forall p \in P$" means "for any $p$ in the set $P$." Why is it true that "not $(\exists p \forall q \ldots)$" is equivalent to "$\forall p \exists q$ (not...)"?

$\alpha$ does the vector $w$ lie halfway between $x$ and $y$? For what value of $\alpha$ does the vector $w$ lie at the center of gravity of a mass of $\frac{1}{3}$ at $x$ and a mass of $\frac{2}{3}$ at $y$?

23. Draw a diagram that shows the vectors $(1,1)$, $(4,2)$, $(2,4)$, and $(3,3)$ in $\mathbb{R}^2$. Indicate the convex hull $H$ of the set consisting of these four vectors. Why is $(3,3)$ a convex combination of $(4,2)$ and $(2,4)$? Indicate in your diagram the vectors $\frac{2}{3}(1,1)+\frac{1}{3}(4,2)$ and $\frac{1}{3}(1,1)+\frac{1}{3}(4,2)+\frac{1}{3}(3,3)$.

review

24. Sketch the following sets in $\mathbb{R}^2$. Which are convex? What are their convex hulls?

review

(a) $\{x : x_1^2+x_2^2=4\}$ (b) $\{x : x_1^2+x_2^2 \leq 4\}$

(c) $\{x : x_1=4\}$ (d) $\{x : x_1=4 \text{ or } x_2=4\}$

25. Let $x$, $y$, and $z$ be three points in $\mathbb{R}^2$. Let $u=ax+by$ $(a+b=1)$ be an affine combination of $x$ and $y$. Geometrically, $u$ lies on the straight line through $x$ and $y$. Why is $v=(1-\gamma)u+\gamma z$ located $\gamma$ of the distance along the line that joins $u$ to $z$? Using the proportional division theorem of Euclidean geometry or otherwise, deduce that the locus of the point $w=\alpha x+\beta y+\gamma z$ when $\gamma=\pi_3$ and $\alpha+\beta+\gamma=1$ is a straight line. (See Figure 6.13(b).)

math

26. Using Figure 6.14(b) as a guide, represent the set $P\times Q$ of all pairs of mixed strategies for the $2\times 3$ bimatrix game of Figure 6.20 as a prism. Sketch player I's reaction curve as a three-dimensional graph within $P\times Q$. Do the same for player II's reaction curve. Where do the reaction curves cross? What is the unique Nash equilibrium? Who gets how much when this is played?

math

27. Verify that the function $f:\mathbb{R}^2\rightarrow\mathbb{R}^2$ defined by $(y_1,y_2)=f(x_1,x_2)$ if and only if

review

$$y_1 = x_1+2x_2+1$$
$$y_2 = 2x_1+x_2+2$$

is affine. Indicate the points $f(1,1)$, $f(2,4)$, and $f(4,2)$ on a diagram.

28. Draw the cooperative and noncooperative payoff regions for the Australian Battle of the Sexes of Figure 6.21(a). Locate the Nash equilibrium outcomes on the latter diagram, and draw their convex hull.

29. Draw the cooperative and noncooperative payoff regions for the game of Figure 6.21(b). Locate the Nash equilibrium outcomes on the latter diagram, and draw their convex hull.

30. Verify that the set of all correlated equilibrium outcomes in the version of Chicken given in Figure 6.15(a) are as shown in Figure 6.19(a).

math

| | | | | | |
|---|---|---|---|---|---|
| | 3 | | 0 | | 2 |
| 5 | | 12 | | 2 | |
| | 0 | | 2 | | 1 |
| 6 | | 6 | | 9 | |

Figure 6.20 The game for Exercise 6.9.26.

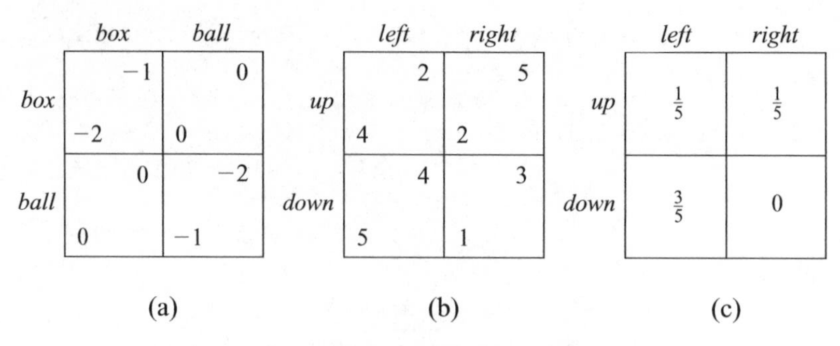

Figure 6.21 Tables for Exercise 6.9.28, 6.9.29, and 6.9.31.

31. Show that there is a correlated equilibrium for the game of Figure 6.21(b) in which the referee observes a chance move that selects one of the cells of the payoff table with the probabilities shown in Figure 6.21(c). He tells Adam to play the row and Eve to play the column in which the cell occurs. Your task is to verify that it is then optimal for Adam and Eve to follow their instructions. Confirm that the payoff pair that Adam and Eve get by playing the correlated equilibrium lies in the convex hull of the set of all the game's Nash equilibrium outcomes (Exercise 6.9.29).

math

32. Find all correlated equilibrium outcomes for the game of Figure 6.21(b).
33. If Adam and Eve play a particular Nash equilibrium in a game, then each pure strategy pair $(s, t)$ will be played with some probability $p(s, t)$. If a referee always tells Adam and Eve to play $s$ and $t$ with probability $p(s, t)$, why is the result necessarily a correlated equilibrium? If the referee begins by choosing the Nash equilibrium at random from those available, why does the result remain a correlated equilibrium? Why does the set of correlated equilibrium outcomes of a game contain the convex hull of its Nash equilibrium outcomes?

math

34. Show that the game of Figure 6.22(a) has a unique Nash equilibrium in which Alice plays *down* with probability $\frac{4}{5}$ and Bob plays *right* with probability $\frac{2}{3}$. Each outcome is then played with the probabilities given in Figure 6.22(b). Show that there are no correlated equilibria for the game other than that in which the referee acts according to the probabilities of Figure 6.22(b).

| | *left* | *right* |
|---|---|---|
| *up* | 1<br>5 | 5<br>1 |
| *down* | 4<br>3 | 3<br>2 |

(a)

| | *left* | *right* |
|---|---|---|
| *up* | $\frac{1}{15}$ | $\frac{2}{15}$ |
| *down* | $\frac{4}{15}$ | $\frac{8}{15}$ |

(b)

Figure 6.22 Tables for Exercise 6.9.33.

35. Alice and Bob participate in an all-pay, sealed-bid auction in which the winner receives a dollar bill and the loser receives nothing—but *both* players must pay what they bid (Section 21.2). If only positive bids in whole numbers of cents are allowed, find a mixed equilibrium in which every bid of less than a dollar is made with positive probability. The players are risk neutral, and both receive nothing if there is a tie.

econ

36. Philosophers sometimes mention correlated equilibria when trying to argue that it is rational to cooperate in the Prisoners' Dilemma. Explain why a correlated equilibrium can never require a player to use a strongly dominated strategy.

phil

37. Other things being equal, a rational person can never be made worse off by becoming better informed. In particular, a rational player can't be harmed in a game by learning something—provided that the other players' information remains unchanged. But it isn't true that everybody will necessarily be better off if *everybody* learns some new piece of information. Use the correlated equilibrium calculated in Section 6.6.2 to explain why both Adam and Eve will suffer if they both learn everything that the referee knows. What will happen if Adam learns what the referee knows but Eve learns only that Adam has learned this information?

phil

38. Exercise 1.13.30 asks what the categorical imperative requires in the case of *Scientific American*'s Million Dollar Game. Assume that the readers are all risk neutral.

math

   a. If the readers can coordinate their choices, why might they randomly select exactly one of their number to enter?
   b. If they must randomize independently, what is the probability that $n$ readers will enter, if each enters with probability $p$? What is the expected payoff to a reader?
   c. Estimate the optimal value of $p$. What is the probability that no prize is then awarded at all?
   d. Why does neither interpretation of the categorical imperative generate a Nash equilibrium?

39. In a simple version of the Ellsberg Paradox, a ball is chosen at random from one of two urns that contain only red or blue balls (Section 13.6.2). Adam wins if he guesses the color of the chosen ball correctly. Urn $A$ is transparent, and Adam can see that it contains an equal number of red and blue balls, Urn $B$ is opaque, and so Adam can't see what mix of balls it contains. Laboratory studies show that most people in Adam's situation prefer that the ball be chosen from Urn $A$.

   If faced with Urn $B$, Adam can always toss a fair coin to decide which color to guess. Given this option, is it possible that a rational agent would be willing to pay some money to have Urn $B$ replaced by Urn $A$?

phil

40. The laboratory evidence in the previous exercise is sometimes explained by saying that Adam may feel that using Urn $B$ confronts him with a version of Newcomb's Paradox with the experimenter in the role of Eve (Exercise 1.13.23). She would then be able to predict his choice before he makes it and so have arranged the mix of balls in Urn $B$ to his disadvantage.

   The situation can be modeled as the game Peeking Pennies. This game is the same as Matching Pennies, except that Eve receives a signal after Adam's

phil

choice, which says "Adam chose *heads*" or "Adam chose *tails*." It is common knowedge that the message is correct with probability $h$ when Adam chooses *heads* and with probability $t$ when he chooses *tails*. If $h > t$ and $h + t > 1$, show that there is a Nash equilibrium in which Eve always chooses *tails* when she hears the message "Adam chose *tails*," but the players otherwise mix their strategies. Confirm that Adam's probability of winning in this equilibrium is less than his probability $\frac{1}{2}$ of winning in regular Matching Pennies.

a. Why is Peeking Pennies relevant to the Ellsberg Paradox?
b. What happens when we erode Eve's predictive power by allowing $h$ and $t$ to approach $\frac{1}{2}$?
c. What happens if we try to instantiate the Newcomb's Paradox of the philosophical literature by taking $h = t = 1$? Why is it impossible to construct a game that incorporates the standard philosophical assumption that Eve can accurately predict Adam's choice *before* he has made it, without dispensing with the standard assumption in game theory that players are free to make any choice they like from their strategy sets?

# 7 Fighting It Out

## 7.1 Strictly Competitive Games

This chapter returns to the special case of strictly competitive games, in which two players have diametrically opposed preferences. The good news is that we can push the study of such zero-sum games quite a long way forward. The bad news is that we make more fuss than usual over the necessary mathematics. Some readers may therefore prefer just to skim the chapter.

Von Neumann and Morgenstern devoted the first half of *Games and Economic Behavior* to zero-sum games because they are simpler than other games. For the same reason, popular accounts of game theory sometimes fail to mention other kinds of games at all. As a consequence, critics often reject game theory altogether on the grounds that "life isn't a zero-sum game."

It is true that life isn't usually a zero-sum game, but anyone who thinks that they are going to solve the Game of Life without first learning to solve simpler games isn't being very realistic. Nor does the rarity of zero-sum games diminish their importance when they do occur. The game played between a pilot and the programmer of an air-to-air missile is one of many possible military applications. But since critics regard such military examples as proof that game theorists are a bunch of Dr. Strangeloves, I have hidden further mention of missiles at the end of the chapter.

→ 7.2

### 7.1.1 Shadow Prices

At what price should Alice sell her little firm to Mad Hatter Enterprises? Alice's plant is worthless, but she owns an $m \times 1$ vector $b$ of raw materials for which Mad

Hatter Enterprises is the only possible purchaser. However, Alice can also process the raw materials and sell the finished products.

To produce the $n \times 1$ vector $x$ of processed goods, Alice requires the $m \times 1$ vector of raw materials given by

$$z = Ax,$$

where $A$ is her $m \times n$ input-output matrix. The processed goods can be sold at fixed prices given by the $n \times 1$ vector $c$. Alice's revenue from such a sale is the inner product $c^\top x = c_1x_1 + c_2x_2 + \cdots + c_nx_n$.

Mad Hatter Enterprises can quote any $m \times 1$ vector $y$ of prices for the raw materials. Once $x$ and $y$ have been determined, the value of Alice's firm is

$$L(x, y) = c^\top x + y^\top(b - Ax).$$

Alice wants to choose $x \geq 0$ to maximize $L(x, y)$. Mad Hatter Enterprises wants to choose $y \geq 0$ to minimize $L(x, y)$. Valuing Alice's firm therefore reduces to solving a strictly competitive game.

The vector of prices $y$ assigned to Alice's stock of raw materials by the solution to the game will be chosen at the lowest level consistent with her being able to process the stock into finished goods that sell at price $c$. Economists say that the coordinates of $y$ are then the *shadow prices* for her stock. They help a manager make decisions by telling her how much the intermediary goods produced during a manufacturing process are worth.

## 7.2 Zero-Sum Games

A *zero-sum* game is a game in which the payoffs always sum to zero. For two players, we need that

$$u_1(\omega) + u_2(\omega) = 0,$$

for each $\omega$ in the set $\Omega$ of pure outcomes, where $u_1 : \Omega \to \mathbb{R}$ and $u_2 : \Omega \to \mathbb{R}$ are the players' Von Neumann and Morgenstern utility functions.

**Theorem 7.1** *A two-player game has a zero-sum representation if and only if it is strictly competitive.*

*Proof* A two-player game is strictly competitive when the players have diametrically opposed preferences over all pairs of outcomes of the game. Thus, $\mathbf{L} \preceq_1 \mathbf{M} \iff \mathbf{L} \succeq_2 \mathbf{M}$ for all lotteries $\mathbf{L}$ and $\mathbf{M}$ whose prizes are the pure outcomes of a strictly competitive game. It follows that

$$\mathcal{E}u_1(\mathbf{L}) \leq_1 \mathcal{E}u_1(\mathbf{M}) \Leftrightarrow \mathbf{L} \succeq_2 \mathbf{M},$$

and so $-u_1$ is a Von Neumann and Morgenstern utility function that represents player II's preference relation $\preceq_2$. Theorem 4.1 then tells us that $u_2 = Au_1 + B$ for some constants $A > 0$ and $B$. To make the game zero sum, we choose $A = -1$ and $B = 0$.

To prove that a two-player, zero-sum game $G$ is strictly competitive is even easier. If $u_2 = -u_1$, then

$$\begin{aligned} \mathbf{L} \preceq_1 \mathbf{M} &\Leftrightarrow \mathscr{E}u_1(\mathbf{L}) \leq \mathscr{E}u_1(\mathbf{M}) \\ &\Leftrightarrow -\mathscr{E}u_1(\mathbf{L}) \geq -\mathscr{E}u_1(\mathbf{M}) \\ &\Leftrightarrow \mathscr{E}u_2(\mathbf{L}) \geq \mathscr{E}u_2(\mathbf{M}) \quad \Leftrightarrow \quad \mathbf{L} \succeq_2 \mathbf{M}. \end{aligned}$$

*Interpersonal Comparison?* It is sometimes wrongly thought that studying zero-sum games commits us to making interpersonal comparisons of utility (Section 4.6.3). But the fact that a gain of one util by one player is balanced by a loss of one util by the other doesn't at all imply that the players feel victory or defeat equally keenly.

We chose $A = -1$ and $B = 0$ in the proof of Theorem 7.1, but we could equally well have taken $A = -2$ and $B = 3$ or $A = -1$ and $B = 1$. The latter choice yields a *constant-sum* representation of our game.

For example, Duel and Russian Roulette are strictly competitive games that were presented in previous chapters as unit-sum games. To convert them into entirely equivalent zero-sum games, just pick a player and subtract one from all of his payoffs.

*Attitudes to Risk?* Sometimes the attitudes that players have to taking risks are overlooked when modeling situations as zero-sum games. For example, games like poker and backgammon are thought to be automatically zero sum because any sum of money won by one player is lost by the others. But this isn't enough to ensure that backgammon or poker are zero-sum games. They certainly won't be if all the players are strictly risk averse.[1]

When games like poker or backgammon are analyzed as zero-sum games, it is implicitly understood that the players are *risk neutral,* so that a player's Von Neumann and Morgenstern utility function $u : \mathbb{R} \to \mathbb{R}$ for money can be chosen to satisfy

$$u(x) = x.$$

We know from studying the St. Petersburg paradox that risk neutrality is unlikely to be a good assumption about people's preferences in general. But assuming risk neutrality may not be too bad an approximation when, as in neighborhood poker games, the sums of money that change hands are small.

### 7.2.1 Matrix Games

The bimatrix game of Figure 7.1(a) is the strategic form of a zero-sum game because the payoffs in each cell sum to zero. The payoff matrices $A$ and $B$ therefore satisfy

[1]In a zero-sum game, $u_1 = -u_2$, and so one player's utility function is strictly concave if and only if the other's is strictly convex. This was one reason for restricting our attention in earlier chapters to win-or-lose games. Only when consideration is restricted to lotteries with just two possible prizes can one deduce from the fact that players have opposing preferences over prizes that they necessarily have opposing preferences over lotteries.

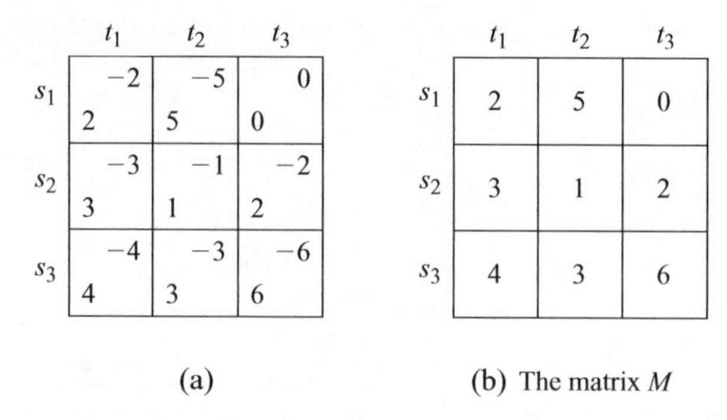


(a) (b) The matrix $M$

Figure 7.1 A zero-sum strategic form.

$A+B=0$. Since $B=-A$, it is redundant to write down player II's payoffs. Instead, the strategic form of a zero-sum game is usually represented by player I's payoff matrix alone, as in Figure 7.1(b). One must remember that such a matrix records only player I's payoffs. It is easy to forget that player II seeks to *minimize* these payoffs.

## 7.3 Minimax and Maximin

Von Neumann's minimax theorem of 1928 is the key to solving zero-sum games. This section prepares the ground by looking at the case of pure strategies.

### 7.3.1 Computing Minimax and Maximin Values

Player I's set $S$ of pure strategies in the game of Figure 7.1(a) corresponds to the rows in the payoff matrix $M$ of Figure 7.1(b). Player II's set $T$ of pure strategies corresponds to the columns of $M$. We denote the entry in row $s$ and column $t$ of the matrix $M$ by $\pi(s,t)$ (rather than $\pi_1(s,t)$ as in Section 5.2).

The largest entries in each column of $M$ are 4, 5, and 6. As usual, these entries are circled in Figure 7.2(a). The smallest entries in each row are 0, 1, and 3. These are enclosed in a square in Figure 7.2(b). For example,

$$\max_{s \in S} \pi(s, t_3) = 6 \quad \text{and} \quad \min_{t \in T} \pi(s_1, t) = 0.$$

The minimax value $\overline{m}$ and the maximin value $\underline{m}$ of the matrix $M$ are given by

$$\overline{m} = \min_{t \in T}\left\{\max_{s \in S} \pi(s,t)\right\} = \min\,\{3,6,4\} = 4,$$

$$\underline{m} = \max_{s \in S}\left\{\min_{t \in T} \pi(s,t)\right\} = \max\,\{0,0,2\} = 3.$$

These quantities are shown with both a circle and a square in Figure 7.2.

The next theorem explains why the minimax value $\overline{m}$ of a matrix $M$ is written with an overline and the maximin value $\underline{m}$ with an underline.

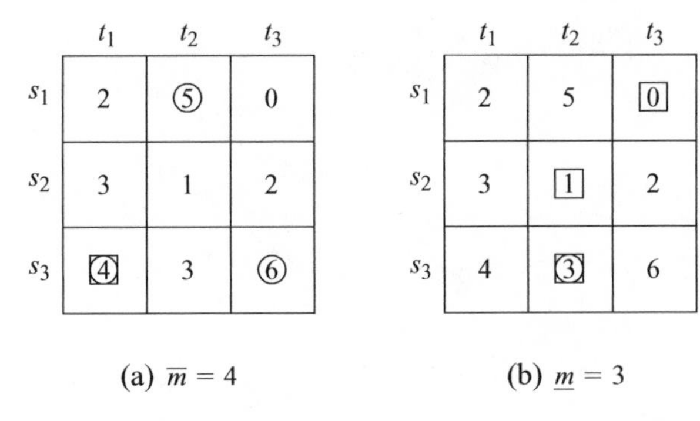


Figure 7.2 Minimax and maximin values for the matrix $M$.

THEOREM 7.2 $\overline{m} \geq \underline{m}$.

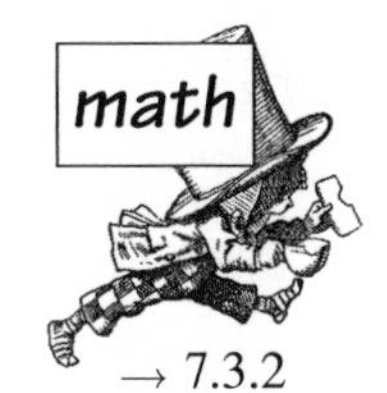


*Proof* For any particular $t \in T$, $\pi(s,t) \geq \min_{t \in T} \pi(s,t)$. It follows that

$$\max_{s \in S} \pi(s,t) \geq \max_{s \in S} \min_{t \in T} \pi(s,t) = \underline{m}.$$

Now apply this inequality with the particular value of $t \in T$ that minimizes the left-hand side to obtain $\overline{m} \geq \underline{m}$.

### 7.3.2 Saddle Points

We have seen that the maximin value of a matrix can be strictly smaller than its minimax value, but the interesting case arises when the two values are equal since we shall see that the matrix then has a saddle point.

A pair $(\sigma, \tau)$ is a *saddle point* for the matrix $N$ of Figure 7.3 when $\pi(\sigma, \tau)$ is largest in its column and smallest in its row (Section 2.8.2). Since the entry in row $s_2$ and column $t_2$ of Figure 7.4(a) gets both a circle and a square, it follows that $(s_2, t_2)$ is a saddle point of $N$.

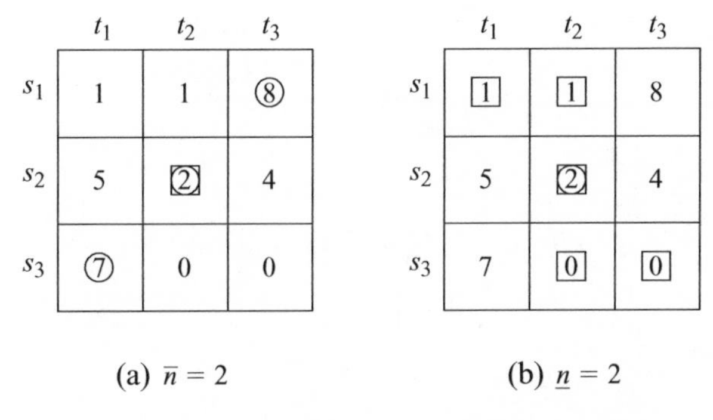


Figure 7.3 Minimax and maximin values for the matrix $N$.

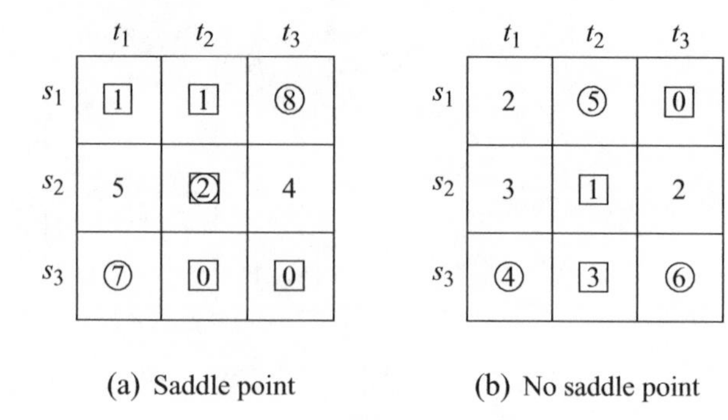


Figure 7.4 Finding saddle points.

The height of the obelisk in row $s_1$ and column $t_3$ of Figure 7.5(a) is 8 because $\pi(s_1, t_3) = 8$ in the matrix $N$ of Figure 7.3(a). The picture is meant to explain why the pair $(s_2, t_2)$ is called a saddle point of $N$, although the saddle drawn would admittedly not be very comfortable to sit on.

Figure 7.5(b) looks more like a real saddle. It shows a saddle point $(\sigma, \tau)$ for a continuous function $\pi : S \times T \to \mathbb{R}$ when $S$ and $T$ are closed intervals of real numbers. For $(\sigma, \tau)$ to be a saddle point, we need that, for all $s$ in $S$ and all $t$ in $T$,

$$\pi(\sigma, t) \geq \pi(\sigma, \tau) \geq \pi(s, \tau). \tag{7.1}$$

Our use of circles and squares probably makes it obvious why matrices have saddle points if and only if their maximin and minimax values are equal, but the next theorem provides a formal proof.

→ 7.3.3

THEOREM 7.3 *A necessary and sufficient condition that $(\sigma, \tau)$ be a saddle point is that $\sigma$ and $\tau$ are given by*

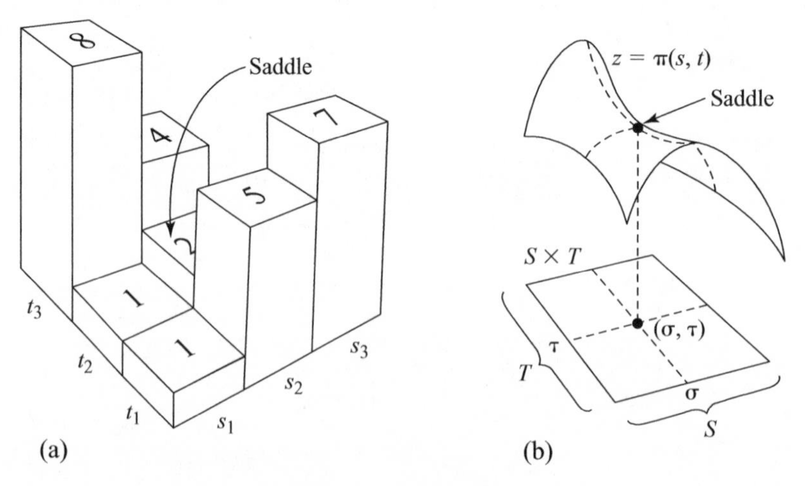


Figure 7.5 Saddle points.

$$\min_{t \in T} \pi(\sigma, t) = \max_{s \in S} \min_{t \in T} \pi(s, t) = \underline{m}, \tag{7.2}$$

$$\max_{s \in S} \pi(s, \tau) = \min_{t \in T} \max_{s \in S} \pi(s, t) = \overline{m}, \tag{7.3}$$

*and* $\underline{m} = \overline{m}$. *When* $(\sigma, \tau)$ *is a saddle point,* $\underline{m} = \pi(\sigma, \tau) = \overline{m}$.

*Proof* A proof that something is necessary and sufficient is usually split into two halves. The first step proves necessity, and the second sufficiency.

**Step 1.** If $(\sigma, \tau)$ is a saddle point, then $\pi(\sigma, t) \geq \pi(\sigma, \tau) \geq \pi(s, \tau)$ for all $s$ in $S$ and $t$ in $T$. Thus $\min_{t \in T} \pi(\sigma, t) \geq \pi(\sigma, \tau) \geq \max_{s \in S} \pi(s, \tau)$, and so

$$\underline{m} = \max_{\sigma \in S} \min_{t \in T} \pi(\sigma, t) \geq \min_{t \in T} \pi(\sigma, t) \geq \max_{s \in S} \pi(s, \tau) \geq \min_{\tau \in T} \max_{s \in S} \pi(s, \tau) = \overline{m}.$$

But Theorem 7.2 says that $\underline{m} \leq \overline{m}$, and so all the $\geq$ signs in the preceding expression may be replaced by $=$ signs.

**Step 2.** Next suppose that $\underline{m} = \overline{m}$. It must then be shown that a saddle point $(\sigma, \tau)$ exists. Choose $\sigma$ and $\tau$ to satisfy (7.2) and (7.3). Then, given any $s$ in $S$ and $t$ in $T$,

$$\pi(\sigma, t) \geq \min_{t \in T} \pi(\sigma, t) = \underline{m} = \overline{m} = \max_{s \in S} \pi(s, \tau) \geq \pi(s, \tau).$$

Taking $s = \sigma$ and $t = \tau$ in this inequality shows that $\underline{m} = \pi(\sigma, \tau) = \overline{m}$. The requirement for $(\sigma, \tau)$ to be a saddle point is therefore satisfied.

→ 7.4

### 7.3.3 Dicing with Death Again

We located a Nash equilibrium for the game of Duel in Section 5.2.1 by identifying a saddle point of Tweeddledum's payoff matrix. We now offer an alternative analysis of the game that uses minimax and maximin values.

We have previously admitted only a finite number of values of $d$ at which a player might open fire in the game of Duel, but each player will now be allowed to choose any $d$ in the closed interval $[0, D]$. The $6 \times 5$ table of Figure 5.3 is therefore replaced by an infinite table, but we will take it for granted that a saddle point continues to exist.

Theorem 7.3 then tells us that, in a Nash equilibrium, Tweedledum will fire his pistol at distance $\delta$ from Tweedledee, where $\delta$ is the value of $d$ at which the maximum is attained in

$$\underline{m} = \max_{d} \inf_{e} \pi(d, e). \tag{7.4}$$

The fact that we have an infinite number of values of $d$ to consider creates two small technical problems. The first is the need to write "inf" instead of "min" in the formula for $\underline{m}$ because $\pi(d, e)$ needn't have a smallest value.[2] The other small

[2]For example, the open interval $(2, 3)$ has no minimum element. Everything in the set $(2, 3)$ is larger than 1, so 1 is a lower bound for the set $(2, 3)$. Its largest lower bound is 2, but 2 isn't the minimum element of the set $(2, 3)$ because 2 isn't even an element of $(2, 3)$. Mathematicians say that the largest lower bound of a set is its *infimum*. The infimum of a set is the same as its minimum when the latter exists. The smallest upper bound of a set is its *supremum*. The supremum of a set is equal to its maximum when the latter exists.

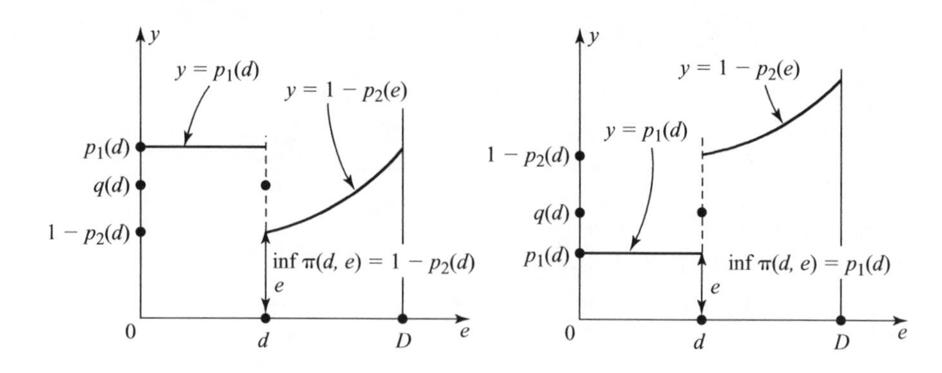


(a) The graph of $y = \pi(d, e)$ for a fixed $d$ when $p_1(d) > 1 - p_2(d)$.

(b) The graph of $y = \pi(d, e)$ for a fixed $d$ when $p_1(d) < 1 - p_2(d)$.

Figure 7.6 Plotting payoffs in Duel.

problem concerns what happens if both players fire at precisely the same instant. We assume that a chance move then selects one of the players to get his shot in just before the other, so that Tweedledum survives with some probability $q(d)$ between $p_1(d)$ and $1 - p_2(d)$.

Figure 7.6 shows how to use the formula for $\pi(d, e)$ given in equation (5.1) to determine $m(d) = \inf_e \pi(d, e)$ for differing values of $d$. (We can't write $m(d) = \min_e \pi(d, e)$ because of the discontinuity in $\pi(d, e)$ at $e = d$. So we write $m(d) = \inf_e \pi(d, e)$ instead, accepting that we can do no better than get arbitrarily close to $m(d)$ by taking values of $e$ sufficiently near to $d$.)

We now plot the graph of $y = m(d)$ in Figure 7.7. The maximum we require for equation (7.4) occurs at the point $d = \delta$, where

$$p_1(d) + p_2(d) = 1,$$

which is reassuringly the same conclusion that we reached in Section 3.7.2 using an entirely different method.

Tweedledee also fires his pistol at distance $\delta$ because swapping $p_1(d)$ and $p_2(d)$ over in the preceding analysis leaves the final result unchanged. Since they fire simultaneously at time $\delta$, the probability that Tweedledum will survive is then $q(\delta) = p_1(\delta) = 1 - p_2(\delta)$.

This analysis of Duel focuses on the fact that it is a Nash equilibrium for both players to fire their pistols when they are distance $\delta$ apart. But more is always true in the special case of a strictly competitive game. A Nash equilibrium then corresponds to a saddle point $(\sigma, \tau)$ of player I's payoff matrix. Theorem 2.2 then tells us that the game has a value. Whatever player II may be planning to do, player I can ensure a payoff of at least $\pi(\sigma, \tau)$ for himself by playing $\sigma$. Whatever player I may be planning to do, player II can ensure that player I gets a payoff of no more than $\pi(\sigma, \tau)$ by playing $\tau$.

In particular, no matter when the other player may be planning to fire, player $i$ can guarantee surviving in Duel with probability at least $p_i(\delta)$ by firing when the players are distance $\delta$ apart.

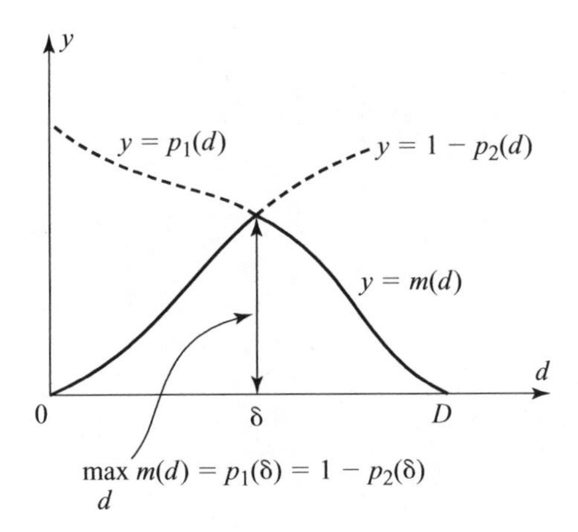


Figure 7.7 The maximin value in Duel.

## 7.4 Safety First

The payoff $p_1(\delta)$ is Tweedledum's security level in Duel. If Tweedledum plays his security strategy of firing when the players are $\delta$ apart, nothing Tweedledee can do will reduce Tweedledum's probability of survival below $p_1(\delta)$.

The next item on the agenda is to extend the idea of a security level to more general games. This will usually involve the use of mixed strategies. People sometimes ask how it can possibly be safe to randomize your choice of strategy, but we already know that Adam's security strategy in Matching Pennies is to play *heads* and *tails* with equal probability (Section 2.2.2). Any other behavior would risk a negative average loss.

### 7.4.1 Security Levels

Adam's security level in a game is the largest expected payoff he can *guarantee,* no matter what the other players do. To compute his security level, Adam therefore has to carry out a worst-case analysis, in which he proceeds on the assumption the other players will predict his strategy choice and then act to *minimize* his payoff. A strategy that guarantees Adam his security level under this paranoid hypothesis is called a *security strategy.*

Adam is player I and Eve is player II in the bimatrix game of Figure 7.8(a). Adam's payoff matrix in this game is the matrix of Figure 7.3. To work through a worst-case scenario, Adam reasons as follows.

If Eve guesses that Adam will choose $s_1$, she can hold his payoff down to 1 by choosing $t_1$ or $t_2$. If she guesses that he will choose $s_2$, then she can hold his payoff down to 2 by choosing $t_2$. If she guesses that he will choose $s_3$, then she can hold his payoff down to 0 by choosing $t_2$ or $t_3$. A worst-case analysis therefore places Adam's payoff in the set $\{1, 2, 0\}$ of payoffs enclosed in squares in the diagram of

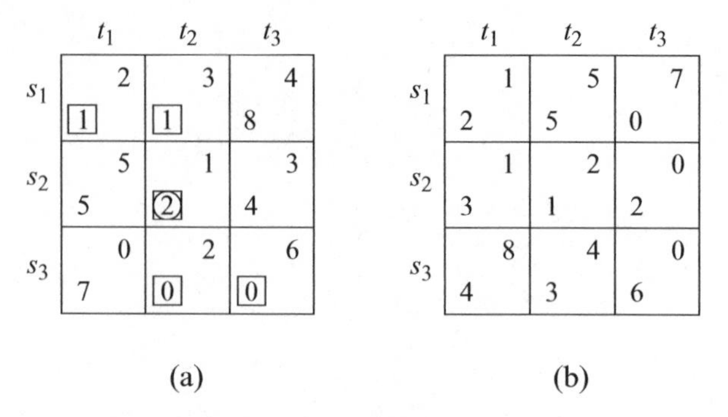


Figure 7.8 Two bimatrix games.

Figure 7.8(a). Since the best payoff in this set is the circled payoff of 2, Adam can guarantee a payoff of at least 2 by using pure strategy $s_2$.

This reasoning mimics the circling and squaring of payoffs in the matrix of Figure 7.3(b) we used to show that $\underline{m} = 2$. The same reasoning shows that Adam can *always* guarantee a payoff at least as good as the maximin value $\underline{m}$ of his payoff matrix. When does this imply that $\underline{m}$ is his security level?

THEOREM 7.4 *If player I's payoff matrix has a saddle point* $(\sigma, \tau)$, *then his security level is* $\underline{m} = \pi_1(\sigma,\ \tau) = \overline{m}$, *and* $\sigma$ *is one of his security strategies.*

*Proof* The worst-case scenario we use when computing player I's security level is equivalent to treating the situation as a strictly competitive game. Player I retains his payoff matrix $A$ in this game, but player II is assigned the payoff matrix $-A$. The proof of the theorem then reduces to observing that $(\sigma, \tau)$ is a solution of this new game (Theorem 2.2). □

Since Adam's payoff matrix $N$ in the game of Figure 7.8(a) has a saddle point, Theorem 7.4 says that his security level is $\underline{n} = 2$ and that $s_2$ is a security strategy. Since Adam's payoff matrix $M$ in the game of Figure 7.8(b) doesn't have a saddle point, Theorem 7.4 doesn't say that his security level is $\underline{m} = 3$. As we show next, his security level is actually $3\frac{1}{2}$.

### 7.4.2 Securing Payoffs with Mixed Strategies

We show that Adam can guarantee a payoff of at least $3\frac{1}{2}$ in the bimatrix game of Figure 7.8(b) by playing his mixed strategy $p = (\frac{1}{4}, 0, \frac{3}{4})$. We then show that Eve can ensure that he gets no more than $3\frac{1}{2}$ by playing her mixed strategy $q = (\frac{1}{2}, \frac{1}{2}, 0)$. It follows that $3\frac{1}{2}$ must be Adam's security level.

*Adam Plays Safe.* Adam will never use his pure strategy $s_2$ because it is strongly dominated by $s_3$. Our first step is therefore to delete row $s_2$, leaving Adam with the payoff matrix shown in Figure 7.9(a).

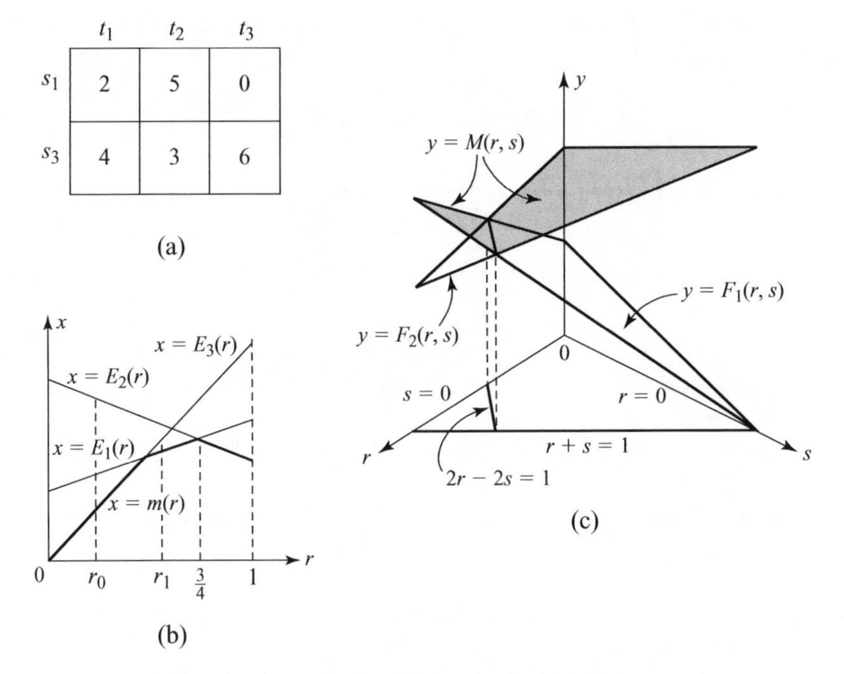


Figure 7.9 Computing mixed security strategies.

We next work out the expected payoff $x = E_k(r)$ that Adam will get if Eve uses her pure strategy $t_k$ and he uses the mixed strategy $(1-r, r)$ in the reduced game. We have that

$$E_1(r) = 2(1-r)+4r = 2+2r;$$

$$E_2(r) = 5(1-r)+3r = 5-2r;$$

$$E_3(r) = 0(1-r)+6r = 6r.$$

The lines $x = E_1(r)$, $x = E_2(r)$, and $x = E_3(r)$ are graphed in Figure 7.9(b).

Adam's paranoic assumption in computing his security level is that Eve will predict his choice of mixed strategy and then choose her strategy so as to assign him whichever of $E_1(r)$, $E_2(r)$, or $E_3(r)$ is smallest.[3] Adam therefore anticipates an expected payoff of

$$m(r) = \min\{E_1(r), E_2(r), E_3(r)\}.$$

The graph of $x = m(r)$ is shown with a bold line in Figure 7.9(b). For example, when $r = r_0$, $m(r) = E_3(r)$. When $r = r_1$, $m(r) = E_1(r)$.

[3]An even worse scenario would be if Eve were able to predict how a tossed coin will land, or what card will be drawn from a shuffled deck. But an analysis that attributed such superhuman powers to Eve wouldn't be very interesting. Alert readers will want to know why Eve neglects her mixed strategies. The reason is that, for each $r$, she can always minimize Adam's payoff by using one of her pure strategies.

Adam must choose $r$ to make the best of this worst-case scenario. His payoff with the optimal choice of $r$ is

$$\underline{v} = \max_r m(r) = \max_r \min_k E_k(r).$$

Figure 7.9(b) reveals that the value of $r$ satisfying $0 \leq r \leq 1$ at which $m(r)$ is largest occurs where the lines $x = E_1(r)$ and $x = E_2(r)$ cross. Since the solution to the equation

$$2 + 2r = 5 - 2r$$

is $r = \frac{3}{4}$, Adam can secure an expected payoff of at least

$$\underline{v} = m(\tfrac{3}{4}) = E_1(\tfrac{3}{4}) = 2 + 2 \times \tfrac{3}{4} = 3\tfrac{1}{2}$$

by using the mixed strategy $p = (\frac{1}{4}, 0, \frac{3}{4})$ in the original game of Figure 7.8(b).

→ 7.4.3

*Eve Plays to Injure Adam.* The next step is to show that Eve can be sure of holding Adam's payoff down to $3\frac{1}{2}$ if she gives up trying to maximize her own payoff and tries to minimize his payoff instead. We therefore treat Eve as player II in the zero-sum game with the payoff matrix of Figure 7.8(a). Recall that the payoffs in this matrix are *losses* to Eve.

We first work out Eve's expected loss $y = F_k(r, s)$ if Adam plays his pure strategy $s_k$ and Eve uses the mixed strategy $q = (1 - r - s, r, s)$. We have that

$$F_1(r, s) = 2(1 - r - s) + 5r + 0s = 2 + 3r - 2s;$$

$$F_2(r, s) = 4(1 - r - s) + 3r + 6s = 4 - r + 2s.$$

The two planes $y = F_1(r, s)$ and $y = F_2(r, s)$ are graphed in Figure 7.9(c).[4]
As in the case of Adam, we look at what happens when Eve adopts the paranoic assumption that Adam will predict her choice of mixed strategy and then choose his strategy so as to assign her whichever of $F_1(r, s)$ or $F_2(r, s)$ represents the larger loss to her. Eve therefore anticipates an expected loss of

$$M(r, s) = \max\{F_1(r, s), F_2(r, s)\}.$$

The graph of $y = M(r, s)$ is shaded in Figure 7.9(c).
Eve now chooses $r$ and $s$ to make the best of this worst-case scenario. Her loss with the optimal choices of $r$ and $s$ is

$$\overline{v} = \min_{(r, s)} M(r, s) = \min_{(r, s)} \max_k F_k(r, s).$$

[4]In Figure 7.9(b), we considered only values of $r$ satisfying $0 \leq r \leq 1$. Here we consider only pairs $(r, s)$ for which $r \geq 0$, $s \geq 0$, and $r + s \leq 1$. Such pairs lie in the triangle bounded by the lines $r = 0$, $s = 0$, and $r + s = 1$.

Figure 7.9(c) reveals that the pair $(r, s)$ at which $M(r, s)$ is smallest occurs where the planes $y = F_1(r, s)$ and $y = F_2(r, s)$ intersect. We therefore examine those pairs $(r, s)$ for which $F_1(r, s) = F_2(r, s)$. This equation reduces to

$$2 + 3r - 2s = 4 - r + 2s$$
$$2r - 2s = 1.$$

Which of the pairs $(r, s)$ lying on this line make $M(r, s)$ smallest?

There are two candidates. The first is the point $(\frac{1}{2}, 0)$ at which the line $2r - 2s = 1$ meets $s = 0$. The second is the point $(\frac{3}{4}, \frac{1}{4})$ at which $2r - 2s = 1$ meets $r + s = 1$.

Since $M(\frac{1}{2}, 0) = F_1(\frac{1}{2}, 0) = 3\frac{1}{2}$, and $M(\frac{3}{4}, \frac{1}{4}) = F_1(\frac{3}{4}, \frac{1}{4}) = 3\frac{3}{4}$, the pair $(r, s)$ that minimizes $M(r, s)$ is $(\frac{1}{2}, 0)$. The minimum value is $\overline{v} = 3\frac{1}{2}$.

*Minimax Equals Maximin?* We have just looked at a case of a two-person zero-sum game in which

$$\underline{v} = \overline{v} = 3\tfrac{1}{2}.$$

Can it *always* be true that the maximin and minimax values of a matrix game are the same when we allow mixed strategies?

If the answer to this question is *yes,* then we can generalize all the conclusions about strictly competitive games of perfect information derived from the existence of saddle points in such games. All our theoretical problems with two-person zero-sum games of imperfect information will then evaporate.

The famous mathematician Emile Borel studied mixed strategies in gambling games some years ahead of Von Neumann. Borel asked himself whether it could always be true that $\underline{v} = \overline{v}$ but guessed the answer was probably *no.* Fortunately, Von Neumann knew nothing of Borel's earlier work when he later proved that the answer is *yes.* Otherwise he mightn't have made the attempt!

However, before we can tackle Von Neumann's minimax theorem, we need to restate the results of Section 7.3.1 to allow for mixed strategies.

### 7.4.3 Minimax and Maximin with Mixed Strategies

Player I's payoff function $\Pi : P \times Q \to \mathbb{R}$ is given by

$$\Pi(p, q) = p^\top A q,$$

where $A$ is his payoff matrix (Section 6.4.3). The minimax value $\overline{v}$ and the maximin $\underline{v}$ value of his payoff function are defined by

$$\underline{v} = \max_{p \in P} \min_{q \in Q} \Pi(p, q) = \min_{q \in Q} \Pi(\tilde{p}, q), \tag{7.5}$$

$$\overline{v} = \min_{q \in Q} \max_{p \in P} \Pi(p, q) = \max_{p \in P} \Pi(p, \tilde{q}), \tag{7.6}$$

where $\tilde{p}$ is the mixed strategy $p$ in $P$ for which $\min_{q \in Q} \Pi(p,q)$ is largest, and $\tilde{q}$ is the mixed strategy $q$ in $Q$ for which $\max_{p \in P} \Pi(p,q)$ is smallest.[5]

A saddle point for the payoff function $\Pi$ is a pair $(\tilde{p}, \tilde{q})$ of mixed strategies such that, for all $p$ in $P$ and all $q$ in $Q$,

$$\Pi(\tilde{p},\, q) \geq \Pi(\tilde{p}, \tilde{q}) \geq \Pi(p,\, \tilde{q}).$$

If one thinks of $\Pi(p,q)$ as being the entry in row $p$ and column $q$ of a generalized "matrix," then the following theorems are natural. Their proofs can be copied from those of Theorems 7.2, 7.3, and 7.4.

THEOREM 7.5 $\underline{v} \leq \overline{v}$.

THEOREM 7.6 *A necessary and sufficient condition that* $(\tilde{p},\, \tilde{q})$ *be a saddle point is that* $\tilde{p}$ *and* $\tilde{q}$ *are given by (7.5) and (7.6) and* $\underline{v} = \overline{v}$. *When* $(\tilde{p},\, \tilde{q})$ *is a saddle point,* $\underline{v} = \Pi(\tilde{p}, \tilde{q}) = \overline{v}$.

THEOREM 7.7 *If player I's payoff function* $\Pi$ *has a saddle point* $(\tilde{p},\, \tilde{q})$, *then his security level is* $\underline{v} = \Pi(\tilde{p},\, \tilde{q}) = \overline{v}$, *and* $\tilde{p}$ *is one of his security strategies.*

### 7.4.4 Minimax Theorem

→ 7.5

The following proof of Von Neumann's minimax theorem is loosely based on an inductive argument of Guillermo Owen. His proof doesn't appeal to any deep theorems, but it does require some heavy algebra. In the argument given below, the algebra will still trouble beginners, but it has been reduced to some playing around with maxima and minima. However, simplifying the algebra in this way makes it necessary to sketch an argument that uses transfinite numbers.

Everyone is familiar with the finite ordinals $0, 1, 2, \ldots,$ which we use for counting finite sets. They need to be supplemented with the transfinite ordinals when counting infinite sets. When we have used up all the ordinals we have constructed so far, we invent a new ordinal to count the next member of a well-ordered set.[6] For example, if we run out of finite ordinals when counting an infinite set, we count its next element with the first transfinite ordinal, which mathematicians denote by $\omega$. However, all that matters for the proof is that for any set there is an ordinal too large to be reached by counting its elements.

THEOREM 7.8 (Von Neumann) *For any finite game,*

$$\underline{v} = \overline{v}.$$

*Proof* We will show that the assumption $\underline{v} < \overline{v}$ implies a contradiction. The minimax theorem then follows from the fact that $\underline{v} \leq \overline{v}$ (Theorem 7.5).

[5]The $\underline{v}$ and $\overline{v}$ defined here are the same as in Section 7.4.2 because the maximum on the right of 7.5 and the minimum on the right of 7.6 are attained at pure strategies.

[6]Every nonempty subset of a well-ordered set has a minimum element. The Well-Ordering Principle says that every set can be well ordered.

The proof requires the construction of a zero-sum game for each ordinal $\alpha$ that has convex and *nonempty* strategy sets $P_\alpha$ and $Q_\alpha$, but the same payoff function as the original game. The first of these games is identical with our original game, so that $P_0 \times Q_0 = P \times Q$. Later games get progressively smaller, in the sense that $\alpha < \beta$ implies $P_\beta \times Q_\beta \subset P_\alpha \times Q_\alpha$, where it is important for the inclusion to be *strict.*

The reason that this construction leads to the desired contradiction is that $P_\gamma \times Q_\gamma$ must be empty if $\gamma$ is a sufficiently large ordinal because one cannot count more points of $P \times Q$ than it contains.

The idea of the construction is to replace $P_\alpha \times Q_\alpha$ by $P_\beta \times Q_\beta$ so that

$$\overline{v}_\beta - \underline{v}_\beta \geq \overline{v}_\alpha - \underline{v}_\alpha. \tag{7.7}$$

We first explain how this is done for the case $\alpha = 0$ and $\beta = 1$.

**Step 1.** If $\underline{v} \geq \Pi(\tilde{p}, \tilde{q})$ and $\Pi(\tilde{p}, \tilde{q}) \geq \overline{v}$, then $\underline{v} \geq \overline{v}$. It follows that our assumption that $\underline{v} < \overline{v}$ implies that either $\underline{v} < \Pi(\tilde{p}, \tilde{q})$ or $\Pi(\tilde{p}, \tilde{q}) < \overline{v}$. The former inequality will be assumed to hold. If the latter inequality holds, a parallel argument is necessary in which it is $P$ that shrinks rather than $Q$, as assumed below.

**Step 2.** Take $Q_1$ to be the *nonempty,* convex set of all $q$ in $Q$ for which

$$\Pi(\tilde{p}, q) \leq \underline{v} + \varepsilon, \tag{7.8}$$

where $0 < \varepsilon < \Pi(\tilde{p}, \tilde{q}) - \underline{v}$. Then $Q_1$ is strictly smaller than $Q$ because it doesn't contain $\tilde{q}$. Let $P_1 = P$.

**Step 3.** With $\tilde{p}_1$ and $\tilde{q}_1$ defined in the obvious way, consider the convex combinations $\hat{p} = \alpha\tilde{p} + \beta\tilde{p}_1$ and $\hat{q} = \alpha\tilde{q} + \beta\tilde{q}_1$. Observe that

$$\begin{aligned} \overline{v} = \min_{q \in Q} \max_{p \in P} \Pi(p, q) &\leq \max_{p \in P} \Pi(p, \hat{q}) \\ &= \max_{p \in P} \{\alpha\Pi(p, \tilde{q}) + \beta\Pi(p, \tilde{q}_1)\} \\ &\leq \alpha \max_{p \in P} \Pi(p, \tilde{q}) + \beta \max_{p \in P_1} \Pi(p, \tilde{q}_1) \\ &= \alpha\overline{v} + \beta\overline{v}_1. \end{aligned} \tag{7.9}$$

**Step 4.** An inequality for $\underline{v}$ requires more effort. Note to begin with that

$$\begin{aligned} \min_{q \in Q_1} \Pi(\hat{p}, q) &\geq \alpha \min_{q \in Q_1} \Pi(\tilde{p}, q) + \beta \min_{q \in Q_1} \Pi(\tilde{p}_1, q) \\ &\geq \alpha \min_{q \in Q} \Pi(\tilde{p}, q) + \beta \min_{q \in Q_1} \Pi(\tilde{p}_1, q) \\ &= \alpha\underline{v} + \beta\underline{v}_1. \end{aligned} \tag{7.10}$$

$$\begin{aligned} \inf_{q \notin Q_1} \Pi(\hat{p}, q) &\geq \alpha \inf_{q \notin Q_1} \Pi(\tilde{p}, q) + \beta \inf_{q \notin Q_1} \Pi(\tilde{p}_1, q) \\ &\geq \alpha(\underline{v} + \varepsilon) + \beta c. \end{aligned} \tag{7.11}$$

To derive the last line, note that, if $\Pi(\tilde{p}, q) \leq \underline{v}+\varepsilon$, then $q$ lies in the set $Q_1$ by (7.8). The constant $c$ is simply an abbreviation for $\inf_{q \notin Q_1} \Pi(\tilde{p}_1, q)$.

**Step 5.** We want (7.10) to be smaller than (7.11). To arrange this, $\alpha = 1-\beta$ and $\beta$ have to be carefully chosen. By taking $\beta$ to be very small, (7.10) can be made as close to $\underline{v}$ as we choose. Similarly (7.11) can be made as close to $\underline{v}+\varepsilon$ as we choose. Thus, if $\beta$ is chosen to be sufficiently small, then (7.10) is less than (7.11). However, it is important that $\beta$ isn't actually equal to zero.

**Step 6.** An inequality for $\underline{v}$ is now possible:

$$\begin{aligned}
\underline{v} = \max_{p \in P} \min_{q \in Q} \Pi(p, q) \geq \min_{q \in Q} \Pi(\hat{p}, q) \\
&= \min\left\{ \min_{q \in Q_1} \Pi(\hat{p}, q), \inf_{q \notin Q_1} \Pi(\hat{p}, q) \right\} \\
&\geq \min\left\{\alpha\underline{v}+\beta\underline{v}_1, \alpha(\underline{v}+\varepsilon)+\beta c\right\} \\
&= \alpha\underline{v}+\beta\underline{v}_1. \qquad (7.12)
\end{aligned}$$

**Step 7.** The desired inequality (7.7) now follows from (7.12) and (7.9).

**Step 8.** It remains to explain how we carry through the construction to ordinals other than $\beta = 1$. There is no difficulty when $\beta$ has an immediate predecessor $\alpha$, but what happens when $\beta$ is an ordinal like $\omega$, which doesn't? In this case, we simply take $P_\beta$ to be the intersection of all $P_\alpha$ with $\alpha < \beta$ and $Q_\beta$ to be the intersection of all $Q_\alpha$ with $\alpha < \beta$.

**Step 9.** The continuity of the payoff function then ensures that (7.7) holds whenever $\alpha < \beta$. The fact that each $P_\alpha$ and $Q_\alpha$ is nonempty, convex, and compact ensures that the same is true of $P_\beta$ and $Q_\beta$. It is also true that the inclusion $P_\beta \times Q_\beta \subset P_\alpha \times Q_\alpha$ is strict when $\alpha < \beta$.
This concludes the construction. The proof of the minimax theorem follows.

### 7.4.5 Security and Equilibrium

The minimax theorem tells us that Adam's security level in any game is the maximin value $\underline{v}$ of his payoff function. He can guarantee at least $\underline{v}$ by playing the security strategy $\tilde{p}$ of (7.5). Eve can hold him to $\underline{v} = \overline{v}$ by playing the security strategy $\tilde{q}$ of (7.6).

In any game, Adam must receive at least his security level $\underline{v}$ at a Nash equilibrium. Otherwise he wouldn't be making a best reply since he could always get more by switching to one of his security strategies. However, the example of the Battle of the Sexes shows the players needn't get more than their security levels. Nor need their equilibrium strategies be secure.

→ 7.5

Recall that mixed strategies in the Battle of the Sexes were represented as line segments in Figure 6.17(b). As explained in Section 6.6.1, the line segment corresponding to $\tilde{p} = \frac{1}{3}$ is horizontal. The line segment corresponding to $\tilde{q} = \frac{2}{3}$ is vertical. Eve therefore always gets the same payoff when Adam plays $\tilde{p} = \frac{1}{3}$, and Adam always gets the same payoff when Eve plays $\tilde{q} = \frac{2}{3}$. It follows that the pair $(\tilde{p}, \tilde{q})$ is a mixed Nash equilibrium.

Similar reasoning can locate Adam's and Eve's security strategies in this special case. The line segment $l$ corresponding to $\hat{p} = \frac{2}{3}$ is vertical. Whatever Eve does, Adam therefore gets the same payoff when he plays $\hat{p} = \frac{2}{3}$. All the other line segments corresponding to Adam's mixed strategies cross $l$ and hence contain points that lie to the left of $l$. The worst possible outcome for Adam when one of these other mixed strategies is used is therefore worse for Adam than the worst possible outcome when he plays $\hat{p} = \frac{2}{3}$. Thus, his security strategy in the Battle of the Sexes is $\hat{p} = \frac{2}{3}$. Similarly, Eve's security strategy is $\hat{q} = \frac{1}{3}$, which corresponds to a horizontal line segment in Figure 6.17(b).

The Nash equilibrium $(\tilde{p}, \tilde{q}) = (\frac{1}{3}, \frac{2}{3})$ and the profile $(\hat{p}, \hat{q}) = (\frac{2}{3}, \frac{1}{3})$ of security strategies correspond to the same pair of line segments in Figure 6.17(b). The players therefore receive the same payoff of $\frac{2}{3}$ at each profile. It follows that Adam and Eve both get their security levels of $\frac{2}{3}$ at the mixed Nash equilibrium, although neither equilibrium strategy is secure.

## 7.5 Solving Zero-Sum Games

It is usually irrational for Adam to proceed on the paranoic assumption that Eve is intent on doing him harm. If Eve is rational, she will seek to maximize her own payoff rather than minimizing his. But paranoia is entirely rational in zero-sum games because Eve's interests are then diametrically opposed to Adam's. Maximizing her payoff is then the same as minimizing his payoff.

### 7.5.1 Values of Two-Player, Zero-Sum Games

In Section 2.8.1, the value $v$ of a strictly competitive game was defined to be an outcome with the property that player I has a strategy $\sigma$ that forces a result that is at least as good for him as $v$, while player II simultaneously has a strategy $\tau$ that forces a result that is at least as good for her as $v$. Things are no different here, except that we now take the value $v$ of a two-player, zero-sum game to be a *payoff* to player I, rather than an outcome.

Theorem 7.9 *Any finite two-player, zero-sum game has a value $v = \underline{v} = \overline{v}$. To ensure that he gets an expected payoff of at least $v$, player I can use any of his security strategies $\tilde{p}$. To ensure that player I gets no more than $v$, player II can use any of her security strategies $\tilde{q}$.*

*Proof* The minimax theorem implies that player I's payoff function always has a saddle point $(\tilde{p}, \tilde{q})$. Theorem 7.7 then applies.

Theorem 7.9 focuses on the value $v$ of a two-person, zero-sum game from the point of view of player I. However, everything is the same for player II, except that her security level is $-v$. In formal terms,

$$\begin{aligned} \max_{q \in Q} \min_{p \in P} \{-\Pi(p, q)\} &= \max_{q \in Q} \{-\max_{p \in P} \Pi(p, q)\} \\ &= -\{\min_{q \in Q} \max_{p \in P} \Pi(p, q)\} = -\overline{v} = -v. \end{aligned}$$

So player II can ensure a payoff of at least $-v$ for herself by using any of her security strategies $\tilde{q}$. To ensure that player II gets no more than $-v$, player I can use any of his security strategies $\tilde{q}$.

### 7.5.2 Equilibria in Two-Player, Zero-Sum Games

It is only necessary to quote the relevant theorem and to give some examples.

THEOREM 7.10 *In a finite two-player, zero-sum game, $\tilde{p}$ is a security strategy for player I and $\tilde{q}$ is a security strategy for player II if and only if $(\tilde{p}, \underline{\tilde{q}})$ is a Nash equilibrium.*

*Proof* The two conditions are equivalent to the existence of a saddle point.

*Rock-Scissors-Paper* Every child knows this game. Adam and Eve simultaneously make a hand signal that represents one of their three pure strategies: *rock, scissors, paper.* The winner is determined by the rules:

| | | |
|---|---|---|
| *rock* | blunts | *scissors* |
| *scissors* | cut | *paper* |
| *paper* | wraps | *rock.* |

If both players make the same signal, the result is a draw. We assume that both players regard a draw as being equivalent to the lottery in which they win or lose with equal probability, so that the game is zero sum. Adam's payoff matrix can then be taken to be

$$A = \begin{bmatrix} 0 & 1 & -1 \\ -1 & 0 & 1 \\ 1 & -1 & 0 \end{bmatrix}$$

The rows and the columns of the payoff matrix $A$ all contain the same numbers shuffled into different orders. It follows that, if Adam and Eve play each of their pure strategies with the same probability, then their opponent will get the same payoff from each pure strategy. It is therefore a Nash equilibrium for both players to use the mixed strategy $(\frac{1}{3}, \frac{1}{3}, \frac{1}{3})^\top$. Theorem 7.10 then tells us that the same mixed strategy is a security strategy for each player.

We can confirm that $(\frac{1}{3}, \frac{1}{3}, \frac{1}{3})^\top$ is a security strategy for both players by observing that they get a payoff of zero from its use, whatever strategy the opponent plays. The value of the game is therefore zero—as it must be for all symmetric, two-player, zero-sum games.

*O'Neill's Card Game.* Section 6.4.5 shows that $(\tilde{p}, \underline{\tilde{q}})$ is a Nash equilibrium for O'Neill's Card Game when $\tilde{p} = \tilde{p} = (\frac{2}{5}, \frac{1}{5}, \frac{1}{5}, \frac{1}{5})^\top$. Theorem 7.10 implies that $\tilde{p}$ and $\tilde{p}$ are therefore security strategies for this strictly competitive game. Unlike the case of Rock-Scissors-Paper, player I enjoys an advantage in O'Neill's game because its value is positive. In fact,

$$v = \tilde{p}^\top A \underline{\tilde{q}} = \tfrac{2}{5}.$$

### 7.5.3 Equivalent and Interchangeable Equilibria

When a game has multiple Nash equilibria, which should count as its solution? Von Neumann and Morgenstern evaded this equilibrium selection problem by focusing on two-player, zero-sum games, in which Theorem 7.10 shows that all pairs of Nash equilibria are interchangeable and equivalent.

Two equilibria $(p, q)$ and $(p', q')$ are *interchangeable* if $(p, q')$ and $(p', q)$ are also Nash equilibria. The equilibria are *equivalent* if $\Pi_1(p, q) = \Pi_1(p', q')$ and $\Pi_2(p, q) = \Pi_2(p', q')$. Since both players then get the same payoff at each equilibrium, neither will then care which gets selected.

If the Nash equilibria of a game are equivalent and interchangeable, then the selection problem disappears. Even if Von Neumann had written a book recommending the equilibrium $(p, q)$, and Morgenstern had written a rival book recommending $(p', q')$, their failure to agree wouldn't trouble the players at all. If Adam follows Von Neumann, he will play $p$. If Eve follows Morgenstern, she will play $\underline{q}'$. The result will be the Nash equilibrium $(p, \underline{q}')$, which assigns both players exactly the payoff they were anticipating.

### 7.5.4 When to Play Maximin

→ 7.6

Some authors say that it is prudent to use maximin strategies in all risky situations, but such folks are irrational in their extreme caution.

As in the case of the Battle of the Sexes, if both players use their security strategies in a general game, then neither is likely to be making a best reply to the strategy choice made by the other (Section 7.4.5). Nor is there any reason why rational players should settle for as little as their security levels in most games. For example, both the pure Nash equilibria in the Battle of the Sexes yield much higher payoffs than the players' security levels.

Theorem 7.10 is therefore definitely only a theorem about two-player, *zero-sum* games, but even when playing in a two-player, zero-sum game, you would be ill advised to use a maximin strategy when you have good reason to suppose that your opponent will play poorly. Playing your security strategy will certainly guarantee you your security level however the opponent plays, but you ought to be aiming for more than your security level against a bad player. You should be probing the opponent's play for systematic weaknesses and deviating from your security strategy in order to exploit these weaknesses. You will be taking a risk in doing so, but it is *irrational* to be unwilling to take a calculated risk when the odds are sufficiently in your favor.

But what if you are playing a good player in a zero-sum game? Evidence gathered by observing strategic situations in professional sport is surprisingly supportive of Von Neumann's theory. The data on how penalty kicks are taken in soccer fit the theory that players mix according to the maximin criterion especially well.

## 7.6 Linear Programming

Mathematical programming consists of finding the maximum or minimum of an objective function $f(x)$ subject to a set of constraints on the values that $x$ is allowed to

→ 7.7

take. Linear programming is the special case in which the objective function and the functions used to specify the constraints are all linear.

This section shows the relevance of zero-sum games to the duality theorem of linear programming. We look only at a special case of a result that is considerably more general.

### 7.6.1 Duality

In Section 6.4.4, we learned that Adam can secure a payoff of $\alpha$ by playing a mixed strategy $p$ that satisfies the inequality $p^\top A \geq \alpha e^\top$. (Recall that $e$ denotes a vector whose entries are all one.)

The problem of finding Adam's security level therefore reduces to locating a vector $p$ that maximizes $\alpha$ subject to the constraints listed on the left below. (The constraints $p^\top e = 1$ and $p^\top \geq 0$ just say that the entries of $p$ must be probabilities.) Eve's security level similarly reduces to locating a vector $q$ that maximizes $\beta$ subject to the constraints listed on the right:

$$\begin{array}{rl} p^\top A \geq \alpha e^\top & \quad Bq \geq \beta e \\ p^\top e = 1 & \quad e^\top q = 1 \\ p^\top \geq 0 & \quad q \geq 0 \end{array}$$

In the case of a zero-sum game, Eve's payoff matrix is $B = -A$. If we are to express everything in terms of Adam's payoffs as usual, we must also write $\gamma = -\beta$. Eve then seeks to minimize $\gamma$ rather than maximize $\beta$. Its minimum value is the negative of Eve's security level, which is equal to Adam's security level by von Neumann's minimax theorem.

We therefore have two problems with the same solution. The maximum value of $\alpha$ subject to the constraints on the left below is the same as the minimum value of $\gamma$ subject to the constraints on the right:

$$\begin{array}{rl} p^\top A \geq \alpha e^\top & \quad Aq \leq \gamma e \\ p^\top e = 1 & \quad e^\top q = 1 \\ p^\top \geq 0 & \quad q \geq 0 \end{array}$$

Rewriting our two problems, we obtain a version of the duality theorem of linear programming. Take $p = \alpha y$ in Adam's problem, so that $\alpha^{-1} = e^\top y$. Assuming that $\alpha > 0$, Adam therefore wants to *minimize* $e^\top y$. His problem therefore reduces to that shown on the *right* below. Writing $q = \gamma x$ similarly reduces Eve's problem to that shown on the *left*.

$$\begin{array}{ll} \text{maximize} & \text{minimize} \\ e^\top x & y^\top e \\ \text{subject to} & \text{subject to} \\ Ax \leq e & y^\top A \geq e^\top \\ x \geq 0 & y \geq 0 \end{array}$$

| (a) Primal program | (b) Dual program |
|---|---|
| maximize $c^\top x$ | minimize $y^\top b$ |
| subject to | subject to |
| $Ax \leq b$ | $y^\top A \geq c^\top$ |
| $x \geq 0$ | $y \geq 0$ |

Figure 7.10 A primal linear programming problem and its dual. If one of the programs is feasible, then both optima exist and are equal.

These two linear programs are said to be dual to each other. This implies, in particular, that they both have the same solution. A more general formulation of a primal program and its dual is given in Figure 7.10.

The duality theorem of linear programming takes as its hypothesis that one of the two programs is feasible. This means that there is at least one vector that satisfies its constraints. The conclusion is then that both programs have a solution and that the maximum in the primal problem is equal to the minimum in the dual problem.

### 7.6.2 Shadow Prices Again

The Lagrangian of the primal problem of Figure 7.10(a) is defined as

$$L(x,y) = c^\top x + y^\top (b - Ax).$$

→ 7.7

Recall that this is the payoff function of the game played between Alice and Mad Hatter Enterprises in Section 7.1.1. The duality theorem tells us that $L(x,y)$ has a saddle point $(\tilde{x}, \tilde{y})$, where $\tilde{x}$ and $\tilde{y}$ solve the primal and dual problems of Figure 7.10 respectively.

To see this, observe that Mad Hatter Enterprises can make $L(x,y)$ as small as it likes if the vector $b - Ax$ has a negative coordinate. Alice will therefore ensure that $Ax \leq b$. The best that Mad Hatter enterprises can then do in minimizing $L(x,y)$ is to choose $y$ so that $y^\top (b - Ax) = 0$. Alice then faces the primal problem of Figure 7.10(a). Thus

$$\max_{x \geq 0} \min_{y \geq 0} L(x,y) = c^\top \tilde{x}.$$

Since $L(x,y) = y^\top b + (c^\top - y^\top A)x$, we can now repeat the argument with the roles of the players reversed. Alice can make $L(x,y)$ as big as she likes if the vector $c^\top - y^\top A$ has a positive coordinate. Mad Hatter Enterprises will therefore ensure that $y^\top A \geq c^\top$. The best that Alice can then do in maximizing $L(x,y)$ is to choose $x$ so that $(c^\top - y^\top A)x = 0$. Mad Hatter Enterprises then faces the dual problem of Figure 7.10(b). Thus

$$\min_{y \geq 0} \max_{x \geq 0} L(x,y) = \tilde{y}^\top b.$$

However, the duality theorem says that $c^\top \tilde{x} = \tilde{y}^\top b$, and so $(\tilde{x}, \tilde{y})$ is a saddle point of $L(x, y)$ by Theorem 7.3.

We learn that Alice can compute the shadow prices of her stock by solving the dual problem of Figure 7.10(b). She should also note that

$$\tilde{y}^\top (b - A\tilde{x}) = 0,$$

which says that Mad Hatter Enterprises will assign a zero price to goods in stock that Alice doesn't use up in producing $\tilde{x}$. The value of her stock is therefore $c^\top \tilde{x} = \tilde{y}^\top b = \tilde{y}^\top A\tilde{x}$.

## 7.7 Separating Hyperplanes

The theorem of the separating hyperplane has important applications. It is used, for example, in proving the existence of clearing prices in general equilibrium models of the economy. The use to which the theorem of the separating hyperplane is put in this section reflects the fact that most proofs of the minimax theorem depend on it.

### 7.7.1 Hyperplanes

→ 7.7.2

Hyperplanes sound like something out of *Star Trek*, but they aren't exciting enough to get into a television script. A *hyperplane* with normal $n \neq 0$ is simply the set of all $x$ that satisfy the equation

$$n^\top x = c. \tag{7.13}$$

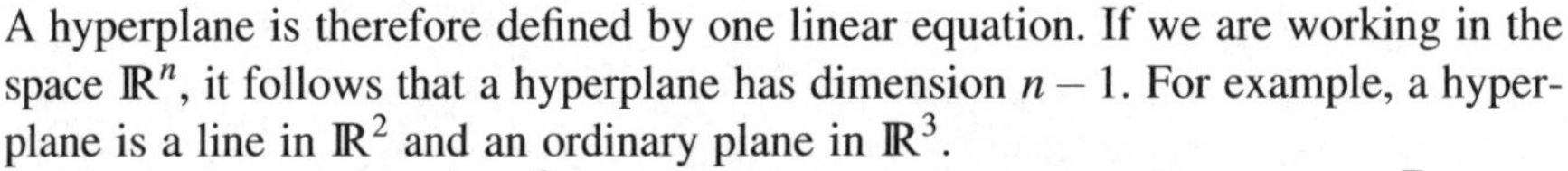

A hyperplane is therefore defined by one linear equation. If we are working in the space $\mathbb{R}^n$, it follows that a hyperplane has dimension $n-1$. For example, a hyperplane is a line in $\mathbb{R}^2$ and an ordinary plane in $\mathbb{R}^3$.

Consider the plane in $\mathbb{R}^3$ that passes through the point $\xi = (3,\ 2,\ 1)^\top$ and is orthogonal to the vector $n = (3,\ 1,\ 1)^\top$. Figure 7.11(a) shows that the point $x$ lies in the plane if and only if the vector $x - \xi$ is orthogonal to the vector $n$. But two vectors are orthogonal if and only if their inner product is zero (Section 6.4.2). The equation of the plane is therefore $n^\top (x - \xi) = 0$, which we can express in the form (7.13) by taking $c = n^\top \xi = 12$. To get a less abstract formulation, simply expand the inner product in (7.13) to obtain

$$3x_1 + x_2 + x_3 = 12.$$

The line in $\mathbb{R}^2$ that passes through the point $\xi = (2,\ 1)^\top$ and is orthogonal to the vector $n = (3,\ 4)^\top$ is a hyperplane in $\mathbb{R}^2$. Figure 7.11(b) shows why the equation of the line is $n^\top (x - \xi) = 0$, which we can express in the form (7.13) by taking $c = n^\top \xi = 10$. Expanding the inner product in (7.13) yields the standard linear equation

$$3x_1 + 4x_2 = 10.$$

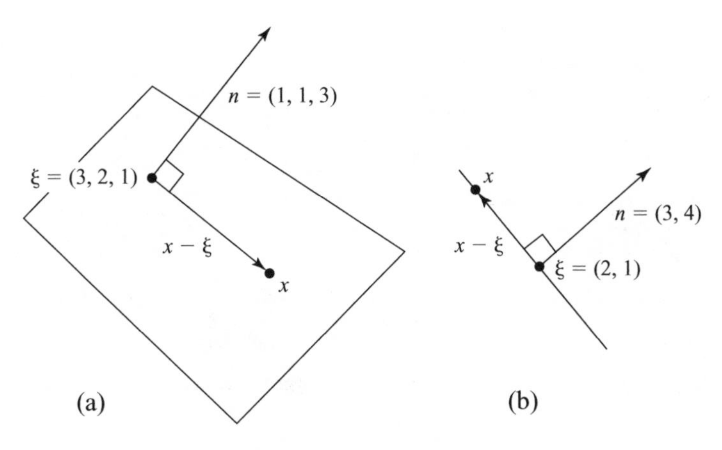


Figure 7.11 Hyperplanes.

Any vector that is orthogonal to a hyperplane will serve as a normal to the hyperplane. We can therefore always adjust the length of a normal to something convenient by multiplying by a suitable scalar. For example, if we want a normal to the line $3x_1 + 4x_2 = 10$ of unit length, we can simply divide through by 5 to obtain the new normal $n = (\frac{3}{5}, \frac{4}{5})^\top$.

### 7.7.2 Separation

Euclid's geometry is commonly thought to be the ultimate in deductive reasoning, but David Hilbert pointed out that some of Euclid's proofs depend on ideas that his axioms neglect. Separation is one of these ideas.

A hyperplane $n^\top x = c$ splits $\mathbb{R}^n$ into two half spaces. Any line joining two points in different half spaces necessarily passes through the hyperplane.

The half space "above" the hyperplane is the set of all $x$ for which $n^\top x \geq c$. This is the half space into which the vector $n$ points. The half space "below" the hyperplane is the set of all $x$ for which $n^\top x \leq c$. To say that the set $G$ lies above the hyperplane therefore means that $n^\top g \geq c$ for each $g$ in $G$. To say that the set $H$ lies below the hyperplane means that $n^\top h \leq c$ for each $h$ in $H$.

Two sets $G$ and $H$ are *separated* by a hyperplane if one lies above the hyperplane and the other lies below. Figure 7.12(a) shows two convex sets $G$ and $H$ in $\mathbb{R}^2$ separated by the hyperplane $n^\top x = c$, which is just a line in this case. Figure 7.12(b) shows a degenerate case, in which the set $H$ consists of a single boundary point $\xi$ of $G$.

A useful version of the theorem of the separating hyperplane is quoted below. Notice that it allows $G$ and $H$ to have boundary points in common.

Theorem 7.11 (Theorem of the Separating Hyperplane) *Let $G$ and $H$ be convex sets in $\mathbb{R}^n$. Suppose that $H$ has interior points but that none of these lie in $G$. Then there exists a hyperplane $n^\top x = c$ that separates $G$ and $H$.*

### 7.7.3 Separation and Saddle Points

Consider a two-person, zero-sum game with matrix $A$. The minimax theorem says that we can always find mixed strategies $\tilde{p}$ and $\tilde{q}$ for the two players that satisfy

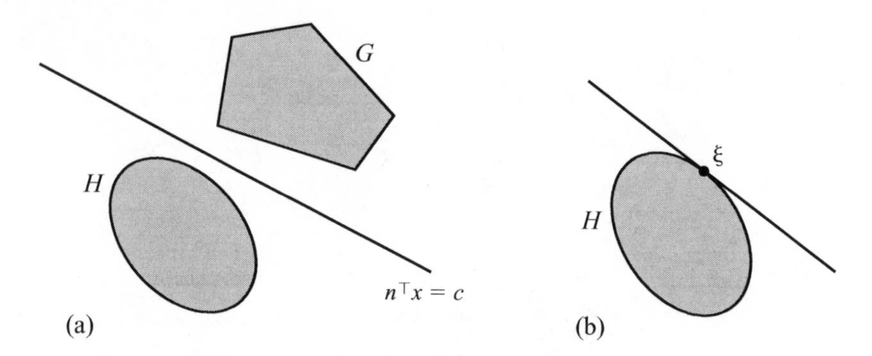


Figure 7.12 Separating hyperplanes.

$\tilde{p}^\top Aq \geq \tilde{p}^\top A\tilde{q} \geq p^\top A\tilde{q}$. Rewriting this saddle point condition in terms of the value $v = \tilde{p}^\top A\tilde{q}$ of the game yields the inequalities

$$\tilde{p}^\top Aq \geq v \geq p^\top A\tilde{q}. \tag{7.14}$$

The theorem of the separating hyperplane allows a geometric interpretation. We construct two convex sets $G$ and $H$ that are separated by a hyperplane $\tilde{p}^\top x = v$, whose normal is player I's security strategy $\tilde{p}$. Player II's security strategy $\tilde{q}$ can be found using the fact that the point $A\tilde{q}$ lies in the set $G \cap H$.

→ 7.7.4

We illustrate the construction using the matrix of Figure 7.9(a):

$$A = \begin{bmatrix} 2 & 5 & 0 \\ 4 & 3 & 6 \end{bmatrix} \tag{7.15}$$

We already know that the value of the game with matrix $A$ is $v = 3\frac{1}{2}$, which is secured by the mixed strategies $\tilde{p} = (\frac{1}{4}, \frac{3}{4})^\top$ and $\tilde{q} = (\frac{1}{2}, \frac{1}{2}, 0)^\top$ (Section 7.4.2).

We take the set $G$ in the theorem of the separating hyperplane to be the convex hull of the columns of the matrix $A$. In Figure 7.13(a), $G$ is a triangle with vertices $(2,4)^\top, (5,3)^\top$, and $(0,6)^\top$.

The points $g$ in $G$ are convex combinations of the columns of $A$. It follows that $G = \{Aq : q \in Q\}$ because, for each $g$ in $G$, there is a $q$ in $Q$ such that

$$\begin{aligned} g &= q_1 \begin{bmatrix} 2 \\ 4 \end{bmatrix} + q_2 \begin{bmatrix} 5 \\ 3 \end{bmatrix} + q_3 \begin{bmatrix} 0 \\ 6 \end{bmatrix} \\ &= \begin{bmatrix} 2 & 5 & 0 \\ 4 & 3 & 6 \end{bmatrix} \begin{bmatrix} q_1 \\ q_2 \\ q_3 \end{bmatrix} = Aq. \end{aligned}$$

The set $H$ of Figure 7.13(b) is defined by

$$H = \{h : h \leq ve\},$$

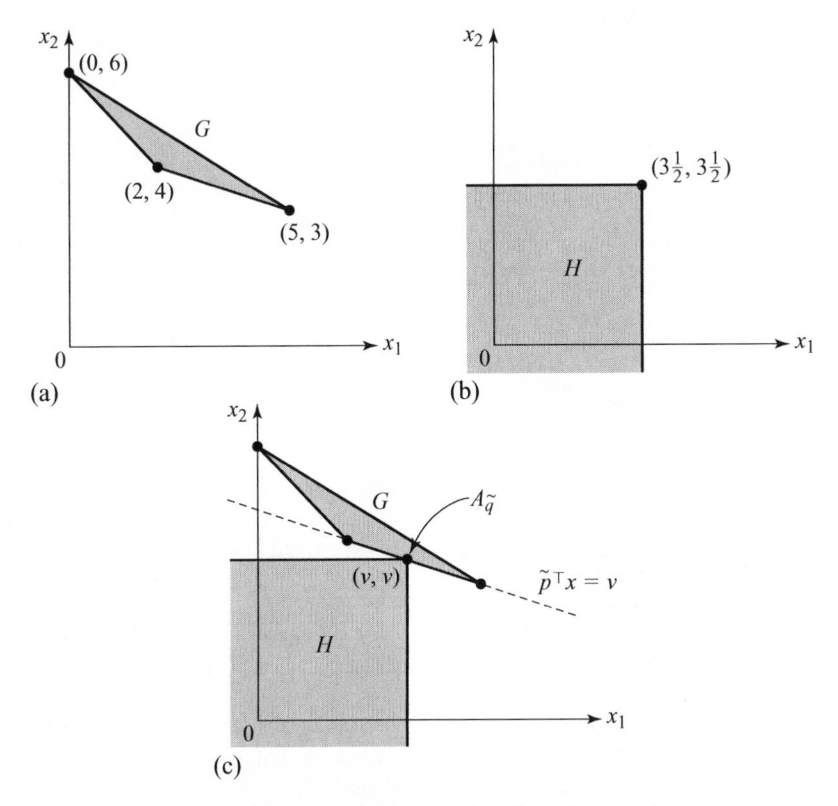


Figure 7.13 A geometric representation of security strategies.

where $v = 3\frac{1}{2}$ is the value of the game. Note that[7] $h$ lies in $H$ if and only if, for all $p$ in $P$,

$$p^\top h \leq v. \tag{7.16}$$

The hyperplane $\tilde{p}^\top x = v$ separates $G$ and $H$. It is immediate that $H$ lies below the hyperplane because we can take $p = \tilde{p}$ in (7.16). To see that $G$ lies above the hyperplane, we need the left half of (7.14). This says that $\tilde{p}^\top Aq \geq v$ for all $q$ in $Q$. On writing $g = Aq$, it follows that, for all $g$ in $G$,

$$\tilde{p}^\top g \geq v.$$

The right half of (7.14) has not yet been used. This says that $p^\top A\,\tilde{q} \leq v$ for all $p$ in $P$. Thus, $A\,\tilde{q}$, which we already know to lie in $G$, must also lie in $H$ by (7.16). That is, the set $G \cap H$ of all points common to $G$ and $H$ contains $A\,\tilde{q}$. Although $G$ and $H$ are separated by the hyperplane $\tilde{p}^\top x = v$, they therefore still have the point $A\,\underline{\tilde{q}}$ in common, as illustrated in Figure 7.13(c).

[7]If $h \leq ve$, then $p^\top h \leq vp^\top e = v$. If $p^\top h \leq v$, for all $p$ in $P$, we can show that $h \leq ve$ by taking $p = e_i$ for each $i$.

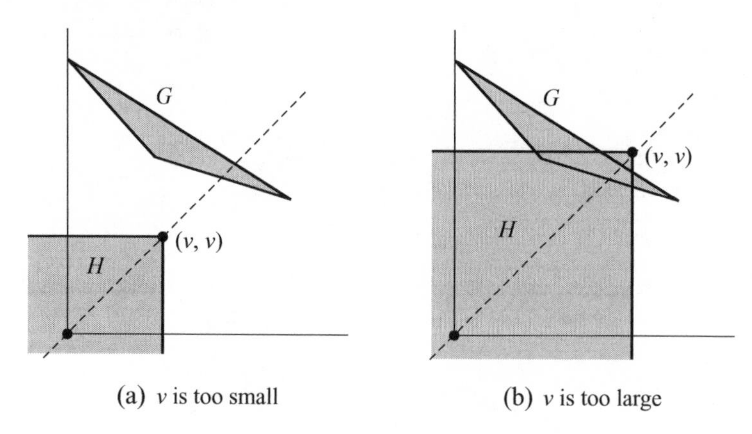


Figure 7.14 Choosing the number $v$.

### 7.7.4 Solving Games Using Separation

We have seen how the minimax theorem can be interpreted geometrically. We now use the geometry to solve some two-player, zero-sum games. The method works for any payoff matrix with only two rows.

*Example 1.* Nobody would choose to analyze a two-person, zero-sum game by the method of Section 7.4.2 with anything more complicated than the payoff matrix $A$ of Figure 7.9(a). A better method is to proceed by turning the argument of the preceding section on its head.

**Step 1.** Mark the location of the columns $(2, 4)^\top$, $(5, 2)^\top$, and $(0, 6)^\top$ of the matrix $A$ on a piece of graph paper. Then draw their convex hull $G$ as in Figure 7.13(a).

**Step 2.** Draw the line $x_1 = x_2$. The point $(v, v)^\top$ on this line determines the set $H$ shown in Figure 7.13(b). We need to choose $v$ to be the *smallest* value such that $G$ and $H$ have at least one point in common.[8] Figure 7.14(a) shows a case where $v$ has been chosen too small, with the result that $G$ and $H$ have no points in common. Figure 7.14(b) shows a case where $v$ has been chosen too large. It could be made a little smaller, and the sets $G$ and $H$ would still have points in common.

**Step 3.** Draw the separating line $\tilde{p},{}^\top x = v$, as in Figure 7.13(c).

**Step 4.** Find player I's security level $\tilde{p}$. This is a normal to the separating line. Often it can be found without the need to calculate, but most people would find it necessary to write down the equation of the separating line in this case. Since the separating line passes through $(2, 4)^\top$ and $(5, 3)^\top$, it has equation

$$\frac{x_2 - 4}{x_1 - 2} = \frac{3 - 4}{5 - 2} = \frac{-1}{3},$$

[8]The sets $G$ and $H$ must have a point in common because $A\tilde{q}$ belongs to both. But their intersection must contain as few other points as possible because the theorem of the separating hyperplane requires that $G$ contain no interior point of $H$.

which may be rewritten as $x_1 + 3x_2 = 14$. The coefficients 1 and 3 in this equation are the coordinates of a normal vector to the separating hyperplane (Section 7.7.1). But we need a normal $\tilde{p}$ that satisfies $p_1 \geq 0$, $p_2 \geq 0$, and $p_1 + p_2 = 1$ and hence lies in the set $P$. The normal $(1,3)^\top$ is therefore replaced by the normal $\tilde{p} = (\frac{1}{4}, \frac{3}{4})^\top$, which is player I's security strategy.

**Step 5.** Find the value $v$ of the game by looking at the point $(v,v)^\top$ where the lines $x_1 = x_2$ and $x_1 + 3x_2 = 14$ meet. Solving these equations, we find that $v + 3v = 14$, and so $v = 3\frac{1}{2}$.

**Step 6.** Find player II's security strategy $\tilde{q}$ using the fact that $A\tilde{q}$ lies in the set $G \cap H$. In the current example, $G \cap H$ consists of the single point $(v,v) = (3\frac{1}{2}, 3\frac{1}{2})$. Thus,

$$\begin{bmatrix} 2 & 5 & 0 \\ 4 & 3 & 6 \end{bmatrix} \begin{bmatrix} \tilde{q}_1 \\ \tilde{q}_2 \\ \tilde{q}_3 \end{bmatrix} = \begin{bmatrix} 3\frac{1}{2} \\ 3\frac{1}{2} \end{bmatrix}.$$

You can solve the system of three simultaneous linear equations created by adding the requirement that $\tilde{q}_1 + \tilde{q}_2 + \tilde{q}_3 = 1$ if you like, but it is usually easier to proceed as follows.

Recall that $G$ is the convex hull of the columns of $A$. Thus $A\tilde{q}$ is a convex combination of the columns of $A$. In fact, $A\tilde{q}$ lies at the center of gravity of weights $\tilde{q}_1$, $\tilde{q}_2$, and $\tilde{q}_3$ located at the points $(2,4)^\top$, $(5,3)^\top$, and $(0,6)^\top$ (Section 6.5.1). In Figure 7.13(c), $(v,v)^\top = A\tilde{q}$ looks as though it is halfway along the line segment joining $(2,4)^\top$ and $(5,3)^\top$. If so, then the appropriate weights must be $\tilde{q}_1 = \frac{1}{2}$, $\tilde{q}_2 = \frac{1}{2}$, and $\tilde{q}_3 = 0$. To verify this, observe that

$$\frac{1}{2}\begin{bmatrix} 2 \\ 4 \end{bmatrix} + \frac{1}{3}\begin{bmatrix} 5 \\ 3 \end{bmatrix} + 0\begin{bmatrix} 0 \\ 6 \end{bmatrix} = \begin{bmatrix} 3\frac{1}{2} \\ 3\frac{1}{2} \end{bmatrix}.$$

Without calculating very much, we have therefore shown that player II has a unique security strategy, $\tilde{q} = (\frac{1}{2}, \frac{1}{2}, 0)^\top$.

*Example 2.* The two-player, zero-sum game with matrix

$$B = \begin{bmatrix} 1 & 2 & 3 \\ 4 & 5 & 4 \end{bmatrix}$$

yields the configuration of Figure 7.15(a). The separating line has equation $x_2 = 4$ and hence $\tilde{p} = (0,1)^\top$. The value of the game is $v = 4$. The set $G \cap H$ consists of all points on the line segment $l$ joining $(1,4)^\top$ and $(3,4)^\top$. If $A\tilde{q}$ lies on $l$, then $\tilde{q}$ is a security strategy for player II. If weights $\tilde{q}_1$, $\tilde{q}_2$, and $\tilde{q}_3$ are placed at $(1,4)^\top$, $(2,5)^\top$, and $(3,4)^\top$, when will their center of gravity lie on $l$? The only restriction necessary is that $\tilde{q}_2 = 0$. Thus, any $\tilde{q}$ in $Q$ with $\tilde{q}_2 = 0$ is a security strategy for player II.

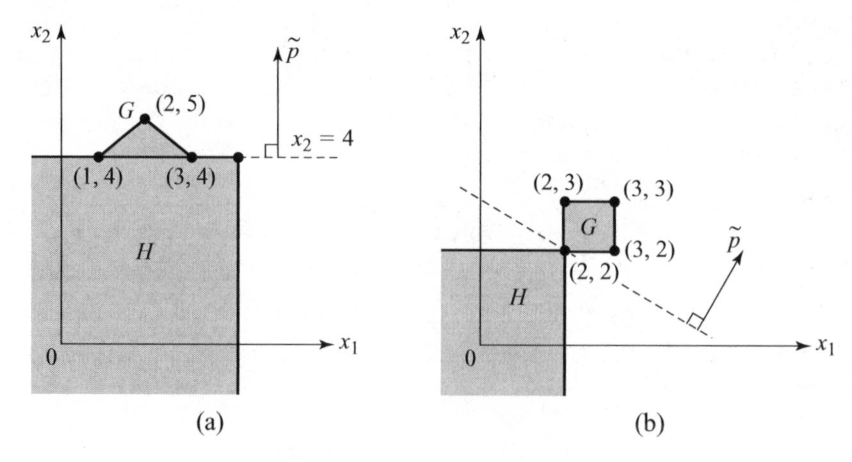


Figure 7.15 Two more examples.

*Example 3.* The two-player, zero-sum game with matrix

$$C = \begin{bmatrix} 2 & 2 & 3 & 3 \\ 2 & 3 & 2 & 3 \end{bmatrix}$$

yields the configuration of Figure 7.15(b). There are many separating lines, of which three have been drawn: the two extremal cases with $\tilde{p}' = (1,0)^\top$ and $\tilde{p}'' = (0,1)^\top$, and an intermediate case $\tilde{p} = (1-r,r)^\top$. Any $\tilde{p}$ with $0 \le r \le 1$ is therefore a security strategy for player I. The value of the game is $v = 2$. The set $G \cap H$ consists of the single point $(2,2)^\top$. For $A\tilde{q}$ to be equal to $(2,2)^\top$, all the weight must be assigned to the single column $(2,2)^\top$, and so player II has a unique security strategy $\tilde{q} = (1,0,0,0)^\top$.

### 7.7.5 Simplifying Tricks

The method of the separating hyperplane always solves two-person, zero-sum games, but it is useful as a practical tool only when the payoff matrix has only two rows or two columns.[9] Larger games can often be reduced in size by various tricks. If not, then linear programming always works (Section 7.6.1).

The following tricks for reducing big games are most useful if you care only about finding the value of a two-player, zero-sum game and at least one security strategy for each player. If you want to find *all* security strategies for the players, you usually have to work harder.

[9]In the latter case, switch the roles of players I and II. The rows and columns of the payoff matrix $A$ then have to be switched. This yields the transpose matrix $A^\top$. The signs of all the payoffs in this matrix then need to be reversed, so that they become the payoffs of the new player I (who is the old player II) rather than the payoffs of the old player I (who is the new player II). The new game therefore has payoff matrix $-A^\top$. After analyzing the new game, security strategies $\tilde{p}, \tilde{q}$, and a value $v$ will be found. The *old* game then has value $-v$. A security strategy for the *old* player I is $\tilde{q}$. A security strategy for the *old* player II is $\tilde{p}$.

- The first trick is simply to check whether the payoff matrix has a saddle point. If it does, we don't need to mess with mixed strategies at all.
- The second trick is to look for symmetries. The example coming up in Section 7.8 shows how these can sometimes be used to simplify things.
- The third trick is even cruder. It consists of deleting dominated strategies as described in Section 5.4.1. For example, we could evade calculating at all in the case of the matrix $B$ of Section 7.7.4.

## 7.8 Starships

In a game once popular with kids, two players secretly mark a number of battleships on a piece of paper. They then alternate in calling out a grid reference they wish to bomb on the other player's piece of paper. The aim is to be the first to eliminate the enemy's fleet. This section analyzes a highly simplified and asymmetric version of the game set in the far future.

fun
→ 7.9

*Hide-and-Seek.* Captain Kirk is trying to save the Starship *Enterprise* from a crazed Mr. Spock, who wants to blow it up with a bunch of atomic missiles he has stolen from Starfleet Command. Spock's aim is to destroy the starship as quickly as possible. Kirk's aim is to delay the destruction of his starship for as long as possible in the hope that rescue willl come.

Kirk hides his starship on a $4 \times 1$ board representing a nebula. The starship occupies two adjacent squares. The diagrams of Figure 7.16(a) show Kirk's three pure strategies, corresponding to the three possible hiding places in the nebula. One by one, in any order he chooses, Spock targets the squares that make up the nebula. He knows when he makes a hit because of the resulting explosion. *Both* squares occupied by the starship must be targeted by Spock's missiles for it to be destroyed.

The diagrams of Figure 7.16(b) represent Spock's pure strategies. The symbols $\circ$ or $\star$ indicate the target of his first missile. The symbol $\circ$ is used to indicate that, if the

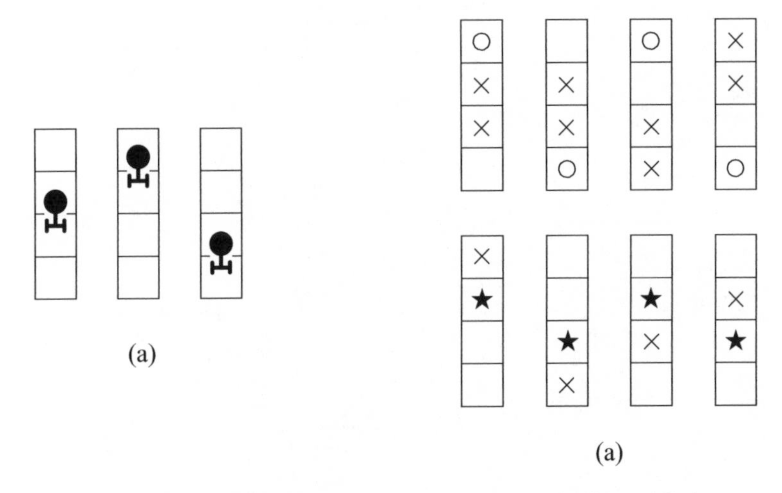


Figure 7.16 Strategies for Captain Kirk and Mr. Spock in Hide-and-Seek.

first missile misses, then the second and third targets are the squares marked with $\times$. The symbol $\star$ indicates that, if the first missile is a strike, then the second target is the square marked with $\times$. What should Spock do under other contingencies? For example, if the symbol $\circ$ is used and the first missile is a strike, what should Spock's second target be? All such questions are answered by considering only strategies that don't require him to make a foolish mistake. For example, if the symbol $\star$ is used and the first missile misses, then Spock knows the location of the battleship precisely, and it would be unwise for him not to target the second and third missile so as to destroy it.

Figure 7.17(a) shows Kirk's payoff matrix for this two-player, zero-sum game. For example, the entry 2 in row 2 and column 3 is calculated by observing that, if Kirk uses row 2 and Spock uses column 3, then Spock's first missile will be a strike. He then knows the location of the remainder of the starship and so uses his second missile to complete its destruction. Thus the game ends after only two missiles have been fired.

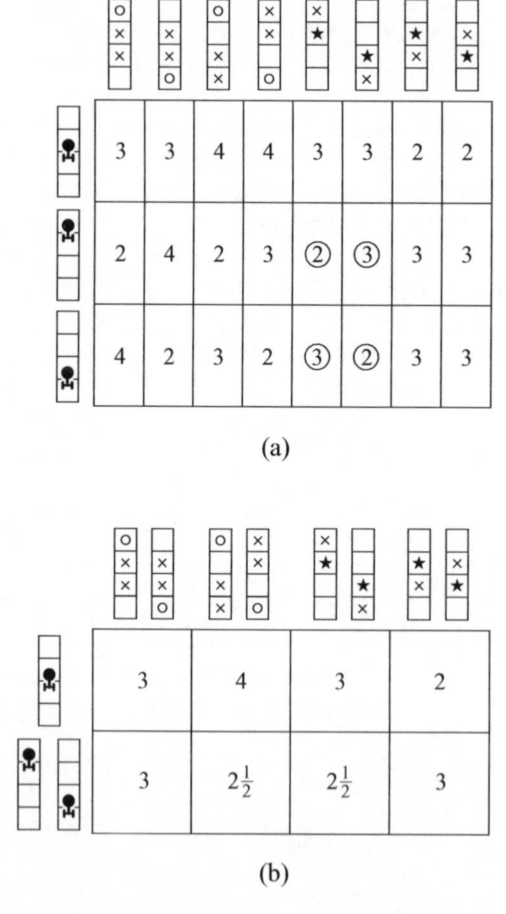


Figure 7.17 Payoff matrices for Captain Kirk in Hide-and-Seek.

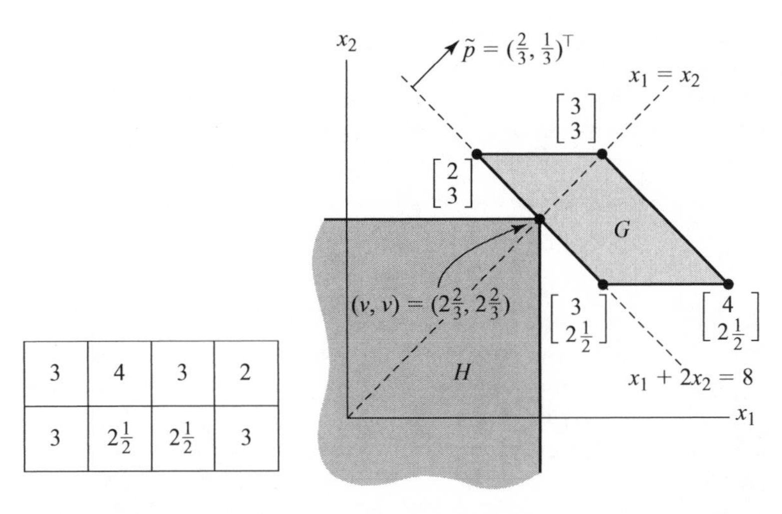


| 3 | 4 | 3 | 2 |
|---|---|---|---|
| 3 | $2\frac{1}{2}$ | $2\frac{1}{2}$ | 3 |

Figure 7.18 The method of separating hyperplanes in Hide-and-Seek.

The $3 \times 8$ payoff matrix in Figure 7.17(a) takes no account of various stupid pure strategies that Spock might use, but it is still too complicated to solve using the method of separating hyperplanes. A further simplification will therefore be made. We assume that if two pure strategies are the same except that north is swapped with south, then each will be used with equal probability. Kirk therefore uses row 2 and row 3 with equal probability. Spock similarly uses columns 7 and 8 with equal probability. This reduces Kirk's payoff matrix to the $2 \times 4$ matrix of Figure 7.17(b).

For example, the entry $2\frac{1}{2}$ in row 2 and column 3 of Figure 7.17(b) arises when Kirk uses each of rows 2 and 3 in Figure 7.17(a) with probability $\frac{1}{2}$, and Spock uses each of columns 5 and 6 with probability $\frac{1}{2}$. Each of the circled payoffs of Figure 7.17(a) then occurs with probability $\frac{1}{4} = \frac{1}{2} \times \frac{1}{2}$. So the expected payoff to Kirk is $\frac{1}{4}(2+3+2+3) = 2\frac{1}{2}$.

*Separating Hyperplanes.* Figure 7.18 shows how to apply the method of separating hyperplanes to the $2 \times 4$ simplified version of Kirk's payoff matrix. The separating line is $x_1 + 2x_2 = 8$. A normal whose coordinates sum to one is $\tilde{p} = (\frac{1}{3}, \frac{2}{3})^\top$.

The set $G \cap H$ consists of just $(2\frac{2}{3}, 2\frac{2}{3})^\top$, which can be found by solving $x_1 + 2x_2 = 8$ simultaneously with $x_1 = x_2$. The value of the game is $v = 2\frac{2}{3}$.

The point $(2\frac{2}{3}, 2\frac{2}{3})$ is one-third of the way along the line segment that joins $(3, 2\frac{1}{2})^\top$ and $(2, 3)^\top$. So $\tilde{q}$ assigns a weight of $\frac{2}{3}$ to column 3 and a weight of $\frac{1}{3}$ to column 4. Columns 1 and 2 get zero weight.[10] Thus $\tilde{q} = (0, 0, \frac{2}{3}, \frac{1}{3})^\top$.

*Conclusion.* How should Hide-and-Seek be played? Taking for granted that the original game has equilibria in which symmetric strategies are used with equal probabilities, Kirk should use the mixed strategy $(\frac{1}{3}, \frac{1}{3}, \frac{1}{3})^\top$ in the $3 \times 8$ game of Figure 7.17(a) (because it assigns equal probabilities to rows 2 and 3 that sum to

[10]We could have eliminated columns 3 and 4 earlier on the grounds that they are weakly dominated by column 1.

$\tilde{p} = \frac{2}{3}$). Spock should use the mixed strategy $(0,0,0,0,\frac{1}{3},\frac{1}{3},\frac{1}{6},\frac{1}{6})^\top$. The average number of missiles needed to destroy the starship will then be $v = 2\frac{2}{3}$.

Even Captain Kirk might guess that he should use each of his three possible hiding places with equal probability, but Mr. Spock will need to use all of his celebrated Vulcan intellect to work out his less obvious optimal strategy.

## 7.9 Roundup

Game theory began with Von Neumann's study of two-person, zero-sum games. These are strictly competitive games in which the players' utility functions are calibrated so that the payoffs always sum to zero. The strategic form of such a game is sometimes called a matrix game because it is necessary only to specify player I's payoff matrix.

The maximin $\underline{m}$ and minimax $\overline{m}$ values of a payoff matrix always satisfy $\underline{m} \leq \overline{m}$. Equality arises if and only if the matrix has a saddle point $(\sigma, \tau)$. The pure strategy $\sigma$ is then a security strategy for player I. Its play guarantees his security level $\underline{m}$.

When player I's payoff matrix lacks a saddle point, his security strategy is mixed. When maximin $\underline{v}$ and minimax $\overline{v}$ values are calculated using mixed strategies, Von Neumann's theorem says that it is always true that $\underline{v} = \overline{v}$. In a two-person, zero-sum game, it follows that any pair of security strategies for the players is a Nash equilibrium. The payoff $v = \underline{v} = \overline{v}$ that player I gets in equilibrium is called the value of the game.

Finding a security strategy for player I in a two-person zero-sum game is a linear programming problem. Player II's problem is its dual. The duality theorem of linear programming is therefore closely related to von Neumann's minimax theorem. Even when a linear programming problem isn't derived from a game, it is often helpful to think of a program and its dual as a game. The solution of the dual problem then has a ready interpretation in terms of shadow prices in the original problem.

The theorem of the separating hyperplane provides a convenient way of solving certain two-player, zero-sum games. Before resorting to this method, first confirm that the game doesn't have a saddle point. If you don't care about finding all the solutions of a game, eliminate dominated strategies before doing anything else. Exploit any symmetries you can find.

## 7.10 Further Reading

*The Compleat Strategyst*, by J. D. Williams: Dover, New York, 1954. This is a delightful collection of simple two-person zero-sum games.

## 7.11 Exercises

1. If $A$ and $B$ are finite sets of real numbers, then[11]

$$A \subseteq B \Rightarrow \max A \leq \max B.$$

[11]Recall that $A \subseteq B$ means that each element of the set $A$ is also an element of the set $B$. The notation $\max A$ means the largest element of $A$.

2. Explain why

math

$$\max\{a_1+b_1, a_2+b_2, \ldots, a_n+b_n\} \leq \max\{a_1, a_2, \ldots, a_n\} + \{\max\{b_1, b_2, \ldots, b_n\}.$$

Give an example with $n=2$ in which the inequality is strict.

3. Explain why

math

$$\max\{-a_1, -a_2, \ldots, -a_n\} = -\min\{a_1, a_2, \ldots, a_n\}$$
$$\min\{-a_1, -a_2, \ldots, -a_n\} = -\max\{a_1, a_2, \ldots, a_n\}.$$

4. Find the maximin and minimax values of the following matrices:

$$A = \begin{bmatrix} 1 & 2 \\ 3 & 4 \end{bmatrix}; \qquad B = \begin{bmatrix} 1 & 3 \\ 4 & 2 \end{bmatrix};$$

$$C = \begin{bmatrix} 2 & 4 & 6 & 3 \\ 6 & 2 & 4 & 3 \\ 4 & 6 & 2 & 3 \end{bmatrix}; \quad D = \begin{bmatrix} 3 & 2 & 2 & 1 \\ 2 & 3 & 2 & 1 \\ 2 & 2 & 3 & 1 \end{bmatrix}.$$

For which matrices is it true that $\underline{m} < \overline{m}$? For which is it true that $\underline{m} = \overline{m}$?

5. Show that, for any matrix $A$, maximin $(-A^\top) = -$minimax $(A)$.
6. Find all saddle points for the matrices of Exercise 7.11.4.
7. For each matrix of Exercise 7.11.4, find all values of $s$ that maximize $\min_{t \in T} \pi(s,t)$ and all values of $t$ that minimize $\max_{s \in t} \pi(s,t)$, where $\pi(s,t)$ denotes the entry of the matrix that lies in row $s$ and column $t$. What do your answers have to do with Exercise 7.11.6?

math

8. Explain why all $m \times 1$ and $1 \times n$ matrices necessarily have a saddle point.

math

9. Explain why the open interval $(1,2)$ consisting of all real numbers $x$ that satisfy $1 < x < 2$ has no maximum and no minimum element. What are the supremum and infimum of this set?

math

10. Let $M$ be player I's payoff matrix in a game. Show that, if $M$ is $A$ or $D$ in Exercise 7.11.4, then player I has a pure security strategy. Find his security level in each case and all his pure security strategies. Decide in each case what player II should do in order to guarantee that player I gets no more than his security level.
11. Repeat Exercise 7.11.10 but with the roles of player I and player II reversed. (You may or may not find Exercise 7.11.5 helpful.)
12. Section 7.4.2 shows that $\underline{m} = p_1(\delta) = 1 - p_2(\delta)$. Employ a similar methodology to show also that $\overline{m} = p_1(\delta) = 1 - p_2(\delta)$, where

math

$$\overline{m} = \min_e \sup_d \pi(d,e).$$

Why does this confirm that firing at distance $\delta$ is a security strategy for Tweedledee?

13. Player I's payoff matrix in a game is

$$\begin{bmatrix} 1 & 2 & 3 & 4 & 5 \\ 9 & 7 & 5 & 3 & 1 \end{bmatrix}.$$

The matrix has no saddle point, and hence player I's security strategies are mixed. Find player I's security level in the game and a mixed security strategy for player I.

14. Why is any mixed strategy a security strategy for player I if his payoff matrix is $D$ in Exercise 7.11.4? What is player I's security level?

15. Explain why the use of the mixed strategy $p = (\frac{1}{3}, \frac{1}{3}, \frac{1}{3})^\top$ by player I guarantees him an expected utility of at least 3 if his payoff matrix is $C$ in Exercise 7.11.4. Show that the use of player II's fourth pure strategy guarantees that player I gets at most 3. What is player I's security level? What is a security strategy for player I?

16. Find player I's security strategies when his payoff matrix is $B$ in Exercise 7.11.4.

math

17. Let $p = (1-x, x)^\top$ and $q = (1-y, y)^\top$, where $0 \le x \le 1$ and $0 \le y \le 1$. If player I's payoff matrix is $B$ in Exercise 7.11.4, show that his expected utility if he uses mixed strategy $p$ and player II uses mixed strategy $q$ is

$$\Pi_1(p,q) = f(x,y) = 1 + 3x + 2y - 4yx.$$

Find the values of $(x, y)$ for which $\partial f / \partial x = \partial f / \partial y = 0$. Explain why these are saddle points of the function $f : [0,1] \times [0,1] \to \mathbb{R}$. Relate this conclusion to your answer for Exercise 7.11.16.

18. Players always get their maximin values or more when they play a Nash equilibrium (Section 7.4.6). By Von Neumann's theorem, they also get their minimax values or more. If they play a *pure* Nash equilibrium, show that they get at least their minimax values in pure strategies.

19. Use the method of Section 7.4.6 to show that the players get only their security levels by playing the mixed equilibrium in the game of Figure 6.15(b). Why are their equilibrium strategies not secure?

20. Adam and Eve simultaneously announce whether or not they will bet on the outcome of an election in which only a Republican and a Democrat are running. If they both bet, Adam pays Eve \$10 if the Republican wins, and Eve pays Adam \$10 if the Democrat wins. Otherwise neither pays anyone anything.

    a. If both are risk neutral and attach the same probability to the event that the Republican will win, explain why the game is zero-sum.

    b. If both are risk neutral but Adam thinks the Democrat will win with probability $\frac{5}{8}$ and Eve believes the Republican will win with probability $\frac{3}{4}$, explain why the game isn't zero sum.

    c. If both attach the same probability to the event that the Republican will win and both are strictly risk averse, explain why the game isn't zero sum.

math

21. Player I's payoff matrix in a zero-sum game is $A$. Why would he be equally happy to be player II in a zero-sum game with payoff matrix $-A^\top$? A matrix $A$ is skew-symmetric if $A = -A^\top$. Why does a symmetric matrix game have a

skew-symmetric payoff matrix? Show that the value of such a game is necessarily zero.

22. Find the values of the zero-sum games that have the following payoff matrices using the method of Section 7.4.3. Confirm that the method of Section 7.7.4 yields the same answers.

$$\text{(a)} \quad \begin{bmatrix} 9 & -5 & 7 & 1 & -3 \\ -10 & 4 & -8 & -6 & 2 \end{bmatrix} \qquad \text{(b)} \quad \begin{bmatrix} 1 & 2 & 3 & 4 & 5 \\ 5 & 4 & 3 & 2 & 1 \end{bmatrix}.$$

Find all security strategies for both players. What are the Nash equilibria for these games?

23. Find the values and all security strategies of the following matrix games using the method of Section 7.4.3.

$$\text{(a)} \quad \begin{bmatrix} 1 & 0 & 2 \\ 3 & 1 & 1 \end{bmatrix} \qquad \text{(b)} \quad \begin{bmatrix} 0 & 1 & 3 \\ 3 & 1 & 0 \end{bmatrix} \qquad \text{(c)} \quad \begin{bmatrix} -2 & 0 \\ -2 & 1 \\ -4 & -3 \end{bmatrix}$$

24. Find the value and at least one security strategy for each player in each of the following matrix games:

$$\text{(a)} \quad \begin{bmatrix} 7 & 2 & 1 & 2 & 7 \\ 2 & 6 & 2 & 6 & 2 \\ 5 & 4 & 3 & 4 & 5 \\ 2 & 6 & 2 & 6 & 2 \\ 7 & 2 & 1 & 2 & 7 \end{bmatrix} \qquad \text{(b)} \quad \begin{bmatrix} 1 & 3 & 2 & 5 \\ 0 & -1 & 6 & 7 \\ 3 & 4 & 2 & 3 \\ -7 & 2 & 2 & 1 \end{bmatrix}$$

25. A $2 \times 2$ matrix $A$ has no saddle point. If $A$ is player I's payoff matrix in a zero-sum game, show that:
    a. A player who uses a security strategy will get the same payoff whatever the opponent does.
    b. A player will get the same payoff whatever he or she does, provided the opponent uses a security strategy.

math

26. A $2 \times 2$ matrix $A$ has no saddle point. If $A$ is player I's payoff matrix in a zero-sum game, show that the value of the game is given by $v = \{e^{\top} A^{-1} e\}^{-1}$, where $e = (1, 1)^{\top}$.

math

27. Alice's input-output matrix in Section 7.1.1 is

econ

$$A = \begin{bmatrix} 1 & 3 \\ 4 & 2 \end{bmatrix}.$$

Her stock of raw materials is $b = (3, 2)^{\top}$. The prices at which she can sell the finished goods are given by $c = (1, 1)^{\top}$. What are the shadow prices for her raw materials?

28. Suppose that the dual problem of Figure 7.10 has a *unique* solution $\tilde{y}$. Explain geometrically why a small change in $b$ will leave $\tilde{y}$ unchanged. The Alice of Section 7.1.1 can buy small amounts of her raw materials at prices specified by the vector $p$. When is this a good idea?

econ

29. Find the values of the following matrix games by exploiting any symmetries you can find.

$$\text{(a)} \begin{bmatrix} 1 & 2 & 3 \\ 3 & 1 & 2 \\ 2 & 3 & 1 \end{bmatrix} \quad \text{(b)} \begin{bmatrix} 1 & 2 & 3 & 0 \\ 3 & 1 & 2 & 0 \\ 2 & 3 & 1 & 0 \\ 0 & 0 & 0 & 1 \end{bmatrix} \quad \text{(c)} \begin{bmatrix} 1 & 2 & 4 & 1 \\ 2 & 1 & 1 & 4 \\ 3 & 1 & 1 & 0 \\ 1 & 3 & 0 & 1 \end{bmatrix}$$

fun

30. Colonel Blotto has four companies that he can distribute among two locations in three different ways: (3, 1), (2, 2) or (1, 3).[12] His opponent, Count Baloney, has three companies that he can distribute among the same two locations in two different ways: (2, 1) or (1, 2). Suppose that Blotto sends $m_1$ companies to location 1 and Baloney sends $n_1$ companies to location 1. If $m_1 = n_1$, the result is a standoff, and each commander gets a payoff of zero for location 1. If $m_1 \neq n_1$, the larger force overwhelms the smaller force without loss to itself. If $m_1 > n_1$, Blotto gets a payoff $n_1$, and Baloney gets a payoff of $-n_1$ for location 1. If $m_1 < n_1$, Blotto gets a payoff $-m_1$, and Baloney gets a payoff of $m_1$ for location 1. Each player's total payoff is the sum of his payoffs at both locations.

    Find the strategic form of this simultaneous-move game. Show that it has no saddle point. Determine a mixed-strategy Nash equilibrium.

fun

31. Repeat the previous exercise for the case when Blotto has five companies and Baloney has four companies. (You may want to use the trick from Section 7.8 by means of which Figure 7.17(a) was reduced to Figure 7.17(b).)

fun

32. Analyze the game of Hide-and-Seek from Section 7.8 on the assumption that Mr. Spock was able to steal only three atomic missiles from Starfleet Command. His aim is to destroy the starship before his missiles are exhausted. Captain Kirk's aim is to survive the bombardment.

math

33. The Inspection Game of Section 2.2.1 becomes zero sum if the players get a payoff of +1 when they win and –1 when they lose. Explain why the value $v_n$ of the $n$-day version of this zero-sum game is also the value of the matrix game of Figure 7.19(a) when $n > 1$. Hence show that

$$v_n = \frac{1 + v_{n-1}}{3 - v_{n-1}}.$$

    Solve this difference equation with the boundary condition $v_1 = -1$, and hence show that $v_n = 1 - 2/n$. (The substitution $v_n = 1 - w_n^{-1}$ will ease your task.) Check the answer against your solution of the five-day version of the Inspection Game given in Exercise 2.12.22.

math

34. The $n$-day Inspection Game of the previous problem is modified so that the agency may inspect on *two* days, freely chosen from the $n$ days on which the river might be polluted. The firm still chooses just *one* of the $n$ days on which to pollute the river. If the value of this game is $u_n$, show that, for $n \geq 3$,

$$u_n = \frac{u_{n-1} + v_{n-1}}{2 - u_{n-1} + v_{n-1}},$$

[12] This isn't the Colonel Blotto we met in Exercise 5.9.11.

|  | act | wait |
|---|---|---|
| act | $-1$ | $1$ |
| wait | $1$ | $v_{n-1}$ |

Figure 7.19 The $n$-day Inspection Game.

where $v_k = 1 - 2/k$. Find $u_4$, and determine the probability with which the agency should inspect on the first day when $n = 4$.

35. Colonel Blotto has to match wits with Count Baloney in yet another military situation. This time Blotto commands two companies, and Baloney commands only one. Each tries to succeed in capturing the enemy camp without losing his own. Every day, each commander sends however many companies he chooses to attack the enemy camp. If the defenders of a camp are outnumbered by the attackers, then the camp is captured. Otherwise the result is a standoff. This continues for a period of $n$ days unless someone is victorious in the interim. Anything short of total victory counts for nothing. Each army then abandons any gain it may have made and retreats to its own camp until the next day.

    Counting a defeat as $-1$, a victory as $+1$, and a standoff as 0, determine optimal strategies for the two players, and compute Blotto's expected payoff if the optimal strategies are used.

fun

36. Odd-Man-Out is a three-player, zero-sum game. Three risk-neutral players simultaneously choose heads or tails. If all choose the same, no money changes hands. If one player chooses differently from the others, he must pay the others one dollar each. What is a security strategy for a player in this game? Find a Nash equilibrium in which no player uses his security strategy. Why does the existence of such a Nash equilibrium contrast with the situation in the two-player case?
37. Use a computer to solve these matrix games by linear programming:

$$A = \begin{bmatrix} 0 & 5 & -2 \\ -3 & 0 & 4 \\ 6 & -4 & 0 \end{bmatrix}; \qquad B = \begin{bmatrix} 4 & 3 & 1 & 4 \\ 2 & 5 & 6 & 3 \\ 1 & 0 & 7 & 0 \end{bmatrix}.$$

# 8 *Keeping Your Balance*

## 8.1 Introduction

Libra is the sign of the zodiac that represents the scales used in classical times for weighing things. So *equilibrium* means something like "equally balanced." For example, in a Nash equilibrium the players' strategy choices are "in balance" because neither would wish to deviate after learning the other's choice.

This chapter explores the idea of a Nash equilibrium in depth. The chapter isn't about how to do computations, but the concepts discussed require quite a lot of mathematics. Readers who don't care why the theorems are true may therefore prefer to skip through the chapter quickly.

Nash equilibria occur where the players' reaction curves cross. But what happens if they don't cross? Nash showed that this problem can't arise in a finite game in which mixed strategies are allowed. His proof ultimately depends on Brouwer's important fixed-point theorem. It is therefore pleasing that Brouwer's theorem can be deduced from the fact that Hex can't end in a draw.

What if the reaction curves of a game cross several times, so that the game has multiple Nash equilibria? Game theorists are still struggling with the problem of determining principles to govern the selection of one of these equilibria as the solution of the game. This chapter begins the study of this equilibrium selection problem by reviewing some of the difficulties.

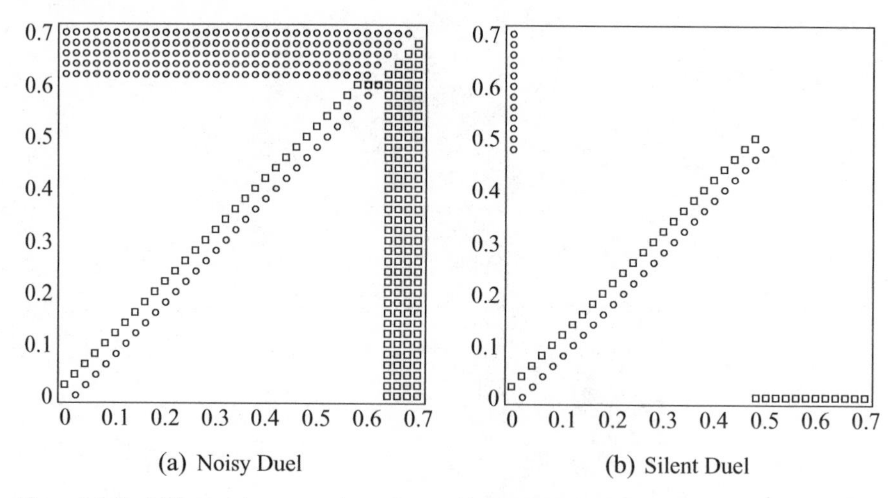


(a) Noisy Duel (b) Silent Duel

Figure 8.1 Duel. The reaction curves for Noisy Duel cross twice in Figure 8.1(a), and so the game has two Nash equilibria in pure strategies. The reaction curves for Silent Duel shown in Figure 8.1(b) don't cross at all, and so the game has no Nash equilibria in pure strategies.

## 8.2 Dueling Again

This section studies two variants of the game of Duel. In the first variant, the reaction curves cross twice. In the second, they fail to cross at all. But the chief lesson is that drawing reaction curves needn't be a trivial task.

*Noisy Duel.* Our first variant of Duel differs from earlier versions only in the details of the mathematical model used to represent it. We call it Noisy Duel to emphasize that Tweedledum and Tweedledee can hear when a shot is fired. After hearing a shot, a player knows that his opponent's pistol is empty, and he can safely walk up to point-blank range before firing himself.

The changes in the mathematical model of Duel alter the reaction curves of Figure 5.3 to those of Figure 8.1(a).[1] Tweedledum and Tweedledee still start out distance $D=1$ apart. We also continue to take $p_1(d)=1-d$ and $p_2(e)=1-e^2$. But now the players are allowed to fire whenever the distance between them is a multiple of $\varepsilon=0.02$. As in Section 7.4.2, they can therefore fire simultaneously. Tweedledum is then assumed to survive with probability $q(d)=\frac{1}{2}\{p_1(d)+1-p_2(d)\}$.

The reaction curves of Figure 8.1(a) cross at $(d,e)=(0.6,0.6)$ and $(d,e)=(0.62,0.6)$. The game therefore has two Nash equilibria in pure strategies.

The existence of multiple equilibria creates serious selection problems in some games, but the appearance of two Nash equilibria when $\varepsilon=0.02$ is an accident without significance in this example. All that really matters in Noisy Duel is that we

[1]Confusion can arise when these two figures are compared. When describing the entries in a matrix, player I's pure strategies correspond to rows and player II's to columns. When presenting the same information using Cartesian axes, player I is assigned the horizontal axis and player II is assigned the vertical axis. Player I's pure strategies then correspond to columns, and Player II's to rows.

can make all the equilibria as close to $(d,e)=(\delta,\delta)$ as we like by taking $\varepsilon$ sufficiently small—where $\delta=(\sqrt{5}-1)/2=0.62$ is the solution of the equation $p_1(d)+p_2(d)=1$ (Section 3.7.2). For example, when $\varepsilon=0.001$, the reaction curves cross only where $(d,e)=(0.618,0.618)$.

Why don't we proceed as in Section 7.3.3 by allowing the players to fire when they are an arbitrarily small distance $d$ apart? The answer is that best replies then sometimes fail to exist. If Tweedledee plans to fire when the players are distance 0.24 apart, then Tweedledum wants to fire a little bit sooner. But if Tweedledum fires when they are distance $0.24+\varepsilon$ apart, he will always wish that $\varepsilon$ were smaller. We can't manage as in Section 7.3.3 by replacing maxima by suprema because we would then end up with a version of Figure 8.1(a) in which the reaction curves sit on top of each other.

Such problems are often handled by first making the gap between the allowed values of $d$ equal to some small $\varepsilon>0$. The limits as $\varepsilon\to 0$ of the equilibria of this discrete game are then treated as the equilibria of the continuous game. However, as in Section 3.7.2, the fact that such a two-step procedure is implicitly being used is seldom made explicit. One eventually learns to take the necessary hand waving in stride, but beginners are advised to work through the two-step procedure whenever they come across it until it ceases to be puzzling. This is one of the reasons we often use Duel as an example when seeing how new ideas work out in practice.

*Silent Duel.* In Noisy Duel, a player can hear when his opponent fires his pistol. In Silent Duel, the only way a player can learn that his opponent has fired is by getting shot.

In the case we will study, sibling rivalry has reached such a pitch that neither Tweedledum nor Tweedledee can bear the prospect of living if their brother also survives. Each therefore assigns a payoff of one to the event that he lives and his brother dies and zero to all other possibilities. The probability $\pi(d,e)$ that Tweedledum attaches to the former event is $p_1(d)$ when $d>e$, and $p_1(d)(1-p_2(e))$ when $d<e$.

Silent Duel is a game of imperfect information that isn't strictly competitive. It therefore differs from Noisy Duel in important ways. We study it here to illustrate that a game's reaction curves can fail to cross even when the strategy spaces are continuous. Unlike its noisy cousin, Silent Duel therefore has no Nash equilibrium in pure strategies.

To keep things simple, we take $D=1$ and make the game symmetric by choosing both hit probabilities to be $p_1(d)=p_2(d)=1-$ d. Tweedledum's payoff function in Silent Duel is then:

$$\pi_1(d,e)=\begin{cases}1-d, & \text{if} \quad d>e,\\ \frac{1}{2}(1-d^2), & \text{if} \quad d=e,\\ e(1-d), & \text{if} \quad d<e.\end{cases}$$

With this information, it is easy to draw the reaction curves of Figure 8.1(b). Their failure to cross is possible because they jump discontinuously from one place to another. The discontinuity isn't caused by restricting $d$ to a grid with separation $\varepsilon=0.02$. The same jump survives no matter how small $\varepsilon$ is made.

→ 8.5

## 8.3 When Do Nash Equilibria Exist?

When reaction curves in pure strategies failed to cross in Chapter 6, we looked for Nash equilibria in mixed strategies. But who says that reaction curves in mixed strategies need to cross? Fortunately, John Nash proved that this problem can't arise in a finite game.

*Nicely Behaved Correspondences.* A mixed strategy $(1-p,p)^\top$ in a $2\times 2$ bimatrix game is determined by naming a real number $p$ in the interval $I=[0,\ 1]$. In such games, we can take the players' sets of mixed strategies to be $P=Q=I$. In the version of Chicken of Figure 6.3(c), player I's payoff function is then given by

$$\Pi_1(p,q)=2+2p-2q-3pq.$$

It is therefore not only a continuous function; it is also an *affine* function of $p$ for each fixed value of $q$ (Section 6.5.1).

Affine functions are simultaneously both convex and concave. Their concavity is the reason that Nash's proof always works in finite games. More generally, his proof works whenever the players' payoff functions $\Pi_i : P\times Q \to \mathbb{R}$ satisfy the following conditions:

- Each strategy set is convex and compact.[2]
- Each payoff function is continuous.
- Each payoff function is concave when the other players' strategies are held constant.

*Kakutani's Fixed-Point-Theorem.* A long time ago, the Japanese mathematician Kakutani asked me why so many economists had attended the lecture he had just given. When I told him that he was famous because of the Kakutani fixed-point theorem, he replied, "What is the Kakutani fixed-point theorem?" I hope I explain his theorem better now than I did then!

We need the conditions on the game listed above to ensure that its best-reply correspondences are nicely behaved. A correspondence $R:X\to Y$ is nicely behaved in the sense that will be needed if it satisfies the following properties when $X$ and $Y$ are convex, compact sets:

- For each $x\in X$, the set $R(x)$ is nonempty and convex.
- The graph of $R:X\to Y$ is a closed subset of $X\times Y$.

Figure 8.2(a) shows the graph $G$ of a nicely behaved correspondence $R:X\to Y$ when both $X$ and $Y$ are compact intervals.

Figure 8.2(b) shows a nicely behaved correspondence $F:X\to X$ that maps $X$ back into itself. Kakutani's fixed-point theorem says that such correspondences always have at least one *fixed point*. This is a point $\tilde{x}$ for which

[2]To be compact, a set in $\mathbb{R}^n$ must be both closed and bounded. To be closed, it must contain all its boundary points. Thus, the compact interval $[0,1]$ is closed because it contains both its boundary points 0 and 1. The interval $(0,1)$ is open because it contains neither of its boundary points 0 and 1.

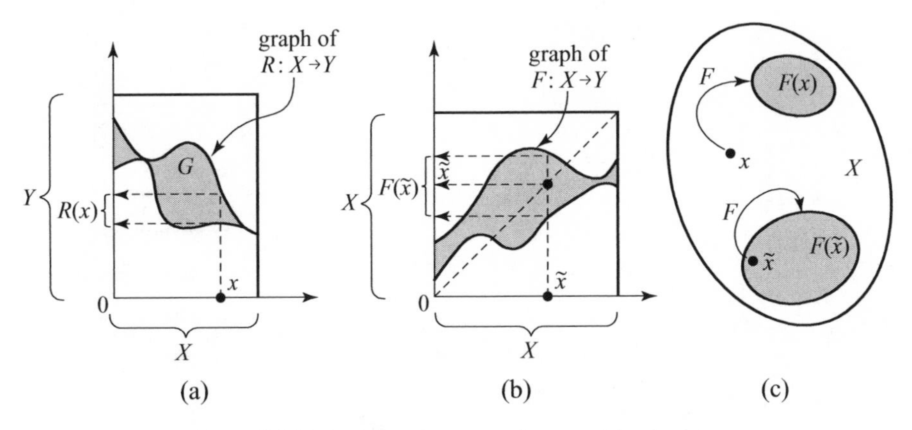


Figure 8.2 Nicely behaved correspondences and fixed points.

$$\tilde{x} \in F(\tilde{x}).$$

As Figure 8.2(b) shows, Kakutani's theorem is trivial when $X$ is a compact interval, but it isn't at all obvious for the case of an arbitrary, nonempty, convex, compact set $X$ like that shown in Figure 8.2(c). However, we leave this subject for the moment while we use the theorem to prove Nash's theorem.

THEOREM 8.1 (Nash) *Every finite game has at least one Nash equilibrium when mixed strategies are allowed.*

*Proof* The steps in the proof are sketched only for the two-player case.

**Step 1.** Confirm that the players' best-reply correspondences $R_i : P \to Q$ are nicely behaved in finite games. Properties of strategy sets and payoff functions that guarantee this conclusion are listed above, but the linking algebra is omitted, even though it isn't very difficult.

**Step 2.** Construct a correspondence $F : P \times Q \to P \times Q$ to which Kakutani's fixed-point theorem can be applied. For each $(p, q)$ in $P \times Q$, define

$$F(p, q) = R_1(q) \times R_2(p)$$

(so that $F(p, q)$ is a *set* in $P \times Q$). The definition is illustrated in Figure 8.3(a) for the $2 \times 2$ bimatrix game case, when $P = Q = I$.

**Step 3.** Deduce that $F$ is nicely behaved using the fact that the same is true of $R_1$ and $R_2$. Again, the not-very-difficult algebra is omitted.

**Step 4.** Apply Kakutani's fixed-point theorem. As illustrated in Figure 8.3(b), the theorem proves the existence of a fixed point $(\tilde{p}, \tilde{q})$ satisfying

$$(\tilde{p}, \tilde{q}) \in F(\tilde{p}, \tilde{q}) = R_1(\tilde{q}) \times R_2(\tilde{p}).$$

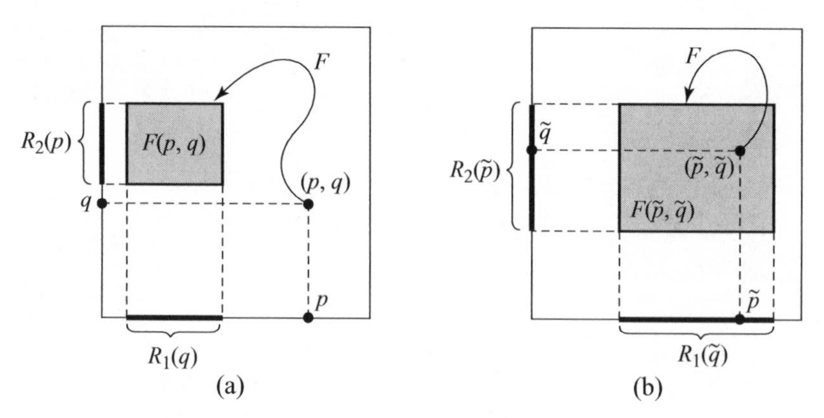


Figure 8.3 The correspondence $F$ in Nash's theorem.

**Step 5.** Notice that $(\tilde{p}, \tilde{q})$ is a Nash equilibrium. The mixed strategy $\tilde{p}$ is a best reply to $\tilde{q}$ because $\tilde{p} \in R_1(\tilde{q})$. The mixed strategy $\tilde{q}$ is a best reply to $\tilde{p}$ because $\tilde{q} \in R_2(\tilde{p})$ (Section 6.2.1). □

### 8.3.1 Symmetric Games

Most of the games we have studied have been symmetric (Section 5.3.1). The Prisoners' Dilemma and Chicken are typical examples. Such games look the same to both players.

In a symmetric equilibrium of a symmetric game, all the players use the same strategy. Since (*dove*, *hawk*) and (*hawk*, *dove*) are Nash equilibria of Chicken, finite symmetric games can certainly have asymmetric equilibria, but the next theorem says that they always have symmetric equilibria as well.

THEOREM 8.2 *Every symmetric finite game has at least one symmetric Nash equilibrium when mixed strategies are allowed.*

*Proof* This proof for the two-player case uses the the fact that $R_1 = R_2 = R$ in a symmetric game. Replace $R_1(q)$ by $R(q)$ and $R_2(p)$ by $\{p\}$ in the proof of Nash's theorem. The fixed point $(\tilde{p}, \tilde{q})$ then satisfies $\tilde{p} \in R(\tilde{q})$ and $\tilde{q} = \tilde{p}$.

Since $\tilde{p} \in R(\tilde{p})$, the mixed strategy $\tilde{p}$ is a best reply to itself, and so $(\tilde{p}, \tilde{p})$ is a symmetric Nash equilibrium of the game.

## 8.4 Hexing Brouwer

→ 8.5

Fixed-point theorems are particularly important for economists because of their need to locate the equilibria of economic systems. Our proof of Nash's theorem illustrates the standard method by means of which fixed-point theorems are used to demonstrate the existence of such equilibria.

Brouwer's fixed-point theorem is the big daddy of the family of fixed-point theorems. Von Neumann used Brouwer's theorem in his original proof of the minimax theorem.[3] Kakutani told me that it was while listening to Von Neumann describing this proof that he thought of his own fixed-point theorem (which can be proved by taking $f(x)$ in Brouwer's theorem to be the center of gravity of the convex set $F(x)$ in Kakutani's theorem).

THEOREM 8.3 (Brouwer) *Suppose that $X$ is a nonempty, compact, convex set in $\mathbb{R}^n$. If the function $f: X \to X$ is continuous, then a fixed point $\tilde{x}$ exists satisfying $\tilde{x} = f(\tilde{x})$.*

David Gale has shown that Brouwer's theorem follows from the fact that Hex can't end in a draw. His argument is a curiosity from the mathematical point of view, but it is too much fun to pass over, in a book on game theory, especially since the version of Hex to be used was invented by Nash. But first we need to learn a little about continuity and compactness.

### 8.4.1 Continuity

We will now be talking about *functions* rather than correspondences, as in the previous section. A function $f: X \to Y$ assigns a unique element $y = f(x)$ in the set $Y$ to each $x$ in the set $X$. A function differs from a correspondence in that $f(x)$ is an element of $Y$ rather than a subset of $Y$. In what follows, $X$ and $Y$ will be subsets of $\mathbb{R}^n$ and $\mathbb{R}^m$ respectively.

As with all important mathematical ideas, the language an author chooses to use in discussing a function depends on the use to which the concept is to be put. In our context, it is perhaps most useful to regard a function as a process that somehow changes $x$ into $f(x)$. This way of thinking is often signaled by calling a function an *operator*, a *transformation*, or a *mapping*.

For example, the continuous function $f: X \to X$ in Brouwer's theorem can be envisaged as a stirring of a tank of water. The stirring will shift a droplet located at point $x$ in the tank to a new location $f(x)$. However the water is stirred, Brouwer's theorem says that at least one droplet will always be returned to its initial location. This metaphor helps explain why $X$ is taken to be convex in Brouwer's theorem. For example, if $X$ were a car's inner tube, we could fill it with water, which could then be rotated a few degrees without any droplet returning its starting point.

To say that a function $f: X \to Y$ is *continuous* means that $f(x_k) \to f(x)$ as $k \to \infty$ whenever $x_k \to x$ as $k \to \infty$.[4] If water is shifted around by a continuous process, sets of droplets that are neighbors at the beginning will still be neighbors at the end. Discontinuities like those created by Moses when he parted the waters of the Red Sea are therefore forbidden.

Our definition of continuity focuses on a point $x$ that is assumed to be a neighbor of the set $S = \{x_1, x_2, \ldots\}$. After the water has been stirred, the requirement for continuity can then be interpreted as saying that the droplet of water that started at $x$ should still be a neighbor of the set of droplets of water that were initially located in $S$. Figure 8.4(a) provides a schematic representation of the idea.

[3]Which perhaps explains von Neumann's dismissive remark when Nash showed him his theorem: "Oh yes, a fixed-point argument."

[4]To say that $y_k \to y$ as $k \to \infty$ means that we can make the distance $\|y_k - y\|$ between $y_k$ and $y$ as small as we like by taking $k$ to be sufficiently large.

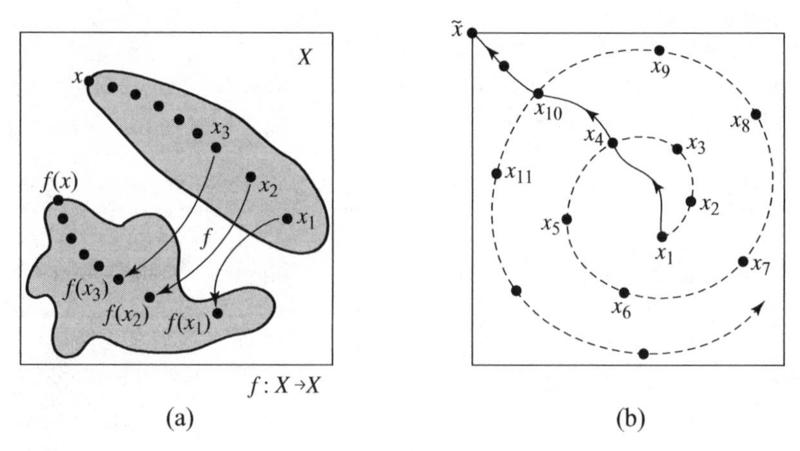


Figure 8.4 Continuity and compactness.

## 8.4.2 Compactness

A compact set in $\mathbb{R}^n$ is closed and bounded. Compact sets are important because any sequence of points chosen from such a set necessarily has a *convergent* subsequence.[5] It isn't easy to appreciate why this property matters until one has seen it being used repeatedly in the proofs of important theorems.

For example, when proving Brouwer's theorem, we will show that, for each natural number $k$, a vector $x_k$ in the compact set $X$ can be found that satisfies

$$\|x_k - f(x_k)\| < \frac{1}{k}. \tag{8.1}$$

We then deduce the existence of a fixed point $\tilde{x}$ satisfying $\tilde{x} = f(\tilde{x})$. How do we use the continuity of the function $f : X \to X$ to get to this conclusion?

The function $g : X \to \mathbb{R}$ defined by $g(x) = \|x - f(x)\|$ is continuous when the same is true of $f$. So if $x_k \to \tilde{x}$ as $k \to \infty$, then $g(x_k) \to g(\tilde{x})$ as $k \to \infty$. But (8.1) implies that $g(x_k) \to 0$ as $k \to \infty$. Thus $g(\tilde{x}) = 0$, as required.

The problem with this argument is that nothing guarantees that the sequence $x_1, x_2, x_3, \ldots$ converges to anything at all. If $X$ weren't compact, this might be an insuperable obstacle, but all we need do when $X$ is compact is to throw away the original sequence and replace it by a convergent subsequence. In the case illustrated in Figure 8.4(b), the convergent subsequence consists of the terms $x_1, x_4, x_{10}, x_{17} \ldots$.

## 8.4.3 Proof of Brouwer's Theorem

This outline of a proof will be confined to the two-dimensional case in which $X$ is the unit square $I^2 = [0, 1] \times [0, 1]$. The extension to the general case isn't difficult, but the details aren't sufficiently interesting to be worth describing.

[5]This nontrivial theorem is attributed to the mathematicians Bolzano and Weierstrass.

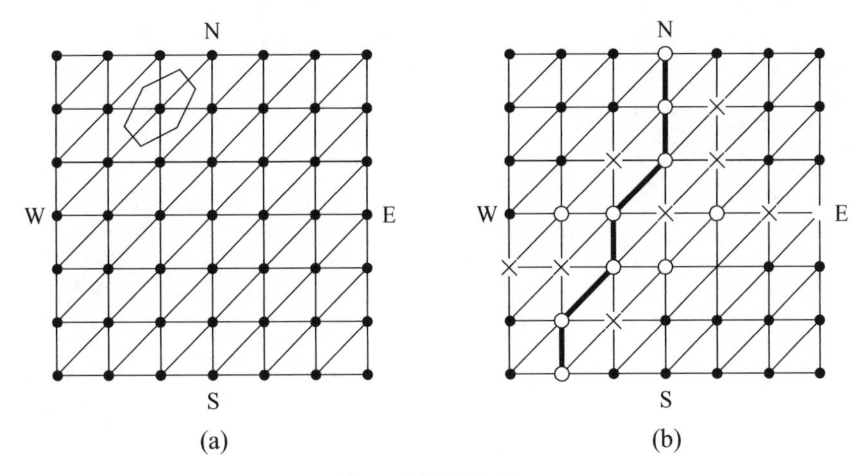


Figure 8.5 Nash's Hex.

Nash's version of Hex is described in Exercise 2.12.13. The board is reproduced in Figure 8.5(a). The hexagon superimposed on the board clarifies why Nash's Hex is equivalent to the conventional version of Section 2.7.1.

The board of Figure 8.5(b) shows a winning configuration for Circle in Nash's Hex. All the nodes on a route linking N and S are labeled with circles. Cross would have won if all the nodes on a route linking W and E were labeled with crosses. Since the game is equivalent to regular Hex, it can't end in a draw. In fact, if all the nodes on the board are labeled with either a circle or a cross, then either Circle or Cross must have won.[6]

**Step 1.** Choose some $d > 0$. Take $O_S$ to be the set of all $x$ in $I^2$ that $f$ shifts a distance of more than $d$ toward the south. Take $X_W$ to be the set of all $x$ in $I^2$ that $f$ shifts a distance of more than $d$ toward the west. Define the sets $O_N$ and $X_E$ in a similar way. Figure 8.6(a) shows what these sets might look like. The unshaded set $S$ in the diagram is the set of all $x$ in $I^2$ that belong to *none* of the four sets $O_N$, $O_S$, $X_E$, or $X_W$.

**Step 2.** If $S$ isn't empty, then we can find at least one $x$ in $I^2$ that is "nearly" fixed because its image $f(x)$ lies in a square of side $2d$ centered at $x$. If such an approximate fixed point always exists no matter how small we take $d$, then we can always find an $x_k$ that satisfies (8.1). But we have seen that the compactness of $X$ and the continuity of $f$ then imply the existence of an *exact* fixed point $\tilde{x}$.

**Step 3.** We must now show that $S$ is never empty. We proceed by assuming that $S$ is empty for some $d > 0$, and seeking a contradiction from the fact that each $x$ in $I^2$ then lies in one of the two sets $O = O_N \cup O_S$ or $X = X_E \cup X_W$.

**Step 4.** Cover $I^2$ with a Hex grid of tiny mesh, as shown in Figure 8.6(b). Label each node on this grid with a circle or a cross depending on whether it lies in $O$ or $X$. (If it lies in both sets, label it at random.) One of the players must have won the Hex position created in this way (Section 2.7.1). Suppose that the winner is Cross.

[6]The fact that *both* players can't win can be used to prove the Jordan curve theorem!

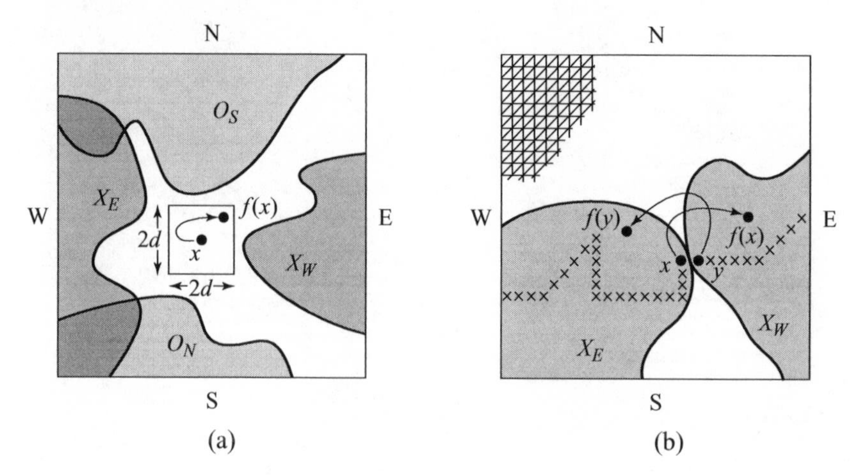


Figure 8.6 Proving Brouwer's theorem.

**Step 5.** The most westerly node on Cross's winning route must lie in $X_E$. The most easterly node must lie in $X_W$. Somewhere in between the route must pass from $X_E$ to $X_W$. Where this happens, we will find a pair of adjacent nodes, $x$ and $y$, one of which lies in $X_W$ and the other in $X_E$.

**Step 6.** The function $f$ shifts the point $x$ more than $d$ to the west and simultaneously shifts the adjacent point $y$ more than $d$ to the east. Since the distance between $x$ and $y$ can be made as small as we please by taking the mesh of the Hex grid sufficiently tiny, this implies the contradiction that the continuous function $f$ has a discontinuity.[7]

## 8.5 The Equilibrium Selection Problem

The equilibrium selection problem is perhaps the greatest challenge facing modern game theory. As soon as one goes beyond the toy models of this book to games that begin to capture the richness of real life, one is deluged with vast numbers of Nash equilibria. Which of these should be selected?

### 8.5.1 Rational Solutions?

Can we always find one equilibrium that is somehow more rational than the others, so that we can identify it as the unequivocal solution of the game?

It is perhaps because Von Neumann and Morgenstern thought their business was to identify unambiguous rational solutions of games that formulating the idea of an

[7]We have shown that, for each sufficiently large natural number $k$, $x_k$ and $y_k$ can be found so that $\|x_k - y_k\| < 1/k$ but $\|f(x_k) - f(y_k)\| \geq d$. If $x_k \to \xi$ as $k \to \infty$, then it follows that $y_k \to \xi$ as $k \to \infty$. Also, since $f$ is continuous, $f(x_k) \to f(\xi)$ as $k \to \infty$, and $f(y_k) \to f(\xi)$ as $k \to \infty$. But this implies that $0 = \|f(\xi) - f(\xi)\| \geq d$, which is a contradiction. But what if the sequence $x_1, x_2, x_3, \ldots$ doesn't converge? The compactness of $X$ then comes to the rescue since we can always pass to a *subsequence* that does converge.

equilibrium was left to Nash. Von Neuman and Morgenstern would probably have denied that the best-reply criterion should be taken to be fundamental in defining the rational solution of a noncooperative game. They would have said, on the contrary, that the best-reply criterion should follow from an independent definition of a rational solution, as it does in the case of two-person, zero-sum games.

John Harsanyi explicitly argued that rational players in the same situation will necessarily make the same decisions. Nowadays, the claim is jokingly referred to as the Harsanyi doctrine, but the joke wouldn't be thought amusing if game theorists hadn't lost faith in the idea that there must be a uniquely rational way of solving games. It only looks that way in two-person, zero-sum games because all their Nash equilibria are equivalent and interchangeable (Theorem 7.10). It then doesn't matter which of the Nash equilibria of a game the players regard as its solution, and so the equilibrium selection problem evaporates.

*Collective Rationality?* If there is no uniquely correct way to write the great book of game theory, then what is the source of its authority? It is sometimes argued that we should conceive of the book as being the product of a hypothetical rational agreement among the citizens of a society. The notion of collective rationality can then be rescued from ignominy and recycled as a possible approach to the problem of equilibrium selection (Section 1.7).

In the new story, everybody knows that only self-policing agreements are viable, and so only equilibria are available for selection (Section 6.6.2). But not all equilibria are equally acceptable. For example, perhaps we can agree not to use an equilibrium if there is a second equilibrium that makes everybody better off. The inferior equilibrium is then said to be *Pareto dominated.*

Figure 8.7(a) illustrates Pareto domination using a version of the Stag Hunt Game (Section 1.9). The Nash equilibrium (*dove*, *dove*) Pareto dominates the Nash equilibrium (*hawk*, *hawk*) because both players get larger payoffs at the first equilibrium.

But what of the Battle of the Sexes reproduced in Figure 8.7(c)? The mixed equilibrium is Pareto dominated by both the pure equilibria (Section 6.6.2). But any argument that favors selecting one of the pure equilibria is an equally good argument for selecting the other. If we can't jointly toss a coin to decide between the pure equilibria, aren't we then stuck with the mixed equilibrium? (Exercise 8.8.9) It isn't even always clear what to do when there is a unique Pareto-dominant equilibrium since this equilibrium may be weakly dominated in the strategic sense (Section 5.4.5).

### 8.5.2 Evolutionary Equilibrium Selection

The knotty philosophical problems that arise when equilibria are interpreted as the end product of the thinking processes of rational players disappear when we turn to the evolutionary interpretation. If equilibria are selected by the inexorable forces of biological or social evolution, we know in principle how to solve the equilibrium selection problem. Just model the dynamics of the relevant evolutionary process, and see where it goes!

However, the kind of questions we would like answered remain intractable. Will evolution always pick out one specific equilibrium in preference to the others if given long enough? Or are the equilibria that we find ourselves playing just a function of

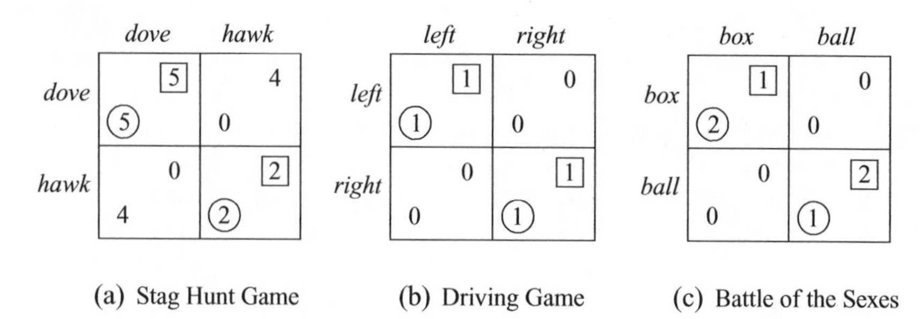


(a) Stag Hunt Game (b) Driving Game (c) Battle of the Sexes

Figure 8.7 Equilibrium selection problems.

the accidents of our evolutionary history? If the latter, which accidents were significant, and which made no difference in the long run?

We can't answer such questions most of the time because the practical problems of modeling social and biological evolutionary processes are way beyond our capacity to solve. In fact, just as we wouldn't need to worry much about equilibria if we knew the "rational solution" of every game, we also wouldn't need to emphasize the role of equilibria in characterizing the long-run behavior of evolutionary processes if we could model the dynamics of such processes adequately. Ending up at an equilibrium is just one of the possibilities for an arbitrary dynamic process.

*Risk Dominance.* Nothing guarantees that we will like the answer when the equilibrium selection problem is solved by evolution. The biologist Sewell-Wright used the landscape metaphor to make this point.[8] Think of evolution as a ball rolling down a valley to an equilibrium at the bottom. This equilibrium may give everybody a low payoff, but how are we to get out of the valley once we are trapped inside?

The Stag Hunt Game of Figure 8.7(a) epitomizes the problem. Imagine an evolutionary game in which pairs of animals are chosen at random from a single population to play the Stag Hunt Game. The points on the line in Figure 8.8 represent all the possible population states. In this simple case, a state is just the proportion $p$ of the population that are currently playing *hawk*.

The three Nash equilibria of the game correspond to the polymorphic equilibria $p=0$, $p=\frac{1}{3}$, and $p=1$ (Section 6.2.3). The arrows show the direction in which evolution will move if animals that play whatever is currently optimal gradually replace those that don't. The mixed equilibrium is unstable, but we might end up at either of the pure equilibria.

The immediate point is that the Pareto-dominant equilibrium (*dove*, *dove*) has the smaller basin of attraction. We are therefore more likely to get trapped in the basin of attraction of the Pareto-dominated equilibrium (*hawk*, *hawk*).

As we saw long ago in Section 1.9, this problem is reflected in its being riskier to play *dove* than *hawk* when there is doubt about which equilibrium should be selected. For this reason, the Nash equilibrium with the larger basin of attraction in such cases is said to be *risk dominant*.

[8]The landscape metaphor is dangerous in game theory because the landscape can be like an Escher picture, in which you keep climbing down but end up higher than you started!

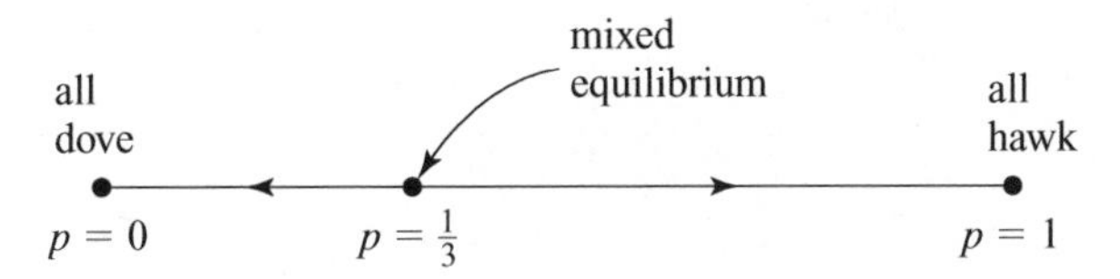


Figure 8.8 Basins of attraction in the Stag Hunt Game.

## 8.6 Conventions

David Hume was the first to draw attention to the importance of evolutive processes in selecting equilibria in the games of everyday life. For example, the words in this book have meaning only by convention. Money is valuable only because it is conventional to regard it as valuable. The house in which I live and the car that I drive are mine only because it is conventional to regard certain exchanges of paper as signifying ownership.

### 8.6.1 Group Selection

The bundle of all the conventions that operate in a society might be thought of as representing its social contract—its collective choice of which equilibrium to follow in the game of life its citizens play.

→ 8.6.2

But does it make sense to speak of collective choice? Game theorists go bananas when told that collective rationality will ensure cooperation in the one-shot Prisoners' Dilemma. Biologists are even less tolerant of the equivalent claim that mutations will be favored that benefit the species rather than the mutated gene (Section 1.7).

Just as collective rationality ceases to be stupid when discussing equilibrium selection in games, so group selection ceases to conflict with the selfish gene paradigm when *equilibria* are competing for survival (Section 1.6.1).

The scope for selection among the social contracts of the small human societies of prehistory was especially great. To see how such selection would work, imagine that everybody in Lilliput plays *dove* in a multiplayer Stag Hunt Game, so that the fitness of each citizen is high. If everybody in Blefuscu plays *hawk*, the fitness of each citizen is low. The population of Lilliput will therefore grow faster than that of Blefuscu. If excess population emigrates to found new colonies that preserve the social contract of the parent society, we can then deploy the standard evolutionary argument to the populations of villages operating the two competing social contracts. Where such group selection arguments apply, it would be surprising to see a Pareto-dominated social contract survive.

Of course, the argument won't work for social contracts that aren't *equilibria* in the game of life, but the selfish gene paradigm tells us that such social contracts aren't stable anyway.

### 8.6.2 Focal Points

Buridan's ass is famous for dying of starvation because it could find no rational reason for preferring one bale of hay to another. The Driving Game of Figure 8.7(b) exemplifies the games of pure coordination in which this problem can't be avoided.

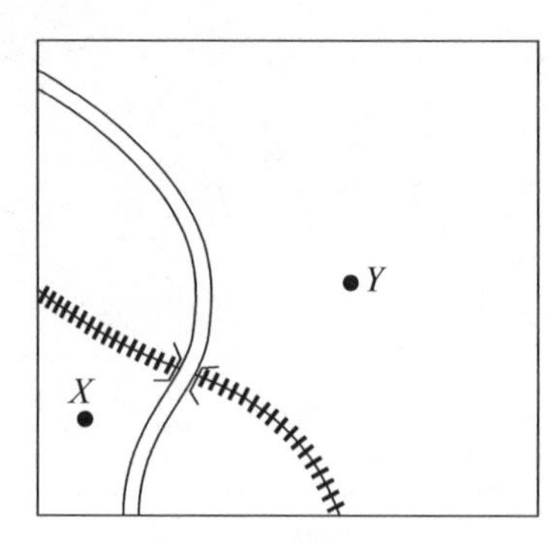


Figure 8.9 Looking for focal points.

There is no reason why either of the equilibria (*left*, *left*) and (*right*, *right*) should be preferred to the other, but social evolution has made it conventional to use the first equilibrium in Britain and the second in France. But conventions aren't always the product of historical accidents. For example, Sweden deliberately switched from driving on the left to driving on the right on 1 September 1967.

Thomas Schelling refers to the mundane conventions that we use to solve such coordination problems in everyday life as *focal points.*[9] In the Driving Game, nobody cares which convention we use, but things are more difficult in a game of impure coordination like the Battle of the Sexes, in which different players would like different equilibria to be focal points. But Schelling pointed out that we are nevertheless rather good at identifying focal points when faced with a new coordination game.

To illustrate this point, we repeat some of Schelling's examples in a slightly doctored form. In each case, ask yourself what choice you would make if you were playing the game. Most people are surprised both at their success in locating focal points and at the arbitrary nature of the contextual cues to which they appeal. An important lesson is that the context in which games appear—the way a game is *framed*—can make a big difference to how real people play them.

1. Two players independently call *heads* or *tails.* They win nothing unless both say the same, in which case each wins $100. What would you call?
2. You are to meet someone in New York tomorrow, but no arrangements have been made about where or when the meeting is to take place. Where will you go? At what time?
3. You are one of a number of saboteurs unexpectedly separated when parachuted into enemy territory. Where will you go in attempting to meet up with your team? Figure 8.9 is a map of the terrain.
4. Alice, Bob, and Carol must each independently write down the letters *A, B,* and *C* in some order. They all get nothing unless they choose the same order, in which case the player whose initial is first gets $300, the player whose initial is second gets $200, and the player whose initial is third gets $100. What would you do if you were Carol?

[9]Thomas Schelling was awarded a Nobel Prize in 2005.

5. Adam and Eve are each given one of two cards. One card is blank, and the other is marked with a cross. A player can put a cross on the first card or erase the cross on the second. Nobody wins anything unless there is one and only one cross on the two cards when they are handed in. In this case, the player who hands in the card with the cross wins $200, and the player who hands in the blank card wins $100. What would you do if given the blank card?
6. Two armies are located at points $X$ and $Y$ on the map in Figure 8.9. It is common knowledge that each commander wishes to occupy as much territory as possible without provoking the conflict that would follow if both commanders attempted to occupy overlapping territories. What area would you attempt to occupy if you were the commander of the army at $X$?
7. A philanthropist donates $100 to Adam and Eve—provided they can agree on how to divide it. Each player is independently required to claim a share. If the shares sum to more than $100, nobody gets anything. Otherwise each player receives the amount that he or she claimed. How much would you claim?
8. Alice loses $100 and Bob finds it. Bob is too honest to spend the money but is unwilling to return it unless suitably rewarded. An argument ensues that is terminated by Carol, who insists that they settle the argument by using the mechanism described in the previous example. What reward would you offer to Bob if you were Alice? What reward would you offer if Bob had already refused $20? What reward would you offer if Alice and Bob had watched a television program together the previous evening on which some guru announced that the fair split in such circumstances is for Bob to get a reward of one-third of the total amount?

Most people say *heads* in Example 1 because it is conventional to say *heads* before *tails* when both are mentioned. How well people do in Example 2 depends on their familiarity with New York. Schelling asked New Englanders, who strongly favored Grand Central Station at noon. In Example 3, the bridge is strongly focal, even in Schelling's more complicated map. In Example 4, Carol usually recognizes that alphabetical order is so focal that she has to say *ABC,* although she will then get the lowest payoff of the three players. In Example 5, the status quo is focal, and most people therefore choose to do nothing. In Example 6, the road or the railway is nearly always chosen as a boundary. The road is chosen more often than the railway, presumably because the territorial split is then slightly less unequal. In Example 7, a fifty-fifty split is almost universal. Example 8 is more challenging. People usually manage to coordinate effectively only after hearing about the guru, in which case they nearly always take his advice.

## 8.7 Roundup

Nash equilibria occur where the players' reaction curves cross. Reaction curves can be complicated. Even when the space of pure strategies is continuous, the reaction curves may be discontinuous and jump over each other. When this happens, the game has no Nash equilibria, but Nash showed that this problem goes away in finite games when mixed strategies are allowed. A finite game always has at least one Nash equilibrium. If the game is symmetric, it has at least one symmetric Nash equilibrium.

Nash's theorem is proved using Kakutani's fixed-point theorem, which is deduced in turn from Brouwer's fixed-point theorem. Such fixed-point theorems are widely used in economics and elsewhere, but they usually have difficult proofs. Our use of the fact that Hex can't end in a draw to prove the Brouwer fixed-point theorem is just a piece of fun, but the accompanying discussion of *compactness* and *continuity* will be found useful in a wide variety of circumstances. (A set in $\mathbb{R}^n$ is compact if it is both closed and bounded. A function $f$ is continuous if it is always true that $x_k \to x$ as $k \to \infty$ implies that $f(x_k) \to f(x)$ as $k \to \infty$.)

When a game's reaction curves cross several times, the game has multiple Nash equilibria. One is then faced with the *equilibrium selection problem,* for which no satisfactory solution is yet known. The reason may be that there is something self-defeating in formulating our difficulties in this way. If we knew everything we need to know to solve the equilibrium selection problem, perhaps we wouldn't want equilibria to be our central concept any more.

In practice, we solve many coordination games by appealing to focal points that are determined by the context in which a game appears. For example, people drive on the left in Japan and on the right in the United States. Such conventions are usually the result of historical accidents, but not always.

## 8.8 Further Reading

*The Game of Hex and the Brouwer Fixed-Point Theorem*, by David Gale: *American Mathematical Monthly* 86 (1979), 818–827.

*Essays on Game Theory*, by John Nash: Edward Elgar, Cheltenham, UK, 1996. The fourth essay contains Nash's theorem on the existence of equilibria in finite games.

*A General Theory of Equilibrium Selection in Games*, by John Harsanyi and Reinhard Selten: MIT Press, Cambridge, MA, 1988. Two Nobel laureates find the equilibrium selection problem hard to solve.

*The Strategy of Conflict*, by Thomas Schelling: Harvard University Press, Cambridge, MA, 1960. Schelling once bravely told a large audience of game theorists that game theory had contributed nothing whatever to the theory of focal points—except perhaps the idea of a payoff table!

## 8.9 Exercises

1. For the three-player game of Exercise 6.9.3 based on the Canadian National Lottery:
   a. Find the strategic form of the game, and locate all its Nash equilibria in pure strategies.
   b. Why is the game symmetric? Explain why the pure Nash equilibria are asymmetric, and deduce that there must be at least one symmetric Nash equilibrium in mixed strategies.
   c. Are there any symmetric Nash equilibria other than that located in Exercise 6.9.3?

2. On the assumption that the gap between the allowed values of $d$ is $\varepsilon = 0.001$, draw the reaction curves for the game of Noisy Duel of Section 8.2 in the region surrounding $(0.62, 0.62)$. Confirm that the reaction curves cross at $(0.618, 0.618)$. Do they also cross elsewhere?
3. Draw an extensive form for the game of Silent Duel of Section 8.2 in the case when we allow only $d = 0$, $d = \frac{1}{2}$, or $d = 1$.
4. Repeat the analysis of Silent Duel of Section 8.2 on the assumption that Tweedledum and Tweedledee are so fond of each other that they would rather not live if their brother dies. They therefore assign a payoff of one to the event that they both survive and a payoff of zero to all events in which one of them dies.
5. Explain why a Nash equilibrium strategy never calls for a strongly dominated strategy to be used with positive probability. Give an example of a game in which a Nash equilibrium strategy is weakly dominated. Explain why every finite game has at least one Nash equilibrium in which no weakly dominated strategy is used with positive probability.[10]
6. A *completely mixed* strategy assigns positive probability to each of a player's pure strategies. If each player's payoff matrix in a bimatrix game is nonsingular, show that the game can have at most one Nash equilibrium in which both players use completely mixed strategies.

   math
7. Let $\Pi_i : P \times Q \to \mathbb{R}$ be player $i$'s payoff function in a bimatrix game in which player I's set of mixed strategies is $P$ and player II's set of mixed strategies is $Q$. Show that, for any Nash equilibrium $(\tilde{p}, \tilde{q})$,

   math

$$\max_{p \in P} \min_{q \in Q} \Pi_1(p, q) \leq \min_{q \in Q} \max_{p \in P} \Pi_1(p, q) \leq \Pi_1(\tilde{p}, \tilde{q}).$$

   What is the corresponding inequality for player II's payoff function? Why do the two inequalities imply that neither player can get less than their security level at a Nash equilibrium? Can you think of a way of seeing why this must be true without calculating at all?
8. Exercise 6.9.29 asked for the cooperative and noncooperative payoff regions of the game of Figure 6.21(b). Find its unique Nash equilibrium. Confirm that player II should use her second pure strategy with probability $\frac{2}{3}$ and receive an expected payoff of $3\frac{2}{5}$ when this equilibrium is played. Show that her security level is also $3\frac{2}{5}$, which she secures by playing her second pure strategy with probability $\frac{3}{5}$. Discuss the relevance of this example to the claim that a unique Nash equilibrium of a game should necessarily be regarded as its rational solution.
9. If the Battle of the Sexes of Figure 6.15(b) is played without any preplay communication and no symmetry-breaking convention is available, explain why the pure Nash equilibria are unavailable as candidates for the rational solution of the game. Show that each player gets an expected payoff of $\frac{2}{3}$ when the mixed Nash equilibrium is used. Show that each player's security level in

[10]First apply Nash's theorem on the existence of Nash equilibria in finite games to the game obtained by deleting all weakly dominated strategies.

the Battle of the Sexes is also $\frac{2}{3}$ but that the players' security strategies aren't the same as their mixed equilibrium strategies.[11] Cast doubt on identifying the mixed equilibrium as the rational solution of the game by asking why the players don't switch to their security strategies since they then get a payoff of $\frac{2}{3}$ for sure. Why would player I profit by sticking to his mixed equilibrium strategy if player II were to switch to her security strategy?

10. The game of Figure 6.21(a) is called the Australian Battle of the Sexes because its cooperative and noncooperative payoff regions are "upside-down" versions of those for the Battle of the Sexes. Follow through an argument like that of Exercise 8.8.9, but show that player I suffers by sticking with his mixed equilibrium strategy if player II switches to her security strategy.
11. Locate the risk-dominant and Pareto-dominant equilibria in the game of Figure 5.10(a).
12. Find a $2 \times 2$ symmetric bimatrix game with two symmetric pure Nash equilibria in which one of the equilibria is both risk dominant and Pareto dominant.

phil

13. Why is money valuable only by convention?
14. In the Boston of Henry James, a lady and a gentleman approach a new-fangled revolving door. In the variant of Chicken with which they are confronted, there are two pure strategy Nash equilibria: the lady can wait for the gentleman to go first, or the gentleman can wait for the lady. Which of these equilibria is focal?
15. Two players have disks divided into five equal sectors. Working around the circle, the sectors are colored red, red, green, red, green. Each disk is now spun like a roulette wheel, so that its orientation is randomized. If each player independently chooses the same sector, both win \$100. Otherwise nobody wins anything. Which sector do you choose? How confident are you that your opponent will choose the same sector?

math

16. A firm's output consists of a commodity bundle chosen from a compact and strictly convex production set $Y$ in $\mathbb{R}^n$. The output bundle is chosen to maximize profit $p^\top y$, where $p$ is the price vector.[12] Because $Y$ is *strictly* convex, there is always a unique profit-maximizing output $y = s(p)$ for each price vector $p$. The function $s : \mathbb{R}^n_+ \to Y$ is then the firm's *supply function*. Answer the parenthetical questions in the following "proof" that the supply function is continuous, and point to a flaw in the argument. What can be done to patch up the proof?[13]

    Let $p_k \to p$ as $k \to \infty$. Write $y_k = s(p_k)$. Then, for any $z$ in $Y$, $p_k^\top z \le p_k^\top y_k$. (Why?) If $y_k \to y$ as $k \to \infty$, it follows that, for any $z$ in $Y$, $p^\top z \le p^\top y$. (Why?) Hence $y = s(p)$. (Why?) Thus, $s(p_k) \to s(p)$ as $k \to \infty$, and so $s$ is continuous.

math

17. The equilibria of economic theory aren't always the equilibria of some game. It may be, for example, that the $i$th player's strategy set is $S_i$ but that some constraint prevents a free choice from all the strategies in $S_i$. Often the subset

[11]The mixed equilibrium calls for player I to use his first pure strategy with probability $\frac{2}{3}$ and for player II to use her second pure strategy with the same probability. Player I's security strategy calls for him to use his first pure strategy with probability $\frac{1}{3}$ and player II's security strategy calls for her to play her second pure strategy with probability $\frac{1}{3}$.

[12]Some of the coordinates of $y$ may be negative and thus represent *inputs*. It isn't therefore being assumed that production is costless.

[13]A sequence $y_1, y_2, y_3, \ldots$ of points in a compact set $Y$ converges to $y$ if and only if all its convergent subsequences converge to $y$. (Proof?)

$T_i$ to which the player is confined depends on the vector $s$ of *all* the players' choices.[14] That is, $T_i = G_i(s)$, where $G_i$: $S_1 \times S_2 \times \ldots \times S_n \to S_i$.

a. Use Kakutani's fixed-point theorem to outline a proof that there is at least one $\tilde{s}$ for which $\tilde{s}_i \in G_i(\tilde{s})$ $(i = 1, 2, \ldots, n)$. List the mathematical assumptions that your proof takes for granted.
b. Soup up your argument to obtain a version of Debreu's "social equilibrium theorem." This asserts that $\tilde{s}$ can be found for which it isn't only true that (a) holds but also that $\tilde{s}_i$ is player $i$'s *optimal* choice from the set $G_i(\tilde{s})$.

18. Game theorists operate on the assumption that rationality is the same for everybody. Immanuel Kant thought he had deduced his categorical imperative from the same principle (Section 1.10). Can you find a reformulation of the categorical imperative that is consistent with the play of Nash equilibria in games?

phil

19. Wonderland has two political parties: the Formalists and the Idealists. They both care only about power and so choose a platform with the sole aim of maximizing their vote at the next election. The voters care only about matters of principle and hence are devoid of party loyalties. For simplicity, the opinions a voter might hold are identified with the real numbers $x$ in the interval $[0, 1]$. Someone with the opinion $x = 0$ believes society should be organized like an anthill, while someone with the opinion $x = 1$ thinks it should be organized like a pool of sharks. Each party chooses its platform somewhere along the political spectrum and isn't able to shift its position later. The voters then cast their votes for the party whose position is nearest to their own.

fun

a. Why is the median voter significant?
b. The parties enter the political arena simultaneously. Why will each party locate its platform at $x = \frac{1}{2}$, thus splitting the vote fifty-fifty?
c. Suppose a new party called the Intuitionists chooses a platform after the Idealists and the Formalists. Show that it is now an equilibrium for the Idealists and the Formalists to locate at $x = \frac{1}{4}$ and $x = \frac{3}{4}$, with the Intuitionists at $x = \frac{1}{2}$. Each of the original parties will get $\frac{3}{8}$ of the vote. The Intuitionists will pick up only $\frac{1}{4}$.
d. Why should the Intuitionists enter the political arena at all if they are doomed to lose? What happens if the Intuitionists think it worthwhile to form a party only if they anticipate receiving more than 26% of the vote?
e. Do we learn anything about why political platforms in two-party systems aren't always the same?

[14]This happens in a simple exchange economy. Economic activity in such an economy is restricted to trading of the players' initial endowments of goods. Each player can be envisaged as selling his or her endowment at the market prices. The sum realized then imposes a *budget constraint* on what the player can then buy with the money. However, the market prices are determined by supply and demand in the market as a whole. That is, they depend on how *everybody* chooses to spend their money. What each player *can* choose is therefore a function of what everybody actually *does* choose.

# 9 *Buying Cheap*

→ 11.1

## 9.1 Economic Models

Buy cheap and sell dear is the classic recipe for making money. How is game theory relevant to this enterprise? We look first at the polar cases of perfect competition and monopoly, on which economic theorists focused almost exclusively before the advent of game theory. The intermediate cases of imperfect competition are left until the next chapter.

Students of economics will be tempted to skip the current chapter since perfect competition and monopoly remain the staple diet of most economic courses from the most elementary to the most advanced. However, I have tried to offer a new angle on the material by evaluating it from a game-theoretic perspective. It will also be a fruitful source of examples in future chapters.

→ 9.3

## 9.2 Partial Derivatives

Every economist knows that a monopolist maximizes her profit by setting marginal revenue equal to marginal cost. Mathematicians prefer to say that profit is maximized where its derivative is zero. Both statements mean the same thing because finding the marginal value of a continuous variable is the same as differentiating it.

Economists typically define a quantity like marginal utility as the increase in utility gained by consuming one more unit of a commodity, without continually explaining that they intend the units in which the variables are measured to become arbitrarily small. In this chapter and the next, it would be easy to be led astray on this point because some of the commodities to be discussed, like apples or hats, naturally come in discrete units. However, we treat all commodities as though they were continuous variables in order to keep the mathematics simple. Even in the case of apples, Eve's marginal utility for a commodity is therefore obtained by differentiating her utility function *partially* with respect to whatever commodity we are talking about.

To find the partial derivative of a function, differentiate it with respect to the variable in question, pretending that all the other variables are constant. For example, if $f:\mathbb{R}^2 \to \mathbb{R}$ is defined by $f(x_1, x_2) = x_1^2 x_2$, then

$$\frac{\partial f}{\partial x_1} = 2x_1x_2; \quad \frac{\partial f}{\partial x_2} = x_1^2.$$

The *gradient* of a differentiable function $f:\mathbb{R}^{\underline{n}} \to \mathbb{R}$ at a point $\xi$ is the $1 \times n$ row vector $\nabla f(\xi)$ of all its partial derivatives evaluated at $\xi$. In our example, $\nabla f(1,3) = (6, 9)$. Geometrically, the vector $\nabla f(\xi)$ points in the direction in which $f(x)$ is increasing fastest at $\xi$. Its modulus or length $|\nabla f(\xi)|$ is the rate of increase of $f(x)$ at $\xi$ in this direction.

Since $f(x)$ doesn't change at all as $x$ moves along one of its contours, it is no surprise that $\nabla f(\xi)$ always points in a direction orthogonal to the contour $f(x) = f(\xi)$. It is therefore a *normal* to the tangent hyperplane to the contour. From Section 7.7.1, we know that the equation of the tangent hyperplane can therefore be written as the inner product

$$\nabla f(\xi)(x - \xi) = 0.$$

For example, the tangent line to the contour $x_1^2 x_2 = 3$ at $x = (1, 3)^\top$ is $6(x_1 - 1) + 9(x_2 - 3) = 0$.

## 9.3 Preferences in Commodity Spaces

An economist observing Adam in the Garden of Eden would have used a Von Neumann and Morgenstern utility function $u$ to describe his preferences over different bundles of fig leaves and apples. Since Adam assigns three utils to each commodity bundle $(f, a)$ on the contour $u(f, a) = 3$, he is indifferent between all such bundles. Economists therefore call $u(f, a) = 3$ an *indifference curve*.[1]

Throughout this chapter and the next, we will keep things simple by assuming that Adam always wants more of everything, so that $u$ is strictly increasing.[2] We will also assume that $u$ is concave, which implies that Adam likes a physical mixture of two bundles on the same indifference curve at least as much as either bundle on its own (Section 6.5.1). Where convenient, we also assume that $u$ can be differentiated as many times as we like.

[1]The equation $u(f, a) = 3$ actually does represent a curve in most examples, but it need not. For example, if Adam is indifferent between all bundles, his only indifference "curve" is the whole commodity space.

[2]Recall that a strictly increasing function has the property that $x > y \Rightarrow f(x) > f(y)$. The meaning of $x > y$ when $x$ and $y$ are vectors is explained in Section 5.3.2.

None of these assumptions about Adam's preferences will be true for all commodities. People usually don't want lots of garbage. Nor is Adam likely to prefer an evening spent with two girlfriends, each giving him half her attention, to an evening alone with one or the other giving him all her attention. Some discretion is therefore necessary in applying the standard model of a consumer to the real world.

If $u$ is strictly increasing and concave on a two-dimensional commodity space, then Adam's indifference curves look something like those shown in Figure 9.1(a). Since Adam has a concave Von Neumann and Morgenstern utility function, he is risk averse. But it would be a mistake to argue that the shape of his indifference curves in Figure 9.1(a) is *caused* by his disliking gambling. As explained in Section 4.5.4, someone to whom the Von Neumann and Morgenstern axioms apply is neutral to the actual act of gambling. A rational person is risk averse partly because of the configuration of his indifference curves in commodity space, rather than the reverse.

### 9.3.1 Prices

It often makes sense to model some or all of the players in a market game as *price takers*. The mechanics of a market somehow determine a price that price takers are unable to alter. Their problem then ceases to be strategic. They simply have to solve a one-person decision problem: How much do I buy or sell at the current prices?

When prices are central, the commodity plotted on the vertical axis will be taken to be the *numeraire,* which is the quantity in which prices are quoted. The numeraire might be dollars or gold, but apples are the numeraire in our stories from the Garden of Eden.

If Adam is a price taker initially endowed with $A$ apples and Eve is willing to buy and sell fig leaves at a fixed price of $p$ apples per fig leaf, then $pf + a = A$ is Adam's budget line. By using some of his endowment of apples to buy fig leaves, Adam can acquire any bundle on this line.

As shown in Figure 9.1(a), the bundle at which Adam's utility is maximized subject to his budget constraint occurs where one of his indifference curves touches

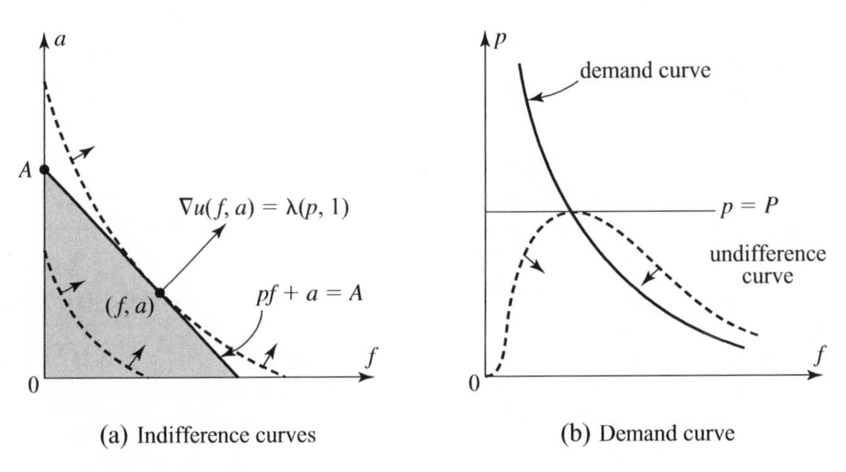


Figure 9.1 Indifference and demand. Indifference curves are drawn with broken lines. Arrows show the direction of increasing preference.

his budget line. One can therefore find the maximizing bundle by noting that the gradient vector $\nabla u(f, a)$ must point in the same direction as the vector $(p, 1)$, which is normal to the budget line $pf + a = A$. Hence, for some $\lambda$, $\nabla u(a, f) = \lambda(p, 1)$.

In the Cobb-Douglas[3] case when $u(a, f) = f^2 a$, we obtain the equations

$$\frac{\partial u}{\partial f} = 2fa = \lambda p, \qquad \frac{\partial u}{\partial a} = f^2 = \lambda,$$

from which it follows that $2a = fp$. Since the solution must also lie on the budget line $pf + a = A$, we find that Adam will choose the bundle $(f, a)$ with $f = 2A/3p$ and $a = A/3$.

The equations $f = 2A/3p$ and $a = A/3$ determine Adam's demand for fig leaves and apples. They specify how many fig leaves and apples Adam will demand when the price of a fig leaf is pegged at $p$ apples.

It is sometimes convenient to draw a diagram like Figure 9.1(b), in which price replaces the numeraire on the vertical axis. A point $(f, p)$ in this diagram corresponds to Adam's buying $f$ fig leaves at a price of $p$ apples per fig leaf. His indifference curves therefore have equations of the form $u(f, A - pf) = c$, where $c$ is a constant.

If the price of a fig leaf is fixed at $P$ apples, then Adam's budget line in this diagram is simply $p = P$. As before, his optimal bundle occurs where an indifference curve touches his budget line. Adam's *demand curve* for fig leaves is therefore the locus of the highest point on each of his indifference curves.

### 9.3.2 Quasilinear Utility

Adam is said to have a *quasilinear* utility function[4] when

$$u(f, a) = a + w(f).$$

With such a utility function, a util is the same thing as an apple—which is our correlate of money in the Garden of Eden. The quantity $w(f)$ is simply the most that Adam would be willing to pay to get $f$ fig leaves. It is standard to assume that $w$ is strictly increasing and concave.

Since Adam's demand for fig leaves at a fixed price $p$ is obtained by differentiating $u(f, A - pf) = A - pf + w(f)$ partially with respect to $f$, the equation for his demand curve takes the particularly simple form:

$$p = w'(f).$$

Because $w$ is assumed to be concave, its derivative $w'$ decreases. The demand curve of a consumer with quasilinear utility therefore slopes downward.

One can recover a quasilinear utility function from the demand curve by integrating (Section 21.3.2). Thus, if Adam uses some of his initial endowment to buy $f$ fig leaves at a fixed price of $p$ per fig leaf, then his increase in utility is the shaded area in Figure 9.2(a). Since utils and money are the same thing for quasilinear preferences, the shaded area also represents how much more than $pf$ Adam would actually be willing to pay to get $f$ fig leaves.

[3]Such a utility function has the form $u(f, a) = f^\alpha a^\beta$, where $\alpha$ and $\beta$ are positive constants.

[4]It is linear in $a$ and $w(f)$ and so is said to be quasilinear in $a$ and $f$.

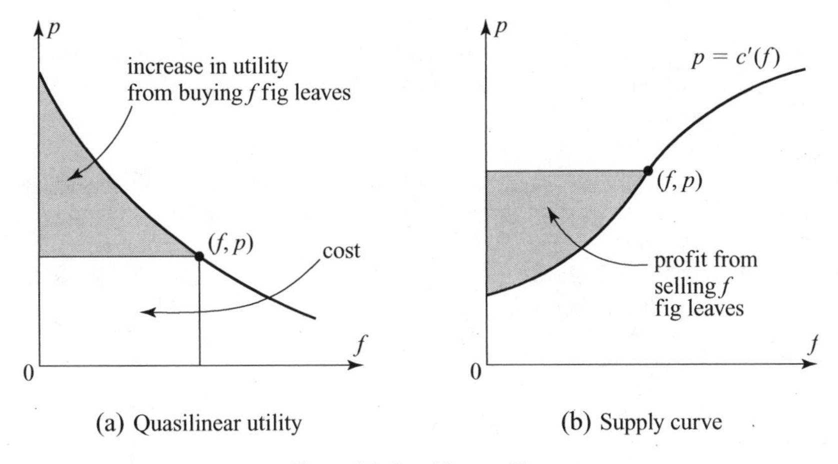


Figure 9.2 Quasilinear utility.

When it rains, why do the rich ride in taxicabs while the poor get wet? The economist Paul Samuelson famously explained that rich people value the cab fare less. Such consumers don't have quasilinear utility functions because Adam's attitude to exchanging apples and fig leaves remains completely unchanged no matter how rich or poor he may become. His indifference curves are simply vertical displacements of one another. For example, $u(f,a)=3$ is the same as $u(f,a-3)=0$.

Attributing quasilinear preferences to consumers is therefore not very realistic, but we are on safer ground when we turn our attention to producers operating in a market that makes them price takers. The reason is that companies arguably have a duty to their shareholders to maximize expected profit.

If Adam pays $a$ apples to Eve for supplying him with $f$ fig leaves that cost her $c(f)$ apples to gather, then her profit from the transaction is

$$\pi(f,a) = a - c(f).$$

If each fig leaf costs more to produce than the last, then $c$ is convex and so $-c$ is concave. Thus $\pi$ satisfies our requirements for a quasilinear utility function. The contours or isoprofit curves of this function can therefore be regarded as Eve's indifference curves.

Because Eve is supplying fig leaves to Adam rather than consuming them herself, we obtain a supply curve instead of a demand curve when we differentiate $\pi = pf - c(f)$ partially with respect to $f$ to find Eve's optimal production of fig leaves at a fixed price $p$. The supply curve is given by

$$p = c'(f),$$

which says that a price taker like Eve equates price and marginal cost when deciding how much to supply.[5]

[5]Economists explain this equation by saying that Eve will produce fig leaves until the extra cost of producing one more fig leaf rises above what it can be sold for.

Since $c$ is convex, Eve's supply curve slopes upward as shown in Figure 9.2(b). Assuming that $c(0)=0$, the shaded area shows the increase in utility (or profit) that Eve derives from producing $f$ fig leaves and then selling them to Adam at a fixed price of $p$ per fig leaf.

If fig leaves were the numeraire instead of apples, we would have drawn a demand curve for Eve and a supply curve for Adam, instead of the other way around. This parallel between consumers and producers is sometimes stressed by explaining consumers' preferences in terms of *opportunity costs*. For example, the opportunity cost to Adam of trading two apples for a fig leaf is the loss of utility he will derive from not being able to eat the apples himself.

## 9.4 Trade

Economics got started when Eve joined Adam in the Garden of Eden. If he has an initial endowment of $A$ apples, and she has an initial endowment of $F$ fig leaves, they both have the opportunity to improve their lot by doing some kind of deal.

The Edgeworth box of Figure 9.3(a) is used to represent their trading opportunities.[6] The box $\mathscr{E}$ is of width $F$ and height $A$. A point $(f, a)$ in the box represents the possible trade in which Adam gets the bundle $(f, a)$ and Eve gets the bundle $(F-f, A-a)$. If Adam and Eve fail to reach an agreement, Adam will be left with the bundle $(0, A)$. Eve will be left with the bundle $(F, 0)=(F-0, A-A)$. The pair $e=(0, A)$ is therefore called the *endowment* point. It represents the empty trade in which no goods are exchanged.

Figure 9.3(b) shows some of Adam's indifference curves $u_1(f, a)=c$ when his utility function $u_1$ satisfies the assumptions made in Section 9.3. Eve's utility function $u_2$ satisfies the same assumptions as Adam's, but her indifference curves have a different shape in Figure 9.3(b) because we have to plot the graph of $u_2(F-f, A-a)=c$ rather than $u_2(f, a)=c$.

### 9.4.1 Bargaining

What deal will Adam and Eve make? The answer depends on a whole raft of issues that will be addressed in later chapters. For example, what do the players know about each other's preferences? Who can make what commitments? How costly is delay? If we know the answers to all such questions, we can model Adam and Eve's bargaining problem as a noncooperative game. The Nash equilibria of this game then correspond to the rational deals available to Adam and Eve.

Knowing the Edgeworth box isn't enough. The Edgeworth box isn't even a game since it tells us nothing about the bargaining *strategies* available to the players. Nevertheless, knowing the Edgeworth box and a few other facts can help us make educated guesses about the deal that Adam and Eve will make.

Edgeworth's educated guess anticipated by some seventy years a result that economists call the Coase theorem. Unless some friction in the bargaining game they play intervenes, rational players will make a Pareto-efficient deal. In Figure 9.3(c),

[6]The Edgeworth box was apparently invented by Pareto!

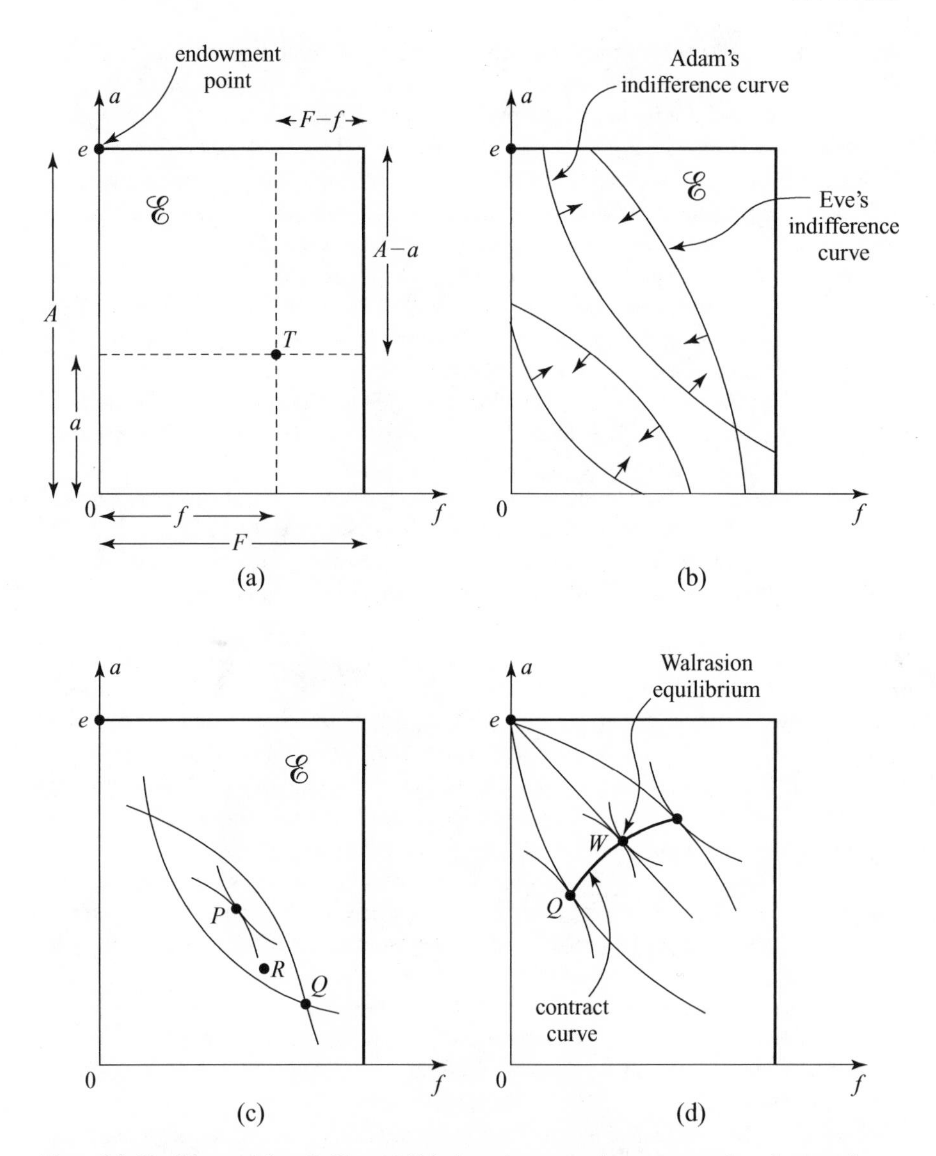


Figure 9.3 The Edgeworth box. In Figure 9.3(a), the endowment point $e$ corresponds to the no-trade outcome in which Adam retains the bundle $(0, A)$ and Eve retains the bundle $(F, 0)$. In the trade $T$, Adam gets $(f, a)$ and Eve gets $(F - f, A - a)$. The arrows in Figure 9.3(b) indicate the direction of the players' preferences. The trade $Q$ on the contract curve in Figure 9.3(d) results when Eve is a fully discriminating monopolist. The trade $W$ is the Walrasian equilibrium that arises under perfect competition.

which shows both Adam's and Eve's indifference curves, the Pareto-efficient trades are easy to spot. No interior point $Q$ of the Edgeworth box $\mathscr{E}$ at which Adam and Eve's indifference curves cross can be Pareto efficient. As indicated by the arrows in Figure 9.3(c), both players would prefer to move from $Q$ to any point $R$ inside the canoe-shaped region bounded by the two indifference curves through $Q$. Adam's and

Eve's indifference curves must therefore touch at any interior point $P$ of $\mathcal{E}$ that corresponds to a Pareto-efficient trade.

Edgeworth also observed that the players won't agree on a deal that makes them worse off than if they hadn't traded at all. Any rational deal must therefore not only be Pareto efficient, it must also lie between the two indifference curves that pass through the endowment point $e$. Our candidates for a rational deal are then reduced to those that lie on the *contract curve* indicated in Figure 9.3(d).

To make a more precise guess about the deal on which Adam and Eve will agree requires making further assumptions. Only one case is relatively straightforward. As in the Stackelberg model of Section 5.5.1, imagine that Eve can open the bargaining game by committing herself to a particular strategy for the remainder of the negotiation. If this strategy is to refuse any deal that gives her less utility than the trade $P$, then the only subgame-perfect equilibrium calls for her to set $P = Q$ in Figure 9.3(d). Adam can then take it or leave it. In equilibrium, he takes it.[7]

Eve's power in the bargaining game therefore guarantees that she will get her best possible outcome on the contract curve. Economists say that she has full monopoly power. Nothing restricts her ability to exploit Adam—short of her actually taking his endowment by force. Adam's helplessness correspondingly results in his getting his worst outcome on the contract curve.[8]

Monopolists are seldom as powerful as Eve in the preceding analysis. The classical assumption is that Eve's monopoly power allows her only to set a price $p$ below which she won't trade. To buy $f$ fig leaves at a price of $p$ apples per fig leaf will cost Adam $pf$ apples. He will then be left with $a = A - pf$ apples. The trades in the Edgeworth box at which fig leaves are exchanged for apples at a fixed price $p$ therefore lie on the straight line $a = A - pf$ through the endowment point $e = (0, A)$, as shown in Figure 9.4(a).

If Eve sets the price $p$, Adam is forced to choose the trade $P$ he likes best on the line $a = A - pf$. As Figure 9.4(a) shows, $P$ lies where one of Adam's indifference curves touches this line. The locus of such points is indicated by a broken curve in Figure 9.4(a). In standard Stackelberg style, Eve can choose $p$ to obtain the trade $M$ that she likes best on this curve. Since $M$ lies where this curve is touched by one of Eve's indifference curves, it is evident that $M$ will be Pareto efficient only by an unlikely accident. The deal reached in a classical monopoly is therefore wasteful, as well as unfair.

Figure 9.4(b) shows a diagram more like those usually drawn to illustrate a classical monopoly. Eve maximizes profit at the point $M$, where one of her isoprofit curves touches Adam's demand curve. We know from Figure 9.1(b) that tangents to Adam's indifference curve at points on his demand curve are horizontal. It follows that Adam's and Eve's indifference curves will touch at $M$ only in pathological cases, and so we have shown again that a classical monopoly isn't normally efficient.

[7]But see Section 19.2.2 for the experimental evidence of how people actually behave in the laboratory when playing such ultimatum games.

[8]Adam may well complain that this isn't fair since Eve appropriates the entire surplus. Nor will he be comforted if we explain that none of the available surplus is wasted, and so the outcome is Pareto efficient. He may even get angry at being treated like a gullible fool if an economist tries to persuade him that his complaint is antisocial because some textbooks say that any Pareto-efficient outcome is "socially optimal."

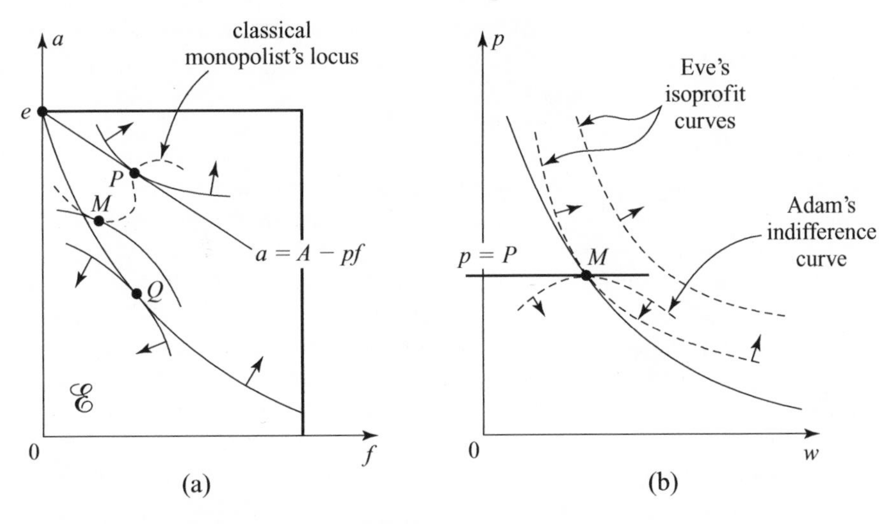


Figure 9.4 Classical monopoly. If she can fix the price, Eve can force a trade on any line $a = A - pf$ in Figure 9.4(a). Adam's optimal reply is $P$. The broken curve is the locus of all such optimal replies. The monopoly point $M$ is Eve's preferred trade on this locus. Since $M$ isn't on the contract curve, it isn't Pareto efficient. Figure 9.4(b) tells the same story in terms of Adam's demand curve. Since Adam's and Eve's indifference curves don't touch at $M$, it isn't a Pareto efficient point.

## 9.5 Monopoly

Economists seldom use the Edgeworth box when discussing a classical monopoly. A more familiar analysis goes like this. Dolly is the only producer of wool in Wonderland. Each ounce costs her \$$c$ to make. The demand curve for wool is given by $w + p = K$, where $K$ is a much larger than $c$.[9] (In Section 5.5.1, we took $c = 3$ and $K = 15$.)

Dolly would be foolish to produce more wool than she can sell at the price she proposes to set. If she produces $w$ ounces, she will therefore sell each ounce at a price of $p = K - w$ because this is the greatest price at which all her wool will be sold.

Dolly's profit is the difference between the revenue she obtains by selling what she produces and the cost of making it. Her profit is therefore

$$\pi(w) = pw - cw = (p - c)w = (K - w - c)w.$$

To find the output $\tilde{w}$ that maximizes profit, she sets marginal revenue equal to marginal cost. That is to say, she differentiates $\pi(w)$ and sets the derivative equal to zero. Since

$$\frac{d\pi}{dw} = K - c - 2w,$$

[9]When dealing with a so-called linear demand curve like $w + p = K$, we implicitly assume that the equation applies only when $w > 0$ and $p > 0$. When $w = 0$, any price $p \geq K$ is also on the curve. When $p = 0$, any quantity $w \geq K$ is on the curve.

profit is maximized when $\tilde{w} = \frac{1}{2}(K - c)$. The price is then $\tilde{p} = \frac{1}{2}(K + c)$. The maximum profit is $\pi = \{\frac{1}{2}(K - c)\}^2$.

### 9.5.1 The Source of Monopoly Power

What is the source of Dolly's monopoly power in the preceding story? How come she is a price maker and Alice is a price taker? The simplest answer is that Dolly is able to make a *commitment* to the price at which she can sell. But why doesn't she then use her commitment power to move away from $M$ in Figure 9.4(a) to some point nearer $Q$?

We leave such commitment questions until Section 9.5.2 and ask instead what features of the economic environment in which Dolly is operating would allow her to act as a price-making monopolist, without attributing unexplained powers of commitment to her enterprise.

The first observation is that a monopolist in economic applications usually has a large number of small customers, rather than one large customer. Economists say that the model of Section 9.4, in which Adam and Eve trade apples for fig leaves, is an exercise in *bilateral* monopoly, thereby recognizing that both the buyer and the seller may have the power to influence the price.

To cope with many consumers isn't difficult in theory. The simplest case arises when a single consumer is replicated many times. Our monopolist Dolly is named after the sheep that was the first mammal to be cloned artificially, but here it is Alice who will be cloned.

Instead of one big Alice demanding $W = K - p$ ounces of wool when the price is $p$, we introduce $N$ small copies of Alice who each demand $W = (K - p)/N$ ounces. Their total demand is then $w = NW = K - p$ ounces of wool, and so the market demand curve is the same as when we were only considering one Alice. We can therefore repeat our monopoly story, telling ourselves that each individual copy of Alice is now too small to be able to exercise any significant market power. If any doubts arise, we can proceed to the limiting case when $N \to \infty$.

But this story is too quick. Suppose, for example, that Dolly has to sell her wool from door to door, confronting each copy of Alice one at a time. Why is her position at each front door then any different from what it was before we split Alice up into lots of small copies? In fact, Section 18.6.2 shows that she is no better off at all. In particular, if each copy of Alice somehow has monopsony power on her own doorstep,[10] then Dolly will get zilch from Alice's fragmentation.

For this reason, economists usually implicitly assume that Dolly is more like a stallholder at a farmer's market than a door-to-door salesperson. She posts a price on her stall, and her customers cluster around competing to buy an ounce of wool when she sets a price that makes the demand for her wool exceed the amount she is able to supply.

### 9.5.2 Price Discrimination

Previous chapters have been scathing about attempts to attribute commitment power to players without explaining the source of this power. A major reason why it

[10]A monopsonist is a *buyer* with monopoly power.

sometimes does make sense to assume that a player can make commitments is that she values her *reputation* for being tough.

To model reputation properly in the case of an aspiring monopolist, one can begin by constructing a *repeated* game in which Dolly sells wool over and over again to an everchanging body of customers, but the analysis of such a model is beyond the scope of this book. We will instead simply observe that equilibria exist in such games that result in Dolly sticking to her posted price because the money she could make today by selling a few more ounces of wool cheaply counts for nothing against the money she would lose by revealing that she is the kind of person who sometimes lowers her price.

When Dolly can make credible price commitments, she may be able to sell different ounces of wool at different prices. Such *price discrimination* can be engineered in various ways. In the most familiar kind of discrimination, Dolly offers different prices to different customers. For example, students can buy airline tickets cheaper than professors. Quantity discounts similarly favor large customers over small.

The ultimate in price discrimination is to sell each ounce of wool at the maximum price that some customer is willing to pay for it. This is what Dolly needs to do to achieve her ideal point $Q$ in Figure 9.3(d). If she must trade ounces of wool one at a time, she should commit herself to refusing to sell any further wool until she has sold the ounce she currently has in her window at negligibly less than the maximum price that someone is willing to pay for it.

If Dolly's only customer is Alice, each ounce of wool is sold at successively lower prices, chosen so as to move Alice's commodity bundle from $e$ in Figure 9.3(d), along her indifference curve through $e$, to the trade $Q$. With each sale of an ounce of wool, Dolly thereby squeezes everything from Alice that there is to be squeezed at that stage. When Alice has quasilinear preferences, we know that the total amount that Dolly can squeeze from Alice can be calculated from the area under Alice's demand curve (Section 9.3.2). The rest of this section looks into this feature of quasilinear preferences more closely.

→ 9.6

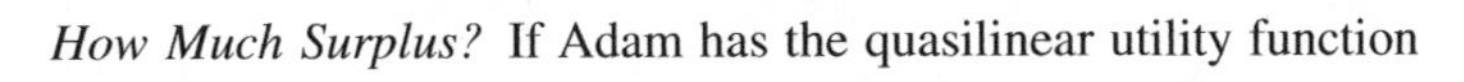

*How Much Surplus?* If Adam has the quasilinear utility function

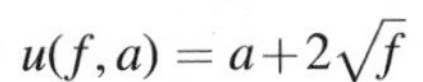

$$u(f,a) = a + 2\sqrt{f}$$

for apples and fig leaves, then his demand curve for fig leaves is given by $p = 1/\sqrt{f}$, (Section 9.3.2). We assume in this model that his initial endowment is the bundle $(F, A)$, where $F < 1$.

Eve is a profit-maximizing producer of fig leaves, who incurs a cost of one apple for each fig leaf that she produces. Her marginal cost of producing a fig leaf is therefore always one apple. Eve has no initial endowment but contracts with Adam to supply him with $f - F \geq 0$ fig leaves, for which he pays her $A - a$ of his apples in advance. Adam ends up with the bundle $(f, a)$. Eve ends up making a profit of $\pi = (A - a) - (f - F)$.

Figure 9.5(a) shows a kind of Edgeworth box. Notice that Adam's indifference curves are vertical displacements of each other. To find where Adam's indifference curves touch Eve's isoprofit curves, we set $\nabla u\ (f,a) = \lambda \nabla \pi(f,a)$ and find that the contract curve lies on the vertical line $f = 1$. The fact that Adam and Eve

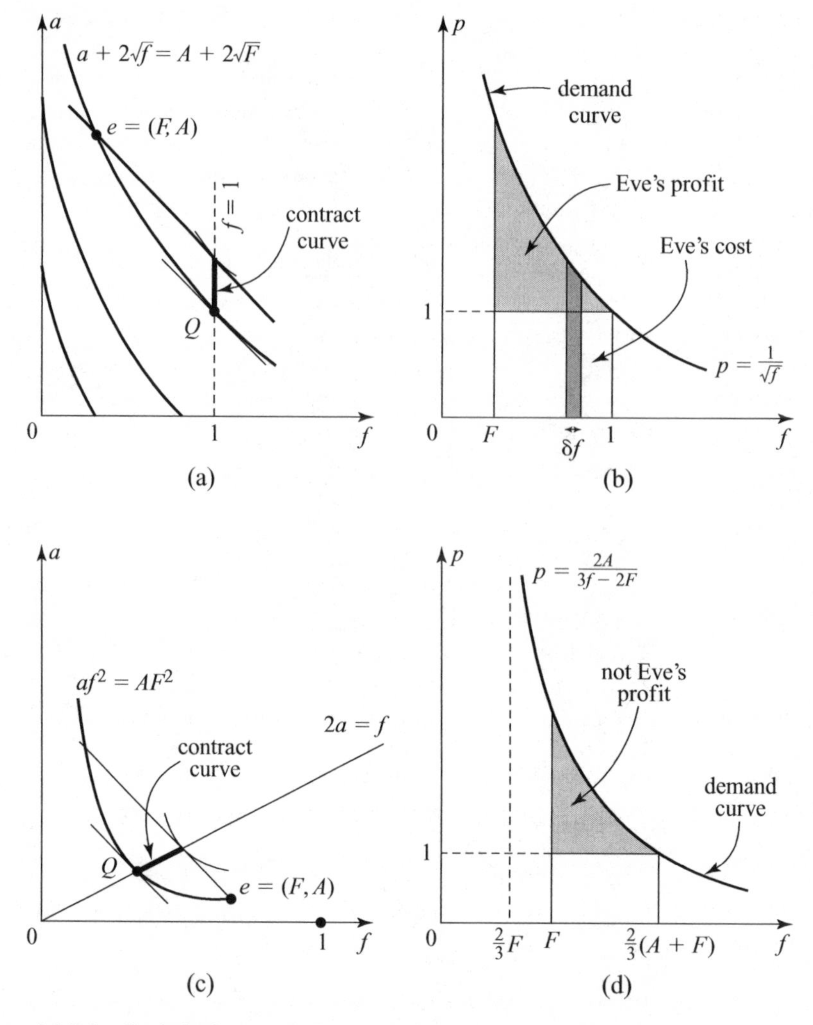


Figure 9.5 Price-discriminating monopolists. When Eve operates as a fully discriminating monopolist, she forces the trade $Q$ in Figures 9.5(a) and (c). Her profit is the shaded area under the demand curve in Figure 9.5(b) when Adam has quasilinear preferences, but not in Figure 9.5(d) when she doesn't. In the latter case, Adam's demand for more fig leaves depends on how many fig leaves he has so far and what he paid for them.

agree on the same number of fig leaves regardless of Adam's wealth in apples reflects the fact that they both have quasilinear preferences.

If Eve is a fully discriminating monopolist, she will secure the trade $Q$. This is located at the point $(f, a)$ on the contract curve where the line $f = 1$ cuts the indifference curve $a + 2\sqrt{f} = A + 2\sqrt{F}$ on which Adam's utility is lowest. Thus $A - a = 2(1 - \sqrt{F})$ and $f - F = 1 - F$. It follows that Eve's profit from acting as a fully discriminating monopolist is

$$\pi = (A - a) - (f - F) = 1 - 2\sqrt{F} + F.$$

How does one get the same answer from looking at the area under Adam's demand curve?

When acting as a fully discriminating monopolist, Eve sells fig leaves to Adam at lower and lower prices. The area of the thin column in Figure 9.5(b) shows how much Eve makes by selling $\delta f$ more fig leaves to Adam at the maximum price $p = 1/\sqrt{f}$ he is willing to pay when he has $f$ fig leaves already.

Eve stops serving Adam as soon as the price $p$ of a fig leaf gets down to her marginal cost of producing a fig leaf. Since $f = 1$ when $p = 1$, Eve serves Adam until his bundle of fig leaves has increased from $f = F$ to $f = 1$. Allowing $\delta f \to 0$, we find that Eve's total revenue $R$ is the area under Adam's demand curve between $f = F$ and $f = 1$. That is,

$$R = \int_F^1 \frac{df}{\sqrt{f}} = 2(1 - \sqrt{F}).$$

To find Eve's profit, we must subtract her cost $1 - F$ of producing $1 - F$ fig leaves. We then obtain that $\pi = 1 - 2\sqrt{F} + F$, as before.

The method of computing a fully discriminating monopolist's profit using the area under the market demand curve is widely used even when it gives the wrong answer. It works when Adam has quasilinear preferences because his attitude toward buying more fig leaves doesn't change as he becomes less wealthy. However, like most of us, I become more careful with my money as I get nearer the bottom of my piggybank. Adam and I might both be willing to pay \$2 per ounce for 10 ounces of wool, but if Dolly makes us pay \$4 per ounce for the first 5 ounces, I won't line up with Adam for a second batch of 5 ounces at \$2 an ounce. Dolly's price will have to come down before I bite.

To illustrate this point, we repeat the above analysis on the assumption that Adam has the Cobb-Douglas utility function $u(f, a) = af^2$ of Section 9.3.1. His demand curve for fig leaves is then given by $p = 2A/(3f - 2F)$. Notice that Adam's initial endowment of $(F, A)$ appears explicitly in this formula. We assume that $2A \geq F$ to keep things simple.

The contract curve lies on the line $2a = f$ as illustrated in Figure 9.5(c). The trade $Q$ is located at the point $(f, a)$ on the contract curve where the line $2a = f$ cuts the indifference curve $af^2 = AF^2$ on which Adam's utility is lowest. Working out $(f, a)$ and substituting in $\pi = (A - a) - (f - F)$, we find that the profit of a fully discriminating monopolist is

$$\pi = A + F - 3\{\tfrac{1}{4}AF^2\}^{\frac{1}{3}}. \tag{9.1}$$

To verify that this isn't the same as the area shaded in Figure 9.5(d), compute

$$\int_F^{\frac{2}{3}(A+F)} \frac{2A}{3f - 2F}\, df = \tfrac{2}{3}A \ln\left(\frac{2A}{F}\right),$$

which is equal to (9.1) only when $2A = F$. Otherwise the integral is larger.

What has gone wrong is that Adam's demand for fig leaves changes with his wealth. Suppose that Adam has paid Eve $b(f)$ apples for $f - F$ fig leaves up to now, leaving him with $a(f) = A - b(f)$ apples. At this stage, Eve offers him a further $\delta f$ fig leaves. Since Adam's current endowment is $(f, a(f))$, his demand curve at this stage is given by $p = 2a(f)/(3(f + \delta f) - 2f)$. Eve can therefore persuade him to pay only an extra

$$b(f + \delta f) - b(f) = \frac{2a(f)}{f + 3\delta f}\delta f$$

for $\delta f$ extra fig leaves.

Allowing $\delta f \to 0$, we are led to the differential equation

$$-\frac{da}{df} = \frac{2a}{f},$$

which has the general solution $af^2 = c$. The constant $c$ of integration is found using the boundary condition $a = A$ when $f = F$. Thus the number $a$ of apples that can be extracted from Adam by a fully discriminating monopolist in return for $f$ fig leaves is given by $af^2 = AF^2$. But this is the equation of the indifference curve on which Adam's utility is lowest. When Eve decides how large to make $f$ by maximizing $\pi = (A - a) - (f - F)$ subject to $af^2 = AF^2$, she will therefore simply be redoing the calculations that led us first to $Q$ and then to the formula (9.1).

*No Income Effects.* Economists say that cases in which a fully discriminating monopolist can't extract the area under the demand curve are caused by "income effects." A leading case without income effects arises when Dolly has many potential customers who each want at most one ounce of wool. We can still end up with the same market demand curve as before because some consumers are likely to be willing to pay more than others to secure an ounce of wool. However, the changes in attitudes that such consumers experience when made to pay more or less for an ounce of wool are irrelevant to our model because they vanish from our sight after being served.

### 9.5.3 Modeling Monopolies

We have looked briefly at several models of monopoly. The first is the classic model in which Dolly is a price maker who chooses the price she likes best and succeeds in serving all the demand at that price. This model can be challenged in various ways. For example, if her customers don't believe Dolly's claim that her price won't be lowered later, she may be forced into the position of a price-taker, as in Section 9.6.1. At the other extreme, she will sometimes have so much price-setting power that she will be able to charge different prices for different ounces of wool.

In seeking to model a monopoly in differing circumstances, it turns out that a lot depends on matters of detail. It can matter how impatient Dolly's customers are. It can matter whether we are talking about a durable good like hats or a perishable good like freshly caught fish. The question of who knows what can be especially important. For example, how does a price-discriminating monopolist know who is willing to pay what? How does a customer know a monopolist's marginal cost?

Even if a price-discriminating monopolist is well informed, what prevents customers to whom she is willing to sell wool cheaply, undercutting the higher price at which she plans to sell hats to others? Perhaps Dolly can get her customers to sign a contract forbidding resale. If so, she may be able to get them to accept other contracts. For example, Section 1.10.3 described how Medicare insisted on a most-favored-customer contract, which guarantees a customer that nobody else will be offered a better price. Dolly's customers may well be pleased to sign such a contract, but the final effect will be to allow Dolly to commit to a price. Since Dolly can't offer wool beyond the monopoly quantity at lower than the classic monopoly price without offering a rebate to the customers she has already served, it now becomes credible that she won't be lowering the monopoly price at all.

If game theory were fully developed, it would provide different models for all the different kinds of market conditions a monopolist could face. However, as things stand, the problem of modeling a monopoly will merely be a source of instructive examples in later chapters.

## 9.6 Perfect Competition

Monopoly and perfect competition are the two classical paradigms of economic theory. We boo the former and cheer the latter. One reason is that perfectly competitive economies are Pareto efficient and classical monopolies are not.

### 9.6.1 The Invisible Hand

Adam Smith was the first economist to draw attention to the virtues of perfectly competitive economies. As he explained, although each of us may be selfishly promoting our own private interests, the market can provide an *invisible hand,* which ensures that goods are distributed efficiently. For game theorists, Adam Smith's invisible hand is a metaphor for the process of trial and error by means of which real people get to the equilibrium of a game.

*Coase Conjecture.* The Coase conjecture isn't the same as the Coase theorem of Section 9.4.1. It is discussed here to illustrate why even a monopolist needs to pay attention to the workings of Adam Smith's invisible hand.

Dolly is a monopolist without commitment power. Each of her many potential customers wants only one ounce of wool. Dolly can produce as much wool as she likes at a constant marginal cost of $1 per ounce, and so her supply curve is $p = 1$.[11]

Dolly's supply curve in this case is labeled $S_1$ in Figure 9.6(a). The market demand curve is labeled $D$. Coase pointed out that no consumer will pay a price $p > 1$ for an ounce of wool if he understands that Dolly has an incentive to make and sell more wool at a lower price $q$ after serving all the consumers who are willing to buy at price $p$. To obtain customers, Dolly will therefore be forced to lower her price all the way down to $p = 1$ per ounce, and so her profit will be zero. The supply and demand for her product can then be read off from Figure 9.6(a) by locating the point $W_1$ at which the market demand curve $D$ and the market supply curve $S_1$ cross.

[11] If she is forced to be a price taker, she will make and sell as much wool as she can at a price $p > 1$. She will make no wool at all at a price $p < 1$. When $p = 1$, she is indifferent between the two possibilities.

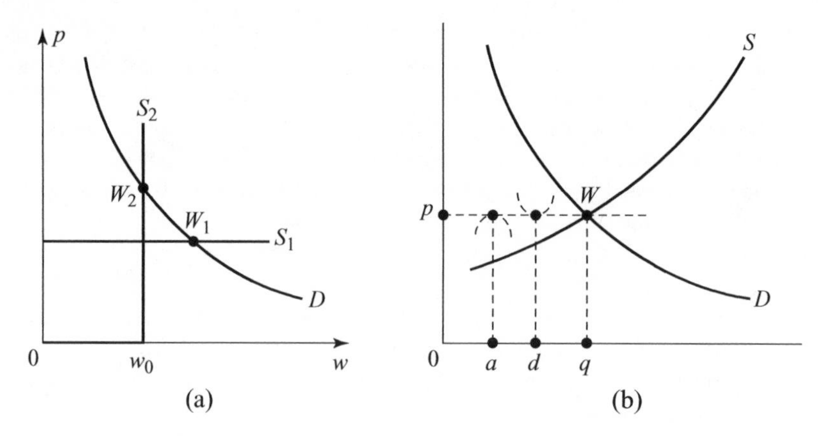


Figure 9.6 Equilibrium where the supply and demand curves cross.

Although Dolly is the only seller of wool, Adam Smith's invisible hand makes her into a price taker.

This is the gloomiest scenario that a monopolist might face. It arises, for example, if Dolly is forced to sell her wool using an auction in which the price rises until the number of customers still bidding is equal to the amount of wool that Dolly is willing to sell at that price. Since prospective customers will progressively drop out of the auction as the auction price reaches their willingness to pay, the result of the auction is $W_1$ in Figure 9.6(a).

How can Dolly evade the Coase scenario? One possibility is for her to adopt an expedient previewed in Section 5.5.2. She can publicly destroy her capacity to sell more wool than the monopoly quantity. To do so may be as easy as restricting the stock she chooses to take to market with her or as painful as firing her shearer. It is to this trick that economists are referring when they criticize monopolists for jacking up the price by restricting supply.

To see how the strategem works, suppose that Dolly produces $w_0$ ounces of wool and then irrevocably fires the only shearer in town, so that no further wool can be produced. Her new supply curve is then labeled $S_2$ in Figure 9.6(a).[12] The horizontal part of $S_2$ arises because Dolly's marginal cost of taking an extra ounce of wool out of stock is assumed to be zero when the demand is $w < w_0$. Her marginal cost of obtaining another ounce of wool when $w = w_0$ is assumed to be infinite, and so the remainder of $S_2$ is vertical.

As illustrated in Figure 9.6(a), an auction will now lead to the point $W_2$, where the market demand curve $D$ crosses the market supply curve $S_2$. The invisible hand is therefore at work even when wicked monopolists force up the price by restricting supply—although this point is usually downplayed so that attention can concentrate on Dolly's profit-maximizing choice of $w_0$, which she chooses just like a classical monopolist.

[12]The marginal cost of producing an ounce of wool is irrelevant to the shape of $S_2$ because the fact that Dolly paid $\$w_0$ to produce her stock of wool has no bearing on what it will sell for. Dolly *sank* this cost when she decided to stock $w_0$ ounces of wool *in advance* of the operation of the market.

*Competitive Pricing.* A monopolist like Dolly might thoughtlessly plan to sell wool for a price $p$ at which demand exceeds supply, but consumers with a high willingness to pay who find themselves near the end of the line that would form would then have an incentive to offer a higher price to her. The resulting informal auction would then save Dolly from the consequences of her folly.

Economists attribute the power of the invisible hand to such informal auctions. The mechanism is particularly effective in a classical perfectly competitive economy, in which there are a large number of small producers, as well as a large number of small consumers. The auctioning process that animates the invisible hand then operates on *both* sides of the market. When the price is high enough to make supply exceed demand, the producers undercut each other in seeking a buyer. When the price is low enough to make demand exceed supply, the consumers overbid each other in seeking a seller. A stable price is therefore possible only when supply and demand are the same.

Figure 9.6(b) is the diagram that economists draw to illustrate such a perfectly competitive economy. The competitive price $p$ and the competitive quantity $q$ of wool traded can be read off from the diagram by locating the point $W$ at which the market demand curve $D$ crosses the market supply curve $S$. At this point, demand equals supply.

*Pareto Efficiency.* If the producers are $M$ small copies of Dolly and the consumers are $N$ small copies of Alice, each Dolly will sell $d$ ounces of wool, and each Alice will buy $a$ ounces of wool, where $Md = Na = q$. Figure 9.1(b) explains why Alice's and Dolly's indifference curves touch the horizontal line in Figure 9.6(b) corresponding to the competitive price $p$.

To make an Alice better off, we have to assign her a bundle below her indifference curve in Figure 9.6(b). The sum of such bundles will therefore lie beneath the horizontal line through $W$. To make a Dolly better off, we have to assign her a bundle above her indifference curve in Figure 9.6(b). The sum of such bundles will therefore lie above the horizontal line through $W$. It follows that no Pareto improvement on the competitive outcome is possible because the two sums need to be equal for the market to clear. We therefore have a justification of Adam Smith's insight that the invisible hand will engineer an efficient outcome in a perfectly competitive market.

### 9.6.2 Walrasian Equilibrium

Walras anticipated game theory by formulating an equilibrium notion that captures the essence of a perfectly competitive economy. However, a Walrasian equilibrium isn't an equilibrium in the sense that game theorists use the term. All consumers and producers are assumed to choose their optimal consumption and production vectors for each possible set of prices. A *Walrasian equilibrium* arises at prices that make the resulting market supply for each commodity adequate to meet the market demand for that commodity.

We return to the bilateral monopoly of Section 9.5.1 to show what a Walrasian equilibrium looks like in an Edgeworth box. Recall that Adam and Eve have the opportunity to trade apples for fig leaves. Figure 9.3(d) shows their contract curve. The Walrasian equilibrium $W$ occurs at a point where a price line is simultaneously

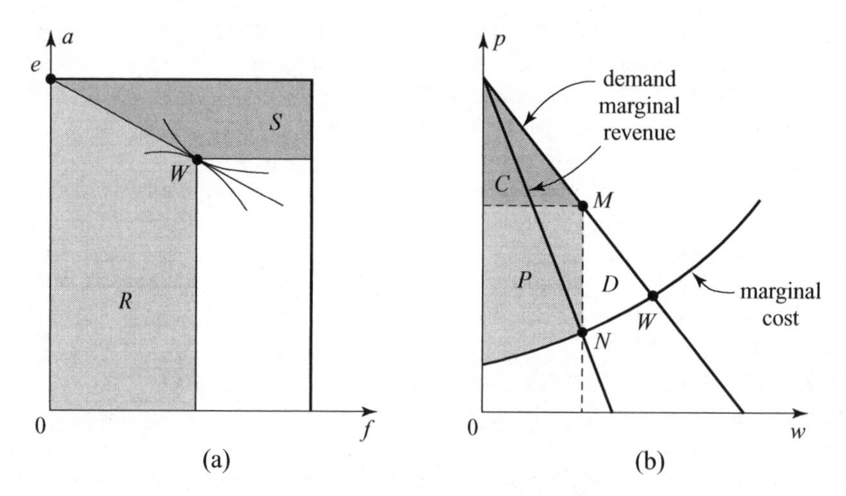


Figure 9.7 Bilateral and classical monopoly.

touched by one of Adam's indifference curves and one of Eve's. If the price is $p$ and $W = (f, a)$, then Adam will demand and Eve will supply $f$ fig leaves. Eve will demand and Adam will supply $A - a$ apples. Demand and supply are therefore equal for both apples and fig leaves. So the market clears, and we have found a Walrasian equilibrium.

The immediate point is that Adam and Eve's indifference curves not only touch the Walrasian price line at $W$, they also touch each other. We are therefore able to confirm that the Walrasian equilibrium $W$ is Pareto efficient—unlike the monopoly point $M$. Economists refer to the general version of this result as the first welfare theorem.[13]

### 9.6.3 Trading Games

A Walrasian equilibrium is Pareto efficient under certain circumstances, but when can we count on the invisible hand taking us there? Game theorists approach this question by trying to model the trading process as a game. One can then ask whether Nash equilibria in this trading game are Walrasian.

Figure 9.7(a) shows a Nash equilibrium for a trading game in which Adam and Eve simultaneously act as (bilateral) monopolists. Both commit themselves to a price and a quantity. Adam's price is the lowest at which he will sell fig leaves. Eve's price is the highest at which she will buy fig leaves. Adam's quantity is the most fig leaves he will exchange for apples. Eve's quantity is the most apples she will exchange for fig leaves. Adam thereby restricts himself to a region $R$ like that shown in Figure 9.7(a), and Eve to a region $S$. In the Nash equilibrium shown, they trade at the Walrasian equilibrium $W$.

But this trading game isn't very realistic because there is no good reason why Adam and Eve should be restricted to trading at a fixed rate of so many apples per fig

[13]The second welfare theorem says that we can make any Pareto-efficient point into a Walrasian equilibrium by choosing the endowment point suitably.

leaf. Indeed, when we come to study bargaining games in later chapters, we will find that a bilateral monopoly is far from the ideal setting in which to apply the concept of a Walrasian equilibrium. The following model, in which there are large number of small buyers and sellers, is a much more favorable environment because nobody is able to exercise any market power.

*Matching and Bargaining.* Consider a market in which each trader wants to buy or sell a particular kind of house. On entering the market, buyers and sellers search for a partner with whom to bargain. If the costs of searching and bargaining are negligible, then all houses will be sold at the same price $p$ (Say's Law). Otherwise, a buyer willing to pay more or a seller willing to accept less would be swamped with offers from players hoping to pick up a bargain.

Suppose that the daily influx of potential buyers and sellers is determined by a demand function $D$ and a supply function $S$. This means that $S(p)$ sellers have an outside option of no more than $p$, and so $S(p)$ house owners will enter the market if they expect to sell their house there at price $p$. Similarly, $D(p)$ potential buyers will enter if they expect to buy a house at price $p$.

Once a deal is reached between a matched pair, they leave the market together. To maintain a steady state, it is therefore necessary that the number of buyers and sellers who enter the market each day be equal. Thus $S(p) = D(p)$, and so we are at a Walrasian equilibrium.

But the costs of searching and bargaining aren't negligible in real life. A major challenge for game theory is therefore to determine how much the outcome deviates from a Walrasian equilibrium when such costs aren't assumed away (Section 18.6.2).

*Walrasian Tâtonnement.* Organized markets present less of a challenge. In such markets, both buyers and sellers participate in a formal "double auction," whose rules are a lot simpler than informal matching and bargaining games.

Walras called the auctioning process he saw in use at the Paris Bourse a *tâtonnement.* The price of gold is fixed twice daily at Rothschild's Bank in London by the same process. Opening prices at the New York Stock Exchange are sometimes determined in much the same way.

Consider the case in which each of a number of traders wishes to buy or sell one gold bar. An auctioneer announces a price, after which the traders simultaneously say whether they are willing to trade at this price or not. If the numbers of buyers and sellers willing to trade are equal, the market closes at this price. If not, the auctioneer adjusts the price upward or downward, depending on whether there are more buyers or sellers willing to trade at the previous price.

If there is a unique Walrasian equilibrium, then it is a Nash equilibrium in this trading game for all players to say that they are willing to trade at any price at which they wouldn't take a loss. They thereby ensure that the tâtonnement can stop only at the unique Walrasian price (where the number of players who say they are willing to sell equals the number who say they are willing to buy). Adam may be able to make the tâtonnement stop at some other price by deviating from the equilibrium strategy, but it won't do him any good. If he stops the process by saying that he is willing to trade when he shouldn't, then he will suffer a loss. If he stops the process by saying that he isn't willing to trade when he should, then he will end up with nothing.

Such results fuel the enthusiasm of commentators who like to attribute magical powers to free markets, but one doesn't need to tweak the preceding example very much to generate Nash equilibria in which players lie about their trading position to manipulate the clearing price in their favor. For example, if there is more than one Walrasian price, Adam may have a strategic incentive to remain silent when he could make a profit by trading at the current price because he expects the auctioneer will then shift the price in his favor (Exercise 9.10.21). If the traders are uncertain about the state of supply and demand, there is no guarantee that the outcome will even be Walrasian.

The moral is that we can't always rely on an invisible hand at the tiller to steer us to a safe haven. Markets with large numbers of small buyers and sellers are relatively immune to manipulation, but game theory tells us how some traders may be able to fix the clearing price in other contexts. When such price fixing gets out of hand, as in the notorious California Power Exchange, game theory has the potential to propose new market mechanisms that aren't so easy to manipulate. With increasing computerization, the demand for expertise in this new area of market design can only increase.

## 9.7 Consumer Surplus

Perfect competition generates Pareto-efficient outcomes. So does a fully discriminating monopoly. But a classic monopoly, in which each fig leaf is sold at the same price, is generally inefficient. Figure 9.7(b) shows how this fact is commonly illustrated in economic textbooks using supply and demand curves.

To find the monopoly quantity of fig leaves, Eve looks for the point $N$ in Figure 9.7(b) at which her marginal revenue curve crosses her marginal cost curve. She then trades at the point $M$ on Adam's demand curve. If Adam has quasilinear preferences, the area marked $C$ is Adam's gain in utility (measured in apples) from trading at $M$ rather than not trading at all (Section 9.3.2). Economists therefore call $C$ the *consumer surplus* generated by the trade $M$. Eve's profit $P$ is called the *producer surplus* generated by $M$.

If Adam and Eve were to trade at the Walrasian point $W$ instead of $M$, the sum of consumer and producer surplus would increase by the area marked $D$. Economists call this area the *deadweight loss* due to monopoly. Since $D > 0$, operating at $M$ must be Pareto inefficient because both Adam and Eve could get larger payoffs by dividing $D$ between them.

Some economists proceed as though the proper aim of government should always be to maximize total surplus. The obvious objection is that what really matters to Adam is his gain in utility, which isn't the same as his consumer surplus when he doesn't have quasilinear preferences. A do-gooder who maximizes Adam's consumer surplus rather than his utility will therefore not be unreservedly welcome. As we know from Section 9.5.2, consumer surplus may not even be the money that Adam saves from what he would have to pay a fully discriminating monopolist. Even if it were, Adam is unlikely to be pleased at the do-gooder's implicit assumption that a dollar saved for a rich man is to be counted the same as a dollar saved for a poor man like himself.

In spite of these failings, consumer surplus will be used in the next chapter as a rough-and-ready measure of the welfare of the consumers under various forms of imperfect competition.

## 9.8 Roundup

This chapter has presented the two polar examples of market organization on which economic textbooks concentrate. The first step was to introduce the standard model of a consumer with convex preferences. The market demand curve is often thought adequate to summarize the properties of a bunch of such consumers, but this chapter includes a number of examples that show that knowing the market demand curve isn't always enough. Only with quasilinear preferences can we recover a consumer's utility function by finding the area under his demand curve.

There is a parallel between consumers and producers that is sometimes worth bearing in mind. A consumer seeks to maximize his utility function, while a producer seeks to maximize her profit function. An isoprofit curve can therefore be thought of as a producer's indifference curve. Even the difference between a consumer's demand curve and a producer's supply curve is only a matter of the point of view one adopts. A producer's supply curve is the same thing as her marginal cost curve, but even a consumer who merely trades some of his endowment can be thought of in these terms by introducing his opportunity cost, which is how much he loses as a consequence of parting with some of his stock instead of keeping it to use for other purposes.

The Edgeworth box allows a geometric interpretation of the deals available to two traders when there are only two commodities. We simplify discussions of the Edgeworth box by always counting the commodity on the vertical axis as the numeraire. The numeraire is the commodity in which prices are quoted.

The contract curve in the Edgeworth box is the set of Pareto-efficient deals that give both players at least as much utility as they would get by not trading at all. Which of these deals results if the players bargain rationally? The answer depends on the details of the game that governs the bargaining process.

When all the power in the bargaining game rests with one player, she is said to be a fully discriminating monopolist. She gradually lowers the price she offers to the consumer to move him along his indifference curve through the endowment point to the point on the contract curve that she likes best.

Monopolists in real life more commonly sell their product at a fixed price. The result is seldom Pareto efficient. Coase asked how a monopolist can commit herself to not undercutting her own price after selling as much as she can at that price. One way she can make her commitment credible is by not stocking more than she can sell at the high price. For this reason, economists often explain monopolists as people who jack up the price by restricting supply.

The outcome of a perfectly competitive market is called a Walrasian equilibrium. It arises when the prices adjust to a level at which the market supply for each commodity meets the market demand for that commodity. Unlike fixed-price monopolies, perfectly competitive markets are Pareto efficient. In the Edgeworth box, a Walrasian equilibrium corresponds to a point on the contract curve at which the common tangent to the indifference curves that touch there passes through the endowment point. In a diagram with supply and demand curves, it corresponds to the point where the two curves cross.

Adam Smith's invisible hand is a metaphor for the process that takes a trading game to one of its Nash equilibria. Such a Nash equilibrium will coincide with a Walrasian equilibrium of the underlying market only if the conditions are right. Even in a Walrasian tâtonnement, when traders respond to the price calls of an auctioneer, the outcome needn't always be Walrasian. The design of organized markets that are maximally robust against attempts by traders to manipulate the clearing price is an increasingly important area of application for game theory.

Consumer surplus is a rough measure of how much the consumers lose or gain under different types of market organization. Maximizing the sum of consumer and producer surplus is sometimes proposed as the proper aim of an enlightened government. There are worse things a government could do, but the proposal lacks any proper justification in the general case.

## 9.9 Further Reading

*Intermediate Microeconomics: A Modern Approach*, by Hal Varian: Norton, New York, 1990. This book is the most popular text for a second course in microeconomics for undergraduates.

*A Course in Microeconomic Theory*, by David Kreps: Princeton University Press, Princeton, NJ, 1990. This is an unusually thoughtful textbook for graduate students of economics.

## 9.10 Exercises

fun

1. The picture that heads up this chapter shows Alice in Dolly's store. The sheep is explaining that one egg costs $5\frac{1}{4}$ pennies. Two eggs cost 2 pennies—but you have to eat them both! Alice buys one egg. What standard assumption does she thereby violate?

review

2. Differentiate the following expressions partially with respect to $a$:

$$\text{(a) } 3a+2f; \qquad \text{(b) } a^2 f; \qquad \text{(c) } \ln(f+2\sqrt{a}).$$

review

3. Find $\nabla u(f,a)$ when $u(f,a)=a^2 f$. Write down the equation of the tangent plane to the curve $a^2 f = A^2 F$ at the point $(F,A)^\top$.

review

4. The functions $u : \mathbb{R}^2 \to \mathbb{R}$ and $v : \mathbb{R}^2 \to \mathbb{R}$ are defined by $u(f,a)=af^2$ and $v(f,a)=a^2 f$. Find the points $(f,a)$ at which $\nabla u(f,a)=\lambda \nabla v(f,a)$, for some $\lambda$. Why are these the points at which contours of the two functions touch?

econ

5. Profit is maximized when marginal revenue equals marginal cost. Why is this the same as setting the derivative of profit equal to zero? What is the relation between marginal revenue and marginal cost when profit is minimized?

econ

6. Adam's utility function $u : \mathbb{R}^2_+ \to \mathbb{R}$ is given by $u(f,a)=af^2$. If his endowment is $(0,A)$ and the price of fig leaves is $p$ apples, find the equation of one of Adam's indifference curves in $(f,p)$ space (Figure 9.1(b)). Sketch the curve and confirm that his demand curve $f=2A/3p$ is the locus of points where $p$ is maximized on such curves.

7. Bob's utility function $u : \mathbb{R}^2_+ \to \mathbb{R}$ is given by $u(f, a) = a^2 f$. The prices of fig leaves and apples are $p$ and $q$, where the numeraire is dollars. Bob has \$$M$, with which he can buy any bundle $(f, a)$ of fig leaves and apples for which $pf + qa \leq M$. Why does Bob demand $f = M/3p$ fig leaves and $a = 2M/3q$ apples? How many fig leaves and apples will $N$ copies of Bob demand? **econ**
8. If Alice is a monopoly seller of apples in a market consisting of $N$ copies of Bob from the previous exercise, show that her revenue is always the same, no matter what price she fixes. If her unit cost of producing an apple is positive, show that she will want to achieve the Wonderland solution of selling no apples at an infinite price. **fun**
9. When Adam has a utility function $u : \mathbb{R}^2_+ \to \mathbb{R}$ defined by $u(f, a) = f + 2a$, fig leaves and apples are said to be perfect substitutes. When $u(f, a) = \min\{f, 2a\}$, fig leaves and apples are said to be perfect complements. Explain this terminology. Sketch the indifference curves in both cases and find Adam's demand for fig leaves when his endowment is $(0, A)$ and the price of fig leaves is $p$ apples. **econ**
10. Adam's utility function $u : \mathbb{R}^2_+ \to \mathbb{R}$ is defined by $u(f, a) = a + \ln f$. **econ**
    a. Sketch the indifference curves of this quasilinear utility function. Verify that these are vertical translations of each other.
    b. Find Adam's demand for fig leaves when his endowment is $(F, A)$ and the price of fig leaves is $p$ apples.
    c. If Adam ends up with $f$ fig leaves, shade an area under his demand curve that equals his utility gain. Integrate his demand to confirm the equality. What goes wrong when $F = 0$?
11. Adam's endowment is $(0, A)$ and Eve's is $(F, 0)$. Draw an Edgeworth box and find the contract curve when Adam and Eve both have the utility functions $u : \mathbb{R}^2_+ \to \mathbb{R}$ defined by: **econ**

$$\text{(a) } u(f, a) = af^2; \text{ (b) } u(f, a) = (f + 1)^2(a + 2).$$

Find the Walrasian equilibria in each case. What trades will Eve enforce if she is a fully discriminating monopolist?

12. Draw a version of Figure 9.4(a) when Adam's utility function is given in Exercise 9.10.10. Comment on the shape of the classical monopolist's locus and the location of the monopoly point $M$. **econ**
13. Repeat Exercise 9.10.10 for the utility functions of Exercise 9.10.9. (Don't expect the results to resemble the diagrams in the text.) **econ**
14. Section 9.5.2 shows that the surplus extracted from Adam by a fully discriminating monopolist is equal to a certain area under his demand curve when his utility function is quasilinear. The same isn't true for other utility functions. Repeat the analysis of Section 9.5.2 that shows this fact using the Cobb-Douglas utility function $u : \mathbb{R}^2_+ \to \mathbb{R}$ defined by $u(f, a) = a^2 f$. **math**
15. Dolly owns the only hardware store in a small Midwestern town. She has stocked her usual supply of snow shovels for the winter, but the demand for shovels increases sharply after an unexpectedly heavy snowfall cuts the town off from the outside world. When Dolly raises the price at which she sells snow shovels, Alice complains that the new price is unfair because Dolly paid no **phil**

more for the shovels that she is selling at the new price than she paid for the shovels she was selling at their old price.

a. Draw demand and supply curves for the old and new situations.
b. Suppose Dolly sells her shovels at the old price. Is this fair to customers who would have bought a shovel at the old price but find that Dolly is out of shovels by the time they get to the store?
c. One might argue that Dolly shouldn't sell on a first-come-first-served basis but ration the shovels instead on a most-needy-first-served basis. But how is she to determine who is the most needy? As the widespread abuse of reserved parking for the disabled shows, she would be unwise to trust her customers' own assessments of their need. What proposals do you have for use in a town big enough that everybody doesn't know everybody else's business?
d. Economists sometimes argue that a person's need for something is reflected by the amount they are willing to pay to get it. If so, then Dolly could determine who is in most need by auctioning her snow shovels to the highest bidders. Show the outcome of running such an auction on your supply and demand diagrams, both before and after the snowfall. If her customers regard it as fair for the price to be determined in this way before the snowfall, why should they regard it as unfair to use the same process after the snowfall?
e. Comment on willingness to pay as a measure of need in health care.

phil

16. Some of the issues raised by the previous exercise are replayed every time OPEC, the oil-producers' cartel, seeks to exercise monopoly power by restricting supply to force up the price. The price at the pump then rises immediately, even though filling stations have their reserve tanks full of gasoline bought at the old price. Explain the backward induction argument that leads to the immediate rise in price. (It is based on the fact that nobody would wish to sell something today if they can sell it for more tomorrow.) To what extent are critics justified in characterizing the immediate price hike as unfair exploitation?

econ

17. In a market for $n$ used cars, a fraction $f$ of the owners are willing to sell their cars for $\$l$ or more. The remaining owners are willing to sell for $\$p$ or more. If $l < p$, draw the supply curve for cars on the assumption that car owners are price takers. The supply curve is made up of horizontal and vertical segments. If the demand curve in a perfectly competitive market cuts the supply curve in a horizontal segment, explain why some owners who are willing to sell at the equilibrium price sell their cars and some do not. If the demand curve cuts the supply curve in a vertical segment, how many cars are sold in equilibrium? Describe the informal auction that drives the price above what car owners who sell at the equilibrium price would be willing to accept.

econ

18. The reason that some owners are willing to sell for less than others in the previous exercise is that they own lemons (which are always breaking down) rather than peaches (which run well). The demand comes from used-car dealers, who are price takers like the owners. Although the dealers kick tires and the like, they actually can't tell a lemon from a peach until after they have bought it, but they must comply with the law that requires them to describe cars accurately when reselling.

a. The dealers are risk neutral. Their demand for used cars is therefore determined by the *expected* resale price. There are $M > n$ potential buyers willing to pay a dealer \$$L$ for a lemon and \$$P$ for a peach ($P > p > L > l$). Explain why the expected resale price for a car bought by a dealer is $LF + P(1 - F)$, where $F$ is the fraction of the $N$ cars bought by dealers that turn out to be lemons.

b. Draw the dealers' demand curve when they all believe that all $n$ used cars will be sold, so that $N = n$ and $F = f$. If $f < (P - p)/(P - L)$, show that the dealers have rational expectations, in that all cars actually are traded at the Walrasian equilibrium. If the inequality is reversed, confirm that the dealers' expectations are irrational, and hence the Walrasian equilibrium isn't viable in the long run.

c. Draw the dealers' demand curve when they all believe that only lemons will be sold, so that $N = nf$ and $F = 1$. Show that the dealers then always have rational expectations.

d. If the fraction of lemons owned isn't too small, confirm Akerlof's result that only lemons will be traded. If the fraction of lemons is small enough, confirm that both belief regimes are consistent with a Walrasian analysis.[14]

19. The closing paragraph of Section 9.6.1 sketches a proof of the first welfare theorem in the case of a market with $M$ clones of Dolly and $N$ clones of Alice. Augment Figure 9.6(b) by indicating the supply and demand curves for each individual Dolly and Alice. Show a pair $(A, \alpha)$ consisting of a quantity $A$ of wool and a price $\alpha$ that Alice would prefer to the Walrasian allocation. Do the same for Dolly and the pair $(D, \delta)$. Why is such a Pareto improvement impossible for both sides of the market unless $MD \geq NA$ and $MD\delta \leq NA\alpha$? Why can't these inequalities both hold when $\alpha < \delta$? Why must the latter inequality hold for Pareto improvement on a Walrasian allocation?

math

20. Build on the previous exercise to obtain a general proof of the first welfare theorem for a pure exchange economy. (Recall the Theorem of the Separating Hyperplane of Section 7.7.2.)

math

21. Ten gold brokers want to buy one gold bar each. A different ten brokers want to sell one gold bar each. Assign reserve prices to each broker so that the demand and supply curves overlap in a vertical line segment. Why are there multiple Walrasian equilibria? If the supply and demand curves are common knowledge, show that it is a Nash equilibrium in a Walrasian tâtonnement for one side of the market always to tell the truth about its willingness to pay and for the other side to remain silent until the tâtonnement reaches the Walrasian price that favors it the most.

econ

22. A leading philosophy journal offers the following story in support of the claim that it can make sense to have intransitive preferences.You always feel worse off if you are tortured a little bit less, provided that the lessened torture must be endured for a sufficiently longer period. By reducing the torture a little at a time and increasing the period that it must be endured, a person with transitive preferences must therefore prefer being tortured severely for two years to suffering the slight discomfort of a hangnail forever. But nobody would choose the former over the latter, and therefore intransitive preferences are reasonable.

phil

[14] What happens in the market will then depend on the expectations of the traders, whose prophecies therefore become self-fulfilling.

Show that the argument is wrong by examining the implications of maximizing the utility function:

$$u(x,t) = -\frac{xt}{1+t},$$

where $x$ represents the intensity of torture, and $t$ represents the length of the period it must be endured. Draw an indifference curve for this utility function through a point $(X_1, T_1)$ that represents being tortured severely for two years. Indicate the direction of preference by drawing appropriate arrows. Show a point $(X_2, T_2)$ that represents suffering a hangnail for a very long time.

Use your diagram to identify the mistake in the argument as a version of Zeno's paradox (in which Achilles runs faster than the tortoise he is racing but supposedly never overtakes it).

# 10

# *Selling Dear*

→ 11.1

## 10.1 Models of Imperfect Competition

In the picture that heads up this chapter, the Mad Hatter says he won't take less than half a guinea for his hat,[1] but the March Hare thinks he can get it for less. His chances would improve if a second hatter were competing for his business. But what prices would the two hatters then charge?

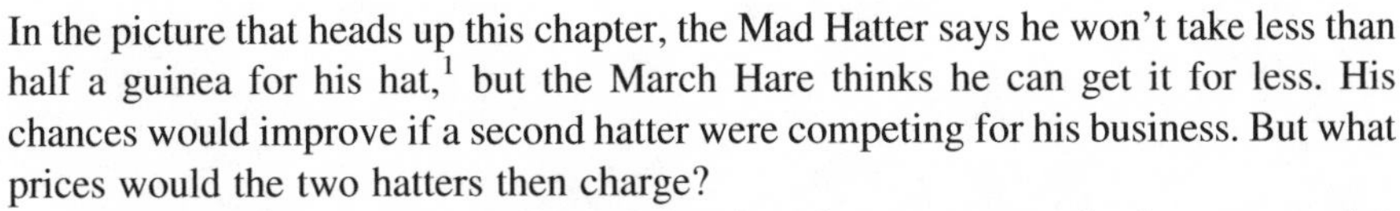

The game played when small numbers of producers compete in the same market is called an oligopoly. Demand curves were studied in the previous chapter so that we could keep things simple here by treating only the producers as players. We can't abstract away the producers in the same way by modeling them as supply curves because we need a large number of small producers to justify using the methods of perfect competition.

## 10.2 Cournot Models

The plan is to work systematically through the cases of principal interest, using the setting of Section 5.5.1. Recall that hats are produced in Wonderland at a cost of $\$c$ each. The demand equation is $h + p = K$, where $K$ is a much larger number than $c$.

[1]There were once twenty shillings in a British pound and twelve pennies in a shilling. Upscale stores priced clothing in the still more ancient guinea, worth twenty-one shillings. Half a guinea is therefore ten shillings and sixpence, written 10/6.

The number of hats that can be sold at a price of $\$p$ each is therefore $h = K - p$. In Section 5.5.1, we took $c = 3$ and $K = 15$.

### 10.2.1 Monopoly

An *oligopoly* is an industry with a small number $n$ of producers, each of appreciable size. An oligopoly with $n = 1$ is called a monopoly.

A price-making monopolist produces $\tilde{h} = \frac{1}{2}(K - c)$ hats and sells them at a price of $\tilde{p} = \frac{1}{2}(K + c)$ per hat (Section 9.5). This output generates her maximum profit of $\pi = \{\frac{1}{2}(K - c)\}^2$. As we will see, the lot of the consumer can be greatly improved by introducing a little competition into the market.

### 10.2.2 Duopoly

An oligopoly with $n = 2$ is called a *duopoly*. In Section 9.5, Alice was one of Dolly's customers, but now she and Bob will be the two producers.

In Cournot's model, both producers choose their output in ignorance of the choice of the other. The price at which hats are sold is then determined by the demand equation. That is, the price adjusts until supply equals demand. If Alice produces $a$ hats and Bob produces $b$ hats, the supply is simply the total number $h = a + b$ of hats produced. The demand for hats when the price is $p$ is $h = K - p$. Thus the price at which hats are sold satisfies

$$p = K - a - b.$$

Alice and Bob play a simultaneous-move game in which they choose $a$ or $b$ from the interval $[0, K]$. Since payoffs are identified with profits, the payoff functions are

$$\pi_1(a, b) = (p - c)a = (K - c - a - b)a,$$
$$\pi_2(a, b) = (p - c)b = (K - c - a - b)b.$$

The game is infinite because each player's strategy set is infinite. Our study of Duel shows that problems can sometimes arise in such games, but it can also happen that things are made a lot simpler. In this case, we can use calculus to find the unique Nash equilibrium $(\tilde{a}, \tilde{b})$ without much hassle.

To find her best replies to Bob's choice of $b$, Alice need only differentiate her profit function partially with respect to $a$ and set the derivative equal to zero. Since

$$\frac{\partial \pi_1}{\partial a} = K - c - 2a - b,$$

Alice's unique best reply to $b$ is

$$a = R_1(b) = \tfrac{1}{2}(K - c - b).$$

Alice's and Bob's reaction curves are shown in Figure 10.1. The equation of Bob's reaction curve is obtained simply by swapping $a$ and $b$ in the formula $a = R_1(b)$. Thus Bob's unique best reply to the choice of $a$ by Alice is

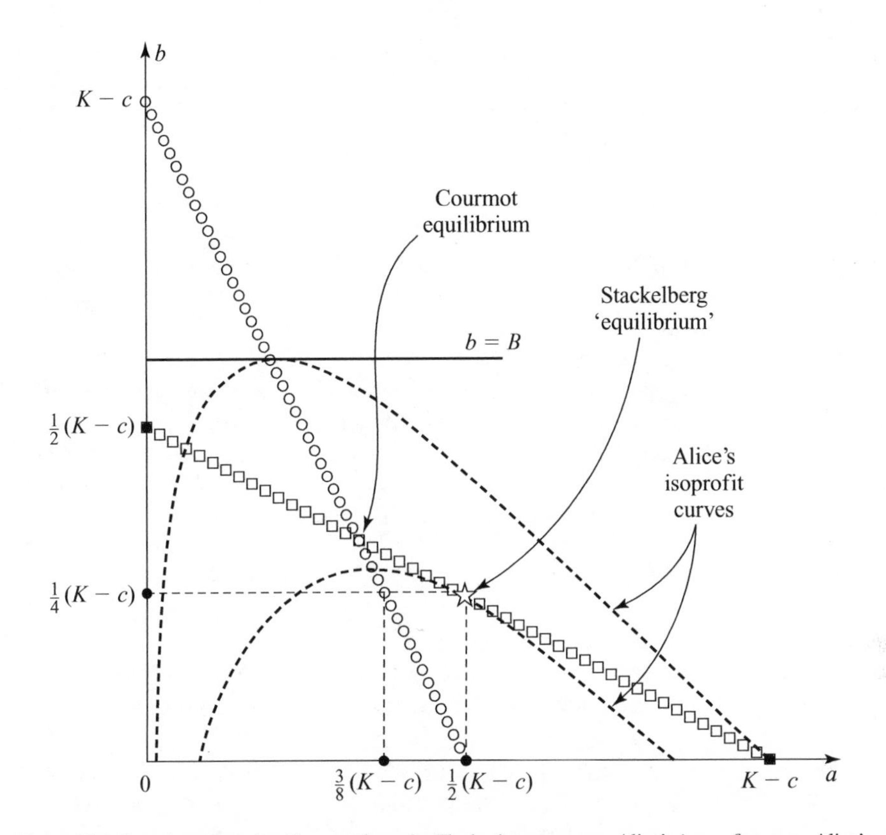


Figure 10.1 Reaction curves in a Cournot duopoly. The broken curves are Alice's *isoprofit curves*. Alice's profit along such curves is constant. For example, $\pi_1(a, b) = 3$ is the isoprofit curve on which Alice's profit is 3. (It has equation $(K - c - a - b)a = 3$, and hence is a hyperbola with asymptotes $a + b = K - c$ and $b = 0$.) Note that each horizontal line $b = B$ is tangent to an isoprofit curve where $a = R_1(B)$. This is because, in computing a best reply to $b = B$, Alice finds the point on $b = B$ at which her profit is largest. The Stackelberg outcome when Alice is the leader and Bob is the follower is marked with a star. It occurs where Alice's isoprofit curve touches Bob's reaction curve, because a Stackelberg leader maximizes profit on the assumption that the follower will make a best reply to her production choice.

$$b = R_2(a) = \tfrac{1}{2}\,(K - c - a).$$

A Nash equilibrium $(\tilde{a}, \tilde{b})$ occurs where the reaction curves cross. To find $\tilde{a}$ and $\tilde{b}$, the equations $a = R_1(b)$ and $b = R_2\ (a)$ must be solved simultaneously. The two equations are:

$$2\tilde{a} + \tilde{b} = K - c,$$
$$\tilde{a} + 2\tilde{b} = K - c,$$

and so $\tilde{a} = \tilde{b} = \frac{1}{3}(K - c)$.

Thus, in the Cournot model of duopoly, there is a unique Nash equilibrium in which each player produces $\frac{1}{3}(K - c)$ hats. The total number of hats produced is therefore $\frac{2}{3}(K - c)$, and so the price at which they are sold is $\tilde{p} = K - \frac{2}{3}(K - c) = \frac{1}{3}K + \frac{2}{3}c$. Each player's profit is $\{\frac{1}{3}(K - c)\}^2$.

These conclusions confirm Section 5.5.1's analysis of the special case when $c = 3$ and $K = 15$. In equilibrium, Alice and Bob each produce four hats and make a profit of \$16.

### 10.2.3 Collusion

The profit a monopolist makes is more than the sum of the profits that two duopolists would make by operating in the same market. Alice and Bob therefore have an incentive to *collude* by agreeing that each will restrict production to reduce total output to the monopoly level of $\frac{1}{2}(K-c)$ (Section 1.7.1).

In such a collusive agreement, who gets what market share will depend on how Alice and Bob bargain behind the scenes (Section 16.7). The simplest case arises when Alice and Bob agree to split the market fifty-fifty, so that each makes $\frac{1}{4}(K-c)$ hats, as shown in Figure 10.1. Each will then make half the monopoly profit. Since

$$\tfrac{1}{2}\{\tfrac{1}{2}(K-c)\}^2 > \{\tfrac{1}{3}(K-c)\}^2,$$

both players prefer their collusive deal to operating a Cournot duopoly.

The consumers suffer from such a collusive deal because they have to pay more for fewer hats. Collusion is therefore commonly illegal. This doesn't stop duopolists from trying to collude, but it does make it harder for them to succeed. No collusive deal worth making is a Nash equilibrium in this context, and so somebody always has an incentive to cheat on the deal. For example, Figure 10.1 shows that if Bob produces $\frac{1}{4}(K-c)$ in accordance with his agreement with Alice, then her best reply isn't to keep the agreement by producing $\frac{1}{4}(K-c)$ herself but to produce $\frac{3}{8}(K-c)$ instead. If she cheats by overproducing, what can Bob do about it? He can't sue Alice because their collusive agreement was illegal to begin with.

The fact that collusive deals are unstable in a Cournot duopoly looks good for the consumer, but Section 1.8 explains that things can be very different when Alice and Bob play the same Cournot duopoly over and over again. In the repeated game that results, worthwhile collusive deals become available as *equilibrium* outcomes since Bob can now punish Alice if she deviates from their agreement by refusing to collude with her in the future (Section 11.3.3).

### 10.2.4 Oligopoly

Cournot's duopoly story can be told again, but with $n$ players instead of only two. Player I's profit function is then

$$\pi_1(h_1, h_2, \ldots, h_n) = (K - c - h_1 - h_2 - \cdots - h_n)h_1.$$

A Nash equilibrium is found by solving the equations

$$\begin{aligned}
2\tilde{h}_1 + \tilde{h}_2 + \cdots + \tilde{h}_n &= K - c, \\
\tilde{h}_1 + 2\tilde{h}_2 + \cdots + \tilde{h}_n &= K - c, \\
&\vdots \\
\tilde{h}_1 + \tilde{h}_2 + \cdots + 2\tilde{h}_n &= K - c.
\end{aligned}$$

These have the unique solution

$$\tilde{h}_1 = \tilde{h}_2 = \cdots = \tilde{h}_n = \frac{1}{n+1}(K-c).$$

Suppose, for example, that $n=9$. Then each firm produces $\frac{1}{10}(K-c)$ hats. The total number of hats produced is therefore $\frac{9}{10}(K-c)$, and so the price at which they are sold is $\tilde{p} = K - \frac{9}{10}(K-c) = \frac{1}{10}K + \frac{9}{10}c$. Each player's profit is $\{\frac{1}{10}(K-c)\}^2$.

### 10.2.5 Perfect Competition

The firms in a perfectly competitive industry are price takers. They don't believe that they can affect the price at which hats sell. Section 9.6.2 explained why one should expect to observe a Walrasian equilibrium in such a market. This can be found by observing where the market supply curve and the market demand curve cross. If this argument is right, then a Cournot oligopoly should approach a perfectly competitive market when we reduce the market power of each producer to zero by allowing $n \to \infty$.

When $n \to \infty$ in a Cournot oligopoly with $n$ firms, the number of hats produced converges to $K-c$, and the price at which they are sold converges to $\tilde{p} = c$. Each firm makes zero profit. To see that this is also what would happen under perfect competition, note that the market supply curve is simply $p = c$ because all the firms have constant marginal cost $c$. The market demand curve is $p + h = K$. The supply and demand curves therefore cross where $\tilde{h} = K - c$ and $\tilde{p} = 1$. Each firm makes zero profit because it sells each hat at marginal cost.

The table of Figure 10.2 goes a long way toward explaining why economists like competition so much. Notice how things get better for the consumers as the industry becomes more competitive. The price of hats goes down, and the number of hats produced goes up.

| | Total output | Price | Total profit | Consumer surplus |
|---|---|---|---|---|
| Monopoly | $\frac{1}{2}(K-c)$ | $\frac{1}{2}K + \frac{1}{2}c$ | $\frac{1}{4}(K-c)^2$ | $\frac{1}{8}(K-c)^2$ |
| Duopoly | $\frac{2}{3}(K-c)$ | $\frac{1}{3}K + \frac{2}{3}c$ | $\frac{2}{9}(K-c)^2$ | $\frac{2}{9}(K-c)^2$ |
| Oligopoly | $\frac{n}{n+1}(K-c)$ | $\frac{1}{n+1}K + \frac{n}{n+1}c$ | $\frac{n}{(n+1)^2}(K-c)^2$ | $\frac{n^2}{2(n+1)^2}(K-c)^2$ |
| Competition | $K-c$ | $c$ | 0 | $\frac{1}{2}(K-c)^2$ |
| Stackelberg | $\frac{3}{4}(K-c)$ | $\frac{1}{4}K + \frac{3}{4}c$ | $\frac{3}{16}(K-c)^2$ | $\frac{9}{32}(K-c)^2$ |

Figure 10.2 Comparing different market structures. The entries in the *consumer surplus* column are a measure of how well off the consumers are under differing regimes.

## 10.3 Stackelberg Models

We met Stackelberg's model of a duopoly in Section 5.5.1. It differs from Cournot's model only in its timing. Alice leads by deciding how many hats to produce. Bob observes Alice's production decision and then follows by deciding how many hats he will produce. A pure strategy for Bob is therefore a *function* $f\colon [0, K] \to [0, K]$. When Alice chooses $a$, Bob's output is $b = f(a)$.

From our study of the Cournot model, we know that Bob has a unique best reply $b = R_2(a)$ to each possible choice of $a$ by Alice. His optimal pure strategy is therefore the function $R_2$. Alice knows that Bob will select $R_2$ and hence chooses the value $a = \tilde{a}$ that maximizes her profit of

$$\pi_1(a,\, R_2(a)).$$

The pair $(\tilde{a}, R_2)$ to which this argument leads is a subgame-perfect equilibrium of the Stackelberg game. The play of the game that results when this equilibrium is used is $[\,\tilde{a},\, \tilde{b}\,]$, where $\tilde{b} = R_2(\tilde{a})$. This outcome is marked with a star in Figure 10.1. Recall from Section 5.5.1 that economists like to call $[\,\tilde{a},\, \tilde{b}\,]$ a Stackelberg "equilibrium," although it is better described as a subgame-perfect play of a Stackelberg game.

We know from the Cournot model that $b = R_2(a) = \frac{1}{2}(K - c - a)$ and $\pi_1(a, b) = (K - c - a - b)a$. Alice therefore has to maximize

$$(K - c - a - R_2(a))a = \tfrac{1}{2}(K - c - a)a.$$

Her problem is easy in this special case because the expression for a Stackelberg leader's profit turns out to be exactly half what a monopolist who produced $a$ would get. Alice will therefore make the same output decision $\tilde{a} = \frac{1}{2}(K - c)$ as a monopolist.

Bob's output is $\tilde{b} = R_2(\tilde{a}) = \frac{1}{4}(K - c)$. Total production is $\frac{3}{4}(K - c)$. Hats are there fore sold at price $\tilde{p} = \frac{1}{4}K + \frac{3}{4}c$. Figure 10.2 explains why consumers prefer a Stackelberg duopoly to a Cournot duopoly.

Section 5.5.1 studied the special case in which $c = 3$ and $M = 15$. The analysis here confirms that Alice produces six hats and Bob produces three hats.

### 10.3.1 Monopoly with a Competitive Fringe

One can think of a market in which one large producer competes with many small rivals as a monopoly with a competitive fringe.

We model the large producer as a Stackelberg leader with unit cost $c$, who produces $l$ hats. She opens the game by publicly committing herself to selling at most $L < K$ hats. If she has no further commitment power, we know from Section 9.6.1 that we can then model her side of the market in the absence of a competitive fringe using a supply curve like that labeled $S_2$ in Figure 9.6(b). When the price $p$ at which hats sell exceeds $c$, the leader's supply curve therefore has equation $l = L$.

The firms in the fringe are assumed to have higher unit costs than the leader and thus don't produce at all when $p \leq c$. When $p > c$, we assume that the total of $f$ hats produced by the competitive fringe is determined by the supply curve $f = s(p-c)$, where $s > 0$ is a small constant.

The Walrasian equilibrium for the market is found by locating the point $W$ at which the market demand curve $p+h=K$ crosses the market supply curve. When $p > c$, the equation of the latter is $h = l + f = L + s(p-c)$. The equilibrium price is therefore $\tilde{p} = (K + sc - L)/(s+1)$, at which price $\tilde{h} = ((K-c)s + L)/(s+1)$ hats are sold. The leader's profit is

$$\pi = \frac{(K-c-L)L}{s+1},$$

which is maximized when $L = \frac{1}{2}(K-c)$. As in the pure Stackelberg model, the leader therefore chooses the same output as a monopolist without any rivals.

## 10.4 Bertrand Models

The time has now come to discuss strategic price setting. For this purpose, we will stay with our Wonderland duopoly, but Alice and Bob will now be selling strawberries at a farmers' market. Strawberries differ from hats in being perishable. In our model, they don't deteriorate at all unless kept overnight, after which they become unsaleable. They are therefore worth nothing at all if not sold on the day of the market.

As before, Alice's and Bob's unit costs are $\$c$ per basket. This isn't the cost of getting a basket to the market in the morning, which we will assume to be negligible. Nor is it the cost of getting an extra basket to the market during the day, which we assume to be infinite. It is the cost of the labor and other factors involved in selling a basket of strawberries. The demand equation continues to be $a+b+p=K$.

In a Cournot duopoly, Alice and Bob choose $a$ and $b$. For the reasons outlined in Section 9.6.1, their entire production is then sold at the highest price $p$ that someone is willing to pay for the last basket sold, so that $p = K-a-b$. The idea is that no customer will pay a high price early in the day, when they know that they can get a lower price by waiting until later.

Cournot's model of imperfect competition was challenged by his countryman Joseph Bertrand, who argued that Cournot had neglected the fierce competition in prices that is a feature of some markets. Instead of Alice and Bob choosing *quantities* and leaving the market to determine the price, Bertrand argued that Alice and Bob should be envisaged as committing themselves to *prices,* leaving the market to determine the quantity that each should supply.

In Section 5.5.2 and elsewhere, we have pointed out the necessity of questioning the credibility of a trader who claims to be offering a take-it-or-leave-it price. An antique dealer who made such a claim wouldn't be taken seriously anywhere in the world. However, take-it-or-leave-it prices are the norm in industries in which traders sell the same good under the same conditions over long periods. For example, you would look pretty foolish if you tried to bargain over the price of basket of strawberries at the checkout desk of a supermarket. However, no Italian housewife would willingly pay the posted price on a basket of strawberries offered for sale at a street market. In brief, the plausibility of the assumption that a trader can commit to a take-it-or-leave-it price depends on the special circumstances of the market under study.

Analyzing a Bertrand duopoly is easy if we assume that customers always buy from the cheaper vendor (and split their demand equally when two vendors offer the same price). The game then reduces to an auction in which both players try to undercut their rival's price so as to grab all the customers. The undercutting stops only when neither Alice nor Bob can cut any more without selling below cost. In equilibrium, the selling price is therefore equal to the players' marginal cost $\$c$. Although Alice and Bob are operating a duopoly, the outcome turns out to be the same as under perfect competition.

It is instructive to draw the players' reaction curves in the case when $c = 3$ and $K = 15$. With these values, a monopolist would set a price of \$9.

If Bob chooses a price $q > 9$ under Bertrand competition, then Alice should ignore him and simply trade at the monopoly price of $p = 9$. Since she is offering a lower price than Bob, the whole market will come to her, and Bob will be left out in the cold. If Bob chooses a price in the range $3 < q \leq 9$, then Alice should undercut him by a tiny amount so as to grab the whole market. If $q \leq 3$, Alice shouldn't undercut Bob because she would then make a loss by selling at less than her unit cost. Any reply $p \geq 3$ is optimal because Alice's profit is zero whatever she does.

As in the analysis of Duel in Section 8.2, some caution is necessary when "tiny amounts" appear on the scene. If prices must be quoted in whole pennies in the Bertrand model, then Alice isn't allowed to reply to Bob's choice of $q = 3.01$ with $p = 3.009$. Nor is it optimal for her to reply with $p = 3.00$ since her profit then becomes zero. Her best reply is $p = 3.01$, even though she then has to split the market with Bob. If we are careful about this detail, we are led to reaction curves of the type shown in Figure 10.3(a). When prices have to be stated in multiples of a cent, these reaction curves cross where $(p, q) = (3, 3)$ and $(p, q) = (3.01, 3.01)$.

However, the size of the smallest coin is usually an irrelevant distraction. We therefore focus on what happens when the value $\varepsilon > 0$ of the smallest coin decreases to zero. Both equilibria $(p, q) = (3, 3)$ and $(p, q) = (3 + \varepsilon, 3 + \varepsilon)$ then converge on (3, 3). Our claim that (3, 3) is the unique equilibrium of the continuous game therefore survives a more careful analysis.

### 10.4.1 Price Leadership

→ 10.6

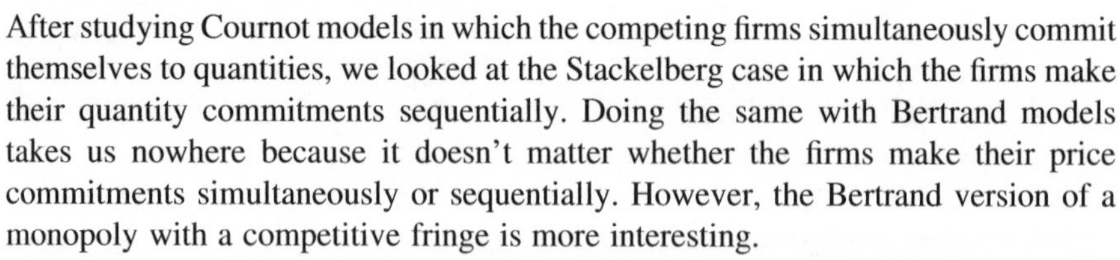

After studying Cournot models in which the competing firms simultaneously commit themselves to quantities, we looked at the Stackelberg case in which the firms make their quantity commitments sequentially. Doing the same with Bertrand models takes us nowhere because it doesn't matter whether the firms make their price commitments simultaneously or sequentially. However, the Bertrand version of a monopoly with a competitive fringe is more interesting.

We proceed as in Section 10.3.1, except that the leader now makes a price commitment rather than a quantity commitment. Economists are interested in such models as a step toward understanding markets in which all but one of the firms seem to play follow-the-leader when making price changes.

The leader won't commit herself to a price $P$ that exceeds the Walrasian price that would result if she weren't present in the market because she would then sell nothing. Equally, the competitive fringe will sell nothing unless they match her price

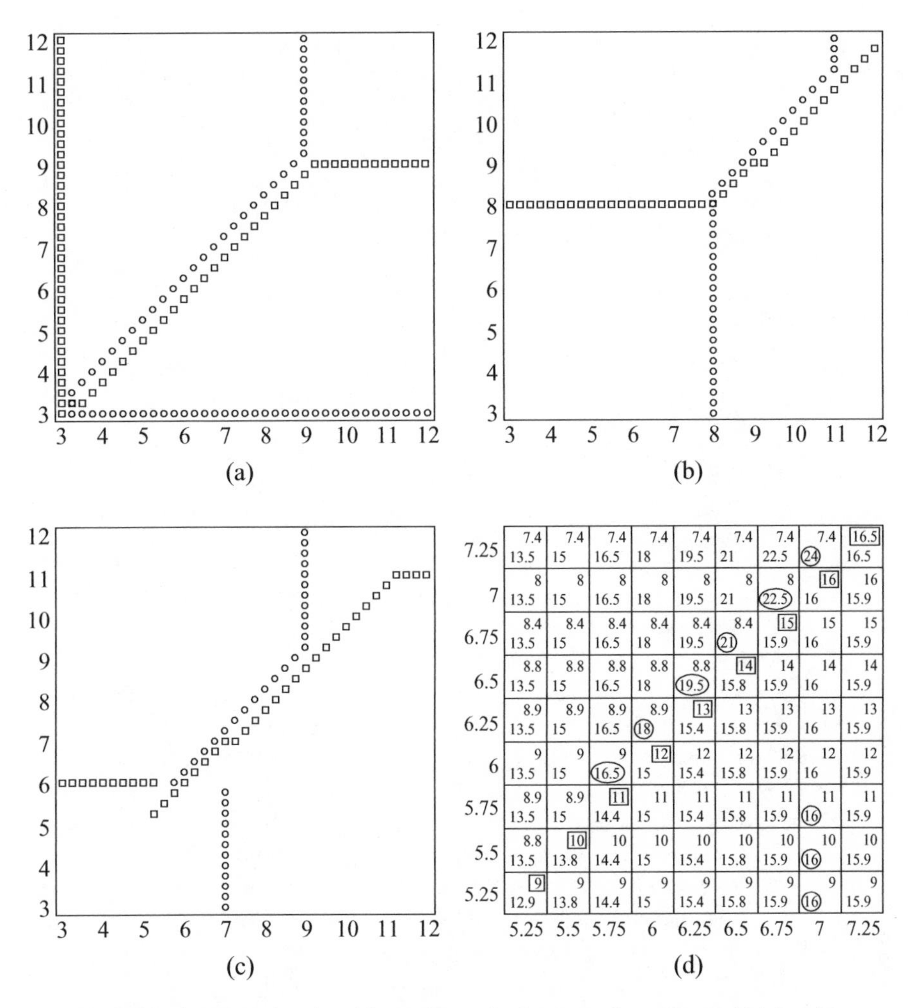

| | 5.25 | 5.5 | 5.75 | 6 | 6.25 | 6.5 | 6.75 | 7 | 7.25 |
|---|---|---|---|---|---|---|---|---|---|
| 7.25 | 7.4 / 13.5 | 7.4 / 15 | 7.4 / 16.5 | 7.4 / 18 | 7.4 / 19.5 | 7.4 / 21 | 7.4 / 22.5 | 7.4 / 24 | 16.5 / 16.5 |
| 7 | 8 / 13.5 | 8 / 15 | 8 / 16.5 | 8 / 18 | 8 / 19.5 | 8 / 21 | 8 / 22.5 | 16 / 16 | 16 / 15.9 |
| 6.75 | 8.4 / 13.5 | 8.4 / 15 | 8.4 / 16.5 | 8.4 / 18 | 8.4 / 19.5 | 8.4 / 21 | 15 / 15.9 | 15 / 16 | 15 / 15.9 |
| 6.5 | 8.8 / 13.5 | 8.8 / 15 | 8.8 / 16.5 | 8.8 / 18 | 8.8 / 19.5 | 14 / 15.8 | 14 / 15.9 | 14 / 16 | 14 / 15.9 |
| 6.25 | 8.9 / 13.5 | 8.9 / 15 | 8.9 / 16.5 | 8.9 / 18 | 13 / 15.4 | 13 / 15.8 | 13 / 15.9 | 13 / 16 | 13 / 15.9 |
| 6 | 9 / 13.5 | 9 / 15 | 9 / 16.5 | 12 / 15 | 12 / 15.4 | 12 / 15.8 | 12 / 15.9 | 12 / 16 | 12 / 15.9 |
| 5.75 | 8.9 / 13.5 | 8.9 / 15 | 11 / 14.4 | 11 / 15 | 11 / 15.4 | 11 / 15.8 | 11 / 15.9 | 11 / 16 | 11 / 15.9 |
| 5.5 | 8.8 / 13.5 | 10 / 13.8 | 10 / 14.4 | 10 / 15 | 10 / 15.4 | 10 / 15.8 | 10 / 15.9 | 10 / 16 | 10 / 15.9 |
| 5.25 | 9 / 12.9 | 9 / 13.8 | 9 / 14.4 | 9 / 15 | 9 / 15.4 | 9 / 15.8 | 9 / 15.9 | 9 / 16 | 9 / 15.9 |

(a) (b) (c) (d)

Figure 10.3 Reaction curves in prices. The smallest unit of currency is a quarter, which is quite large. It therefore sometimes pays to match your opponent's price rather than undercutting it. Figure 10.3(d) includes the payoffs for a $9 \times 9$ chunk of Figure 10.3(c). (Don't get confused by the fact that Alice's strategies correspond to *columns* and Bob's to *rows* in this final figure.)

of $P$ per hat. However, the invisible hand will ensure that they don't sell hats at significantly below $P$. It follows that they will supply $f = s(P-c)$ hats at a price negligibly less than $P$. Since the total demand at price $P$ is $K-P$ hats, the leader is then left to meet the residual demand of $K-P-s(P-c)$ hats. Her profit from meeting the residual demand is

$$\pi = (K + sc - (s+1)P - c)P,$$

which is maximized by taking

$$P = (K-(s-1)c)/2(s+1).$$

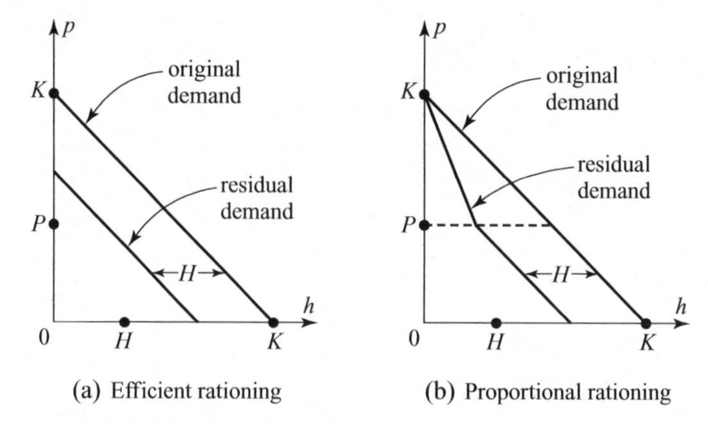


Figure 10.4 Residual demand curves. The original market demand curve has equation $p+h=K$. A group of $H$ customers is now served at price $P<K-H$. To obtain the residual demand curve under efficient rationing, throw out the $H$ consumers who are willing to pay a price $p>K-H$. Then shift the original demand curve a distance $H$ to the left. For the residual demand curve under proportional rationing, we continue to shift the segment of the original demand curve that lies in the range $0 \leq p \leq P$ a distance $H$ to the left, but the top point of the shifted segment is then joined by a straight line to the top of the original demand curve.

*Residual Demand.* One reason for taking an interest in the price leadership model is that it introduces the idea of *residual demand.* The original demand curve is $p+h=K$. What is the new demand curve after $H$ hats have been sold at price $P$? This is one of those questions that can't be answered unless we know something more about the consumers than the shape of their market demand curve.

The most interesting case is probably that in which the market demand is found by aggregating the demands of large numbers of consumers who want only one hat each. At price $P$, $K-P$ of these consumers will be demanding a hat, but only $H$ of them will be served by the competitive fringe. Who will the lucky customers be? Economists call the method that determines who gets served a *rationing* scheme.

Textbooks often proceed as though it were unproblematic that the rationing scheme will be efficient. Under *efficient rationing,* the customers served first are those who value a hat most.[2] One can imagine that the consumers who are the most eager to buy are the most forceful in pushing their way to the head of the line at Alice's store. But if customers actually join the line at random. we obtain the case of *proportional rationing* (provided there are enough tiny consumers to justify applying the law of large numbers). Of the consumers who are willing to pay Alice's price of $P$ for a hat, each willingness-to-pay category then contributes in proportion to its size to the lucky group of $H$ consumers who succeed in buying a hat from the competitive fringe.

Figure 10.4(a) shows the residual demand curve after $H$ customers have been served at price $P$ with efficient rationing. Figure 10.4(b) shows the residual demand curve with proportional rationing. Since the demand at price $P$ is the same in both

[2]Efficient rationing maximizes consumer surplus, but proportional rationing is no less Pareto efficient.

cases, the rationing scheme doesn't affect our analysis of the price leadership model, but it can make a big difference in other models.

## 10.5 Edgeworth Models

Consumers would like to live in a world in which Bertrand's model of duopoly were correct because a Bertrand duopoly is just like a perfectly competitive market in that the price is forced down to unit cost. The firms would prefer a world in which Cournot's model were correct because they make zero profit in Bertrand's model.

Which is the right model? Economists still dispute this question today, but game theorists agree that there is no "right" model of imperfect competition. Tolstoy famously said that all happy families are the same but that each unhappy family is unhappy in its own way. Similarly, all perfectly competitive markets are alike, but each imperfectly competitive market requires a model tailored to its own special circumstances.

*Capacity Constraints.* Even when fierce price competition is a feature of a market, it is seldom true that Bertrand's model can be uncritically applied. Francis Edgeworth pointed out the importance of the capacity constraints that duopolists typically face when they compete on price. Even when Alice and Bob can make price commitments, they will still take only a limited number of baskets of strawberries to the market as in a Cournot model. But now we can no longer call upon the invisible hand to tell us what price will prevail.

If Alice takes one basket and Bob takes ten, he can afford to laugh when she undercuts his price. Once Alice has sold her basket, Bob will act as a monopolist in serving the residual demand that remains after Alice's satisfied customers have departed. Bob's profit then depends on the shape of the residual demand curve, which depends in turn on the rationing scheme that decides which consumers Alice serves. For the moment, we shall assume that the rationing scheme is efficient (Section 10.4.1).

Edgeworth modeled the strategic realities of Alice's and Bob's problem as a two-stage game:

**Stage 1. Capacity choice.** Alice and Bob first simultaneously decide how many baskets to bring to market.

**Stage 2. Price setting.** Alice and Bob then simultaneously commit themselves to a price at which to sell for the rest of the day.

Since Alice and Bob are each assumed to observe the capacity choice of the other before committing themselves to a price, we can solve the game by backward induction.

Each possible capacity pair leads to a price-setting subgame, for which we need to find a Nash equilibrium. We then repeat the Cournot analysis, but with the equilibrium profits for each subgame replacing the Cournot profits. A Nash equilibrium for this replacement of the Cournot game then corresponds to a subgame-perfect equilibrium of the whole Edgeworth game. The restricted Cournot payoff table of Section 5.5.1 is shown in Figure 10.5(a). Figure 10.5(b) shows the new table

(a) Cournot

| | $b=3$ | $b=4$ |
|---|---|---|
| $a=4$ | 20, 15 | 16, 16 |
| $a=6$ | 18, 9 | 12, 8 |

(b) Edgeworth

| | $b=3$ | $b=4$ |
|---|---|---|
| $a=4$ | 20, 15 | 16, 16 |
| $a=6$ | $20\frac{1}{4}$, $8\frac{7}{8}$ | 16, $10\frac{2}{3}$ |

Figure 10.5 Edgeworth competition. The Cournot payoff table, which is repeated from Figure 5.11(c), shows only four of the possible pairs of capacity choices. The Edgeworth payoff table shows how the Cournot table changes when the players' quantity choice is followed by Bertrand competition in prices with efficient rationing.

that results from replacing the Cournot payoffs by the equilibrium profits in the four price-setting subgames that follow the four pairs of capacity choices.

The notable feature of Figure 10.5 is that the Cournot equilibrium remains an equilibrium after the payoffs have been changed to allow for Bertrand competition in prices.[3] At this equilibrium, Alice and Bob choose the Cournot quantities of $a=b=4$, and then both set their prices equal to the Cournot price of \$7. So Bertrand competition in prices needn't have any effect at all on the outcome of the game!

→ 10.5.1

We next sketch the argument used by Kreps and Scheinkman to show that this result is no accident.

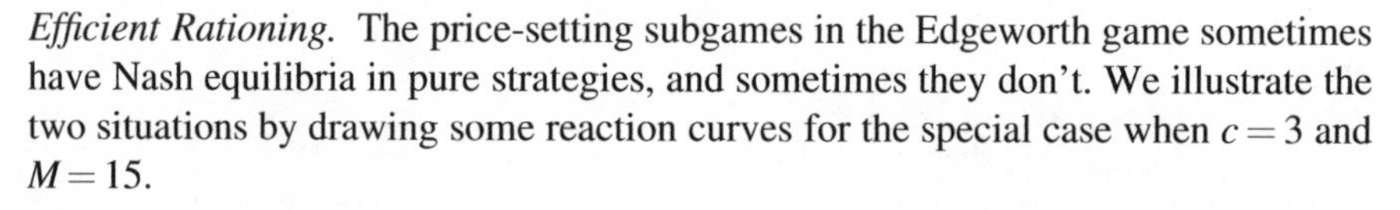

*Efficient Rationing.* The price-setting subgames in the Edgeworth game sometimes have Nash equilibria in pure strategies, and sometimes they don't. We illustrate the two situations by drawing some reaction curves for the special case when $c=3$ and $M=15$.

**The case $(\mathbf{a}, \mathbf{b}) = (\mathbf{3}, \mathbf{4})$.** Figure 10.3(b) shows the players' reaction curves in pure strategies for the price-setting subgame that follows the capacity choice $(a, b) = (4, 3)$. They differ from the reaction curves for a Bertrand duopoly since Alice and Bob can't meet demands that exceed their capacity.

It remains true that Alice and Bob will wish to undercut each other when the price is high enough, but the existence of capacity constraints prevents this phase from continuing all the way down to unit cost. Once the price gets low enough, Alice will be happy to let Bob undercut her. All of the customers will then want to buy their strawberries from Bob, but he has only three baskets to sell. After Bob's baskets are sold, the customers will have to buy their strawberries from Alice at her higher price.

With Kreps and Scheinkman's assumption that rationing is efficient, Bob will sell his three baskets of strawberries to the customers whose valuations are the highest. The residual demand left for Alice is then given by $a=12-p$ (instead of the demand of $a=15-p$ that she would face if she were acting as a monopolist, without Bob having creamed off the most valuable customers.)

[3]Alice's strategy in the Cournot equilibrium $(4, 4)$ of Figure 10.5(b) is weakly dominated, but this phenomenon disappears when we allow all capacity choices.

With her residual monopoly, Alice makes a profit of $\pi = (p-3)(12-p)$, which reaches a maximum when $p = 7\frac{1}{2}$. But to obtain this monopoly profit, Alice would need to sell $12-p = 4\frac{1}{2}$ baskets, which is more than the 4 baskets she has to sell. The nearest she can come to her monopoly profit is therefore to sell all 4 baskets at the most they will go for, namely $p = 12-4 = 8$. Once Bob's price $q \leq 8$, Alice will therefore cease to undercut him. Her optimal reply is then simply to stick with $p = 8$.

We can go through exactly the same story for Bob. Once $p \leq 8$, he will cease to undercut Alice. His optimal reply is also $q = 8$ because this is the price that a monopolist with only three baskets to sell is able to charge the customers that Alice was unable to satisfy at her lower price.

Since the players' reaction curves cross where $(p, q) = (8, 8)$, it is a Nash equilibrium for both players to commit themselves to a price of \$8. It is significant that this is the Cournot price when seven baskets are sold. The equilibrium profits that Alice and Bob receive in the price-setting subgame that arises when $(a, b) = (4, 3)$ are therefore identical to the Cournot profits when $(a, b) = (4, 3)$.

**The case $(\mathbf{a}, \mathbf{b}) = (\mathbf{6}, \mathbf{4})$.** Figure 10.3(c) shows the reaction curves for the price-setting subgame of the Edgeworth game that follows the capacity choice $(a, b) = (6, 4)$. The curves fail to cross, and hence there is no Nash equilibrium in pure strategies. The failure is possible because the reaction curves jump discontinuously from one place to another.

Alice's reaction curve jumps because she is no longer capacity constrained when acting as a residual monopolist. When facing a residual demand of $a = 11-p$, Alice maximizes her profit of $\pi = (p-3)(11-p)$ by setting $p = 7$. She then sells $a = 11-7 = 4$ baskets, which is less than her capacity of 6 baskets. Her profit is $\pi = \$16$. When $q \leq 5\frac{2}{3}$, this is better than she would get by fractionally undercutting Bob. By undercutting, she will sell her entire capacity at a profit of just less than $(q-3)6$, but $(q-3)6 \leq 16$ when $q \leq 5\frac{2}{3}$. As $q$ falls through $5\frac{2}{3}$, Alice's best reply $p$ therefore jumps from a fraction less than $q$ to $p = 7$.

Bob's situation is similar. As $p$ falls through 7, Bob's best reply $q$ jumps from a fraction less than $p$ to $q = 6$. As Figure 10.3(c) shows, the jumps are badly placed for the existence of a pure Nash equilibrium. Only mixed Nash equilibria are therefore possible.

Finding the mixed equilibria of a complicated game is seldom easy. A good beginning is to determine the support of the mixed strategies used in the equilibrium. The *support* of a mixed strategy is the set of pure strategies that are played with positive probability when it is used. As in Section 6.1.1, the supports we are looking for in this example are found by successively deleting dominated strategies, but one isn't always so lucky.

Figure 10.3(d) shows a 9×9 payoff table, with Alice as the *column* player and Bob as the *row* player. Notice that we lose the first and last rows and columns by successively deleting strongly dominated strategies, leaving us with a 7×7 table that covers prices between \$5.50 and \$7 inclusive. We would have ended up with the same 7×7 table if we had started with the whole payoff table. Any Nash equilibrium for the whole payoff table must therefore also be a Nash equilibrium for our 7×7 bimatrix game.

Since no pure equilibrium exists for the 7×7 bimatrix game, we look for an equilibrium in which Alice and Bob use mixed strategies, $\alpha$ and $\beta$. Without forgetting that Bob is player I and Alice is player II in our current formulation, we denote Alice's payoff matrix by $A$ and Bob's by $B$.

The vector $\beta^\top A$ lists the payoffs that Alice gets with each of her pure strategies when Bob plays $\beta$ (Section 6.4.4). If $\alpha$ calls for Alice to use each price between 5.5 and 7 with positive probability, then each such price must be equally profitable. This equilibrium profit is \$16 because all the entries in the last column of $A$ are 16. Thus

$$\beta^\top A = 16e^\top, \tag{10.1}$$

where $e$ is the $7\times 1$ vector whose entries are all 1. This vector equation expands into a system of seven linear equations in seven unknowns that can be solved for $\beta$ by pressing the right buttons on a computer—but one would need to recompute Figure 10.3(d) to a much greater degree of accuracy before placing much reliance on the answer.

In formal terms, the solution to (10.1) is $\beta = 16e^\top A^{-1}$, where $A^{-1}$ is the inverse matrix to $A$. The matrix $A$ has a simple structure in which the entry corresponding to price $(q,p)$ is $(11-p)(p-3)$ when $q \le p$, and $6(p-3)$ when $p > q$. As a consequence, many of the entries of $A^{-1}$ are zero, and so it is unusually easy to work out $A^{-1}$.

However, nobody inverts even an easy matrix if it can be avoided. As in Section 6.1.1, we therefore short-circuit the difficulties by passing to the continuous case and using the fact that the players must be indifferent between each pure strategy that they use with positive probability. Suppose that the equilibrium probability with which Bob uses a price $q \le p$ is $Q(p)$. Then Alice's profit when she uses a price $p$ with positive probability is

$$(11-p)(p-3)Q(p)+6(p-3)(1-Q(p)) = 16.$$

The equilibrium probability with which Bob uses a price $q \le p$ is therefore

$$Q(p) = \frac{6(p-5\frac{2}{3})}{(p-3)(p-5)},$$

which increases from 0 at $p = 5\frac{2}{3}$ to 1 at $p=7$.

The equilibrium probability $P(q)$ with which Alice uses a price $p<q$ can be somewhat more painfully calculated as

$$P(q) = \frac{4(q-5\frac{2}{3})}{(q-3)(q-5)},$$

which increases from 0 at $q = 5\frac{2}{3}$ to $\frac{2}{3}$ at $q=7$. Alice's equilibrium strategy therefore has an atom of mass $\frac{1}{3}$ at $q=7$. Each particular price is used with zero probability, except for \$7, which is used with probability $\frac{1}{3}$.

*Edgeworth Payoffs.* The preceding discussion tells us more than we need to know about Bertrand competition in two subgames of the Edgeworth game. The two cases typify what happens in general.

The pair $(3,4)$ of capacity choices typifies the points in the set $R$ that lie on or below *both* reaction curves in Figure 10.1. The price-setting subgame that follows such a pair $(a,b)$ of capacity choices has a pure equilibrium in which both players set the Cournot price and then sell their entire output. The Edgeworth payoffs that follow such capacity choices are therefore identical to the Cournot payoffs.

The pair $(6,4)$ of capacity choices typifies the points outside the set $R$. These pairs lie above one or the other of the two reaction curves of Figure 10.1. The price-setting subgame that follows such a pair $(a,b)$ of capacity choices has a mixed equilibrium. The player who makes the larger capacity choice at the equilibrium gets an expected payoff equal to the payoff he or she would receive as the follower in a Stackelberg game. In the case $(a,b)=(6,4)$, the player with the larger payoff is Alice, and her payoff is \$16, which is what she would get in a Stackelberg game, if she chose her capacity after observing Bob's choice of $b=4$.

These results allow us to confirm Kreps and Scheinkman's discovery that the Cournot outcome remains a subgame-perfect equilibrium of the Edgeworth game. If Alice's payoff matrices in Figure 10.3 included all capacity choices, the row corresponding to $a=4$ in Figure 10.3(b) would be identical to the row in Figure 10.3(a) for columns corresponding to $b \leq 4$. For columns corresponding to $b>4$, the entries would all be 16. Since the game is symmetric, similar observations apply to Bob's payoffs in the column corresponding to $b=4$. It follows that $(4,4)$ remains a Nash equilibrium in Figure 10.3(b), even when the payoff table is expanded to include all pairs of capacity choices.

### 10.5.1 Proportional Rationing

Kreps and Scheinkman's result shows that fierce price competition doesn't necessarily eliminate the high prices and low production typical of a Cournot duopoly. However, this doesn't imply that the laurels of victory should be awarded to Cournot in his posthumous debate with Bertrand. For example, we get a different result if we follow Beckmann in working with proportional rationing (Section 10.4.1).

As Figure 10.4 shows, a monopolist will then have an easier time when confronted with the residual demand curve. In particular, Alice and Bob are less likely to be capacity constrained when operating a residual monopoly, and so their reaction curves are more likely to jump. With proportional rationing, we should therefore expect to see mixed strategies in the price-setting subgame, even when Alice and Bob have chosen their capacities optimally. As Bertrand predicted, we will also see lower prices and higher production than in the Cournot case.[4]

*Package Holidays.* How realistic are models in which duopolists roll dice to decide what price to set? When mixed strategies are interpreted in such a naive way, the answer is: not at all. But we have seen that a player's choice of strategy may be effectively unpredictable without any need for dice to be rolled (Section 6.3).

Hal Varian plausibly explains sales at which goods are sold at knock-down prices as a way of implementing mixed strategies in practice. One can see the same phenomenon in action simply by walking around a fruit market at the end of the day and observing the wide variation in prices offered by vendors trying to unload their

[4]Davidson and Deneckere have confirmed these expectations.

stock. But a marketing executive working for Alice would think you were crazy if you asked what random device was used to decide when and where a sale should be held. Such decisions are commonly made by committees of experts who believe that their experience tells them exactly the right time and place for each sale. But Bob's experts have access to similar experience. If they can't predict what Alice's experts will decide, then Alice might as well be rolling dice for all that they can tell!

My own small experience in this area comes from consulting for a large package holiday business accused of anticompetitive activity by the European Commission. Package holidays perhaps fit the assumptions we have been making about strawberries better than real strawberries do. A successful firm has to book capacity far ahead of the holiday season, but whenever an airplane leaves with an empty seat, the corresponding package holiday is lost forever. On the other hand, empty seats don't decay at all during the booking season. When package holiday companies book more capacity than turns out to be in demand, they are therefore in the same position as strawberry sellers trying to unload their stock at the end of the day. Since proportional rationing seems to fit the realities of the package holiday business reasonably well, mixed equilibria in the price-setting subgame should therefore be observed.

Do we observe mixed equilibria in the package holiday business? Its executives are certainly no more inclined to roll dice than the executives of other industries, but the observed dispersion in prices offered late in the season for similar holidays is much too large to be attributed to cost or demand differences between rival firms. Trial-and-error learning has taught the marketing executives to be a lot more rational than they realize!

## 10.6 Roundup

In this chapter, some standard models of imperfect competition were considered for their own sake rather than to make some game-theoretic point.

In Cournot models, the firms simultaneously choose how much to produce. The price at which they can sell is then determined by the demand equation. Cournot oligopolies with $n$ firms cover a whole range of possibilities, from the case of monopoly when $n = 1$ to the case of perfect competition when $n \to \infty$. As $n$ increases, the consumers benefit as more is sold at a cheaper price. Stackelberg models differ only in that the firms make their production decisions sequentially.

Mixed strategies can arise in models of imperfect competition when price setting is modeled. In Bertrand competition, the players commit themselves to a price and then meet all the demand at that price. Since it always pays to undercut an opponent who sets prices above unit cost, the only equilibrium is for both players to sell at unit cost. Edgeworth competition introduces an earlier stage at which the players choose their capacities. Kreps and Scheinkman showed that the equilibria of simple models of Edgeworth competition reproduce the Cournot outcome, even though pricing is conducted à la Bertrand.

More realistic models generate results intermediate between the Bertrand and Cournot outcomes. For this chapter, the most significant feature of such models is that they typically require the use of mixed strategies for the price-setting phase of the game. Marketing executives will deny that they are using mixed strategies, but unexplained price dispersion sometimes provides evidence that they may have unconsciously purified a mixed equilibrium.

## 10.7 Further Reading

*Theory of Industrial Organization*, by Jean Tirole: MIT Press, Cambridge, MA, 1988. This popular book surveys a large number of models of imperfect competition, including a general version of the Edgeworth-Bertrand model. An appendix provides a quick introduction to a variety of game-theoretic tools.

*Game Theory with Economic Applications*, by Scott Bierman and Luís Fernández: Addison-Wesley, Reading, MA, 1998. Many economic models are studied without any fancy mathematics. The chapter on oligopoly is particularly relevant.

## 10.8 Exercises

1. If Alice and Bob bargain about which collusive deal to operate in the Cournot Game of Section 10.2.2, they will presumably agree on an outcome that is Pareto efficient for them (ignoring the interests of the consumers). Explain why the Pareto-efficient output pairs occur where Alice's and Bob's isoprofit curves touch. Deduce that the Pareto-efficient pairs lie on the straight line segment that joins the points corresponding to a monopoly by Alice and a monopoly by Bob. Why should this have been obvious straight away? Confirm that the Nash equilibrium of the game isn't Pareto efficient. **econ**
2. In the Cournot Game of Section 10.2.2, Alice and Bob have the same unit cost $c > 0$. Suppose instead that $0 < c_1 < c_2 < \frac{1}{2}K$. Show that **econ**
   a. The reaction curves are given by $q_1 = R_1(q_2) = \frac{1}{2}(K - c_1 - q_2)$ and $q_2 = R_2(q_1) = \frac{1}{2}(K - c_2 - q_1)$.
   b. The Nash equilibrium outputs are $q_1 = \frac{1}{3}K - \frac{2}{3}c_1 + \frac{1}{3}c_2$ and $q_2 = \frac{1}{3}K - \frac{2}{3}c_2 + \frac{1}{3}c_1$.
   c. The equilibrium profits are $\pi_1 = \frac{1}{9}(K - 2c_1 + c_2)^2$ and $\pi_2 = \frac{1}{9}(K - 2c_2 + c_1)^2$.
3. Sketch the isoprofit curves for the previous exercise. **math**
   a. Show the players' reaction curves in your diagram, together with the Nash equilibrium of the game.
   b. Show the equilibrium outputs of the Stackelberg version of the game in which Alice is the leader and Bob the follower.
   c. Indicate the curve of Pareto-efficient output pairs that are potential collusive agreements. Show that the curve has equation

$$2(q_1 + q_2)^2 - (2q_1 + q_2)(K - c_2) - (2q_2 + q_1)(K - c_1) + (K - c_1)(K - c_2) = 0.$$

   Confirm that the monopoly outcomes of the game lie on this curve but that the Nash equilibrium outcome doesn't.
4. In Section 10.2.2, all firms manufacture the same product. Consider instead the case when the goods are *differentiated*. Perhaps Alice produces widgets at unit cost $c_1$, but Bob produces wowsers at unit cost $c_2$. If $q_1$ widgets and $q_2$ wowsers are produced, the respective prices for the two goods are determined by the **econ**

demand equations $p_1 = K - 2q_1 - q_2$ and $p_2 = K - q_1 - 2q_2$. Adapt Cournot's duopoly model to this new situation and find:

a. the players' reaction curves
b. the quantities produced in equilibrium and the prices at which the goods are sold
c. the equilibrium profits

econ

5. Repeat Exercise 10.8.4 with the demand equations $p_1 = K - 2q_1 + q_2$ and $p_2 = K + q_1 - 2q_2$. Comment on how the consumers' view of the products must have changed to yield these new demand equations.

econ

6. In the $n$-player Cournot oligopoly game of Section 10.2.4:
   a. Modify the game so that each firm has to pay a fixed cost of $F$ regardless of the quantity it produces in order to enter the hat industry. Explain why nobody's behavior changes if the fixed cost $F$ is less than each player's equilibrium profit.
   b. If the fixed cost exceeds the equilibrium profit with $n$ players, then at least one firm would have been better off if it hadn't entered the hat industry. Assuming there are no barriers to entry other than payment of the fixed entry cost of $F$, determine the number of firms that will end up producing hats. What happens as $F \to 0$?

econ

7. Section 10.4 studied Bertrand's model when both firms have the same unit cost $c$, but now Alice's and Bob's unit costs differ, so that $c_1 > c_2 > 0$. Show that only Bob sells strawberries at price $p = c_1$. Alice therefore doesn't enter the market, but the possibility that she might determines the price at which Bob is able to sell his product.

econ

8. Repeat Exercises 10.8.4 and 10.8.5 for the case of a Bertrand duopoly.

econ

9. Widget consumers are located with uniform density[5] $\rho$ along a single street of length $l$. Each consumer has a need for at most one widget. A consumer will buy the widget he needs from whatever source costs him the least.[6] In calculating costs, he considers not only the price at which a widget is sold at an outlet but also his transportation expenses. It costs a consumer $\$tx^2$ to travel a distance $x$ and back again.

   In Hotelling's model, two widget firms are to open outlets on a street. Each firm independently decides where to locate its outlet. After their outlets have been opened, they engage in Bertrand competition. The unit cost to a firm is always $\$c > 0$. There are no fixed costs.

   a. Alice locates her outlet a distance $x$ from the west end of the street, and Bob locates his outlet a distance $X$ from the east end of the street. If Bob now sets price $P$, determine the number of customers Alice will get if she sets price $p$. What will her profit be?
   b. After $x$ and $X$ have been chosen, the subgame that ensues is a simultaneous-move game in which the pure strategies for Alice and Bob are their prices $p$ and $P$. Find the unique Nash equilibrium of this subgame for all values of $x$ and $X$. What profits will the players make if this Nash equilibrium is played?

[5]This means that there are $\rho x$ consumers in any segment of the street of length $x$.
[6]His reserve price for a widget is so high that it needn't be considered.

c. Consider the simultaneous-move game in which the locations $x$ and $X$ are chosen. Take for granted that a Nash equilibrium will be played in the price-fixing game that follows. What is the unique Nash equilibrium?

d. Comment on the relevance of the idea of a subgame-perfect equilibrium to the preceding analysis.

e. Where do the firms locate in equilibrium? What prices do they set? What are their profits?

10. Repeat the oligopoly analysis of Section 10.2.4 on the assumption that the firms play follow-the-leader instead of moving simultaneously. Player I first chooses the quantity $q_1$ that he will produce. Player II chooses her quantity $q_2$ second, after having observed player I's choice. Then player III chooses $q_3$ after having observed $q_1$ and $q_2$, and so on. What is a "Stackelberg equilibrium" for this game? Show that the equilibrium outcome approaches perfect competition as $n \to \infty$. **econ**

11. Analyze the $n$-player oligopoly model of Section 10.2.4 again but without the assumption that the players all move simultaneously. Assume instead that player I chooses the quantity $q_1$ first. After observing his choice, all the remaining players then choose how much to produce simultaneously. What happens as $n \to \infty$? **econ**

12. In the Hotelling model of Exercise 10.8.9, show that the conclusion is unchanged if one firm acts as a leader by locating first, provided that everything else remains the same. **econ**

13. We sometimes see the same product being sold at widely different prices. A possible explanation of such price dispersion is that the pricing game has a mixed equilibrium. Even Bertrand duopolies can have mixed equilibria. Consider the case in which both players face a constant unit cost of $c > 0$, and the demand equation is $q = p^{-l}$ $(0 < l < 1)$. Show that, for each $a > c$, there is a symmetric mixed equilibrium in which a player's price $p$ exceeds $P \geq a$ with probability **econ**

$$\text{prob}\,(p > P) = \left(\frac{a-c}{P-c}\right)\left(\frac{P}{a}\right)^{l}.$$

14. One reason for neglecting the mixed equilibrium of the previous exercise when studying a Bertrand duopoly is that it requires the use of arbitrarily large prices with positive probability. This possibility is excluded when $l > 1$ because the monopoly price $p^* = lc/(l-1)$ is then finite. Confirm that any price $p > p^*$ is strongly dominated in the Bertrand game. **econ**

    Let $c < a < b \leq p^*$. Confirm that there is no symmetric Nash equilibrium in which all prices in the interval $[a, b)$ are played with positive probability and all prices outside are played with zero probability.

15. For each $\varepsilon > 0$, find a mixed $\varepsilon$-equilibrium (Section 5.6.1) for a Bertrand duopoly under the assumptions of the previous exercise. (Take $a$ close to $c$ and $b = p^*$.) Sketch a graph showing the probability density function of a mixed strategy in your $\varepsilon$-equilibrium. In what sense does this strategy approach the traditional equilibrium strategy (each player chooses $p = c$) as $\varepsilon \to 0$? What happens to the players' payoffs as $\varepsilon \to 0$? **math**

# 11 *Repeating Yourself*

## 11.1 RECIPROCITY

With no external means of *enforcing* preplay agreements, rational players must forego the fruits of cooperation in games like the Prisoners' Dilemma when they are played just once. One might say that rational players need a police officer to help them cooperate in such one-shot games. However, cooperation can become available as an *equilibrium* outcome when the game is played repeatedly.

For example, Alice and Bob may be duopolists looking for a way to cooperate in the Prisoners' Dilemma. In the one-shot case, no agreement they make will last because collusion between duopolists is illegal, and so neither Alice nor Bob will have legal recourse if the other cheats. But it is a Nash equilibrium in a *repeated* version of the game if both players use the GRIM strategy (Section 1.8). At this equilibrium, Alice and Bob always cooperate—but not because they have ceased to be money-grubbing misfits. They cooperate because their partner will give them hell in the future if they don't!

Everybody understands that such *self-policing* or *incentive-compatible* arrangements are important in ordinary life. People provide a service to others expecting to get something in return. As the saying goes, I'll scratch your back if you'll scratch mine. If the service a person provides *isn't* satisfactorily reciprocated, then the service will be withdrawn. Sometimes, some disservice will be offered instead.

The philosopher David Hume argued that this type of *reciprocity* is the glue that holds human societies together. When we cease to reciprocate adequately, those around us apply a little discipline to bring us back into line. Not much is usually needed. A half-turned shoulder or an almost imperceptible pout are usually enough to indicate that further social exclusions will follow if you keep straying from the approved equilibrium path. But everything up to and including the electric chair is available for those who refuse to fit in at all.

Although we all play our part in maintaining a complex network of reciprocal arrangements with those around us, we understand how the system works no better than the physics we use when riding a bicycle. Game theory offers some insight into the nuts and bolts of such self-policing agreements. How do they work? Why do they survive? How much cooperation can they support?

## 11.2 Repeating a Zero-Sum Game

What happens when Adam and Eve play Matching Pennies twice? The zero-sum game $Z$ of Figure 11.1(a) has player II's payoff matrix from Section 6.2.2. Its value is $v = \frac{1}{2}$. The players' security strategies are both $(\frac{1}{2}, \frac{1}{2})$.

When $Z$ is played twice by the same players, it becomes the *stage game* of the repeated game $Z^2$. (If the stage games aren't all the same, the game obtained by playing them one after the other is called a *supergame*.)

For this example, we assume that the players don't discount the future. Their payoffs in the repeated game $Z^2$ are obtained simply by adding up the payoffs in each stage game. For example, if the strategy pair $(s_1, t_2)$ is used at the first stage and the strategy pair $(s_2, t_2)$ is used at the second stage, then Adam gets $0 + 1 = 1$ in the repeated game $Z^2$.

*The Repeated Game Isn't M.* The strategic form of $Z^2$ is often confused with the matrix game $M$ of Figure 11.1(b). The error becomes apparent when we try to use a security strategy from one game in the other.

The mixed strategy $(0, \frac{1}{2}, \frac{1}{2}, 0)$ is a security strategy for Adam in the game with matrix $M$. It guarantees him an expected payoff of exactly $+1$. He can't guarantee getting more than $+1$ because the mixed strategy $(0, \frac{1}{2}, \frac{1}{2}, 0)$ similarly guarantees Eve an expected payoff of exactly $-1$.

But suppose Eve knows that Adam will toss a fair coin to decide which of $s_1s_2$ and $s_2s_1$ to play. If Adam uses $s_i$ at stage one, Eve will then reply with $t_i$ at stage two.

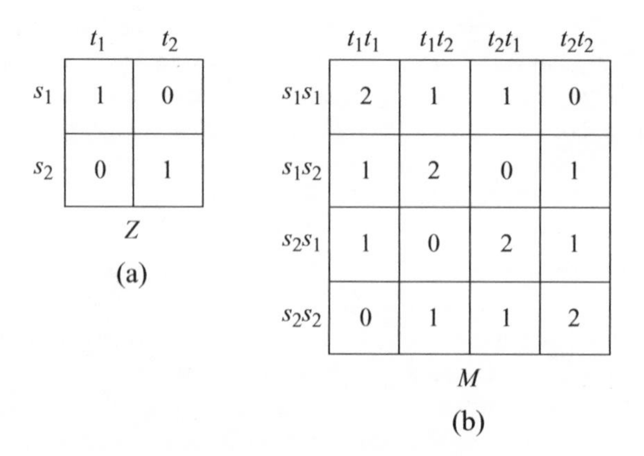


Figure 11.1 Two zero-sum games.

Since she always gets 0 at the second stage by playing this way, her total expected payoff becomes $-\frac{1}{2}+0=-\frac{1}{2}$. Thus, Adam gets only $+\frac{1}{2}$, which is less than the supposedly secure $+1$.

The reason for this anomaly is that the pure strategies of $M$ don't allow the players to make their behavior at the second stage *contingent* on what happened at the first stage.

*Making Actions Contingent on the History of Play.* The set $S=\{s_1, s_2\}$ of Adam's pure strategies in the stage game $Z$ are called *actions* so as not to confuse them with pure strategies in the repeated game $Z^2$. The set of actions for Eve in the stage game $Z$ is $T=\{t_1, t_2\}$.

The set of possible outcomes at the first stage of $Z^2$ is $H=S\times T$. The four elements of the set $H$ are therefore the possible *histories* of play at the second stage. For example, the history $h_{21}=(s_2,t_1)$ means that Adam used action $s_2$ and Eve used action $t_1$ at the first stage.

A pure strategy for Adam in $Z^2$ is a pair $(s,f)$, in which $s$ is an action in $S$ to be used at the first stage and $f: H\to S$ is a *function.* If Eve uses action $t$ at the first stage, then the history of the game at the second stage will be $h=(s,t)$, and so his pure strategy demands that Adam take the action $f(h)=f(s,t)$ at the second stage. His play at the second stage is therefore contingent on what happened at the first stage.

*How Many Pure Strategies?* The fact that Adam and Eve don't forget what has happened so far when deciding what action to take in the next stage game has the unpleasant consequence that the number of pure strategies in a repeated game quickly gets very large.

The 16 possible functions $f: H\to S$ are shown as tables in Figure 11.2(a). Since Adam has 2 choices for $s$ and 16 choices for $f$, he has $2\times 16$ choices of pure strategy in $Z^2$. Eve has the same number of pure strategies, and so the strategic form of $Z^2$ is represented by the $32\times 32$ matrix of Figure 11.3(a).

This strategic form isn't so monstrous as it first appears because each row and column is repeated four times. If each distinct row and column is written down only once, we obtain the $8\times 8$ matrix of Figure 11.3(b). This $8\times 8$ matrix is a *reduced* strategic form in which the pure strategies included are just those in which a player's behavior at the second stage is contingent only on what the *opponent* did at the first stage.

A pure strategy for an Eve who ignores what she did at the first stage is a pair $(t, G)$ in which $t$ is an action in $T$ and $G: S\to T$ is a function. If Adam uses action $s$ at the first stage, then Eve will use action $t$ at the first stage and action $G(s)$ at the second stage. The four possible functions $G: S\to T$ are shown as tables in Figure 11.2(b).

*Solving $Z^2$.* It is obvious that one solution of a repeated two-person, zero-sum game is for both players always to play their security strategies for the stage game independently at every repetition. However, it is instructive to see that this isn't the only security strategy available to the players.

For example, it is a security strategy for Adam to use each of his pure strategies in the zero-sum game of Figure 11.3(b) with probability $\frac{1}{8}$. His expected payoff is then exactly $+1$, whatever Eve does. Eve similarly guarantees an expected payoff of exactly $-1$ by using each of her pure strategies with probability $\frac{1}{8}$. Another security strategy calls for Adam to choose each of $(s_1,F_{12})$, $(s_1,F_{21})$, $(s_2,F_{12})$, and $(s_2,F_{21})$ with probability $\frac{1}{4}$. Alternatively, he can choose each of $(s_1,F_{11})$, $(s_1,F_{22})$, $(s_2,F_{11})$,

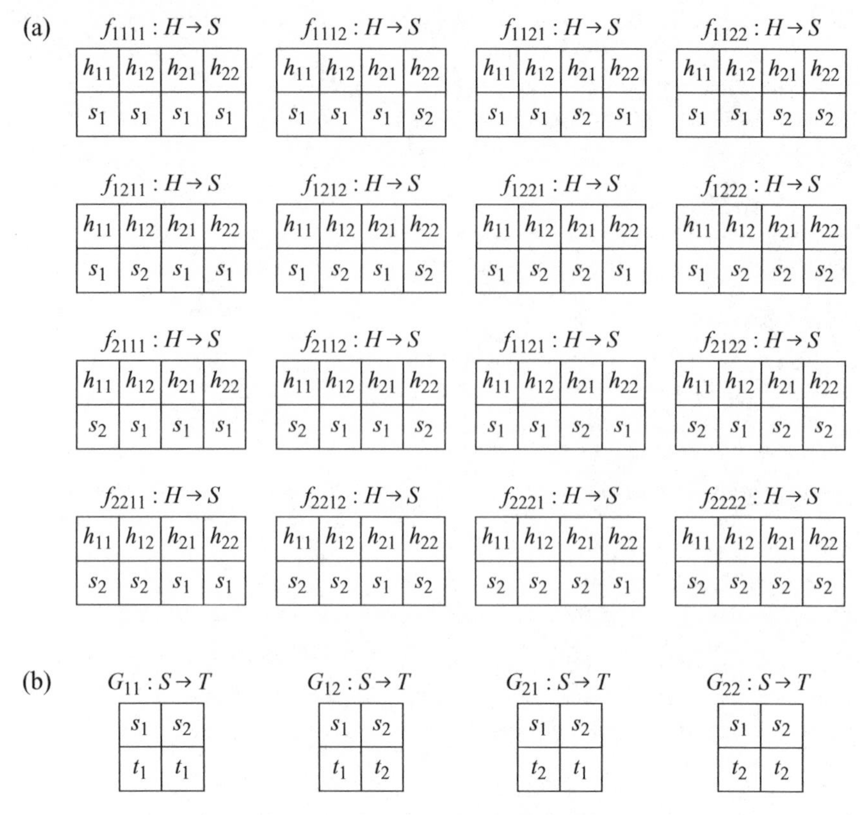


Figure 11.2 Some functions.

and $(s_2, F_{22})$ with probability $\frac{1}{4}$. It is this last security strategy that corresponds to his always playing the stage-game security strategy independently at every repetition.

## 11.3 Repeating the Prisoners' Dilemma

We now study the game obtained by repeating the Prisoners' Dilemma of Figure 11.4(a) $n$ times. If $n = 10$, each player then has $2^{349,525}$ pure strategies (Exercise 11.9.3), but it is still easy to analyze. There is a unique subgame-perfect equilibrium in which each player always chooses *hawk*.

The reason is simple. Before the last stage of the repeated game, it is possible that Adam might be deterred from choosing *hawk* because of the fear that Eve will retaliate later in the game. But, at the *final* stage, no later retaliation is possible. Since *hawk* dominates *dove* in the one-shot Prisoners' Dilemma, both players will therefore choose *hawk* at the final stage, whatever the history of play may have been.

Now consider the last stage but one. Nobody can be punished for playing *hawk* at this stage because the worst punishment the opponent could inflict at the final stage for such bad behavior is to play *hawk*. But the opponent is planning to use *hawk* at the final stage anyway, no matter what happens now. Both players will therefore use *hawk* at the last stage but one.

<table>
<tr><th></th><th>$(t_1, g_{1111})$</th><th>$(t_1, g_{1112})$</th><th>$(t_1, g_{1121})$</th><th>$(t_1, g_{1122})$</th><th>$(t_1, g_{1211})$</th><th>$(t_1, g_{1212})$</th><th>$(t_1, g_{1221})$</th><th>$(t_1, g_{1222})$</th><th>$(t_1, g_{2111})$</th><th>$(t_1, g_{2112})$</th><th>$(t_1, g_{2121})$</th><th>$(t_1, g_{2122})$</th><th>$(t_1, g_{2211})$</th><th>$(t_1, g_{2212})$</th><th>$(t_1, g_{2221})$</th><th>$(t_1, g_{2222})$</th><th>$(t_2, g_{1111})$</th><th>$(t_2, g_{1112})$</th><th>$(t_2, g_{1121})$</th><th>$(t_2, g_{1122})$</th><th>$(t_2, g_{1211})$</th><th>$(t_2, g_{1212})$</th><th>$(t_2, g_{1221})$</th><th>$(t_2, g_{1222})$</th><th>$(t_2, g_{2111})$</th><th>$(t_2, g_{2112})$</th><th>$(t_2, g_{2121})$</th><th>$(t_2, g_{2122})$</th><th>$(t_2, g_{2211})$</th><th>$(t_2, g_{2212})$</th><th>$(t_2, g_{2221})$</th><th>$(t_2, g_{2222})$</th></tr>
<tr><td>$(s_1, f_{1111})$</td><td rowspan="8" colspan="8">2</td><td rowspan="8" colspan="8">1</td><td rowspan="4" colspan="4">1</td><td rowspan="4" colspan="4">0</td><td rowspan="4" colspan="4">1</td><td rowspan="4" colspan="4">0</td></tr>
<tr><td>$(s_1, f_{1112})$</td></tr>
<tr><td>$(s_1, f_{1121})$</td></tr>
<tr><td>$(s_1, f_{1122})$</td></tr>
<tr><td>$(s_1, f_{1211})$</td><td rowspan="4" colspan="4">0</td><td rowspan="4" colspan="4">1</td><td rowspan="4" colspan="4">0</td><td rowspan="4" colspan="4">1</td></tr>
<tr><td>$(s_1, f_{1212})$</td></tr>
<tr><td>$(s_1, f_{1221})$</td></tr>
<tr><td>$(s_1, f_{1222})$</td></tr>
<tr><td>$(s_1, f_{2111})$</td><td rowspan="8" colspan="8">1</td><td rowspan="8" colspan="8">2</td><td rowspan="4" colspan="4">1</td><td rowspan="4" colspan="4">0</td><td rowspan="4" colspan="4">1</td><td rowspan="4" colspan="4">0</td></tr>
<tr><td>$(s_1, f_{2112})$</td></tr>
<tr><td>$(s_1, f_{2121})$</td></tr>
<tr><td>$(s_1, f_{2122})$</td></tr>
<tr><td>$(s_1, f_{2211})$</td><td rowspan="4" colspan="4">0</td><td rowspan="4" colspan="4">1</td><td rowspan="4" colspan="4">0</td><td rowspan="4" colspan="4">1</td></tr>
<tr><td>$(s_1, f_{2212})$</td></tr>
<tr><td>$(s_1, f_{2221})$</td></tr>
<tr><td>$(s_1, f_{2222})$</td></tr>
<tr><td>$(s_2, f_{1111})$</td><td rowspan="2" colspan="2">1</td><td rowspan="2" colspan="2">0</td><td rowspan="2" colspan="2">1</td><td rowspan="2" colspan="2">0</td><td rowspan="2" colspan="2">1</td><td rowspan="2" colspan="2">0</td><td rowspan="2" colspan="2">1</td><td rowspan="2" colspan="2">0</td><td>2</td><td>1</td><td>2</td><td>1</td><td>2</td><td>1</td><td>2</td><td>1</td><td>2</td><td>1</td><td>2</td><td>1</td><td>2</td><td>1</td><td>2</td><td>1</td></tr>
<tr><td>$(s_2, f_{1112})$</td><td>1</td><td>2</td><td>1</td><td>2</td><td>1</td><td>2</td><td>1</td><td>2</td><td>1</td><td>2</td><td>1</td><td>2</td><td>1</td><td>2</td><td>1</td><td>2</td></tr>
<tr><td>$(s_2, f_{1121})$</td><td rowspan="2" colspan="2">0</td><td rowspan="2" colspan="2">1</td><td rowspan="2" colspan="2">0</td><td rowspan="2" colspan="2">1</td><td rowspan="2" colspan="2">0</td><td rowspan="2" colspan="2">1</td><td rowspan="2" colspan="2">0</td><td rowspan="2" colspan="2">1</td><td>2</td><td>1</td><td>2</td><td>1</td><td>2</td><td>1</td><td>2</td><td>1</td><td>2</td><td>1</td><td>2</td><td>1</td><td>2</td><td>1</td><td>2</td><td>1</td></tr>
<tr><td>$(s_2, f_{1122})$</td><td>1</td><td>2</td><td>1</td><td>2</td><td>1</td><td>2</td><td>1</td><td>2</td><td>1</td><td>2</td><td>1</td><td>2</td><td>1</td><td>2</td><td>1</td><td>2</td></tr>
<tr><td>$(s_2, f_{1211})$</td><td rowspan="2" colspan="2">1</td><td rowspan="2" colspan="2">0</td><td rowspan="2" colspan="2">1</td><td rowspan="2" colspan="2">0</td><td rowspan="2" colspan="2">1</td><td rowspan="2" colspan="2">0</td><td rowspan="2" colspan="2">1</td><td rowspan="2" colspan="2">0</td><td>2</td><td>1</td><td>2</td><td>1</td><td>2</td><td>1</td><td>2</td><td>1</td><td>2</td><td>1</td><td>2</td><td>1</td><td>2</td><td>1</td><td>2</td><td>1</td></tr>
<tr><td>$(s_2, f_{1212})$</td><td>1</td><td>2</td><td>1</td><td>2</td><td>1</td><td>2</td><td>1</td><td>2</td><td>1</td><td>2</td><td>1</td><td>2</td><td>1</td><td>2</td><td>1</td><td>2</td></tr>
<tr><td>$(s_2, f_{1221})$</td><td rowspan="2" colspan="2">0</td><td rowspan="2" colspan="2">1</td><td rowspan="2" colspan="2">0</td><td rowspan="2" colspan="2">1</td><td rowspan="2" colspan="2">0</td><td rowspan="2" colspan="2">1</td><td rowspan="2" colspan="2">0</td><td rowspan="2" colspan="2">1</td><td>2</td><td>1</td><td>2</td><td>1</td><td>2</td><td>1</td><td>2</td><td>1</td><td>2</td><td>1</td><td>2</td><td>1</td><td>2</td><td>1</td><td>2</td><td>1</td></tr>
<tr><td>$(s_2, f_{1222})$</td><td>1</td><td>2</td><td>1</td><td>2</td><td>1</td><td>2</td><td>1</td><td>2</td><td>1</td><td>2</td><td>1</td><td>2</td><td>1</td><td>2</td><td>1</td><td>2</td></tr>
<tr><td>$(s_2, f_{2111})$</td><td rowspan="2" colspan="2">1</td><td rowspan="2" colspan="2">0</td><td rowspan="2" colspan="2">1</td><td rowspan="2" colspan="2">0</td><td rowspan="2" colspan="2">1</td><td rowspan="2" colspan="2">0</td><td rowspan="2" colspan="2">1</td><td rowspan="2" colspan="2">0</td><td>2</td><td>1</td><td>2</td><td>1</td><td>2</td><td>1</td><td>2</td><td>1</td><td>2</td><td>1</td><td>2</td><td>1</td><td>2</td><td>1</td><td>2</td><td>1</td></tr>
<tr><td>$(s_2, f_{2112})$</td><td>1</td><td>2</td><td>1</td><td>2</td><td>1</td><td>2</td><td>1</td><td>2</td><td>1</td><td>2</td><td>1</td><td>2</td><td>1</td><td>2</td><td>1</td><td>2</td></tr>
<tr><td>$(s_2, f_{2121})$</td><td rowspan="2" colspan="2">0</td><td rowspan="2" colspan="2">1</td><td rowspan="2" colspan="2">0</td><td rowspan="2" colspan="2">1</td><td rowspan="2" colspan="2">0</td><td rowspan="2" colspan="2">1</td><td rowspan="2" colspan="2">0</td><td rowspan="2" colspan="2">1</td><td>2</td><td>1</td><td>2</td><td>1</td><td>2</td><td>1</td><td>2</td><td>1</td><td>2</td><td>1</td><td>2</td><td>1</td><td>2</td><td>1</td><td>2</td><td>1</td></tr>
<tr><td>$(s_2, f_{2122})$</td><td>1</td><td>2</td><td>1</td><td>2</td><td>1</td><td>2</td><td>1</td><td>2</td><td>1</td><td>2</td><td>1</td><td>2</td><td>1</td><td>2</td><td>1</td><td>2</td></tr>
<tr><td>$(s_2, f_{2211})$</td><td rowspan="2" colspan="2">1</td><td rowspan="2" colspan="2">0</td><td rowspan="2" colspan="2">1</td><td rowspan="2" colspan="2">0</td><td rowspan="2" colspan="2">1</td><td rowspan="2" colspan="2">0</td><td rowspan="2" colspan="2">1</td><td rowspan="2" colspan="2">0</td><td>2</td><td>1</td><td>2</td><td>1</td><td>2</td><td>1</td><td>2</td><td>1</td><td>2</td><td>1</td><td>2</td><td>1</td><td>2</td><td>1</td><td>2</td><td>1</td></tr>
<tr><td>$(s_2, f_{2212})$</td><td>1</td><td>2</td><td>1</td><td>2</td><td>1</td><td>2</td><td>1</td><td>2</td><td>1</td><td>2</td><td>1</td><td>2</td><td>1</td><td>2</td><td>1</td><td>2</td></tr>
<tr><td>$(s_2, f_{2221})$</td><td rowspan="2" colspan="2">0</td><td rowspan="2" colspan="2">1</td><td rowspan="2" colspan="2">0</td><td rowspan="2" colspan="2">1</td><td rowspan="2" colspan="2">0</td><td rowspan="2" colspan="2">1</td><td rowspan="2" colspan="2">0</td><td rowspan="2" colspan="2">1</td><td>2</td><td>1</td><td>2</td><td>1</td><td>2</td><td>1</td><td>2</td><td>1</td><td>2</td><td>1</td><td>2</td><td>1</td><td>2</td><td>1</td><td>2</td><td>1</td></tr>
<tr><td>$(s_2, f_{2222})$</td><td>1</td><td>2</td><td>1</td><td>2</td><td>1</td><td>2</td><td>1</td><td>2</td><td>1</td><td>2</td><td>1</td><td>2</td><td>1</td><td>2</td><td>1</td><td>2</td></tr>
</table>

(a)

| | $(t_1, G_{11})$ | $(t_1, G_{12})$ | $(t_1, G_{21})$ | $(t_1, G_{22})$ | $(t_2, G_{11})$ | $(t_2, G_{12})$ | $(t_2, G_{21})$ | $(t_2, G_{22})$ |
|---|---|---|---|---|---|---|---|---|
| $(s_1, F_{11})$ | 2 | 2 | 1 | 1 | 1 | 1 | 0 | 0 |
| $(s_1, F_{12})$ | 2 | 2 | 1 | 1 | 0 | 0 | 1 | 1 |
| $(s_1, F_{21})$ | 1 | 1 | 2 | 2 | 1 | 1 | 0 | 0 |
| $(s_1, F_{22})$ | 1 | 1 | 2 | 2 | 0 | 0 | 1 | 1 |
| $(s_2, F_{11})$ | 1 | 0 | 1 | 0 | 2 | 1 | 2 | 1 |
| $(s_2, F_{12})$ | 1 | 0 | 1 | 0 | 1 | 2 | 1 | 2 |
| $(s_2, F_{21})$ | 0 | 1 | 0 | 1 | 2 | 1 | 2 | 1 |
| $(s_2, F_{22})$ | 0 | 1 | 0 | 1 | 1 | 2 | 1 | 2 |

(b)

Figure 11.3 Some big matrices.

| | $d$ | $h$ |
|---|---|---|
| $d$ | 2 / 2 | 3 / −1 |
| $h$ | −1 / 3 | 0 / 0 |

(a) Prisoners' Dilemma

| | $d$ | $h$ |
|---|---|---|
| $d$ | $2 + y(h)$ / $2 + x(h)$ | $3 + y(h)$ / $-1 + x(h)$ |
| $h$ | $-1 + y(h)$ / $3 + x(h)$ | $0 + y(h)$ / $0 + x(h)$ |

(b) Final stage

Figure 11.4 Repeating the Prisoners' Dilemma a finite number of times.

Now apply the same argument at the last stage but two, and so on.

THEOREM 11.1 *The finitely repeated Prisoners' Dilemma has a unique subgame-perfect equilibrium in which both players plan always to use* hawk.

→ 11.3.1

*Proof* For a formal proof, we need to appeal to the principle of induction. To this end, we take $P(n)$ to be the proposition that the theorem is true for the $n$-times repeated Prisoners' Dilemma.

We know that $P(1)$ is true because this is just the one-shot case. To deduce the theorem it remains to show that $P(n) \Rightarrow P(n+1)$ for each $n = 1, 2, \ldots$. For this purpose, we assume that $P(n)$ holds for some particular value of $n$ and try to deduce that $P(n+1)$ holds as well.

Suppose the last stage of the $(n+1)$-times repeated game has been reached after a history $h$ of play. If the play at the $k$th stage resulted in a payoff of $x_k$ to Adam, then his total payoff by the time the final stage is about to be played is $x(h) = x_1 + x_2 + \cdots + x_n$. Eve will similarly have accumulated a payoff of $y(h)$. The final stage game shown in Figure 11.4(b) is therefore strategically identical to the Prisoners' Dilemma of Figure 11.4(a) since adding a constant to each of a player's payoffs makes no strategic difference to a game. In particular, *hawk* strongly dominates *dove*, and so the final stage game has the unique Nash equilibrium (*hawk*, *hawk*).

The game of Figure 11.4(b) is a smallest subgame of the $(n+1)$-times repeated Prisoners' Dilemma. Backward induction requires replacing each such smallest subgame by a leaf labeled with a payoff pair that results from using a Nash equilibrium in the subgame. As (*hawk*, *hawk*) is the only Nash equilibrium in Figure 11.4(b), the required payoff pair is $(0 + x(h),\ 0 + y(h))$.

The new game obtained by this reduction is precisely the same as the $n$-times repeated Prisoners' Dilemma. Since $P(n)$ is being assumed, *hawk* will therefore always be used by both players. We already know that they play *hawk* at the final stage of the $(n+1)$-times repeated Prisoners' Dilemma, and so they *always* play *hawk* in this game. Thus $P(n+1)$ is true.

→ 11.3.2

### 11.3.1 Rational Fools?

Critics who regard playing *hawk* in the one-shot Prisoners' Dilemma as the act of a "rational fool" think that the same applies doubled when the Prisoners' Dilemma is repeated. Surely game theory must be nonsensical if it claims that rational people can't cooperate even in an ongoing relationship.

In countering this kind of criticism, it is important to recognize how different the repeated case is from the one-shot case. It is best for Eve to choose *hawk* in the one-shot Prisoners' Dilemma, whatever may or may not be known about Adam's rationality because *hawk* strongly dominates *dove*. But to get a similar result in the finitely repeated Prisoners' Dilemma, it isn't even enough that it be common knowledge that both players are rational. We need their beliefs on this subject to be so firmly rooted that nothing that happens in the game can ever lead to the beliefs being abandoned (Section 2.9.4). No matter how often Adam may behave irrationally, Eve must continue to attribute his behavior to some transient influence that won't persist into the future (Section 5.6.2).

Such an idealizing assumption is very unrealistic. Toward the end of a long repeated game, what real person is going to believe that an opponent with an unbroken history of irrationality is likely to behave rationally in the future? When the finitely repeated Prisoners' Dilemma is analyzed with more realistic assumptions, different conclusions follow. In particular, equilibria exist that call for the play of *dove* (Exercise 5.9.22).

One step toward more realism involves looking at repetitions of the Prisoners' Dilemma that don't have a definite time horizon. Of course, nobody lives forever, and so Adam knows his relationship with Eve will end eventually, but he is unlikely to be able to tie down the precise date of their final meeting.

### 11.3.2 An Infinite Horizon Example

What happens when the Prisoners' Dilemma is repeated an *indefinite* number of times? We start with the case when the probability that the game will continue to the next stage is always $\frac{2}{3}$.

The repeated game doesn't have a finite horizon. The probability that the game won't be over after the $N$th stage is $(\frac{2}{3})^N$, and so there is no value of $N$ for which the game is certain to be over after the $N$th stage. It is true that $(\frac{2}{3})^N \to 0$ as $N \to \infty$, and hence the probability that the game will literally go on forever is zero. But it is nevertheless a game with an infinite horizon.

The GRIM strategy calls for *dove* to be played as long as the opponent reciprocates by playing *dove* also (Section 1.8). If the opponent ever fails to do so, GRIM calls for *hawk* always to be played thereafter. Any deviation will therefore be well and truly punished, but if both players stick to GRIM, no occasion for punishment will arise. The players will cooperate forever.

Each player's expected payoff will then be

$$C = 2+2(\tfrac{2}{3})+\cdots+2(\tfrac{2}{3})^{N-1}+2(\tfrac{2}{3})^{N}+2(\tfrac{2}{3})^{N+1}+2(\tfrac{2}{3})^{N+2}+\cdots.$$

Suppose a player deviates from GRIM by playing *hawk* for the first time at the $(N+1)$st stage. The deviant will then get a payoff of three at this stage but no more than zero thereafter. If the other player sticks with GRIM, the most the deviant can get from switching is therefore

$$D = 2+2(\tfrac{2}{3})+\cdots+2(\tfrac{2}{3})^{N-1}+3(\tfrac{2}{3})^{N}+0(\tfrac{2}{3})^{N+1}+0(\tfrac{2}{3})^{N+2}+\cdots.$$

It is unprofitable to deviate if $C \geq D$. We therefore consider

$$\begin{aligned} C - D &= (2-3)(\tfrac{2}{3})^N + (2-0)(\tfrac{2}{3})^{N+1} + (2-0)(\tfrac{2}{3})^{N+2} + \cdots \\ &= (\tfrac{2}{3})^N\{-1 + 2\times\tfrac{2}{3}(1 + \tfrac{2}{3} + (\tfrac{2}{3})^2 + \cdots)\} \\ &= (\tfrac{2}{3})^N\left\{-1 + \tfrac{4}{3}\left(\frac{1}{1-\frac{2}{3}}\right)\right\} = 3(\tfrac{2}{3})^N > 0. \end{aligned}$$

It follows that a player who deviates from GRIM loses if the opponent sticks with GRIM. Thus, (GRIM, GRIM) is a Nash equilibrium whose play results in the players cooperating all the time in the infinite horizon game.

This story explains why rational cooperation can be viable in a repeated Prisoners' Dilemma with an infinite horizon. It is such a good story that we will repeat it every time we meet a new repeated game!

### 11.3.3 Collusion in a Repeated Cournot Duopoly

→ 11.4

It is difficult for Alice and Bob to collude in a one-shot Cournot Duopoly Game because someone always has an incentive to cheat on any deal that isn't a Nash equilibrium. But duopolists almost never play just once. They usually play day after day without any definite view about when their interaction will come to an end. Such a repeated environment is much more favorable for sustaining collusive deals than the harsh one-shot environment we considered in Section 10.2.3. To see why, we need only copy the argument of Section 11.3.2 that shows cooperation to be feasible in an indefinitely repeated version of the Prisoners' Dilemma.

In the Cournot duopoly of Section 10.2.2, the firms would jointly extract the most from the consumers if they colluded in restricting their joint production to $\tilde{h} = \frac{1}{2}(K-c)$ hats, which is the output of a profit-maximizing monopolist. In the repeated version to be studied now, suppose they agree that Alice will produce $a$ hats in each period and that Bob will produce $b$ hats, where $a + b = \tilde{h}$. If this agreement holds up, Alice makes a profit of $A$ per period, and Bob makes a profit of $B$. But what if someone cheats?

In the one-shot case, this consideration destroys their prospects of colluding successfully. But, in the indefinitely repeated case, Alice and Bob can build a provision into their agreement about what action should be taken if someone cheats. The simplest provision is that the partnership is then dissolved, and both play their one-shot Nash equilibrium strategies in all succeeding periods.

Is it a Nash equilibrium in the *repeated* game if Alice and Bob play this way? The answer depends on how Alice and Bob evaluate the stream of payoffs they will receive while playing the repeated game. Economists usually proceed by computing the *present value* of such an income stream (Exercise 19.11.19).

For example, if the yearly interest rate is fixed at $r$%, then the present value of an IOU promising to pay \$$X$ three years from now is $Y = X/(1+r)^3$. More generally, the present value of an income stream $X_0, X_1, X_2, \ldots,$ in which \$$X_t$ is to be received $t$ years from now, is simply $X_0 + \delta X_1 + \delta^2 X_2 + \cdots$, where $\delta = 1/(1+r)$ is the *discount factor* associated with the fixed interest rate $r$.

If Alice's discount factor is $\delta$, where $0 < \delta < 1$, then she will evaluate the income stream she gets when neither player deviates from their collusive agreement as being worth

$$C = A + A\delta + A\delta^2 + \cdots + A\delta^N + \cdots .$$

If Bob sticks to the agreement but Alice deviates, how much will Alice get?

If Alice deviates for the first time at the $(N+1)$st stage, she gets

$$D = A + A\delta + \cdots + A\delta^{N-1} + Z\delta^N + E\delta^{N+1} + E\delta^{N+2} + \cdots ,$$

where $Z$ is the bonanza that Alice enjoys from cheating on Bob at the $(N+1)$st stage and $E$ is the profit per period that each firm receives when each plays the one-shot Nash equilibrium strategy.

Alice will cheat if $C < D$. We therefore consider

$$\begin{aligned} C - D &= \delta^N\{(A - Z) + (A - E)\delta + (A - E)^2\delta^2 + \cdots\} \\ &= \delta^N\{(A - Z) + (A - E)\delta/(1 - \delta)\}, \end{aligned}$$

which is nonnegative when

$$\delta \geq \frac{Z - A}{Z - E}.$$

This inequality holds when the discount factor $\delta$ is sufficiently large because the right-hand side is less than 1 when $E < A < Z$.[1] A similar inequality holds for Bob under similar circumstances, and so collusion is indeed compatible with the players' incentives in the repeated Cournot Duopoly Game, provided that the players don't discount the future too heavily.

*Colluding in the Dark.* The preceding argument shows that a range of collusive deals can be sustained as Nash equilibria when a Cournot duopoly is modeled as a repeated game with an infinite horizon—provided that the players care sufficiently about their future income streams.

Is collusion therefore endemic in oligopolistic situations? Many cases of blatant collusion have come to light, and the documented cases are doubtless only the tip of a large iceberg, but one must remember that the model we have been studying neglects many important issues.

In particular, our definition of a repeated game assumes that Alice and Bob know for certain what action the other took at all previous stages of the game. It is then easy for them to monitor whether the other is sticking to the deal. But collusion in the real world is more like a game of Blindman's Buff played in a room where someone keeps shifting the furniture around at random.

[1] If $a = b$ as in Section 10.2.3, then $A = B = \frac{1}{8}(K - c)^2$ and $E = \frac{1}{9}(K - c)^2$. The optimal deviation for Alice at the $N$th stage is $R_1(b) = \frac{3}{8}(K - c)$, for which the corresponding profit is $Z = \{\frac{3}{8}(K - c)\}^2$.

If Bob doesn't have a spy in Alice's factory, how does he know how many hats she is producing? If his profit falls below what he should be making, he may suspect that Alice has cheated, but she will put the blame on some external glitch over which she has no control. Should he punish her anyway? If he punishes her when she is innocent, he will be needlessly wrecking their cozy arrangement. If he fails to punish her when she is guilty, she will continue to take advantage of him in the future.

There are no easy answers to this kind of problem, and so there is probably little or no collusion in industries like the package holiday business, where the terms of trade fluctuate a great deal in an unpredictable way.

## 11.4 Infinite Repetitions

The strategy sets in infinitely repeated games are huge and complicated. As the first of several simplifications, we therefore restrict our attention to those strategies that can be represented by finite automata.

### 11.4.1 Finite Automata

An automaton is an idealized computing machine. When strategies are represented by automata, a player's choice of strategy can therefore be regarded as a decision to delegate the play of the game to a suitably programmed computer. A *finite* automaton can remember only a finite number of things, and so it can't keep track of all possible histories in a long repeated game. Confining attention to strategies that can be represented by finite automata is therefore a real restriction.

The kind of finite automata suitable for playing repeated games respond to what Eve does at the $n$th stage by choosing an action for Adam at the $(n+1)$st stage. Figure 11.5 shows little pictures of various finite automata capable of playing the repeated Prisoners' Dilemma. The circles represent possible states the machines may be in. The letter inside each circle says what action the machine will take in that state. The arrows show how a machine shifts from one state to another according to what the opponent did in the previous stage game. The arrow that comes from nowhere indicates the state in which the machine starts the game.

The machine labeled TIT-FOR-TAT gets its name because it always does next time what its opponent did last time. If it is in the state in which it outputs $h$ for *hawk*, it will stay in the same state if it receives the input $h$. If it receives the input $d$ for *dove*, it switches to the state in which it outputs $d$.

Because it begins by playing *dove*, TIT-FOR-TAT is said to be a nice machine. By contrast, TAT-FOR-TIT is nasty because it begins by playing *hawk* in an attempt to exploit its opponent. It then stays in its current state when the opponent plays *dove* and shifts states when the opponent plays *hawk*.

Figure 11.6 shows what happens when TAT-FOR-TIT plays TIT-FOR-TAT and when it plays itself. In both cases, the two machines end up by cycling through the same sequence of states forever. In Figure 11.6(a), the cycle is three stages long and begins immediately. In Figure 11.6(b), the cycle is only one stage long, and it begins only after some preliminary jostling at stage one.

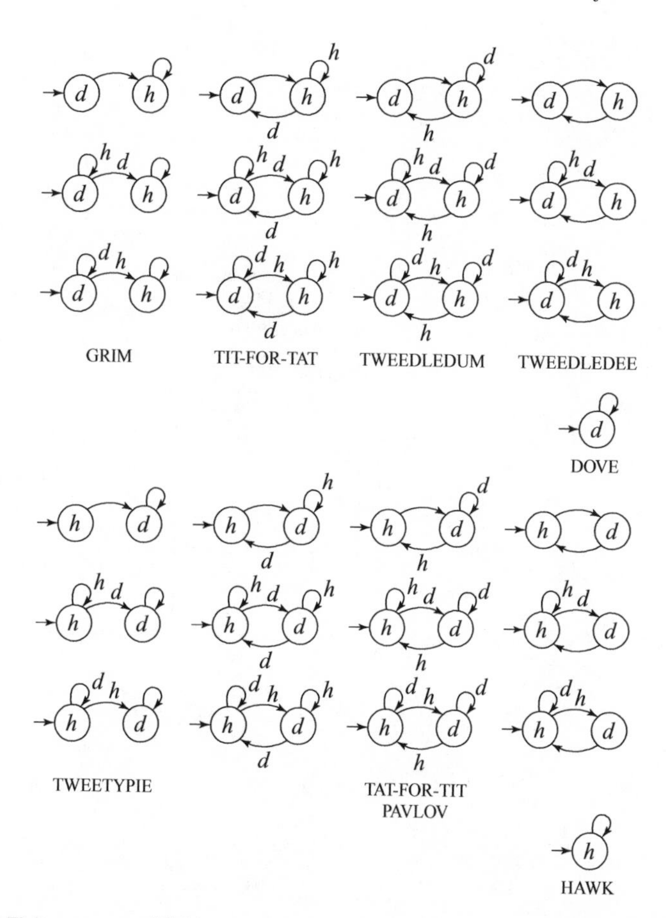


Figure 11.5 Finite automata. All 26 one-state and two-state finite automata capable of playing the Prisoners' Dilemma are listed. Each circle represents a possible state of the machine. The letter written within the circle is the output the machine offers in that state. The arrows indicate transitions. Each machine has one arrow that comes from nowhere, which indicates the machine's initial state. Unlabeled transitions are made independently of what the opponent does at the previous stage. The machines at the top that start by cooperating are said to be 'nice.' Those at the bottom are 'nasty.'

*Any* two finite automata playing each other in a repeated game will eventually end up cycling through the same sequence of states over and over again.[2] This makes it easy to work out their total payoffs in the repeated game.

## 11.4.2 Patient Players

What is Adam's payoff in a repeated game when he uses strategy $a$ and Eve uses strategy $b$? If Adam and Eve choose actions $s_n$ and $t_n$ at the $n$th stage of the game,

→ 11.4.3

[2]If $a$ has $m$ states and $b$ has $n$ states, then there are only $mn$ pairs of states. Thus, after $mn$ stages, the two machines *must* return to a situation identical to one they have jointly experienced previously. They are then doomed to reiterate their past behavior.

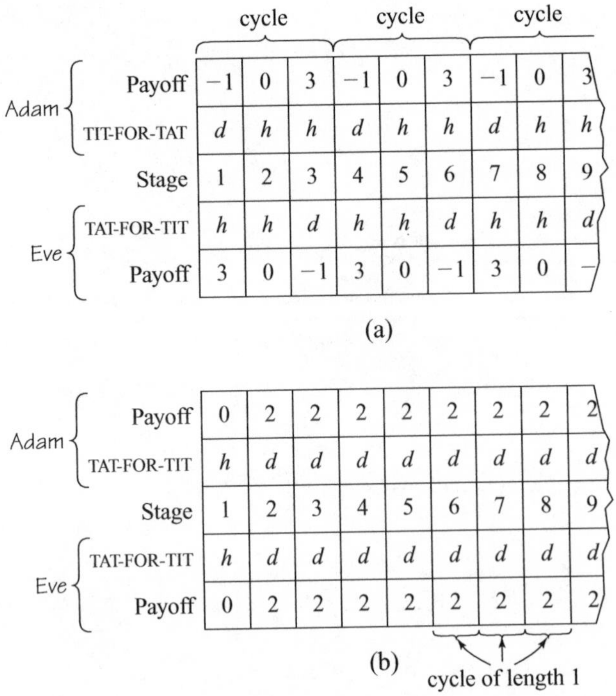

| | | cycle | | | cycle | | | cycle | | |
|---|---|---|---|---|---|---|---|---|---|---|
| Adam | Payoff | −1 | 0 | 3 | −1 | 0 | 3 | −1 | 0 | 3 |
| | TIT-FOR-TAT | *d* | *h* | *h* | *d* | *h* | *h* | *d* | *h* | *h* |
| | Stage | 1 | 2 | 3 | 4 | 5 | 6 | 7 | 8 | 9 |
| Eve | TAT-FOR-TIT | *h* | *h* | *d* | *h* | *h* | *d* | *h* | *h* | *d* |
| | Payoff | 3 | 0 | −1 | 3 | 0 | −1 | 3 | 0 | − |

(a)

| | | | | | | | | | | |
|---|---|---|---|---|---|---|---|---|---|---|
| Adam | Payoff | 0 | 2 | 2 | 2 | 2 | 2 | 2 | 2 | 2 |
| | TAT-FOR-TIT | *h* | *d* | *d* | *d* | *d* | *d* | *d* | *d* | *d* |
| | Stage | 1 | 2 | 3 | 4 | 5 | 6 | 7 | 8 | 9 |
| Eve | TAT-FOR-TIT | *h* | *d* | *d* | *d* | *d* | *d* | *d* | *d* | *d* |
| | Payoff | 0 | 2 | 2 | 2 | 2 | 2 | 2 | 2 | 2 |

(b) cycle of length 1

Figure 11.6 Computer wars.

then Adam's payoff at the $n$th stage is $\pi_1(s_n, t_n)$. To find his payoff in the repeated game as a whole, he must evaluate the income stream

$$\pi_1(s_1, t_1),\ \pi_1(s_2, t_2),\ \pi_1(s_3, t_3),\ \ldots.$$

As in Section 11.3.3, the players seek to maximize a discounted sum of such an income stream. Adam's payoff function $U_1 : S \times T \to \mathbb{R}$ in the repeated game then takes the form

$$U_1(a,b) = \pi_1(s_1,t_1) + \delta\pi_1(s_2,t_2) + \delta^2\pi_1(s_3,t_3) + \cdots,$$

where $\delta$ is his *discount factor.*

Adam's income stream in Figure 11.6(a) is $-1, 0, 3, -1, 0, 3, -1, 0, 3, \ldots$. If $a$ is TIT-FOR-TAT and $b$ is TAT-FOR-TIT, Adam would therefore then get a payoff in the repeated game equal to

$$\begin{aligned} U_1(a,b) &= -1 + 0\delta + 3\delta^2 - 1\delta^3 + 0\delta^4 + 3\delta^5 - 1\delta^6 + 0\delta^7 + \cdots \\ &= (-1+3\delta^2) + (-1+3\delta^2)\delta^3 + (-1+3\delta^2)\delta^6 + \cdots \\ &= (-1+3\delta^2)(1+\delta^3+\delta^6+\cdots) \\ &= (-1+3\delta^2)/(1-\delta^3) \\ &= (-1+3\delta^2)/(1-\delta)(1+\delta+\delta^2). \end{aligned}$$

The plan is to focus on very patient players, but we can't simply set $\delta = 1$ as in Section 11.2 because the series obtained when $\delta = 1$ won't converge. For example, the series $-1+0+3-1+0+3-1+0+3\cdots$ diverges to $+\infty$. A little fancy footwork is therefore required.

The utility functions $U_1$ and $AU_1 + B$ represent the same preferences (Section 4.6.1). Thus $U_1$ can be replaced by $(1-\delta)U_1$ without changing the strategic situation. We then take the limit as $\delta \to 1$. In Adam's case,

$$\lim_{\delta\to 1}(1-\delta)U_1(a,b) = \lim_{\delta\to 1}\left(\frac{-1+3\delta^2}{1+\delta+\delta^2}\right) = \frac{-1+3}{3} = \tfrac{2}{3},$$

which is simply what Adam gets on average as his stage-game payoffs cycle through the values $-1$, 0, and 3.

One of the advantages of working with finite automata is that this trick always works. When two finite automata play each other in a repeated game, they will eventually end up cycling through a fixed sequence of states. Each player will then be assumed to evaluate the income stream he or she obtains by taking the average of the payoffs they receive *during this cycle*.[3]

Figure 11.6(b) provides a second example. Adam and Eve both evaluate their income streams as being worth two utils. Notice that the initial jockeying for position at the very beginning of the game is ignored in this evaluation. The players are assumed to care only about what happens *in the long run*.

### 11.4.3 Nash Equilibria

From now on, it will be taken for granted that the players in a repeated game evaluate their income streams in terms of their long-run average payoffs. We already know that two GRIM strategies then make up a Nash equilibrium for the infinitely repeated Prisoners' Dilemma (Section 1.8). What other Nash equilibria can we find?[4]

In this chapter, we use the version of the Prisoners' Dilemma given in Figure 11.4(a). Figure 11.7 then shows the strategic form of the game that would result if the players were restricted to choosing from the finite automata given names in Figure 11.5.

This strategic form reveals that we must expect lots of Nash equilibria in an infinitely repeated game. When we allow *all* finite automata, the number of Nash equilibria becomes infinite. But, for the moment, we will look at only 4 of the 22 Nash equilibria shown in Figure 11.7.

[3]Evaluating an income stream this way is equivalent to using the utility function

$$V_1(a,b) = \lim_{N\to\infty}\frac{1}{N}\sum_{n=1}^{N}\pi_1(s_n,t_n).$$

It is therefore often referred to as the limit-of-the-means criterion. One reason for confining our attention to strategies representable by finite automata is that the limit of the means needn't exist in the general case.

[4]Except for the sketchy remarks of Section 11.4.5 concerning subgame-perfect equilibria, our attention is confined to the case of Nash equilibria to keep things reasonably simple.

| | DOVE | HAWK | GRIM | TIT-FOR-TAT | TAT-FOR-TIT | TWEEDLEDUM | TWEEDLEDEE | TWEETYPIE |
|---|---|---|---|---|---|---|---|---|
| DOVE | 2, 2 | −1, [3] | (2), 2 | (2), 2 | −1, [3] | (2), 2 | (2), 2 | −1, [3] |
| HAWK | (3), −1 | (0), [0] | 0, [0] | 0, [0] | $1\frac{1}{2}$, $-\frac{1}{2}$ | $1\frac{1}{2}$, $-\frac{1}{2}$ | $1\frac{1}{2}$, $-\frac{1}{2}$ | (3), −1 |
| GRIM | 2, [2] | (0), 0 | (2), [2] | (2), [2] | 0, 0 | (2), [2] | (2), [2] | (3), −1 |
| TIT-FOR-TAT | 2, [2] | (0), 0 | (2), [2] | (2), [2] | $\frac{2}{3}$, $\frac{2}{3}$ | (2), [2] | (2), [2] | 2, [2] |
| TAT-FOR-TIT | (3), −1 | $-\frac{1}{2}$, $1\frac{1}{2}$ | 0, 0 | $\frac{2}{3}$, $\frac{2}{3}$ | (2), [2] | (2), [2] | (2), [2] | 2, [2] |
| TWEEDLEDUM | 2, [2] | $-\frac{1}{2}$, $1\frac{1}{2}$ | (2), [2] | (2), [2] | (2), [2] | (2), [2] | (2), [2] | 2, [2] |
| TWEEDLEDEE | 2, [2] | $-\frac{1}{2}$, $1\frac{1}{2}$ | (2), [2] | (2), [2] | (2), [2] | (2), [2] | (2), [2] | 2, [2] |
| TWEETYPIE | (3), −1 | −1, [3] | −1, [3] | (2), 2 | (2), 2 | (2), 2 | (2), 2 | 2, 2 |

Figure 11.7 A restricted strategic form.

*Hawk versus Hawk.* If Eve knows that Adam is planning to play *hawk* at every repetition of the Prisoners' Dilemma, she may sigh at losing the opportunity to cooperate, but her best reply is to play *hawk* all the time as well. So (HAWK, HAWK) is a Nash equilibrium in the repeated game.

This fact illustrates a general result. Whenever $(s, t)$ is a Nash equilibrium of a one-shot game, it is also a Nash equilibrium in the repeated game if Adam always plays $s$ and Eve always plays $t$.

*Grim versus Grim.* As in Section 11.3.2, it is a Nash equilibrium when GRIM plays itself. The outcome is that both players cooperate all the time.

If GRIM weren't a best reply to itself, there would be some other machine DEVIANT that got a bigger payoff than 2 when playing GRIM. So DEVIANT couldn't always use *dove* when playing GRIM. Eventually, it would have to play *hawk*. But, as soon as DEVIANT plays *hawk*, GRIM retaliates by switching to a state in which it plays *hawk* itself. Thus, when DEVIANT plays GRIM, the latter will be using *hawk* and only *hawk* in the long run. The best that DEVIANT can then do is to play *hawk* as well in the long run. Thus DEVIANT will get a payoff of 0, which is a lot worse than the payoff of at least 2 it was supposed to get.

*Tit-for-Tat versus Tit-for-Tat.* The GRIM strategy offers no opportunity for repentance to a deviant who defects at some stage. Any transgression condemns the deviant to an eternity of punishment. The TIT-FOR-TAT strategy isn't so fierce. It punishes a transgression enough to make the deviation unprofitable but forgives the offender if he starts to cooperate again.

Why are two TIT-FOR-TATS a Nash equilibrium? Two TIT-FOR-TATS cooperate when they play each other, and so both get a payoff of 2. Is there a DEVIANT machine that can get more than 2 when playing TIT-FOR-TAT?

The DEVIANT machine would have to play *hawk* eventually, but TIT-FOR-TAT then retaliates by playing *hawk* until DEVIANT plays *dove* again. The DEVIANT machine therefore gains nothing. For each stage at which it gets a payoff of 3 by playing *hawk* when TIT-FOR-TAT plays *dove*, it suffers a countervailing payoff of −1 when it plays *dove* to persuade TIT-FOR-TAT to return to cooperating.

*Tat-for-Tit versus Tat-for-Tit.* This pair of strategies is a Nash equilibrium for much the same reason as two TIT-FOR-TATS are a Nash equilibrium. Notice that TAT-FOR-TIT is a nasty machine that defects at the first stage. But when it plays itself, both machines then switch to cooperating all the time. Since only the long-run outcome matters, both players therefore still get the cooperative payoff of 2.

### 11.4.4 Folk Theorem

The one-shot Prisoners' Dilemma is shown yet again in Figure 11.8(a). Its cooperative payoff region $X$ is shaded in Figure 11.8(b) (Section 6.6.1). We have seen that the infinitely repeated version of the game has many Nash equilibria, but the full count is enormous. Every point in the deeply shaded part of $X$ is a Nash equilibrium outcome of the infinitely repeated game.

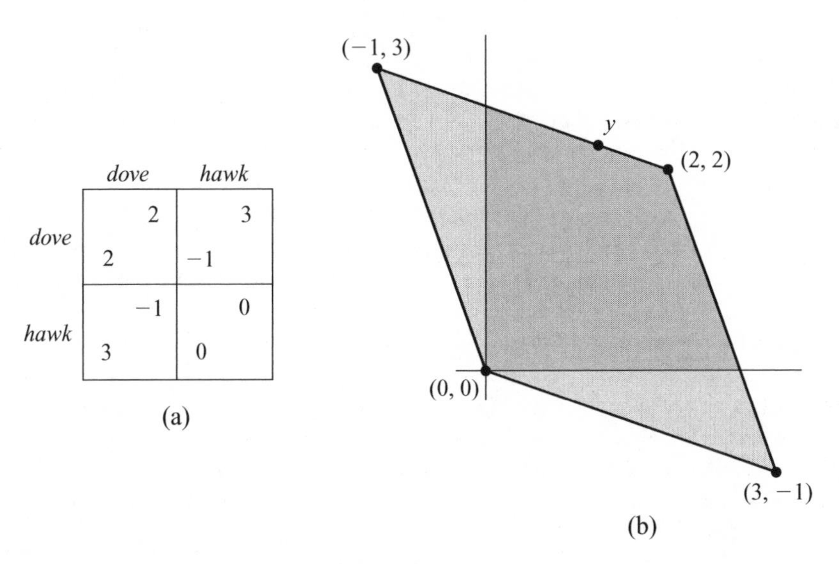


Figure 11.8 The folk theorem. The lightly shaded part of Figure 11.8(b) is the cooperative payoff region of the one-shot Prisoners' Dilemma of Figure 11.8(a). The deeply shaded part is the set of all Nash equilibrium outcomes in the infinitely repeated game.

The general version of this result is called the *folk theorem*, where "folk" is as in "folklore." In the early days of game theory, it seems that everybody knew the theorem, but nobody was willing to claim credit as its author. However, Bob Aumann was among the first to recognize its full significance.[5] It says that:

> The set of all Nash equilibrium outcomes of an indefinitely repeated game consists of all points in the cooperative payoff region of the stage game at which all players get their security levels or more.

The folk theorem is of fundamental importance for political philosophy. Without an external enforcement agency to deter contract violations, most of the outcomes in the cooperative payoff region of a one-shot game lie outside our reach (Section 11.1). But when we consider cooperation in society as a whole, there is no external enforcement agency to which we can appeal. All earthly sources of authority—kings, presidents, judges, policemen, and the like—are themselves but players in the game of life. They too must be incentivized if they are to carry out their specified roles properly. The only stable agreements available to society as a whole must therefore police themselves.

Political philosophers before David Hume saw no solution to this conundrum. Even today, philosophers trying to get around the problem vainly invent reasons why it is rational to cooperate in the one-shot Prisoners' Dilemma. But a society doesn't play a one-shot game. It plays a repeated game, in which the folk theorem tells us that we need lose none of the fruits of cooperation by restricting ourselves to agreements on *equilibria* in the game of life.[6] Any contract that rational players might sign in the presence of an external enforcement agency in the one-shot case is also available as a *self-policing* agreement in the infinitely repeated case.

So why don't we all live together in amity and peace? One of many reasons is that our formulation of a repeated game assumes that history is common knowledge, so nobody can cheat without being found out. The standard folk theorem therefore better fits small village societies in which secrets are hard to keep than the large anonymous societies of today. Variants of the theorem in which information is restricted in various ways show that it is sometimes still possible to maintain a substantial measure of rational cooperation even when cheating is hard to detect, but this is one of many areas in game theory that aren't properly understood as yet.

→ 11.5

*The Game $G^{\#}$.* It is easy to prove a simple version of the folk theorem, but we need to get ready for the proof by generalizing some ideas already introduced for the infinitely repeated Prisoners' Dilemma.

In what follows, the role previously played by the Prisoners' Dilemma will be taken over by a general finite game $G$. This will be the stage game for an infinitely repeated game $G^{\infty}$. Adam's pure strategy set $S$ for the one-shot game $G$ is the set of actions available to him at each stage of $G^{\infty}$. Eve's pure strategy set $T$ for $G$ is the set of actions available to her at each stage of $G^{\infty}$.

The set of finite automata that input actions from the set $T$ and output actions from the set $S$ is denoted by $\mathscr{A}$. The set of finite automata that input actions from the set $S$

[5] His role was recognized in 2005 by the award of a Nobel Prize.

[6] Nobody will sign a contract that gives them less than their security level.

and output actions from the set $T$ is denoted by $\mathscr{B}$. The sets $\mathscr{A}$ and $\mathscr{B}$ are the pure strategy sets for a game $G^\#$ that is to be the final object of study. A player's choice of a strategy for $G^\#$ can be regarded as a decision to delegate responsibility for playing $G^\infty$ to a suitably chosen computing machine.

If Adam chooses $a$ in $\mathscr{A}$ and Eve chooses $b$ in $\mathscr{B}$, then the two automata will eventually cycle through the same sequence of states forever (as in Figure 11.6). If the pairs of actions through which the machines cycle are $(s_1, t_1), (s_2, t_2), \ldots, (s_N, t_N)$, then player $i$'s payoff in $G^\#$ is

$$V_i(a,b) = \frac{1}{N}\sum_{n=1}^{N} \pi_i(s_n, t_n). \tag{11.1}$$

So a player's payoff in $G^\#$ is what the player gets on average during the cycle into which play finally settles.

For example, the one-shot game $G$ in Figure 11.6(a) is the Prisoners' Dilemma. The automaton $a$ is TIT-FOR-TAT, and the automaton $b$ is TAT-FOR-TIT. The length of a cycle is $N=3$, and $(s_1, s_2)=(d,h)$, $(s_2,t_2)=(h,h)$, $(s_3,t_3)=(h,d)$. Thus,

$$(V_1(a,b), V_2(a,b)) = \tfrac{1}{3}(-1,3)+\tfrac{1}{3}(0,0)+\tfrac{1}{3}(3,-1) = (\tfrac{2}{3}, \tfrac{2}{3}).$$

Notice that the payoffs that result when two automata play the repeated Prisoners' Dilemma can only ever be rational numbers.[7] In proving a folk theorem in which strategies are represented by finite automata, the best we can therefore hope for is to get a result that says that Nash equilibrium outcomes are *dense* in some part of the cooperative payoff region of the stage game.[8]

LEMMA 11.1 *Any outcome of $G^\#$ lies in the cooperative payoff region of the one-shot game $G$.*

*Proof* If $(s,t)$ is a pure strategy pair for $G$, then $(\pi_1(s,t),\ \pi_2(s,t))$ is the pair of payoffs that goes in the $s$th row and $t$th column of the strategic form of $G$. The cooperative payoff region of $G$ is the convex hull of all such payoff pairs (Section 6.6.1). From (11.1),

$$(V_1(a,b), V_2(a,b)) = \frac{1}{N}\sum_{n=1}^{N}(\pi_1(s_n,t_n), \pi_2(s_n,t_n)),$$

and hence the outcome $(V_1(a,b), V_2(a,b))$ of the game $G^\#$ is a convex combination of payoff pairs in the strategic form of $G$ (Section 6.5.1).

*Minimax Point.* The folk theorem quoted in Section 11.4.4 takes for granted that mixed strategies are allowed, but the proof we are working up to applies only to pure strategies.

[7]A rational number is a fraction $m/n$ in which $m$ and $n \neq 0$ are integers.

[8]The rational numbers are dense in the set of all real numbers because each real number can be approximated arbitrarily closely by rational numbers. For example, $\pi = 3.14159\ldots$ is approximated to within an accuracy of 0.0005 by the rational number 3142/1000.

Instead of being able to show that each $x \geq \overline{v} = \underline{v}$ in the cooperative payoff region of $G$ is a Nash equilibrium outcome, we will be able to show this only for $x \geq \overline{m}$.

The maximin point for $G$ is $\underline{m} = (\underline{m}_1, \underline{m}_2)$, but it is the *minimax point* $\overline{m}$ that matters here. When mixed strategies are allowed, the distinction between maximin and minimax disappears because Von Neumann's minimax theorem says that $\underline{v} = \overline{v}$, but $\underline{m} < \overline{m}$ unless both payoff matrices have saddle points (Theorems 7.2 and 7.3).

In the one-shot Prisoners' Dilemma of Figure 11.8(a), $\underline{m} = \overline{m} = (1, 1)$. Figure 11.9(b) shows the cooperative payoff region of the game of Figure 11.9(a) together with the location of $\underline{m} = (2, 2)$ and $\overline{m} = (3, 2)$ (neither of which need appear in the payoff matrix).

Let $r_1(t)$ be one of Adam's best replies in $S$ to Eve's choice of a pure strategy $t$ in $T$. Then

$$\overline{m}_1 = \min_{t \in T} \max_{s \in S} \pi_1(s, t) = \min_{t \in T} \pi_1(r_1(t), t) \tag{11.2}$$

because the maximum in the middle term is achieved where $s = r_1(t)$. It follows that any Nash equilibrium $(\sigma, \tau)$ in pure strategies of the one-shot game $G$ necessarily assigns the players their minimax values or more. The reason is simple. Since $\sigma$ is a best reply to $\tau$,

$$\pi_1(\sigma, \tau) = \pi_1(r_1(\tau), \tau) \geq \min_{t \in T} \pi_1(r_1(t), t) = \overline{m}_1.$$

Similarly, the fact that $\tau$ is a best reply to $\sigma$ implies that $\pi_2(\sigma, \tau) \geq \overline{m}_2$.

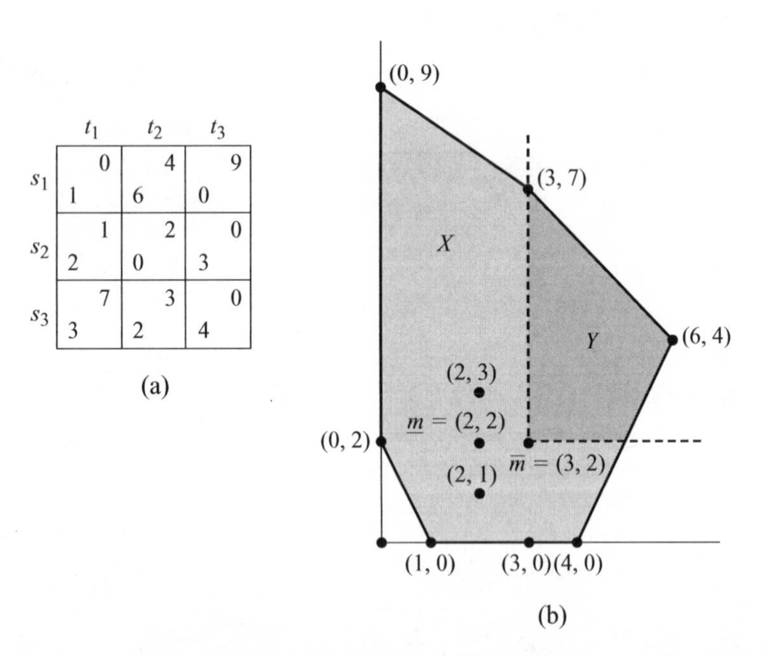


Figure 11.9 A minimax point. Imagine that Eve wants to punish Adam after he has deviated in a repeated game. If she uses a pure strategy for this purpose, she knows he will respond with his best reply. So the worst she can do to Adam is to hold him to his minimax payoff.

The following lemma says something that is superficially very similar. But remember that $G^{\#}$ is a very different game from $G$. The pure strategies in $G^{\#}$ are automata that play the *repeated* game $G^{\infty}$.

LEMMA 11.2 *Any Nash equilibrium of $G^{\#}$ assigns the players at least their minimax values in the one-shot game $G$.*

*Proof* If $V_1(a,b) < \overline{m}_1$, we show that Adam has a better reply to $b$ than $a$, and hence $(a,b)$ can't be a Nash equilibrium for $G^{\#}$. The better reply is easy to find. Simply take an automaton $c$ in $\mathscr{A}$ that makes a best one-shot reply to $b$ at every stage of the repeated game. If $\pi_1(s_n, t_n)$ is the very worst stage-game payoff that $c$ ever gets in playing $b$, then

$$\begin{aligned} V_1(c,b) &\geq \pi_1(s_n, t_n) \\ &= \pi_1(r_1(t_n), t_n) \\ &\geq \min_{t \in T} \pi_1(r_1(t), t) = \overline{m}_1. \end{aligned}$$

The strategy $c$ isn't necessarily a *best* reply to $b$, but it is a better reply than $a$ when $V_1(a,b) < \overline{m}_1$. It follows that, if $(a,b)$ is a Nash equilibrium for $G^{\#}$, then $V_1(a,b) \geq \overline{m}_1$. Similarly, $V_2(a,b) \geq \overline{m}_2$. □

The cooperative payoff region $X$ of the game $G$ of Figure 11.9(a) is shown in Figure 11.9(b). Lemma 11.2 says that the Nash equilibria of $G^{\#}$ lie in the set $Y$. One equilibrium is easy to identify. Since $(s_3, t_1)$ is a Nash equilibrium for the *one-shot* game $G$, it must be a Nash equilibrium in $G^{\#}$ for Adam and Eve to choose automata that always play $s_3$ and $t_1$ respectively. Thus $(3, 7)$ is a Nash equilibrium outcome for $G^{\#}$. But this is only one Nash equilibrium outcome. The folk theorem tells us about *all* Nash equilibrium outcomes.

THEOREM 11.2 (folk theorem) *Let $X$ be the cooperative payoff region of a finite one-shot game $G$, and let $\overline{m}$ be its minimax point. Then the outcomes corresponding to Nash equilibria in pure strategies of the game $G^{\#}$ are dense in the set*

$$Y = \{x : x \in X \text{ and } x \geq \overline{m}\}.$$

*Proof* The idea of the proof is almost ridiculously simple. How do we make $y$ in Figure 11.9(b) into a Nash equilibrium outcome of the repeated game? If Adam deviates from whatever is necessary to implement $y$, Eve punishes him by switching permanently to whatever strategy holds him to his minimax payoff $\overline{m}_1$. Since $y_1 \geq \overline{m}_1$, he therefore won't deviate.

**Step 1.** Suppose that $x_1, x_2, \ldots, x_K$ are payoff pairs that appear in the strategic form of $G$. Let $q_1, q_2, \ldots, q_K$ be nonnegative rational numbers satisfying $q_1 + q_2 + \cdots + q_K = 1$. Then

→ 11.4.5

$$y = q_1 x_1 + q_2 x_2 + \cdots + q_K x_K$$

is a convex combination of $x_1, x_2, \ldots, x_K$ and hence lies in $X$. The set of all such $y$ is dense in $X$. We show that, if $y \geq \overline{m}$, then $y$ is a Nash equilibrium outcome of $G^{\#}$.

**Step 2.** The fractions $q_1, q_2, \ldots, q_K$ can be written with a common denominator $N$, so that $q_k = n_k/N$ ($k = 1, 2, \ldots K$), where $n_k$ is a nonnegative integer. We then have that $n_1 + n_2 + \cdots + n_K = N$.

**Step 3.** Let the action pairs that generate the outcomes $x_1, x_2, \ldots, x_K$ of $G$ be $(s_1, t_1)$, $(s_2, t_2)$, ..., $(s_K, t_K)$. To achieve the outcome $y$ of $G^{\#}$, two automata $a$ and $b$ will be constructed that perpetually cycle through a sequence of $N$ action pairs. First they play $(s_1, t_1)$ for $n_1$ stages, then $(s_2, t_2)$ for $n_2$ stages, then $(s_3, t_3)$ for $n_3$ stages, and so on. After they complete the cycle by playing $(s_K, t_K)$ for $n_K$ stages, the cycle begins again.

**Step 4.** The payoff pair that results when $a$ plays $b$ is $y$ because

$$\frac{1}{N}\sum_{k=1}^{K} n_k \pi(s_k, t_k) = \sum_{k=1}^{K} q_k x_k = y.$$

*Example.* We now put the proof on hold while we work through an example for the case when $G$ is the Prisoners' Dilemma of Figure 11.8(a) and $y$ is the point shown in Figure 11.8(b). Since

$$y = \tfrac{3}{4}(2,2) + \tfrac{1}{4}(-1,3),$$

implementing $y$ as an equilibrium outcome in the repeated game requires running through the cycle generated by the action pairs $(s_1, t_1) = (d, d)$, $(s_2, t_2) = (d, d)$, $(s_3, t_3) = (d, d)$, and $(s_4, t_4) = (d, h)$. But this is what the four states at the top of the diagrams representing HUMPTY and DUMPTY in Figure 11.10 are wired up to do.

The state at the bottom of the diagrams representing HUMPTY and DUMPTY in Figure 11.10 is included to ensure that HUMPTY and DUMPTY are best replies to each other. Any deviation from the cycle that generates $y$ is punished by the opponent's switching permanently to the bottom state in which *hawk* is always played. The same argument that shows (GRIM, GRIM) is a Nash equilibrium therefore also works for (HUMPTY, DUMPTY).

**Step 5.** We now use HUMPTY and DUMPTY as patterns to complete the construction of the automata $a$ and $b$.

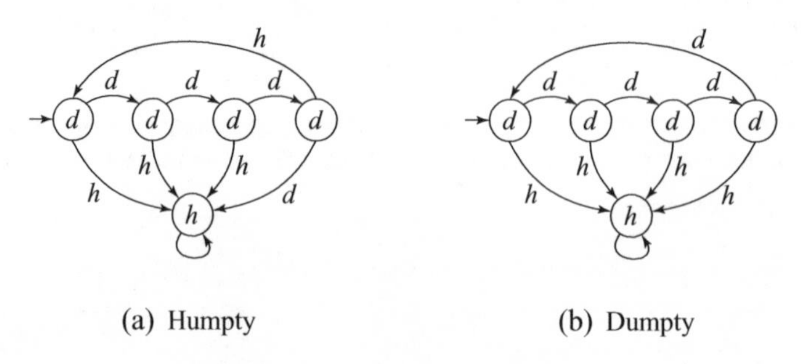


Figure 11.10 Humpty and Dumpty.

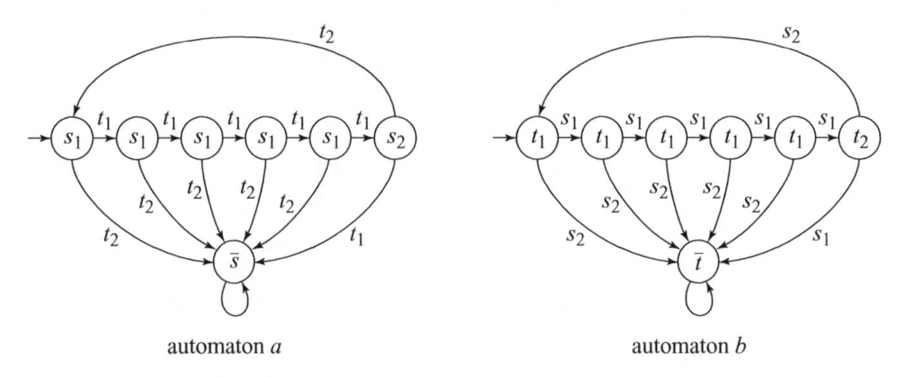


Figure 11.11 Folk automata. The equilibrium cycle in this example requires the automata to play $(s_1, t_1)$ for five stages and $(s_2, t_2)$ for one stage.

Figure 11.11 shows their final structure. The states at the top of the diagram are wired up to ensure that the two machines cycle through the action pairs necessary to implement the outcome $y$. The states at the bottom of the diagrams are included to ensure that $(a, b)$ is a Nash equilibrium. But what determines the punishment actions $\bar{s}$ and $\bar{t}$?

**Step 6.** The significant feature of the punishments $\bar{s}$ and $\bar{t}$ is that they *minimax* the opponent. Thus $\bar{s}$ is chosen so that

$$\pi_2(\bar{s}, r_2(\bar{s})) = \min_{s \in S} \pi_2(s, r_2(s)) = \overline{m}_2.$$

So even if Eve makes a best reply $r_2(\bar{s})$ to Adam's choice of $\bar{s}$, she still gets no more than her minimax value. It follows that $\overline{m}_2$ is the worst payoff that Adam can inflict on Eve if she knows what he is doing.

**Step 7.** Provided that $y \geq \overline{m}$, any deviation by a player from the cycle that implements $y$ triggers a permanent transition by the opponent into a punishment state in which the opponent gets no more than his or her minimax value in $G$. So neither player can gain from replacing their current machine by a DEVIANT machine because any attempt by DEVIANT to improve on $y$ will only make things worse. Thus $(a, b)$ is a Nash equilibrium, and so $y$ is an equilibrium outcome as the folk theorem requires. □

### 11.4.5 Who Guards the Guardians?

→ 11.5

The reasons for introducing subgame-perfect equilibria in Section 2.9.3 apply with even greater force for repeated games. In the folk theorem, we studied Nash equilibria in which players are deterred from departing from cooperative play by the prospect of being punished. If they were to deviate, they believe that their opponent will retaliate by minimaxing them. So they never *actually* deviate, and the punishment is never *actually* inflicted.

But do the beliefs we have been attributing to the players make sense? If Eve were to deviate, is it really credible that Adam would then minimax her relentlessly thereafter, no matter how damaging this may be to him? Not if he pays attention to his incentives!

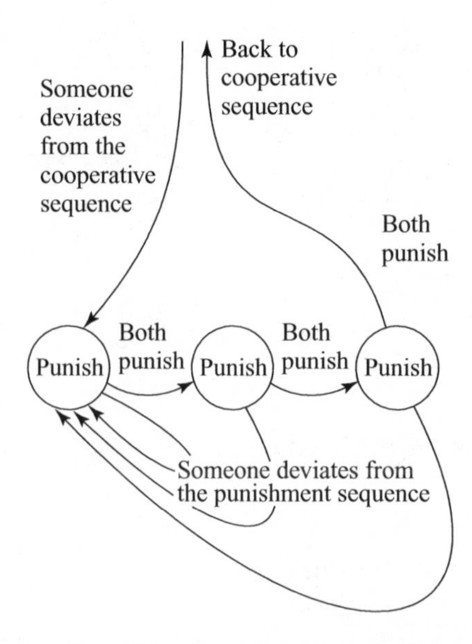


Figure 11.12 Guarding the guardians. Three stages of punishment are taken to be adequate to deter deviation from the equilibrium cycle. Any failure to punish when punishment is due will itself be punished.

So the question arises: Can equilibrium strategies be found in which the planned punishments *are* always credible? The answer is *yes*. That is to say, a version of the folk theorem holds with Nash equilibria replaced by subgame-perfect equilibria.

A formal proof for such an improved version of the folk theorem is too fussy to be worth reproducing, but the idea is very simple. Figure 11.12 shows a punishment scheme that will support a suitable subgame-perfect equilibrium.

Any player who deviates from the cooperative sequence is punished for however many stages are necessary to render the deviation unprofitable, whereupon both players return to their cooperative phase.[9] But what if players fail to punish when the equilibrium says that they should punish? Then this failure is punished. And if someone fails to punish someone who has failed to punish when punishment is called for, then this failure is punished also.

Such constructions provide a formal answer to a perennial question that is usually posed by quoting some politically incorrect lines from Juvenal:

> Pone seram; cohibe:
> Sed *quis custodiet* ipsos *custodes*?
> Cauta est, et ab illis incipit uxor.

The phrase in italics translates as "Who guards the guardians?" The game theory answer is that they guard each other.

[9]In the story told here, *both* players then switch into the punishment schedule. This means that the automata would need to input not only what the opponent did but also what they did themselves last time.

## 11.5 Social Contract

→ 11.6

What is the glue that holds a society together? Philosophers have traditionally tried to frame explanations in terms of a "social contract"—a tacit agreement to which we are all party that somehow regulates our dealings with each other.

The word "contract" is far from ideal. It suggests that we consciously signed on to the agreement and that some external enforcement agency polices our observance of its terms. But neither of these features of a legal contract applies in the case of a *social* contract. In particular, if we want to envisage a social contract as the organizing principle of a society, we have to explain why people honor its terms when there is no possibility of their being sued if they don't.

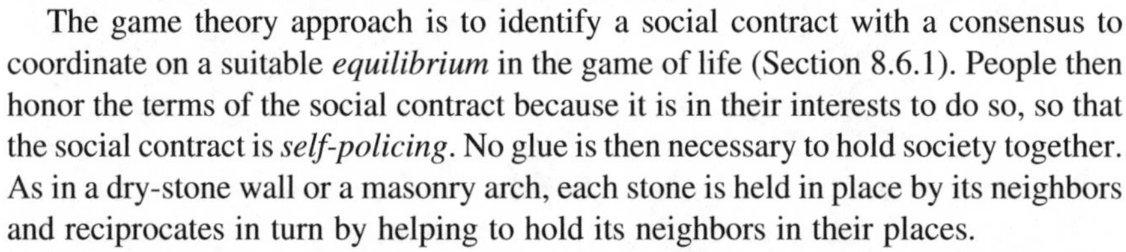

The game theory approach is to identify a social contract with a consensus to coordinate on a suitable *equilibrium* in the game of life (Section 8.6.1). People then honor the terms of the social contract because it is in their interests to do so, so that the social contract is *self-policing*. No glue is then necessary to hold society together. As in a dry-stone wall or a masonry arch, each stone is held in place by its neighbors and reciprocates in turn by helping to hold its neighbors in their places.

David Hume first made this argument more than two hundred years ago, but it remains unpopular because critics reject it as "reductive." Do love and duty count for nothing? Are mutual trust and respect to be thrown out of the window? Not at all! Game theorists love their neighbors as much as anyone else. But we aren't ready to say that this is just the way things happen to be. We want to know *why*.

An experiment with apes may clarify the point. Some bananas were hung in the apes' cage, but whenever an ape tried to take a banana, the whole group was thoroughly hosed down. After a while, individual apes that approached the bananas were punished by the other apes. Eventually, the bananas remained untouched. They continued to remain untouched, even after the hosing policy had been abandoned, and all the apes had gradually been replaced by new apes who had never observed any hosing. If they could talk, perhaps the apes left in the cage would tell each other that nobody must touch the bananas because this is what is right and proper in ape societies—just as we say similar things about the various taboos that operate in human societies. But to say something of this kind doesn't *explain* a social contract; it merely describes it.

### 11.5.1 Trust

We met the holdup problem in Section 5.6.2. Alice delivers a service to Bob, trusting him to reciprocate by making a payment in return. But why should he pay up if nothing will happen to him if he doesn't? Sociologists model the holdup problem using the toy game of Figure 11.13(a), which we call the Trust Minigame. The game has a unique subgame-perfect equilibrium in which Alice doesn't deliver the service because she predicts that Bob won't pay.

But people mostly do pay their bills. When asked why, they usually say that they have a duty to pay and that they value their reputation for honesty. Game theorists agree that this is a good description of how our social contract works, but we want to know why it embodies such virtues. We therefore look at the infinitely repeated version of the Trust Minigame.

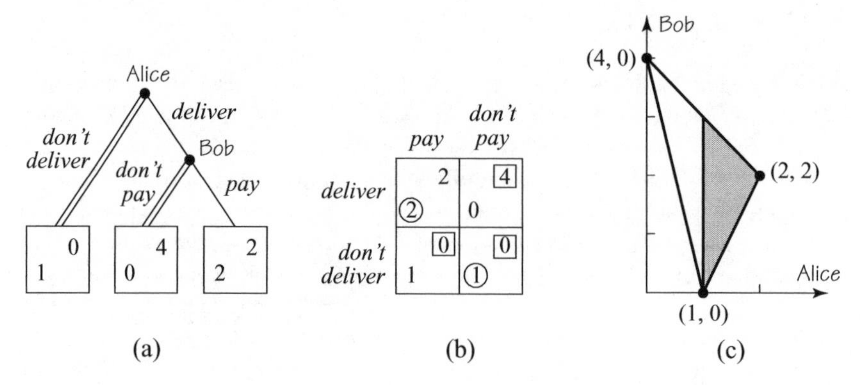


Figure 11.13 The Trust Minigame.

The folk theorem says that all points in the shaded region of Figure 11.13(c) are equilibrium outcomes of the repeated game, including the payoff pair $(2, 2)$ that arises when Alice always delivers and Bob always pays. We explain this equilibrium in real life by saying that Bob can't afford to lose his reputation for honesty by cheating on Alice because she will then refuse to provide any service to him in the future. In practice, Alice will usually be someone new, but the same equilibrium works just as well because nobody will be any more ready than Alice to trade with someone with a reputation for not paying.

Critics argue that people still pay up, even in one-shot games, where their reputation for honesty is irrelevant. But game theorists see no problem here. When the one-shot game is rare, there is little to gain in having a special strategy different from the one you use in the repeated version.

As for commonly encountered one-shot games, it simply isn't true that people are particularly virtuous.[10] Experiments on how people play the one-shot Prisoners' Dilemma are sometimes quoted in an attempt to refute this banal observation about human nature. It is true that about half the subjects cooperate at first, but as they gain experience in playing against a new opponent each time, the frequency with which they defect climbs relentlessly upward until about 90% of subjects have learned to defect.

### 11.5.2 Authority

Immanuel Kant is one of many philosophers who have argued that duty is the cement that holds societies together. His story is that we have a duty to obey those in authority and that societies must therefore have a big boss who is the ultimate source of all authority. Otherwise we would get into an infinite regress when we tried to trace who was responsible to whom.

But the subgame-perfect version of the folk theorem explicitly *closes* the chains of responsibility. The guardians guard each other. Some societies get along fine with no

[10]Tipping in restaurants you are unlikely to visit again is widely quoted as a counterexample. Having worked as a waiter in my youth, I get a warm glow from tipping generously myself, but the amounts are a negligible fraction of my income.

bosses at all—as in the hunter-gatherer societies that still survive in odd corners of the world. Even in authoritarian societies, Kant's story doesn't help much because it doesn't explain *why* the big boss has authority.

For example, the Queen of Hearts is the big boss in Wonderland, but why does anyone obey her? Alice obeys because she believes that the queen will order the executioner to cut off her head if she doesn't. He obeys because he believes that she will order someone else to cut off his head if he doesn't. And the same goes for everybody else in Wonderland. When we look for the source of the queen's authority in this equilibrium, we find that she has power over her subjects only because they think she has.

Such a bossy social contract needs more than two players to make it work. The secret remains *reciprocity*, but now it is no longer necessary that the punishment for cheating on the social contract should be administered by the injured party. As David Hume pointed out more than two hundred years ago, the punishment that deters cheating in a multiplayer repeated game commonly comes from a *third party*.

### 11.5.3 Altruism

From a Humean perspective, bosses like the Queen of Hearts are simply coordinating mechanisms for an equilibrium in a repeated game. But if modern hunter-gatherer societies are any guide, the human societies of prehistory got by with no bosses at all, using *fairness* as a coordinating mechanism.

To see how this might work, imagine a toy world in which only a mother and a daughter are alive at any time. Each player lives for two periods. The first period is her youth, and the second her old age. When young, a player bakes two (large) loaves of bread. She then gives birth to a daughter and immediately grows old. Old players are too feeble to produce anything.

One equilibrium requires each player to consume both her loaves of bread in her youth. Everyone will then have to endure a miserable old age, but everyone will be optimizing, given the choices of the others. All players would prefer to consume one loaf in their youth and one loaf in their old age. But this "fair" outcome can be achieved only if the daughters all give one of their two loaves to their mothers because bread perishes if not consumed when baked.

Since mothers can't retaliate if their daughters are selfish, it is a little surprising that the fair outcome can be sustained as an equilibrium. In this fair equilibrium, a *conformist* is a player who gives her mother a loaf if and only if her mother was a conformist in her youth. Conformists therefore reward other conformists and punish nonconformists.

To see why a daughter gives her mother a loaf, suppose that Alice, Beatrice, and Carol are mother, daughter, and granddaughter. If Beatrice neglects Alice, she becomes a nonconformist. Carol therefore punishes Beatrice to avoid becoming a nonconformist herself. If not, she will be punished by her daughter—and so on. If the first-born player is deemed to be a conformist, it is therefore a subgame-perfect equilibrium for everybody to be a conformist.

In real life, daughters commonly look after their aged mothers because they love them. But the model teaches us that, even if all daughters were stonyhearted egoists, their aged mothers wouldn't necessarily be neglected.

## 11.6 The Evolution of Cooperation

The game theorists who proved versions of the folk theorem in the early fifties knew nothing of David Hume. The biologist Robert Trivers was equally unaware of their work when he rediscovered the idea fifteen years later. He referred to the mechanism that makes the folk theorem work as *reciprocal altruism.* Some twelve years later, the word was finally disseminated to the world at large by Bob Axelrod's *Evolution of Cooperation.*

The folk theorem says that infinitely repeated games have immense numbers of equilibria. It therefore looks like we are faced with the equilibrium selection problem in a particularly acute form. However, the fact that the equilibria are all packed close together means that it isn't easy for evolution to get trapped in the basin of attraction of a Pareto-inferior equilibrium (Section 8.5.2). Axelrod's contribution was to run computer simulations that suggest that one should normally expect evolution to select a Pareto-efficient equilibrium.

*Axelrod's Olympiad.* Axelrod invited various social scientists to submit computer programs for a competition in which each entry would be matched against every other entry in the indefinitely repeated Prisoners' Dilemma. After learning the outcome of a pilot round, contestants submitted computer programs that implemented sixty-three of the possible strategies of the game. For example, TIT-FOR-TAT was submitted by the psychologist Anatole Rapaport. The GRIM strategy was submitted by the economist James Friedman.

In the Olympiad, TIT-FOR-TAT was the most successful strategy. Axelrod then simulated the effect of evolution operating on his sixty-three strategies using an updating rule which ensures that strategies that achieve a high payoff in one generation are more numerous in the next. The fact that TIT-FOR-TAT was the most numerous of all the surviving programs at the end of the evolutionary simulation clinched the question for Axelrod, who then proceeded to propose tit-for-tat as a suitable paradigm for human cooperation across the board. In describing its virtues, he says:

> What accounts for TIT-FOR-TAT's robust success is its combination of being nice, retaliatory, forgiving and clear. Its niceness prevents it from getting into unnecessary trouble. Its retaliation discourages the other side from persisting whenever defection is tried. Its forgiveness helps restore mutual cooperation. And its clarity makes it intelligible to the other player, thereby eliciting long-term cooperation.

As a consequence of Axelrod's claims, a whole generation of social scientists has grown up who believe that TIT-FOR-TAT embodies everything that they need to know about how reciprocity works.

But it turns out that TIT-FOR-TAT wasn't so very successful in Axelrod's simulation.[11] Nor is the limited success it does enjoy robust when the initial population of entries is varied. The unforgiving GRIM does extremely well when the initial pop-

[11]The successful strategy was a mixture of six entries. TIT-FOR-TAT was the strategy played most frequently, but its probability was only a little more than one-sixth.

ulation of entries consists of all twenty-six finite automata with at most two states (Figure 11.5). Nor does evolution generate nice machines that are never the first to defect, when some small fraction of suckers worth exploiting is allowed to flow continually into the system. As for clarity, for cooperation to evolve, it is only necessary that a mutant be able to recognize a copy of itself. All that is then left on Axelrod's list is the requirement that a successful strategy be retaliatory. But this is a lesson that applies only in *pairwise* interactions.

For example, it is said that reciprocity can't explain the evolution of friendship. It is true that the offensive-defensive alliances of chimpanzees can't be explained with a tit-for-tat story. If Adam needs help because he is hurt or sick, his allies have no incentive to come to his aid because he is now unlikely to be useful as an ally in the future. Any threat he makes to withdraw his cooperation will therefore be empty. But it needn't be the injured party who punishes a cheater in multiperson interactions (Section 11.5). The rest of the band will be watching if Adam is abandoned to his fate, and they will punish his faithless allies by refusing to form alliances with them in the future. Who wants to make an alliance with someone with a reputation for abandoning friends when they are in trouble?

I think that the enthusiasm for TIT-FOR-TAT survives for the same reason that people invent reasons why it is rational to cooperate in the one-shot Prisoners' Dilemma. They want to believe that human beings are essentially nice. But the real lesson to be learned from Axelrod's Olympiad and many later evolutionary simulations is much more reassuring.

Although the claims for TIT-FOR-TAT are overblown, the conclusion that evolution is likely to generate a cooperative outcome seems to be genuinely robust. We therefore don't need to pretend that we are all Doctor Jekylls in order to explain how we manage to get along with each other fairly well much of the time. Even a society of Mr. Hydes will eventually learn to coordinate on a Pareto-efficient equilibrium in an indefinitely repeated game!

## 11.7 Roundup

Sages from Confucius on have identified reciprocity as the key to human cooperation. Reciprocity can't arise in one-shot games, and so its study requires looking at repeated games.

If a game $G$ is repeatedly played by the same players, it is said to be the stage game of a repeated game. The strategies of $G$ then become the available actions at each stage of the repeated game, but it isn't true that a strategy for the repeated game consists simply of naming an action for each stage of the game. We must allow the action chosen at any stage to be contingent on the previous history of the game. It is sometimes unrealistic to assume that the history of the game so far is common knowledge among the players, but this chapter lives with this defect.

When the Prisoners' Dilemma is repeated ten times, the only subgame-perfect equilibrium calls for both players always to plan to play *hawk*. But when the Prisoners' Dilemma is repeated indefinitely often, playing *dove* all the time can be supported as an equilibrium outcome—provided that the players are sufficiently patient, and the probability that the next game will be the last is always small. The same holds for collusion in a Cournot duopoly. The general result is called the folk theorem. It says that the set of all Nash equilibrium outcomes of an indefinitely repeated game consists of all points in the cooperative payoff region of the stage game at which all players get their security levels or more.

The proof of the folk theorem generalizes the observation that it is a Nash equilibrium for Adam and Eve both to play the GRIM strategy in the infinitely repeated Prisoners' Dilemma. Nobody ever dares to play anything but *dove* because anyone who cheats will be relentlessly punished by the other player switching permanently to *hawk*.

The version of the folk theorem proved in the text is restricted to pure strategies that can be represented as finite automata. When two such automata play each other, they eventually start cycling through the same sequence of action pairs over and over again. We capture the idea that the players are very patient by making their payoffs in the repeated game equal to their average payoffs during the cycle. Such limit-of-the-means payoffs correspond to first computing the discounted sum of a player's income stream and then taking the limit as the discount factor $\delta \to 1$.

To prove our folk theorem, first find a cycle that generates payoffs for the players close to any particular outcome $x$ in the cooperative payoff region of the stage game. Players can then be deterred from deviating from this cycle by building appropriate punishments into the strategies. But this trick works only when $x \geq \overline{m}$ because Eve can't do worse to Adam than minimax him, if he knows what she is doing.

Who guards the guardians? This question arises when we ask why players should stick to their strategy and punish a deviant opponent when it is costly to administer the punishment. The answer is that the folk theorem still holds for subgame-perfect equilibria because one can build in the proviso that failures to punish when punishment is due must themselves be punished. This closing of the chains of responsibility explains why some political philosophers choose to model the social contracts that form the organizing principle of particular societies as different subgame-perfect equilibria in a repeated game of life. We then have an opportunity to try to understand why concepts like reputation and trust matter so much in human societies.

Axelrod popularized the idea of reciprocity in repeated games by highlighting the strategy TIT-FOR-TAT. It is an equilibrium in the infinitely repeated Prisoners' Dilemma if both players use this strategy, which requires playing *dove* at the outset of the game and then copying what the opponent did at the previous stage. But the evolutionary arguments offered in support of TIT-FOR-TAT could equally well be made for many other strategies. It certainly doesn't embody everything that matters about reciprocity in repeated games. It is particularly poor at capturing reciprocal behavior in games with more than two players, where an attempt by Adam to cheat Eve will often be punished by a *third* player. However, Axelrod's basic claim that evolution is likely to generate Pareto-efficient equilibria in indefinitely repeated games seems to be genuinely robust.

## 11.8 Further Reading

*Evolution of Cooperation*, by Bob Axelrod: Basic Books, New York, 1984. This book sold the world on the idea that reciprocity matters, but the claims it makes for TIT-FOR-TAT are overblown.

*Game Theory*, by Drew Fudenberg and Jean Tirole: MIT Press, Cambridge, MA, 1991. Look here for the details of fancier folk theorems.

*Game Theory and the Social Contract.* Vol. 2: *Just Playing,* by Ken Binmore. MIT Press, Cambridge, MA, 1998. Axelrod's claims for TIT-FOR-TAT are reviewed in Chapter 3.
*Social Evolution*, by Bob Trivers: Cummings, Menlo Park, CA, 1985. Reciprocity and much more in animal societies.

## 11.9 Exercises

1. The twice-repeated game $Z$ of Figure 11.1(a) is studied under the assumption that a player's payoff in the repeated game $Z^2$ is $x+y$, where $x$ and $y$ are the player's payoffs at the first and second stages. What matrix would replace Figure 11.3(b) if the payoffs in $Z^2$ were

$$\text{(a)}\, x+\tfrac{1}{2}y \qquad \text{(b)}\, xy?$$

2. The set $H$ in Section 11.2 is the set of possible histories of play just before $Z$ is played for the second time. How many elements does $H$ have? How many elements would $H$ have if $Z$ were a $3\times 4$ matrix game? How many elements would $H$ have if it were the set of histories of play just before $Z$ was played for the fifth time?
3. Show that the $n$-times-repeated Prisoners' Dilemma has

math

$$2^{4^0}\times 2^{4^1}\times 2^{4^2}\times\cdots\times 2^{4^{n-1}} = 2^{(4^n-1)/3}$$

pure strategies. Give an estimate of how many decimal digits it takes to write down the number of pure strategies in the ten-times-repeated Prisoners' Dilemma.
4. A repeated game $G^n$ results when $G$ is played precisely $n$ times in succession. The payoffs in $G^n$ are obtained by adding the payoffs in each stage game. If $G$ has a unique Nash equilibrium, show that $G^n$ has a unique subgame-perfect equilibrium and that this requires each player to plan always to use his or her Nash equilibrium strategy at every stage.
5. The game Chicken of Figure 1.13(a) has three Nash equilibria. Deduce that the game obtained by repeating Chicken twice has at least nine subgame-perfect equilibria.
6. Theorem 11.1 shows that, when the Prisoners' Dilemma is repeated a finite number of times, there is a unique subgame-perfect equilibrium in which each player always plans to play *hawk.* Prove that all Nash equilibria also lead to *hawk* always actually being played but that Nash equilibria exist in which players plan to use *dove* under certain contingencies that never arise when the equilibrium is used.

math

7. Theorem 11.1 shows that, when the Prisoners' Dilemma is repeated a finite number of times, there is a unique subgame-perfect equilibrium in which each player always plans to play *hawk.* Use a similar formal argument to prove the conclusion of Exercise 5.9.17(b) for the finitely repeated Chain Store Game.

econ

8. Section 11.3.2 studies a version of the repeated Prisoners' Dilemma in which the probability $p$ that any particular repetition will be the last is given by $p=\frac{1}{3}$. What is the largest value of $p$ for which a pair of GRIM strategies constitutes a Nash equilibrium?

phil

9. Exercise 5.9.22 considers one way in which imperfect rationality can lead to cooperation in the finitely repeated Prisoners' Dilemma. In the current exercise, the players are perfectly rational, but they can choose only finite automata as strategies that have at most 100 states.[12] Why can't such a machine count up to 101? Why does it follow that the pair (GRIM, GRIM) is a Nash equilibrium in the automaton-selection game when the Prisoners' Dilemma is to be repeated 101 times?[13]
10. Section 6.6 contains diagrams of various payoff regions for the versions of Chicken and the Battle of the Sexes given in Figure 6.15. Locate their minimax points in mixed strategies and hence draw the set of payoff pairs that can be sustained as equilibria when the games are played repeatedly by very patient players. (Appeal to the general form of the folk theorem given in Section 11.4.4)
11. Repeat the previous exercise for the Stag Hunt Game of Figure 8.7(a).

math

12. The finite automata studied in this chapter are called Moore machines. Given an input set $T$ and an output set $S$, a Moore machine is formally a quadruple $\langle Q, q_0, \lambda, \mu \rangle$ in which $Q$ is a set of states, $q_0$ is the initial state, $\lambda : Q \to S$ is an output function, and $\mu : Q \times T \to Q$ is a transition function. Which of the machines of Figure 11.5 is determined by the following specifications?

$$
\begin{aligned}
S = T &= \{d, h\} \\
q_0 &= d \\
\lambda(d) &= d; \lambda(h) = h \\
\mu(d, d) &= d; \mu(d, h) = h; \mu(h, d) = d; \mu(h, h) = h
\end{aligned}
$$

phil

13. Explain why a computer with no access to external storage is a finite automaton in which each state consists of all possible sets of memories the computer could be holding. If we deny the computer access to an external clock or a calculator, does its complexity "really" represent the complexity of the strategy it implements?

econ

14. The interest rate is fixed at 10%. You are offered an asset that pays $1,000 from now until eternity at yearly intervals. You find its present value by calculating the sum of the discounted annual payments in the income stream secured by the asset. What discount factor will you use? Assuming no uncertainties, at what price will the asset be traded?

econ

15. To borrow $1,000, you must pay back twelve monthly installments of $100.
    a. It cost you $200 to borrow $1,000 for a year. Why is your yearly interest rate not equal to 200/1,000 = 20%?
    b. What is the present value of the income stream 1,000, −100, −100, ..., −100 if the *monthly* interest rate is *m*? Find the approximate monthly

[12]A kibitzer would then think the players are boundedly rational because it would seem that the players were incapable of solving computational problems whose resolution requires a finite automaton with more than hundred states.

[13]Neyman has shown that cooperation remains possible as a Nash equilibrium outcome even when the number of states allowed is very large compared with the number of times the Prisoners' Dilemma is to be repeated.

interest rate $\mu$ you are paying by determining the value of $m$ that makes this present value equal to zero.

c. What *yearly* interest rate corresponds to the monthly interest rate?

16. Obtain a version of the folk theorem that concerns mixed strategy equilibria. Assume that each player can directly observe the randomizing devices employed by the opponent in the past and not just the actions that the opponent actually used. Why does this assumption matter?

math

17. Suppose it is common knowledge that the players in a repeated game always jointly observe the toss of a coin before each stage is played. Give an example to show why this might be relevant.

phil

18. Pandora can choose any amount between zero and one dollar for herself. If this one-player game is repeated infinitely often and Pandora is very patient, explain why a subgame-perfect equilibrium like that considered in Section 11.4.5 can't be found in which she disciplines herself not to take the whole dollar all the time.

fun

19. In Exercise 5.9.19, Alice is an incumbent monopolist in the finitely repeated Chain Store Game and is unable to establish a reputation for being tough by fighting early entrants into her markets. This exercise concerns the infinitely repeated case. Assume that Alice evaluates her income stream using a discount factor $\delta$ satisfying $0 < \delta < 1$.

    Consider a strategy $s$ for Alice that calls for her to fight an entrant if and only if she has never acquiesced to an entry in the past. Consider a strategy $t_i$ for the $i$th potential entrant that calls for entering the market if and only if Alice has acquiesced to an entry in the past. Is this strategy profile a Nash equilibrium? Is it subgame perfect?

econ

20. The Ultimatum Game has been the object of extensive laboratory studies (Section 19.2.2). In one version, Adam can offer any share of four dollars to Eve. If she accepts, she gets her share and Adam gets the rest. If she refuses, both get nothing. The Ultimatum Minigame shown in Figure 11.14 is a simplified version in which Adam can make only a fair offer to split the money evenly or an unfair offer in which in which he gets three times as much as Eve. Eve is assumed to accept the fair offer for sure but can say *yes* or *no* to the unfair offer.

    a. Explain why the doubled lines in Figure 11.14(a) show the unique subgame-perfect equilibrium of the game. Confirm that the strategic form of the game is as shown in Figure 11.14(b). Confirm that the cooperative payoff region is the shaded part of Figure 11.14(c)

    b. Find all pure and mixed Nash equilibria of the one-shot game.

    c. Show that each outcome in the deeply shaded part of Figure 11.14(c) can be sustained as a Nash equilibrium in the repeated game, provided that the players are sufficiently patient.

21. In laboratory studies, real people don't play the subgame-perfect equilibrium in the Ultimatum Game of the previous exercise. The Humean explanation is that people are habituated to playing the fair equilibrium in *repeated* versions of the game. Use the Ultimatum Minigame to comment on how people would use the words *fairness, reputation,* and *reciprocity* if the Humean explanation were correct. Why would this explanation be difficult to distinguish from the claim

phil

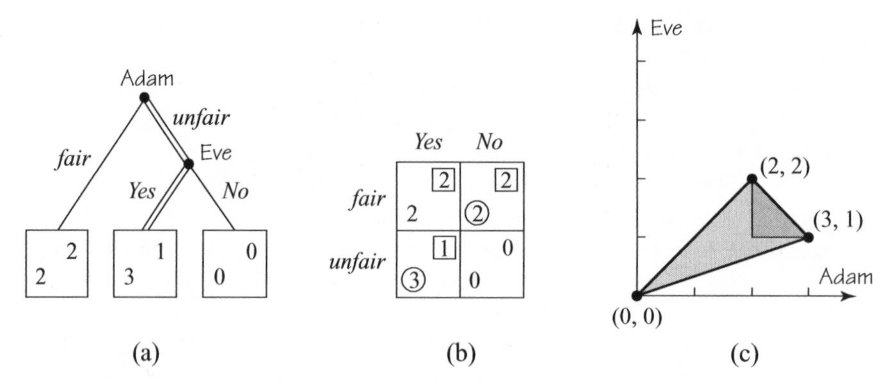


Figure 11.14 The Ultimatum Minigame.

that people have a taste for a good reputation, fairness, or reciprocity built into their utility functions?

phil

22. Suppose the Queen of Hearts takes the role of Eve in a new version of the Battle of the Sexes of Figure 6.15(b). Adam is replaced by all the rest of the cards in the pack. In this multiplayer coordination game, everybody must make the same strategy choice, or else everybody gets a payoff of zero. If everybody chooses *box,* the queen gets a payoff of 1, and everybody else gets 2. If everybody chooses *ball,* the queen gets 2 and everybody else gets 1.
    a. If everybody sees the queen move first, explain why the outcome will be that everybody plays her preferred strategy.
    b. If moves are made simultaneously, show that everybody will play the queen's preferred strategy if it is common knowledge that everybody *believes* the queen will play this strategy herself.

    Relate this conclusion to the discussion of authority in Section 11.5.2.

fun

23. Hans Christian Andersen tells the story of an emperor who was deceived by two tricksters into believing that they had woven a suit of clothes for him that were visible only to the pure in heart. They then pretended to dress the emperor in the nonexistent new clothes for a big parade through the town. Although the emperor was naked, everybody pretended otherwise. Use the story to explain how the folk theorem can explain how false assertions that everybody knows to be false can nevertheless be treated as true in a social context.

phil

24. In an overlapping generations model, there are always three persons alive at any time. Every so often, two are matched to play the Prisoners' Dilemma while the other looks on. Currently, Alice, Bob, and Carol are alive. They sustain a social contract in which everybody cooperates. But Carol dies and is replaced by the youthful Dan, who doesn't know the ropes. Dan is matched for the first time with Alice, who is tempted to exploit his inexperience. Describe an equilibrium in which such bad behavior is prevented by the threat of punishment from Bob.
25. The Prisoners' Dilemma is played infinitely often by pairs of anonymous players drawn at random each time from a finite population. If the players are sufficiently patient and forward looking, explain why it is a Nash equilibrium of this multiplayer repeated game if everyone uses the GRIM strategy. Cooperation is therefore achieved even though it isn't possible to identify cheaters.

26. In the previous exercise, the innocent are knowingly punished for the crimes of the guilty. Why is the mechanism called "contagion"? Is this a case where the end justifies the means? What of the similar equilibria in which cooperation is sustained by responding to a crime committed by a member of an outsider group by punishing anyone in the outsider group who happens to be available?

phil

27. Explain why pairwise reciprocal altruism can't explain the altruism of the model of Section 11.5.3.

phil

28. The version of Chicken given in Figure 6.15(a) is repeated 100 times. The repeated game payoffs are just the sum of the stage-game payoffs. Consider a strategy $s$ that tells you always to choose *slow* up until the 100th stage and to use *slow* and *speed* with equal probabilities at the 100th stage—*unless* the two players have failed to use the same actions at every preceding stage. If such a coordination failure has occurred in the past, $s$ tells a player to look for the first stage at which differing actions were used and then always to use whatever action that person *didn't* play at that stage.
    a. Why is $(s, s)$ a Nash equilibrium?
    b. Prove that $(s, s)$ is a subgame-perfect equilibrium.
    c. Give some examples of income streams other than 2, 2, 2, ... 2, 1 that can be supported as equilibrium outcomes in a similar way.
    d. What is it about Chicken that allows such folk theorem results to be possible in the finitely repeated case?
29. The version of the Battle of the Sexes given in Figure 6.15(b) has two Nash equilibria in pure strategies and one in mixed strategies. Explain why the one-shot game poses an equilibrium selection problem if there is no way to break the symmetry.

    Now suppose that the Battle of the Sexes is repeated $n$ times. The repeated game payoffs are just the sum of the stage-game payoffs.

    Consider a strategy $s$ that tells you always to play the mixed strategy of the one-shot game until your choice coincides with that of the opponent at some stage. If the latter eventuality occurs, $s$ requires you to continue by alternating between *box* and *ball* to the end of the game. Explain why $(s, s)$ is a symmetric Nash equilibrium.

# 12
# *Getting the Message*

## 12.1 Knowledge and Belief

The tradition in philosophy is that knowledge is justified true belief, but game theorists make a sharp distinction between knowledge and belief. This chapter looks at how we treat knowledge. Belief is studied in the next chapter.

### 12.1.1 Decision Problems

A decision problem is determined by a function $f : A \times B \to C$, where $A$ is the set of available actions, $B$ is the set of possible states of the world, and $C$ is the set of possible consequences or outcomes (Section 3.2).

Pandora chooses an action $a$ in the set $A$, but what happens next depends also on what state $b$ the world happens to be in. The consequence $c = f(a, b)$ therefore depends on both Pandora's action $a$ and the state $b$.

A player may be faced with many decision problems as a game proceeds. At each stage, players *know* what decision problem they are facing, but they don't usually know what the state of the world is. On this subject, they have to rely on their *beliefs* (Section 3.3.2). Beliefs are therefore defined on the set $B$ of states of the world.

What a player knows in a game changes as the game is played. For example, after Alice trumps your ace in bridge, you now know that she no longer holds that trump in her hand. Von Neumann saw that one can keep track of what a player knows during a game simply by introducing *information sets* (Section 2.2.1). Although this idea is now taken for granted, it seems to me another tribute to Von Neumann's genius that he should have realized that something that looks so complicated should admit such a simple resolution.

Once Pandora learns that she has reached a particular information set, then she knows what decision problem she has to solve. How she solves the problem will depend on her preferences over the possible consequences and her beliefs over the states of the world.

Each time play reaches a new information set, she will need to *update* her beliefs to take account of her new knowledge. The next chapter discusses how players condition their probabilities for the possible states of the world on the knowledge that they have reached a particular information set (Section 3.3). The current chapter is about the information sets themselves.

→ 12.3

## 12.2 Dirty Faces

The next section makes such a big fuss about the knowledge operator that you will surely wonder whether such care is really necessary. Mostly it isn't, but we shall use the following ancient conundrum to illustrate how easy it can sometimes be to get confused without a proper mathematical model.

Alice, Beatrice, and Carol are three very proper Victorian ladies traveling together in a railway carriage. Each has a dirty face, but nobody is blushing, even though a Victorian lady who was conscious of appearing in public with a dirty face would surely do so. It follows that none of the ladies knows that her own face is dirty, although each can clearly see the dirty faces of the others.

Victorian clergymen always told the whole truth and nothing but the truth, and so the ladies pay close attention when a local minister enters the carriage and announces that one of the ladies has a dirty face.

After his announcement, one of the ladies blushes. How come? Didn't the minister simply tell the ladies something they knew already?

To explain what the minister added to what the ladies already knew, we need to look carefully at the chain of reasoning that leads to the conclusion that one of the ladies must blush. If neither Beatrice nor Carol blushes, Alice would reason as follows:

> *Alice:* Suppose that my face were clean. Then Beatrice would reason as follows:
>
> > *Beatrice:* I see that Alice's face is clean. Suppose that my face were also clean. Then Carol would reason as follows:
> >
> > > *Carol:* I see that Alice's and Beatrice's faces are clean. If my face were clean, nobody's face would be dirty. But the minister's announcement proves otherwise. So my face is dirty, and I must blush.
> >
> > *Beatrice:* Since Carol hasn't blushed, my face is dirty. So I must blush.
>
> *Alice:* Since Beatrice hasn't blushed, my face is dirty. So I must blush.

This argument shows that *someone* will blush—not that *everyone* will blush, which is the claim that is usually mistakenly made.

So what did the the minister add to what the ladies already knew? Everybody knew that someone had a dirty face, but he made this fact *common knowledge*. The idea of common knowledge has been touched upon several times in previous chapters, but this is one of the issues that will be tied down once and for all in the current chapter.

## 12.3 Knowledge

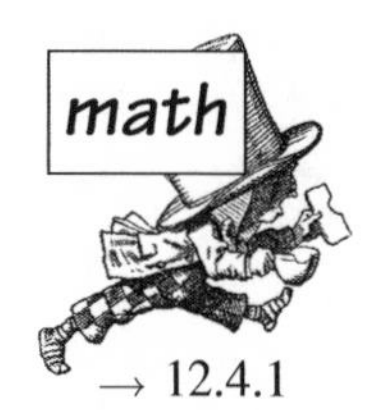


→ 12.4.1

The philosophy of knowledge is called *epistemology*. In this context, the humble sample space $\Omega$ of Section 3.2 often gets called the set of possible states of the world. We shall inflate its importance even more by calling $\Omega$ our *universe of discourse*. But a subset $E$ of $\Omega$ will still just be called an event.

In the case of our Victorian ladies, the universe of discourse contains the eight states listed as the columns in Figure 12.1. For example, in the state of the world $\omega = 8$, all three ladies have dirty faces. If $\omega = 8$ is the true state of the world, then any event that contains $\omega$ is said to have occurred—for example, the event $D_{\text{B}} = \{3, 5, 7, 8\}$ that Beatrice has a dirty face.

### 12.3.1 Knowledge Operators

Pandora's knowledge can be specified with the help of a *knowledge operator* $\mathscr{K}$. For each event $E$, the set $\mathscr{K}E$ is the set of states of the world in which Pandora knows that $E$ has occurred. That is to say, $\mathscr{K}E$ is the event that Pandora knows $E$.

For example, when playing poker, Pandora might be sure that her full house is the winning hand, provided that Olga isn't hiding two fives in her hand to go with the two fives showing on the table. If $E$ is the event that Pandora's hand is better, then $\mathscr{K}E$ is the event that Pandora has seen one of the fives that Olga might be holding being dealt to someone else.

The properties that game theorists assume about knowledge are listed in Figure 12.2 for a finite universe of discourse.

Properties (K0) and (K1) are bookkeeping assumptions. Property (K2) says that Pandora can't know something unless it actually happens.

Property (K3) is really redundant because it can be deduced from (K2) and (K4). Since $\mathscr{K}^2E = \mathscr{K}(\mathscr{K}E)$, property (K3) says that Pandora can't know something

| | 1 | 2 | 3 | 4 | 5 | 6 | 7 | 8 |
|---|---|---|---|---|---|---|---|---|
| Alice | Clean | Dirty | Clean | Clean | Dirty | Dirty | Clean | Dirty |
| Beatrice | Clean | Clean | Dirty | Clean | Dirty | Clean | Dirty | Dirty |
| Carol | Clean | Clean | Clean | Dirty | Clean | Dirty | Dirty | Dirty |

Figure 12.1 Victorian states of the world.

| | | | |
|---|---|---|---|
| (K0) | $\mathcal{K}\Omega = \Omega$ | (P0) | $\mathcal{P}\emptyset = \emptyset$ |
| (K1) | $\mathcal{K}(E \cap F) = \mathcal{K}E \cap \mathcal{K}F$ | (P1) | $\mathcal{P}(E \cup F) = \mathcal{P}E \cup \mathcal{P}F$ |
| (K2) | $\mathcal{K}E \subseteq E$ | (P2) | $\mathcal{P}E \supseteq E$ |
| (K3) | $\mathcal{K}E \subseteq \mathcal{K}^2 E$ | (P3) | $\mathcal{P}E \supseteq \mathcal{P}^2 E$ |
| (K4) | $\mathcal{P}E \subseteq \mathcal{K}\mathcal{P}E$ | (P4) | $\mathcal{K}E \supseteq \mathcal{P}\mathcal{K}E$ |

Figure 12.2 Knowledge and possibility.

without knowing that she knows it. Game theory thereby finesses an old worry: How do you know that you know that you know that you know something?[1] If you don't know all these knowings, then you know nothing at all!

Property (K4) introduces the *possibility operator* $\mathscr{P}$. Not knowing that something didn't happen is the same as thinking it possible that it did happen. So we define the possibility operator by $\mathscr{P}E = \sim \mathscr{K} \sim E$, where $\sim F$ means the complement of the set $F$. Property (K4) then says that, if Pandora thinks something is possible, then she knows that she thinks it possible.

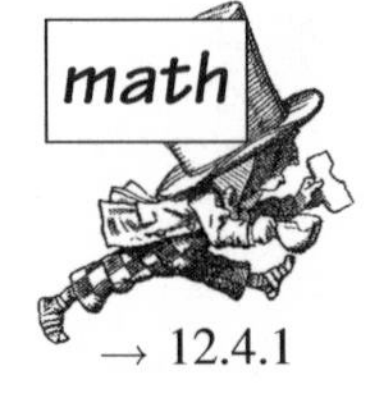


→ 12.4.1

*Notes.* The properties (P0)–(P4) for the possibility operator $\mathscr{P}$ given in Figure 12.2 are equivalent to (K0)–(K4). We could equally well have started with (P0)–(P4) and defined $\mathscr{K}$ by $\mathscr{K}E = \sim \mathscr{P} \sim E$.

Since $E \subseteq F$ implies that $E \cap F = E$ and $E \cup F = F$, we can deduce from (K1) and (P1) that

$$\left.\begin{array}{l} E \subseteq F \Rightarrow \mathscr{K}E \subseteq \mathscr{K}F \\ E \subseteq F \Rightarrow \mathscr{P}E \subseteq \mathscr{P}F \end{array}\right\} \qquad (12.1)$$

→ 12.3.2

It follows that $\subseteq$ can be replaced by $=$ in (K3), (K4), (P3), and (P4).

*Small Worlds.* Assumptions (K0)–(K4) are too strong to be generally applicable to all situations in which we talk about knowledge.[2] They make good sense only when the universe of discourse is sufficiently small that all possible implications of all possible events can be explored in minute detail. The statistician Leonard Savage called this proviso on the type of universe of discourse to be considered a *small-world* assumption (Section 13.6.2).

The axiom that makes the necessity of restricting attention to small worlds most apparent is (P4). This can be rewritten as $\mathscr{K}E = \sim \mathscr{K} \sim \mathscr{K}E$, which says that, if Pandora doesn't know that she doesn't know something, then she knows it (Exercise 12.12.2).

This assumption is inevitable in the small world of a game. For example, suppose that Pandora doesn't know that she doesn't know she has been dealt the queen of hearts. Then it isn't true that she knows she doesn't know she has been dealt the queen of hearts. But she would know she hadn't been dealt the queen of hearts if

[1]Thomas Hobbes addressed this exotic complaint to René Descartes in 1641.

[2]The axioms correspond to what philosophers call the modal logic S-5. Other modal logics are controversially said to be more suitable in large worlds.

she had been dealt some other card. So she knows that she wasn't dealt some other card.

But the world of everyday life isn't so cut and dried. For example, I was surprised yesterday by my mother-in-law's coming to stay for the weekend, although I certainly didn't know that I didn't know she was coming to stay. The moral is that large worlds contain possibilities of which we fail even to conceive.

### 12.3.2 Truisms

Although it is not a standard usage, we define a *truism* for Pandora to be something that can't be true without her knowing it. So $T$ is a truism if and only if $T \subseteq \mathcal{K}T$. By (K2), we then have $T = \mathcal{K}T$.

If we regard a truism as capturing the essence of what happens when making a direct observation, it can be argued that all knowledge necessarily derives from truisms. The following theorem expresses this formally. It isn't a deep result, but its proof will provide some practice in using the knowledge operator.

THEOREM 12.1 *Pandora knows that E has occurred if and only if a truism T that implies E has occurred.*

*Proof* The proof of necessity and sufficiency is split into two steps:

**Step 1.** If the true state $\omega$ lies in a truism $T$ with $T \subseteq \mathcal{K}E$, we show that Pandora knows that $E$ has occurred. But if $\omega \in T \subseteq \mathcal{K}E$, then $\omega \in \mathcal{K}E$, whether or not $T$ is a truism.

**Step 2.** If Pandora knows that $E$ has occurred, we show that a truism $T$ has occurred with $T \subseteq E$. This is easy because we can just take $T = \mathcal{K}E$. The event $T$ is a truism because (K3) says that $T \subset \mathcal{K}T$. The truism $T$ must have occurred because to say that Pandora knows that $E$ has occurred means that the true state $\omega \in \mathcal{K}E = T$. □

## 12.4 Possibility Sets

A possibility set $P(\omega)$ is the set of all states that Pandora thinks are possible when the true state is $\omega$. We can therefore define it by requiring that

$$\omega_2 \in P(\omega_1) \Leftrightarrow \omega_1 \in \mathcal{P}\{\omega_2\}.$$

It doesn't matter that there is a risk of confusing the two sets $P(\omega)$ and $\mathcal{P}\{\omega\}$ because the next theorem implies that they are the same.

THEOREM 12.2 $\omega_1 \in \mathcal{P}\{\omega_2\} \Leftrightarrow \omega_2 \in \mathcal{P}\{\omega_1\}$.

*Proof* Assume to the contrary that $\omega_1 \in \mathcal{P}\{\omega_2\}$ but $\omega_2 \notin \mathcal{P}\{\omega_1\}$.

**Step 1.** Rewrite $\omega_1 \in \mathcal{P}\{\omega_2\}$ as $\{\omega_1\} \subseteq \mathcal{P}\{\omega_2\}$. If we can show that $\omega_2 \notin \mathcal{P}\{\omega_1\}$ implies $\mathcal{P}\{\omega_2\} \subseteq \sim\{\omega_1\}$, we will then have the contradiction we need since only the empty set can be a subset of its complement.

**Step 2**. Rewrite $\omega_2 \notin \mathscr{P}\{\omega_1\}$ as $\{\omega_2\} \subseteq \sim\mathscr{P}\{\omega_1\} = \mathscr{K}\sim\{\omega_1\}$. Then,

$$\mathscr{P}\{\omega_2\} \subseteq \mathscr{P}\mathscr{K}\sim\{\omega_1\} \subseteq \mathscr{K}\sim\{\omega_1\} \subseteq \sim\{\omega_1\},$$

where we have appealed successively to (12.1), (P4), and (K2).

COROLLARY 12.1 $\zeta \in P(\omega) \Rightarrow P(\zeta) = P(\omega)$.

*Proof* $\zeta \in P(\omega) \Rightarrow \{\zeta\} \subseteq \mathscr{P}\{\omega\} \Rightarrow \mathscr{P}\{\zeta\} \subseteq \mathscr{P}\{\omega\} \Rightarrow P(\zeta) \subseteq P(\omega)$ by (12.1) and (P3). But Theorem 12.2 implies that $\omega \in P(\zeta)$, and so we also have that $P(\omega) \subseteq P(\zeta)$.

THEOREM 12.3 *The smallest truism containing $\omega$ is $P(\omega)$.*

*Proof* Property (P2) implies that $\omega \in \mathscr{P}\{\omega\}$. Property (K4) implies that $\mathscr{P}\{\omega\}$ is a truism. Why is $\mathscr{P}\{\omega\}$ the *smallest* truism containing $\omega$? If $T$ is another truism containing $\omega$, we need to show that $\mathscr{P}\{\omega\} \subseteq T$. But, by (P1) and (P4), $\{\omega\} \subseteq T = \mathscr{K}T$ implies that

$$\mathscr{P}\{\omega\} \subseteq \mathscr{P}T = \mathscr{P}\mathscr{K}T \subseteq \mathscr{K}T = T.$$

COROLLARY 12.2 *Pandora knows that $E$ has occurred in state $\omega$ if and only if $P(\omega) \subseteq E$.*

*Proof* If $P(\omega) \subseteq E$, then Theorem 12.3 tells us that Pandora knows $E$ in state $\omega$ because $P(\omega)$ is a truism that contains $\omega$. On the other hand, if Pandora knows that $E$ has occurred, there must be a truism $T$ such that $\omega \in T \subseteq E$. But $P(\omega)$ is the smallest truism containing $\omega$. So $\omega \in P(\omega) \subseteq T \subseteq E$.

### 12.4.1 Knowledge Partitions

To partition a set $S$ is to break it down into a collection of subsets so that each element of $S$ belongs to one and only one subset in the collection.

For example, in Section 15.2, we look at a toy model of poker in which Alice and Bob are each dealt one card from a deck containing only the king, queen, and jack of hearts. The card dealt to Alice from the top of the deck then defines a partition of the set

$$\Omega = \{KQJ,\ KJQ,\ QKJ,\ QJK,\ JKQ,\ JQK\}$$

of all possible ways the cards could be shuffled. The collection of subsets that make up the partition is

$$\{\{KQJ,\ KJQ\},\ \{QKJ,\ QJK\},\ \{JKQ,\ JQK\}\}. \tag{12.2}$$

Our theorems on possibility sets can be summarized by saying that they partition Pandora's universe of discourse into units of knowledge. When the true state is

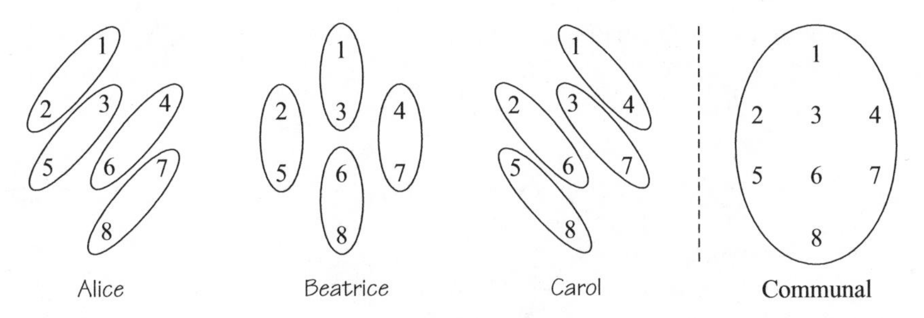


Figure 12.3 Possibility sets *before* the minister speaks.

determined, Pandora will necessarily learn that one and only one of these units of knowledge has occurred. Everything else she knows can then be deduced from this fact.

For example, in the toy poker model, it may be that the cards are shuffled so that the true state is $\omega = QKJ$. Alice is then dealt the queen of hearts from the top of the deck. She then can't help but know that the event $P(\omega) = \{QKJ, QJK\}$ from her knowledge partition (12.2) has occured.

*Dirty Possibilities.* What are the possibility sets in the story of the dirty-faced ladies? Figure 12.3 shows possibility sets for each lady *before* the minister makes his announcement. (Ignore the fourth column for the moment.)

For example, whatever Alice sees when she looks at the faces of her companions, it remains possible for Alice that her own face is clean or dirty. Thus, writing $P_A$ to indicate that we are discussing what Alice thinks is possible, $P_A(1) = P_A(2) = \{1, 2\}$.

Figure 12.4 shows possibility sets for the ladies *after* the minister's announcement but *before* any blushing takes place. When Alice sees two clean faces, she can now deduce the state of her own face from whether or not the minister says anything. Thus $P_A(1) = \{1\}$ and $P_A(2) = \{2\}$.

### 12.4.2 Refining Your Knowledge

Some possibility partitions can be compared. A partition $\mathscr{C}$ is a *refinement* of a partition $\mathscr{D}$ if each set in $\mathscr{C}$ is a subset of a set in $\mathscr{D}$. Under the same circumstances, $\mathscr{D}$ is said to be a *coarsening* of $\mathscr{C}$. For example, Alice's partition in Figure 12.4 is a

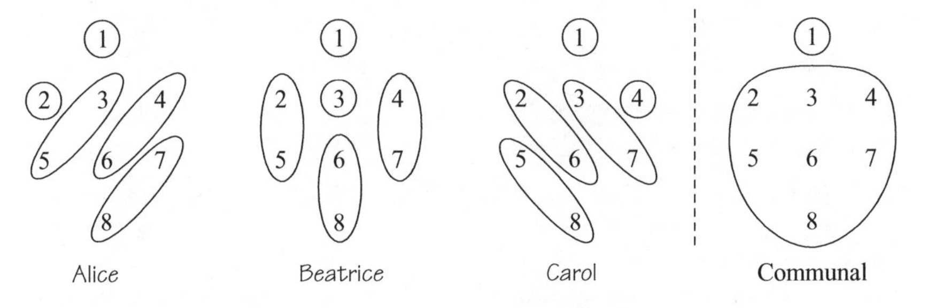


Figure 12.4 Possibility sets *after* the minister speaks, *before* blushing begins.

refinement of her partition in Figure 12.3. Equivalently, her partition in Figure 12.3 is a coarsening of her partition in Figure 12.4. This reflects the fact that she is better informed in the latter case.

*Blushing in Rotation.* If a lady blushes on discovering that her face is dirty, the other players will use what they thereby learn about her knowledge to refine their own knowledge partitions.

The following sequence of events follows from the assumption that the opportunity to blush rotates among the three ladies, starting with Alice. Figure 12.5(a) illustrates how the ladies' knowledge partitions evolve.

**Step 1.** Before the minister has had a chance to speak, the knowledge situation is as shown in Figure 12.3.

**Step 2.** After the minister has had a chance to speak, the knowledge situation is as shown in Figure 12.4. This diagram is repeated as the first row of Figure 12.5(a), but with the states in which a lady has a dirty face indicated by the addition of shading. (Ignore the fourth column of the figure for now.)

**Step 3.** Alice (but not Beatrice or Carol) now has the opportunity to blush. She will blush only in state 2 because this is the only state in which she knows her face is dirty. Alice's own information is unchanged whether she blushes or not. However, Beatrice and Carol learn something from her behavior. If Alice blushes, the true state must be $\omega = 2$. This allows Bearice to split her possibility set $\{2,5\}$ into two subsets $\{2\}$ and $\{5\}$.

As with the dog that *didn't* bark in the Sherlock Holmes story, observing that Alice doesn't blush is just as informative for Beatrice when her possibility set is $\{2,5\}$ as observing that Alice does blush. The fact that Alice doesn't blush excludes the possibility that the true state is $\omega = 2$. It must therefore be that $\omega = 5$.

Carol makes similar inferences and so splits her possibility set $\{2,6\}$ into $\{2\}$ and $\{6\}$. The result is shown in the second row of Figure 12.5(a).

**Step 4.** Beatrice (but not Carol or Alice) now has the opportunity to blush. She blushes only in states 3 and 5. This is very informative for Carol, whose new possibility partition becomes as refined as it can possibly get. Alice, however, learns nothing. In particular, her possibility set $\{3,5\}$ can't be refined because Beatrice will blush both in state 3 and in state 5. The result is shown in the third row of Figure 12.5(a).

**Step 5.** Carol (but not Alice or Beatrice) now has the opportunity to blush. She blushes in states 4, 6, 7, and 8. However, neither Alice nor Beatrice can refine their possibility partitions on the basis of this information.

**Step 6.** Alice now has the opportunity to blush again. She blushes only in state 2. This helps neither Beatrice nor Carol.

**Step 7.** Beatrice now has the opportunity to blush again. She blushes only in states 3 and 5. This helps neither Alice nor Carol.

No further steps need be examined since steps 5, 6, and 7 will just repeat over and over again. The final informational situation is therefore as recorded in the third row of Figure 12.5(a).

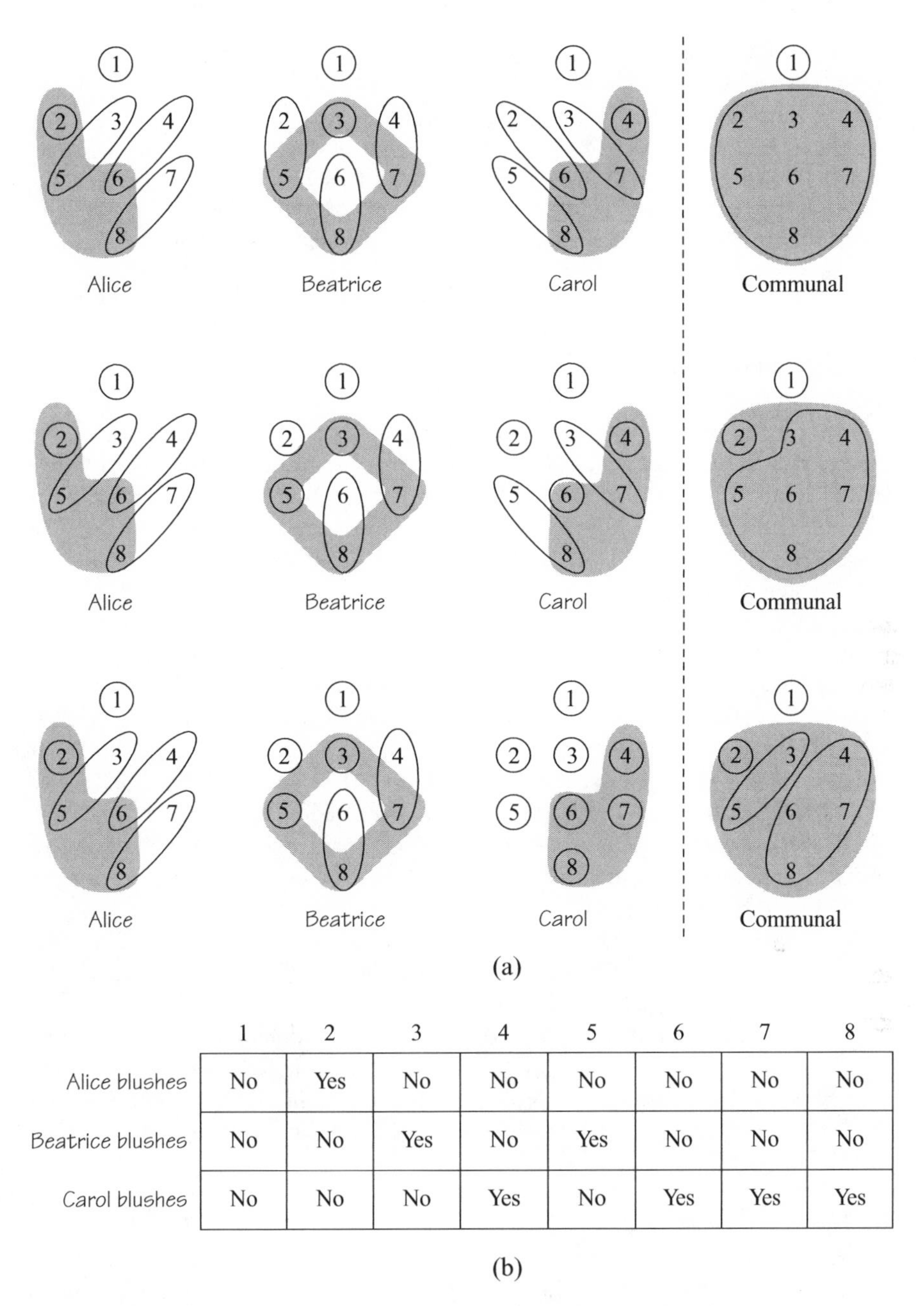


| | 1 | 2 | 3 | 4 | 5 | 6 | 7 | 8 |
|---|---|---|---|---|---|---|---|---|
| Alice blushes | No | Yes | No | No | No | No | No | No |
| Beatrice blushes | No | No | Yes | No | Yes | No | No | No |
| Carol blushes | No | No | No | Yes | No | Yes | Yes | Yes |

Figure 12.5 Blushing in rotation.

*Who Blushes?* The blushing table of Figure 12.5(b) can now be constructed using the third row of Figure 12.5(a) on the assumption that any lady who knows that her face is dirty necessarily blushes.

For example, Beatrice's possibility set when $\omega = 8$ is $P_B(8) = \{6, 8\}$. The event that she has a dirty face is $D_B = \{3, 5, 7, 8\}$. It is therefore false that $P_B(8) \subseteq D_B$.

Hence, by Corollary 12.2, Beatrice doesn't blush when the true state is $\omega = 8$. However, $P_C(8) = \{8\}$ and $D_C = \{4, 6, 7, 8\}$. Thus $P_C(8) \subseteq D_C$, and therefore Carol blushes when the true state is $\omega = 8$.

However, the story of blushing in rotation is only one of several stories that could have been told that are consistent with the informational specifications given in the tale of the dirty-faced ladies. Other possibilities are explored in Exercises 12.12.14 and 12.12.15. Someone always blushes, but who it is depends on how the blushing mechanism works.

## 12.5 Information Sets

In principle, the states of the world in a game are all of its possible plays. As the game proceeds, Pandora will update her knowledge partition as she learns things about the preceding history of play. However, it is too clumsy to draw pictures like those of Figure 12.5(a), in which the players' knowledge partitions of the set $\Omega$ of possible plays become more and more refined with each successive move. It is more convenient to summarize the properties of the players' knowledge partitions that we need by drawing *information sets* (Section 2.2.1).

Information sets aren't possibility sets, but they inherit many of the properties of the possibility sets that they determine. The most important property is that Pandora's information sets must *partition* her set of decision nodes. In particular, her information sets mustn't overlap.

For example, the Monty Hall Game of Figure 3.1 is a game of imperfect information in which there are four nodes at which Alice might have to make a decision. These decision nodes are partitioned into two information sets, which become possibility sets if we restrict the states of the world to be the four possible histories of play: [13], [23], [21], and [31].

*Properties of Information Sets.* One can't partition a player's set of decision nodes any old way and expect to obtain a game in which the information sets make sense. In particular, neither of the situations of Figure 12.6 is admissible if $\{x, y\}$ is to be interpreted as an information set. In Figure 12.6(a), Adam could tell which decision node he was at by counting the choices available to him. In Figure 12.6(b), he could deduce where he was from the labels used to describe his choices.

### 12.5.1 Perfect Recall

In a game of perfect recall, nobody ever forgets something they once knew because the information sets are drawn in such a way that it is always possible to deduce anything that you knew in the past from the fact that you have arrived at a particular information set.

A game of perfect information is necessarily a game of perfect recall because all information sets in a game of perfect information contain only one decision node. Thus, everybody always knows everything about the history of play in the game so far. But a game of perfect recall may have imperfect information, as in the Monty Hall Game of Figure 3.1.

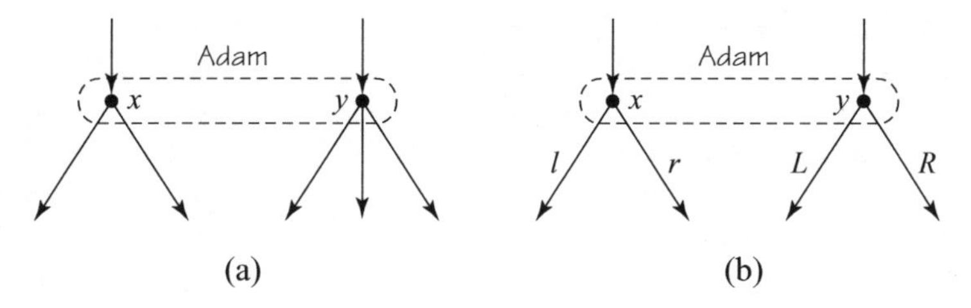


Figure 12.6 Illegal information sets.

*Absent-Minded Drivers.* Terence Gorman was a much-loved economist well known for being absent minded. In the Mildly Forgetful Driver's Game of Figure 12.7(a), Terence's home is on the opposite corner of the block to his office. He can get home by taking either two left turns or two right turns. If he does anything else he is hopelessly lost. But when he comes to make the second turn, Terence can't remember whether the first turn he took was a right or a left. His forgetfulness is represented in the game tree by including both nodes $x$ and $y$ in an information set $I$ to indicate that he doesn't know whether the history of play that brought him to $I$ is $[l]$ or $[r]$.

In the Seriously Forgetful Driver's Game of 12.7(b), Terence needs to make a right turn and then a left turn to get home. But in this game he can't even remember whether he has made a turn already when he gets to the second turn. The information set that represents his forgetfulness now indicates that he doesn't know whether the history of play that brought him to $I$ is $[\emptyset]$ or $[r]$. This is a much more serious form of imperfect recall because we now have an information set that contains two decision nodes *on the same play.*

Terence could escape the problems that both these one-player games of imperfect recall create for him by taking notes of things as they happen in the game and referring to his notebook when in doubt. Since we allow him to consult the great book of game theory free of charge, it would be unreasonable to make him pay for taking notes. In the idealized world inhabited by game theorists, perfect recall should therefore always be taken for granted unless something is said to the contrary.

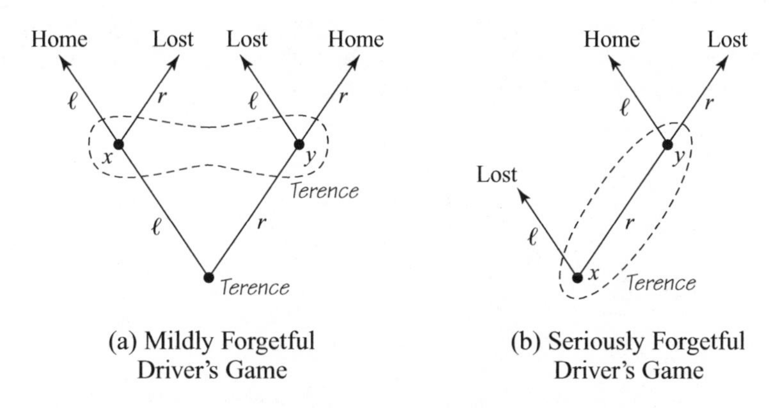


Figure 12.7 Absent-minded drivers.

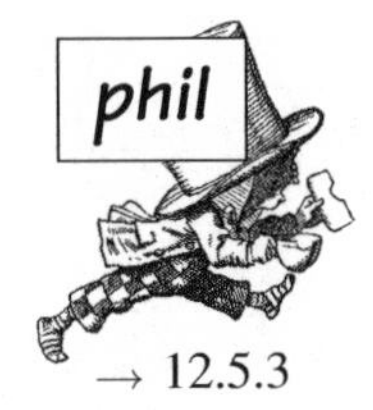


→ 12.5.3

*Perfect Recall and Knowledge.* The relative seriousness of the two violations of perfect recall in our Forgetful Driver Games are illustrated by Figure 12.8. In these diagrams, the states of the world are all possible plays of the game. The possibility sets shown refer to what Terence thinks is possible after he has just made a decision. There are therefore two rows in Figure 12.8(a) because Terence is aware of making one decision after another.

What goes wrong in the case of the Mildly Forgetful Driver's Game is simply that the second possibility partition isn't a refinement of the first. But things are much worse in the case of the Seriously Forgetful Driver's Game because the possibility sets overlap—which is as serious a violation of our knowledge requirements as it is possible to make.

### 12.5.2 Agents

Games like the Seriously Forgetful Driver's Game seem unlikely ever to be useful as models because they generate incoherent knowledge structures. However, models in which there is some forgetfulness can sometimes be useful. Bridge is an example.

One may study bridge as a four-player game. It will then be a game of imperfect information with perfect recall. North and South will be two separate players who happen to have identical preferences. Sometimes such a set of players is called a *team*. East and West will also be a team but with diametrically opposed preferences to the North-South partnership.

Alternatively, one may study bridge as a two-player, zero-sum game between Adam and Eve. Adam is then a manager for the North-South partnership. North and South act as puppets who simply follow his instructions, given in detail before the game begins. We say that North and South are Adam's *agents*. Similarly, East and West are agents for Eve.

The latter may seem the simpler formulation because two-player games are easier than four-player games. But if bridge is formulated according to the second model, it becomes a game of imperfect recall. It would make nonsense of the game if, when Adam puts himself into South's shoes, he were able to remember what cards North had when Adam was occupying his shoes a moment before.

### 12.5.3 Behavioral Strategies

A pure strategy specifies a particular action for each of a player's information sets. For example, when $n = 10$, Tweedledum has five (singleton) information sets in the

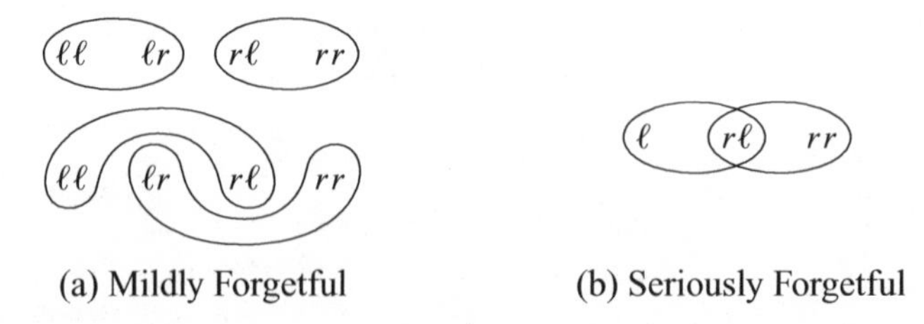


Figure 12.8 Violating the knowledge requirements. In the Mildly Forgetful Game, the second possibility partition over plays of the game isn't a refinement of the first. In the Seriously Forgetful Game, the possibility sets aren't even a partition.

game Duel of Figure 3.14. At each information set, he has two choices, so he has a total of $2^5 = 32$ pure strategies.

A mixed strategy $p$ is a vector whose coordinates correspond to the pure strategies of a game (Section 6.4.2). Tweedledum's use of the mixed strategy $p$ results in his $i$th pure strategy being played with probability $p_i$. Since Duel has thirty-two pure strategies, its mixed strategies are very long vectors.

A *behavioral strategy* resembles a pure strategy in that it specifies how players are to behave at each of their information sets. But instead of selecting a particular action at each information set, a behavioral strategy assigns a *probability* to each of the available actions. In Duel, a behavioral strategy is therefore determined by only five probabilities, rather than the thirty-two probabilities required for a mixed strategy.

A player using a behavioral strategy can be thought of as decentralizing the decision process to a bunch of agents, one for each of the player's information sets. Each agent is given a piece of paper saying with what probability he should select each of the available actions at the information set the agent is responsible for. Each agent then acts *independently* of all the others.

When using a mixed strategy, Tweedledum does all his randomizing *before* the game begins. When using a behavioral strategy, he rattles a dice box or spins a roulette wheel only *after* reaching an information set.

Although they seem so different, the next result says that the two types of strategy are effectively the same in games of perfect recall. This fact is useful because behavioral strategies are so much simpler than mixed strategies.

Proposition 12.1 (Kuhn) *Whatever mixed or behavioral strategy $s$ that Pandora may choose in a game of perfect recall, she has a strategy $t$ of the other type with the property that, however the opponents play, the resulting lottery over the outcomes of the game is the same for both $s$ and $t$.*

We offer only an illustration of how Kuhn's theorem works for the simple game of Figure 12.9.

Eve's pure strategy *LLR* is shown in Figure 12.9(a), and her pure strategy *RRL* in Figure 12.9(b). Our aim is to find a behavioral strategy $b$ that has the same effect as the mixed strategy $m$ that assigns probability $\frac{1}{3}$ to *LLR* and $\frac{2}{3}$ to *RRL*. To specify such a behavioral strategy, we need to determine the probabilities $q_1, q_2$, and $q_3$ with which Eve's agents use the action $R$ at each of her three information sets.

The randomization specified by $m$ leads to the use of either *LLR* or *RRL*. So $L$ will get played at Eve's first information set with probability $\frac{1}{3}$, and $R$ will get played with probability $\frac{2}{3}$. To mimic this behavior with the behavioral strategy $b$, take $q_1 = \frac{2}{3}$.

Eve's second information set won't be reached at all if the randomizing specified by $m$ leads to the use of *LLR*. If her second information set is reached, the randomizing called for by $m$ must therefore have led to the use of *RRL*. So $R$ will be played for certain at Eve's second information set. To mimic this behavior with $b$, take $q_2 = 1$.

Eve's third information set can't be reached at all when $m$ is used. So $q_3$ can be chosen to be anything.

## 12.6 Common Knowledge

Every so often in the previous chapters, we heard that something or other must be common knowledge. The philosopher David Lewis said that something is common

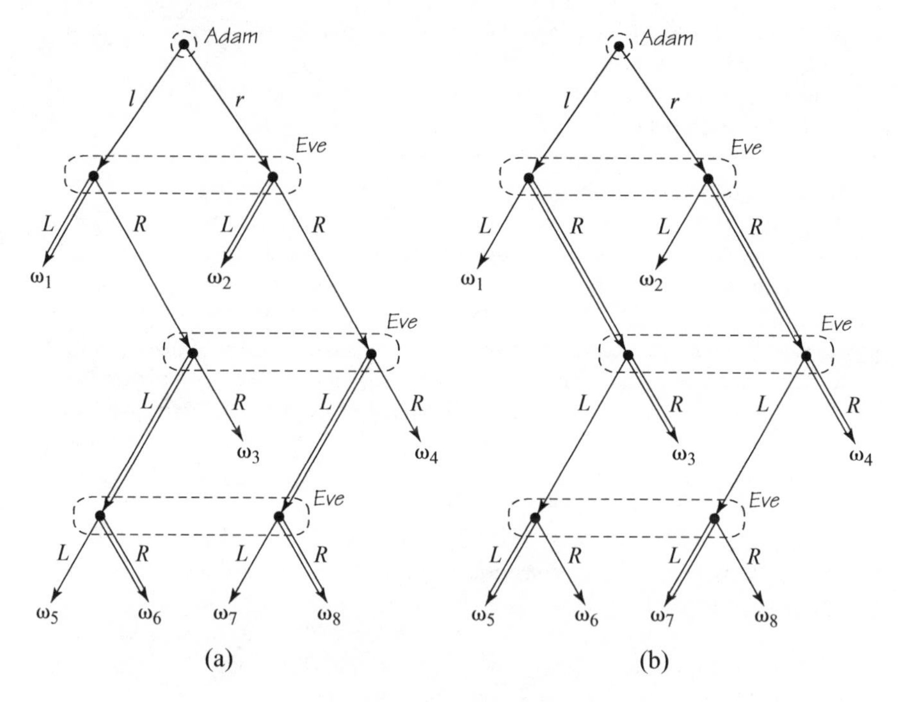


Figure 12.9 Kuhn's theorem.

knowledge if everybody knows it, everybody knows that everybody knows it, everybody knows that everybody knows that everybody knows it, and so on. But how do you know whether all the statements in such an infinite regress are true? This section adapts the story of the dirty-faced ladies to explain how Bob Aumann made common knowledge into a useful tool by answering this question.

### 12.6.1 Meeting of Minds

The common knowledge operator turns out to satisfy the same set of axioms as the individual knowledge operator $\mathscr{K}$. In particular, it has a dual operator $\mathscr{M}$ that registers what the community of players as a whole think possible. By the common knowledge version of Corollary 12.2, $E$ is common knowledge when $\omega$ is the true state of the world if and only if

$$M(\omega) \subseteq E.$$

If we can get a grip on the communal possibility sets $M(\omega)$, we will therefore have solved the problem of determining when an event $E$ is common knowledge. Aumann pointed out that $M(\omega)$ is simply the *meet* of the possibility sets of each individual player.[3]

[3]Some authors prefer to say *join* rather than *meet*. Since these terms represent dual concepts in lattice theory, this is a bit confusing for mathematicians.

*Finding the Meet.* Just as it is hard for something to be common knowledge, so it is easy for something to be communally possible. It is enough for something to be communally possible if Alice thinks it possible. But it is also enough if Beatrice thinks it possible that Alice thinks it possible. Or if Carol thinks it possible that Beatrice thinks it possible that Alice thinks it possible. And so on.

It is easy to keep track of these possibility chains in a diagram. Figure 12.10 shows how this is done. The possibility partitions for Alice, Beatrice, and Carol are those of the third row of Figure 12.5(a). Their meet is another partition consisting of the communal possibility sets shown in the fourth column.

To find the meet, join two states with a line if they belong to the same possibility set for at least one individual. For example, 4 and 7 get linked because they are both included in one of Beatrice's possibility sets. When all such links have been drawn, two states belong to the same communal possibility set if and only if they are connected by a chain of linkages. For example, 4 and 8 belong to the same communal possibility set because 4 gets linked to 7 and 7 gets linked to 8.

With this technique in our pocket, it is easy to trace the evolution of what becomes common knowledge as time passes in the story of the dirty-faced ladies. The fourth columns of Figures 12.3, 12.4, and 12.5(a) show how the communal possibility sets change as information percolates through the community. The event that someone has a dirty face is $D = \{2,3,4,5,6,7,8\}$. This becomes common knowledge in Figure 12.4 because $M(8) \subseteq D$. The event that Carol has a dirty face is $D_C = \{4,6,7,8\}$. This becomes common knowledge in the third row of Figure 12.5(a). Only then does it become true that $M(8) \subseteq D_C$.

*Public Events.* The chain of reasoning that leads to more and more becoming common knowledge is sparked by the minister's announcement that someone in the carriage has a dirty face. An implicit understanding is that it is common knowledge that he will always speak up when he sees a dirty face and remain silent otherwise.

Such an understanding makes $D$ into a *public event.* This means that $D$ is a common truism and so can't occur without everybody knowing it. As we know from the analogue of Theorem 12.1, an event $E$ becomes common knowledge if and only if it is implied by a public event.

How should we interpret the idea of a public event in general? Just as a truism is to be understood as representing what an individual directly observes, so a public event represents what a community observes when everybody is present together observing that everybody else is observing it, too. This is perhaps why we attach so much importance to eye contact. When looking into another person's eyes, the messages we thereby exchange become common knowledge between us.

### 12.6.2 Mutual Knowledge

→ 12.7

We turn again to the story of the dirty-faced ladies in explaining how the common knowledge operator is defined.

Different people often know different things. For the story of the dirty-faced ladies we therefore need three knowledge operators, $\mathscr{K}_A$, $\mathscr{K}_B$, and $\mathscr{K}_C$.

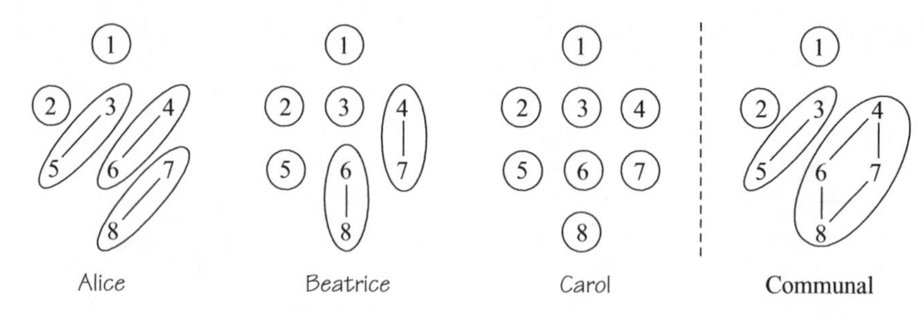


Figure 12.10 Communal possibility sets.

Something is *mutual knowledge* if everybody knows it. More precisely, if the relevant individuals are Alice, Beatrice, and Carol, then the "everybody knows" operator is defined by

$$\text{(everybody knows)}E = \mathscr{K}_A E \cap \mathscr{K}_B E \cap \mathscr{K}_C E.$$

Thus $E$ is mutual knowledge when the true state of the world is $\omega$ if and only if $\omega \in$ (everybody knows)$E$.

For example, before the minister made his announcement, it was mutual knowledge that someone in the railway carriage has a dirty face. To see this, recall that $D_A = \{2,5,6,8\}$ is the event that Alice's face is dirty. Similarly, $D_B = \{3,5,7,8\}$ and $D_C = \{4,6,7,8\}$ are the events that Beatrice and Carol have dirty faces. The event that someone has a dirty face is therefore $D = D_A \cup D_B \cup D_C = \{2,3,4,5,6,7,8\}$. Notice that $\mathscr{K}_A D = \{3, 4, 5, 6, 7, 8\}$, $\mathscr{K}_B D = \{2, 4, 5, 6, 7, 8\}$, and $\mathscr{K}_C D = \{2, 3, 5, 6, 7, 8\}$. Hence

$$\text{(everybody knows)}\, D = \mathscr{K}_A D \cap \mathscr{K}_B D \cap \mathscr{K}_C D = \{5, 6, 7, 8\}.$$

The true state of the world is actually $\omega = 8$. Thus, $D$ is mutual knowledge because $8 \in$ (everybody knows)$D$.

Mutual knowledge is what we need to define a public event $E$. As with a truism, the criterion is

$$E \subseteq \text{(everybody knows)}E.$$

### 12.6.3 Common Knowledge Operator

Because the (everybody knows) operator satisfies (K2) of Figure 12.2:

$$\begin{aligned} E &\supseteq \text{(everybody knows)}E \\ &\supseteq \text{(everybody knows)}^2 E \\ &\supseteq \text{(everybody knows)}^3 E \\ &\;\vdots \\ &\supseteq \text{(everybody knows)}^N E \\ &= \text{(everybody knows)}^{N+1} E \\ &= \text{(everybody knows)}^{N+2} E \end{aligned}$$

Why do the inclusions become identities after the $N$th step? The reason is that the finite set $\Omega$ contains only $N$ elements, and so we will run out of things that can be discarded from (everybody knows)$^n E$ to make it a strictly smaller set on or before the $N$th step.

When the universe of discourse is finite, we can therefore define the *common knowledge* operator by taking

$$\text{(everybody knows)}^{\infty} E = \text{(everybody knows)}^{N} E$$

for a large enough value of $N$. Lewis's criterion for an event $E$ to be common knowledge when the true state is $\omega$ then becomes

$$\omega \in \text{(everybody knows)}^{\infty} E.$$

*Properties of Common Knowledge.* The mutual knowledge operator fails to satisfy all the axioms of Figure 12.2. It satisfies (K0), (K1), and (K2) but not (K3). For example, in state 5 of Figure 12.3, everybody knows that someone has a dirty face, but Beatrice thinks state 2 is possible. In state 2, Alice thinks state 1 is possible. Since everybody has a clean face in state 1, it is therefore false that everybody knows that everybody knows someone has a dirty face in state 5.

However, such problems disappear when we turn to the common knowledge operator, which satisfies all the axioms of Figure 12.2. It follows that analogues exist for all the results obtained for the individual knowledge operator $\mathscr{K}$, provided that we define the communal possibility operator $\mathscr{M}$ by

$$\mathscr{M}E = \sim\text{(everybody knows)}^{\infty}\sim E$$

## 12.7 Complete Information

Strictly speaking, everything in the description of a game must be common knowledge among the players. This includes the rules, the players' preferences over the possible outcomes of the game, and the players' beliefs about the chance moves of the game. We then say that information is *complete.*

It will be obvious that we don't always need so much to be common knowledge. For example, the players in the one-shot Prisoners' Dilemma need to know only that *hawk* strongly dominates *dove* to figure out their optimal strategy. However, other games can be much more tricky.

The best way to see why one needs strong knowledge requirements in general is to look at what can go wrong when the complete information requirement is relaxed. We therefore leave this issue until Chapter 15, which is about situations in which information is incomplete.

## 12.8 Agreeing to Disagree?

→ 12.9

Can rational people genuinely agree to disagree? This was the issue that first led Robert Aumann to study common knowledge. The version of his approach given here is due to Michael Bacharach.

### 12.8.1 Elementary, My Dear Watson

One of Alice, Beatrice, and Carol is guilty of a crime. The only available clues are the state of their faces in the railway carriage. Sherlock Holmes and Hercule Poirot are engaged to solve the mystery. The size of their fees limits the time each is able to devote to the case. They therefore agree that Sherlock will pursue one of two possible lines of inquiry and Hercule will investigate another.

At the end of the inquiry, each detective will have reduced the state space $\Omega = \{1,2,3,4,5,6,7,8\}$ to one of a number of possibility sets. However, Sherlock's possibility partition won't be the same as Hercule's because they will have received different information during their separate investigations. It may be, for example, that Sherlock's and Hercule's possibility partitions will be as in Figure 12.11(a) after their inquiries are concluded.

Each possibility set $P(\omega)$ in Figure 12.11 is labeled with one of the suspects. This is the person that the investigator will accuse if the true state is $\omega$. Thus, if the true state is $\omega = 8$, Sherlock will accuse Carol because $P_S(\omega) = \{6,8\}$.

It is important for the story that Sherlock and Hercule *reason* in the same way. Perhaps they both went to the same detective school (or read the same game theory book). Thus it is given that, if Sherlock and Hercule arrive at the same possibility set, they will both accuse the same person. For example, $P_S(\omega) = P_H(\omega) = \{6,8\}$ when $\omega = 8$. Thus Sherlock and Hercule will both accuse Carol if $\omega = 8$.

Now suppose that Sherlock and Hercule discuss the case *after* both have completed their inquiries but *before* reporting their findings. Each simply tells the other

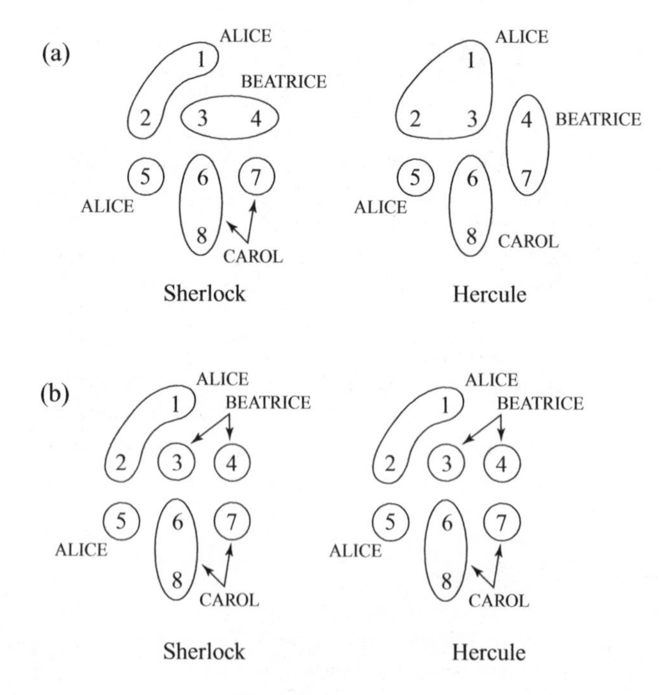


Figure 12.11 Whodunit?

whom they plan to accuse on the basis of their current evidence. Can they agree to disagree? For example, if the true state is $\omega = 3$, will Sherlock persist in accusing Beatrice, while Hercule points his finger at Alice?

In the circumstances of Figure 12.11(a), the answer is *no*. Suppose that the true state is $\omega = 3$, and Sherlock and Hercule simultaneously name the suspect they would accuse if they got no further information. Thus Sherlock names Beatrice, and Hercule names Alice. Such a naming of suspects is very informative for both Sherlock and Hercule. They use this new information to refine their possibility partitions. The new partitions are shown in Figure 12.11(b). These partitions are the *same* for both Sherlock and Hercule. Thus, the investigators will now accuse the *same* person. In Figure 12.11(b), the person accused is taken to be Beatrice.

The point here is that Sherlock, for example, would be foolish not to react to Hercule's conclusion. Hercule reasons exactly as Sherlock would reason if he had Hercule's information. Thus, when Hercule reports his conclusion, this conclusion is just as much a piece of hard evidence for Sherlock as the evidence he collected himself.

### 12.8.2 Reaching a Consensus

The conclusion of the preceding story holds in general if we make appropriate assumptions, of which the most important is that Sherlock's and Hercule's preliminary conclusions become *common knowledge*.

To see why, suppose that both detectives have completed their investigations. Not only this, but they have also met, and it is now common knowledge between them whom each plans to accuse. Can each now finger a different person?

Imagine that Sherlock's final possibility partition of $\Omega$ is

$$\{\text{ALICE}, \text{BEATRICE}_1, \text{BEATRICE}_2, \text{BEATRICE}_3, \text{CAROL}\},$$

where, for example, $\text{BEATRICE}_2$ represents a possibility set in which Sherlock will accuse Beatrice. Suppose that it is common knowledge that Sherlock will accuse Beatrice when the true state is $\omega$, so that

$$M(\omega) \subseteq \text{BEATRICE}_1 \cup \text{BEATRICE}_2 \cup \text{BEATRICE}_3.$$

But the partition $M$ is a coarsening of Sherlock's possibility partition. Thus, for example, either $\text{BEATRICE}_2 \subseteq M(\omega)$ or $\text{BEATRICE}_2 \subseteq \sim M(\omega)$. Similar inclusion relations hold for Sherlock's other possibility sets. It follows that $M(\omega)$ must be the *union* of some of the possibility sets in which Sherlock accuses Beatrice. It may be, for example, that

$$M(\omega) = \text{BEATRICE}_2 \cup \text{BEATRICE}_3. \tag{12.3}$$

*Umbrella Principle.* We now need the weak rationality assumption that we met when discussing the case of Professor Selten's umbrella (Section 1.4.2).

In their detective school, Sherlock and Hercule were both trained how to decide who should be accused under all possible contingencies. If a detective's investigations lead him to the conclusion that the set of possible states of the world is $E$, his training will therefore tell him the right person to accuse. Denote this person by $d(E)$. For example, when $E = \text{ALICE}$, the person a detective will accuse is $d(E) = \text{Alice}$.

Let $E$ and $F$ be two events that can't both occur. The detectives' decision rule will then be required to have the following property:

$$d(E) = d(F) \Rightarrow d(E \cup F) = d(E) = d(F).$$

If a detective's decision rule violates this requirement, he would sometimes find himself in court replying to the defense attorney as follows:

> Did you accuse my client Beatrice?—*Yes.*
> When you accused her, what did you know about the state of Alice's face?—*Nothing.*
> Whom would you have accused if you had known Alice's face was dirty?—*Carol.*
> Whom would you have accused if you had known Alice's face was clean?—*Carol.*
> Are you not using an irrational decision rule?—*I guess so.*

Since Sherlock accuses Beatrice in $\text{BEATRICE}_2$ and $\text{BEATRICE}_3$, the Umbrella Principle tells us that (12.3) implies

$$d(M(\omega)) = \text{Beatrice}. \tag{12.4}$$

Hercule must therefore also be accusing Beatrice in state $\omega$ because applying the same argument to him must also lead to (12.4).

The result is general. With the Umbrella Principle, we have the following proposition—provided everybody uses the same rule of inference:

PROPOSITION 12.2 *If it is common knowledge that everybody knows something different in state $\omega$, then the different things they know must all be consistent on $M(\omega)$.*

*The Speculation Paradox.* Aumann used a version of the preceding proposition to show that players can't agree to disagree about probabilities (Exercise 13.10.28), but the economic version is more fun. It says that speculation is impossible for rational players.

In the crudest version of the paradox, Alice and Bob are playing a zero-sum game, but they don't know what the payoffs are. Alice asks Bob to sign a binding contract in which the players agree to switch from their old strategies to some new strategies. Should Bob agree? Obviously not, since Alice wouldn't propose the contract unless she were expecting to gain. But in a zero-sum game, what Alice wins, Bob loses.

In terms of what the players know, the act of signing the contract makes it common knowledge that both players expect to gain. But these views are necessarily inconsistent in a zero-sum game.

Paul Milgrom and Nancy Stokey offer a more elaborate version of the paradox. A market has traded to a Pareto-efficient outcome. Since the traders' world is risky, this means that nobody can improve the expected utility of their holding by trading any further. But some traders then get insider information. Will there now be more trading, as they try to exploit their knowledge?

In Milgrom and Stokey's idealized world, the answer is *no.* The signing of a trading contract would make it common knowledge that there is an event $E$ in which all the signatories expect to be better off. But if this is so, we would have been better off in the first place by writing a contract that specified that the new trading arrangements would operate if $E$ were to occur. This result is sometimes called the Groucho Marx theorem after his joke that he wouldn't want to belong to a club that would have him as a member.

So how come speculation survives? The paradox assumes that all people have the same inference rule. Many authors have claimed that this is necessarily true of rational beings. Harsanyi was one such, and so Aumann refers to the claim as the Harsanyi doctrine (Section 13.5.1). But why should there be only one way of being rational? This certainly isn't true in Bayesian decision theory, where the inference rules the players use are the same only if they all begin with the same prior beliefs (Exercise 13.10.28). As for actual speculators on the stock market, they laugh at people like us who think that rationality is relevant to making money.

## 12.9 Coordinated Action

David Lewis introduced his definition of common knowledge while writing about conventions, which we met in Section 8.6 when discussing equilibrium selection. For example, the Driving Game that we play every morning on the way to work has two Pareto-efficient equilibria. In France, convention demands the use of the equilibrium in which everyone drives on the right. In Britain, the convention is that everyone drives on the left.

Lewis argues that conventions must be common knowledge in order to work. Others have said the same thing about any Nash equilibrium at all. But such claims are obviously wrong. All that is necessary for it to be optimal to play a particular Nash equilibrium is that all the players believe that the other players will play their equilibrium strategies with a high enough probability.

It is fortunate that coordinated action doesn't require common knowledge among the players of an agreement to act together since such a requirement would often make coordinated action impossible! To see why, we look at the paradox of the Byzantine generals from computer science literature.

*Beware of Greeks Bearing Gifts.* The Greeks of the Byzantine empire were so sneaky that they didn't even trust each other. The following story supposedly shows that they therefore couldn't ever coordinate on anything.

In this story, two Byzantine generals occupy adjacent hills, with the enemy in the valley between. If both generals attack together, victory is certain, but if only one general attacks, he will suffer badly. The first general therefore sends a messenger to the second general proposing an attack. Since there is a small probability that any messenger will be lost while passing through the enemy lines, the second general sends a messenger back to the first general confirming the plan to attack. But when this messenger arrives, the second general doesn't know that the first general knows that the second general received the first general's message proposing an attack. The first general therefore needs to send another messenger confirming the arrival of the second general's messenger. But when this messenger arrives, the first general doesn't know that the second general knows that the first general knows that the second general received the first general's message.

The fact that an attack has been proposed is therefore not common knowledge because, for an event $E$ to be common knowledge, *all* statements of the form (everybody knows that)$^n E$ must be true. Further messengers may be shuttled back and forward until one of them is picked off by the enemy, but no matter how many confirmations each general receives before this happens, it *never* becomes common knowledge that an attack has been proposed.

If it were really true that rational coordinated action is impossible in such stories, then computer scientists who work on distributed systems would be in serious trouble since automated agents in different locations would never be able to act together! Nor would Sweden have been able to switch from driving on the left to driving on the right on 1 September 1967.

### 12.9.1 The Email Game

Rubinstein's E-mail Game is a formal version of the Byzantine paradox. It is based on the Stag Hunt Game of Figure 8.7(a). The game has two Nash equilibria in pure strategies: (*dove*, *dove*) and (*hawk*, *hawk*). The first is Pareto dominant and the second is risk dominant (Section 8.5.2). We first discussed a version of the Stag Hunt Game in Section 1.9 as an example of a case in which it might be difficult for the players to persuade each other to move from the risk-dominant equilibrium to the Pareto-dominant equilibrium.

In the E-mail Game, Alice and Bob must independently choose between DOVE and HAWK. Their payoffs are then determined by whether Chance has made DOVE correspond to *dove* and HAWK to *hawk* in the Stag Hunt Game or whether she has reversed these correspondences. It is common knowledge that the former happens with probability $\frac{2}{3}$.

Only Bob learns what decision Chance has made. He would like to communicate this information to Alice, so that they can coordinate on the equilibrium they both prefer, but their only contact is by e-mail. The sending of messages is automatic. On the understanding that the default action is DOVE, a message goes to Alice that says "Play HAWK" whenever Bob learns that DOVE corresponds to *hawk*. Alice's machine confirms receipt of the message by bouncing it back to Bob's machine. Bob's machine confirms that the confirmation has been received by bouncing the message back again, and so on.

*Who Knows What?* The (everybody knows)$^n$ operator becomes applicable with ever-higher values of $n$ as confirmation after confirmation is received. So if the players could wait until infinity before acting, Chance's choice would become common knowledge.[4]

[4]If the first message takes one second and each subsequent message takes half as long as the one before, then the waiting time will be only two seconds!

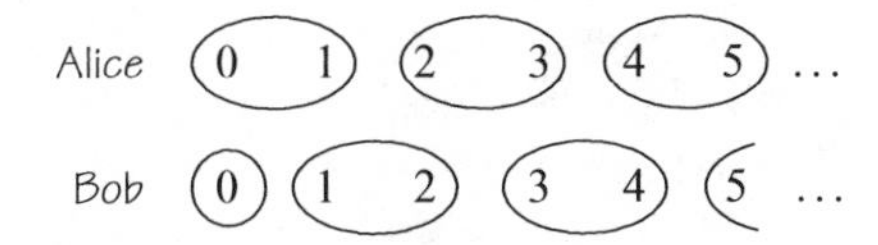


Figure 12.12 Possibility sets in the E-mail Game.

However, the E-mail Game is realistic to the extent that the probability of any given message failing to arrive is some very small $\varepsilon > 0$. The probability of Chance's choice becoming common knowledge is therefore zero. But we can still ask whether coordinated action is possible for Alice and Bob. Is there a Nash equilibrium in which they do better than always playing their default action of DOVE? We will find that the answer is *no*.

Figure 12.12 shows possibility sets for Alice and Bob in the E-mail Game. The possible states of the world are the number of messages that could get sent. For example, $P_A(3) = \{2, 3\}$ and $P_B = \{3, 4\}$. To see why $P_A(3) = \{2, 3\}$, observe that if the fourth message goes astray, then Alice thinks it is also possible that the third message (sent by Bob's machine) wasn't sent because the second message (sent by her machine) didn't arrive.

*Finding the Equilibrium.* As always, a pure strategy names an action (either DOVE or HAWK in the E-mail Game) for each of a player's information sets. The only Nash equilibrium consistent with Bob's choosing DOVE when he learns that DOVE corresponds to *dove* requires both players to choose DOVE at *all* their information sets—even though both players know that DOVE corresponds to *hawk* at all information sets not containing the state 0.

The proof is by induction. We first show that if Alice plays the default action DOVE at $\{0, 1\}$, then it is optimal for Bob to play DOVE at $\{1, 2\}$. On reaching this possibility set, Bob believes it more likely that the state of the world is 1 rather than 2.[5] Can it then be optimal for him to play HAWK? The most favorable case is when each state is equally likely and Alice is planning to play DOVE at $\{2, 3\}$. Bob might as well then be playing against someone playing each strategy in the ordinary Stag Hunt Game with equal probability, so his optimal reply is *hawk*, which he knows corresponds to DOVE at $\{1, 2\}$.

Similarly, Bob's playing DOVE at $\{1, 2\}$ implies that Alice plays DOVE at $\{2, 3\}$, and so on. Thus DOVE is *always* played in a Nash equilibrium of the E-mail Game. Although Lewis's claims for the necessity of common knowledge are mistaken, it nevertheless looks like the Byzantine generals are still in trouble!

*Byzantium Saved!* The E-mail Game is a nice exercise in handling knowledge problems, but its paradoxical conclusion disappears when the model is made more realistic by making communication both purposeful and costly. Many Nash equilibria appear when we allow the players to choose whether to send and receive messages, given that both activities involve a small cost.

[5]Because the second message can go astray only if the first message is received.

In the most pleasant equilibrium, both players play HAWK whenever Bob proposes doing so and Alice says OK—as when friends agree to meet in a coffee shop. But there are other equilibria in which the players settle on HAWK only after a long exchange of confirmations of confirmations. Hosts of polite dinner parties suffer from this equilibrium when the guests start moving infinitely slowly toward the door at the end of the evening, stopping every so often to exchange meaningless sentiments of good will.[6]

## 12.10 Roundup

A decision problem can be modeled as a function $f: A \times B \to C$. Pandora chooses an action $a$ in the set $A$, but the consequence $c = f(a, b)$ also depends on the state $b$ of the world. Since Pandora knows what decision problem she is solving, she *knows* the set $B$ of all currently possible states of the world. She doesn't know which of the states in $B$ is the true states of the world, but her choice of action will be guided by her *beliefs* about which states are more or less likely than others.

In small worlds, the knowledge operator $\mathscr{K}$ satisfies a number of useful axioms that we wouldn't be entitled to assume in general. In game theory, the possibility operator $\mathscr{P} = \sim\mathscr{K}\sim$ is often more useful. The event $\mathscr{P}\{\omega\}$ in which Pandora thinks that the state $\omega$ is possible is the same as the possibility set $P(\omega)$, which is the set of states Pandora thinks possible when the true state is $\omega$. These possibility sets partition Pandora's universe.

All that matters about what the players know in a game is captured by its information sets, which determine what the players think is possible when it is their turn to move. Game theorists dig deeper into epistemology only when considering how knowledge assumptions limit the way information sets can legitimately be defined in a game.

Unless something is said to the contrary, games should always be assumed to have perfect recall. This means that players never forget anything. By looking at games played by forgetful drivers, we found that perfect recall imposes important restrictions on legitimate information sets. In particular, two nodes on the same play can't belong to the same information set.

Kuhn's theorem says that we can forget about mixed strategies in games of imperfect recall and work instead with behavioral strategies. A behavioral strategy simply specifies the probability with which Pandora plans to use each action at each of her information sets. She might then be said to decentralize her choice of strategy by delegating responsibility to separate agents at each of her information sets.

An event $E$ is common knowledge when the true state is $\omega$ if and only if

$$\omega \in (\text{everybody knows})^n E$$

for all values of $n$. Events that are common knowledge are implied by $M(\omega)$, which is the set of states that the players as a whole think possible when $\omega$ occurs. It is easy

[6]Will social evolution eventually eliminate such long goodbyes? The prognosis isn't good. Only the unique equilibrium of the original Email Game—in which HAWK is never played—fails to pass an appropriate evolutionary stability test (Binmore and Samuelson, *Games and Economic Behavior* 35 (2001), 6–30).

to find $M(\omega)$ because the communal possibility partition is simply the meet of the players's individual possibility partitions.

Players who are rational enough to honor the Umbrella Principle can't agree to disagree if their decision rules are identical. They may have different private information, but they will all necessarily make the same choice if their planned choices are common knowledge. Rational speculation then becomes impossible if it is common knowledge that someone must lose from trading.

The paradox of the Byzantine generals is based on the claim that coordinated action is impossible unless the plan to act together becomes common knowledge. An analysis of the E-mail Game shows that this conclusion holds water only under unduly restrictive circumstances.

## 12.11 Further Reading

*A Mathematician's Miscellany*, by J. E. Littlewood: Cambridge University Press, Cambridge, 1953. I was a schoolboy when I first came across the paradox of the dirty-faced ladies in this popular work by one of the great mathematicians.

*Conventions: A Philosophical Study*, by David Lewis: Harvard University Press, Cambridge, MA, 1969. The author is generous in acknowledging his debt to David Hume and Thomas Schelling.

## 12.12 Exercises

1. What subsets of $\Omega$ in Figure 12.1 correspond to the following events? Which of these events occur when the true state of the world is $\omega = 3$?
   a. Beatrice has a dirty face.
   b. Carol has a clean face.
   c. Precisely two ladies have dirty faces.
2. The Oracle at Delphi puzzled the philosopher Socrates by naming him the wisest man in Greece. He finally decided that it must be because he was the only man in Greece who knew he was ignorant. Everybody else didn't know that they didn't know any secrets of the universe. **phil**
   Show that the properties (K0)–(K4) of Section 12.3.1 imply that $(\sim \mathcal{K})^2 E = \mathcal{K} E$. Deduce that Socrates thought he was living in a large world.
3. Use the knowledge properties (K0)–(K4) of Section 12.3.1 to prove **math**
   a. $E \subseteq F \Rightarrow \mathcal{K}E \subseteq \mathcal{K}F$
   b. $\mathcal{K}E = \mathcal{K}^2 E$
   c. $(\sim \mathcal{K})^2 E \subseteq \mathcal{K}E$
   Offer an interpretation of each of these statements.
4. Show that (K0) – (K4) of Section 12.3.1 are equivalent to (P0)–(P4). **math**
5. Write down properties of the possibility operator $\mathcal{P}$ that are analogous to those given in Exercise 12.12.3. Interpret these properties. **math**
6. In the story of the dirty-faced ladies of Section 12.2, it is true that everybody has a dirty face. Why isn't this a truism for Alice before the minister speaks?
7. Show that an event $T$ is a truism if and only if $T = \mathcal{K}T$. Show that the same is true of a public event $T$ when $\mathcal{K}$ is replaced by the common knowledge operator.

math

8. Show that, for any event $E$, all of the following are truisms:

$$\text{(a) } \mathcal{K}E \qquad \text{(b) } \sim\mathcal{K}E \qquad \text{(c) } \mathcal{P}E \qquad \text{(d) } \sim\mathcal{P}E$$

math

9. Show that $\sim S$, $S \cap T$, and $S \cup T$ are truisms when the same is true of $S$ and $T$.

math

10. Explain why

$$\bigcap_{\omega \in \mathcal{K}E} \mathcal{K}E \subseteq \bigcap_{\omega \in \mathcal{K}E} E \subseteq \bigcap_{\omega \in \mathcal{K}(\mathcal{K}E)} \mathcal{K}E = \bigcap_{\omega \in \mathcal{K}E} \mathcal{K}E.$$

Use Theorem 12.2 and Exercise 12.12.7 to deduce that

$$\mathcal{P}\{\omega\} = \bigcap_{\omega \in \mathcal{K}E} E.$$

math

11. Use Theorem 12.3 to prove that

$$\mathcal{K}E = \{\omega : \mathcal{P}\{\omega\} \subseteq E\}.$$

12. Suppose that the minister in the story of the dirty-faced ladies of Section 12.2 no longer announces that somebody has a dirty-face whenever this is true. Instead, he announces that there are at least two dirty-faced ladies if and only if this is true. Assuming that the ladies know the minister's disposition, draw a diagram showing the ladies' possibility sets after the minister has had the opportunity to make an announcement.
13. Continue the preceding exercise by drawing diagrams like those of Figure 12.5(a) to show how the ladies refine their possibility partitions if the opportunity to blush rotates among them as in Section 12.4.2.
14. Suppose that the dirty-faced ladies no longer take turns in having the opportunity to blush as in Section 12.4.2. Instead, all three ladies have the opportunity to blush precisely one second after the minister's announcement and then again precisely two seconds after the announcement and so on. Draw diagrams to show how the ladies' possibility partitions get refined as time passes. Who will blush in this story? How many seconds after the announcement will the first blush occur?
15. Find a blushing story that leads to a final configuration of possibility sets that is different from those obtained in Section 12.4.2 and Exercise 12.12.14.
16. For the game of Figure 12.9:
    a. Find a mixed strategy for Eve that always leads to the same lottery over outcomes as the behavioral strategy in which she assigns equal probabilities to each action at each information set.
    b. Find a behavioral strategy for Eve that always leads to the same lottery over outcomes as the mixed strategy in which *RLR* is used with probability $\frac{2}{3}$ and *LRL* with probability $\frac{1}{3}$.
17. Explain why the game of Figure 5.16 has imperfect information but perfect recall. Find a behavioral strategy for player II that always leads to the same

lottery over outcomes as the mixed strategy in which she uses $dD$ with probability $\frac{2}{3}$ and $uU$ with probability $\frac{1}{3}$.

18. In the Mildly Forgetful Driver's Game of Figure 12.7(a), find a mixed strategy that leads to the same lottery over outcomes as the behavioral strategy in which $r$ is chosen at Terence's first information set with probability $p$ and at his second information set with probability $P$. Show that no behavioral strategy results in the same lottery over outcomes as the mixed strategy that assigns probability $\frac{1}{2}$ to the play $[ll]$ and probability $\frac{1}{2}$ to the play $[rR]$. Why doesn't Kuhn's theorem apply?
19. In the Seriously Forgetful Driver's Game of Figure 12.7(b), what outcome does Terence get for each of his two pure strategies? Deduce that all his mixed strategies lead to his getting lost, but find a behavioral strategy that yields a payoff of $\frac{1}{4}$. Why doesn't Kuhn's theorem apply?
20. Prove that the $\mathscr{K}$ = (everybody knows) operator of Section 12.6.2 satisfies properties (K0), (K1), and (K2) of Figure 12.2. An example is given in Section 12.6.2 to show that everybody can know something without everybody knowing that everybody knows it. Give another example. **math**
21. How should the operator $\mathscr{K}$ = (somebody knows) be defined in formal terms? Why does this operator not satisfy (K1) of Figure 12.2? **math**
22. Why does the common knowledge operator $\mathscr{K}$ = (everybody knows)$^{\infty}$ satisfy (K3) of Figure 12.2 as claimed in Section 12.6.3? **math**
23. Return to Exercises 12.12.13 and 12.12.14. In each case, find the communal possibility partitions at each stage of the blushing process. Eventually, it is common knowledge that Beatrice and Carol both have dirty faces when this is true. Explain why. In the case of Exercise 12.12.13, why does it never become common knowledge that Beatrice and Carol both have clean faces when this is true?
24. It is common knowledge that Gino and Polly always tell the truth. The state space is $\Omega = \{1,2,3,4,5,6,7,8,9\}$. The players' initial possibility partitions are shown in Figure 12.13(a). The players alternate in announcing how many elements their current possibility set contains.
    a. Why does Gino begin by announcing three in all states of the world?
    b. How does Gino's announcement change Polly's possibility partition?
    c. Polly now makes an announcement. Explain why the possibility partitions afterward are as in Figure 12.13(b).
    d. Continue updating the players' possibility partitions as announcements are made. Eventually, Figure 12.13(c) will be reached. Why will there be no further changes?
    e. In Figure 12.13(c), the event $E$ that Gino's possibility set contains two elements is $\{5,6,7,8\}$. Why is this common knowledge when the true state is $\omega = 5$? Is $E$ a public event?
25. In the previous exercise, it is now common knowledge that Gino and Polly think each element of $\Omega$ is equally likely. Instead of announcing how many elements their current possibility set contains, they announce their current conditional probability for the event $F = \{3,4\}$.
    a. In Figure 12.13(a), explain why the event that Gino announces $\frac{1}{3}$ is $\{1,2,3,4,5,6\}$ and the event that he announces 0 is $\{7,8,9\}$.

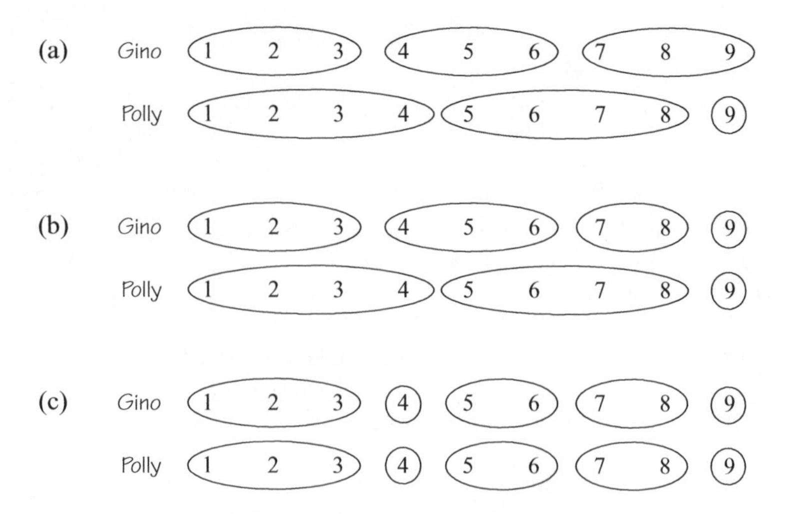


Figure 12.13 Reaching consensus.

b. What is Polly's possibility partition after Gino's initial announcement? Explain why the event that Polly now announces $\frac{1}{2}$ is $\{1,2,3,4\}$ and the event that she announces 0 is $\{5,6,7,8,9\}$.
c. What is Gino's new possibility partition after Polly's announcement? Explain why the event that Gino now announces $\frac{1}{3}$ is $\{1,2,3\}$, the event that he announces 1 is $\{4\}$, and the event that he announces 0 is $\{5,6,7,8,9\}$.
d. What is Polly's new possibility partition? Explain why the events that Polly will now announce $\frac{1}{3}$, 1, or 0 are the same as in (c).
e. Explain why each player's posterior probability for the event $F$ is now common knowledge, whatever the true state of the world.
f. In Figure 12.13(a), why is it true that no player's posterior probability for $F$ is common knowledge in any state?
g. What will the sequence of announcements be when the true state of the world is $\omega = 2$?

26. Alice's, Beatrice's, and Carol's initial possibility partitions are as shown in Figure 12.14. It is common knowledge that their common prior attaches equal probability to each state. The table on the right of Figure 12.14 shows Alice's, Beatrice's, and Carol's initial posterior probabilities for $F$ for each state and also the average of these probabilities. Each player now *privately* informs a kibitzer of her posterior probability for the event $F = \{1,2,3\}$. The kibitzer computes the average of these three probabilities and announces the result of his computation *publicly*. Beatrice and Carol update their probabilities for $F$ in the light of this new information. They then privately report their current posterior probabilities to the kibitzer, who again publicly announces their average, and so on.
    a. Draw Figure 12.14 again, but modify it to show the situation *after* the kibitzer's first announcement.
    b. Repeat (a) for the kibitzer's second announcement.
    c. Repeat (a) for the kibitzer's third announcement.

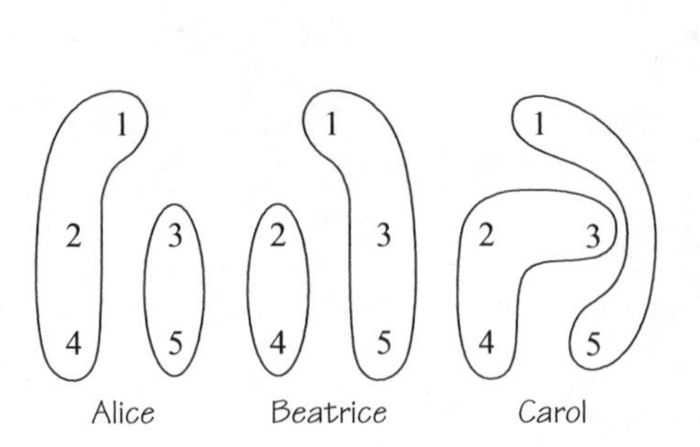


| State | Alice | Beatrice | Carol | Average |
|---|---|---|---|---|
| 1 | $\frac{2}{3}$ | $\frac{2}{3}$ | $\frac{1}{2}$ | $\frac{11}{18}$ |
| 2 | $\frac{2}{3}$ | $\frac{1}{2}$ | $\frac{2}{3}$ | $\frac{11}{18}$ |
| 3 | $\frac{1}{2}$ | $\frac{2}{3}$ | $\frac{2}{3}$ | $\frac{11}{18}$ |
| 4 | $\frac{2}{3}$ | $\frac{1}{2}$ | $\frac{2}{3}$ | $\frac{11}{18}$ |
| 5 | $\frac{1}{2}$ | $\frac{2}{3}$ | $\frac{1}{2}$ | $\frac{5}{9}$ |

Figure 12.14 Reaching consensus again.

d. How many announcements are necessary before consensus is reached on the probability of $F$?
e. What will the sequence of events be when the true state of the world is $\omega = 1$?
f. If the true state of the world is $\omega = 1$, does this ever become common knowledge?
g. If $\omega = 5$ isn't the true state, at which stage will this fact become common knowledge?
h. If $\omega$ is even, at what stage does this become common knowledge?
i. Consensus is reached when everybody reports the same probability for $F$ to the kibitzer. Why is it common knowledge that consensus has been reached as soon as it happens?

27. Explain why rational players are necessarily playing a Nash equilibrium in a game if the strategy choice of each player is mutual knowledge. **phil**
28. Alice is playing poker with Bob. The cards are dealt, and Alice takes a peek at her hand without letting Bob see. She now proposes a bet. If she doesn't hold the queen of hearts, she pays him one dollar. If she does, he pays her one dollar. Why should Bob refuse to bet? **phil**

    What if Alice asks Bob to bet against her being able to prove that time travel is possible? Remember that she might be a time traveler herself!

# 13 *Keeping Up to Date*

## 13.1 Rationality

What is rationality? Game theorists have tried as hard as anybody to pin down the concept, but nobody would claim to have all the answers. Perhaps rationality is a concept like life that will turn out not to have sharp boundaries. But just as philistines know a great work of art when they see one, so most of us think we can smell an irrational argument when it is thrust under our noses.

However, the myth of the wasted vote is a cautionary tale (Section 1.3.3). People think that democracy would collapse if it were true that each individual voter might as well stay at home on a rainy election night for all the difference a single vote makes to the outcome of the election. Since they like living in a democracy, they therefore argue that no vote cast for a party that stands a chance of winning can be "wasted." The error they make is to allow their *preferences* to influence their *beliefs*.

This chapter is devoted to the contrary principle that rationality demands separating your beliefs from your preferences. Bayesian decision theory is the embodiment of this principle within game theory.

## 13.2 Bayesian Updating

As players encounter information sets while playing a game, they learn something about the choices made by Chance in the past. For example, if East plays the queen of hearts in bridge, then Chance can't have chosen to give the queen of hearts to North at the opening move that represents the shuffling and dealing of the cards.

However, players don't necessarily learn something for sure. They mostly learn only that some events have become more or less probable. For example, if an opponent at bridge turns out to have no spades at the first trick, then it becomes more likely that she has the queen of hearts, rather. But how much more likely?

The method used to answer such questions is called *Bayesian updating.* This section gives the gist of how it works.

→ 13.2.3

### 13.2.1 Bayes's Rule

If $E$ and $F$ are independent events, then $\text{prob}(E \cap F) = \text{prob}(E)\,\text{prob}(F)$. But what of the probability of $E \cap F$ when the events $E$ and $F$ aren't independent? In Section 3.3, we learned that we must then introduce the conditional probability $\text{prob}(E|F)$, which quantifies your new belief about $E$, given that you now know that $F$ has occurred.

A fair die is rolled. You win in the event $E$ that the dice shows more than 3. What is your probability of winning conditional on the event $F$ that the result is even?

The scientific way of answering this question is to record the outcomes when the die is rolled $6n$ times. If $n$ is large enough, it is very likely that each number on the dice will appear in the record about $n$ times. If we now cross out all the odd numbers, we will be left with a record containing about $3n$ even numbers. You would lose when one of these numbers is 2 and win when it is 4 or 6. The number of times that the latter event occurs is about $2n$. The frequency with which you win when the die shows an even number is therefore about $2n/3n$. For this reason, we say that $\text{prob}(E \mid F) = \frac{2}{3}$.

This counting is summarized in the formula

$$\text{prob}(E \cap F) = \text{prob}(E|F)\,\text{prob}(F)$$

that we used to define a conditional probability in Section 3.3. (In the dice example, $\text{prob}(E \cap F) = \frac{1}{3}$ and $\text{prob}(F) = \frac{1}{2}$.)

The defining equation for a conditional probability leads immediately to Bayes's rule, which says that

$$\text{prob}(E|F) = \frac{\text{prob}(F|E)\,\text{prob}(E)}{\text{prob}(F)}.$$

The denominator can also be expressed in terms of conditional probabilities. Since $\text{prob}(F) = \text{prob}(E \cap F) + \text{prob}(\sim E \cap F)$, we have

$$\text{prob}(F) = \text{prob}(F|E)\,\text{prob}(E) + \text{prob}(F|\sim E)\,\text{prob}(\sim E),$$

but it is often possible to escape without bothering with this equation.

Bayes's rule follows immediately from the fact that

$$\text{prob}(E|F)\,\text{prob}(F) = \text{prob}(E \cap F) = \text{prob}(F|E)\,\text{prob}(E)$$

and thus is no more than a minor reshuffling of the definition of a conditional probability. However, since the latter simply records an arithmetical relationship between the frequencies with which events occur, we will need to think again about our reasons for believing in Bayes's rule when we broaden the scope of the probabilities we consider from the objective variety derived from observed frequencies to the subjective variety to be introduced in Section 13.3.

### 13.2.2 Guessing in Examinations

The candidates in a multiple-choice test have to choose among $m$ answers. Each candidate is either entirely ignorant and simply chooses an answer at random or else is omniscient and knows the right answer for sure. If the proportion of omniscient candidates is $p$, what is the probability that a candidate who got the answer right was guessing?

We need to compute prob (ignorant | right). Bayes's rule tells us that

$$\text{prob}\,(\text{ignorant}\,|\text{right}) = \frac{\text{prob}\,(\text{right}\,|\,\text{ignorant})\,\text{prob}\,(\text{ignorant})}{\text{prob}\,(\text{right})}.$$

Since ignorant candidates choose at random, prob (right | ignorant) $= 1/m$. We are given that prob (ignorant) $= 1 - p$. What of prob (right)?

One can avoid calculating the denominator directly using the following trick. Write $c = 1/\text{prob}\,(\text{right})$. Then

$$\text{prob}\,(\text{ignorant}\,|\,\text{right}) = c(1-p)/m.$$

The same mode of reasoning also shows that prob (omniscient | right) $= cp$ because prob (right | omniscient) $= 1$ and prob (omniscient) $= p$. We can therefore work out $c$ from the formula

$$\text{prob}\,(\text{ignorant}\,|\,\text{right}) + \text{prob}\,(\text{omniscient}\,|\,\text{right}) = 1.$$

We learn that $c(1-p)/m + cp = 1$, and so $c = m/(1-p+pm)$. Thus,

$$\text{prob}\,(\,\text{ignorant}\,|\,\text{right}) = \frac{1-p}{1-p+pm}.$$

If there are three answers to choose from and only one person in a class of hundred is omniscient, then $m = 3$ and $p = 0.01$. The probability that a person who got the answer right was guessing is then 0.971.

### 13.2.3 Monty Hall's Last Show

We return to the Monty Hall Game of Section 3.1.1 to expand on the brief discussion of Bayesian updating in a game of imperfect information offered in Section 3.3.3.

Figure 13.1(a) shows the information set $R$ at which Alice arrives after the Mad Hatter opens *Box 1* to show that it is empty. Alice then knows that the game

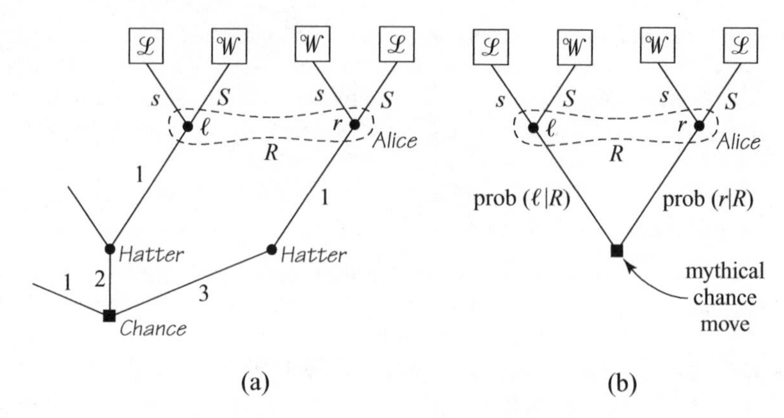


Figure 13.1 Updating at the right information set in the Monty Hall Game. A subgame can be rooted only in a singleton information set, but Figure 13.1(b) shows how to create a mythical chance move in which to root a subgame when using backward induction in games of imperfect information.

has reached one of the two nodes in $R$. Either she is at the left node $l$ or the right node $r$.

Alice doesn't know whether she is at $l$ or $r$, so she works out the probabilities $\text{prob}\,(l \mid R)$ and $\text{prob}\,(r \mid R)$ that represent her beliefs on arriving at $R$. She can appeal directly to the definition of a conditional probability, but most people prefer to use Bayes's rule:

$$\begin{aligned} \text{prob}\,(l\,|R) &= c\,\text{prob}\,(R \mid l)\,\text{prob}\,(l) &= c\,\text{prob}\,(l), \\ \text{prob}\,(r\,|R) &= c\,\text{prob}\,(R \mid r)\,\text{prob}\,(r) &= c\,\text{prob}\,(r), \end{aligned}$$

where $\text{prob}\,(R \mid l) = \text{prob}\,(R \mid r) = 1$ because Alice is certain to be at the information set if she is at one of the nodes $l$ or $r$. The constant $c$ is found by observing that $\text{prob}\,(l \mid R) + \text{prob}\,(r \mid R) = 1$. Hence $c = 1/(\text{prob}\,(l) + \text{prob}\,(r))$.

Working out the unconditional probabilities $p(l)$ and $p(r)$, we find that[1]

$$\begin{aligned} \text{prob}\,(l\,|R) &= \frac{\text{prob}\,(l)}{\text{prob}\,(l) + \text{prob}\,(r)} = \frac{p}{1+p} \\ \text{prob}\,(r\,|R) &= \frac{\text{prob}\,(r)}{\text{prob}\,(l) + \text{prob}\,(r)} = \frac{1}{1+p}, \end{aligned}$$

where $p$ is Alice's prior subjective probability that the Mad Hatter will open *Box 1* on those occasions when Chance puts the prize in *Box 2*.

Figure 13.1(b) shows that Alice's posterior probabilities for the nodes $l$ and $r$ in the information set $R$ can be thought of as the probabilities at an invented chance

[1]The game reaches $l$ if and only if Chance first puts the prize in *Box 2* and the Mad Hatter opens *Box 1*. Since the first of these events occurs with probability $1/3$, $\text{prob}\,(l) = p/3$. The game reaches $r$ if and only if Chance puts the prize in *Box 3* since the Mad Hatter must then open *Box 1* for sure. Thus, $\text{prob}\,(r) = 1/3$.

move that opens a mythical subgame in which Alice decides between switching and staying after being shown that *Box 1* is empty.

We can now proceed as in a game of perfect information when looking for a subgame-perfect equilibrium. To find Alice's optimal behavior at $R$, we treat the mythical subgame we have created just like any other subgame (Section 14.3). Alice maximizes her probability of winning the prize when $l$ is more likely than $r$ by playing $S$ (and thus staying with *Box 2*). She maximizes her probability of winning the prize when $r$ is more likely than $l$ by playing $s$ (and so switching from *Box 2* to *Box 3*).

But $\text{prob}(l \mid R) < \text{prob}(r \mid R)$ whenever $p < 1$. Alice therefore always prefers to switch boxes at $R$ unless $p = 1$, when she is indifferent.

### 13.2.4 Wasted Votes

→ 13.3

The probability of a vote being pivotal in a national election is infinitesimal. Democracy has nevertheless not collapsed because the prospect of being pivotal has little to do with why people vote. I certainly don't go to the polling booth because I think that the probability that my vote will be pivotal is high enough to justify the nuisance of my making the trip. Like most other people, I go to the polling booth because I like being part of the democratic process. But once having sunk the cost of making the trip to the polling booth, I try to maximize the effectiveness of my vote. This means conditioning my beliefs on the highly unlikely event that I will be pivotal since only if this very low-probability event occurs will my vote make any difference.

To show how a game theorist in the polling booth might reason, consider an election in which the candidates are Alice and Bob. Pandora is one of five voters. Two of the other voters are Alice's ma and pa. They can be counted on to vote for Alice no matter what. Pandora and the other two voters want to see the better candidate elected. How should Pandora vote?

Since it doesn't matter how Pandora votes unless she is pivotal, she should cast her vote on the assumption that the other free voters went for Bob. If she thinks Bob is the better candidate, she should therefore join them. But what if she thinks Alice is the better candidate? Instead of simply casting her vote for Alice, she should ask herself *why* the other free voters went for Bob. Unless she has reason to think that her sources of information are better than theirs, she may then want to vote for *Bob* with some probability $p$.

To illustrate this point with a simple model, assume that Chance first chooses either $A$ or $B$ with probability $\frac{1}{2}$. Alice is the better candidate in event $A$ and Bob in event $B$.

The voters learn something about the quality of the candidates, but their information may be wrong. In event $A$, a voter is sent message $a$ with probability $\frac{2}{3}$ and message $b$ with probability $\frac{1}{3}$. In event $B$, a voter is sent message $b$ with probability $\frac{2}{3}$ and message $a$ with probability $\frac{1}{3}$. Each of these messages is independent of the others.

Assuming that the other free voters always vote for Bob when they receive $b$ and continue to vote for Bob with probability $p$ when they receive $a$, how should Pandora vote when she gets the message $a$?

If $\beta$ is the event that a voter goes for Bob, the event that Pandora's vote is pivotal after receiving the message $a$ can be represented as $a\beta\beta$. To make her decision, Pandora needs to use Bayes's rule to find the larger of the conditional probabilities:[2]

$$\text{prob}\,(A \mid a\beta\beta) = c\,\text{prob}\,(a\beta\beta \mid A)\,\text{prob}\,(A) = c\,\tfrac{2}{3}\left(\tfrac{2}{3}p + \tfrac{1}{3}\right)^2 \tfrac{1}{2}\,;$$
$$\text{prob}\,(B \mid a\beta\beta) = c\,\text{prob}\,(a\beta\beta \mid B)\,\text{prob}\,(B) = c\,\tfrac{1}{3}\left(\tfrac{1}{3}p + \tfrac{2}{3}\right)^2 \tfrac{1}{2}.$$

We consider two cases. In the first, Pandora knows that the other free voters won't notice that their vote can matter only when they are pivotal. They therefore simply vote for whichever candidate is favored by their own message. Thus $p = 0$, and so $\text{prob}\,(A \mid a\beta\beta) < \text{prob}\,(B \mid a\beta\beta)$. It follows that Pandora should vote for Bob all the time—even when her own message favors Alice! If this outcome seems paradoxical, reflect that Pandora will be pivotal in favor of Alice only when the two other free voters have received messages favoring Bob. The messages will then favor Bob by two to one.

The second case arises when it is common knowledge that all the free voters are game theorists. To find a symmetric equilibrium in mixed strategies we simply set $\text{prob}\,(A \mid a\beta\beta) = \text{prob}\,(B \mid a\beta\beta)$, which happens when $p \approx 0.32$ (Exercise 13.10.8.). Pandora will then vote for Bob slightly less than a third of the time when her own message favors Alice.

Critics of game theory don't care for this kind of answer. Strategic voting is bad enough, but randomizing your vote is surely the pits! However, Immanuel Kant is on our side for once. If everybody except Alice's parents votes like a game theorist, the better candidate is elected with a probability of about 0.65. If everybody except Alice's parents votes for the candidate favored by their own message, not only is the outcome unstable, but the better candidate is elected with a probability of about 0.63 (Exercise 13.10.9).

→ 13.4

## 13.3 Bayesian Rationality

If Bayesian decision theory consisted of just updating probabilities using Bayes's rule, there wouldn't be much to it. But it also applies when we aren't told what probabilities to attach to future events. This section explains how Von Neumann and Morgenstern's theory can be extended to cover this case.

### 13.3.1 Risk and Uncertainty

Economists say they are dealing with *risk* when the choices made by Chance come with objectively determined probabilities. Spinning a roulette wheel is the arche-

[2] Note that $\text{prob}\,(a\beta\beta \mid A) = \text{prob}\,(a \mid A)\{\text{prob}\,(\beta \mid A)\}^2$. Also, $\text{prob}\,(a \mid A) = \frac{2}{3}$ and $\text{prob}\,(\beta \mid A) = \text{prob}\,(\beta \mid a \mid A)\,\text{prob}\,(a \mid A) + \text{prob}\,(\beta \mid b \mid A)\,\text{prob}\,(b \mid A) = p \times \frac{2}{3} + 1 \times \frac{1}{3}$. We don't need to find $c$. If we did, we could use the fact that $c^{-1} = \text{prob}\,(a\beta\beta) = \text{prob}\,(a\beta\beta \mid A) + \text{prob}\,(a\beta\beta \mid B)$ or the equation $\text{prob}\,(A \mid a\beta\beta) + \text{prob}\,(B \mid a\beta\beta) = 1$.

typal example. On a standard wheel, the ball is equally likely to stop in any of one of thirty-seven slots labeled $0, 1, \ldots, 36$. The fact that each slot is equally likely can be verified by observing the frequency with which each number wins in a very large number of spins. These frequencies are the data on which we base our estimates of the objective probability of each number. For example, if the number seven came up fifty times in one hundred spins, everybody would become suspicious of the casino's claim that its probability is only $\frac{1}{37}$.

Economists speak of *uncertainty* when they don't want to claim that there is adequate objective data to tie down the probabilities with which Chance moves. Sometimes they say that such situations are *ambiguous* because different people might argue in favor of different probabilities. Betting on horses is the archetypal example.

One can't observe the frequency with which Punter's Folly will win next year's Kentucky Derby because the race will be run only once. Nor do the odds quoted by bookies tell you the probabilities with which different horses will win. Even if the bookies knew the probabilities, they would skew the odds in their favor. Nevertheless, not only do people bet on horses, but they also go on blind dates. They change their jobs. They get married. They invest money in untried technologies. They try to prove theorems. What can we say about rational choice in such uncertain situations?

Economists apply a souped-up version of the theory of revealed preference described in Section 4.2. Just as Pandora's purchases in a supermarket can be regarded as revealing her preferences, so also can her bets at the racetrack be regarded as revealing both her preferences *and* her beliefs.

### 13.3.2 Revealing Preferences and Beliefs

A decision problem is a function $f: A \times B \to C$ that assigns a consequence $c = f(a, b)$ in $C$ to each pair $(a, b)$ in $A \times B$ (Section 12.1.1). If Pandora chooses action $a$ when the state of the world happens to be $b$, the outcome is $c = f(a, b)$. Pandora *knows* that $B$ is the set of states that are currently possible. Her *beliefs* tell her which possible states are more or less likely.

Let $a$ be the action in which Pandora bets on Punter's Folly in the Kentucky Derby. Let $E$ be the event that Punter's Folly wins and $\sim E$ the event that it doesn't. The consequence $\mathscr{L} = f(a, \sim E)$ represents what will happen to Pandora if she loses. The consequence $\mathscr{W} = f(a, E)$ represents what will happen if she wins.

All this can be summarized by representing the action $a$ as a table:

$$a = \begin{array}{|c|c|} \hline \mathscr{L} & \mathscr{W} \\ \hline \sim E & E \\ \hline \end{array} \qquad (13.1)$$

Such betting examples show why an act $a$ can be identified with a function $G: B \to C$ defined by $c = G(b) = f(a, b)$. When thinking of an act in this way, we call it a *gamble*.

Von Neumann and Morgenstern's theory doesn't apply to horse racing because the necessary objective probabilities for the states of the world are unavailable, but the theory can be extended from the case of risk to that of uncertainty by replacing the top line of Figure 4.6 by:

$$\mathbf{G} = \begin{array}{|c|c|c|c|c|} \hline w_1 & w_2 & w_3 & \cdots & w_n \\ \hline E_1 & E_2 & E_3 & \cdots & E_n \\ \hline \end{array} \sim \begin{array}{|c|c|c|c|c|} \hline w_1 & w_2 & w_3 & \cdots & w_n \\ \hline p_1 & p_2 & p_3 & \cdots & p_n \\ \hline \end{array} \qquad (13.2)$$

The new line simply says that Pandora treats any gamble **G** as though it were a lottery **L** in which the probabilities $p_i = \text{prob}\,(E_i)$ are Pandora's *subjective* probabilities for the events $E_i$.

If Pandora's subjective probabilities $p_i = \text{prob}\,(E_i)$ don't vary with the gamble **G**, we can then follow the method of Section 4.5.2 and find her a Von Neumann and Morgenstern utility function $u : \Omega \to \mathbb{R}$. Her behavior can then be described by saying that she acts *as though* maximizing her expected utility

$$Eu(\mathbf{G}) = p_1 u(\omega_1) + p_2 u(\omega_2) + \cdots + p_n u(\omega_n)$$

relative to a subjective probability measure that determines $p_i = \text{prob}\,(E_i)$.

*Bayesian rationality* consists in separating your beliefs from your preferences in this particular way. Game theory assumes that all players are Bayesian rational. All that we need to know about the players is therefore summarized by their Von Neumann and Morgenstern utilities for each outcome of the game and their subjective probabilities for each chance move in the game.

→ 13.4

### 13.3.3 Dutch Books

Why would Pandora behave as though a gamble **G** were equivalent to a lottery **L**? How do we find her subjective probability measure? Why should this probability measure be the same for all gambles **G**?

To appeal to a theory of revealed preference, we need Pandora's behavior to be both *stable* and *consistent*. Consistency was defended with a money-pump argument in Section 4.2.1. When bets are part of the scenario, we speak of Dutch books rather than money pumps.

For an economist, making a Dutch book is the equivalent of an alchemist finding the fabled philosopher's stone that transforms base metal into gold. But you don't need a crew of nuclear physicists and all their expensive equipment to make the "economist's stone." All you need are two stubborn people who differ about the probability of some event.

Suppose that Adam is quite sure that the probability of Punter's Folly winning the Kentucky Derby is $\frac{3}{4}$. Eve is quite sure that the probability is only $\frac{1}{4}$. Adam will then accept small enough bets at any odds better than $1:3$ against Punter's Folly winning. Eve will accept small enough bets at any odds better than $1:3$ against Punter's Folly losing.[3] A bookie can now make a Dutch book by betting one cent with Adam at odds of $1:2$ and one cent with Eve at odds of $1:2$. Whatever happens, the bookie loses one cent to one player but gets two cents from the other.

[3]Assuming they have smooth Von Neumann and Morgenstern utility functions.

This is the secret of how bookies make money. Far from being the wild gamblers they like their customers to think, they bet only on sure things.

*Avoiding Dutch Books.* To justify introducing subjective probabilities in Section 13.3.2, we need to assume that Pandora's choices reveal full and rational preferences over a large enough set of gambles (Section 4.2.2).

Having full preferences will be taken to include the requirement that Pandora never refuses a bet—provided that she gets to choose which side of the bet to back, which means that she chooses whether to be the bookie offering the bet or the gambler to whom the bet is offered. Being rational will simply mean that nobody can make a Dutch book against her.

We follow Anscombe and Aumann in allowing our gambles to include all the lotteries of Section 4.5.2. We then have the Von Neumann and Morgenstern theory of rational choice under risk at our disposal. This makes equation (13.2) meaningful and also allows us to introduce notional poker chips that each correspond to one util on Pandora's Von Neumann and Morgenstern utility scale. We can then admit compound gambles denominated in poker chips.

Compound gambles represent bets about which consequence will arise in a simple gamble of the form:

$$G = \begin{array}{|c|c|c|c|c|} \hline w_1 & w_2 & w_3 & \cdots & w_n \\ \hline E_1 & E_2 & E_3 & \cdots & E_n \\ \hline \end{array} .$$

An example is the bet in which a bookie offers the gamblers odds of $x:1$ against the event $E$ occurring. For each such bet, Pandora chooses whether to be the gambler or the bookie. If she chooses to be the bookie when $x = a$ and the gambler when $x = b$, then we must have $a \leq b$ since the kind of Dutch book we made against Adam and Eve in Section 13.3.3 could otherwise be made against Pandora.

If Pandora doesn't choose to be the bookie or the gambler all the time,[4] then we can find odds $c:1$ such that Pandora chooses to be the bookie when $x < c$ and the gambler when $x > c$. She is then acting as though she believes that the probability of $E$ is $p = 1/(c+1)$. We then say that $p$ is her subjective probability for $E$.

When the state $E$ arises in other gambles, Pandora must continue to behave as though its probability were $p$; otherwise a Dutch bookie will exploit the fact that she sometimes assigns one probability to $E$ and sometimes another. Nor must Pandora neglect to manipulate her subjective probabilities according to the standard laws of probability lest further Dutch books be made against her.

Our assumptions therefore ensure that Pandora is Bayesian rational.

### 13.3.4 Priors and Posteriors

Among the laws of probability that Pandora must honor if she is to be immune to Dutch books are those that govern the manipulation of conditional probabilities. Her

[4]If she does, then her subjective probability for $E$ is $p = 0$ when she chooses to be the bookie all the time and $p = 1$ when she chooses to be the gambler all the time.

subjective probabilities must therefore obey Bayes's rule. It is for this reason that Bayesian rationality is named after the Reverend Thomas Bayes.[5] People rightly think that making sensible inferences from new information is one of the most important aspects of rational behavior, and Bayesian updating is how such inferences are made in Bayesian decision theory.

The language of *prior* and *posterior* probabilities is often used when discussing such inferences. When economists ask for your prior, you are being invited to quantify your beliefs before something happens. Your posterior quantifies your beliefs after it has happened.

→ 13.4

*Tossing Coins.* A weighted coin lands heads with probability $p$. Your prior probabilities over the possible values of $p$ are prob $(p = \frac{1}{3}) = 1 - q$ and prob $(p = \frac{2}{3}) = q$. (Values of $p$ other than $\frac{1}{3}$ and $\frac{2}{3}$ are impossible.) What are your posterior probabilities after observing the event $E$ in which heads appears $m$ times and tails $n$ times in $N = m + n$ tosses? From Bayes's rule:[6]

$$\text{prob}\,(p = \tfrac{2}{3} \mid E) = c\,\text{prob}\,(E \mid p = \tfrac{2}{3})\,\text{prob}\,(p = \tfrac{2}{3}) = \frac{2^m q}{2^m q + 2^n(1 - q)};$$

$$\text{prob}\,(p = \tfrac{1}{3} \mid E) = c\,\text{prob}\,(E \mid p = \tfrac{1}{3})\,\text{prob}\,(p = \tfrac{1}{3}) = \frac{2^n(1 - q)}{2^m q + 2^n(1 - q)}.$$

What happens if $m \approx \frac{2}{3}N$ and $n \approx \frac{1}{3}N$, so that the frequency of heads is nearly $\frac{2}{3}$? If $N$ is large, we would regard this as evidence that the objective probability of the coin landing heads is about $\frac{2}{3}$. Your posterior probability that $p = \frac{2}{3}$ is correspondingly close to one because

$$\text{prob}\,(p = \tfrac{2}{3} \mid E) \approx \frac{q}{q + (1 - q)2^{-N/3}} \to 1 \text{ as } N \to \infty.$$

This example illustrates the relation between subjective and objective probabilities. Unless your prior assigns zero probability to the true value of a probability $p$, your posterior probability for $p$ will be approximately one with high probability after observing enough independent trials (Exercise 13.10.15).

→ 13.5

## 13.4 Getting the Model Right

The arguments offered in defense of consistency in Section 4.8.3 become even harder to sustain when the criteria include immunity against Dutch books. However, critics of the consistency requirements of Bayesian decision theory often miss their target by attacking applications of the theory that fail—not because the consistency requirements are unreasonable but because the decision problem was wrongly modeled.

[5]He would be amazed that a whole theory of rational decision making was named in his honor centuries after his death. The theory was actually put together over the years by a number of researchers, including Frank Ramsey and Leonard Savage.

[6]The binomial distribution tells us that the probability of exactly $m$ heads in $m + n$ tosses when heads lands with probability $p$ is $(m + n)!p^m(1 - p)^n/m!n!$

*Miss Manners.* Amartya Sen tells us that people never take the last apple from a bowl. They are therefore inconsistent when they reveal a preference for no apples over one apple when offered a bowl containing only one apple but reverse this preference when offered a bowl containing two apples.

The data supporting this claim must have been gathered in some last bastion of good manners—and this is relevant when modeling Pandora's choice problem. Pandora's belief space $B$ must allow her to recognize that she is taking an apple from a bowl in a society that subscribes to the social values of Miss Manners rather than those of Homer Simpson. Her consequence space $C$ must allow her to register that she cares more about her long-term reputation than the transient pleasure to be derived from eating an apple right now. Otherwise, we won't be able to model the cold shoulders she will get from her companions if they think she has behaved rudely.

Pandora's apparent violation of the consistency postulates of revealed preference theory then disappears like a puff of smoke. She likes apples enough to take one when no breach of etiquette is likely, but not otherwise.

*Sour Grapes.* Sen's example shows the importance of modeling a choice problem properly before applying Bayesian decision theory. The reason is that its consistency assumptions essentially assert that rational players faced with a choice problem $f: A \times B \to C$ won't allow what is going on in one of the domains $A$, $B$, or $C$ to affect their treatment of the other domains.

For example, the fox in Aesop's fable is irrational in judging the grapes to be sour because he can't reach them. He thereby allows his beliefs in domain $B$ to be influenced by the actions available in domain $A$. If he decided that chickens must be available because they taste better than grapes, he would be allowing his assessment of what actions are available in domain $A$ to be influenced by his preferences in domain $C$. The same kind of wishful thinking may lead him to judge that the grapes he can reach must be ripe because ripe grapes taste better than sour grapes or that sour grapes taste better than ripe grapes because the only grapes that he can reach are sour. In both cases, he fails to separate his beliefs in domain $B$ from his preferences in domain $C$.

Such irrationalities are inevitable if $A$, $B$, and $C$ are chosen in a way that links their content. As an example of a possible linkage between $A$ and $C$, suppose that Pandora refuses a draw when playing chess but then loses. If she is then unhappier than she would have been if no draw had been offered, we made a mistake if we took $C = \{\mathscr{L}, \mathscr{D}, \mathscr{W}\}$. At the very least, we should have distinguished between losing-having-refused-a-draw and losing-without-having-refused-a-draw. That is to say, where necessary, the means by which an end is achieved must be absorbed into the definition of an end.

Linkages between $A$ and $B$ and between $B$ and $C$ can cause similar problems. For example, suppose that an umbrella and an ice cream cone are among the prizes available at a county fair and the possible states of the world are *sunny* and *wet*. It wouldn't then be surprising if Pandora's preferences over the prizes were influenced by her beliefs about the state of the world. If so, the prizes themselves mustn't be taken to be the objects in $C$. If we did, Pandora would seem to be switching her preference between umbrellas and ice cream cones from day to day, and we wouldn't have the *stable* preferences we need to apply revealed preference theory. In such cases, we identify $C$ with Pandora's *states of mind.* Instead of an umbrella being a consequence, we use the states of mind that accompany having an umbrella-on-a-sunny-day or having an umbrella-on-a-wet-day as consequences.

When such expedients are employed, our critics accuse us of reducing the theory to a bunch of tautologies. However, as noted at the end of Section 1.4.2, this is a puzzling accusation. What could be safer than to be defending propositions that are true by definition?

*Warning.* If we model an interaction between Alice and Bob as a game in strategic form, then Alice's consequence space $C$ is the set of cells in the payoff table. Her action space $A$ is the set of rows. Since she doesn't know what Bob is planning to do, her belief space $B$ is the set of columns.

If we want to be able to appeal to orthodox decision theory, the interaction between Alice and Bob must involve no linkages between $A$, $B$, and $C$ that aren't modeled within the game. If such unmodeled linkages exist, it is a good idea to look around for a more complicated model of the interaction that doesn't have such linkages.

For example, Figure 5.11(c) isn't the right strategic form for the Stackelberg model because it doesn't take into account the fact that Bob sees Alice's move before moving himself. Economists get around this problem by inventing the non-standard idea of a Stackelberg equilibrium (Section 5.5.1), but game theorists prefer the model of Figure 5.12(a), in which the strategy space assigned to Bob recognizes the linkage neglected in Figure 5.11(c). Only then are we are entitled to appeal to the standard theory.

→ 13.6

## 13.5 Scientific Induction?

We have met objective and subjective probabilities. Philosophers of science prefer a third interpretation. A *logical* probability is the degree to which the evidence supports the belief that a proposition is true.

An adequate theory of logical probability would solve the age-old problem of scientific induction. Does my boyfriend really love me? Is the universe infinite? Just put the evidence in a computer programmed with the theory, and out will come the appropriate probability.

Bayesianism is the creed that the subjective probabilities of Bayesian decision theory can be reinterpreted as logical probabilities without any hassle. Its adherents therefore hold that Bayes's rule is the solution to the problem of scientific induction.

### 13.5.1 Where Do Priors Come From?

If Bayes's rule solves the problem of scientific induction, then upating your beliefs when you get new information is simply a matter of carrying out some knee-jerk arithmetic. But what of the prior probabilities with which you begin? Where do they come from?

*Harsanyi Doctrine.* Rational beings are sometimes said to come with priors already installed. John Harsanyi even advocates a mind experiment by means of which we can determine these rational priors. You imagine that a veil of ignorance conceals all the information you have ever received. Harsanyi thinks that ideally rational folk in this state of sublime ignorance would all select the same prior. Such claims are fondly known among game theorists as the Harsanyi doctrine (12.8.2). But even if Harsanyi were right, how are we poor mortals to guess what this ideal prior would be? Since nobody knows, priors are necessarily chosen in more prosaic ways.

*The Principle of Insufficient Reason.* Bayesian statisticians use their experience of what has worked out well in the past when choosing a prior. Bayesian physicists prefer whatever prior maximizes entropy. Otherwise, an appeal is usually made to Laplace's principle of insufficient reason. This says that you should assign the same probability to two events if you have no reason to think one more likely than the other. But the principle is painfully ambiguous.

What prior should we assign to Pandora when she knows nothing at all about the three horses running in a race? Does the principle of insufficient reason tell us to give each horse a prior probability of $\frac{1}{3}$? Or should we give a prior probability of $\frac{1}{2}$ to Punters' Folly because Pandora has no reason to think it more likely that Punters' Folly will win than lose?

## 13.6 Constructing Priors

When objective probabilities are unavailable, how do we manage in the absence of a sound theory of logical probability? We use *subjective* probabilities instead.

We commonly register our lack of understanding of how Pandora converts her general experience of the world into subjective beliefs by saying that the latter reflect her "gut feelings." But she would be irrational to treat the rumblings of her innards as an infallible oracle. Our gut feelings are usually confused and inconsistent. When they uncover such shortcomings in their beliefs, intelligent people modify the views about which they are less confident in an attempt to bring them into line with those about which they are more confident.

Savage thought that his theory would be a useful tool for this purpose. His response to Allais mentioned in Section 4.8 illustrates his attitude. When Allais pointed out an inconsistency in his choices, Savage recognized that his gut had acted irrationally and modified his behavior accordingly. Similarly, if you were planning to accept $96 \times 69$ dollars in preference to $87 \times 78$ dollars, you would revise your plan after realizing that it is inconsistent with your belief that $96 \times 69 = 6{,}624$ and $87 \times 78 = 6{,}786$ (Section 4.8.3).

So how would Savage form a prior? He would test any snap judgments that came to mind by reflecting that his gut is more likely to get things right when it has more evidence rather than less. For each possible future course of events, he would therefore ask himself, "What subjective probabilities would my gut come up with *after* experiencing these events?" In the likely event that these posterior probabilities were inconsistent with each other, he would then massage his initial snap judgments until consistency was achieved.[7] Only then would he feel that he had done justice to what his gut had to tell him.

Although Savage's consistency axioms are considerably more sophisticated than our story of Dutch books, he was led to the same theory. In particular, consistency demands that all posterior probabilities can be derived from the *same* prior using Bayes's rule. After massaging his original snap judgments until they became consistent, Savage would therefore act as a Bayesian—but for reasons that are almost the opposite of those assumed by Bayesianism. Instead of mechanically deducing his posterior probabilities from a prior chosen when he was in a maximal state of ignorance, Savage would have used his judgement to derive a massaged prior from the unmassaged posterior probabilities that represented his first stab at quantifying his gut feelings.

Savage was under no illusions about the difficulty of bringing such a massaging process to a successful conclusion. If the set of possible future histories that have to be taken into account is sufficiently large, the process obviously becomes impractical. He therefore argued that his theory was only properly applicable in what he called a small world.

### 13.6.1 Small Worlds

Savage variously describes the idea that one can use Bayesian decision theory on the grand scale required by Bayesianism as "ridiculous" and "preposterous." He insists that it is sensible to use his theory only in the context of a *small world*. Even the theory of knowledge on which we base our assumptions about information sets makes sense only in a small world (Section 12.3.1).

For Savage, a small world is a place where you can always "look before you leap." Pandora can then take account *in advance* of the impact that all conceivable future pieces of information might have on the inner model that determines her gut feelings. Any mistakes built into her original model that might be revealed in the future will then *already* have been corrected, so that no possibility remains of any unpleasant surprises.

In a large world, one can "cross certain bridges only when they are reached." The possibility of an unpleasant surprise that reveals some factor overlooked in the original model can't then be discounted. Knee-jerk consistency is no virtue in such a world. If Pandora keeps backing losers, she may be acting consistently, but she will lose a lot more money in the long run than if she temporarily lays herself open to a Dutch book while switching to a strategy of betting on winners.

Perhaps Pandora began by choosing her prior in a large world as Bayesianism prescribes, but, after being surprised by a stream of unanticipated data, wouldn't she be foolish not to question the basis on which she made her initial choice of prior? If her doubts are sufficient to shake her confidence in her previous judgment, why not

[7]Much of the wisdom of Luce and Raiffa's *Games and Decisions* has been forgotten (see Section 4.10). On this subject they say, "Once confronted with inconsistencies, one should, so the argument goes, modify one's initial decisions so as to be consistent. Let us assume that this jockeying—making snap judgments, checking up on their consistency, modifying them, again checking on consistency etc—leads ultimately to a bona fide, prior distribution."

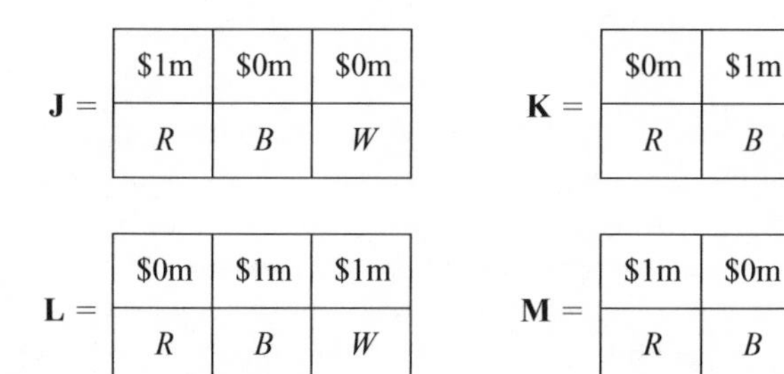


Figure 13.2 Lotteries for Ellsberg's Paradox. The prizes are given in millions of dollars to dramatize the situation.

abandon her old prior and start again with a new prior based on better criteria? I can think of no good reason why not. But Pandora will then have failed to update using Bayes's rule.

*Ellsberg's Paradox.* An urn contains 300 balls, of which 100 are known to be red. The other 200 balls are black or white, but we don't know in what proportions. A ball is drawn at random, generating one of three possible events labeled $R$, $B$, or $W$, depending on the color of the ball. You are asked to consider your preferences over the gambles of Figure 13.2.

A Bayesian who takes the conditions of the problem to imply that $\text{prob}\,(R) = \frac{1}{3}$ and $\text{prob}\,(B) = \text{prob}\,(W)$ would express the preferences $\mathbf{J} \sim \mathbf{K}$ and $\mathbf{L} \sim \mathbf{M}$. However, most people express the preferences $\mathbf{J} \succ \mathbf{K}$ and $\mathbf{L} \succ \mathbf{M}$, thereby exposing themselves to a Dutch book. They can't be assessing the three events using subjective probabilities because $\mathbf{J} \succ \mathbf{K}$ is the same as $\text{prob}\,(R) > \text{prob}\,(B)$ and $\mathbf{L} \succ \mathbf{M}$ is the same as $\text{prob}\,(B) > \text{prob}\,(R)$.

People presumably prefer **J** to **K** because $\text{prob}\,(R)$ is objectively determined, but $\text{prob}\,(B)$ isn't. Similarly, they prefer **L** to **M** because $\text{prob}\,(B \cup W)$ is objectively determined, but $\text{prob}\,(R \cup W)$ isn't. The paradox is therefore said to be an example of uncertainty aversion.

My own view is that some uncertainty aversion is entirely reasonable for someone making decisions in a large world. Who knows what dirty work may be going on behind the scenes? (Exercise 13.10.23) It is true that the Ellsberg paradox itself is arguably a small-world problem, but people are unlikely to see the distinction when put on the spot. Their answers are simply gut responses acquired from living all their lives in a very large world indeed.

## 13.7 Bayesian Rationality in Games

The toy models we use in game theory are small worlds almost by definition. Thus we can use Bayesian decision theory without fear of being haunted by Savage's ghost, telling us that it is ridiculous to use his theory in a large world. However, we have to be wary when enthusiasts apply the theorems we have derived for the small worlds of game theory to worlds that the players perceive as large.

→ 13.8

### 13.7.1 Subjective Equilibria

From an evolutionary viewpoint, mixed equilibria summarize the objective frequencies with which different strategies can coexist in large populations. But mixed equilibria aren't so easy to justify on rational grounds. If you are indifferent between two pure strategies, why should you care which you choose?

For this reason, Section 6.3 suggests interpreting mixed equilibria as a statement about what rational players will *believe*, rather than a prediction of what they will actually *do*. When an equilibrium is interpreted in this way, it is called a *subjective equilibrium*. But what is an equilibrium in beliefs?

I think this is another of those questions that will properly be answered only when we are nearer a solution to the problem of scientific induction, but naive Bayesians don't see any problem at all. When playing Matching Pennies, so the story goes, Adam's gut feelings tell him what subjective probabilities to assign to Eve's choosing *heads* or *tails*. He then chooses *heads* or *tails* to maximize his own expected utility. Eve proceeds in the same way. The result won't be an equilibrium, but so what?

But it isn't so easy to escape the problems raised by sentences that begin: "Adam thinks that Eve thinks..." In forming his own subjective beliefs about Eve, Adam will simultaneously be trying to predict how Eve will form her subjective beliefs about him. While using something like the massaging process of Section 13.6, he will then not only have to massage his own probabilities until consistency is achieved but also have to simulate Eve's similar massaging efforts. The end product will include not only Adam's subjective probabilities for Eve's choice of strategy but also his prediction of her subjective probabilities for his choice of strategy. The two sets of subjective probabilities must be consistent with the fact that both players will optimize on the basis of their subjective beliefs. If so, we are looking at a Nash equilibrium. If not, a Dutch book can be made against Adam.

### 13.7.2 Common Priors?

We have always assumed that the probabilities with which Chance moves are objective, but what if we are playing games at a race track rather than a casino?

We then have to build the players' subjective beliefs about Chance into the model. The argument justifying subjective equilibria still applies, but if Adam is to avoid a Dutch book based on his predictions of everybody's beliefs, his massaging efforts must generate a *common* prior from which each player's posterior beliefs can be deduced by conditioning on their information.

But why should Eve be led to the *same* common prior as Adam? In complicated games, one can expect the massaging process to converge on the same outcome for all players only if their gut feelings are similar. But we can expect the players to have similar gut feelings only if they all share a common culture and so have a similar history of experience. Or to say the same thing another way, only when the players of a game are members of a reasonably close-knit community can they be expected to avoid leaving themselves open to a Dutch book being made against their group as a whole.

This isn't a new thought. Ever since Section 1.6, we have kept returning to the idea that it is common knowledge that all players read the same authoritative game theory book. What we are talking about now is how Von Neumann—or whoever else the author may be—knows what to say when offering advice on how to play each particular game. If he decides to assume that it is common knowledge that all players have the same common prior, then he is proceeding as though the players all share a common culture.

Some authors deny that a common culture is necessary to justify the common prior assumption. They appeal to the Harsanyi doctrine of Section 13.5.1 in arguing that a shared *rationality* is all that is necessary for common knowledge of a common prior. However, I feel safe in making this assumption only when the players determine their priors *objectively* by consulting social statistics or other data that everybody sees everybody else consulting.

*Correlated Subjective Equilibrium.* Bob Aumann claims a lot more for subjective equilibrium by making the truly heroic assumption that the whole of creation can be treated as a small world in which a state specifies not only things like how decks of cards get dealt but also what everybody is thinking and doing. If Alice is Bayesian rational, she then behaves just like her namesake in Section 6.6.2 when operating a correlated equilibrium in Chicken. The referee is now the entire universe, which sends a signal that tells her to take a particular action. She then updates her prior to take account of the information in the signal. Because she is Bayesian rational, the action she then takes is optimal given her posterior beliefs. Aumann's idea of a correlated equilibrium therefore encompasses everything!

The result isn't a straightforward correlated equilibrium, which would require that the players all share a common prior. An implicit appeal to the Harsanyi doctrine is therefore usually made to remove the possibility that the players may agree to disagree about their priors.

## 13.8 Roundup

Bayes's rule says that

$$\text{prob}\,(F|E) = \frac{\text{prob}\,(E|F)\,\text{prob}\,(F)}{\text{prob}\,(E)}.$$

It is so useful in computing conditional probabilities at information sets in games that the process is called Bayesian updating. Your probability measure over possible states of the world before anything happens is called your *prior.* The probability measure you get from Bayesian updating after observing an event $E$ is called a *posterior.*

We sometimes need to calculate many conditional probabilities of the form $\text{prob}\,(F_i|E)$ at once. If one and only one of the events $F_1, F_2, \ldots, F_n$ is sure to happen after $E$ has been observed, we write

$$\text{prob}\,(F_i|E) = c\,\text{prob}\,(E|F_i)\,\text{prob}\,(F_i)$$

and find $c$ using the formula $\text{prob}\,(F_1 \mid E) + \text{prob}\,(F_2 \mid E) + \cdots + \text{prob}\,(F_n \mid E) = 1$.

Bayesian rationality means a lot more than believing in Bayes's rule. Our assumption that players are Bayesian rational implies that they separate their beliefs from their preferences by quantifying the former with a subjective probability measure and the latter with a utility function. When choosing among gambles **G** in which you get the prize $\omega_i$ when the event $E_i$ occurs, Bayesian rational players act as though seeking to maximize their expected utility:

$$\varepsilon u(\mathbf{G}) = p_1 u(\omega_1) + p_2 u(\omega_2) + \cdots + p_n u(\omega_n),$$

where $u(\omega_i)$ is their Von Neumann and Morgenstern utility for the prize $\omega_i$ and $p_i = \text{prob}\,(E_i)$ is their subjective probability for the event $E_i$.

You won't be able to separate your beliefs from your preferences if you are careless in your choice of the sets $B$ and $C$ in which they live. If your preference between an umbrella and an ice cream cone depends on whether the day is rainy or sunny, you can't treat getting an umbrella as one of the possible consequences in your decision problem. Although you will be accused of making the theory tautological, you must think of your possible consequences as getting an umbrella-on-a-rainy-day or getting an umbrella-on-a-sunny-day. Sometimes it is necessary to redefine your actions in a similar way before trying to apply Bayesian decision theory.

What should it mean to say that Pandora reveals full and rational preferences when choosing among gambles? The simplest criterion requires that Pandora's choices should immunize her against Dutch books. A *Dutch book* is a system of bets that guarantee that Pandora will lose whatever happens if she takes them on. Assuming that Pandora is always willing to take one side of every bet, she can be immune to a Dutch book only if she always behaves as though each event has a probability. Since she may have no objective evidence about how likely the events are, we say that the probabilities revealed by her betting behavior are *subjective.* If we also assume that Pandora honors the Von Neumann and Morgenstern theory, we are then led to the conclusion that she must be Bayesian rational.

Leonard Savage came to the same conclusion from a more sophisticated set of criteria. His work is often quoted to justify Bayesianism—the claim that Bayesian updating is the solution to the problem of scientific induction. Savage rejected this idea as "ridiculous" outside the kind of small world in which you are able to evaluate each possible future history before settling on a prior. Fortunately, the models of game theory are small worlds in this sense.

Bayesianism tells you to keep updating the prior with which you started, even when you receive data whose implications reveal that you chose your prior on mistaken principles. The Harsanyi doctrine says that two rational people with the same information will start with the same prior. The principle of insufficient reason says that this prior will assign two events the same probability, unless there is some reason to suppose that one is more likely than the other. All three propositions deserve to be treated with a good measure of skepticism.

Savage envisaged a process in which you massage your original gut feelings into a consistent system of beliefs by the use of the intellect. The same reasoning can be employed to explain *subjective equilibria,* provided that we insist that players massage the beliefs they attribute to other players along with their own. The result will be that all the beliefs they attribute to the players will be derivable from a *common prior.* However, the argument doesn't imply that it will be common knowledge that all players have the *same* common prior, which is a standard assumption in some contexts.

## 13.9 Further Reading

*The Foundations of Statistics*, by Leonard Savage: Wiley, New York, 1954. Part I is the Bayesian bible. Part II is an unsuccessful attempt to create a decision theory for large worlds.

*Notes on the Theory of Choice*, by David Kreps: Westview Press, London, 1988. A magnificent overview of the whole subject.

*A Theory of Probability*, by John Maynard Keynes: Macmillan, London, 1921. An unsuccessful attempt to create a theory of logical probability by one of the great economists of the twentieth century.[8]

## 13.10 Exercises

1. Each of the numbers $0, 1, 2, 3, \ldots, 36$ is equally likely to come up when playing roulette. You have bet a dollar on number 7 at the odds of 35 : 1 offered by the casino. What is your expected monetary gain? As the wheel stops spinning, you see that the winning number has only one digit. What is your expected gain now? **review**
2. Find $\text{prob}\,(x = a \mid y = c)$ and $\text{prob}\,(y = c \mid x = a)$ in Exercise 3.11.8. **review**
3. The $n$ countries of the world have populations $M_1, M_2, \ldots, M_n$. The number of left-handed people in each country is $L_1, L_2, \ldots, L_n$. What is the probability that a left-handed person chosen at random from the world population comes from the first country?
4. A box contains one gold and two silver coins. Two coins are drawn at random from the box. The Mad Hatter looks at the coins that have been drawn without your being able to see. He then selects one of the coins and shows it to you. It is silver. At what odds will you bet with him that the other is gold? At what odds will you bet if the coin that you are shown is selected at random from the drawn pair?
5. In a new version of Gale's Roulette, the players know that the casino has things fixed so that the sum of the numbers shown on the roulette wheels of Figure 3.19 is always 15 (Exercise 3.11.31). Explain the extensive form given in Figure 13.3. **fun**
   a. With what probability does each node in player II's center information set occur, given that the information set has been reached after player I has chosen wheel 2?
   b. What are player II's optimal choices at each of her information sets? Double the branches that correspond to her optimal choices in a copy of Figure 13.3.
   c. Proceeding by backward induction, show that the value of the game is $\mathbf{2/5}$, which player I can guarantee by choosing either wheel 2 or wheel 3.
6. Redraw the information sets in Figure 13.3 to model the situation in which both players know that player I will get to see where wheel 1 stops before picking a wheel and player II will get to see where wheel 2 stops before picking a wheel. Double the branches corresponding to player II's optimal choices at each of her nine information sets. Proceeding by backward induc- **fun**

[8] A version of his illustration of the ambiguity implicit in the principle of insufficient reason appears as Exercise 14.9.21.

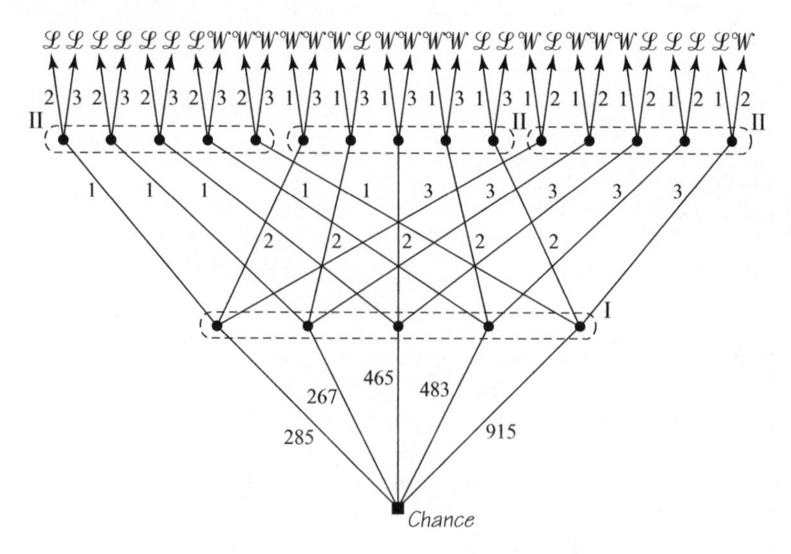


Figure 13.3 An extensive form for Gales' Roulette when both players know that the wheels are rigged so that the numbers on which they stop always sum to 15. The wheels are no longer independent and so are treated as a single entity in the opening chance move.

tion, double the branches corresponding to player I's optimal replies at each of his three information sets. Deduce that the value of the game is **3/5** and that player I can guarantee this lottery or better by always choosing wheel 2.

math

7. Explain why $\text{prob}\,(E) = \text{prob}\,(E \cap F) + \text{prob}\,(E \cap {\sim}F)$. Deduce that

$$\text{prob}\,(E) = \text{prob}\,(E|F)\,\text{prob}\,(F) + \text{prob}\,(E|{\sim}F)\,\text{prob}\,({\sim}F).$$

Find a similar formula for $\text{prob}\,(E)$ in terms of the conditional probabilities $\text{prob}\,(E \mid F_i)$ when the sets $F_1, F_2, \ldots, F_n$ partition $E$.

8. Calculate $\text{prob}(A \mid a\beta\beta)$ and $\text{prob}\,(B \mid a\beta\beta)$ in the discussion of strategic voting in Section 13.2.4. Show that these conditional probabilities are equal when

$$p = \frac{2 - \sqrt{2}}{2\sqrt{2} - 1} \approx 0.32.$$

Why does this value of $p$ correspond to a mixed equilibrium?

9. In the discussion of strategic voting in Section 13.2.4, show that the probability that the better candidate is elected is

$$q = \tfrac{1}{2}\left\{1 - \left(\tfrac{2}{3}p + \tfrac{1}{3}\right)^3 + \left(\tfrac{1}{3}p + \tfrac{2}{3}\right)^3\right\}.$$

Prove that this quantity is maximized when $p$ takes the value computed in the previous problem.

phil

10. Casting your vote on the assumption that it will be pivotal may require you to suppose that large numbers of people will change their current plans on how to

vote. Why does making this assumption not involve you in the Twins' Fallacy of Section 1.3.3?

11. Pundits commonly urge that a vote for a small central party is wasted because the party has no chance of winning. Construct a very simple model in which people actually vote on the assumption that their vote won't be wasted but with the result that *everybody* votes for the central party, even though *nobody* would vote for it if they simply supported the party they liked best.

phil

12. Discuss the problem that a green game theorist faced in the polling booth when deciding whether to vote for Ralph Nader's green party in the presidential election in which George W. Bush finally beat Al Gore by a few hundred votes in Florida.[9] (Nader said that Bush and Gore were equally bad, but most Nader voters would have voted for Gore if Nader hadn't been running.)

phil

13. A bookie offers odds of $a_k$:1 against the $k$th horse in a race being the winner. There are $n$ horses in the race, and

$$\frac{1}{a_1+1}+\frac{1}{a_2+1}+\cdots+\frac{1}{a_n+1}<1.$$

How should you bet to take advantage of the rare opportunity to make a Dutch book against a bookie?

14. Adam believes that the Democrat will be elected in a presidential election with probability $\frac{5}{8}$. Eve believes the Republican will be elected with probability $\frac{3}{4}$. Neither gives third-party candidates any chance at all. They agree to bet $10 on the outcome at even odds. What is Adam's expected dollar gain? What is Eve's?

Make a Dutch book against Adam and Eve on the assumption that they are both always ready to accept any bet that they believe has a nonnegative dollar expectation.

15. In Section 13.3.4, a coin lands heads with probability $p$. Pandora's prior probabilities for $p$ are $\text{prob}\,(p=\frac{1}{3})=1-q$ and $\text{prob}\,(p=\frac{2}{3})=q$. Show that her posterior probabilities after observing the event $E$ in which heads appears $m$ times and tails $n$ times in $N=m+n$ tosses are

math

$$\text{prob}\,(p=\tfrac{2}{3}\,|E) = \frac{2^m q}{2^m q+2^n(1-q)};$$
$$\text{prob}\,(p=\tfrac{1}{3}\,|E) = \frac{2^n(1-q)}{2^m q+2^n(1-q)}.$$

If $q=\frac{1}{2}$, $N=7$, and $m=5$, what is Pandora's posterior probability that $p=\frac{2}{3}$? What is her posterior probability when $q=0$?

16. A coin lands heads with probability $p$. Pandora's prior probabilities for $p$ are $\text{prob}\,(p=\frac{1}{4})=\text{prob}\,(p=\frac{1}{2})=\text{prob}\,(p=\frac{3}{4})=\frac{1}{3}$. Show that her posterior

math

[9] The question actually turned out to be less whether your vote would count than whether it would be counted.

probability for $p = \frac{1}{2}$ after observing the event $E$, in which heads appears $m$ times and tails $n$ times in $N = m + n$ independent tosses, is $\text{prob}(p = \frac{1}{2}|E) = 2^N/(2^N + 3^m + 3^n)$.

Suppose that the value of $p$ is actually $\frac{1}{2}$. We can read off from Figure 3.8 that it is more likely than not that $m = 3$ or $m = 4$ heads will be thrown in $N = 7$ independent tosses. Deduce that it is more likely than not that Pandora's posterior probability for $p = \frac{1}{2}$ exceeds $\frac{1}{2}$.

phil

17. A theater critic gave good first-night reviews to all the Broadway hits a newspaper editor can remember. Why isn't this a good enough reason for the editor to hire the critic?

    Let $H$ be the event that the critic predicts a hit, and let $h$ be the event that the show actually is a hit. Let $F$ be the event that the critic predicts a flop, and let $f$ be the event that the show actually flops. Pandora's prior is that $\text{prob}(h) = \text{prob}(f)$. Unless she receives further information, she is indifferent between attending a performance and staying at home. To be persuaded to see the performance on the advice of the critic, she needs that $\text{prob}(h \mid H) > \text{prob}(f \mid H)$. If she is also not to regret taking the critic's advice to stay away from a performance that later turns out to be a hit, she needs that $\text{prob}(h \mid F) < \text{prob}(f \mid F)$. Will Pandora's criteria necessarily be met if the editor uses the criterion $\text{prob}(H \mid h) = 1$ when deciding whom to hire? If nothing else but being hired were relevant, how would a critic exploit the use of such a criterion?

phil

18. If Alice is dealt four queens in poker, her posterior probability for a queen remaining in the deck is zero. But Bob will still be assigning a positive probability to this event. Alice now offers to bet with Bob that no further queen will be dealt, at odds that seem favorable to him relative to his current subjective probability for this event. Why should Bob treat Alice's invitation to bet as a piece of information to be used in updating his probability? After updating, he will no longer want to bet at the odds she is willing to offer. How do things change if Bob can choose to take either side of any bet that Alice proposes? (Section 13.3.3)

phil

19. Bayesianism can be applied to anything, including the Argument by Design that some theologians still argue is a valid demonstration of the existence of God. The argument is that the observation of organization demonstrates the existence of an organizer.

    Let $F$ be the event that something appears organized. Let $G$ be the event that there is an organizer. Everybody agrees that $\text{prob}(F \mid G) > \text{prob}(F \mid \sim G)$. However, the Argument by Design needs to deduce that $\text{prob}(G \mid F) > \text{prob}(\sim G \mid F)$ if God's existence is to be more likely than not. Explain why people whose priors satisfy $\text{prob}(G) > \text{prob}(\sim G)$ are ready to make the deduction, but others are more hesitant.

20. Large numbers of people claim to have been abducted by aliens. Let $E$ be the event that this story is true and $R$ the event that large numbers of people report it to be true. If $\text{prob}(R \mid E) = 1$ and $\text{prob}(R \mid \sim E) = q < 1$, show that Bayesians will think alien abduction more likely than not when their prior probability $p = \text{prob}(E)$ satisfies $p > q/(1 + q)$.

phil

21. David Hume famously argued that belief in a miracle is never rational because a breach in the laws of nature is always less credible than that the witnesses

| | *dove* | *hawk* |
|---|---|---|
| *dove* | \$2 | \$0 |
| *hawk* | \$3 | \$1 |

(a) Newcomb á la Lewis

| | *correct* | *mistaken* |
|---|---|---|
| *dove* | \$2 | \$0 |
| *hawk* | \$1 | \$3 |

(b) Newcomb á la Ferejohn

Figure 13.4 Attempts to model the Newcomb paradox.

should lie or be deceived. Use the previous exercise to show that a Bayesian's prior probability of a miracle would have to be zero for Hume's argument to hold irrespectively of the supporting evidence that the witnesses might present.

Comment on the implications for science if Hume's argument could be sustained. For example, the laws of quantum physics seem miraculous to me, but I believe physicists when they tell me that they work.

22. We looked at a version of Pascal's Wager in Exercise 4.11.29. God is commonly thought to demand belief in His existence as well as observance of His laws. Is it consistent with Bayesian decision theory to argue that Pandora should attach a high subjective probability to the event that God exists and hence that there is an afterlife because this makes her expected utility large? **phil**

23. As the experimenter in the Ellsberg paradox of Section 13.6.1, you are eager to save money. Against someone who goes for **J** and **L**, you expect to lose \$1 million per subject. If your subjects are Bayesians who are willing to accept **K** and **M** instead, can you lose less by fixing the proportion of black and white balls in the urn?

24. Various approaches to Newcomb's paradox were reviewed in Exercises 1.13.23 onward. In Exercise 1.13.24, the philosopher David Lewis treats Adam as a player in the Prisoners' Dilemma. Figure 13.4(a) then illustrates Adam's choice problem. What is the function $f: A \times B \to C$? What are the sets $A$, $B$, and $C$? **phil**

    The political scientist John Ferejohn suggests modeling Newcomb's paradox as in Figure 13.4(b). The states in $B$ labeled *correct* and *mistaken* now represent Eve's success in predicting Adam's choice. Why does this model provide an example in which $B$ is linked to $A$, and hence Bayesian decision theory doesn't apply? (Section 13.4)

25. The philosopher Richard Jeffries is credited with improving Bayesian decision theory by making it possible for Adam's beliefs about Eve's choice of strategy to depend on his own choice of strategy in the Prisoners' Dilemma. How does this scenario violate the precepts of Section 13.4? **phil**

26. Bob is accused of murdering Alice. His DNA matches traces found at the scene. An expert testifies that only ten people in the entire population of 100 million people come out positive on the test. The jury deduces that the chances of Bob being innocent are one in ten million, but the judge draws their attention to the table of Figure 13.5. The defense attorney says that this implies that there is only **phil**

| | Positive | Negative |
|---|---|---|
| Acquaintance | 1 | 999 |
| Stranger | 9 | |

Figure 13.5 DNA testing. The numbers in the table show how many people in a population of 100 million fall into each category. All but 1,009 people belong in the empty cell.

one chance in ten than Bob is guilty. The prosecuting attorney says that the table implies that Bob is guilty for sure. Assess the reasoning of each party.

phil

27. The fact that there is something wrong with the prosecution's reasoning in the previous exercise becomes evident if we observe that the logic would be the same if the first row of the table gave the results of testing a sample of one thousand people chosen at random from the whole population. Reconstruct the prosecution case on the assumption that convincing evidence can be produced that it is more likely than not that the guilty party knows the victim in this kind of murder.

math

28. Bayesian-rational players make whatever decision maximizes their expected payoff given their current beliefs. Prove that such a decision rule satisfies the Umbrella Principle of Section 12.8.2: If $E \cap F = \emptyset$ and $d(E) = d(F)$, then $d(E \cup F) = d(E) = d(F)$.

    Explain why two Bayesian rational players will have the same decision rule only if they have the same prior.

phil

29. Observing a black raven adds support to the claim that all ravens are black. Hempel's paradox exploits the fact that "$P \Rightarrow Q$" is equivalent to "not $Q \Rightarrow$ not $P$." Observing a pink flamingo therefore also adds support because pink isn't black and flamingos aren't ravens. One way of resolving the paradox is to argue that observing a pink flamingo adds only negligible support because there are so many ways of not being black or a raven. Formulate a Bayesian version of this argument.

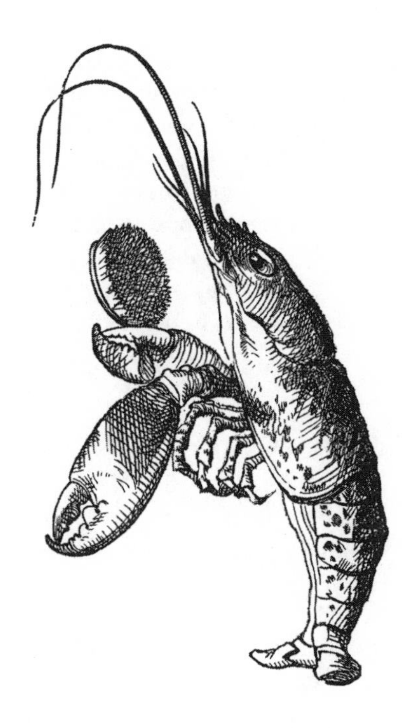

# 14
# *Seeking Refinement*

## 14.1 Contemplating the Impossible

The Red Queen famously told a doubtful Alice that she sometimes believed six impossible things before breakfast. Alice was only seven and a half years old, but she should have known better than to doubt the value of thinking about things that won't happen. Making rational decisions *always* requires contemplating the impossible. Why won't Alice touch the stove? Because she would burn her hand if she did.

Politicians pretend to share Alice's belief that hypothetical questions make no sense. As George Bush Senior put it when replying to a perfectly reasonable question about unemployment benefit, "If a frog had wings, he wouldn't hit his tail on the ground." But far from being meaningless, hypothetical questions are the lifeblood of game theory—just as they ought to be the lifeblood of politics. Players stick to their equilibrium strategies because of what would happen if they didn't. It is true that Alice won't deviate from equilibrium play. However, the reason that she won't deviate is that she predicts that unpleasant things would happen if she did.

Game theory can't avoid subjunctives, but they often fly thicker and faster than is really necessary—especially when we ask how some equilibrium selection problems might be solved by *refining* the idea of a Nash equilibrium.

The refinement approach can't help with the problem of choosing among *strict* Nash equilibria, which we found so difficult in Chapter 8. In such equilibria, each player has only one best reply. Refinement theory works by eliminating some of the alternatives when there are multiple best replies. For example, subgame perfection is a refinement in which we eliminate best replies in which the players aren't planning to optimize in subgames that won't be reached in equilibrium (Section 2.9.3). In the impossible event that such a subgame *were* reached, the players are presumed to reason that the actions chosen there *would* be optimal.

Inventing refinements is properly the domain of social climbers, but game theorists were once nearly as prolific in inventing abstruse reasons for excluding unwelcome equilibria. So many refinements with such different implications were proposed that the profession is now very skeptical about the more exotic ideas. Some authors have even moved in the opposite direction by *coarsening* the Nash equilibrium concept. However, this chapter makes no attempt to survey all the proposals for refining or coarsening Nash equilibria. It focuses instead on the problems that the proposals failed to solve.

## 14.2 Counterfactual Reasoning

→ 14.3

The classic opening line of a mathematical proof is: Suppose $\varepsilon > 0$. But suppose it isn't? Everybody laughs when someone says this in class, but it deserves a proper response.

Theorems consist of material implications of the form "$P \Rightarrow Q$." This means the same as "(not $P$) or $Q$" and so is necessarily true when $P$ is false. Theorems are therefore automatically true when their hypotheses are false.

Mathematicians often think that any sentence with an *if* must be a material implication, but conditional sentences written in the subjunctive often say something substantive when their hypotheses are false. For example, it is true that Alice would burn her hand if she were to touch the stove but false that she will in fact touch the stove. She doesn't touch the stove *because* she knows the subjunctive conditional is true. She therefore reasons counterfactually—drawing a valid conclusion from an implication based on a premise that is factually false.

Alice's counterfactual is easy to interpret. But what of the following example from the Australian philosopher David Lewis?

> If kangaroos had no tails, they would topple over.

Since kangaroos actually do have tails, a sentence that says what would happen if kangaroos had no tails can be of interest only if it is meant to apply in some fictional world different in some respect from the actual world.

In one possible world, it might be that a particular kangaroo survives after its tail has been severed, but everything else is as before. Such an unfortunate kangaroo would indeed topple over if it stood on its feet, but one can also imagine a possible world in which some crucial event in the evolutionary history of the kangaroo is changed so that all the marsupials later called kangaroos have no tails. Kangaroos wouldn't then topple over because a species with such a handicap couldn't survive.

The meaning of a counterfactual statement is therefore as much to be found in its context as in its content. Often the context is very clear. For example, Eve will have no trouble understanding Adam if he tells her that he wouldn't have lost this month's mortgage repayment if he had been dealt the queen of hearts rather than the king in last night's poker game. Before the deal, there were many cards that Adam might have drawn, each of which represents a different possible world. But only in the possible world corresponding to the queen of hearts would Adam and Eve retain a roof over their heads.

One can't anticipate such clarity when dealing with more exotic counterfactuals, but the approach we will take is to try to pin down whatever is serving as a substitute for the shuffling and dealing of the cards in Adam's poker story. Only in the presence of such a contextual model can a counterfactual be interpreted unambiguously.

Biological evolution provides one important example. How do we explain how animals behave in circumstances that don't normally arise? If this behavior was shaped by evolution, it was in the world of the past when different sets of genes were competing for survival. When we apply the selfish gene paradigm, the possible world that we use to interpret counterfactuals must therefore be this lost world of the past. The relevant context is then the evolutionary history of the species.

### 14.2.1 Chain Store Paradox

Section 2.5 offers an impeccable defense of backward induction for the case of win-or-lose games. It is often thought that backward induction is equally unproblematic in any game. Nobody claims that rational players will necessarily use their subgame-perfect strategies whatever happens, but it is sometimes argued that the backward induction *play* must be followed when it is common knowledge that the players are rational. Selten's Chain Store paradox explains that such claims can't always be right because they ignore the necessity of interpreting the counterfactuals that keep players on the equilibrium path.

*Chain Store Game.* Alice's chain of stores operates in two towns. If Bob sets up a store in the first town, Alice can acquiesce in his entry or start a price war. If he later sets up another store in the second town, she can again acquiesce or fight. If Bob chooses to stay out of the first town, we simplify by assuming that he necessarily stays out of the second town. Similarly, if Alice acquiesces in the first town, we assume that Bob necessarily enters the second town, and Alice again acquiesces.

This story is a simplified version of the full Chain Store paradox explored in a sequence of exercises in Chapter 5. The doubled lines in Figure 14.1(a) show that backward induction leads to the play [*ia*], in which Bob enters and Alice acquiesces. The same result is obtained by successively deleting (weakly) dominated strategies in Figure 14.1(b).

*Rational Play?* Suppose the great book of game theory says the play [*ia*] is rational. Alice will then arrive at her first move with her belief that Bob is rational intact. To check that the book's advice to acquiesce is sound, she needs to predict what Bob would do at his second move in the event that she fights. But the book says that fighting is irrational. Bob would therefore need to interpret a counterfactual at his second move: If a rational Alice behaves irrationally at her first move, what would she do at her second move?

There are two possible answers to this question: Alice might acquiesce or she might fight. If she would acquiesce at her second move, then it would be optimal for Bob to enter at his second move, and so Alice should acquiesce at her first move. In this case, the book's advice is sound. But if Alice would fight at her second move,

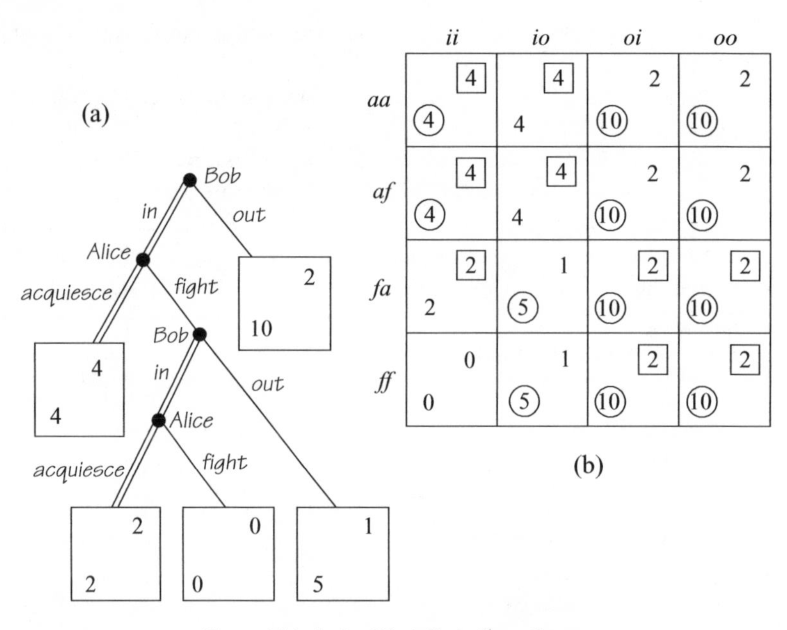


Figure 14.1 A simplified Chain Store Game.

then it would be optimal for Bob to stay out at his second move, and so Alice should fight at her first move. In this case, the book's advice is unsound.

What possible worlds might generate these two cases? In any such world, we must give up the hypothesis that the players are superhumanly rational. They must be worlds in which players sometimes make mistakes. The simplest such world arises when the mistakes are transient errors—like typos—that have no implications for mistakes that might be made in the future. In such a world, Bob still predicts that Alice will behave rationally at her second move, even though she behaved irrationally at her first move. If the counterfactuals that arise in games are always interpreted in terms of this world, then backward induction is always rational.

Lewis argues that the default world in which to interpret a counterfactual is the world "nearest" to our own. He would therefore presumably be happy with the preceding analysis.[1] But when we apply game theory to real problems, we aren't especially interested in the errors that a superhuman player might make. We are interested in the errors that real people make when trying to cope intelligently with complex problems. Their mistakes are much more likely to be "thinkos" than "typos." Such errors do have implications for the future (Section 2.9.4). In the Chain Store Game, the fact that Alice irrationally fought at her first move may signal that she would also irrationally fight at her second move.[2] But if Bob's counterfactual is interpreted in terms of such a possible world, then the backward induction argument collapses.

The Chain Store paradox tells us that we can't always ignore the context in which games are played. Modern economists respond by trying to make the salient features

[1]In the counterfactual event that he were still alive!

[2]Selten repeated the game a hundred times to make this the most plausible explanation after Alice has fought many entrants in the past.

of the context part of the formal model. However, it isn't easy to model all the psychological quirks to which human players are prey!

### 14.2.2 Dividing by Zero?

In Bayesian decision theory, the problem of interpreting a counterfactual arises when one seeks to condition on an event $F$ that has zero probability. Since prob $(E \mid F) = \text{prob}(E \cap F)/\text{prob}(F)$, we are then given the impossible task of dividing by zero.

Kolmogorov's *Theory of Probability* is the bible of probability theory. When you would like to update on a zero probability event $F$, he recommends considering a sequence of events $F_n$ such that $F_n \to F$ as $n \to \infty$ but for which prob $(F_n) > 0$. One can then seek to define prob $(E|F)$ as

$$\lim_{n \to \infty} \text{prob}(E|F_n).$$

However, Kolmogorov warns against using the "wrong" events $F_n$ by giving examples in which the derived values of prob $(E|F)$ make no sense (Exercise 14.9.21). In the geometric problems that Kolmogorov considers, it isn't hard to see what the "right" value of prob $(E|F)$ ought to be, but game theorists aren't so fortunate. So how do they manage?

When Alice tells the Red Queen that she *can't* believe something impossible, she may well be right when talking about an action that nobody would ever take in a world of superhumanly rational players. But a zero probability event $F$ in this ideal world may correspond to an event $F_n$ that occurs with positive probability in some nearby possible world in which people sometimes make mistakes. One may then ask how people would act in this nearby possible world if $F_n$ were to occur and use *these* actions as approximations of how they would act if the impossible event $F$ were to occur in the world of superhumans.

### 14.2.3 Trembling Hands

Selten says that the players in a possible world in which mistakes are like typos have "trembling hands." The players always mean to choose rationally, but chance events outside their control intervene. Metaphorically, their hands tremble when they reach out to press their action buttons (Section 2.9.4). Selten gives the example of a male bird whose genes choose the strategy of helping his female partner to rear their fledglings but is shot by a hunter while on his way back to the nest.

We can build such psychological or physical limitations into the rules of an expanded game by introducing independent chance moves that select each of the $n$ available actions that a player *didn't* choose at an information set with some tiny probability $\varepsilon > 0$, leaving the action that the player *did* choose to be selected with the residual probability of $1 - n\varepsilon$. We then follow Kolmogorov's advice and envisage the world of superhuman players as the limit of this world of trembles as $\varepsilon \to 0$.

In the trembling world, the need to condition on zero probability events vanishes because every information set is reached with positive probability. In any Nash equilibrium, all of the players then necessarily optimize at all information sets. Our problems with backward induction therefore disappear.

The Nash equilibria of the expanded game converge on strategy profiles of the original game. Selten calls these strategy profiles *perfect equilibria,* but they are more commonly known as *trembling-hand equilibria*. When the context in which a game is played justifies confining our attention to such trembling-hand equilibria, we then have a watertight argument for throwing out weakly dominated strategies and Nash equilibria that aren't subgame perfect.[3]

The reason is simple. Weakly dominated strategies become strongly dominated when small enough positive trembles are introduced into a game, so rational players will never use them. Similarly, rational players will automatically play a Nash equilibrium in every subgame that is reached with positive probability. One has to verify that these properties survive in the limit, but the proof isn't very hard.

### 14.2.4 Correlated Trembles

When mistakes are like typos, it is easy to defend backward induction and the successive deletion of weakly dominated strategies. However, the defense depends on each mistake being *independent* of any other.

When mistakes can be thinkos, life becomes more difficult. We can still imagine that chance jogs the elbows of players when they reach for their action buttons, but now we must model the fact that Alice's past errors signal the kind of errors she is likely to make in the future (Section 2.9.4). The chance moves we introduce must therefore allow for correlation between trembles.

What happens when we go to the limit depends very much on what correlations between trembles we build into our expanded game. But if we are sufficiently inventive, we can make almost any Nash equilibrium of the original game emerge as the limit of equilibria of the expanded game—including equilibria that aren't subgame perfect or are weakly dominated.

*The Chain Store Paradox Again.* To illustrate this last point, we return to the Chain Store Game of Figure 14.1 to see how the Nash equilibrium (*fa*, *oi*) might not be eliminated when we go to the limit, although it is weakly dominated and not subgame perfect.

The simple trick is to expand the Chain Store Game by adding a new Chance move as in Figure 14.2(a). This Chance move occasionally replaces Alice with a robot player, who always fights, no matter what.[4] The strategic form of Figure 14.2(b) shows that (*fa*, *oi*) is always a Nash equilibrium of the expanded game and hence survives when we take the limit as $\varepsilon \to 0$.

*Should We Junk Backward Induction?* The preceding argument shows that we can't afford to throw any Nash equilibrium away without considering the context in which a game is played. So what good is backward induction?

[3]Although all subgame-perfect equilibria aren't trembling-hand perfect. They may even involve the use of weakly dominated strategies.

[4]For an expanded game in which all information sets are always visited with positive probability, further trembles must be added. This is why we look at (*fa*, *oi*) instead of (*ff*, *oo*), which would be eliminated if we added extra trembles in the natural way.

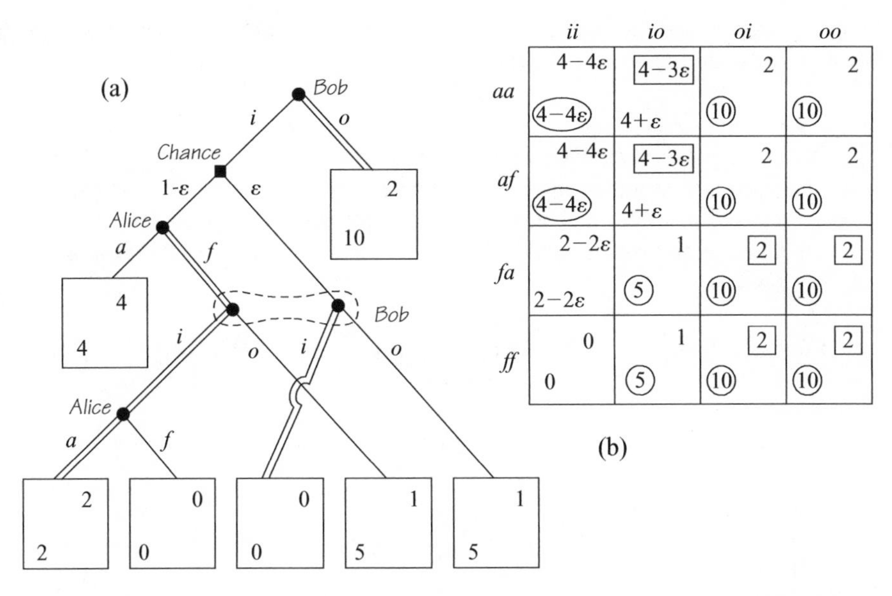


| | $ii$ | $io$ | $oi$ | $oo$ |
|---|---|---|---|---|
| $aa$ | $4-4\varepsilon$, $4-4\varepsilon$ | $4+\varepsilon$, $4-3\varepsilon$ | 10, 2 | 10, 2 |
| $af$ | $4-4\varepsilon$, $4-4\varepsilon$ | $4+\varepsilon$, $4-3\varepsilon$ | 10, 2 | 10, 2 |
| $fa$ | $2-2\varepsilon$, $2-2\varepsilon$ | 5, 1 | 10, 2 | 10, 2 |
| $ff$ | 0, 0 | 5, 1 | 10, 2 | 10, 2 |

Figure 14.2 Correlated trembles in the Chain Store Game. With probability $\varepsilon > 0$, the Chance move of Figure 14.2(a) replaces Alice with a robot that always fights. Figure 14.2(b) shows the strategic form of the game.

There are at least two reasons why it would be stupid to throw backward induction away. Trembling-hand equilibria are often exactly what the context in which a game is played demands. Many toy games, for example, are too short for correlated trembles to arise. Backward induction is then a sensible refinement of the Nash equilibrium concept. Even when backward induction isn't appropriate as a refinement, it can still be immensely useful as a tool for computing Nash equilibria. It is especially important to remember this fact when contemplating the monstrous strategic forms that result when a game is expanded by building in trembles, so that each information set will always be visited with positive probability. Any way of avoiding working out the payoffs in such strategic forms is to be welcomed with open arms!

## 14.3 Backward and Imperfect

How do we carry out backward induction in games of imperfect information? The idea of a subgame-perfect equilibrium is then inadequate because a subgame must be rooted in a single node. We get around this problem by prefixing each information set with a mythical chance move to act as the root of a mythical subgame (Section 13.2.3).

### 14.3.1 Assessment Equilibria

An *assessment* is a pair $(s, \mu)$ in which $s$ is a profile of behavioral strategies and $\mu$ is a profile of beliefs, one for each information set in the game. The idea of a subgame-perfect equilibrium is then generalized by requiring that an assessment satisfy two conditions to be in equilibrium:[5]

- Assuming that $s$ is used in the future, the strategy profile $s$ must require that each player plan to optimize at each information set, given the beliefs assigned to that information set by the belief profile $\mu$.
- Whenever possible, the belief profile $\mu$ must be obtained by Bayesian updating on the assumption that the players have used the strategy profile $s$ in the past.

*The Chain Store Paradox Yet Again.* We first show that the Nash equilibrium (*fa*, *oi*) in the expanded Chain Store Game of Figure 14.2(a) corresponds to an assessment equilibrium.

We focus on the information set at Bob's second move. There is no difficulty in seeing why the doubled lines in Figure 14.2(a) correspond to optimal choices at other moves in the game. Moreover, it is only at Bob's second move that we have a choice for the belief profile $\mu$ that must accompany (*fa*, *oi*), since all the other information sets contain only one node.

If Bob finds himself at his second move, he will consult the great book of game theory, where he will find that Alice fights at her first move. Bayesian updating therefore becomes trivial because Bob learns nothing from discovering that his opponent has chosen to fight. He therefore continues to assign a probability $\varepsilon$ to the right node in his information set. But with this belief, it is optimal for him to enter (because $2(1-\varepsilon) > 1$, when $\varepsilon$ is sufficiently small).

*Backward Induction?* The subgame-perfect equilibrium (*aa*, *ii*) of the original game isn't an assessment equilibrium of the expanded game. If it were, Bob would learn at his second move that his opponent must be the robot. He would then update the probability of the right node in his information set to 1. But with this belief, it is optimal for him to stay out.

However, we don't dispose of backward induction in the original game so easily. We can't find an assessment equilibrium of the expanded game that converges on the subgame-perfect equilibrium of the original game, but we can find an assessment equilibrium that converges on a Nash equilibrium of the original game that results in play moving along the backward induction path [*ia*].

In this assessment equilibrium, Bob enters at his first move, and Alice acquiesces at her second move. Their behavioral strategies at their other moves require that they mix. Alice is therefore indifferent between acquiescing and fighting at her first move. So Bob must stay out at his second move with a probability $q$ that makes $4 = 2(1-q) + 5q$. Thus $q = \frac{2}{3}$.

If it is optimal for Bob to mix, he must be indifferent between entering and staying out, given his beliefs at his second move. So he must think his left and right nodes are equally likely.

[5]David Kreps and Bob Wilson called this idea a sequential equilibrium, but I have thrown away some of their auxiliary assumptions. Sometimes people speak of a perfect Bayesian equilibrium, but an assessment equilibrium is neither perfect nor Bayesian in the sense in which these terms are normally applied to equibria.

We have made it optimal for Alice to fight with some probability $p$ at her first move. After Bob looks this fact up in the great book of game theory, he will update his beliefs at his second move. Proceeding as in the Monty Hall Game, he updates the probability of his right node to $\varepsilon/(p(1-\varepsilon)+\varepsilon)$ (Section 13.2.3). But to keep things in equilibrium, we know that this must be equal to $\frac{1}{2}$. Thus, $p=\varepsilon/(1-\varepsilon)$.

In our new assessment equilibrium, Bob enters for certain at his first move, and Alice acquiesces for certain at her second move. The belief profile makes Bob think each of his nodes is equally likely at his second move, where he stays out with probability $\frac{1}{3}$. Alice fights with probability $p=\varepsilon/(1-\varepsilon)$ at her first move. Allowing $\varepsilon \to 0$, we find that the limiting equilibrium path is [*ia*].

Note that Bob's beliefs at his second move in the limiting equilibrium can't be obtained by Bayesian updating since the strategy profile being used makes it impossible that the second move be reached. It is true that his beliefs were obtained by following Kolmogorov's advice, but our definition of an assessment equilibrium leaves us free to assign any beliefs whatsoever in such cases.

## 14.4 Gang of Four

The idea that we can model possible worlds in which players are subject to various psychological quirks was introduced by a gang of four economists.[6] Until psychology is better understood, using their methodology will remain more of an art than a science, but it has the major advantage that it forces us to build our underlying psychological assumptions into the rules of the games we construct—and hence to come clean on precisely what they are.

As in the expanded Chain Store Game of Figure 14.2(a), the idea is to introduce chance moves that decide who the players will be. With high probability, the individuals whom Chance selects to play are rational, but there is a small probability that one of a number of *irrational* types is chosen instead to fill any particular player's role. The irrationality of these types lies in their being modeled as automata who never question their program. Whether they are programmed with a clever or a stupid strategy, they stick with the strategy no matter what.

Their ideas have mostly been applied to problems in which the traditional models are too simple because they abstract away small imperfections that turn out to matter a lot more than seems likely at first sight. It can then be very misleading to look at the case in which $\varepsilon \to 0$ because the limiting case would be a good approximation to reality only if $\varepsilon$ were a great deal smaller than the already small values one might reasonably expect to see in practice.

The finitely repeated Prisoners' Dilemma is the leading example of such a case, but we focus on the simpler Centipede Game of Figure 5.14(a), so that we can look at more complex kinds of irrational behavior. Recall that Rosenthal's Centipede Game was introduced as a response to the holdup problem of classical economics (Section 5.6.2). How is it possible to trade when people can't trust each other? We continue to argue that the answer lies in assuming some small element of

[6]David Kreps, Paul Milgrom, John Roberts, and Bob Wilson. The real "gang of four" were high-ranking Chinese politicos who made a sensational attempt to flee from Beijing.

irrationality, but we now approach the problem with a less crude instrument than approximate Nash equilibria.

### 14.4.1 Your Reputation for Honesty

Equilibria that call on the players to cooperate in infinitely repeated games can be interpreted in terms of the players' need to sustain a reputation for honesty and fair play (Section 11.5.1). But what of finitely repeated games? In laboratory experiments, subjects do in fact cooperate a great deal when they play the Prisoners' Dilemma or a Cournot duopoly a fixed number of times with the same partner. Only toward the very end of the game do some people start stabbing their partners in the back.

One can explain this behavior by noting that subjects in a laboratory are unlikely to believe that the other subjects are all rational (Section 5.6). Most of the predictions of game theory don't change much if one follows the gang of four by introducing a small probability that an opponent is an irrational robot, but we are now talking about cases in which they do. For example, Adam and Eve will never cooperate in the finitely repeated Prisoners' Dilemma if we eliminate all possibility of irrational behavior (Section 11.3). However, the gang of four shows that if Adam and Eve believe that there is a very small probability that their opponent in the finitely repeated Prisoners' Dilemma is a robot playing TIT-FOR-TAT, then rational play is close to what we actually observe in laboratories.

The reason is very simple. Suppose the great book of game theory told Adam to play *hawk* for certain at some stage of the game at which the robot will play *dove*. If Eve then observes the play of *dove,* she will immediately update her probability that she is playing the robot to one—no matter how small her prior probability for this event may be. It is then optimal for her to play *dove* all the way up to the penultimate stage of the game. But Adam will predict this and so deviate from the book's advice by playing *dove* when the book says to play *hawk*. Eve will then mistake him for the robot and discover her error only when he stabs her in the back by playing *hawk* one stage before she was planning to do the same to him.

An equilibrium in the game must therefore require Adam and Eve to play *dove* with positive probability in the early stages of the game. The gang of four showed that this probability is high for most of the game and hence provided a plausible rationale for the observed behavior of laboratory subjects.

Of course, critics of game theory are impatient with the story. Intelligent critics say that people cooperate because it is important to maintain a reputation for reciprocating favors. But the gang of four doesn't disagree. On the contrary, the gang of four sees its role as explaining *why* it makes sense to behave as though you were committed to responding to each tit with a tat.

### 14.4.2 Centipede Game

Figure 14.3 shows how the gang of four might apply its technique to the Centipede Game of Figure 5.14(a). The initial chance move either chooses Adam to be player I with probability $1 - \varepsilon$ or else fills the role with one of a number of irrational types. These types differ in how long they cooperate before dropping out of the game. The probabilities assigned to the later chance moves reflect the frequency with which

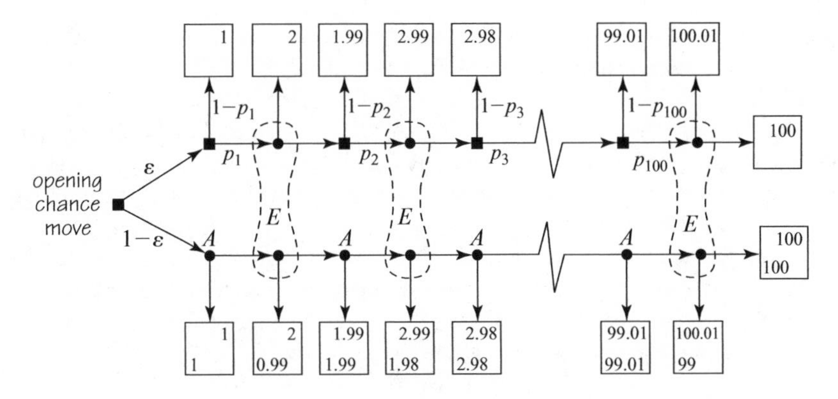


Figure 14.3 An expanded Centipede Game.

different irrational types are present in the population. Eve's information sets are drawn to ensure that she doesn't know whether she is playing Adam or some irrational type.

If the conditions are favorable, an assessment equilibrium of this expanded game can require both Adam and Eve to honor their deal by playing *across* for most of the game.

*An Assessment Equilibrium.* In the Centipede Game, Adam and Eve alternate in giving up one cent of their current payoff in order to increase their partner's payoff by one dollar. We emphasize this structure in Figure 14.4 by writing one cent as $c$ rather than 0.01, as in Figure 14.3.

To simplify our task, we make a heroic simplification by assuming that the probability with which the robot plays *across* is always 1 in the early part of the game, and 0 in the later part of the game. Figure 14.3 shows the robot playing *across* with probability 0 for the first time at the node at which Eve gets a payoff of $x$ when the

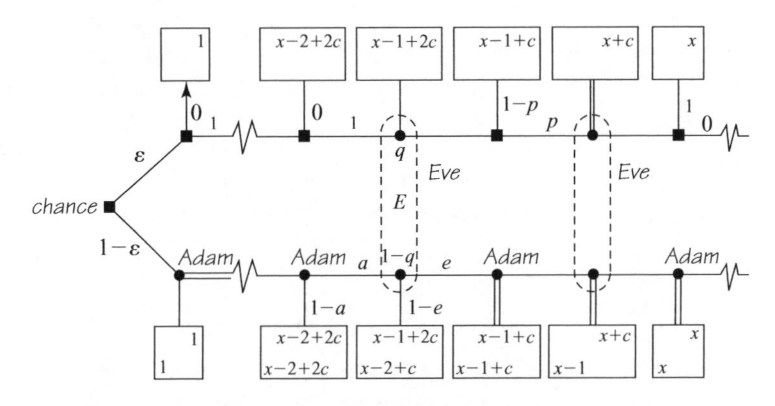


Figure 14.4 Looking for an equilibrium. The diagram shows part of the tree of the expanded Centipede Game of Figure 14.3. The doubled lines show actions that are taken in all the equilibria studied. When $p$ is small, it is necessary for Adam and Eve to mix in equilibrium. They then move with probabilities $a$ and $e$.

robot plays *down*. At the robot's previous move, it plays *across* with probability $p$. At its earlier moves, it plays *across* with probability 1.

The plan is to look for an assessment equilibrium in which Eve plans to play *down* the move before the robot would play *down* if it got the opportunity. Adam will then plan to play *down* the move before that. But if Adam is going to stab Eve in the back at this move, wouldn't it be better for her to deviate by stabbing him in the back one move earlier still?

The answer depends on what Eve believes when she arrives at the information set $E$ in Figure 14.4 that represents this earlier move. The belief profile in the assessment equilibrium we are seeking will assign a conditional probability $q$ to the event that she is playing the robot, given that she has arrived at $E$. This must be obtained by Bayesian updating on the assumption that everybody follows the advice offered in the great book of game theory.

If the book says that Adam will play *down* for sure at one of these earlier moves, then $q=1$ because Eve wouldn't then reach $E$ if she were playing Adam. Her optimal play at $E$ is then *across*. However, Adam would read this fact in the great book of game theory and respond by delaying playing *down* until his first move after $E$. So the book won't say that Adam will play *down* for sure at an earlier move because this behavior can't be in equilibrium.

Going to the other extreme, suppose that the book says that both Adam and Eve should always play *across* at earlier moves. Then the calculation of $q$ is very simple. Eve's arrival at $E$ gives her no more information than she had at the beginning of the game about whom she is playing, and so $q=\varepsilon$. Playing *across* at $E$ is then optimal for her if and only if

$$x-1+c \le q\{(1-p)(x-1-c)+p(x+c)\}+(1-q)\{x-1+c\}, \tag{14.1}$$

which reduces to the requirement that $p \ge c/q = c/\varepsilon$.

If $\varepsilon$ isn't too small and $p$ is sufficiently large, Eve will therefore play *across* at $E$. But then it becomes optimal for both Adam and Eve to play *across* at all previous moves, and so our assumption about what the great book of game theory might say turns out to be in equilibrium.

To summarize, if $\varepsilon$ isn't too small and $p$ is sufficiently large, we have found an assessment equilibrium in which Adam and Eve both cooperate by playing *across* in the early part of the game. Eve is planning to play *down* the move before the robot does this for sure. Adam plans to play *down* the move before Eve plans to play *down*. Eve knows that Adam will stab her in the back at this move but doesn't respond by stabbing him in the back at the move before because she thinks that she will get more on average by proceeding as though her opponent is the robot.

## 14.5 Signaling Games

Signaling games are the classic arena for testing out refinements. The fundamental problem is of great interest: How do words get their meaning?

Game theorists think of this question as a coordination problem. The human game of life has enormous numbers of equilibria. How is it possible that we succeed in coordinating on one of these equilibria rather than another after making noises at each other? How do we know that some noises are significant and others are just cheap talk? (Section 6.6.2)

One answer is that some signals are costly to send and so must convey some information about the intentions of the sender. Zahavi's Handicap Principle is an application of this idea in biology. Why do peacocks have such big tails? Because by sending this costly signal to peahens, they advertise that they are so fit that they can even compete when severely handicapped. Some skylarks similarly sing when pursued by young hawks. Older hawks don't bother because they have learned that the skylarks who sing are signaling that they don't need all their lung capacity to outfly a bird of prey.

Game theory's contribution to this debate is to provide models that pose the issues in a sharp form, but such models typically have many equilibria. Refinements of the Nash equilibrium concept were invented in an attempt to solve such equilibrium selection problems. We look at two such refinements, with a view to giving the flavor of the kind of counterfactual reasoning used in justifying them. I think such reasoning inadequate because it fails to make explicit the nature of the possible worlds used to interpret the counterfactuals, but other game theorists are less critical.

### 14.5.1 Burning Money

The Handicap Principle also applies in economics. Newspapers often complain that strikes that inconvenience the public are irrational. The firm and the workers are going to agree eventually, so why don't they agree before the strike instead of after? Many strikes are doubtless irrational, but there would still be strikes even if everybody were the soul of sweet reason. A strike or a lockout is a costly signal intended to communicate how strong you are.

A crude example of such a signal would be if Alice were to begin negotiating with Bob by taking a hundred dollar bill from her purse and burning it! She thereby conveys to Bob that she is rich enough to afford taking the risk that he won't accept a deal favorable to her.

The following model tries to capture this scenario using the Battle of the Sexes as a primitive model of bargaining.

*Forward Induction.* The honeymoon couple of Section 6.6.1 played the Battle of the Sexes without previously discussing the game. In the coming example, they get a limited opportunity to communicate. Before playing the version of the Battle of the Sexes shown at the top left of Figure 14.5, Adam can send a signal to Eve by taking two dollar bills from his wallet and burning them. The resulting game is shown schematically in Figure 14.5(a). The payoffs are measured in dollars on the assumption that the players are risk neutral.

Adam opens the game by choosing between $D$ (don't burn the money) and $B$ (burn the money). If he chooses the latter, two dollars is subtracted from each of his payoffs in the ensuing Battle of the Sexes. A reduced strategic form is shown in Figure 14.5(b). (The pure strategy $Bt$ for Adam means that he burns the money and then plays $t$ in the Battle of the Sexes. The pure strategy $rl$ for Eve means that she plays $r$ in the Battle of the Sexes if Adam doesn't burn the money and $l$ if he does.)

The Burning Money Game has several Nash equilibria, but all but $(Dt, ll)$ are eliminated by the forward induction refinement. If correct, this would be a remarkable

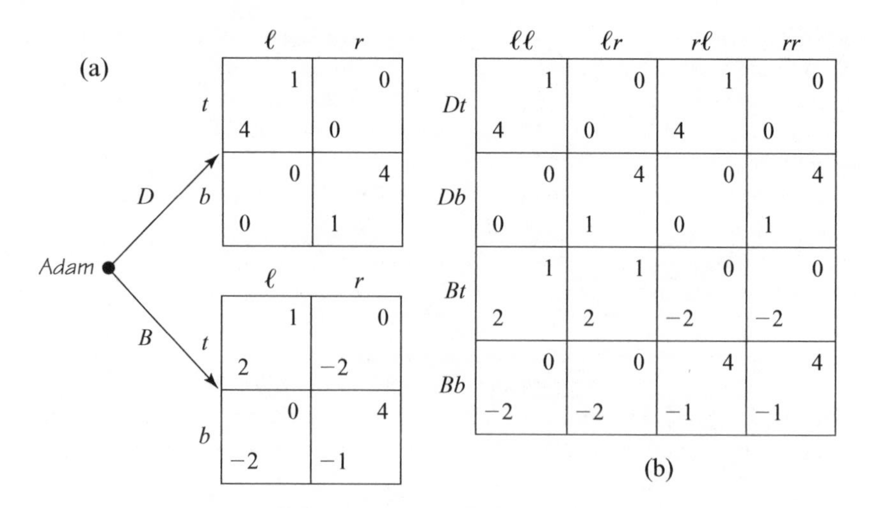


Figure 14.5 The Burning Money Game.

conclusion! When $(Dt, ll)$ is played, Adam's decision *not* to burn the money somehow convinces Eve that the right equilibrium on which to coordinate in the ensuing Battle of the Sexes is his favorite $(t, l)$ rather than her favorite $(b, r)$.

The forward induction argument goes like this:

**Step 1.** If Adam burns the money, he can't be thinking that the equilibrium $(b, r)$ will then be played in the Battle of the Sexes because he can be sure of getting more than $-1$ by not burning the money.

**Step 2.** If Adam burns the money, he must therefore be planning to play $t$ in the expectation that the equilibrium that is then played in the Battle of the Sexes will be $(t, l)$. His payoff would then be 2.

**Step 3.** If Adam *doesn't* burn the money, he can't be thinking that the equilibrium $(b, r)$ will then be played in the Battle of the Sexes because we have seen that he could get a payoff 2 rather than 1 by burning the money.

**Step 4.** If Adam doesn't burn the money, he must therefore be planning to play $t$ in the expectation that the equilibrium played in the Battle of the Sexes will be $(t, l)$. His payoff would then be 4.

**Step 5.** Whether he burns the money or not, we have seen that Adam will then play $t$. Eve's best reply is to play $ll$. Adam therefore does best by not burning the money.

One can also defend the selection of $(Dt, ll)$ by observing that it corresponds to successively deleting weakly dominated strategies in the order *Bb, rr, rl, Db, lr,* and *Bt*. However, we saw in Section 5.4.5 that the claim that it is necessarily irrational to use a weakly dominated strategy doesn't survive close scrutiny. Nor does the forward induction argument.

If the argument is right, then it isn't rational for Adam to burn the money. If he were to burn the money, Eve would therefore have a counterfactual to interpret. Step 2 assumes that she would interpret his counterfactual behavior as a rational act, but why wouldn't she simply deduce that Adam had behaved irrationally—and so is likely to behave irrationally in the future? In brief, the argument works only for possible worlds in which making a mistake consists of consulting game theory books written by authors who subscribe to alternative theories of equilibrium selection. But such an esoteric possible world is very distant from our own!

### 14.5.2 Real Men Don't Eat Quiche

Kreps's game of Quiche shown in Figure 14.6(a) takes us back to the fundamental problem of how signals acquire their meaning.

Chance begins by deciding whether Adam will be tough or wimpish. In either case, Adam then has to face Eve, who may choose to *bully* Adam or *defer* to him. She would defer to him if she knew he were tough, and she would bully him if she knew he were a wimp. But only Adam knows for sure the temperament with which he has been endowed by Nature. However, he can send signals to Eve by acting tough or behaving like a wimp. Here the signals are stylized as drinking *beer* or eating *quiche*. Tough guys prefer beer and wimps prefer quiche, but they won't necessarily consume what they prefer. For example, a wimp may conceal his distaste for beer in the hope of being mistaken for a tough guy.

The chance move chooses tough guys with probability $1-r=\frac{1}{3}$ and wimps with probability $r=\frac{2}{3}$. The information sets labeled TOUGH and WIMP show that Adam knows his own temperament. The information sets labeled QUICHE and BEER show that Eve knows only Adam's signal but not whether he is tough or wimpy. The payoffs are chosen so that Adam gets a bonus of 2 if Eve defers to him, plus a bonus of 1 if he avoids consuming something he dislikes. Eve gets a bonus of 1 for guessing right.

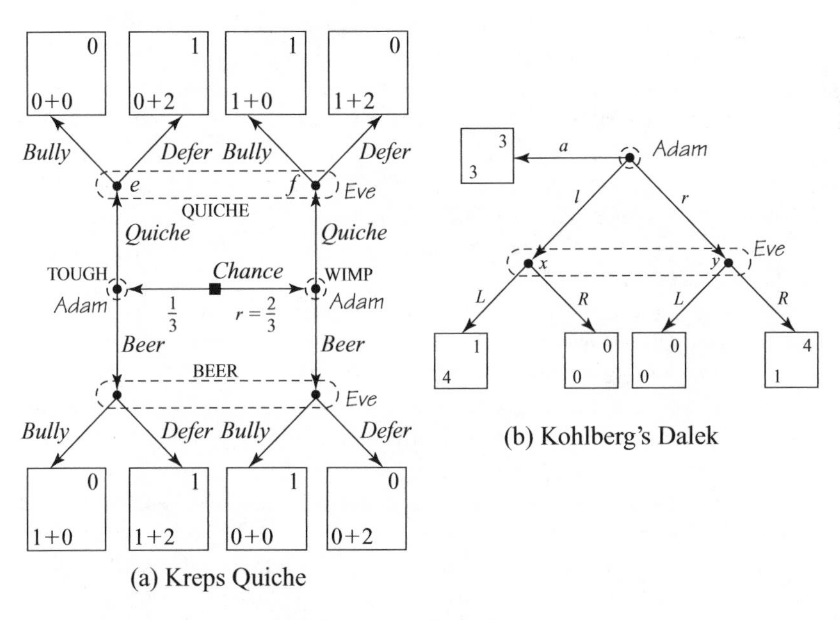


Figure 14.6 Two games of imperfect information.

The game has a unique assessment equilibrium in which Adam drinks beer for certain at TOUGH. Eve bullies for certain at QUICHE. Adam vacillates at WIMP. He eats quiche with probability $\frac{1}{2}$, in accordance with his wimpish nature. But with probability $\frac{1}{2}$, he drinks beer and hopes to be mistaken for a tough guy. This keeps Eve guessing at BEER. In equilibrium, she bullies with probability $\frac{1}{2}$ and defers with probability $\frac{1}{2}$.

To confirm that this behavior is in equilibrium, we first need to figure out Eve's beliefs at her two information sets. The unconditional probability that the game will reach the left node in Eve's information set QUICHE is 0. The unconditional probability of reaching the right node is $\frac{2}{3} \times \frac{1}{2} = \frac{1}{3}$. So the belief profile in the assessment equilibrium must assign a conditional probability of 0 to the left node and 1 to the right node. The unconditional probability that the game will reach the left node in Eve's information set BEER is $\frac{1}{3}$. The unconditional probability of reaching the right node is $\frac{2}{3} \times \frac{1}{2} = \frac{1}{3}$. So the belief profile in the assessment equilibrium must assign a conditional probability of $\frac{1}{2}$ to the left node and $\frac{1}{2}$ to the right node.

With these beliefs, it is optimal for Eve to *bully* at QUICHE. Since she is indifferent between her two choices at BEER, it is also optimal for her to use both with equal probability. It remains to verify that Adam is optimizing at his two information sets. At TOUGH, he gets zero from *quiche* and something positive from *beer*. So he chooses *beer*. At WIMP, the equilibrium requires that he mix. For this to be optimal, Adam must be indifferent between choosing *quiche* and *beer*. This is true because he gets 1 from the former and $\frac{1}{2} \times 0 + \frac{1}{2} \times 2 = 1$ from the latter.

### 14.5.3 Pooling and Separating Equilibria

The signals in the preceding equilibrium convey genuine information. Eve sometimes finds out whether Adam is tough or wimpy. Overgrown teenagers might say that the beer-drinking test "separates the men from the boys." Economists distinguish such *separating equilibria* from the *pooling equilibria* that we will encounter next. In a pooling equilibrium, different types of players always send the same signal and so can't be told apart. The signals then convey no information at all.

In interpreting such results, it is important to forget the significance that drinking beer or eating quiche have in a macho culture. Words and other signals mean what they mean only because a series of historical accidents has led society to coordinate on one equilibrium rather than another. If we wanted to know what the other possible equilibria might have been, we need to separate words from their familiar meanings in English and treat them as abstract signals.

If beer drinking is the "right" signal for a tough guy to send, this conclusion should emerge endogenously from the analysis. *After* an equilibrium has been computed, it may make sense to ask how the players interpret the signals they receive while using the equilibrium strategies. However, *before* the equilibria are computed, one isn't entitled to take for granted that any particular signal will be interpreted in any particular way. When communicating in an ironic or satirical mode, for example, we play by different rules than when simply communicating facts. We don't then expect the words we use to be taken literally. What we say is often the opposite of what we mean, and yet we would be astonished to be misunderstood.

Matters are usually less subtle in economic games, but the essential issues are just the same. What does it mean, for example, if the other firm lowers its price? Is it a sign of strength or weakness? Sometimes a study of the equilibria of the game allows an unambiguous interpretation of such signals to be made, as in the version of Quiche with $r = \frac{2}{3}$. However, when $r = \frac{1}{3}$, matters are much less simple.

### 14.5.4 Unreached Information Sets

When $r = \frac{2}{3}$ in Quiche, we don't have to worry about counterfactuals because all information sets are visited with positive probability in equilibrium. But things are different if $r = \frac{1}{3}$. One then obtains a game that typifies the difficulties that arise in signaling problems.

*Quiche Eaters Are Wimps.* We first look at an assessment equilibrium in which Adam drinks beer, whatever his type. Eve defers if she sees him drink beer and bullies if she sees him eat quiche.

Eve's behavior at BEER is optimal because she learns nothing about Adam's type from seeing him drink beer. Bayesian updating at BEER is therefore trivial, with the left node retaining a probability of $r = \frac{2}{3}$. It is then optimal for Eve to defer because it is more likely that Adam is tough than wimpy.

The problematic information set is QUICHE because this isn't reached in equilibrium. The definition of an assessment equilibrium then leaves us free to assign *any* probabilities to its nodes. We choose to assign a probability of 0 to the left node and 1 to the right node. In the counterfactual event that QUICHE is reached, we therefore assume that Eve will deduce that her opponent is a wimp, in which case it is optimal for her to bully him. But if she is going to bully anyone who eats quiche, then it is optimal for Adam to drink beer whatever his type.

*Beer Drinkers Are Wimps.* The general problem that Quiche was constructed to illustrate emerges when one realizes that we can swap the roles of beer and quiche in the preceding story. We then obtain an assessment equilibrium in which Eve deduces that Adam is a wimp in the counterfactual event that she sees him drinking beer. So both tough and wimpish Adams eat quiche!

*Intuitive Criterion.* The second of these assessment equilibria seems very perverse. The signals have somehow reversed their "natural" meaning.

Kreps proposed a refinement called the intuitive criterion that would eliminate the perverse equilibrium. If the players have the opportunity for an extended period of cheap talk before Quiche is played, he argues that a tough Adam would "refute" a supposedly authoritative game theory book that recommends the perverse equilibrium by making the following speech.

> *Adam:* I am a tough guy. The game theory book says that I am going to eat quiche. However, I plan to drink beer anyway. I'm telling you this so you don't take me for a wimp and bully me.
> *Eve:* How do I know that you aren't a wimp trying to fool me?
> *Adam:* You can tell I'm tough from the fact that I'm trying to persuade you not to follow the book, whereas a wimp would keep his mouth shut. If I succeed in persuading you, I will get a payoff of 3 after drinking beer, whereas I would get a payoff of only 2 if we all follow the book. On the other

> hand, a wimp would have no incentive to drink beer regardless of whether you are persuaded by my argument. According to the book he gets 3, whereas the most he can get after drinking beer is 2.

If I were Eve, I wouldn't be persuaded. My response would go like this:

> *Eve:* You argue that I should believe that you are tough because a wimp would have no grounds for challenging the authority of the game theory book. But if it were right for me to be persuaded by your argument, a wimp wouldn't keep his mouth shut. He would see this as just another opportunity for sending a signal and so make precisely the same speech as a tough guy would make. Any such speech is therefore just cheap talk.

Although this attempt to use a cheap-talk story to refine away the perverse equilibrium fails, it successfully focuses attention on the fact that the *cost* of making mistakes is different for tough guys and wimps. If this fact is significant for the kind of trembles that operate in the possible world that Eve uses to evaluate her counterfactuals, then we have a solid reason for eliminating the perverse equilibrium. If costly mistakes are less likely than cheap mistakes, then Eve's beliefs at BEER in the perverse equilibrium would assign a larger probability to the event that a beer drinker was tough. She would then defer and so destabilize the perverse equilibrium.

## 14.6 Rationalizability

We close this chapter by looking at a coarsening of the Nash equilibrium concept called *rationalizability*. Its proponents ask what can be said if we know only that it is common knowledge that the players are Bayesian rational. Their answer is that any strategy profile that survives the successive deletion of strongly dominated strategies is then possible.[7]

*The Rationalizing Argument.* Alice and Bob are playing a finite game. Each is uncertain about what pure strategy the opponent will finally use. A Bayesian rational player therefore assigns a subjective probability to each of the possible alternatives. The players then choose a strategy to play that maximizes their expected payoffs with respect to these subjective probabilities. Each player therefore behaves as though choosing a best reply to one of the opponent's mixed strategies.

Let Bob's set of mixed strategies be $M$. Then Alice will necessarily choose a strategy from the set $\mathscr{B}M$ of best replies to strategies in $M$. If Bob knows that Alice is Bayesian rational, he will therefore choose a strategy from the set $\mathscr{B}^2M = \mathscr{B}(\mathscr{B}M)$ of best replies to strategies in $\mathscr{B}M$. If Alice knows that Bob knows that she is Bayesian rational, she will therefore choose a strategy from $\mathscr{B}^3M$, and so on in a style that will by now be familiar.

The proof is completed by observing that a mixed strategy is a best reply to some mixed strategy choice of the opponent if and only if it isn't strongly dominated

[7]Sometimes it is argued that one should also be allowed to delete *weakly* dominated strategies, but only at the first round of deletions.

(Exercise 14.9.20). It follows that the strategy profiles that can be played when it is common knowledge that the players are Bayesian rational are those that survive the successive deletion of strongly dominated strategies.

*Discussion.* Some games have no strongly dominated strategies at all. In such games, *all* strategies are rationalizable. Does this mean that we can forget about the Nash equilibria of such games and simply tell the players that anything goes? One would have to answer *yes* to this question if it really were the case that the only piece of common knowledge available about the players was the fact that they are Bayesian rational. However, a theory that assumed only this would be very barren. Real people, even total strangers from different countries, have a great deal more in common than rationalizability assumes. The mere fact that we are all human ensures that we share a common culture to some extent.

Orthodox game theory captures this insight, albeit crudely, by discussing what should be written in the great book of game theory, with the implicit understanding that what is written there will be common knowledge among the players. More generally, the implicit assumption behind much game-theoretic reasoning is that the way to behave in a game—the conventional thing to do—is somehow common knowledge among the players. Attention can then be concentrated on those commonly understood conventions that aren't self-destabilizing. These are the conventions that select *equilibria*. Like all idealizations, the assumption that the convention in use is common knowledge will sometimes be wildly inappropriate. However, I can't think of any real situations at all in which it makes sense to assume common knowledge of Bayesian rationality and nothing else.

## 14.7 Roundup

A subgame-perfect equilibrium is a refinement of the Nash equilibrium concept because not all Nash equilibria are subgame perfect. Like other refinements that have been proposed, it requires that the counterfactuals that arise when a rational player deviates from rational play be interpreted in a particular way.

One way of making sense of such counterfactuals is to expand a game by introducing chance moves or trembles that result in all information sets always being visited with positive probability. Counterfactual beliefs in the original game are then interpreted as the limits of the beliefs in the expanded game as the probability of trembling goes to zero. If the trembles are independent, one can then justify subgame perfection and the iterated deletion of weakly dominated strategies. If different trembles are correlated—as when mistakes are thinkos rather than typos—then a justification can be found for almost any Nash equilibrium. The Chain Store paradox provides an example of a game in which a backward induction analysis leading to a subgame-perfect equilibrium fails to be compelling.

Backward induction nevertheless remains a valuable tool, especially when it is applied to games of imperfect information using the idea of an assessment equilibrium. An assessment is a pair $(s, \mu)$ in which $s$ is a profile of behavioral strategies and $\mu$ is a profile of beliefs. For an equilibrium, $s$ must specify optimal play at every information set, given the beliefs assigned by $\mu$ to the nodes in that information set. Wherever possible, these beliefs must be obtained by Bayesian updating on the assumption that $s$ was used in the past. The belief profile may assign any beliefs at all at any information set that can't be reached if $s$ is used.

The gang of four pointed out that the interesting case sometimes arises before the trembles go all the way to zero. In the case of the Centipede Game, one can then find assessment equilibria in which the players behave as though they trust each other until late in the game. Similar conclusions can be obtained in the finitely repeated Prisoners' Dilemma.

Signaling games are the classic arena for the deployment of refinements. Forward induction supposedly eliminates all but one of the equilibria in the Burning Money Game. The intuitive criterion similarly eliminates the perverse equilibrium in Quiche. Whether such refinements make sense depends on the extent to which the implicit assumptions they make about how counterfactuals are to be interpreted fit the context in which a game is played.

Rationalizability is a coarsening of the Nash equilibrium concept. From the fact that it is common knowledge that the players are all Bayesian rational, we can justify only the successive deletion of strongly dominated strategies. But why should we assume that the players know nothing else?

## 14.8 Further Reading

*Counterfactuals*, by David Lewis: Harvard University Press, Cambridge, MA, 1973. The book successfully popularizes Leibniz's idea of possible worlds but isn't very useful as a guide to using counterfactuals in practise.

*The Situation in Logic*, by Jon Barwise: Center for the Study of Language and Information, Stanford, CA, 1989. Another take on counterfactuals and common knowledge.

*Game Theory for Applied Economists*, by Robert Gibbons: Princeton University Press, Princeton, NJ, 1992. An unfussy introduction to game theory, with an orthodox treatment of refinements.

## 14.9 Exercises

phil

1. In the Surprise Test paradox, anything follows from a contradiction, and so it isn't surprising when the teacher gives the test on Monday (Section 2.3.1). Why isn't the contradiction a counterfactual?
2. Show that more than the subgame-perfect equilibrium strategies survive after successively deleting weakly dominated strategies in the version of the Chain Store Game of Figure 14.1 but that all the surviving strategy profiles result in play moving along the backward induction path.

phil

3. Construct a toy game in which a male bird can either return to the nest with food or abandon the female bird to look for another partner. If he doesn't return to the nest, the female bird can either continue to incubate the eggs without support or seek a new partner. With appropriate payoffs, backward induction will keep the male bird faithful. But birds aren't famous for their capacity to analyze subjunctive conditionals. Is it reasonable to defend backward induction as a trembling-hand equilibrium in such a biological context?

math

4. Figure 5.2(b) shows the strategic form of a three-player game. Which strategies are dominated? What are the two Nash equilibria in pure strategies? Show that

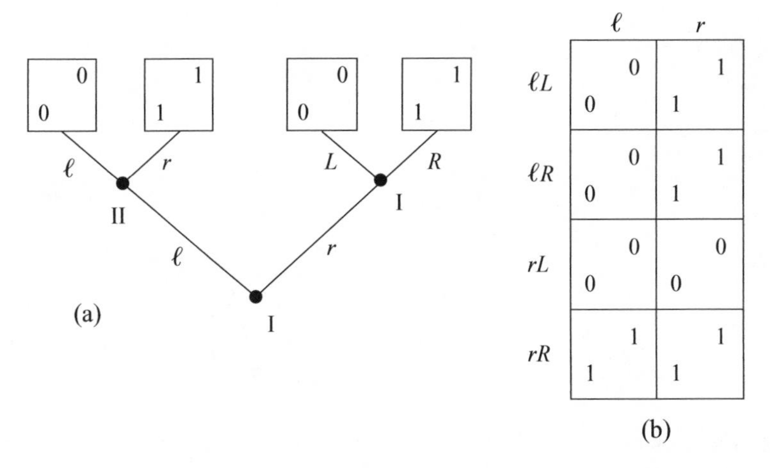


Figure 14.7 How many angels? The trembling-hand equilibria of this extensive form don't coincide with the trembling-hand equilibria of the simultaneous-move game with the same strategic form.

neither requires the use of a weakly dominated strategy but that only one is a trembling-hand equilibrium.

5. Von Neumann and Morgenstern proposed that one should normally discard the extensive form of a game in favor of its strategic form. This amounts to saying that all games are essentially simultaneous-move games. Explore this claim by studying the new game obtained by allowing the players of a given game to begin by simultaneously committing themselves to a strategy. Confirm that the extensive form of Figure 14.7(a) has the strategic form shown in Figure 14.7(b). Write the extensive form of a simultaneous-move game that has the same strategic form. Show that there is a trembling-hand perfect equilibrium in the latter that isn't a trembling-hand perfect equilibrium of the original game.

phil

6. Finish the analysis of the Monty Hall Game by backward induction begun in Section 13.2.3. Express your answer as an assessment equilibrium. Compare this equilibrium with previous analyses of the game.
7. Change Bob's payoff in the bottom left cell of Figure 14.2(a) from 2 to 3. Explain why (*fa*, *oi*) is no longer part of an assessment equilibrium. What of the other assessment equilibrium found in Section 14.3.1?
8. Selten's Horse Game is shown in Figure 14.8. Verify that $(d, A, l)$ and $(a, A, r)$ are Nash equilibria. Why is it obvious that both are also subgame perfect? Show that $(d, A, l)$ can't be part of an assessment equilibrium but $(a, A, r)$ can.
9. In Section 14.4.2, the Centipede Game provides the basis for an example using the gang-of-four methodology. Draw an extensive form in which Chance may also choose an irrational type to be player II.
10. Find an assessment equilibrium of the modified Centipede Game of Section 14.4.2 on the assumption that the robot *always* plays *across* with probability $p$.

math

11. What possible world was implicitly evoked in deciding on the beliefs at unreached information sets in the assessment equilibrium of Section 14.4.2?
12. Find all pure Nash equilibria of the Burning Money Game (Section 14.5.1).

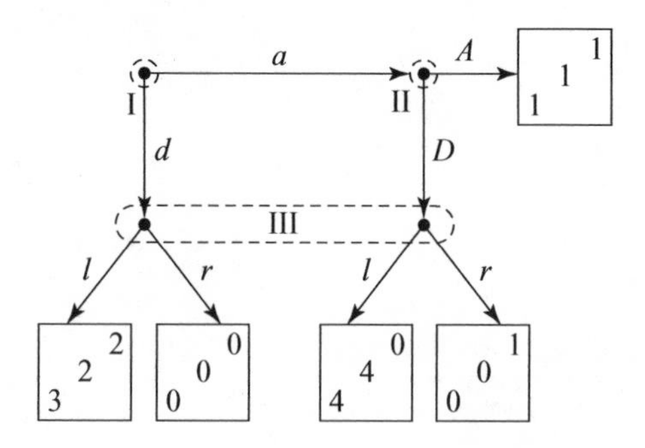


Figure 14.8 Selten's Horse Game.

13. Confirm that the strategic form for the game of Quiche with $r = \frac{2}{3}$ is as shown in Figure 14.9. Use the strategic form to find the unique Nash equilibrium of the game (Section 14.5.2).
14. Kohlberg's Dalek Game is given in Figure 14.6(b). Why is there an equilibrium selection problem? How would the forward induction refinement resolve this problem? (Laboratory experiments show that real player I's end up playing *a*.)
15. Give an example of a $2 \times 2$ bimatrix game in which *every* pair of pure strategies is rationalizable.
16. Obtain a unique pair of pure strategies by applying the rationalizability criterion to the bimatrix game of Figure 14.10. (Don't forget that mixed strategies may be relevant.)

econ

17. In the Cournot Duopoly Game of Section 10.2.2:
    a. Confirm that each player's profit is strictly concave as a function of the player's output. Deduce that mixed strategies are never best replies in the game and hence can be ignored.
    b. Use Figure 10.1 to assist in drawing a large diagram showing both Alice's reaction curve $a = R(b)$ and Bob's reaction curve $b = R(a)$, where $R(x) = \frac{1}{2}(K - c - x)$ for $0 \leq x \leq K - c$.

| | *bb* | *bd* | *db* | *dd* |
|---|---|---|---|---|
| *qq* | $\frac{2}{3}$, $\frac{2}{3}$ | $\frac{2}{3}$, $\frac{2}{3}$ | $\frac{8}{3}$, $\frac{1}{3}$ | $\frac{8}{3}$, $\frac{1}{3}$ |
| *qb* | 0, $\frac{2}{3}$ | $\frac{4}{3}$, 0 | $\frac{2}{3}$, 1 | 2, $\frac{1}{3}$ |
| *bq* | 1, $\frac{2}{3}$ | $\frac{5}{3}$, 1 | $\frac{7}{3}$, 0 | 3, $\frac{1}{3}$ |
| *bb* | $\frac{1}{3}$, $\frac{2}{3}$ | $\frac{7}{3}$, $\frac{1}{3}$ | $\frac{1}{3}$, $\frac{2}{3}$ | $\frac{7}{3}$, $\frac{1}{3}$ |

Figure 14.9 The strategic form of Quiche.

| | | | |
|---|---|---|---|
| 5, 2 | 2, 6 | 1, 4 | 0, 3 |
| 4, 1 | 3, 4 | 2, 1 | 1, 2 |
| 1, 0 | 1, 1 | 1, 5 | 5, 1 |
| 2, 3 | 0, 1 | 0, 2 | 4, 4 |

Figure 14.10 A game to rationalize.

c. Let $x_0 = 0$, and define $x_{n+1} = R(x_n)$ for $n = 0, 1, \ldots$. If $x_n \to \tilde{x}$ as $n \to \infty$, explain why $\tilde{x} = R(\tilde{x})$. Deduce that $(\tilde{x}, \tilde{x})$ is the unique Nash equilibrium for the Cournot Duopoly Game computed in Section 10.2.2.

d. Mark $x_1$ on both players' axes in the diagram drawn for part (b). Explain why it is never a best reply for either player to produce more than $x_1$. Erase the part of your diagram with $a > x_1$ or $b > x_1$.

e. Mark $x_2$ on both players' axes in your diagram. Explain why it is never a best reply for either player to produce less than $x_2$ if it is known that the strategy choices erased in part (d) will never be used. Erase the part of your diagram with $a < x_2$ or $b < x_2$.

f. Mark $x_3$ on both players' axes in your diagram, and then decide which part of your diagram should now be erased.

g. Explain why an output $q$ that never gets erased in the process whose initial three steps are described above satisfies $x_{2n} \le q \le x_{2n+1}$.

h. Deduce that the only *rationalizable* strategy pair for the Cournot Duopoly Game is the unique Nash equilibrium $(\tilde{x}, \tilde{x})$.

18. Player I has payoff matrix $A$ in a finite, two-player game. Explain why his mixed strategy $\tilde{p}$ is a best reply to some mixed strategy for player II if and only if

math

$$\exists q \in Q \; \forall p \in P \; (\tilde{p}^\top A q \ge p^\top A q),$$

where $P$ is player I's set of mixed strategies and $Q$ is player II's set of mixed strategies.[8] Why is the above statement equivalent to

$$\min_{q \in Q} \max_{p \in P_0} p^\top A q \le 0,$$

where $P_0 = \{p - \tilde{p} : p \in P\}$?

[8]The notation "$\exists q \in Q$" means "there exists a $q$ in the set $Q$ such that." The notation "$\forall p \in P$" means "for any $p$ in the set $P$."

math

19. With the notation of the previous exercise, explain why player I's mixed strategy $\tilde{p}$ is strongly dominated (possibly by a mixed strategy) if and only if

$$\exists p \in P \, \forall q \in Q \; (p^\top Aq > \tilde{p}^\top Aq).$$

Deduce that $\tilde{p}$ is *not* strongly dominated if and only if[9]

$$\forall p \in P \; \exists q \in Q \; (p^\top Aq \leq \tilde{p}^\top Aq).$$

Why is the second statement equivalent to

$$\max_{p \in P_0} \min_{q \in Q} p^\top Aq \leq 0?$$

math

20. Use the previous exercises to show that a mixed strategy in a finite, two-player game is a best reply to some mixed-strategy choice by the opponent if and only if it isn't strongly dominated. (You will need to appeal to Von Neumann's minimax theorem with $P$ replaced by $P_0$.)

phil

21. What is the probability $p$ that a randomly chosen chord to a circle is longer than its radius? If the midpoint of the chord is a distance $x$ from the center of the circle, show that the length of the chord is $2\sqrt{r^2 - x^2}$, where $r$ is the radius of the circle. Deduce that

$$p = \text{prob}\left(\frac{x}{r} \leq \frac{\sqrt{3}}{2}\right).$$

   a. The midpoint of a randomly chosen chord is equally likely to lie on any radius. Given that it is on any particular radius, it is equally likely to be at any point on the radius. Show that $p = \sqrt{3}/2$.
   b. The midpoint of a randomly chosen chord is equally likely to be anywhere inside the circle. Show that $p = 3/4$.
   Discuss the relevance of these results to Kolmogorov's advice on how to condition on a zero probability event (Section 14.2.2).

phil

22. Deepen the paradox of the previous exercise by considering the implications of assuming that one endpoint of the chord is equally likely to be anywhere on the circumference of the circle and that the second endpoint is equally likely to be anywhere else on the circumference, given that the first endpoint has been chosen. Discuss the implications for the principle of insufficient reason (Section 13.5.1).

[9]Why is it true that "not ($\exists \, p \, \forall q \ldots$)" is equivalent to "$\forall \, p \, \exists p \, q$ (not...)"?

# 15
# *Knowing What to Believe*

## 15.1 Complete Information

Information is perfect when the players always know everything that has happened so far in the game. Information is complete when everything needed to specify a game is common knowledge among the players—including the preferences and beliefs of the other players.

So far, we have always assumed that information is complete, although such a strong hypothesis isn't always necessary. In the Prisoners' Dilemma, the players need know only that *hawk* is a strongly dominant strategy in order to optimize, but changing the payoffs only a little yields the game of Chicken, in which we certainly do need complete information to get anywhere.

When is it reasonable to assume that information is complete? A game like chess creates no problem, but Russian Roulette in another story. Is it likely that Boris will know exactly how averse Vladimir is to risk? In a Cournot duopoly, both firms maximize profit, but how reliably can one firm estimate the profit of another? In real life, firms keep their costs a closely guarded secret in the hope of making life harder for their competitors.

John Harsanyi's theory of incomplete information is a way of getting a handle on such problems. It is a technique for *completing* a strategic structure in which information is incomplete. The theory leaves a great deal to the judgment of those who use it. It points a finger at what is missing in an informational structure but doesn't say where the missing information is to be found. What it offers is the right questions. Coming up with the right answers is something that Harsanyi leaves to you and me.

If we can answer the questions adequately, the result is a game of *imperfect* information. To say that information is *complete* in this game is superfluous. If the information structure hadn't been completed, it wouldn't be a game. The "games of incomplete information" one reads about in economic textbooks are really games of complete but imperfect information obtained by applying Harsanyi's methodology to an incomplete information structure.

## 15.2 Bluffing

Von Neumann's analysis of poker was what made me into a game theorist. I knew that good poker players had to bluff frequently, but I just didn't believe that it could be optimal to bluff as much as Von Neumann claimed. I should have known better than to doubt the master! After much painful calculation, I not only had to admit that he was right but also found myself hopelessly hooked on game theory ever after.

*Von Neumann's Model.* Von Neumann's second poker model will set the scene for our general approach to the problem of incomplete information. In this model, two risk-neutral players, Alice and Bob, are independently dealt a real number between 0 and 1. All possible deals are equally likely.

Before the deal, each player must put an ante of $1 into the pot. After the deal, there is a round of betting, during which Bob may *fold.* If he folds, then Alice wins the pot, no matter who has the better hand. If Bob doesn't fold, there is a showdown, after which the player with the higher card wins the pot. The showdown occurs when Bob *calls* Alice's last bet by making his total contribution to the pot equal to hers.

To keep things simple, Von Neumann restricted the betting possibilities very severely. In his model, Alice moves first. She can either *check* (by adding $0 to the pot) or *raise* (by adding $1 to the pot). If she checks, Bob must call. If Alice raises, Bob has a choice. He can fold or call.

Figure 15.1(a) shows the players' optimal strategies in Von Neumann's model. Everybody who plays nickel-and-dime poker knows that Alice must sometimes raise with poor hands; otherwise, Bob will learn never to call when she raises with a good hand. Amateurs try to compromise by bluffing with middle-range hands, but Von Neumann has no time for such timidity. If you want to win at poker against good opposition, bluff a lot with really bad hands!

*A Simplified Model.* Von Neumann's model simplifies poker a lot while still capturing the essence of the game. The next model simplifies even more by replacing his infinite deck of cards by one containing only the king, queen, and jack of hearts. However, Figure 15.1(b) shows that optimal play still has the same character as in Von Neumann's model.

The chance move that begins the game tree of the simplified model in Figure 15.2 represents the dealer shuffling the deck into one of six equally likely orders. The top card is then dealt to Alice and the second card to Bob. The rest of the game tree then shows Von Neumann's betting rules in operation with the new deck of cards.

As with Russian Roulette, this two-person, zero-sum game will be solved in two different ways (Sections 4.7 and 5.2.2). We first look for Nash equilibria in the strategic form, and then we attack the extensive form using backward induction. But first

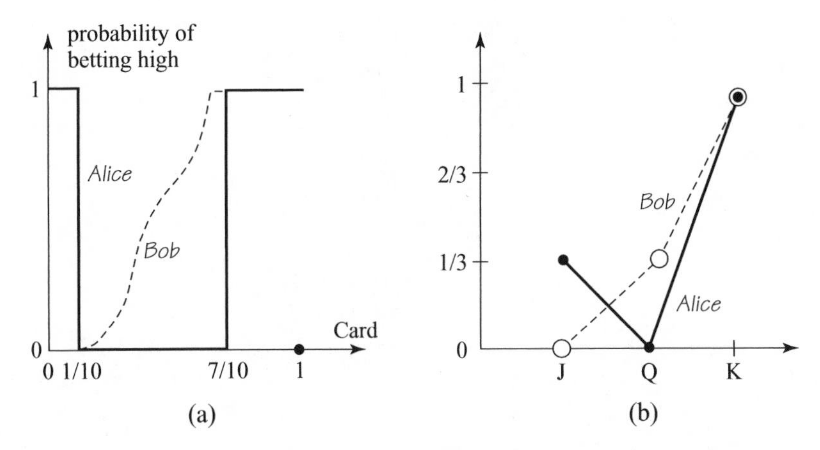


Figure 15.1 Optimal play in Von Neumann's Poker model. Figure 15.1(a) shows how the probability with which Alice or Bob should bet high varies as a function of their hands. (Bob has multiple optimal strategies, of which only one is shown.) Figure 15.1(b) shows that optimal play in the simplified version studied in the text has a similar character.

some obvious reductions can be made. These are indicated by doubling appropriate lines in Figure 15.2. For example, Alice checks when holding the queen because Bob calls only when he has her beaten. Only two decisions then remain in doubt. Does Alice bluff when holding the jack? Does Bob call when holding the queen?

Figure 15.3(a) shows all the pure strategies of the game, but only the shaded part of the strategic form matters (strategies that don't correspond to the shaded part are dominated). Figure 15.3(b) is a close-up of the shaded part. This game has a unique Nash equilibrium in which Alice plays *RCR* (raises when holding the jack) with probability $r = \frac{1}{3}$. Bob plays *CCF* (calls when holding the queen) with probability $c = \frac{1}{3}$. Alice's expected payoff is then $\frac{1}{18}$.

The strategic-form analysis looks simpler than it really is because the computation of the relevant payoffs was suppressed. Such computations can be avoided altogether by tackling the extensive form directly.

Bob must mix when holding the queen, and so he is indifferent between folding and calling. The unconditional probability that the left node of his information set $Q$ is reached is $\frac{1}{6}$. The unconditional probability that the right node is reached is $\frac{1}{6}r$, where $r$ is the probability that Alice raises when holding the jack. The relevant conditional probabilities are therefore $1/(1+r)$ and $r/(1+r)$. So Bob is indifferent between his two actions at $Q$ if

$$-1 = -2/(1+r)+2r/(1+r),$$

which implies that $r = \frac{1}{3}$.

Alice must similarly mix when holding the jack. If Bob calls with probability $c$ when holding the queen, she is indifferent between her two actions at $J$ if

$$-1 = \tfrac{1}{2}(-2)+\tfrac{1}{2}\{(1-c)-2c\},$$

which implies that $c = \frac{1}{3}$.

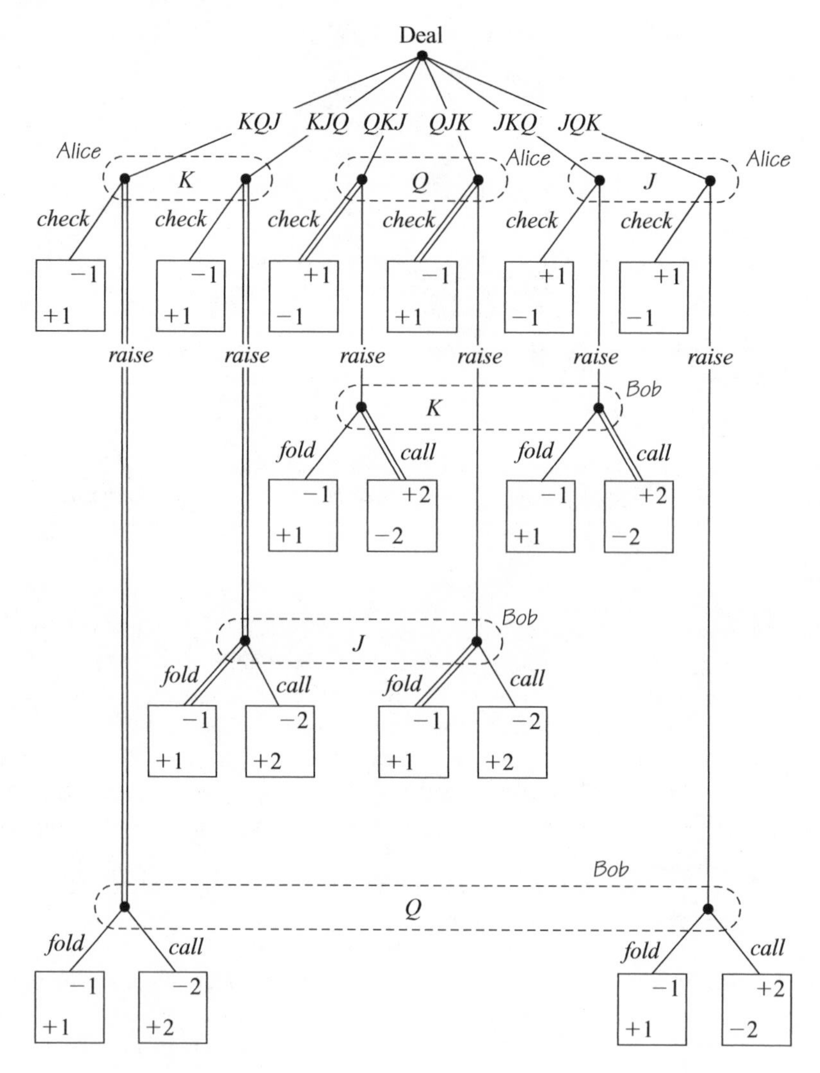


Figure 15.2 A simplified version of Von Neumann's second Poker model.

→ 15.3

*Risking Your Shirt.* Players at the world poker championships in Las Vegas play a lot more like Von Neumann recommends than amateurs like you and me. But you're in for a disappointment if you think that boning up on mathematical models will allow you to emulate the triumphs of legendary players like the great Amarillo Slim. You may work out the security strategy for whatever poker variant the dealer proposes, but playing this strategy will yield only an expected profit of zero in a fair game. To make money at the Poker table, you need to be more adventurous. You must actively seek out and exploit the psychological flaws of your opponents. However, unless you are a natural master of human psychology like Amarillo Slim, your nerdish attempts to exploit the flaws of others are likely to end up with them exploiting yours instead!

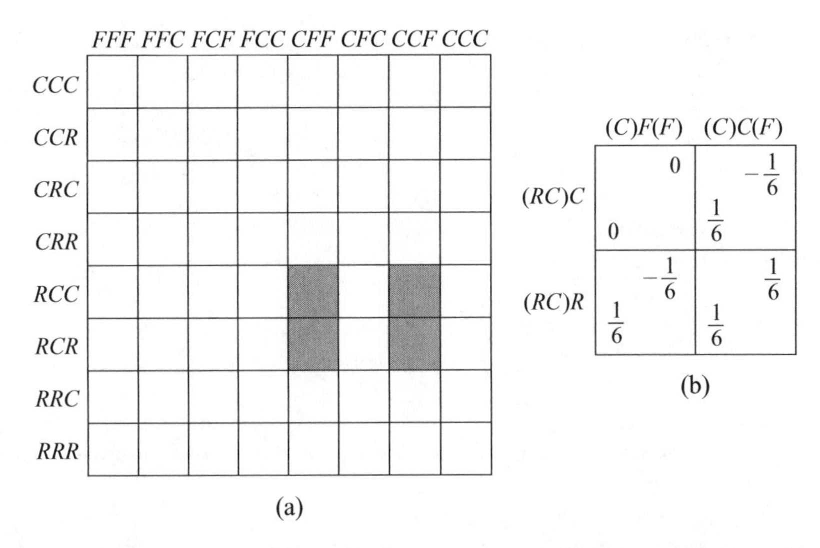


Figure 15.3 The strategic form for the simplified version of Von Neumann's second Poker model. In Figure 15.3(b), Alice chooses between checking or raising when holding the Jack. Bob chooses between folding or calling when holding the Queen.

## 15.3 Incomplete Information

According to the grumpy philosopher Thomas Hobbes, man is characterized by his strength of body, his passions, his experience, and his reason. In game theory, Pandora's strength of body is determined by the rules of the game. Her reason leads her to honor the principles of Bayesian rationality. The remaining properties translate into her preferences and her beliefs. For information to be complete, three of Hobbes's criteria need to be common knowledge:

- the rules of the game
- the players' preferences over the possible outcomes of the game
- the players' beliefs about the chance moves in the game

Harsanyi's method provides a possible approach when information on one or more of these issues is incomplete. We focus on incomplete information over preferences, but it turns out that one can't discuss incomplete information at all without also taking on board incomplete information about beliefs.

### 15.3.1 Typecasting

If we remove any reference to the players' preferences from the description of a game, the result is called a game form or a mechanism, but it brightens things up to use the language of the theater. We can then think of the rules of a game as a *script.* In a two-player game, the script will have *roles* for two *actors.*

A set $\mathscr{M}$ of out-of-work male actors and a set $\mathscr{F}$ of out-of-work female actors turn up to be auditioned. Chance is the casting director. In her casting move, she selects one of the actors from the set $\mathscr{M}$ to fill the role of player I and one of the actors from the set $\mathscr{F}$ to fill the role of player II. An actor who has been cast in one of the roles will know this fact but not who has been cast in the opposing role. Actors must therefore decide how to play their part in ignorance of the identity of their opponent.

The strategy the actors choose depends on their *type*. All of the actors are Bayesian rational, and so a type is determined by an actor's preferences and beliefs:

*Preferences.* An actor's preferences are specified by a Von Neumann and Morgenstern utility function defined over the set of possible outcomes.

*Beliefs.* An actor's beliefs are specified by the subjective probabilities the actor assigns to the choices available to Chance at the casting move.

Specifying the type of each possible actor completes the incomplete informational structure with which we started. The result is a game of imperfect information that we shall call a *Bayesian game*.

The root of the Bayesian game is the casting move. Information sets show that actors know when they are chosen to play but not who the other actors are. The probabilities for the choices available to Chance at the casting move are built into the players' beliefs. It is usual—but not necessary—to assume that these beliefs are all the same (Section 13.7.2). Making such a *common prior* assumption doesn't put all the actors into the same informational state because the first thing they will do after arriving at the information set where they learn that they have been chosen to occupy the role of a player in the game is to use Bayes's rule to *update* their prior.

*Bayes-Nash Equilibrium.* A Nash equilibrium in a Bayesian game is called a *Bayes-Nash equilibrium.* The *Bayes* before the hyphen reminds you that a Bayesian game is involved. You therefore need to find out who the actors are. What are their preferences and beliefs? The *Nash* after the hyphen tells you what calculations then need to be made.

Economists often refer to "Bayesian equilibria in games of asymmetric information" when talking about Bayes-Nash equilibria. Information is said to be *asymmetric* because the actors know different things after they are selected to play. The equilibrium is said to be *Bayesian* because the actors use Bayes's rule to update their prior beliefs. I dislike this terminology for the same reason I dislike talking about "Stackelberg equilibria" when referring to subgame-perfect equilibria in Stackelberg games (Section 5.5.1).

→ 15.3.2

*No Infinite Regress.* The cleverness of Harsanyi's formulation becomes apparent only when one realizes how it avoids an infinite regress. This feat is achieved by making the heroic assumption that information *is* complete in the Bayesian game that one constructs from an incomplete information structure.

To see why an infinite regress may arise, consider Alice and Bob after the deal in poker. Alice doesn't know what hand Bob is holding. Bob doesn't know what Alice believes about the hand he is holding. Alice doesn't know what Bob believes about what she believes about the hand he is holding, and so on. The chain of beliefs about beliefs is closed in poker by assuming that the chance move that represents shuffling and dealing the cards is common knowledge. Harsanyi does the same when information is incomplete by assuming that the casting move is common knowledge.

Although this is a subject that is seldom discussed in the economics literature, the truth is that only sometimes is it realistic to make the big assumption that the casting move is common knowledge. We therefore play safe by illustrating Harsanyi's theory with our poker model.

### 15.3.2 Types in Poker

Imagine an anthropologist from Mars who saw our poker model being played without realizing that the cards dealt to Alice and Bob were significant. The Martian would then apply the theory of incomplete information by asking the following questions:

*1. What Is the Script?* This consists of Von Neumann's betting rules as illustrated in Figure 15.4(a). The payoff boxes are empty, and so information that is necessary for a game-theoretic analysis is missing. What should be written in the payoff boxes depends on the characteristics of the players. In our theatrical terminology, this means that we need to know the set of actors auditioning for each of the two roles.

*2. Who Are the Actors?* If the Martian watches long enough, three types of behavior will emerge for each of the roles of Alice and Bob. We know that the different behaviors arise because Alice and Bob may each be dealt one of three cards, but the Martian may attribute the different behaviors to Alice and Bob each having personalities split into three different types, which we call Alice King, Alice Queen, Alice Jack, Bob King, Bob Queen, and Bob Jack.

*3. What Are Their Preferences?* Figure 15.4(b) shows the preferences when Alice Jack plays Bob Queen. Figure 15.4(c) shows how Bob Queen's preferences over the possible outcomes change when he stops playing Alice Jack and starts playing Alice King. He prefers to stop calling and start folding. It is important for the theory that we have such freedom to make an actor's preferences contingent on the types of the actors cast in other roles.

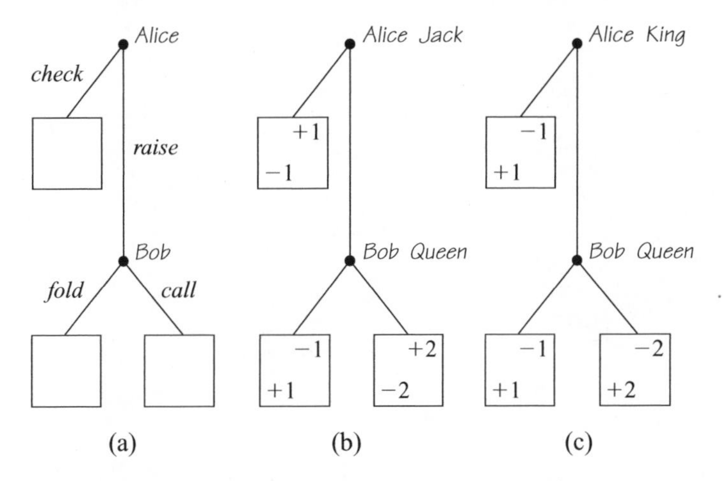


Figure 15.4 Preferences with incomplete information. Figure 15.4(a) shows the script of the Poker model of Figure 15.2. Figures 15.4(b) and 15.4(c) show some of the preferences of some of the actors.

*4. What Are Their Beliefs?* If it were common knowledge which actor were occupying each role, then we would be playing a game of perfect information like Figure 15.4(b), but no actor knows for sure who is occupying the other role. The actors' beliefs on this subject therefore need to be built into the Bayesian game we are constructing.

Figure 15.5(a) shows the probabilities that an actor of each type assigns to the possible types of the actor filling the opposing role. In poker, the actors derive these beliefs by updating the prior probabilities of the possible deals after observing the hand they have been dealt themselves.

*5. What Is the Casting Move?* In poker, the casting move is the deal, which is shown as the chance move that opens the game of Figure 15.2. What the actors know about the casting move is indicated by introducing information sets. For example, Alice Jack uses the fact that the casting move has chosen her to occupy the role of Alice to deduce that the casting move can't have chosen Bob Jack to occupy the role of Bob.

An alternative representation is given in Figure 15.5(b). (The number in the third row and second column is the probability that Alice is dealt the jack and Bob the queen.) This is the prior probability distribution from which the actors filling the roles of Alice and Bob deduce their posterior beliefs after learning their own type. In poker, it is easy to defend the assumption that it should be common knowledge that all actors have the same prior because much objective data is available about shuffling and dealing decks of cards when this is done honestly (Section 13.7.2).

*6. Who Are the Players?* The theory of incomplete information offers two choices when deciding how to analyze the game of imperfect information with which we end up. Do we analyze it like an extensive form or a strategic form?

**(i) Actors as players.** We implicitly took this approach when analyzing our poker model using backward induction in the extensive form. Each actor is treated as a separate player. In our case, Alice King, Alice Queen, and Alice Jack choose at the three upper information sets in Figure 15.2. Bob King, Bob Queen, and Bob Jack choose at the three lower information sets.

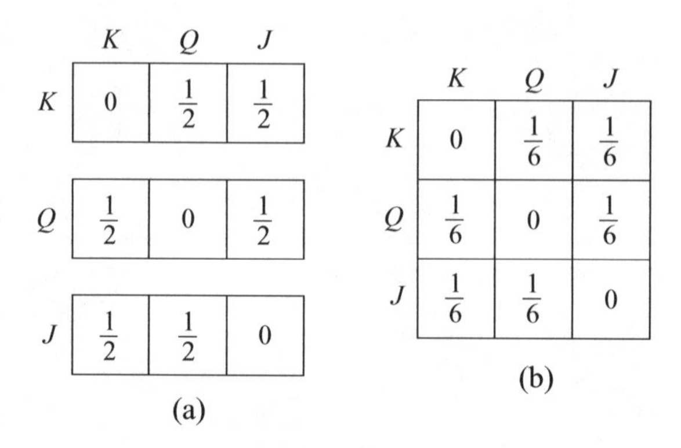


Figure 15.5 Beliefs with incomplete information. Figure 15.5(a) shows the beliefs that each actor has about the actor filling the opposing role. Figure 15.5(b) shows a common prior from which these beliefs can be derived by actors who update on the information that they have been chosen to play.

Since the game is now being treated as a *six*-player game, we really ought to write payoffs for all six actors in the payoff boxes of Figure 15.2. It doesn't matter what payoff is assigned to actors who aren't chosen to play, but it is usual to assign them a payoff of zero. If we follow this convention, the payoff box reached when Alice raises and Bob calls after the deal JQK will contain the payoff vector $(0, 0, -2, 0, 2, 0)$. We should properly make Alice Jack's payoff $-2\times\frac{1}{3}$ rather than $-2$ because she is chosen to play only with probability $\frac{1}{3}$, but life is usually too short for such niceties.[1]

**(ii) Actors as agents.** The second approach retains the *two*-player structure of the original script. We implicitly took this line when we analyzed the strategic form of our poker model.

Imagine that the female actor cast in the role of Alice consults Von Neumann about what strategy she should use. Similarly, the male actor cast in the role of Bob consults Morgenstern. In order to induce Von Neumann and Morgenstern to give optimal advice, their commissions must be arranged so that each has identical preferences to the actor they are advising. Von Neumann will then give advice to the actor that is *contingent* on her type. For example, in poker, Von Neumann will recommend a different strategy to Alice King than to Alice Jack.

This approach reduces the actors to puppets. The actual players in this version of the story are the guys who pull their strings: namely, Von Neumann and Morgenstern. Before Chance makes her casting decisions, Von Neumann and Morgenstern have a complicated game to consider. Each must have instructions in readiness for every actor who might call upon them for advice.

## 15.4 Russian Roulette

In Russian Roulette, everything hinges on how risk averse Boris and Vladimir are, but this isn't likely to be common knowledge. To keep things simple while studying this new problem, we look at a version of the game of Figure 4.4 in which the revolver has only three chambers.

Both players have three outcomes to consider: $\mathscr{L}$ (shooting yourself), $\mathscr{D}$ (chickening out), and $\mathscr{W}$ (winning the fair maiden). Their Von Neumann and Morgenstern utility functions are calibrated so that they both attach a utility of 0 to $\mathscr{L}$ and 1 to $\mathscr{W}$. Boris's and Vladimir's levels of risk aversion are determined by the respective values of $a$ and $b$ they assign to $\mathscr{D}$ (Section 4.7).

Although this problem of incomplete information isn't very hard, it is still worthwhile to be systematic in following Harsanyi's routine:

*1. Script.* A reduced script is shown in Figure 15.6(a), which takes for granted that Boris will chicken out if his second move is reached (because he would otherwise be sure of shooting himself).

*2. Actors.* For each $a$ between 0 and 1, the set $\mathscr{F}$ of actors who might occupy Boris's role contains an actor, whom we call Ms. $a$. (All the actors are skilled male

[1] Just as it doesn't make any difference what payoff we assign to actors who don't get to play, so it doesn't matter if we multiply all an actor's payoffs by the same number when the actor does get to play.

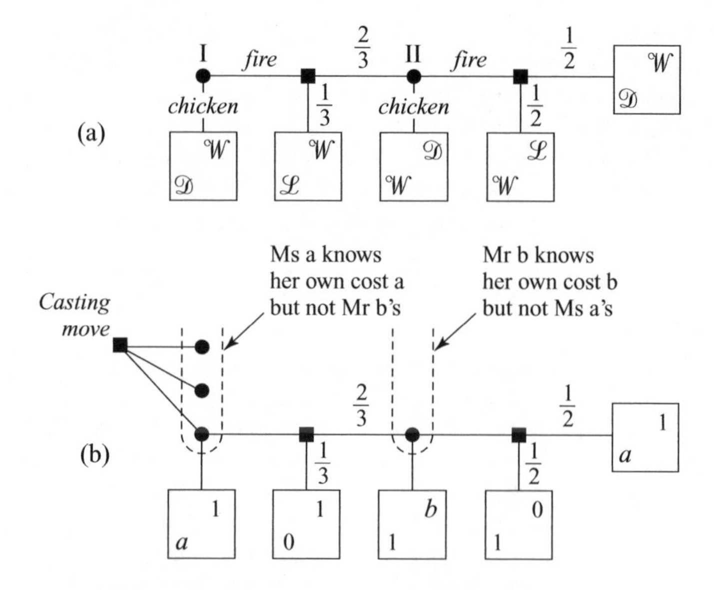


Figure 15.6 Russian Roulette with incomplete information.

impersonators.) For each $b$ between 0 and 1, the set $\mathscr{M}$ of actors who might occupy Vladimir's role contains an actor, whom we call Mr. $b$.

*3. Preferences.* Ms. $a$'s Von Neumann and Morgenstern utility for the outcome $\mathscr{D}$ is $a$. Mr. $b$'s is $b$.

*4. Beliefs.* A substantive assumption is required here. The simplest is that all actors have the *same* belief about the risk aversion of their opponent. We could assign any probabilities to the various possibilities without complicating the model much, but it's easiest to make each possibility equally likely. It is asking a lot that Boris's and Vladimir's beliefs should be common knowledge, but that is what we need in order to apply Harsanyi's approach.

*5. Casting Move.* The beliefs we have assigned the actors are consistent with a casting move that independently chooses $a$ and $b$ from a uniform distribution on the interval [0, 1]. (This means that the probability that $a$ or $b$ lies in any particular subinterval of [0, 1] is equal to the length of the subinterval.) Assuming that $a$ and $b$ are independent simplifies matters a lot because the actors learn nothing about their opponent from the fact that they have been chosen to play, and so no Bayesian updating is necessary.

*6. Actors as Players.* We will treat each actor as a separate player. Figure 15.6(b) then shows part of the Bayesian game.

*7. A Bayes-Nash Equilibrium.* We use backward induction to solve the Bayesian game. If he gets to move, Mr. $b$ chickens out if $b > \frac{1}{2} \times 0 + \frac{1}{2} \times 1 = \frac{1}{2}$. He pulls the trigger if $b < \frac{1}{2}$.

Ms. $a$ doesn't know $b$, but she believes that $b$ is equally likely to be greater or smaller than $\frac{1}{2}$. (The event $b = \frac{1}{2}$ occurs with probability zero.) So Ms. $a$ believes that her opponent will chicken out with probability $\frac{1}{2}$. It follows that Ms. $a$ will chicken out when

$$a > \tfrac{1}{3}\times 0 + \tfrac{2}{3}\times\tfrac{1}{2}\times 1 + \tfrac{2}{3}\times\tfrac{1}{2}\times\tfrac{1}{2}\times 1 + \tfrac{2}{3}\times\tfrac{1}{2}\times\tfrac{1}{2}\times a,$$

which reduces to $a > \frac{3}{5}$.

Muscovite maidens who don't observe the casting move will then be gratified to observe that Russian cavalry officers pull the trigger on the first chamber with probability $\frac{3}{5}$ and on the second chamber with probability $\frac{1}{2}$.

*Olga Intervenes.* We now modify the preceding assumption about the actors' beliefs to give an example where Bayesian updating is necessary.

Olga secretly prefers Boris, so she sneaks him the information that he is less risk averse than Vladimir. Although the source is unreliable, Boris treats it as gospel.

When Ms. $a$ now computes the probability that $b > \frac{1}{2}$, she conditions on the event that b>a. If $a > \frac{1}{2}$, it is then certain that Vladimir will chicken out, so Boris chickens out when $a > \frac{1}{3}\times 0 + \frac{2}{3}\times 1 = \frac{2}{3}$.

If $a < \frac{1}{2}$, the probability that Vladimir chickens out is

$$p = \text{prob}\,(b > \tfrac{1}{2}\,|\, b > a) = \frac{\text{prob}\,(b > \frac{1}{2} \text{ and } b > a)}{\text{prob}\,(b > a)} = \frac{1}{2(1-a)}.$$

The criterion for Ms. $a$ to chicken out is now

$$a > \tfrac{1}{3}\times 0 + \tfrac{2}{3}\times p\times 1 + \tfrac{2}{3}\times(1-p)\times\tfrac{1}{2} + \tfrac{2}{3}\times(1-p)\times\tfrac{1}{2}\times a,$$

which reduces to $4a^2 - 7a + 3 < 0$. This holds only when $\frac{3}{4} < a < 1$, and so Ms. $a$ always pulls the trigger when $a < \frac{1}{2}$.

Muscovite maidens will now be thrilled at seeing the first cavalry officer behaving recklessly more often (since he now pulls the trigger with probability $\frac{2}{3}$ instead of $\frac{3}{5}$). If they don't know that Boris has inside information, they might attribute this change in behavior to his having bolder preferences, but it is Boris's *beliefs* about Vladimir's boldness that have changed.

## 15.5 Duopoly with Incomplete Information

→ 15.6

In real life, firms sometimes take elaborate precautions to prevent their rivals from learning their costs. This section models the case when Alice and Bob are both successful in concealing their costs from each other.

In the Cournot duopoly of Section 10.2.2, it is common knowledge that Alice and Bob both have constant unit cost $c > 0$. We now suppose that Alice and Bob have unit costs $A$ and $B$, but neither knows the unit cost of the other. We therefore need an actor Ms. $A$ for each possible value of $A$ and an actor Mr. $B$ for each possible value of $B$. One can think of these actors as possible managers of Alice's and Bob's firms.

*Actors' Preferences.* Ms. $A$ and Mr. $B$ seek to maximize their firms' profits, given that they have been chosen to manage them. Their profit functions are

$$\begin{aligned}\pi_A(a,b) &= (K-A-a-b)a,\\ \pi_B(a,b) &= (K-B-a-b)b.\end{aligned}$$

*Actors' Beliefs.* A pleasant feature of a Cournot duopoly is that we don't need to be very specific about the probabilities with which the casting move selects different actors to manage Alice's and Bob's firms since everything can be expressed in terms of expectations. For example, with the notation $\mathscr{E}X = \overline{X}$, Bob's expectation of Alice's cost is $\overline{A}$.

In the general case, we will also have to worry about $\overline{\overline{A}}$, which is Alice's expectation of Bob's expectation of Alice's cost. But we can put aside such concerns about beliefs about beliefs—and beliefs about beliefs about beliefs—if we close the model at the first opportunity by making the heroic assumption that it is common knowledge that the casting move assigns costs to Alice and Bob *independently*. This assumption ensures that all the possible managers of a firm have the *same* beliefs about the manager of the other firm.

*Players as Agents.* We now have a game of imperfect information played between Von Neumann and Morgenstern, acting on behalf of Alice and Bob. A pure strategy for Von Neumann in this game is a *function* $\alpha : \mathscr{F} \to \mathbb{R}_+$, where $\mathscr{F}$ is the set of possible managers of Alice's firm. If Ms. $A$ is chosen to run Alice's firm, she tells Von Neumann her type and asks how many hats she should produce. His reply is $\alpha(A)$. Morgenstern chooses a function $\beta : \mathscr{M} \to \mathbb{R}_+$. If Mr. $B$ is chosen to run Bob's firm, Morgenstern's advice to Bob is $\beta(B)$.

*Bayes-Nash Equilibrium.* For a Nash equilibrium, $\alpha$ and $\beta$ must be best replies to each other. Since Von Neumann's advice to each Ms. $A$ must be optimal, given that Morgenstern has chosen $\beta$, the value of $a$ we set equal to $\alpha(A)$ must maximize Ms. $A$'s expected profit:[2]

$$\mathscr{E}\pi_A(a,\beta) = (K-A-a-\overline{b})a,$$

where $\overline{b} = \mathscr{E}\beta(B)$ is Ms. $A$'s expectation of Mr. $B$'s output. We find the maximum by differentiating with respect to $a$, just as in Section 10.2.2. Setting the derivative equal to zero yields

$$K-A-2a-\overline{b} = 0.$$

It follows that Von Neumann's advice function $\alpha$ is a best reply to $\beta$ when

$$\alpha(A) = \tfrac{1}{2}(K-A-\overline{b}). \tag{15.1}$$

[2]Remember that the expectation operator is linear. For example, $\overline{2X+3} = 2\overline{X}+3$.

A similar argument shows that the advice that Morgenstern gives to Mr. $B$ is optimal, given that Von Neumann has chosen $\alpha$ when

$$\beta(B) = \tfrac{1}{2}(K - B - \overline{a}). \tag{15.2}$$

We have been working with the expectations $\overline{a}$ and $\overline{b}$ of $\alpha(A)$ and $\beta(B)$ without knowing their values. But now we can take expectations in (15.1) and (15.2) to work them out:

$$\overline{a} = \tfrac{1}{2}(K - \overline{A} - \overline{b}),$$
$$\overline{b} = \tfrac{1}{2}(K - \overline{B} - \overline{a}).$$

In this calculation, it is important that $\overline{a}$ and $\overline{b}$ remain constant when we take expectations. This follows from our assumption that all possible managers of a firm have the *same* beliefs about the manager of the other firm.

In the symmetric case[3] when $\overline{A} = \overline{B} = c$, we obtain

$$\overline{a} = \overline{b} = \tfrac{1}{3}(K - c).$$

The *average* outputs are therefore the same as in the case when it is common knowledge that Alice and Bob both have unit cost $c$ (Section 10.2.2).

Substituting our values for $\overline{a}$ and $\overline{b}$ in (15.1) and (15.2), we obtain formulas for the advice functions $\alpha$ and $\beta$ that make $(\alpha, \beta)$ a Bayes-Nash equilibrium:

$$\alpha(A) = \tfrac{1}{6}(2K - 3A + c),$$
$$\beta(B) = \tfrac{1}{6}(2K - 3B + c).$$

For example, when it turns out that $A = B = 0$, the market is flooded with hats because Alice's and Bob's expectations about their rival's costs are then badly wrong.

### 15.5.1 Beliefs about Beliefs

What happens in the Cournot model with incomplete information about costs if we don't close the model by assuming it to be common knowledge that costs are assigned independently to Alice and Bob? All is the same until we take expectations in (15.1) and (15.2). However, we must then take into account the fact that different managers may believe different things about their opponent, so that all our expectations now depend on the actor to whom they are attributed. In particular, $\overline{a}$ and $\overline{b}$ are no longer constant, and so

$$\overline{a} = \tfrac{1}{2}(K - \overline{A} - \overline{\overline{b}}),$$
$$\overline{b} = \tfrac{1}{2}(K - \overline{B} - \overline{\overline{a}}).$$

[3]Although the Bayesian game is then symmetric, economists still say that they are dealing with a case of asymmetric information.

But how do we find $\bar{\bar{a}}$ and $\bar{\bar{b}}$ ? Taking expectations again, we get

$$\bar{\bar{a}} = \tfrac{1}{2}(K - \bar{\bar{A}} - \bar{\bar{\bar{b}}}),$$
$$\bar{\bar{b}} = \tfrac{1}{2}(K - \bar{\bar{B}} - \bar{\bar{\bar{a}}}).$$

The term $\bar{\bar{\bar{a}}}$ is as monstrous as it looks. It represents Mr. $B$'s expectation of Ms. $A$'s expectation of Mr. $B$'s expectation of Ms. $A$'s output! And things get worse as we keep on taking expectations in an attempt to eliminate the terms in $a$ and $b$ whose values we don't know. But pressing on intrepidly, we find that the outputs in a Bayes-Nash equilibrium are given by the infinite sums:

$$\alpha(A) = \tfrac{1}{2}(K - A) - \tfrac{1}{4}(K - \bar{B}) + \tfrac{1}{8}(K - \bar{\bar{A}}) - \tfrac{1}{16}(K - \bar{\bar{\bar{B}}}) + \cdots,$$
$$\beta(B) = \tfrac{1}{2}(K - B) - \tfrac{1}{4}(K - \bar{A}) + \tfrac{1}{8}(K - \bar{\bar{B}}) - \tfrac{1}{16}(K - \bar{\bar{\bar{A}}}) + \cdots.$$

These infinite sums pose the fundamental problem of game theory with a vengeance. How do we cope with an infinite regress of the kind: If I think that he thinks that I think ...? In the context of a problem of incomplete information, Harsanyi's answer is to close the model by assuming that sufficient information about the casting move is common knowledge.

*Closing the Model.* We have looked at one way of closing a Cournot duopoly with incomplete information about costs, but there are many others.

A particularly elegant closure assumes that it is common knowledge that actors always believe the expected value of their rival's cost to be equal to their own cost. All expectations about expectations then collapse because

$$\bar{B} = A; \ \bar{\bar{A}} = \bar{B} = A; \ \bar{\bar{\bar{B}}} = \bar{\bar{A}} = A,$$

and so on. Ms. $A$'s Bayes-Nash output therefore becomes

$$\begin{aligned}\alpha(A) &= \tfrac{1}{2}(K - A) - \tfrac{1}{4}(K - A) + \tfrac{1}{8}(K - A) - \tfrac{1}{16}(K - A) + \cdots \\ &= \tfrac{1}{2}(K - A)\{1 - \tfrac{1}{2} + \tfrac{1}{4} - \tfrac{1}{8} + \cdots\} \\ &= \tfrac{1}{3}(K - A).\end{aligned}$$

All of the complication therefore reduces to a very simple result. A firm's output is the same as if it were common knowledge that both firms had the same unit cost as itself (Section 10.2.2).

→ 15.6

### 15.5.2 Agreeing to Disagree?

Actors deduce their beliefs about other agents by updating their prior distribution over the possible choices available to Chance at the casting move. These prior distributions must be common knowledge, but they needn't be the same. If all the actors do have the same prior, their beliefs are said to be *consistent*.

In both our Cournot models with incomplete information, we can find common priors for the actors, and so the actors' beliefs are consistent. But what if each actor believes it is certain that the other firm has such a large unit cost that it will never produce anything at all? Both firms will then produce the monopoly output appropriate to their actual unit cost and thus make a kind of Dutch book against themselves to the benefit of the consumer (Section 13.3.3).

Such examples make economists uncomfortable with the assumption that rational players may "agree to disagree." They therefore commonly invoke the Harsanyi doctrine by requiring that, if there is common knowledge of the players' priors, then the priors are all the same (Section 13.7.2).

## 15.6 Purification

→ 15.7

Like much else in game theory, mixed strategies have more than one interpretation. Sometimes we think of a mixed equilibrium as an equilibrium in actions—as when Alice actively tosses a coin when playing Matching Pennies to decide what to do (Section 2.2.2). Sometimes we interpret a mixed equilibrium as an equilibrium in beliefs (Section 6.3). The latter interpretation is more troublesome since we have to wrap our minds around the idea that mixed equilibria can be purified. Adam may then choose his strategy without randomizing at all, but when Eve looks up his plan in the great book of game theory, she isn't told for sure what he is going to do.

Harsanyi used his theory of incomplete information to provide a precise model that demystifies the idea of a purified mixed equilibrium. His argument is outlined here for the bimatrix game of Figure 6.2(a), which is repeated as Figure 15.7(a). Adam's and Eve's payoff matrices in the game are $A$ and $B$.

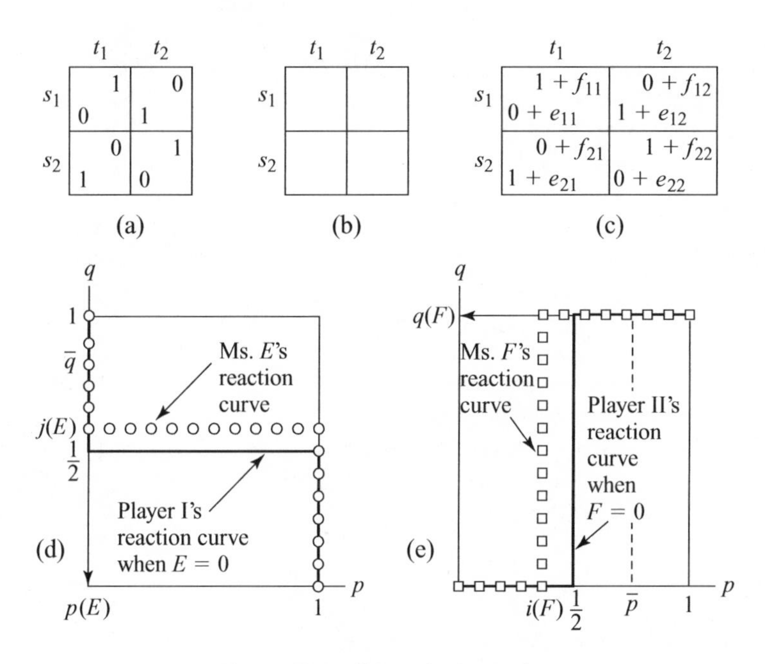


Figure 15.7 Purifying mixed strategies.

Figure 6.2(b) shows both players' reaction curves. They cross just once, showing that the game has a unique Nash equilibrium. This mixed equilibrium would usually be described by saying that both players use each of their pure strategies with probability $\frac{1}{2}$. Our task is to *purify* this mixed equilibrium.

*Fluctuating Payoffs.* Since the players are not to randomize, the random element that creates uncertainty in everybody's mind has to come from somewhere else. Harsanyi makes each player uncertain about the precise nature of their opponents' preferences. For example, people's attitudes toward risk vary from day to day, and so a player is likely to be doubtful about how risk averse an opponent is. Such variations will be reflected in a player's Von Neumann and Morgenstern utility function and hence in the payoffs of the game.

The script for our model is shown in Figure 15.7(b). The fact that the cells in this table are empty indicates that information is incomplete. An actor, Mr. $E$, in the set $\mathcal{M}$ of male actors will be identified with a $2 \times 2$ matrix $E$. The entries of $E$ represent fluctuations from Adam's payoffs in the basic game of Figure 15.7(a). An actor, Ms. $F$, in the set $\mathcal{F}$ of female actors will also be identified with a $2 \times 2$ matrix $F$ representing fluctuations from Eve's payoffs.

If it were common knowledge who had been cast in each role, then the game being played would be that of Figure 15.7(c). However, the actor, Mr. $E$, cast in the role of Adam, knows only that his own payoff matrix is $A+E$. He is uncertain about the payoff matrix of his opponent. Similarly, the actor, Ms. $F$, cast in the role of Eve, knows only that her own payoff matrix is $B+F$.

Specifying the actors' payoff matrices determines their preferences. As regards their beliefs, it is common knowledge that the casting director's selections of Mr. $E$ and Ms. $F$ are made independently. As in the preceding section, we don't need to say exactly what the probability density functions are.

*Bayes-Nash Equilibrium.* Before the casting move, each Mr. $E$ selects a $2 \times 1$ col-1 column vector $P(E)$ to represent his choice of mixed strategy. The second coordinate of $P(E)$ will be denoted by $p(E)$. This is the probability with which $P(E)$ requires Mr. $E$'s *second* pure strategy to be used. If our purification project is successful, the mixed strategy $P(E)$ will actually turn out to be a pure strategy, and so $p(E)$ will be 0 or 1. Eve doesn't know what $E$ is, but she can compute that Adam's expected mixed-strategy choice is $\overline{P} = \mathcal{E}_E\{P(E)\}$. We denote the second coordinate of this $2 \times 1$ column vector by $\overline{p}$.

Ms. $F$ similarly chooses a $2 \times 1$ column vector $Q(F)$ that represents a mixed strategy for her. Its second coordinate is denoted by $q(F)$. Adams calculates that Eve's expected mixed-strategy choice is $\overline{Q} = \mathcal{E}_F\{Q(F)\}$. We denote the second coordinate of this $2 \times 1$ column vector by $\overline{q}$.

Recall from Section 6.4.3 that Mr. $E$'s payoff is $P(E)^\top(A+E)Q(F)$ when he is matched with Ms. $F$. Since he doesn't know who his opponent is, he calculates the expected value of this quantity:

$$\mathcal{E}_F\{P(E)^\top(A+E)Q(F)\} = P(E)^\top(A+E)\mathcal{E}_F\{Q(F)\} = P(E)^\top(A+E)\overline{Q}.$$

If each actor is making a best reply to the choices of the others, as required for a Nash equilibrium, then this equation tells us that $P(E)$ must be a best reply to $\overline{Q}$ in a game in which player I's payoff matrix is $A+E$. Figure 15.7(d) shows player I's reaction curve in such a game.

In seeking a best reply to $\overline{Q}$, Mr. E cares about whether $\overline{q} > j(E)$ or $\overline{q} < j(E)$. In the former case, he takes $p(E)=0$. In the latter case he takes $p(E)=1$. Only when $\overline{q} = j(E)$ is it possible that Mr. E might use a mixed strategy because only then is he indifferent between his two pure strategies.

Similarly, Ms. $F$'s expected payoff is

$$\mathscr{E}_E\{P(E)^\top (B+F)Q(F)\} = \mathscr{E}_E\{P(E)^\top\}(B+F)Q(F) = \overline{P}^\top (B+F)Q(F).$$

In seeking a best reply to $\overline{P}$, Ms. $F$ studies Figure 15.7(e). She then takes $q(F)=0$ if $\overline{p} < i(F)$ and $q(F)=1$ if $\overline{p} > i(F)$. Only when $\overline{p} = i(F)$ might she use a mixed strategy.

*Small Fluctuations.* We haven't yet used the fact that the entries in $E$ and $F$ represent *small* fluctuations in the payoffs of the game of Figure 15.7(a). We need this assumption to ensure that the reaction curves of Figures 15.7(d) and 15.7(e) are close to the reaction curves for the case when $E$ and $F$ are the zero matrix. Both $i(F)$ and $j(E)$ will then be close to $\frac{1}{2}$ for all $E$ and $F$. It follows that both $\overline{p}$ and $\overline{q}$ will be approximately $\frac{1}{2}$.[4]

*Purification Achieved.* What do we learn from this argument?

- All actors use a pure strategy.
- Adam's beliefs about Eve are summarized by $\overline{q}$. This is the probability with which Adam thinks that Eve will play her second pure strategy. Similarly, Eve thinks that Adam will play his second pure strategy with probability $\overline{p}$.
- When the fluctuations become small, $\overline{p}$ and $\overline{q}$ approach $\frac{1}{2}$.

Although the players actually use *pure* strategies, their *beliefs* about what the opponent will do approach the mixed Nash equilibrium of the underlying game as the fluctuations in the payoffs become vanishingly small. The mixed Nash equilibrium has therefore been purified.

## 15.7 Incomplete Information about Rules

→ 15.8

If it is common knowledge that the Prisoners' Dilemma is to be repeated ten times, then Adam and Eve will always play *hawk*. The critics who say that they are therefore "rational fools" fail to recognize how strong the requirement is that information be complete (Section 11.3.1). In particular, the rules of the game must be common knowledge among the players. To see how different things can get when

[4]This conclusion depends on the fact that the original game of Figure 15.7(a) has no pure strategy equilibria. Suppose, for example, that $\overline{p}$ were so much larger than $\frac{1}{2}$ that $i(F) < \frac{1}{2}$ for all $F$. Then $q(F)=1$ for all $F$ and so $\overline{q}=1$. Hence $p(E)=0$ for all $E$, and so $\overline{p}=0$. But this isn't consistent with the assumption that $\overline{p} > \frac{1}{2}$.

the rules aren't common knowledge, we apply Harsanyi's methodology to a case in which nobody knows exactly how many times the Prisoners' Dilemma will be repeated—although it will remain common knowledge that the game will be repeated only a finite number of times.

### 15.7.1 Ignorance Is Bliss

It is sometimes wrongly claimed that rational people can't fail to profit from being better informed (Exercise 5.9.24). In the ten-times repeated Prisoners' Dilemma of Figure 1.3(a), Adam and Eve can become very much better off when they *cease* to know that the tenth repetition will be the last. A little ignorance can therefore sometimes lead to the same happy outcome as a little irrationality when the Prisoners' Dilemma is repeated a finite number of times (Section 14.4.1).

When it is common knowledge that the game is of finite length but the actual length is an open question, we need an actor for each possible finite length of the game. If $n$ is even, we make Mr. $n$ a candidate for the role of Adam. If $n$ is odd, we make Ms. $n$ a candidate for the role of Eve.

The casting move chooses the actual length $N$ of the game. To keep things simple, we use an information structure similar to that of the Email Game (Section 12.9.1). The only actors chosen to play are therefore those corresponding to $N$ and $N+1$. An actor with $n > 1$ who is chosen to play knows that the game is of length $n$ or $n-1$, but nothing else. (The exceptional case is Ms. 1, who will know for sure that the game is of length 1.) Figure 15.8 shows the actors together with their information sets.

The actors' preferences are the same as those of the players whose role they occupy. We assume that an actor with $n > 1$ who is chosen to play believes it twice as likely that the length of the game will be $n$ than $n-1$ (Exercise 15.10.23).

We are interested in what happens when the casting move chooses $N = 10$. Adam and Eve are then played by Mr. 10 and Ms. 11, but Mr. 10 thinks his opponent is as likely to be Ms. 9 as Ms. 11. He must therefore take account of what both Ms. 9 and Ms. 11 would believe if chosen to play. Ms. 9 would believe her possible opponents are Mr. 8 or Mr. 10. Ms. 11 would believe them to be Mr. 10 or Mr. 12. Continuing in this way, we are forced to extend the number of actors who need to be considered until nobody is left out. The game of imperfect information to which we are led therefore has an infinite number of actors, each of whom we will treat as a separate player.

We look for an equilibrium in which each actor uses a "grim-trigger" strategy. This requires that *dove* be played until one of two things happens. The first possibility is that the opponent plays *hawk* on some occasion. Such an opponent is punished with the grim response of always playing *hawk* thereafter, no matter what.

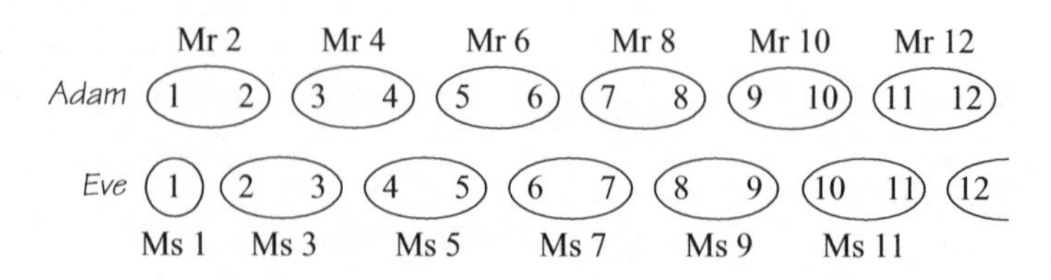


Figure 15.8 How long does this go on for?

The second possibility is that a "trigger" stage of the game is reached. *Hawk* is then played from that stage on, no matter how cooperative the opponent may have been. In our case, the trigger stage is the stage that the actor knows to be the very last stage of the game when it is reached. For example, Ms. 11's trigger stage is the eleventh stage of the game.

If all actors' players use such a grim-trigger strategy, the result is a Bayes-Nash equilibrium. If this equilibrium is used and $N = 10$, then the players play *dove* until stage 9. Only at the tenth and last stage does someone play *hawk*.

Why does no actor ever have an incentive to deviate? Consider Ms. 7. If her opponent is Mr. 6, then the length of the game must be $N = 6$. If both use the grim-trigger strategy, then Ms. 7's income stream is 1, 1, 1, 1, 1, $-1$. If her opponent is Mr. 8, then the length of the game is $N = 7$, and her income stream is 1, 1, 1, 1, 1, 1, 3. Thus, Ms. 7 expects to get $\frac{1}{3} \times 4 + \frac{2}{3} \times 9 = 7\frac{1}{3}$ by sticking to her grim-trigger strategy.

What can she get by deviating? She does best by beating Mr. 6 to the draw. That is, if she is going to deviate, she should plan to play *hawk* at stage 5 and thereafter. This generates the two income streams: 1, 1, 1, 1, 3, 0 and 1, 1, 1, 1, 3, 0, 0. So the most that Ms. 7 can get from deviating is $\frac{1}{3} \times 7 + \frac{2}{3} \times 7 = 7$, which is less than her equilibrium payoff.

## 15.8 Roundup

For something to be a game, information must be complete. This means that the rules of the game, the players' payoffs, and the subjective probabilities they assign to chance moves must be common knowledge. Harsanyi proposed a methodology that completes a scenario with incomplete information. We call the end product a Bayesian game.

Actors are defined by their preferences and beliefs. Bayesian games are games of imperfect information that open with a casting move that assigns an actor to each player role. The first thing actors do on finding that they have been chosen to play is to update their beliefs to take this fact into account.

The basic problem that Harsanyi's methodology tackles is the fundamental problem of game theory. How do we deal with the infinite regress involved in chains of reasoning of the kind: If I think that he thinks that I think ...? In an incomplete information context, Harsanyi's method closes such chains by insisting that the casting move in the Bayesian game is common knowledge. This is sometimes an unreasonably strong assumption.

A Nash equilibrium of a Bayesian game is often said to be a "Bayesian equilibrium of a game of asymmetric information," but I prefer to speak of a Bayes-Nash equilibrium because there isn't really such a thing as a *game* of incomplete or asymmetric information.

The chief reason for invoking Harsanyi's theory of incomplete information is that people are often doubtful about the precise preferences of other people. We then ask the following questions:

What is the script?
Who are the actors?

What are their types?
What is the casting move?
Who are the players in the Bayesian game?

A script consists of the rules of a game, commonly called a mechanism or game form. To state an actor's type is to specify the actor's preferences and beliefs. Even when our focus is on the unknown preferences of the players, we can't avoid taking into account the beliefs of the actors about the casting move. We can analyze a Bayesian game as an extensive form or as a strategic form. In the former case, we make each actor a player. In the latter case, we treat each actor as an agent of the original players. The difference between the two approaches is no more than the difference between working with behavioral strategies and mixed strategies (Section 12.5.3).

Harsanyi's theory of incomplete information introduces no new principles. It is simply a modeling technique, as is clear from the fact that we can apply it directly to a game of imperfect information like poker. When the technique is applied to a Cournot duopoly with incomplete information about costs, we find that Alice's output as a function of her unit cost $A$ is

$$\alpha(A) = \tfrac{1}{2}(K-A) - \tfrac{1}{4}(K-\overline{B}(A)) + \tfrac{1}{8}(K-\overline{\overline{A}}(A)) - \tfrac{1}{16}(K-\overline{\overline{\overline{B}}}(A)) + \cdots,$$

where $\overline{B}(A)$ is Alice's expectation of Bob's cost $B$ and $\overline{\overline{A}}(A)$ is Alice's expectation of Bob's expectation of Alice's cost, and so on. Such formulas make it apparent why economists find it necessary to close their models by insisting that everything becomes common knowledge at some level.

Harsanyi's method can be adapted to the case in which there is incomplete information about the rules of the game. When there is common knowledge that the Prisoners' Dilemma is to be repeated a finite number of times but not of precisely how many times it will be repeated, then cooperation can emerge as a Bayes-Nash equilibrium.

## 15.9 Further Reading

*Game Theory: Analysis of Conflict*, by Roger Myerson: Harvard University Press, Cambridge, MA, 1991. This advanced book on game theory is particularly careful about the application of Bayesian methods.

*The Education of a Poker Player, including Where and How One Learns to Win,* by Herbert Yardley: Jonathan Cape, London, 1959. Don't waste time reading mathematical models if you want to make money playing poker.

## 15.10 Exercises

phil

1. Damon Runyon famously warned against betting with a man who claims he can make the jack of spades jump out of a deck of cards and squirt cider in your ear. If you do, be prepared for an earful of apple juice! Poker experts similarly warn against betting against someone who might know he is on to a

sure thing. Is this really good advice? Use the following model to illustrate your answer.

Alice and Bob are playing a version of poker with no ante. Bob is dealt a card that is *H* or *L* with equal probability. He looks at his card and then either folds or bets one chip. If he bets, Alice may then fold or call. If she calls, the game is over. Bob wins both chips if his card is *H*. Alice wins both chips if his card is *L*. Show that Alice never calls in equilibrium. If Bob deletes his weakly dominated strategies, show that he will bet when dealt *H* and fold when dealt *L*. Now analyze the game on the assumption that Bob gets to see his card only with some small probability $p > 0$ before deciding whether to fold or bet.

2. How come poker is possible if you should never bet with someone who might be sure of holding a better hand? Analyze Exercise 15.10.1 on the assumption that Alice and Bob must ante up one chip each before Bob is dealt his card. Show that Alice will now sometimes bet, even though Bob might know for sure that he has her beaten.

   The moral is that betting with someone who might be betting on a sure thing can be a good idea if it is costly not to bet at all.

3. Alice has raised Bob in the final round of a game of five-card stud. If her hole card is an ace, she knows she has the better hand for sure. In considering whether to fold, Bob reflects on the wisdom offered in the previous exercises. He argues that the cost of folding will be that he will lose all the money he has contributed to the pot so far. Since he can call for far less than this amount, he decides to keep betting.

   econ

   Use the idea of a sunk cost to explain why Bob's analysis is incorrect. However, Bob does incur a cost if he folds. Use the idea of an opportunity cost to explain how this cost arises.

4. After paying an ante of $\$a$, Alice and Bob are each independently dealt one card that is *H* (high) or *L* (low) with equal probability. After looking at their cards, Alice and Bob simultaneously either fold or bet an extra $\$b$. If they both fold or have the same card at a showdown, their antes are returned.

   Write down the $4 \times 4$ strategic form of the game. By successively deleting strongly dominated strategies, show that both players always use the Colonel Blimp strategy of betting with a high hand and folding with a low hand when $b > a > 0$. If $a > b > 0$, show that they always bet.

5. Section 15.3.2 reinterprets a poker model using the language of Harsanyi's theory of incomplete information. Do the same for the game of Quiche from Section 14.5.2, using Figure 15.9 to assist the explanation.

6. Game theorists use the Stag Hunt Game of Figure 8.7(a) to illustrate a story told by Jean-Jacques Rousseau in developing his theory of the social contract. Adam and Eve agree to cooperate in hunting a deer. They know that, after parting company, each might happen to see a hare, but they exchange promises not to be diverted from the hunt for the deer, although the knowledge that a hare is around enhances the attractiveness of abandoning the hunt in favor of attempting to trap a hare.

   phil

   Figure 15.10 models Rousseau's story on the assumption that the probability each player assigns to the other spotting a hare is $\frac{1}{2}$. For example, the payoff table labeled YN applies in the case when Adam spots a hare but Eve doesn't. Solve this game by successively deleting dominated strategies, and hence show

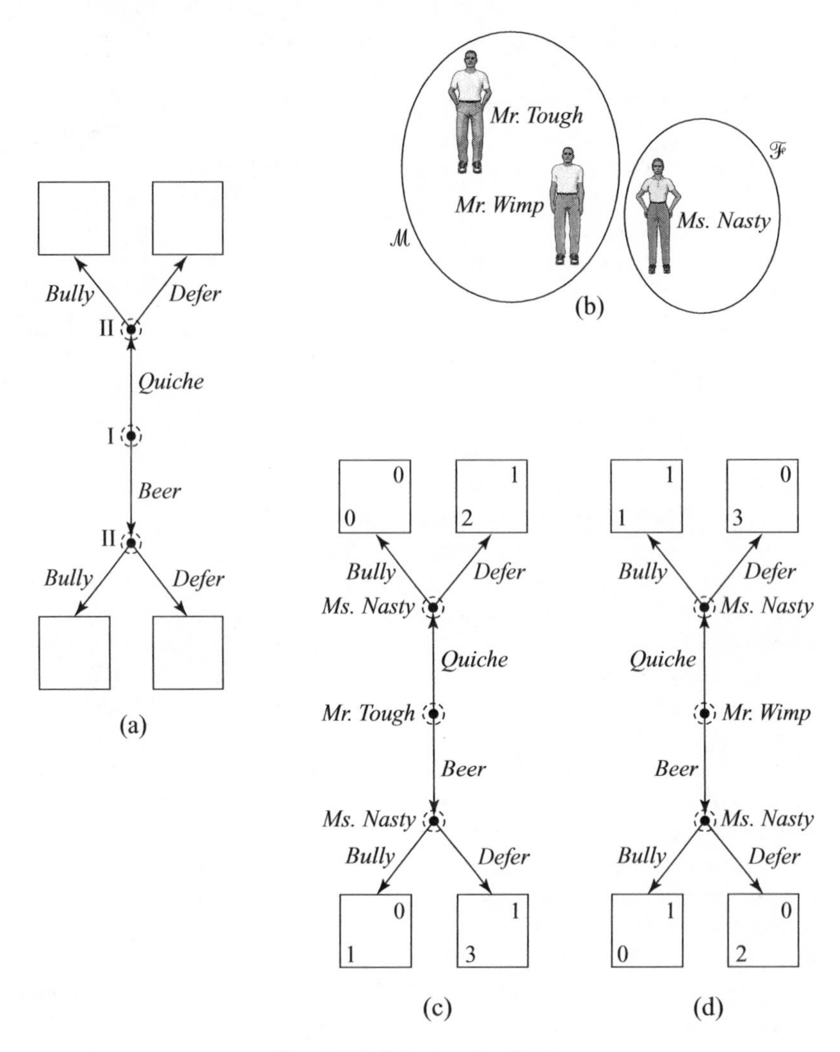


Figure 15.9 Reinterpreting Quiche.

that Rousseau's claim that it is rational to cooperate is no more valid here than in the one-shot Prisoners' Dilemma.

7. Analyze our incomplete information version of Russian Roulette when it is common knowledge that the casting move determines $a$ and $b$ as follows (Section 15.4). First $a$ is drawn from a uniform distribution on $[0, 1]$. Half of the time, $b$ is then chosen equal to $a$. The other half of the time, $b$ is chosen independently from the same distribution.

econ

8. Our first model of a Cournot duopoly with incomplete information about costs reduces to that of Section 10.2.2 when it is common knowledge that $A = B = c$, where $c$ is a constant. Analyze the case when $\overline{A} = c_1$ and $\overline{B} = c_2$. Your answer should reduce to that given in Exercise 10.8.2 when it is common knowledge that $A = c_1$ and $B = c_2$.

| NN | *dove* | *hawk* |
|---|---|---|
| *dove* | 5, 5 | 0, 4 |
| *hawk* | 4, 0 | 2, 2 |

| NY | *dove* | *hawk* |
|---|---|---|
| *dove* | 5, 5 | 0, 8 |
| *hawk* | 4, 0 | 2, 4 |

| YN | *dove* | *hawk* |
|---|---|---|
| *dove* | 5, 5 | 0, 4 |
| *hawk* | 8, 0 | 4, 2 |

| YY | *dove* | *hawk* |
|---|---|---|
| *dove* | 5, 5 | 0, 8 |
| *hawk* | 8, 0 | 4, 4 |

Figure 15.10 Deer hunting.

9. In our model of a Cournot duopoly with incomplete information about costs (Section 15.5), suppose that a player's cost can be only $H$ (high) or $L$ (low). It is common knowledge that the casting move chooses (L, L), (L, H), and (H, L) with probabilities 0.8, 0.1, and 0.1, respectively. Explain why Alice knows that Bob is badly mistaken in his beliefs when her cost is high. Without calculating, explain how her output in a Bayes-Nash equilibrium will reflect her knowledge. **econ**
10. Section 15.5.1 discusses our model of a Cournot duopoly with incomplete information about costs, under the assumption that the actors expect their opponent's cost to equal their own cost. Suppose instead that **econ**

$$\overline{A} = cB; \quad \overline{B} = dA,$$

where $c$ and $d$ are positive constants. In a Bayes-Nash equilibrium, show that

$$\alpha(A) = \tfrac{1}{3}K - A(2-d)/(4-cd),$$
$$\beta(B) = \tfrac{1}{3}K - B(2-c)/(4-cd).$$

11. Show that the beliefs assigned to the players in the previous exercise are consistent only if $cd = 1$. (Note that $\mathcal{E}(X|Y=y) = cy$ implies $\mathcal{E}X = c\mathcal{E}Y$.) **econ**
12. In a new version of the Cournot duopoly model with incomplete information about costs, all Mr. $B$'s have the same unit cost $c_2$ and the same expectation $\overline{A}$ of Alice's cost (Section 15.5.1). If these facts are common knowledge, show that the Bayes-Nash outputs are **econ**

$$\alpha(A) = \tfrac{1}{6}(2K + 2c_2 - 3A - \overline{A}),$$
$$b = \tfrac{1}{3}(K - 2c_2 + \overline{A}).$$

Although Alice's cost may be low compared with Bob's, explain why his lack of knowledge of this fact may result in his producing the most hats.

econ

13. In the previous exercise, Alice suffers if her cost is low, but it is common knowledge that Bob thinks it is high. What happens if she has the opportunity to make her cost common knowledge before outputs are chosen? Explain why there can't be a Bayes-Nash equilibrium in which Ms. $A$ reveals her cost when it is low and conceals it when it is high. (In such an equilibrium, Mr. $B$ will update $\overline{A}$ when he learns that Ms. $A$ is concealing her cost. Consider what is then optimal for Ms. $A$ when her cost is only just too high to be revealed.)

econ

14. In a new version of Exercise 15.10.12, it ceases to be true that all Mr. $B$'s have the same expectation of Alice's cost. Each Mr. $B$ now has his own expectation $\overline{A}(B)$. However, each Ms. $A$ has the same $\overline{\overline{A}}$ expectation of $\overline{A}(B)$. If these facts are common knowledge, show that the Bayes-Nash outputs are

$$\alpha(A) = \tfrac{1}{6}(2K + 2c_2 - 3A - \overline{\overline{A}}),$$
$$\beta(B) = \tfrac{1}{12}(4K - 8c_2 + 3\overline{A}(B) + \overline{\overline{A}}).$$

If $A$ is low compared with $c_2$ and Bob knows this, how might Bob still end up producing most of the hats?

15. Here is a method for completing a scenario with incomplete information:
    a. The script is shown in Figure 15.11(a).
    b. The actors who might be cast in role I are Mr. $A$ and Mr. $B$. The actors who might be cast in role II are Ms. $C$ and Ms. $D$.
    c. Figure 15.11(b) shows the probabilities with which the casting move assigns actors to roles. For example, the probability that Mr. $B$ and Ms. $D$ are cast as player I and player II is 0.1.
    d. The actors' preferences are given in Figure 15.11(c). These diagrams show how the empty payoff boxes of Figure 15.11(a) would be filled if it were common knowledge how the players had been cast. Notice that the payoffs always sum to zero.[5]

    Explain why it isn't true that Chance's choice of who is to be player I is independent of her choice of who is to be player II. Find the probability prob(B | C) that Ms. $C$ will assign to the event that her opponent is Mr. $B$, conditional on Ms. $C$ learning that she has been chosen to be player II. Which actors will know for sure who their opponent is should they be chosen to play?

16. This exercise continues Exercise 15.10.15.
    a. Solve each of the four two-player, zero-sum games shown in Figure 15.11(c). Mark the cell in each payoff table that will result when the solution strategies are used. These four games show the four possible ways Figure 15.11(a) could be completed if it were common knowledge who was occupying what role.
    b. Think of the game of imperfect information described in Exercise 15.10.15 as a four-player game whose players are Mr. $A$, Mr. $B$, Ms. $C$, and Ms. $D$.

[5]The Bayesian game will therefore also be zero sum, but we wouldn't be entitled to draw this conclusion if the actors had inconsistent beliefs.

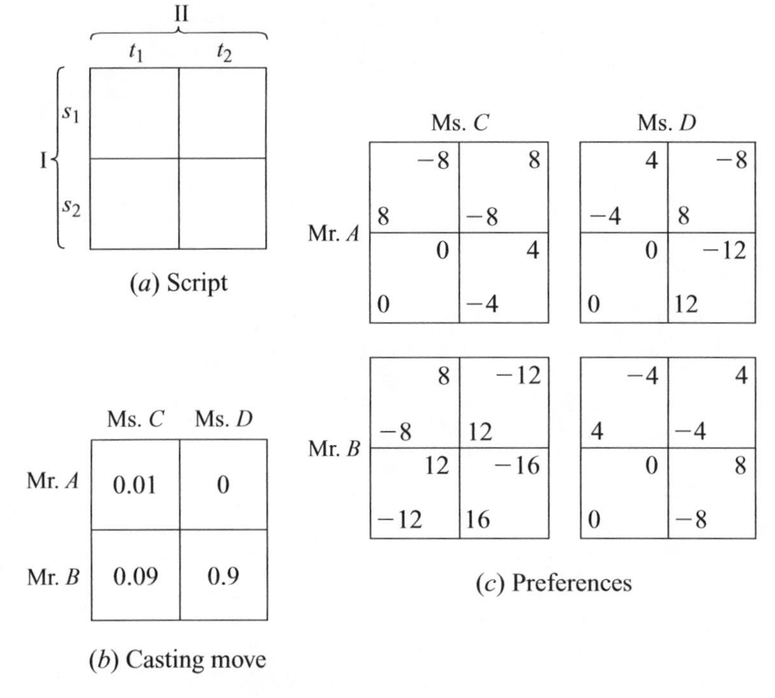


Figure 15.11 Information for Exercise 15.10.15.

Solve this four-player game by successively deleting strongly dominated strategies.

c. What did you assume to be common knowledge in part (b)?

d. Return to Figure 15.11(c), and mark the cell in each payoff table that will result if the actors for that table are chosen to play and each uses the pure strategy you calculated in part (b).

e. Comment on the difference between the cells marked in part (a) and part (d). Explain how Mr. *A* is able to exploit Ms. *C*'s ignorance.

17. Here is a second approach to the problem of Exercise 15.10.15.

a. Section 15.3.2 suggests thinking of player I as a reincarnation of Von Neumann, who gives advice to Mr. *A* and Mr. *B*. Player II may similarly be seen as a reincarnation of Morgenstern, who gives advice to Ms. *C* and Ms. *D*. Figure 15.12 shows the shape of the extensive form of the game that Von Neumann and Morgenstern will then see themselves as playing. Fill in the payoff boxes.

b. What are Von Neumann and Morgenstern's four pure strategies?

c. Find the strategic form of the game between Von Neumann and Morgenstern. Confirm that the game is zero sum.

d. Observe that Von Neumann's payoff matrix has a saddle point. Then solve the game.

e. Confirm that the game can also be solved by the successive deletion of dominated strategies.

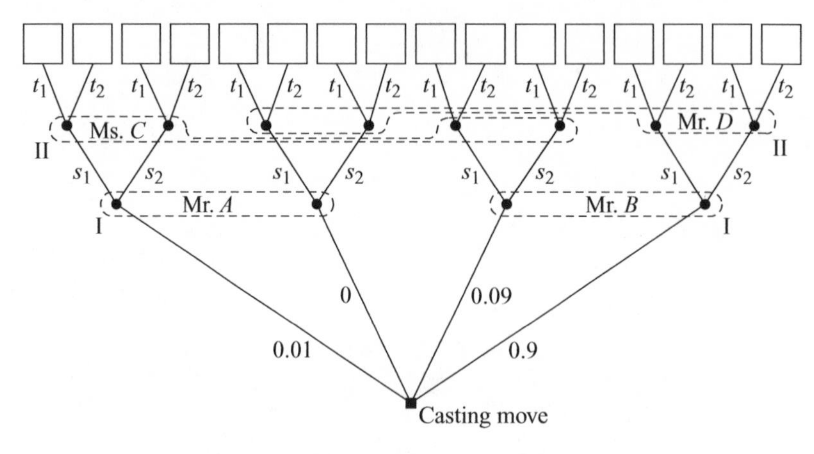


Figure 15.12 The skeleton game tree for Exercise 15.10.17.

phil

18. Von Neumann and Morgenstern get the right answer in the zero-sum game constructed in the previous exercise by applying the maximin principle. Why is the same not true of the actors in Exercise 15.10.16?

phil

19. This exercise asks still more about the problem of Exercise 15.10.15.
    a. How does the tree of Figure 15.12 need to be modified for it to represent the four-player game studied in Exercise 15.10.16?
    b. If the players needed to randomize, explain why the approach of Exercise 15.10.16 would amount to using behavioral strategies where Exercise 11.10.17 would use mixed strategies.
20. This is the final exercise using the problem of Exercise 15.10.15. It is now common knowledge that the actors' beliefs are as shown in Figure 15.13.
    a. Show that the specified beliefs are inconsistent (Section 15.5.2).
    b. Model the situation as a two-player game of imperfect information played between Von Neumann and Morgenstern.
    c. Compute enough of the strategic form to demonstrate that the game is *not* zero sum, even though the payoffs in each cell of Figure 15.11(c) sum to zero.

econ

21. Each of two agents simultaneously decides whether to pay for the provision of a *public good*. The good is said to be public because, if it is made available, an agent who free rides by paying nothing gets just as much pleasure from its enjoyment as an agent who paid for it. Figure 15.14(a) shows what the payoffs would be if all costs and benefits were common knowledge. In this payoff table, $c_i$ represents the cost to player $i$ of ensuring that the public good is available.
    a. Explain why Figure 15.14(a) is a version of Chicken (Figure 6.3(c) and Figure 6.15(a)). Find its three Nash equilibria in the case when $0 < c_i < 4$. If $c_1 = c_2 = 1$ and the mixed Nash equilibrium is used, how often will the public good be provided? Answer the same question when $c_1 = c_2 = 3$.
    b. Now consider the case when the costs $c_i$ aren't common knowledge. Assume instead that each agent's cost can be "high" ($c_i = 3$) or "low" ($c_i = 1$). It is common knowledge that each cost is determined independently and that the

# 16 *Getting Together*

## 16.1 Bargaining

We can achieve a lot more by getting together and cooperating than would be possible if we all just did our own thing. The surplus that we are able to create by pooling our talents and resources is often hugely larger than the sum of its parts. But who will get together to create what surplus? What bargain will we make about how to divide the surplus we create?

These questions are studied in cooperative game theory. This chapter opens the subject by broadening our understanding of the kind of economic bargaining that we looked at when studying the Edgeworth box in Section 9.4.

### 16.2 Cooperative Game Theory

Game theorists are often accused of treating life as a zero-sum game. Can't we see that human beings are social animals for whom cooperation is the norm?

Critics who take this line are unaware that anything has happened in game theory since the publication of Von Neumann and Morgenstern's *Theory of Games and Economic Behavior.* Since nobody knew how to extend Von Neumann and Morgenstern's notion of the rational solution of a noncooperative game beyond the strictly competitive case, the early literature necessarily concentrated on two-person, zero-sum games. This literature was then quoted by self-proclaimed strategic experts pretending to be using game theory while proposing wildly dangerous nuclear adventures in the depths of the cold war. All this was long ago, but the prejudice is renewed every time a game theorist points out the fallacy in whatever attempt to justify rational cooperation in the Prisoners' Dilemma is currently popular.

To set the record straight, not only do game theorists agree that human beings are social animals who cooperate most of the time, we also think that they are usually rational to do so in the games they commonly play. We see no inconsistency in simultaneously denying that rational cooperation is possible in two-person, zero-sum games or the Prisoners' Dilemma because such games arise only rarely in everyday life. If they were representative of the games people commonly play, we simply wouldn't have evolved as social animals.

*Cooperative or Noncooperative Theory?* Why does it make sense for players to trust each other in some situations and not in others? When trying to answer such questions, we have no choice but to use the methods of *noncooperative game theory* to which the previous chapters of this book were devoted. The alternative would be to assume some of what we should be proving by postulating mechanisms that induce people to cooperate without explaining how they work.

Noncooperative game theory is therefore not the study of conflict, as critics of game theory imagine, but the study of games in which any cooperation is fully explained by the choice of strategies the players make. For example, in Chapter 17 we will study noncooperative bargaining games, in which the players' choice of bargaining strategies may or may not lead to an agreement. If the players reach agreement, they will have succeeded in cooperating on a negotiated outcome. If they don't, then cooperation will have broken down.

Cooperative game theory differs from noncooperative game theory in abandoning any pretension at explaining *why* cooperation survives in our species. It postulates instead that the players have access to an unmodeled black box whose contents somehow resolve all the problems of commitment and trust that have worried us periodically since Section 1.7.1. In management jargon, cooperative theory assumes that the problem of how cooperation is sustained is solved "off model" rather than "on model" as in noncooperative theory. In studying ideas from cooperative game theory in this chapter, we can therefore only ask *what* should be expected to happen when rational cooperation is unproblematic, leaving the question of *how* rational cooperation might cease to be a problem for the next chapter.

*Pandora's Box.* When Pandora opened the box containing the gifts the gods intended for the human race, only hope didn't fly away. However, the black box of cooperative game theory must contain something more than hope if the theory is to be relevant to anything in real life.

In economic applications, one can sometimes argue that the black box contains all the apparatus of the legal system. The players then honor their contracts for fear of being sued if they don't. In social applications, the black box may contain the reasons why the players care about the effect that behaving dishonestly in the present may have on their reputation for trustworthy behavior in the future. One can even argue that the black box contains the results of our childhood conditioning or an inborn aversion to immoral behavior.[1]

[1]Although, as in Section 1.4.1, one can then stick with noncooperative theory after suitably modifying the game's payoffs or strategies.

The big mistake is to follow the Pandoras lampooned in Section 1.7 for seeking to furnish the black box of cooperative game theory with nothing more than the fond hope that conflict would disappear if only people would behave rationally. Much conflict in real life is admittedly stupid, but we won't make people less stupid by teaching them that their hearts are more rational than their heads.

## 16.3 Cooperative Payoff Regions

As a bare minimum, the cooperative black box needs to contain a preplay negotiation period. During this negotiation period, the players are free to sign whatever agreements they choose about the strategies to be used in the game they are about to play. Previous chapters have emphasized the difficulty of making commitments stick, but the preplay agreements of cooperative game theory are understood to be *binding*. Once the players have signed an agreement, there is no way they can wriggle out of their contractual obligations should they later prove to be inconvenient.

With binding preplay agreements, the strategic structure of the game becomes largely irrelevant. Nor are the details of the agreements commonly important. All the players really need to know about an agreement is what payoffs each will receive if it is implemented. In cooperative game theory, we therefore transfer our focus from a game's *strategy space* to its *payoff space*.

### 16.3.1 From Strategies to Payoffs

Section 5.5.1 introduces a simplified Cournot game in which Alice can produce either $a=4$ or $a=6$ hats. Bob can produce $b=3$ or $b=4$ hats. The players are assumed to be maximizers of expected profit. Their payoff functions are therefore $\pi_1(a,b)=(p-c)a$ and $\pi_2(a,b)=(p-c)b$, where $c$ is the unit cost of production and $p=K-a-b$ is the price at which hats sell. The strategic form of Figure 5.11(c) tells us everything we need to know about these payoff functions when $c=3$ and $K=15$.

Figure 16.1 is a more complicated representation of the strategic form of the game that separates its strategy and payoff spaces. The payoff functions $\pi_1 : \mathbb{R}^2 \to \mathbb{R}$ and $\pi_2 : \mathbb{R}^2 \to \mathbb{R}$ are represented by arrows linking the two spaces. For example, the arrow linking the strategy pair $(6,4)$ to the payoff pair $(12,8)$ indicates that Alice will get \$12 and Bob will get \$8 when Alice produces 6 hats and Bob produces 4 hats. Mathematicians say that the arrows represent the *vector function* $\pi : \mathbb{R}^2 \to \mathbb{R}^2$ defined by

$$\pi(a,b) = (\pi_1(a,b), \pi_2(a,b)) .$$

The game's unique Nash equilibrium is the strategy pair $(a,b)=(4,4)$, which corresponds to the payoff pair $N=(16,16)$ in Figure 16.1. However, there is no particular reason to anticipate the play of an equilibrium when the players can sign binding preplay agreements. For example, Alice and Bob could conspire to screw the consumers by jointly acting like a monopolist. They would then restrict the total supply of hats to drive up the price. The nearest they can get to imitating a monopolist in our game is to agree on playing the strategy pair $(a,b)=(4,3)$. The corresponding payoff pair is $M=(20,15)$.

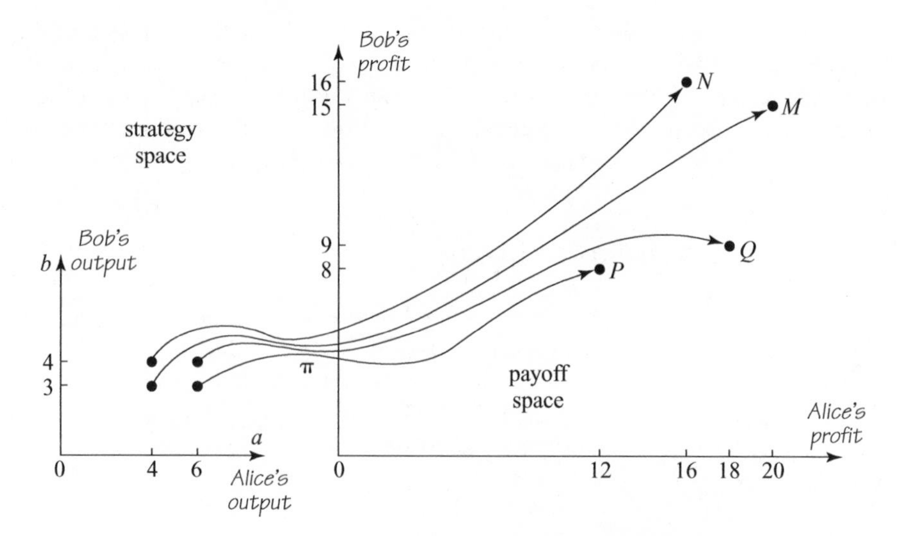


Figure 16.1 Separating payoff and strategy spaces. The diagram is an alternative representation of the strategic form of Figure 5.11(c). If mixed strategies are ignored, the cooperative payoff region of the game is the set $\{M, N, P, Q\}$.

Section 6.6.1 introduced cooperative payoff regions as the set of all payoff pairs that are available to the players if they can write binding preplay agreements on how to play the game. If we don't allow mixed strategies, the players can agree to implement any of the four pairs of pure strategies that are possible in the game. Its cooperative payoff region then consists of the four points *M, N, P,* and *Q* in payoff space.

It isn't usually practical to draw an equivalent to Figure 16.1 when mixed strategies are allowed.[2] However, we know from Section 6.5 that the cooperative payoff region $X$ is then simply the convex hull of the set of pure payoff pairs, as shown in Figure 16.2(a).

### 16.3.2 Free Disposal

In cooperative game theory, the players are allowed to make any binding agreement they choose. What else goes into the cooperative black box depends on the modeler's aims. The ability to dispose of utility at will is usually a harmless addition. If $x$ is a payoff profile on which the players can agree and $y \leq x$, then the players can achieve $y$ by agreeing that each player will burn an appropriate amount of money after $x$ has been implemented.

A player might dispose of utility by burning some money or shooting himself in the foot. But why would a rational player ever want to do such a thing? Section 5.5.2 gives some examples in which Bob might have strategic reasons for making himself worse off, but the chief reason for allowing free disposal is that we shouldn't exclude payoff profiles from the players' feasible set unless they are actually impossible to implement. If some of the profiles included in the feasible set will never be chosen, this fact will emerge when we determine what is optimal for the players.

[2]Exercise 16.12.1 treats the exceptional case of $2 \times 2$ bimatrix games (Section 6.2.2).

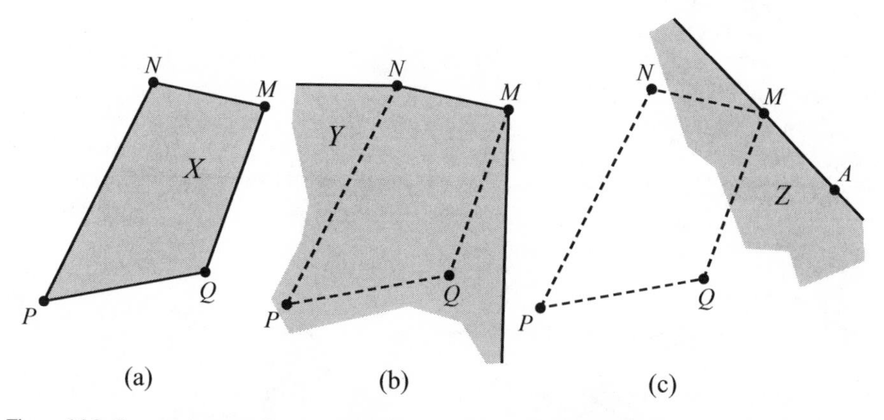


Figure 16.2 Cooperative payoff regions. Figure 16.2(a) shows the cooperative payoff region $X$ of the simplified Cournot game of Figure 16.1 when mixed strategies are allowed. Figure 16.2(b) shows how the cooperative payoff region expands to $Y$ when free disposal is permitted. Figure 16.2(c) shows how the cooperative payoff region expands again to $Z$ when utility can be transferred.

When free disposal is allowed, the cooperative payoff region $X$ must be replaced by a simpler region $Y$. For example, we have seen that Alice and Bob can agree on any payoff pair $x$ in the set $X$ of Figure 16.2(a) when playing the Cournot game of Figure 16.1. With free disposal, they can also agree on any $y$ with $y \leq x$ since this vector inequality means that $y_1 \leq x_1$ and $y_2 \leq x_2$ (Section 5.3.2). The geometric interpretation is that $y$ lies to the southwest of $x$. When we gather together all such $y$ for all $x$ in the set $X$, we obtain the set $Y$ illustrated in Figure 16.2(b).

Cooperative payoff regions are always convex. When free disposal is allowed, they are also *comprehensive*. This means that if $x$ is in the set $Y$ and $y \leq x$, then $y$ is in $Y$ as well.

### 16.3.3 Compensation

In real life, Alice will sometimes persuade Bob to use one strategy rather than another by offering him money to compensate him for the loss he may thereby incur. When such transactions are illegal, the payment is called a bribe.

Compensation can be included in the contents of the cooperative black box by assuming that the players are able to transfer utils to other players as well as to the trash can. Game theorists use the morally neutral term *side payments* when referring to such transfers.

When side payments are allowed, the set $Y$ of Figure 16.2(b) must be replaced by the set $Z$ of Figure 16.2(c). For example, to achieve the payoff pair $R = (15, 20)$, Alice and Bob can agree to implement the payoff pair $M = (20, 15)$, after which Alice gives 5 utils to Bob.

In general, if $y$ is in the set $Y$ and $z = (y_1 - r, y_2 + r)$, then $z$ is in the set $Z$, whatever the value of the side payment $r$. With two players, the set $Z$ therefore always consists of all payoff pairs to the southwest of some straight line of slope $-1$. In our example, the straight line passes through the point $M$.

*Transferable Utility?* Alert readers will recall from Section 4.6.3 that attempting to compare the utils of different players creates problems that get worse if one goes further and asks that utils be transferable as well as comparable. Utils aren't real objects and thus can't really be transferred. Only physical commodities can actually pass from one hand to another.

Transferable utility therefore only makes proper sense in special cases. The leading case arises when both players are risk neutral and their Von Neumann and Morgenstern utility scales have been chosen so that their utility for a sum of money $x$ is simply $u_1(x) = u_2(x) = x$. Transferring one util from Alice to Bob is then just the same as Alice handing Bob a dollar.

## 16.4 Nash Bargaining Problems

John Nash idealized a bargaining problem as a pair $(X, \xi)$. In such a *Nash bargaining problem,* $X$ represents the set of feasible payoff pairs on which two players can agree, and $\xi$ is a payoff pair in $X$ called the *status quo,* which represents the consequences of a disagreement.

In Chapter 17, we will find that Nash's simple model works fine in some contexts but needs to be elaborated in others. The minimal extension that captures the essence of wage bargaining requires that we recognize *two* disagreement points: a breakdown point $b$ and a deadlock point $d$. A bargaining problem then becomes a triple $(X, b, d)$. Nash's original formulation of a bargaining problem as a pair $(X, \xi)$ is treated as the special case $b = d = \xi$.

The Coase theorem tells us that rational players will agree (Section 9.4.1). So why worry about what would happen if they were to disagree? The answer is that counterfactuals always matter when decisions are made (Section 14.1). I never cross the road when a car is coming because I would be run over if I did. Something that isn't going to happen therefore determines *when* I cross the road. Similarly, what would happen if Alice and Bob were to disagree determines *what* agreement they make when they do agree.

### 16.4.1 The Feasible Set

A payoff pair $x$ is feasible if the players can agree on a deal that results in their receiving the payoffs specified by $x$. We always assume that the set $X$ of feasible payoff pairs satisfies three requirements:

The set $X$ is convex.
The set $X$ is closed and bounded above.[3]
Free disposal is allowed.

These conditions are always satisfied when we expand the cooperative payoff region of a finite game by allowing free disposal. As we saw in Section 6.6, it would be necessary to base our model of the feasible set $X$ on a different payoff region if we weren't restricting our attention to contracts that are automatically binding. However, much of the time we won't need to worry about where a particular feasible set $X$ came from at all. It will be enough that it satisfies our three requirements.

[3]A set is closed if it contains all its boundary points. A set $S$ is bounded above if there exists $b$ such that $x \leq b$ for each $x$ in $S$.

### 16.4.2 Divide-the-Dollar

A Beverley Hills mansion is worth \$4 million to its owner and \$5 million to a potential buyer. By getting together and agreeing a sale, the buyer and the seller can create a surplus of \$1 million. How this surplus is divided between them is decided by bargaining. A simple model that captures the essence of this archetypal bargaining problem is traditionally known as Divide-the-Dollar.

The story that goes with the model envisages a philanthropist who offers Alice and Bob the opportunity to share a dollar—provided they can agree on how to divide it between them. If they can't come to an agreement, the philanthropist takes his dollar back again. In this story, the dollar represents the surplus over which two economic agents bargain. The philanthropist's provision that the dollar is available only if Alice and Bob can reach an agreement represents the fact that there will be no surplus unless the agents get together to create it.

What determines who gets how much? Attitudes to risk obviously matter since Alice is likely to get more of the dollar if she is less averse to the risk of the negotiations breaking down than Bob. Figure 16.3 shows how Alice's and Bob's risk attitudes are reflected in the shape of the feasible set $X$.

In money terms, Alice and Bob can agree on any pair $m = (m_1, m_2)$ of dollar amounts in the set $M = \{m : m_1 + m_2 \leq 1\}$ of Figure 16.3(a). To achieve the point $(0.4, 0.6)$, the dollar is split so that Alice gets 40 cents and Bob gets 60 cents. To achieve $(2.4, -1.4)$, they agree to split the dollar 40 : 60, and then Bob pays Alice an additional two dollars from his pocket. To achieve $(-3, -3)$, they can refuse the philanthropist's dollar and then each burn three dollars taken from their own pockets.

Assume that both players care about nothing except how much money they get from the deal. In particular, Alice is interested neither in helping Bob along nor in doing him any injury, unless this results in some financial benefit for herself. Bob feels exactly the same about Alice. Player $i$'s Von Neumann and Morgenstern utility function $u_i : M \to \mathbb{R}$ for deals is then given by

$$u_i(m) = v_i(m_i),$$

where $v_i : \mathbb{R} \to \mathbb{R}$ represents player $i$'s utility for money.[4]

Figure 16.3(b) illustrates the feasible set $X$ for Divide-the-Dollar when Alice and Bob are risk averse, so that $u_1$ and $u_2$ are concave functions. In this case $X = u(M)$ is necessarily convex.[5]

[4]It is taken for granted that $v_i$ is strictly increasing, continuous, bounded above, and unbounded below on $\mathbb{R}$. If $v_i$ is concave, these assumptions guarantee that the feasible set $X$ of payoff pairs satisfies the conditions given in Section 16.4.1.

[5]Just as $m = (m_1, m_2)$, so $u(m) = (u_1(m), u_2(m))$. An argument to show that $u(M) = \{u(m) : m \in M\}$ is convex goes like this. Suppose $x$ and $y$ are in $u(M)$. Then $x = u(m)$ and $y = u(n)$ for some $m$ and $n$ in $M$. To prove that $u(M)$ is convex, we need to show that $ax + by \in u(M)$ for each $a$ and $b$ with $a + b = 1$, $a \geq 0$, and $b \geq 0$. Since $M$ is convex, $am + bn \in M$. Thus $u(am + bn) \in u(M)$. If $u(am + bn) \geq z$, it follows that $z \in u(M)$ because free disposal is allowed. The utility function $u_i$ is concave. Thus $u_i(am + bn) \geq au_i(m) + bu_i(n)$ $(i = 1, 2)$. Hence, $u(am + bn) \geq au(m) + bu(n) = ax + by$. It follows that $z = ax + by \in u(M)$, and so $u(M)$ is convex.

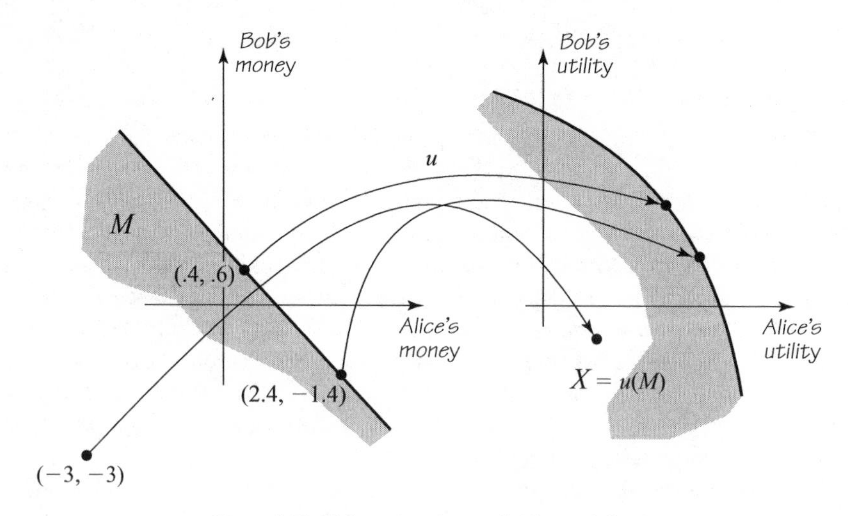


Figure 16.3 Risk-averse players dividing a dollar.

### 16.4.3 The Bargaining Set

A new item now needs to be added to the contents of our cooperative black box. We require that the payoff profile generated by a rational agreement lies in what Von Neumann and Morgenstern called the *bargaining set.* This is the equivalent in payoff space of Edgeworth's contract curve (Section 9.4.1). The bargaining set therefore consists of all Pareto-efficient payoff profiles that assign all players as least as much as they can get without doing a deal at all.

*Coase Theorem.* Economists refer to the Coase "theorem" when claiming that rational agreements will be Pareto efficient (Section 9.4.1). The claim is made only with strong qualifications. For example, nobody would expect the outcome to be Pareto efficient if it were very costly for the bargainers to exchange messages or to find out what agreements are feasible. Economists idealize such frictions away by saying that the Coase theorem applies only when the transaction costs are zero.

Saying that transaction costs are zero is often also understood to sweep away any informational differences between the bargainers, but airbrushing out the informational issues in this way is a bad mistake. For example, if it is common knowledge that the valuations of a potential buyer and a seller of a house are independent and equally likely to be anything between \$4 million and \$5 million, then the result of optimal bargaining is very inefficient indeed. Even when the bargaining process is chosen to maximize the expected surplus that rational bargainers can achieve, the house is sold only when it is worth a quarter million more to the buyer than the seller! (Exercise 20.9.8) It is therefore important to keep in mind that the results of this chapter properly apply only when everybody's preferences are common knowledge among the bargainers.

→ 16.4.4

*Pareto Improvements.* If $x$ and $y$ are payoff profiles, $x \gg y$ means that all players like the payoff they get with $x$ more than the payoff they get with $y$. In such a case, we say that $x$ is a weak Pareto improvement on $y$. To write $x > y$ means that all

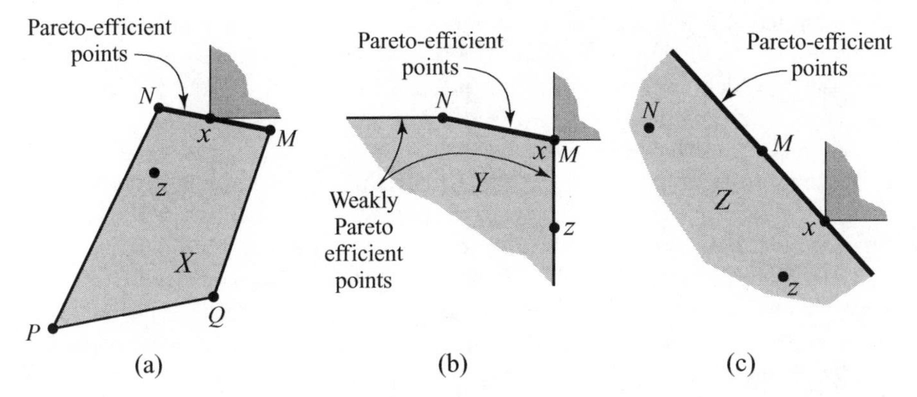


Figure 16.4 Pareto efficiency. The sets $X$, $Y$, and $Z$ are those of Figure 16.2. Notice that some boundary points of $Y$ are weakly Pareto efficient but not (strongly) Pareto efficient.

players like $x$ at least as much as $y$ and that at least one player likes $x$ strictly more than $y$ (Section 5.4.1). We then say that $x$ is a strong Pareto improvement on $y$.

A feasible payoff profile $x$ is *weakly* Pareto efficient if the only payoff profiles that weakly improve $x$ are infeasible (Section 1.7). A feasible payoff profile $x$ is *strongly* Pareto efficient if the only payoff profiles that strongly improve $x$ are infeasible.[6]

The Coase theorem is "proved" by arguing that rational players with negligible transaction costs won't cease bargaining when a Pareto-inefficient proposal is on the table because each will see that both have something to gain by replacing the current proposal with an alternative that they both prefer.

To argue that *every* player will want to continue bargaining rather than sticking with $y$, we need a *weak* improvement of $y$ to be available. So the Coase argument only supports the conclusion that rational bargainers will agree on a *weakly* Pareto-efficient deal. But we will follow the common practice of interpreting Pareto efficiency in the strong sense unless something is said to the contrary. Nearly always, it doesn't matter which sense is adopted.

We use this convention for the first time in Figure 16.4, which shows the Pareto-efficient points of the sets $X$, $Y$, and $Z$ of Figure 16.2. The shaded sets indicate the Pareto improvements on $x$. In each case, $z$ is a feasible point for which $x$ is a Pareto improvement.

### 16.4.4 Maximizing the Cake

Bargaining is typical of human interaction in that both competition and cooperation need to be handled simultaneously. Cooperation is necessary to create a large cake. Competition arises because each player wants a big share. In real life, people often squabble so fiercely over who will get how much of the cake that no cake gets baked

[6]The use of the words *weak* and *strong* in this context can sometimes be confusing. With the definitions used here, a feasible set $X$ has fewer strongly Pareto-efficient points than it has weakly Pareto-efficient points. It is therefore stronger to say that $x$ is strongly Pareto efficient than to say $x$ is weakly Pareto efficient.

at all, but the Coase theorem says that rational players will find some compromise that allows them to get together to produce the biggest of all possible cakes.

Cases in which bargaining separates into the cooperative activity of identifying the largest cake and the competitive activity of deciding how to share it are commonplace. Consider, for example, what happens if we allow side payments or bribes in the simplified Cournot game of Section 16.3.1. When the Coase theorem applies, a rational Alice and Bob will agree on some point $A$ on the boundary of the feasible set $Z$ of Figure 16.2(c). To obtain $A$, Alice and Bob must first jointly use the strategies that implement the payoff pair $M = (20, 15)$. This act corresponds to their maximizing the cake by colluding to act like a monopolist. Their subsidiary agreement on how to share the cake is then determined by the size of the side payment that one player must make to the other to get them from $M$ to $A$.

→ 16.4.5

*Wage Bargaining.* We return to Alice's hat factory as a source of examples of wage bargaining. Bob will be the union representative for Alice's workforce of one mad hatter.

In this chapter, Alice's company makes hats that sell at \$8 each. The production function is given by $h = \sqrt{l}$, wlhere $h$ is the number of hats produced in a day, and $l$ is the total number of hours of labor expended each day by the hatter. When the hourly wage that Alice pays to the hatter is $w$, her daily profit is $\pi = 8\sqrt{l} - wl$. If the hatter regards each hour of leisure as being worth \$1, then his total daily income is $I = wl + (24 - l)$.

What will happen when Alice and Bob bargain over the future wage $w$ and the level of employment $l$? We assume that Alice wants to maximize her profit $\pi$ and Bob wants to maximize the hatter's total income $I$. Both players are risk neutral. The total available surplus is

$$s = \pi + I = 8\sqrt{l} + 24 - l,$$

which is maximized when $l = 16$. If the Coase theorem applies, Alice and Bob will therefore agree that the total level of employment should be $l = 16$ hours a day. This creates a daily surplus of \$40. Their negotiations over the wage rate $w$ then reduce to playing Divide-the-Dollar with a \$40 cake instead of a \$1 cake.

### 16.4.5 Breakdown

Returning to the general case, suppose that Alice and Bob's attempt to come to an agreement breaks down altogether. Both players will then be forced to do as well as they can independently of each other. In Divide-the-Dollar, they will simply return to what they were doing before the philanthropist appeared with his dollar. In the simplified Cournot game of Section 16.3.1, we assume that a breakdown results in the game being played noncooperatively, so that Alice and Bob use their Nash equilibrium strategies $a = 4$ and $b = 4$.

A player's breakdown payoff $b_i$ is the most that he or she can get when acting independently of the other player. The profile $b$ of both players' breakdown payoffs is called the *breakdown point* of the bargaining problem.

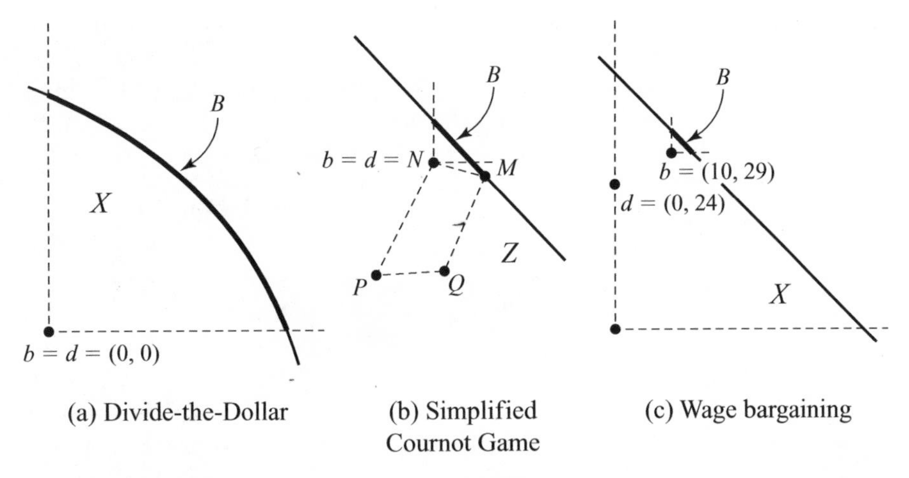


Figure 16.5 Three bargaining problems. It is assumed that $b = d$ in Divide-the-Dollar and the simplified Cournot game with bribes. In the wage bargaining problem, $d = (0, 24)$ and $b = (10, 29)$.

Players would be foolish to sign on to a deal that gives them less than they can get on their own.[7] Agreements in the bargaining set are therefore assumed to consist of all Pareto-efficient points $x$ in $X$ that satisfy $x \gtrsim b$. Taking the breakdown point $b$ to be $(0, 0)$ in Divide-the-Dollar and the Nash equilibrium outcome $(16, 16)$ in our simplified Cournot game with bribes, we obtain the bargaining sets $B$ illustrated in Figures 16.5(a) and 16.5(b).

### 16.4.6 Deadlock

If the bargaining breaks down altogether in our wage bargaining example, the firm and the workers will look around for their next best options. Alice may hire fewer skilled workers or go bankrupt. Bob's workers may find poorer jobs or go on the dole. The breakdown point $b$ is the pair of incomes that Alice and Bob would derive from these outside options.

Economists using the idea of a Nash bargaining problem $(X, d)$ in predicting wage settlements have traditionally identified the disagreement point $d$ with the breakdown point $b$. But the location of $b$ can't be crucial in determining the deal that Alice and Bob will reach. If it were, Alice would always get a better deal when Bob's outside option is reduced. But why should Bob make any new concessions? Alice will threaten to walk off if he doesn't, but her threat isn't credible. Both the payoff he is offering her now and her outside option are exactly the same as before, and she didn't walk off then!

We won't be throwing $b$ away altogether because we need it to define the bargaining set. However, we will add a second disagreement point to our armory by asking what would happen if the negotiations were to get deadlocked, with Alice and Bob glaring at each other across the negotiation table without either giving up and

[7]Players who understand this are traditionally said to be individually rational. The term isn't used much in this book because it is often taken to imply that the players' breakdown payoffs should be identified with their security levels (Section 18.2).

walking off for good. The pair of incomes they would get during such an impasse will be called the *deadlock point d.*

During a strike, for example, Alice's and Bob's deadlock payoffs are likely to be substantially smaller than their breakdown payoffs. When modeling a bargaining problem as a triple $(X, b, d)$, we will therefore always simplify by assuming that $d \leq b$.

*Wage Bargaining 2.* Figure 16.5(c) shows our wage bargaining problem on the assumption that Alice gets no income during the dispute, but the hatter continues to value his leisure at \$1 per hour. The deadlock point is therefore $d = (0, 24)$. Alice has the outside option of selling out and retiring to Palm Beach. The hatter could go and work for the post office. Alice regards each day in Palm Beach as being worth \$10, and so her breakdown payoff is 10. The government would pay the hatter \$2 per hour for a total of five hours of work a day, and so Bob's breakdown payoff is $2 \times 5 + (24 - 5) = 29$. Thus, $b = (10, 29)$.

### 16.4.7 Stocks and Flows

In compressing a real-life bargaining problem into Nash's format, nothing that is strategically relevant must be overlooked. It is particularly important to take note of whether the goods being traded are stocks or flows and whether they are durable or perishable.

The case of bargaining in the Edgeworth box provides an example (Section 9.4.1). If Adam has a stock of $F$ fig leaves and Eve has a stock of $A$ apples, then the result of their negotiation breaking down will simply be that the players consume their initial endowments on their own. Their bargaining set therefore corresponds to the contract curve. On the other hand, none of the information given in an Edgeworth box tells us anything about Adam and Eve's deadlock point. However, if their endowments will dwindle away to nothing during an indefinite standoff, then a deadlock corresponds to their both having no endowments at all.

The breakdown point remains the same if Adam's and Eve's endowments are *flows* of $F$ fig leaves and $A$ apples *per day*. But if we assume that apples and fig leaves perish if not consumed on the day they are delivered, the deadlock point is now the same as the breakdown point because Adam and Eve will be consuming their daily endowments during an indefinite standoff.

→ 16.6

## 16.5 Supporting Hyperplanes

If a point $\xi$ lies outside the interior of a convex set $S$, then we can find a hyperplane that separates $\xi$ and $S$ (Theorem 7.11). In the case when $\xi$ is actually a boundary point of $S$, we obtain a hyperplane $H$ through $\xi$ with the property that $S$ is contained in one of the two half spaces it defines. We then say that $H$ is a *supporting* hyperplane to $S$ at $\xi$. The half space that contains $S$ is similarly a supporting half space.

Figure 16.6 shows some supporting hyperplanes in two-dimensional space, where a hyperplane is just a straight line. Notice that a supporting hyperplane at a point $\xi$ where the boundary is smooth is just the tangent hyperplane at $\xi$. If $\xi$ is located at a corner, there are multiple supporting hyperplanes at $\xi$.

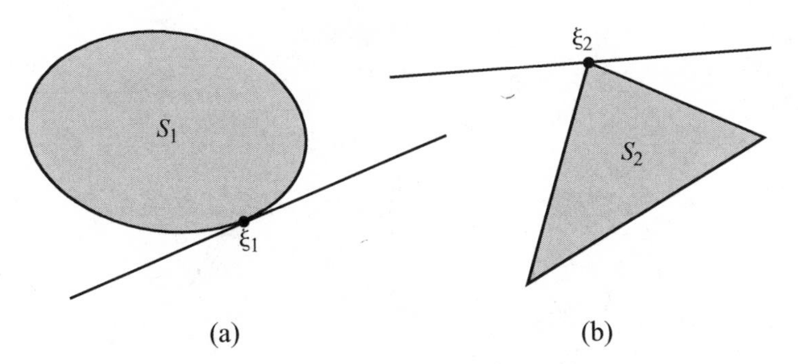


Figure 16.6 Supporting hyperplanes.

## 16.6 Nash Bargaining Solution

Von Neumann and Morgenstern thought that game theory could say no more about rational agreements than that they lie in the bargaining set. Nash took the reverse position by proposing rationality assumptions that pinpoint a unique payoff pair in the bargaining set. This payoff pair is said to be the *Nash bargaining solution* of the problem.

### 16.6.1 Bargaining Solutions

A *bargaining solution* gives the payoff pair on which Alice and Bob will agree, not just for one bargaining problem but for *all* bargaining problems. Mathematically, we take $\mathscr{B}$ to be the set of all bargaining problems and define a bargaining solution to be any function $F : \mathscr{B} \to \mathbb{R}^2$ for which the payoff pair $F(X, b, d)$ always lies in the feasible set $X$.

We are especially interested in the generalized *Nash bargaining solution* $G : \mathscr{B} \to \mathbb{R}^2$. Given two parameters $\alpha \geq 0$ and $\beta \geq 0$, we define $G(X, b, d)$ as the point $s = x$ in $X$ at which the Nash product

$$(x_1 - d_1)^\alpha (x_2 - d_2)^\beta$$

is maximized, subject to the requirement that $x \geq b$.

Figure 16.7(a) shows that $s = G(X, b, d)$ lies in the bargaining set $B$ of the problem. As we allow $\alpha$ to increase relative to $\beta$, $s$ moves through points of $B$ that are successively more favorable to Alice. The parameters $\alpha$ and $\beta$ therefore tell us who is more advantaged by the rules of the bargaining game. For this reason, $\alpha$ and $\beta$ are called Alice's and Bob's *bargaining powers.*

A chess analogy may help in interpreting $\alpha$ and $\beta$. I would lose if I played Black against a grandmaster because grandmasters play well and I play badly. But what if some genius published the solution of chess the night before our game? If it says that White has a winning strategy, I would still lose—but not because I would be playing chess less skillfully than a grandmaster. We would now both play optimally. The grandmaster would win only because the *game* puts me at a disadvantage.

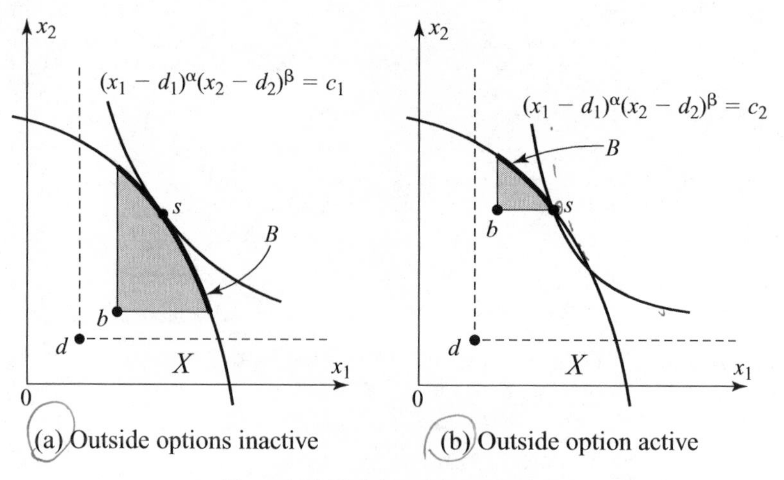


Figure 16.7 Nash bargaining solutions.

Similarly, $\alpha$ and $\beta$ aren't measures of bargaining skill. If Alice and Bob are rational, they will both bargain as well as it is possible to bargain. If they have different bargaining powers, it is because they are advantaged or disadvantaged by their role in the bargaining game. For example, players endowed with preferences that make them anxious for an early agreement will have less bargaining power than players who are willing to wait (Section 17.5.3).

When the the Nash bargaining solution is mentioned without a qualifier, it is always the *regular* or *symmetric* Nash bargaining solution that I have in mind. This arises when $\alpha = \beta$. One might say that the balance of advantage in the bargaining game is then effectively even.

*Divide-the-Dollar 2.* We return to Section 16.4.2, taking Alice's and Bob's utility functions to be $v_1(z) = z^\gamma$ and $v_2(z) = z^\delta$, where $0 < \gamma \leq 1$ and $0 < \delta \leq 1$.[8] The Pareto-efficient utility pairs are $(z^\gamma, (1-z)^\delta)$, where $z$ is Alice's share of the dollar and $1 - z$ is Bob's share. We assume that $b = d = (0, 0)$.

The generalized Nash bargaining solution can be found by maximizing the Nash product:

$$(x_1 - d_1)^\alpha (x_2 - d_2)^\beta = z^{\alpha\gamma}(1-z)^{\beta\delta}.$$

Differentiating with respect to $z$ and setting the result equal to zero, we find that the maximizing value of $z$ is given by

$$z = \frac{\gamma\alpha}{\gamma\alpha + \delta\beta}\,; \qquad 1 - z = \frac{\delta\beta}{\gamma\alpha + \delta\beta}.$$

[8]The utility functions aren't defined when $z < 0$, but we take free disposal for granted. Note that $v_1$ and $v_2$ are strictly increasing and concave, and so the players prefer more money to less and are risk averse.

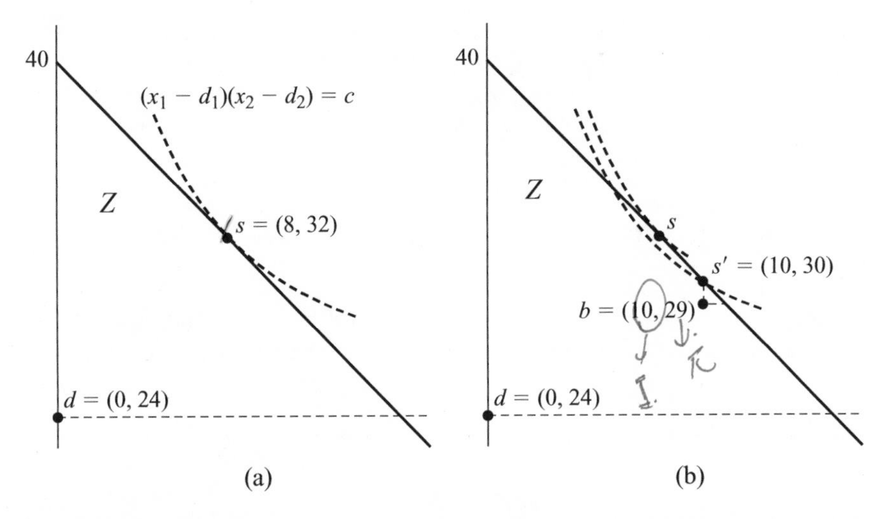


Figure 16.8 Wage bargaining. Figure 16.8(a) shows how to use the Nash bargaining solution of $(Z, d)$ to find the Nash bargaining solution of $(Z, b, d)$ when the constraint $x \geq b$ is active. When the constraint is inactive, the Nash bargaining solutions of $(Z, d)$ and $(Z, b, d)$ coincide.

To obtain the symmetric Nash bargaining solution, set $\alpha = \beta$. The dollar is then split in the ratio $\gamma : \delta$.

The moral is that it is bad to be risk averse in this kind of bargaining situation. The more risk averse you are, the less money you get. If $\gamma = \delta = 1$, so that Alice and Bob are both risk neutral, they split the dollar 50:50. If $\delta = \frac{1}{3}$, so that Bob is seriously risk averse, the split becomes 75:25.

Used car dealers therefore affect lighthearted unconcern at the prospect of losing a sale. But, as the Bible says, although they say it is naught, when they goeth their way, they boasteth (Proverbs 20:14).

*Wage Bargaining 3.* Our wage bargaining example has already been reduced to a variant of Divide-the-Dollar in which a surplus of forty dollars is to be divided. Since Alice and Bob are risk neutral, the feasible set can be identified with $Z$ in Figure 16.8(a). Taking $d = (0, 24)$ and $b = (10, 29)$ as before, we then have a bargaining problem $(Z, b, d)$ to solve.

In the absence of the constraint $x \geq b$, the symmetric Nash bargaining solution calls for the players to split the difference by agreeing on $s = (8, 32)$. Figure 16.8(a) illustrates this conclusion by showing a contour $x_1(x_2 - 24) = C$ of the relevant Nash product that touches the boundary $x_1 + x_2 = 40$ of the feasible set $Z$ at the point $(8, 32)$.

Imposing the constraint $x \geq b$ shrinks the bargaining set to $B$ in Figure 16.8(b). Drawing the contour $x_1(x_2 - 24) = c$ shows that the Nash product attains its maximum value on $B$ at the point $s' = (10, 30)$,[9] which is therefore the symmetric Nash bargaining solution of the problem $(Z, b, d)$. Note that Alice gets no more than her

[9] The constraint $x_1 \geq 10$ is said to be *active* in the maximization because the maximum would be something else if the constraint were absent.

outside option. She may threaten to retire to Palm Beach, but Bob can always make it worth her while to stay in business by offering her a tiny bit more.

Since $l = 16$ hours are worked and Bob's agreed income is $I = 30$, the formula $I = wl + (24 - l)$ tells us that his final wage will be $w = \$1.375$. Without Alice's outside option, it would have been $w = \$1.50$.

### 16.6.2 Finding Nash Bargaining Solutions

Only the relative size of the bargaining powers is significant in determining a generalized Nash bargaining solution $s$. We can therefore normalize by taking $\alpha + \beta = 1$. Instead of maximizing a Nash product, we can then use Figure 16.9(a) to locate $s$, using the following steps:

**Step 1.** Find the convex set $S$ of all points $x$ in $X$ that satisfy $x \geq b$.

**Step 2.** Focus on supporting lines to $S$ at $s$.

**Step 3.** For each supporting line, locate the points $r$ and $t$.

**Step 4.** The point $s$ we are looking for satisfies $s = \alpha r + \beta t$.

As an example of how this geometric method can be used, consider collusion in the simplified Cournot game of Section 16.3.1. When bribes are allowed, the feasible set $Z$ is shown in Figure 16.2(c). We take $d = (16, 16)$ and look for the generalized Nash bargaining solution of the problem $(Z, d)$. In the case when $\alpha = \frac{1}{3}$ and $\beta = \frac{2}{3}$, the solution $s$ is shown in Figure 16.10(a). It lies two-thirds of the way along the line segment joining $r$ to $t$.

Finding the symmetric Nash bargaining solution $s$ geometrically is even easier. Since $s$ lies halfway between $r$ and $s$ in Figure 16.9(b):

> The symmetric Nash bargaining solution $s$ is the point on the boundary of $S$ where the ray from $d$ to $s$ makes the same angle to the horizontal as a supporting line to $S$ at $s$.

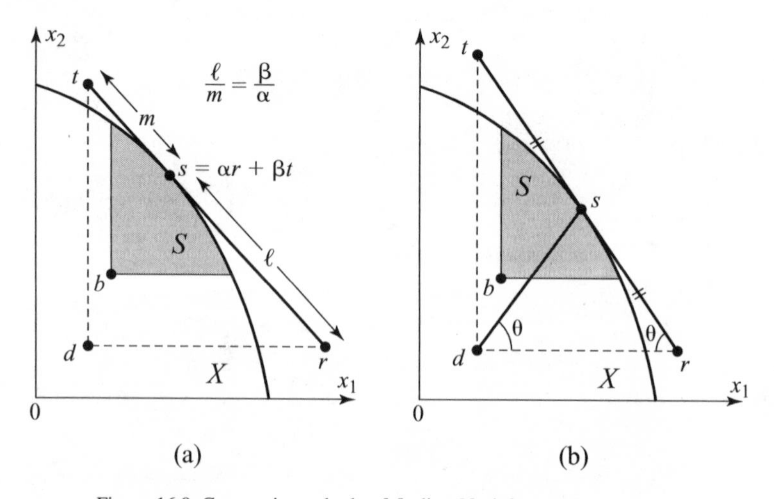


Figure 16.9 Geometric methods of finding Nash bargaining solutions.

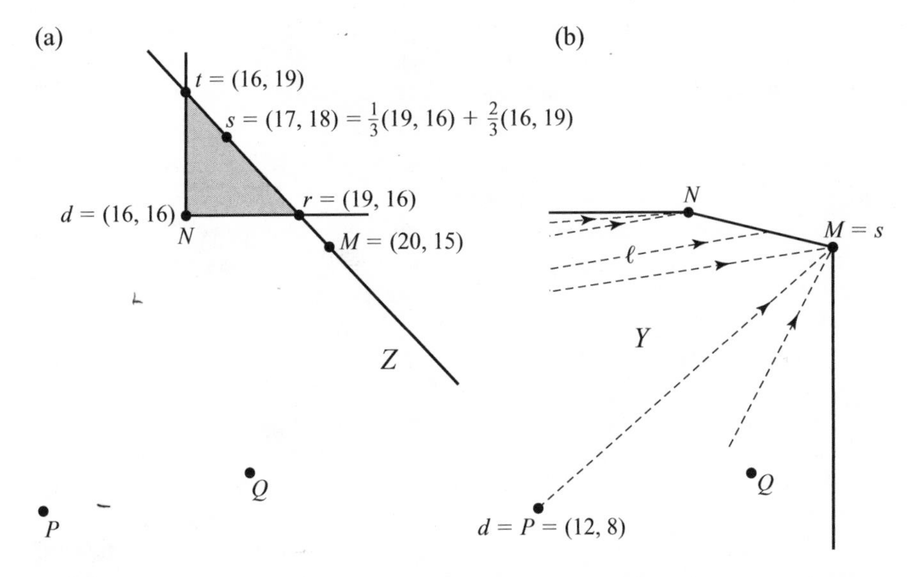


Figure 16.10 Collusion in a simplified Cournot game. Figure 16.10(a) shows the bargaining problem $(Z, d)$, in which the feasible set $Z$ is taken from Figure 16.2(c) and $d = (16, 16)$. The point $s$ is the generalized Nash bargaining solution for $(Z, d)$ with $\alpha = \frac{1}{3}$ and $\beta = \frac{2}{3}$. Figure 16.10(b) shows the bargaining problem $(Y, d)$, in which the feasible set $Y$ is taken from Figure 16.2(b) and $d = (12, 8)$. The point $s$ is the symmetric Nash bargaining solution for $(Y, d)$.

We apply this result to the case of collusion in the simplified Cournot game of Section 16.2.1 when bribes are impossible. The relevant feasible set $Y$ appears in Figure 16.2(b).

If the broken line $l$ in Figure 16.10(b) cuts the boundary of $Y$ at $s$, then $s$ is the symmetric Nash bargaining solution for the problem $(Y, d)$ for each point $d$ in $Y$ that lies on the line $l$. The line $l$ is chosen so that it makes the same angle with the horizontal as a supporting line at $s$. Since there are many supporting lines at the two corners $(16, 16)$ and $(20, 15)$ of $Y$, many broken lines pass through these two points. When you are looking for a Nash bargaining solution, you are therefore quite likely to find it at such a corner.

Since $(16, 16)$ is already a Pareto-efficient point of $Y$, the case $d = (16, 16)$ isn't very interesting. We therefore look instead at the case $d = (12, 8)$, which arises if the players can credibly threaten each other before the negotiation begins that they will maximize production if a collusive agreement can't be reached (Section 17.4.1). Figure 16.10(b) shows that the symmetric Nash bargaining solution is then $s = (20, 15)$.

### 16.6.3 Nash's Bargaining Theorem

What would a rational procedure for settling bargaining problems look like? John Nash proposed the following criteria:

- The agreed payoff pair should always be in the bargaining set.
- The final outcome shouldn't depend on how the players calibrate their utility scales.

- If the players sometimes agree on the payoff pair $s$ when $t$ is feasible, then they never agree on $t$ when $s$ is feasible.
- In symmetric situations, both players get the same.

The first property has already been discussed at length. The second property simply recognizes the fact that the choice of an origin and a unit for a utility scale is arbitrary. The fourth property isn't so much a rationality assumption as a decision to confine attention to bargaining procedures that treat the players symmetrically. It isn't used in proving the theorem that follows. This leaves the third property to be considered.

The third property is a version of the *Independence of Irrelevant Alternatives*. A story told by the venerable game theorist Bob Aumann illustrates how it should work. A committee of the prestigious Econometric Society was to decide which of Mme. Chicken, Dr. Dodo, and Prof. Eagle should be invited to give a fancy lecture. Dr. Dodo was quickly eliminated, but it took a long time to agree that the invitation should go to Mme. Chicken rather than Prof. Eagle. Someone then pointed out that Dr. Dodo couldn't make the event anyway. This observation provoked a renewal of the debate, which ended up with the invitation going to Prof. Eagle. This is a violation of the Independence of Irrelevant Alternatives, which says that the choice between Mme. Chicken and Prof. Eagle should be independent of the availability of Dr. Dodo. He is an irrelevant alternative because he wasn't going to get chosen even if he were available.

→ 16.7

*Nash's Axioms.* In order to prove a theorem, we formulate Nash's criteria as axioms or postulates concerning an abstract bargaining solution $F : \mathscr{B} \to \mathbb{R}^2$. To keep things simple, we restrict our attention to the case of bargaining problems $(X, d)$ with a single disagreement point.

The first axiom says that the bargaining solution lies in the bargaining set.

Axiom 16.1

$$\begin{aligned} &(i) \quad F(X,d) \geq d \\ &(ii) \quad y > F(X,d) \Rightarrow y \notin X. \end{aligned}$$

The second axiom says that it doesn't matter how the utility scales are calibrated. Suppose, for example, that the bargaining solution awards 50 utils to Eve. She now adopts a new utility scale so that an outcome whose old utility was $u$ is assigned a utility of $U = \frac{9}{5}u + 32$ on the new scale. If nothing else has changed, Eve should then be awarded $\frac{9}{5} \times 50 + 32 = 12$ new utils by the bargaining solution.

To express this idea more generally, we need two strictly increasing, affine transformations $\tau_1 : \mathbb{R} \to \mathbb{R}$ and $\tau_2 : \mathbb{R} \to \mathbb{R}$. Recall that a strictly increasing affine transformation is defined by $\tau_i(u) = A_i u + B_i$ where $A_i > 0$. A function $\tau : \mathbb{R}^2 \to \mathbb{R}^2$ can then be constructed from $\tau_1$ and $\tau_2$ by defining

$$\tau(x) = (\tau_1(x_1), \tau_2(x_2)) = (A_1 x_1 + B_1,\ A_2 x_2 + B_2).$$

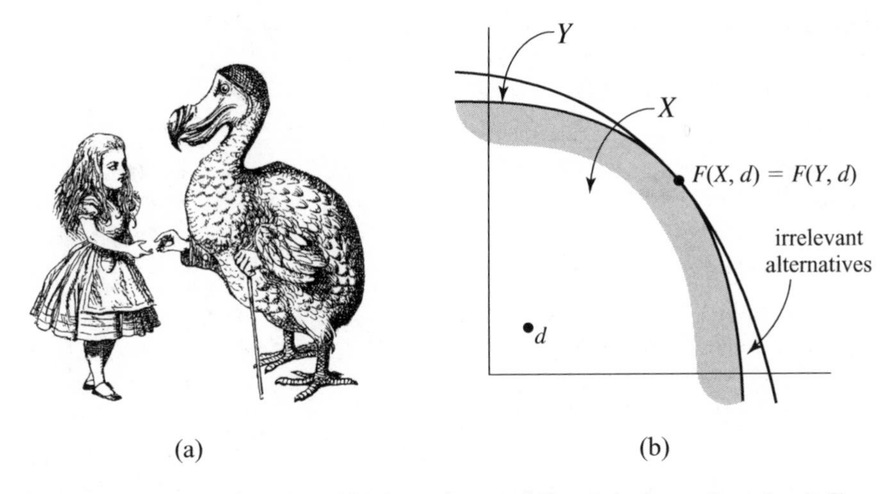


Figure 16.11 Irrelevant Alternatives. Alice is meeting one of Nature's irrelevant alternatives in Figure 16.11(a). In Figure 16.11(b), the points of $X$ not in $Y$ are irrelevant alternatives because they weren't chosen when available.

AXIOM 16.2 *Given any strictly increasing affine transformations* $\tau_1$ *and* $\tau_2$,

$$F(\tau(X), \tau(d)) = \tau(F(X, d)).$$

The third axiom formalizes the Independence of Irrelevant Alternatives. In Figure 16.11(b), the set $Y$ is a subset of $X$ that contains $F(X, d)$. The elements of $X$ that aren't in $Y$ are irrelevant alternatives. If the bargaining solution selects $F(X, d)$ for the bargaining problem $(X, d)$, then it should also select $F(X, d)$ for the bargaining problem $(Y, d)$ because the choice should be independent of the availability or unavailability of irrelevant alternatives.

AXIOM 16.3 *If* $d \in Y \subseteq X$, then

$$F(X, d) \in Y \ \Rightarrow \ F(Y, d) = F(X, d).$$

Any generalized Nash bargaining solution $G : \mathscr{B} \to \mathbb{R}^2$ satisfies Axioms 16.1, 16.2, and 16.3. The next theorem says that these are the only bargaining solutions that satisfy the axioms.[10]

THEOREM 16.1 (Nash) *If* $F : \mathscr{B} \to \mathbb{R}^2$ *satisfies Axioms 16.1–16.3, then* $F$ *is a generalized Nash bargaining solution for some bargaining powers* $\alpha$ *and* $\beta$.

*Proof* Start with the simple bargaining problem $(Z, 0)$ illustrated in Figure 16.12(a). The feasible set $Z$ consists of all payoff pairs $x$ that satisfy $x_1 + x_2 \leq 1$. The disagreement point is the zero vector $0 = (0, 0)$.

[10]The theorem is actually a generalization of Nash's theorem. It can be generalized further by omitting Axiom 16.1(ii), provided that the conclusion is altered to admit the possibility that $F(X, d) = d$ for all $(X, d)$.

**Step 1.** By Axiom 16.1, the solution $s' = F(Z,0)$ of the bargaining problem $(Z,0)$ lies on the line segment joining $r' = (1,0)$ and $t' = (0,1)$. Since $s'$ is therefore a convex combination of $r'$ and $t'$, we can write $s' = \alpha r' + \beta t'$ where $\alpha + \beta = 1$, $\alpha \geq 0$, and $\beta \geq 0$.

**Step 2.** Next consider an arbitrary bargaining problem $(X,d)$, as illustrated in Figure 16.12(c). Let $s = G(X,d)$, where $G$ is the generalized Nash bargaining solution corresponding to the bargaining powers $\alpha$ and $\beta$. Then $s = \alpha r + \beta t$ in Figure 16.12(c). The aim is to prove that $F(X,d) = G(X,d)$.

**Step 3.** Recalibrate Adam's and Eve's utility scales using strictly increasing, affine transformations $\tau_i : \mathbb{R} \to \mathbb{R}$ chosen so that $\tau_1(d_1) = \tau_2(d_2) = 0$ and $\tau_1(r_2) = \tau_2(t_2) = 1$. The affine function $\tau : \mathbb{R}^2 \to \mathbb{R}^2$ then has the property that $\tau(d) = 0$, $\tau(r) = r'$ and $\tau(t) = t'$, as illustrated in Figure 16.12(b).

**Step 4.** Since affine functions preserve convex structures, the image of the line through $r$, $s$, and $t$ remains a supporting line to the image of the set $X$. That is, the line $x_1 + x_2 = 1$ through $r'$, $s'$, and $t'$ is a supporting line to the convex set $X' = \tau(X)$. In particular, since $\tau$ preserves convex combinations, $s' = \tau(s)$. Thus, by Axiom 16.2,

$$F(Z,0) = \tau(G(X,d)). \tag{16.1}$$

**Step 5.** This is the heart of the proof. Since $X' \subseteq Z$, $F(X',0) = F(Z,0)$, by Axiom 16.3.

**Step 6.** Since $X' = \tau(X)$ and $0 = \tau(d)$, we have from (16.1) that

$$F(\tau(X), \tau(d)) = \tau(G(X,d)).$$

If $\tau^{-1} : \mathbb{R}^2 \to \mathbb{R}^2$ is the inverse function to $\tau$, it follows that[11]

$$G(X,d) = \tau^{-1}(F(\tau(X), \tau(d)))\,. \tag{16.2}$$

**Step 7.** Replace $\tau$ by $\tau^{-1}$ in Axiom 16.2. Then apply the result to the right-hand side of (16.2) to obtain

$$G(X,d) = F(\tau^{-1}(\tau(X)), \tau^{-1}(\tau(d))) = F(X,d),$$

which is what had to be proved.

### 16.6.4 Symmetry

The symmetric Nash bargaining solution $N : \mathscr{B} \to \mathbb{R}$ is the special case of a generalized Nash bargaining solution that occurs when the bargaining powers $\alpha$ and $\beta$ are equal. The fact that it treats the players symmetrically can be expressed

[11] A function $f : X \to Y$ has an inverse function $f^{-1} : Y \to X$ if the equation $y = f(x)$ has a unique solution $x = f^{-1}(y)$ in $X$ for each $y$ in $Y$.

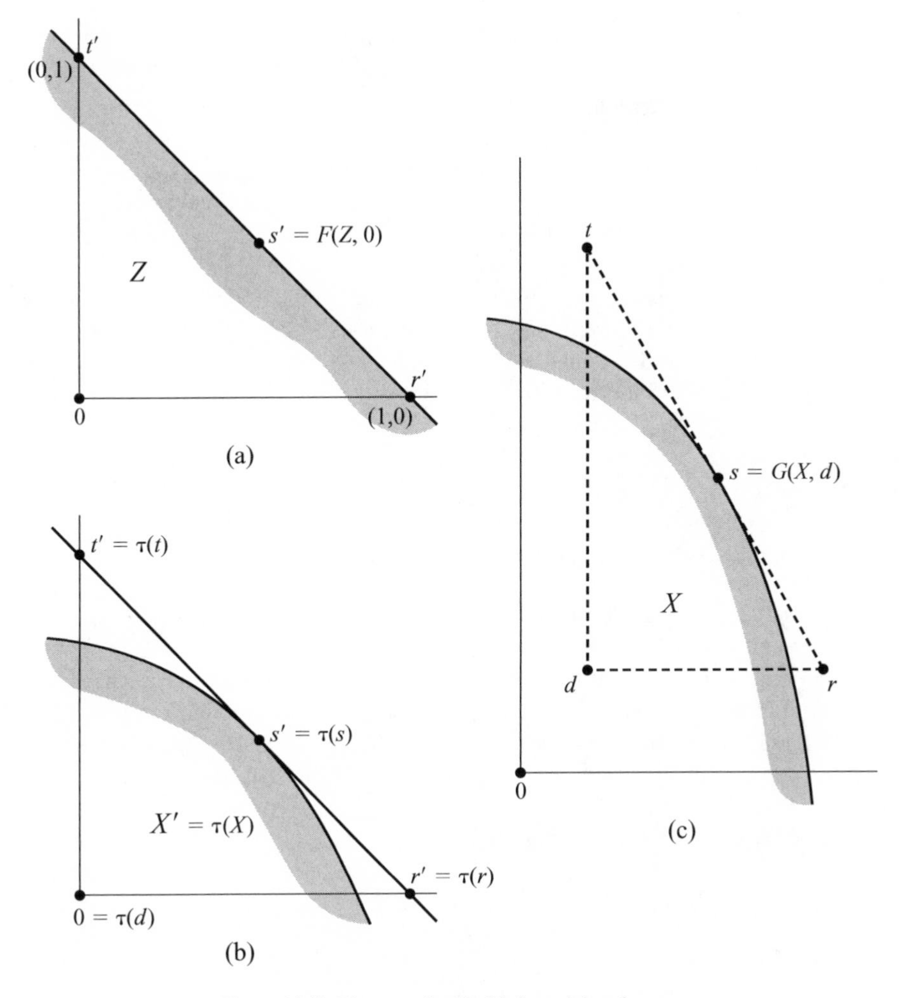


Figure 16.12 Diagrams for Nash's bargaining theorem.

mathematically using the function $\rho : \mathbb{R}^2 \to \mathbb{R}^2$ defined by $\rho\,(x_1, x_2) = (x_2, x_1)$ that simply swaps the players' payoffs.

Nash's symmetry requirement can then be expressed as the axiom:

Axiom 16.4 $F(\rho(X), \rho(d)) = \rho\,(F(X, d))$.

This axiom says that the bargaining solution doesn't care who is labeled player I and who is labeled player II. If the players' labels are reversed, each will still get the same payoff.

Corollary 16.1 (Nash) *If* $F : \mathscr{B} \to \mathbb{R}^2$ *satisfies Axioms* 16.1–16.4, *then F is the symmetric Nash bargaining solution N.*

*Proof* The bargaining problem $(Z, 0)$ in the proof of Theorem 16.1 is symmetric. Its solution is therefore symmetric by Axiom 16.4. Thus $\alpha = \beta$.

→ 16.8

## 16.7 Collusion in a Cournot Duopoly

Firms don't like competition, whether perfect or imperfect. They make more money by getting together in cozy cartels (Section 10.2.3). Sometimes they fix the price at which their product sells, but in Cournot duopolies they bargain over each firm's market share. We have been using a simplified version of the Cournot Game as an example, but now we look at the full version of Exercise 10.8.2, in which Alice's and Bob's unit costs satisfy $0 < c_1 < c_2 < \frac{1}{2}K$.

If bribery is too risky to be practical, we model Alice and Bob's bargaining problem as $(X, d)$ in Figure 16.13(a). The Pareto frontier of the feasible set $X$ satisfies the equation of Exercise 10.8.3($c$). The disagreement point $d$ is the Nash equilibrium of the Cournot Game without collusion. From Exercise 10.8.2, we know that $d_1 = \frac{1}{9}(K - 2c_1 + c_2)^2$ and $d_2 = \frac{1}{9}(K - 2c_2 + c_1)^2$.

When $K = 6$, $c_1 = 1$, and $c_2 = 2$, the symmetric Nash bargaining solution of the problem $(X, d)$ is $t = (4.31, 1.20)$ (Exercise 16.12.22). This outcome is generated by a market-sharing arrangement in which Alice restricts her production to $q_1^* = 1.59$, and Bob to $q_2^* = 0.70$.

The problem is more interesting when Alice and Bob are willing to give and take bribes. The new bargaining problem is then $(Z, d)$, as illustrated in Figure 16.13(b). The Coase theorem tells us that Alice and Bob will collude on a point on the Pareto frontier of $Z$. It isn't efficient for Bob to produce anything at all because Alice has a lower unit cost. A collusive deal therefore consists of Alice's running a monopoly, which she sustains by bribing Bob to stay out of the market. Alice therefore produces $q_1 = \frac{1}{2}(K - c_1)$ and makes a profit of $\frac{1}{4}(K - c_1)^2$. How much of this goes to Bob as a bribe?

The symmetric Nash bargaining solution $s$ of the problem $(Z, d)$ is given by $s_1 = \frac{1}{8}(K - c_1)^2 + \frac{1}{2}(d_1 - d_2)$ and $s_2 = \frac{1}{8}(K - c_1)^2 - \frac{1}{2}(d_1 - d_2)$. The final quantity is Alice's bribe to Bob. In the case when $K = 6$, $c_1 = 1$, and $c_2 = 2$, the disagreement point is $d = (4, 1)$, and Alice's monopoly profit is 6.25. It follows that $s = (4.38, 1.88)$, and so the bribe is 1.88.

The consumer can take some comfort in the fact that the collusive deals we have been considering here are unstable. Talk is cheap when executives meet in smoke-filled hotel rooms because the deals they reach aren't binding. How can you sue someone for breaking an illegal deal? It is therefore likely that any agreement will be broken if one of the firms has an incentive to do so (Section 10.2.3). However, the consumers would be unwise to throw away the antitrust laws that forbid collusion because Alice and Bob will typically be playing the same Cournot Duopoly Game day after day for a long time to come. Such *repeated* games can have many equilibria, including equilibria that sustain the kind of collusive deals we have been talking about here (Section 11.3.3).

## 16.8 Incomplete Information

Nearly everything this book has to say about bargaining assumes that anything of interest to the players is common knowledge before the bargaining begins. Game theorists know that bargaining with incomplete information is much more important

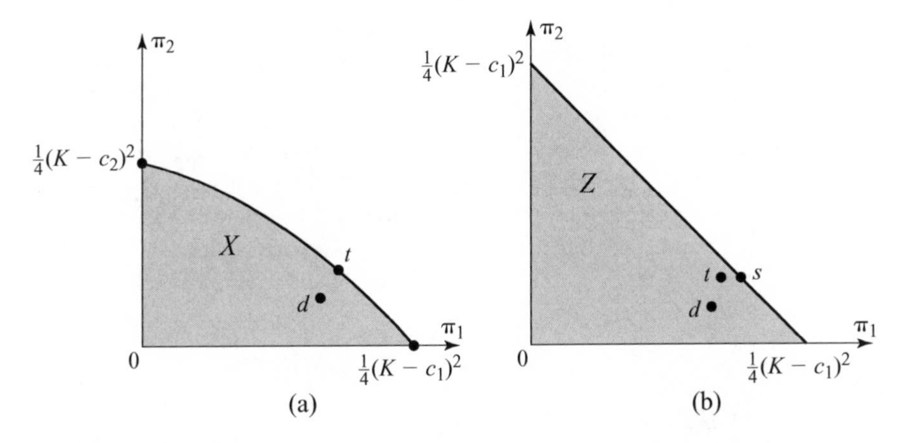


Figure 16.13 Collusion in a full Cournot duopoly. Alice and Bob are engaged in a collusive negotiation before playing a Cournot duopoly game in which Alice's unit cost is lower. Their deadlock point $d$ is the Nash equilibrium of the Cournot game. The Nash bargaining solutions $s$ and $t$ of the problems $(Z, d)$ and $(X, d)$ correspond to the case when bribery is and isn't practical.

in practice, but we don't have a systematic theory to offer. The best we can do at present is to analyze some special cases.

Section 16.4.3 points out that the Coase theorem generally fails when information is incomplete. But what if it doesn't fail? Harsanyi and Selten offer a set of axioms that generalizes the Nash bargaining solution to such cases. We describe their approach with an example.

When analyzing Divide-the-Dollar in Section 16.6.1, we took for granted that Alice's and Bob's utility functions are common knowledge. But are the players really likely to know how risk averse everybody is? (Section 15.4) How do we model the problem when they are sure only of their own utility function? Such problems of incomplete information are tackled by assuming that Alice and Bob may be one of many possible *types,* each of which has a different Von Neumann and Morgenstern utility function (Section 15.3.1).

Suppose it is common knowledge that Alice is type $A_i$ with probability $p_i$ and Bob is type $B_j$ with probability $q_j$. Harsanyi and Selten's theory then predicts that the dollar will be divided so as to maximize the Nash product:[12]

$$(x_1 - d_1)^{p_1} \cdots (x_I - d_I)^{p_I} (y_1 - e_1)^{q_1} \cdots (y_J - e_J)^{q_J}.$$

In this product, $x_i$ is the utility that type $A_i$ would get from the agreement if she were Alice, and $d_i$ is what type $A_i$ would get after a disagreement. Similarly, $y_j$ is the utility that type $B_j$ would get from the agreement if he were Bob, and $e_j$ is what type $B_j$ would get after a disagreement.

Updating the analysis of Divide-the-Dollar of Section 16.6.1 to this new situation, we accordingly look for the split $(z, 1-z)$ of the dollar that maximizes

$$z^{\bar{\gamma}}(1-z)^{\bar{\delta}},$$

[12]The theory incorporates a symmetry axiom, without which we would need to replace the powers $p_i$ and $q_j$ by $\alpha p_i$ and $\beta q_j$.

where $\overline{\gamma} = p_1\gamma_1 + \cdots + p_I\gamma_I$ and $\overline{\delta} = q_1\delta_1 + \cdots + q_J\delta_J$. So when Bob doesn't know the value of $\gamma$ in Alice's utility function $v_1(z) = z^\gamma$, he simply proceeds as though it were certain that $\gamma$ is equal to its expected value $\overline{\gamma}$. Alice similarly proceeds as though it were certain that $\delta$ is equal to its expected value $\overline{\delta}$.

Notice how the result depends on what Alice and Bob *believe* (Section 15.5.1). In particular, if Alice and Bob know nothing beyond the fact that their opponent has been chosen from the population at large, the dollar will be split fifty-fifty. It will be split fifty-fifty even when Alice actually turns out to be risk neutral and Bob turns out to be very risk averse—provided that Bob can avoid giving himself away by sweating too much during the negotiation!

## 16.9 Other Bargaining Solutions

The Nash bargaining solution was the first of many bargaining solutions for which systems of axioms have been proposed. Critics sometimes use this plethora of concepts as an excuse to attack the axiomatic method itself. What is the point of defending a bargaining solution with a system of axioms if *any* bargaining solution can be defended in this way? However, to ask this question is to miss the point altogether. One formulates a system of axioms characterizing a bargaining solution in order to summarize what needs to be defended. The defense begins only when one asks how well the axioms fit the practical context in which the bargaining solution is to be applied.

→ 16.9.2

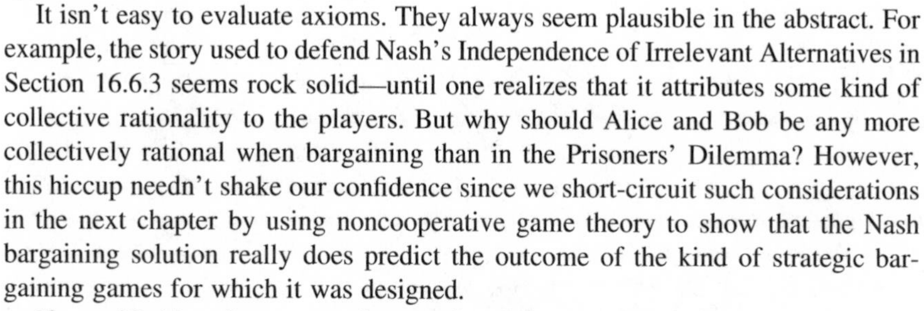

It isn't easy to evaluate axioms. They always seem plausible in the abstract. For example, the story used to defend Nash's Independence of Irrelevant Alternatives in Section 16.6.3 seems rock solid—until one realizes that it attributes some kind of collective rationality to the players. But why should Alice and Bob be any more collectively rational when bargaining than in the Prisoners' Dilemma? However, this hiccup needn't shake our confidence since we short-circuit such considerations in the next chapter by using noncooperative game theory to show that the Nash bargaining solution really does predict the outcome of the kind of strategic bargaining games for which it was designed.

If even Nash's axioms create interpretive difficulties, what about the axioms that characterize alternative bargaining solutions? A bargaining solution proposed by Kalai and Smorodinsky will serve as the basis of a number of exercises that show how cautious one needs to be.[13]

### 16.9.1 Kalai-Smorodinsky Solution

This bargaining solution for the Nash bargaining problem $(X, d)$ is easy to describe geometrically in terms of what one might call the utopian point $U$ of the problem. Alice's utopian payoff $U_1$ is the most she can get, given that Bob is to receive at least his disagreement payoff of $d_1$. Bob's utopian payoff $U_2$ is the most that he can get, given that Alice is to receive at least $d_1$.

[13]Kalai and Smorodinsky proposed their solution partly to illustrate this point, but their followers have been much less circumspect.

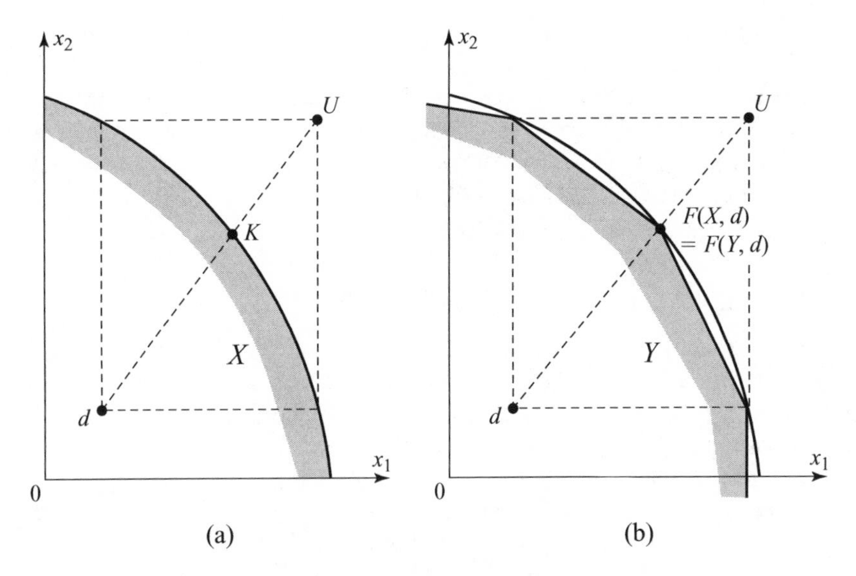


Figure 16.14 The Kalai-Smorodinsky bargaining solution.

It isn't usually possible for both Alice and Bob to get their utopian payoffs simultaneously—a fact reflected in Figure 16.14(a) by placing the utopian point $U$ outside the feasible set $X$. Having located the utopian point $U$, join this to the disagreement point $d$ with a straight line. The Kalai-Smorodinsky bargaining solution $K$ is located where this straight line crosses the Pareto frontier of $X$.

The axioms that characterize the Kalai-Smorodinsky bargaining solution are the same as those that characterize the symmetric Nash bargaining solution, except that the Independence of Irrelevant Alternatives (Axiom 16.3) is replaced by the following axiom, called Individual Monotonicity:[14]

Axiom 16.5 *Suppose that* $d \in Y \subseteq X$, *and the utopian point* $U$ *is the same for both* $(X,d)$ *and* $(Y,d)$. *If* $F(Y,d)$ *is a Pareto efficient point of* $X$, *then*

$$F(X,d) = F(Y,d).$$

The proof that Kalai and Smorodinsky's axioms characterize their solution is so easy that we leave it as an exercise (Exercise 16.12.25).

*Why Individual Monotonicity?* Figure 16.14(b) shows the conditions under which the problems $(X,d)$ and $(Y,d)$ have the same solution according to the preceding axiom. Kalai and Smorodinsky observe that, for each possible outcome $y$ in $Y$, there is an outcome $x$ in $X$ that assigns Bob more than $y_2$ without assigning Alice less than $y_1$. They argue that Bob should therefore get more in $(X,d)$ than in $(Y,d)$. Since the

[14]Monotonicity means different things to different people. To mathematicians, a monotonic function is either increasing or decreasing, but economists usually mean that the function is increasing. Sometimes they mean that it is *strictly* increasing. A monotonic bargaining solution is one that assigns more to both players when the agreement set expands (Exercise 16.12.27). Individual Monotonicity is something yet again.

same goes for Alice, it follows that $f(X,d) \geq f(Y,d)$. But when $f(Y,d)$ is a Pareto-efficient point of $X$, this implies that $f(X,d) = f(Y,d)$.

However, why should we suppose that Bob's bargaining position must necessarily be at least as strong after some change has expanded the agreement set so that, for each payoff Alice might get, there is more left for him? Even if we accept this principle, why should it be applied only when the two situations being compared have the same utopian point? (Exercises 16.12.26)

### 16.9.2 Walrasian Bargaining Solution

→ 16.10

When Nash formulated a bargaining problem as a pair $(X, \xi)$, he took for granted that only utility information is relevant to the bargaining behavior of rational players. But what if the bargaining strategies available to the players are constrained by the institutions within which they are forced to operate?

The bilateral monopoly studied in Section 9.6.3 is a leading example. Figure 16.15 shows representations of Adam and Eve's bargaining problem as an Edgeworth box $\varepsilon$ and also as a Nash bargaining problem $(X, \xi)$. To keep things simple, the endowment point $e$ of the Edgeworth box is identified with a single disagreement point $\xi$ in the corresponding Nash bargaining problem. Recall from Section 16.4.7 that this identification makes sense when a commodity bundle $(f, a)$ represents a flow of perishable goods rather than a stock.

In Section 9.6.3, Adam and Eve are able to communicate only in prices and quantities. In such cases, we must be prepared to consider bargaining solutions that take account not only of the players' utilities but also of the *economic environment* on which their utility functions are defined.

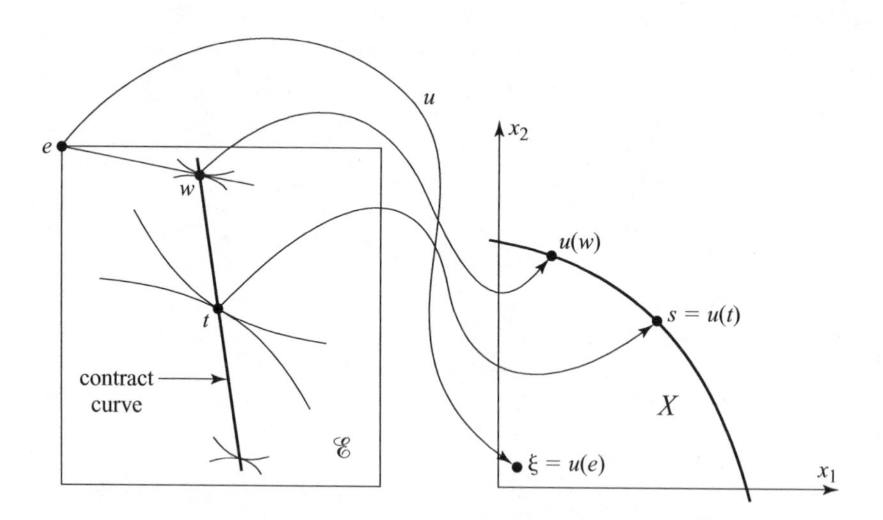


Figure 16.15 The Walrasian bargaining solution. Adam's and Eve's indifference curves in the Edgeworth box on the left have been carefully chosen to ensure that the translation of their bargaining problem to utility space shown on the right is symmetric. The commodities bundles represent *flows* of perishable goods to justify the correspondence between the endowment point $e$ and the disagreement point $d$. Note that Walrasian equilibria need not be at all equitable.

It is natural to begin the study of such informationally rich bargaining solutions by formulating a set of axioms that retain the spirit of Nash's axioms. Somewhat surprisingly, the axioms turn out to characterize the set of Walrasian equilibria of a pure exchange economy, thereby providing some justification for those economic textbooks that treat a bilateral monopoly as just another example of perfect competition. But when using the idea in the context of a bilateral monopoly, I prefer to speak of the Walrasian bargaining solution to make it clear that we aren't pretending that nobody has market power.

Exercise 16.12.29 runs through the argument that leads from a set of Nash-like axioms to the Walrasian bargaining solution $w$. In brief, Axiom 16.1 is replaced by the requirement that $w$ lies on the contract curve. Axiom 16.2 is replaced by an axiom that says that the outcome shouldn't depend on how we calibrate either the utility scales or the commodity scales. Axiom 16.4 is replaced by an axiom that requires that $w$ be symmetric when the whole Edgeworth box is symmetric.

As in the case of the Nash bargaining solution, the vital axiom is Axiom 16.3—the Independence of Irrelevant Alternatives. As with Axiom 16.2, the new axiom needs to be split into two parts, one of which says something about utilities and the other about commodities. Taking the commodity part first, the axiom requires that deleting trades from the Edgeworth box so as to create a new Edgeworth box that still contains both the endowment point $e$ and $w$ should leave the solution invariant. The utility part is more interesting. It requires that the solution remain invariant when each player's utility function is changed so that no possible trade increases in utility, but the utilities of $e$ and $w$ remain the same.

## 16.10 Roundup

Cooperative game theory abstracts away an unmodeled preplay negotiation period, in which *binding* contracts may be signed. Attention can then be transferred from noncooperative strategy spaces to cooperative payoff regions in utility space (Section 6.6.1). Assuming free disposal or transferable utility makes the cooperative payoff region larger.

In a Nash bargaining problem $(X, b, d)$, the set $X$ contains all possible deals. The breakdown point $b$ lists the players' outside options—the most that each can get when acting independently. The deadlock point $d$ arises if the players are unable to reach agreement, but nobody walks off to take up their outside option. We always assume that $b \geq d$. Nash himself considered the special case $(X, \xi)$ in which $\xi = b = d$.

The Coase theorem says that rational bargainers will agree on a Pareto-efficient outcome when transaction costs are zero—which includes the proviso that everything the bargainers might want to know before the bargaining begins is common knowledge. The bargaining set consists of all Pareto-efficient points $x$ in $X$ with $x \geq b$. Von Neumann and Morgenstern argued that rational deals might lie anywhere in the bargaining set, but Nash's axioms characterize a *unique* outcome as the Nash bargaining solution of the problem.

The fundamental axiom is the Independence of Irrelevant Alternatives. It says that players who sometimes agree on $s$ when $t$ is available never agree on $t$ when $s$ is available. Without the Symmetry Axiom, Nash's axioms characterize a generalized Nash bargaining solution determined by bargaining powers $\alpha \geq 0$ and $\beta \geq 0$. It lies where the Nash product

$$(x_1 - d_1)^\alpha (x_2 - d_2)^\beta$$

is maximized subject to the constraints that $x$ is in $X$ and $x \geq b$. We can use the Nash product to show that a player profits from having a larger bargaining power and being less risk averse, but it is usually easier to locate a Nash bargaining solution geometrically.

Information is assumed to be complete in most of the chapter. The players' preferences are therefore common knowledge. Harsanyi and Selten have proposed an extension of the Nash bargaining solution to the case of incomplete information, but it is unclear how broadly it can be applied.

Alternatives to the Nash bargaining solution exist, but their axiomatic characterizations are less persuasive. An exception arises when information that isn't payoff relevant may affect the bargaining outcome. A case can then sometimes be made for Walrasian equilibria as bargaining solutions.

## 16.11 Further Reading

*Game Theory and the Social Contract.* Vol. 2: *Just Playing,* by Ken Binmore: MIT Press, Cambridge, MA, 1998. Bargaining in a richer context.

*Getting to Yes: Negotiating Agreement without Giving In,* 2d ed., by Roger Fisher and William Ury: Houghton Mifflin, Boston, 1992. This best-seller argues that good bargaining consists of insisting on a fair deal. Thinking strategically is dismissed as a dirty trick!

*Everything Is Negotiable*, by Gavin Kennedy: Prentice Hall, Englewood Cliffs, NJ, 1983. His books are sold in airport bookstores but nevertheless give good advice.

*The Art and Science of Negotiation*, by Howard Raiffa: Harvard University Press, Cambridge, MA, 1982. The magnum opus of one of the authors of the classic *Games and Decisions*.

## 16.12 Exercises

1. Draw an equivalent of Figure 16.1 in which contracts that specify joint mixed strategies are allowed. Your picture of the strategy space will then resemble Figure 6.2(b). Draw an arrow that links the mixed-strategy profile in which the pure-strategy profiles $(4, 4)$ and $(6, 4)$ are equally likely to the pair of expected payoffs that then results.

econon

2. In Section 9.3.2, Adam and Eve have quasilinear preferences that allow a util to be identified with an apple. Their contract curve in the Edgeworth box is then a vertical line segment, as shown in Figure 9.5(a). Explain why all Pareto-efficient trades require the same number of fig leaves to be transferred from Eve to Adam but that the number of apples that Adam pays to Eve in return may vary. Relate this conclusion to the concept of transferable utility. Sketch the bargaining set in utility space for the case when Adam and Eve are risk neutral.

3. The Prussian military guru, von Clausewitz, famously said that war is the continuation of diplomacy by other means. Would the Coase theorem then imply that rational players never go to war?

phil

4. What is the symmetric Nash bargaining solution of the wage bargaining problem studied in Section 16.6.1 when Alice's outside option is reduced to $5? What does it become when Bob's outside option is also changed as a consequence of the Post Office's willingness to offer ten hours of labor a day at a wage of $2 an hour?
5. The pseudonymous Marshall Jevons writes whodunits that are solved by a Harvard professor using the eternal verities of neoclassical economics. In *A Deadly Indifference,* the hero explains why he was able to buy a house cheap after the only rival buyer is murdered: "It never fails. Where there is less competition for an asset, the price of that asset falls.'' Why is this wrong? What principle is being misapplied?

fun

6. When does it make sense to identify both the breakdown point and the deadlock point with the endowment point in the problem of Exercise 16.12.2? Assuming that this identification makes sense in the quasilinear trading problem of Section 9.5.2, find the symmetric Nash bargaining solution. How many fig leaves and apples are traded when this is used?

econ

7. Draw supporting lines to the convex set $H$ of Exercise 6.9.23 at the points $(1,1)$, $(2,4)$, $(3,3)$, and $(2,\frac{4}{3})$. Where it is possible to draw more than one supporting line, draw several supporting lines.

review

8. The function $f:\mathbb{R}^2\to\mathbb{R}^2$ is defined by $(y_1,y_2)=f(x_1,x_2)$, where

review

$$y_1 = x_1 + 2x_2 + 1$$
$$y_2 = 2x_1 + x_2 + 2.$$

Why is $f$ affine? Show the points $f(1,1)$, $f(2,4)$, and $f(4,2)$ on a diagram. Also show $f(H)$ for the set $H$ of Exercise 6.9.23 and $f(l)$ for one of the supporting lines $l$ that you drew in the preceding exercise.

9. Find the cooperative payoff region for the game of Figure 16.16 in each of the following cases:
   a. The players can make binding agreements but can neither freely dispose of utils nor transfer them.
   b. Free disposal is permitted, but utils can't be transferred.
   c. Free disposal is permitted, and utils can be transferred from one player to the other.
10. Which of the following values of $y$ satisfy $y>x$ and hence are Pareto improvements on $x=(2,3)$?

(a) $y=(4,4)$ (*b*) $y=(1,2)$ (c) $y=(2,4)$
(d) $y=(3,3)$ (*e*) $y=(2,3)$ (f) $y=(3,2)$

For which of these values of $y$ is it true that $y\geq x$? For which values of $y$ is it true that $y\gg x$?

| | $t_1$ | $t_2$ | $t_3$ |
|---|---|---|---|
| $s_1$ | $-1$ | $3$ | $0$ |
| | $-1$ | $1$ | $3$ |
| $s_2$ | $0$ | $1$ | $3$ |
| | $1$ | $0$ | $0$ |

Figure 16.16 The game for Exercise 16.12.9.

11. Find the Pareto-efficient points for the cooperative payoff regions $Y$ and $Z$ obtained in Exercises 16.12.9(b) and 16.12.9(c). What are the bargaining sets when the breakdown point is $b=(0,1)$? What are the bargaining sets when it is $c=(1,0)$?
12. If $d=(0,-3)$, find the value of the symmetric Nash bargaining solution for each of the problems $(Z,b,b)$, $(Z,c,c)$, $(Z,b,d)$, and $(Z,c,d)$, where $Z$, $b$, and c are given in the previous exercise. Do the same when $Z$ is replaced by $Y$.
13. Find the values of the generalized Nash bargaining solution with bargaining powers $\alpha=\frac{1}{3}$ and $\beta=\frac{2}{3}$ for each of the bargaining problems of the previous exercise. Do the same with $\alpha=\frac{2}{3}$ and $\beta=\frac{1}{3}$.
14. How is Divide-the-Dollar resolved by the generalized Nash bargaining solution with bargaining powers $\alpha=\frac{2}{5}$ and $\beta=\frac{3}{5}$ in the case when Alice's Von Neumann and Morgenstern utility for \$$x$ is $v_1(x)=x^{\gamma}$ and Bob's is $v_2(x)=x^{\delta}$, where $\gamma=\frac{1}{4}$ and $\delta=\frac{3}{4}$? Whose share of the dollar would increase if both $\gamma$ and $\delta$ were changed to $\frac{1}{2}$?
15. If Alice and Bob are both risk loving, explain why the use of the symmetric Nash bargaining solution in Divide-the-Dollar will result in their tossing a fair coin to see who gets the whole dollar.

math

16. Various geometric ways of finding the symmetric Nash bargaining solution $s$ are given in Section 16.6.2. Using the method illustrated in Figure 16.9(b) for the case when the Pareto-efficient points of $X$ lie on a differentiable curve defined by $y=f(x)$, show that

$$f'(s_1) = -\frac{s_2-d_2}{s_1-d_1},$$

when the breakdown constraint $s \geq b$ is inactive. Get the same result by setting the derivative of the relevant Nash product equal to zero.

phil

17. Explain why *every* payoff pair in the bargaining set is the generalized Nash bargaining solution for *some* bargaining powers $\alpha$ and $\beta$. Should we deduce that Theorem 16.1 is devoid of substantive content?

econ

18. In the wage bargaining problem of Section 16.4.4, suppose that Alice and Bob bargain only about the wage $w$. After a wage has been agreed upon, Alice unilaterally sets the number of hours of labor per day for which she is willing to pay. What is the feasible set $X$ in this new situation? Write down the Nash product for the symmetric Nash solution as a function of $w$ in the case when the deadlock point is $b=d=(0,24)$. What wage will be agreed upon? What will

the length of the working day be? Why will the firm say that there is overstaffing when both the wage and the length of the working day are negotiated simultaneously? Why can the worker respond by saying that bargaining only over the wage is inefficient?

19. "Men pay lip service to equal rights in the home while letting women do three-quarters of the household chores." This is a quote from a newspaper article that is typical of the genre. Other things being equal, the fact that wives do more housework than husbands would indeed show that the balance of power within marriages is biased in favor of men, but are other things equal? This exercise considers the implications of taking into account one of many factors that journalists neglect.

    Alice and Bob are getting married. They have no interest in enjoying any of the benefits of marriage other than sharing the housework. In the modern style, they agree on a binding marriage contract that specifies how many hours of housework a week each will contribute.

    In the world in which Alice and Bob live, people think that there is a right and proper number of hours that a married couple should spend doing housework. The Von Neumann and Morgenstern utility of a person who contributes $h$ hours of housework and has a partner who contributes $k$ hours is $u = 1 - ch$ if $h + k \geq C$ and $x = -ch$ if $h + k < C$, but men and women differ on the values of the positive constants $C$ and c. For women, $C = W$ and $c = w$. For men, $C = M$ and $c = m$, where $W > M > 0$ and $m > w > 0$.

    a. How many hours a week will men and women spend on housework when living alone? Why will men live in squalor when $mM > 1$?
    b. Explain why a Pareto-efficient marriage contract will specify that the total number of hours per week they devote to housework will be what Alice thinks is right rather than what Bob thinks is right.
    c. Assuming that everybody's preferences are common knowledge and that men do housework when living alone, describe the Nash bargaining problem that Alice and Bob must solve before getting married. Ignore the possibility that Alice and Bob may have other candidates for their hand in marriage.
    d. Use the symmetric Nash bargaining solution to solve Alice and Bob's bargaining problem. Confirm that if women think that twice as much housework needs to be done as men do then wives will do three times as much housework.

20. If $U$ is the utopian point of a Nash bargaining problem $(X, d)$, explain why the symmetric Nash bargaining solution always assigns a utility of at least $\frac{1}{2}(U_i - d_i)$ to player $i$. **math**

21. Use the symmetric Nash bargaining solution to predict the outcome of collusive negotiation in the Bertrand version of the duopoly of Section 16.7. **econ**

22. Find the symmetric Nash bargaining solution $t$ of the problem $(X, d)$ of Section 16.7. One method is to maximize the Nash product $N = (\pi_1 - d_1)(\pi_2 - d_2)$ by solving the equations $\partial N / \partial q_i = 0$ (i = 1, 2). (When using a computer to solve the equations numerically, remember to reject the stationary point at which $N = 0$.) **math**

23. Find the Kalai-Smorodinsky bargaining solution for each of the bargaining problems of Exercise 16.12.11 in the case when the deadlock point and the breakdown points are the same.

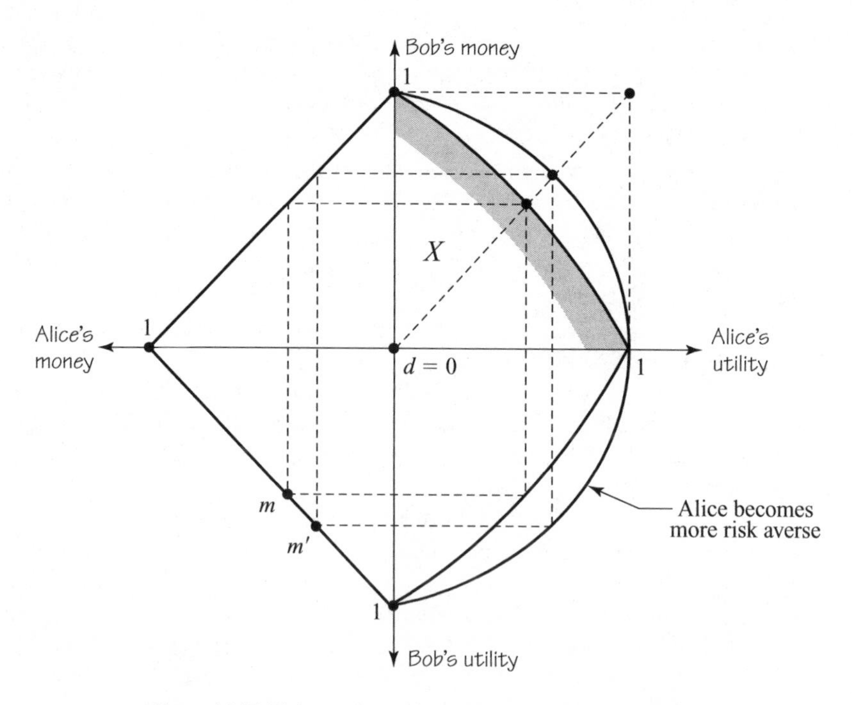


Figure 16.17 Risk aversion with the Kalai-Smorodinsky solution.

math

24. Show that the Kalai-Smorodinsky bargaining solution fails to satisfy Nash's Independence of Irrelevant Alternatives.
25. Show that axioms 16.1, 16.2, 16.4, and 16.5 characterize the Kalai-Smorodinsky bargaining solution.
26. Why should Kalai and Smorodinsky's Individual Monotonicity be applied only to bargaining problems with the same utopian point $U$? What happens if the principle is applied without this constraint?
27. Following up on the previous exercise, what happens if we insist that a bargaining solution be monotonic in the sense that $F(X,d) \leq F(Y,d)$ whenever $X \subseteq Y$? Show that neither the Nash bargaining solution nor the Kalai-Smorodinsky solution is monotonic in this sense.
28. Use Figure 16.17 to show that the Kalai-Smorodinsky bargaining solution of Divide-the-Dollar favors risk-neutral players over risk-averse players.[15]

math

29. A set of axioms for the Walrasian bargaining solution is informally described in Section 16.9.2. Give formal versions of the axioms while following the following steps to show that they do indeed characterize the Walrasian equilibrium in an Edgeworth box when this is unique:
    a. Expand the Edgeworth box so that the endowment point $e$ and the Walrasian equilibrium $w$ lie on its secondary diagonal.
    b. Change the commodity units so that the box becomes square.
    c. Change the utility units so that the payoffs at $e$ and $w$ are $(0,0)$ and $(1,1)$.

[15]Avoid an error I made in my book *Just Playing* by noting that the diagram plots Alice's utility as a function of *Bob's* share of the dollar.

d. Let $\tau$ reflect the Edgeworth box across its secondary diagonal. Replace Adam's utility function $u_1$ by the smallest function $U_1$ that is larger than both $u_1$ and $\tau \circ u_2$ at every point in the box.[16] Do the same for Eve's utility function $u_2$.
e. Apply the symmetry and efficiency axioms.
f. Return to the original configuration using the independence and calibration axioms.

[16]The function $U_1$ may not be concave, but it is always quasiconcave. (To say that $f : \mathbb{R}^n \to \mathbb{R}$ is quasiconcave means that the sets $\{x : f(x) \geq c\}$ are convex for all $c$.)

# 17 *Cutting a Deal*

## 17.1 Noncooperative Bargaining Models

Noncooperative bargaining sounds like a contradiction. Why bargain if you aren't going to cooperate? But we're into counterfactuals again. If disagreement were impossible, there wouldn't be anything to bargain about (Section 16.4). If Alice never says *no* to anything, why not demand everything she's got?

Cooperative game theory sometimes provides simple characterizations of *what* agreement rational players will reach, but we need noncooperative game theory to understand *why* (Section 16.2). When disputes arise about which cooperative bargaining solution to use and how to use it, noncooperative models of bargaining are therefore the place to look for an answer.

## 17.2 The Nash Program

Cooperative game theory presupposes a preplay negotiation period during which the players come to a binding agreement on how a game is to be played. However, all this activity is packed away in a black box during a cooperative analysis (Section 16.2). The *Nash program* invites us to open such black boxes to see whether the mechanism inside really does works in the way the axioms characterizing a cooperative solution concept assume.

Nash observed that any negotiation is itself a species of game, in which the moves are everything the players may say or do while bargaining. If we model any bargaining that precedes a game $\mathcal{G}$ in this way, the result is an enlarged game $\mathcal{N}$.

A strategy for this negotiation game first tells a player how to conduct the preplay negotiations and then how to play $\mathcal{G}$ depending the outcome of the negotiations.

Negotiation games must be studied *without* presupposing preplay bargaining, all preplay activity having already been built into their rules. Analyzing them is therefore a task for noncooperative game theory. This means looking for their Nash equilibria in the hope that the equilibrium selection problem won't prove too difficult (Section 8.5).

When negotiation games can be solved successfully, we have a way of checking up on cooperative game theory. If a cooperative solution concept says that the result of a rational agreement on how to play $\mathcal{G}$ will be $s$, then we should also get $s$ from solving $\mathcal{N}$.

An important proviso is that the outcomes that cooperative game theory assumes to be enforceable as irrevocable contracts should be available as *equilibria* of the game $\mathcal{G}$, but life is usually too short to make a fuss about the legal and economic apparatus that needs to be added to $\mathcal{G}$ to explain why players keep their commitments (Section 5.5.2).

→ 17.2.1

*Implementation Theory.* The Nash program is sometimes wrongly represented as a battle for supremacy between cooperative and noncooperative game theory. But the Nash program *unifies* the two approaches and so allows the weaknesses of one to be buttressed by the strengths of the other. I think the confusion arises partly because of a failure to distinguish adequately between the Nash program and *implementation theory*.

Welfare functions describe the objectives of a government (Section 19.4). To implement a welfare function, the government must invent and enforce an appropriate set of rules. These create a game for its citizens to play. In an ideal world, welfare would be maximized at the equilibria of this game. Implementation theory is therefore about designing games whose equilibria have desirable welfare properties (Section 20.5).

However, we have to pass up the fun of designing games in the Nash program. When we open the cooperative black box, we are stuck with whatever negotiation game we find inside. If its equilibria turn out to have socially undesirable properties, there is nothing we can do about it.

### 17.2.1 What Matters in Bargaining?

Much of what passes as bargaining consists of attempts by one party to exploit the folly of the other. But rational players will ignore all the colorful bombast of traditional bargaining. Insults, flattery, and simulated outrage will all be water off a duck's back. Watching rational players bargain is therefore unlikely to be a bundle of laughs, but we have no grounds for complaint. A negotiation game that was sufficiently general to capture each twist and turn that a real-life negotiation might conceivably take would be complicated beyond all imagining. Even after all the irrelevancies have been suppressed, we will still find it a challenge to analyze the resulting negotiation games.

What are the factors that really matter in rational bargaining? The previous chapter emphasized *risk*. If other things are equal, bold players get more than timid players. In this chapter, we study *commitment* and *threats,* both credible and incredible. Where threats are incredible, *delay* becomes a factor. If other things are equal again, we will find that patient players get more than impatient players. But *information* trumps all other considerations.

*Complete Information.* When Alice tries to sell Bob a used car, she would like to know the most that he would really be willing to pay—but he won't tell her. Nor will she tell him the lowest price she will take. Such informational asymmetries matter enormously. In particular, the Coase theorem fails (Section 16.4.3). There may be a price at which both players would benefit from a sale, but the car will nevertheless sometimes remain unsold. Even when Alice and Bob succeed in agreeing on a price, the deal will depend heavily on who knew what at the outset of the negotiations.

We don't yet know how to tackle such bargaining problems of incomplete information in a satisfactory way. Chapter 15 explains how game theorists deal with the general problem of incomplete information, but attempts to apply the theory to bargaining have mostly proved inconclusive because the models pose the equilibrium selection problem in an intractable form. We will look at some special cases in this chapter, but information should always be assumed to be complete unless otherwise stated.

## 17.3 Commitment in Bargaining

Usually it is a weakness to let things get out of control—but not in bargaining, where it can be "strong to be weak." For example, if Bob opens the bargaining by making a commitment to accept no less than 99 cents in Divide-the-Dollar, then Alice's bargaining position becomes hopeless—provided she believes that Bob's commitment is genuinely irrevocable. She must now choose between one penny or nothing. If she rationally chooses the former,[1] then Bob's strategy will have won him the lion's share of the dollar.

Of course, Bob won't convince Alice just by saying he is committed. Who believes someone who claims he is now making his "last and final offer?" Even the prices posted on expensive items in upscale stores are seldom final. The seller will usually try to make you feel like a cheapskate for challenging the price, but folk wisdom is right for once. Everything is negotiable. Never take no for an answer. Sometimes the seller even turns out to be a pushover. After I was quoted a price at a car rental counter somewhere in Texas, my conversation with the clerk went like this:

> *Me*: What discounts do you offer?
> *Clerk*: 20%.

But how many people think to ask?

It is genuinely hard to establish commitments (Section 5.5). People sometimes make a career of establishing a reputation for being stubborn or stupid for this purpose. Trade unionists occasionally succeed in committing themselves by voting

[1] Experiments show that real people are either irrational or care about things other than money in such Ultimatum Games! (Section 19.2.2)

for intransigent leaders. But aside from such special circumstances, the vocabulary of commitment is usually just so much cheap talk.

Real commitments can't often be made, but what happens when they can? Both players will then rush to get a take-it-or-leave-it demand on the table first. If the negotiation game they are playing doesn't favor one player over the other, their demands will arrive simultaneously. The game that results is the Nash Demand Game.

Studying this game won't only be a useful exercise in applying the techniques of Chapter 8; it will also be our first opportunity to apply the Nash program. Will the solution outcome of the Nash Demand Game turn out to be the Nash bargaining solution? It must have been a good moment for John Nash when the answer turned out to be *yes*.

### 17.3.1 Nash Demand Game

→ 17.5

The Nash Demand Game is a simultaneous-move game based on a Nash bargaining problem $(X, d)$. Alice and Bob simultaneously announce a demand. Their demands $x_1$ and $x_2$ are either compatible or incompatible. They are compatible if the pair $x = (x_1, x_2)$ lies in the set $X$ of feasible payoff pairs. Both players then get their demands. If the demands are incompatible, both players get their disagreement payoffs.

The zero and the unit on a player's von Neumann and Morgenstern utility scale can be chosen to suit our convenience (Section 4.6.2). So let us take $d = 0$ and make the frontier of $X$ passes through the points $(0, 1)$ and $(1, 0)$. Figure 17.1(a) shows the resulting bargaining problem $(X, d)$. As a further simplification, demands will be restricted to the interval $[0, 1]$.

The players' reaction curves are shown in Figure 17.1(b). If Bob makes a demand satisfying $0 < x_2 < 1$, then Alice's best reply is to choose the demand $x_1$ that makes $(x_1, x_2)$ Pareto efficient in $X$. She won't claim less because she would then get less. She won't claim more since the demands would then be incompatible, and so her payoff would be $d_1 = 0$. If Bob makes his maximum demand of $x_2 = 1$, then Alice

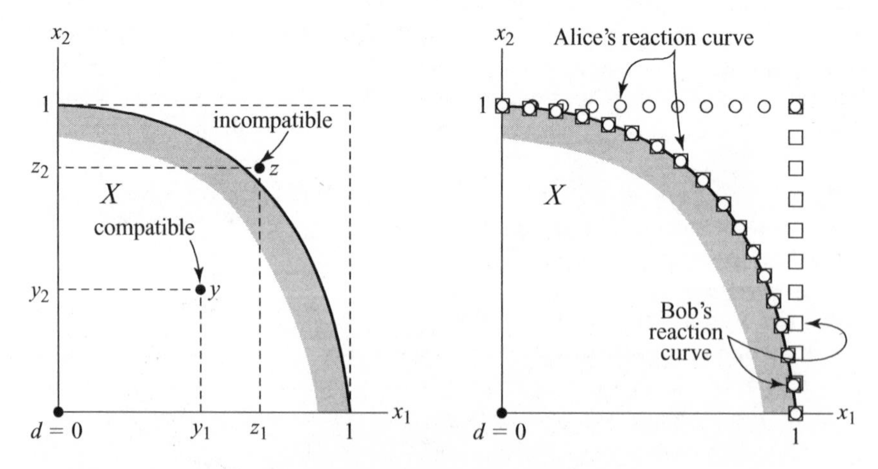


Figure 17.1 The Nash Demand Game.

will get nothing regardless of what she does. Thus any demand is a best reply for her in this case.

Figure 17.1(b) shows that any point in the bargaining set for the problem $(X, d)$ corresponds to a Nash equilibrium for the Nash Demand Game (Section 16.4.3). There is also a "noncooperative" Nash equilibrium that results if both players are greedy enough to make their maximum possible demand. Both then get nothing.

We have found an infinite number of Nash equilibria. This creates a major equilibrium selection problem. Sometimes such problems can be solved by moving to a more realistic model. Since nothing is certain but death and taxes, Nash therefore introduced a little uncertainty into his game.

### 17.3.2 The Smoothed Nash Demand Game

In the smoothed Nash Demand Game, the players aren't quite sure what the feasible set $X$ is. So they don't know in advance whether a pair of demands $(x_1, x_2)$ will prove to be compatible or not.

To model this situation, we assign a probability $p(x_1, x_2)$ to the event that the pair $(x_1, x_2)$ is compatible. Figure 17.2(a) shows some contours for the probability function $p: [0, 1]^2 \to [0, 1]$. For example, if a demand pair $x = (x_1, x_2)$ lies on the contour $p(x) = \frac{1}{3}$, then the probability that the demand pair $x$ will prove to be compatible is $\frac{1}{3}$.

The players know that the frontier of the set $X$ is somewhere in the strip sandwiched between the regions labeled $p(x) = 0$ and $p(x) = 1$. If the amount of uncertainty is small, then the strip will be narrow. Our focus is on what happens as its width becomes vanishingly small.

Player $i$'s payoff function in the game is

$$\pi_i(x_1, x_2) = x_i p(x_1, x_2),$$

because the outcome that results from the players' choosing the pair $(x_1, x_2)$ of demands is a lottery in which player $i$ gets $x_i$ with probability $p(x_1, x_2)$ and zero with probability $1 - p(x_1, x_2)$.

It will be assumed that the function $p: [0, 1]^2 \to [0, 1]$ is sufficiently well behaved that a naive approach to calculating the players' reaction functions for the smoothed demand game will be successful.[2] To find her best replies to Bob's demand of $x_2$, Alice simply differentiates her payoff function partially with respect to $x_1$. She then sets the partial derivative $\partial \pi_1 / \partial x_1$ equal to zero. Bob does the same, and so we get the equations:

$$x_1 p_{x_1}(x_1, x_2) + p(x_1, x_2) = 0, \qquad (17.1)$$

$$x_2 p_{x_2}(x_1, x_2) + p(x_1, x_2) = 0, \qquad (17.2)$$

where $p_{x_1}(x)$ denotes the partial derviative of $p$ with respect to $x_1$ evaluated at the point $x$. Figure 17.2(b) shows the typical shape of these reaction curves.[3]

[2] It is enough if $p$ is differentiable, quasiconcave, and strictly decreasing.

[3] What happens when $x_1 = 1$ or $x_2 = 1$? Depending on what is assumed about $p$, the reaction curves may bend back, as in Figure 17.1(b), and cross at an equilibrium in which both players get nothing.

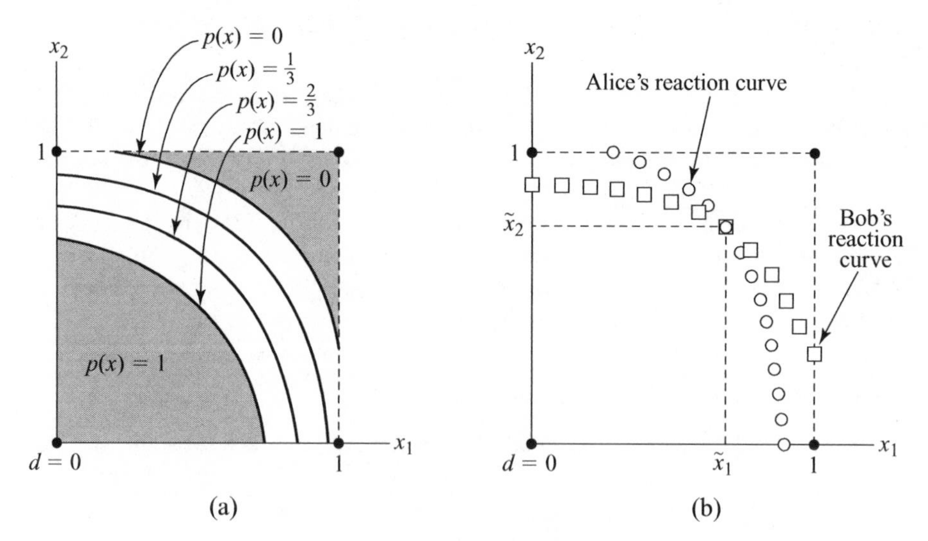


Figure 17.2 The smoothed Nash Demand Game.

A Nash equilibrium $\tilde{x} = (\tilde{x}_1, \tilde{x}_2)$ occurs where these two reaction curves cross. Figure 17.2(b) shows the two reaction curves crossing once. But it may happen that the reaction curves cross several times, so that multiple Nash equilibria exist. However, we shall see that they all approximate the symmetric Nash bargaining solution when the players are fairly certain about what $X$ is.

The tangent to the curve $p(x) = p(\tilde{x})$ at the point $\tilde{x}$ has equation $\nabla p(\tilde{x})(x - \tilde{x}) = 0$ (Section 9.2). When written in full, the equation becomes

$$p_{x_1}(\tilde{x}_1, \tilde{x}_2)(x_1 - \tilde{x}_1) + p_{x_2}(\tilde{x}_1, \tilde{x}_2)(x_2 - \tilde{x}_2) = 0. \tag{17.3}$$

If $\tilde{x} = (\tilde{x}_1, \tilde{x}_2)$ is a Nash equilibrium, it must lie on both Alice's and Bob's reaction curves. Thus (17.1) and (17.2) are both true with $x = \tilde{x}$. This observation allows the partial derivatives in (17.3) to be eliminated, leaving us with the simple equation

$$\frac{x_1}{2\tilde{x}_1} + \frac{x_2}{2\tilde{x}_2} = 1.$$

Figure 17.3 serves as a reminder that this is the equation of the tangent line to $p(x) = p(\tilde{x})$ at the point $\tilde{x}$.

In Figure 17.2, the tangent line cuts the horizontal axis at the point $r = (2\tilde{x}_1, 0)$. It cuts the vertical axis at the point $t = (0, 2\tilde{x}_2)$. Thus, the Nash equilibrium $\tilde{x} = (\tilde{x}_1, \tilde{x}_2)$ lies at the midpoint of the line segment joining $r$ and $t$. If $Y$ is the shaded region in Figure 17.3, it follows that $\tilde{x}$ is the symmetric Nash bargaining solution for the bargaining problem $(Y, 0)$ (Section 16.6.2).

The frontier of the set $Y$ is a probability contour. All such contours converge on the frontier of the set $X$ as the width of the strip shown in Figure 17.2(a) approaches

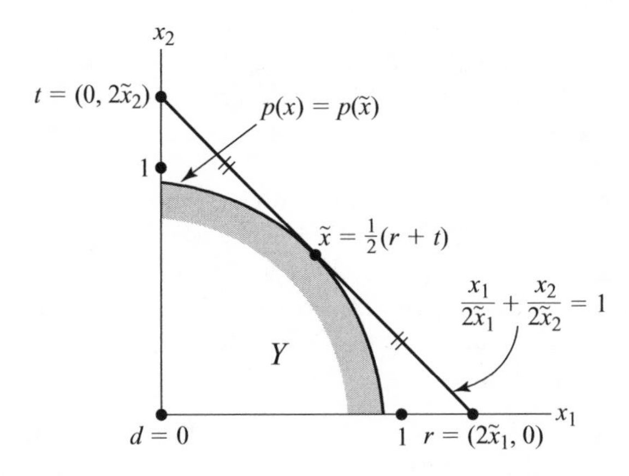


Figure 17.3 Characterizing Nash equilibria in the smoothed game.

zero.[4] It follows that $\tilde{x}$ converges to the symmetric Nash bargaining solution of the bargaining problem $(X, 0)$.

The limiting value of $\tilde{x}$ is a Nash equilibrium of the original Demand Game. It has been selected from all the other equilibria by Nash's technique of looking at the limits of Nash equilibria in the smoothed game. Since this equilibrium selection criterion selects the symmetric Nash bargaining solution from the set of all Nash equilibria for the original Nash Demand Game, we have our first example of a cooperative solution concept surviving the kind of test required by the Nash program—albeit only for the case when the players are free to make whatever commitments strike their fancy.

### 17.3.3 Incomplete Information

→ 17.4

The Nash bargaining solution predicts the equilibrium outcome of the Nash Demand Game when information is complete. But what of Harsanyi and Selten's application of the idea to the case of incomplete information?

Continuing the Divide-the-Dollar example of Section 16.8, suppose that Alice's and Bob's types are independently chosen from the same distribution of possible types. We don't need to know any more to write down Harsanyi and Selten's prediction of the outcome. Since the situation is symmetric, the dollar will be split fifty-fifty, regardless of Alice's and Bob's types.

This result receives some endorsement from the Nash program because it turns out to be a Bayes-Nash equilibrium of the smoothed Nash Demand Game if Alice and Bob always demand approximately half the dollar independently of their type. But the endorsement is only partial because the game usually has other Nash equilibria.

As an example, consider the case of two equally likely types. Each type's utility for money takes the form $v_i(z) = z^{\gamma_i}$, where $\gamma_1 = 1$ and $\gamma_2 = \frac{1}{3}$. The first type is

[4]Mathematicians should note that the "width" of the strip may be taken to be the Hausdorff distance between the set where $p = 0$ and the set where $p = 1$. (The Nash bargaining solution is continuous in the Hausdorff metric.)

therefore risk neutral, and the second is strictly risk averse. In this situation, it is a Bayes-Nash equilibrium if each type plays the smoothed Nash Demand Game as though certain that their opponent will be of the other type. Risk-neutral types then demand approximately $\frac{3}{4}$ of the dollar; risk-averse types demand approximately $\frac{1}{4}$ (Section 16.6.1).

To confirm that we are looking at a Nash equilibrium, we need to check that neither type can profit by switching to the strategy of the other type. The expected payoff for risk-neutral types is $\frac{3}{8} = \frac{1}{2} \times \frac{3}{4}$ because they get nothing when matched with another risk-neutral type. If they imitate a risk-averse type, they always get $\frac{1}{4}$ of the dollar but are worse off because $\frac{3}{8} > \frac{1}{4}$. Similarly, it is bad for risk-averse types to imitate risk-neutral types because

$$\left\{\tfrac{1}{4}\right\}^{\frac{1}{3}} > \tfrac{1}{2} \times \left\{\tfrac{3}{4}\right\}^{\frac{1}{3}}.$$

The first equilibrium is a Pareto improvement on the second, but it doesn't follow that the latter won't be played in practice!

## 17.4 Nash Threat Games

→ 17.5

When the cooperative black box contains unlimited commitment power, our analysis of the Nash Demand Game supports the use of the symmetric Nash bargaining solution. But what if we aren't sure where to put the disagreement point? Nash's answer is that $d$ is determined by the players' *threats* about what they will do if agreement isn't reached.

As in Section 17.2, Alice and Bob have the opportunity to negotiate an agreement on how to play a game $\mathcal{G}$. A simplified version of Nash's threat theory supposes that the players simultaneously open the negotiations by making a commitment to the strategy they will play in $\mathcal{G}$ if the negotiations break down.[5] These threats then determine a disagreement point $d$ for an application of the symmetric Nash bargaining solution. The problem for the players is to choose their threats in a maximally effective manner.

In the Battle of the Sexes of Figure 6.15(b), each player has an infinite number of mixed strategies available as possible threats. Figure 17.4(a) copies the noncooperative payoff region $X$ of the game from Figure 6.17(b). Recall that the line segments labeled $p$ and $q$ correspond to mixed strategies for each player (Section 6.6.1). If $p$ and $q$ are played in the Battle of the Sexes, the outcome is the payoff pair $d(p,q)$ that lies at the point of intersection of the corresponding line segments. If Alice and Bob commit themselves to the threats $p$ and $q$, the final outcome will therefore be the symmetric Nash bargaining solution $s(p,q)$ of the problem $(Y, d(p,q))$.[6]

We have transformed the problem into a *threat game* whose outcome is $s(p,q)$ when Alice and Bob choose $p$ and $q$. Finding its solution is easy for the Battle of the Sexes because $s_1(p,q) + s_2(p,q)$ is always 3, and so the threat game is constant sum

[5]There is no reason why they shouldn't also simultaneously commit themselves to a negotiation strategy, but it turns out to make no difference at which stage the latter commitment is made.

[6]The feasible set has been expanded from $X$ to $Y$ because it is sensible to assume free disposal here.

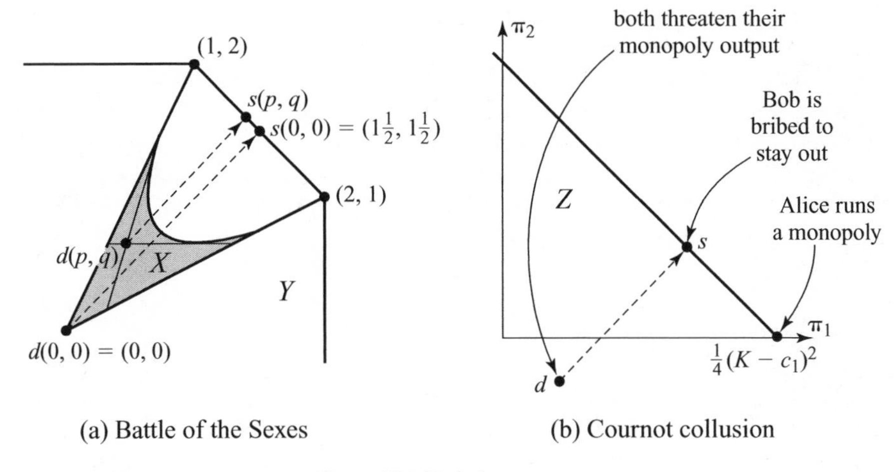


(a) Battle of the Sexes

(b) Cournot collusion

Figure 17.4 Nash threat games.

(Section 7.2). The Minimax Theorem then says that the players' optimal threats are their security strategies (Section 7.4). These can be read from Figure 17.4(a).

If Alice plays $p=0$ in the threat game, her worst possible payoff is $1\frac{1}{2}$. If Bob plays $q=0$, his worst possible payoff is also $1\frac{1}{2}$. In a game where the payoffs always add up to 3, neither can therefore guarantee getting more than $1\frac{1}{2}$. Thus, $p=0$ and $q=0$ are the security strategies we are looking for.

In terms of the Battle of the Sexes, the wife threatens to go the ballet by herself unless an agreement is reached. The husband simultaneously threatens to go to the boxing match. They then toss a fair coin and go off together to the ballet or the boxing match, depending on how it lands.

### 17.4.1 Collusion with Threats

Section 16.7 studied collusion in a Cournot duopoly when bribes are practical. The disagreement point $d$ in the resulting bargaining problem $(Z, d)$ was taken to be the Nash equilibrium of the underlying Cournot Game. This makes sense when threats aren't credible, but what if they are?

We already know that the symmetric Nash bargaining solution to the problem $(Z,d)$ is $\pi_1 = \frac{1}{2}(d_1-d_2)+\frac{1}{8}(K-c_1)$ and $\pi_2 = \frac{1}{2}(d_2-d_1)+\frac{1}{8}(K-c_1)$ for all disagreement points $d$ (Section 16.7). If Alice and Bob threaten to play $q_1$ and $q_2$, their payoffs after a disagreement will be $d_1=(K-c_1-q_1-q_2)q_1$ and $d_2=(K-c_2-q_1-q_2)q_2$. Alice's payoff in the Nash threat game when she threatens $q_1$ and Bob threatens $q_2$ is therefore

$$\tfrac{1}{2}(K-c_1-q_1)q_1 - \tfrac{1}{2}(K-c_2-q_2)q_2 + \tfrac{1}{8}(K-c_1)^2.$$

It follows that it is a strictly dominant strategy for Alice to choose the value of $q_1$ that maximizes $(K-c_1-q_1)q_1$. But the maximizing value $q_1 = \frac{1}{2}(K-c_1)$ is Alice's monopoly output. For the same reason, Bob's monopoly output $q_2 = \frac{1}{2}(K-c_2)$ is strictly dominant for him in the threat game. When each player uses these optimal threats, the agreement that results assigns a payoff of $\frac{1}{4}(K-c_1)^2 - \frac{1}{8}(K-c_2)^2$ to Alice and $\frac{1}{8}(K-c_2)^2$ to Bob.

As shown in Figure 17.4(b), each player therefore threatens to behave like a monopolist if an agreement isn't reached. They then agree that Bob will be paid a bribe of half of what his own monopoly payoff would have been to leave the field clear for Alice to run her monopoly for real. But will Bob keep the deal if he receives the bribe before the outputs are chosen? If the bribe is to be paid after the outputs are chosen, will Alice honor her promise?

## 17.5 Bargaining without Commitment

Inspired by the preceding analysis of the Battle of the Sexes, I threatened to burn our house down if my wife didn't let me watch my favorite television program, but she explained that my threat was incredible because it wasn't consistent with the subgame-perfect equilibrium of our bargaining game.

When threats must be credible to be effective, *time* enters the picture. Patient players can then threaten impatient players with the prospect of a delayed agreement. If the threat is credible, an impatient player will then be forced to concede a larger share. To see how large a share, we need to study the subgame-perfect equilibria of noncooperative bargaining models that allow indefinite numbers of offers and counteroffers. We begin by studying some primitive models with only a few rounds of bargaining.

### 17.5.1 The Ultimatum Game

The Ultimatum Game is based on Divide-the-Dollar. The players prefer more money to less and care about nothing else. We will also assume they are risk neutral, although nothing much would change if they weren't.

The rules specify that Alice first makes a proposal to Bob on how to divide the dollar. He may then accept or refuse. If he accepts, Alice's proposal is adopted. If he refuses, the game ends with both players getting nothing.

The game tree is illustrated in Figure 17.5. The branches at the root are labeled with the amounts that Alice claims for herself. After each such claim, Bob can choose $Y$ or $N$. To choose $Y$ signifies acceptance. To choose $N$ signifies a refusal. After a refusal, both players get nothing.

*Nash Equilibria.* The doubled branches in Figure 17.5(a) indicate one of the many Nash equilibria of the game. The outcome is very counterintuitive. Alice offers everything to Bob, who accepts. How can such an odd outcome result from Nash equilibrium play?

Denote the pure strategy for Alice indicated in Figure 17.5(a) by $s$. Thus $s$ calls upon Alice to offer the entire dollar to Bob. A pure strategy for Bob is much more complicated. For each possible proposal that Alice might make, Bob must say whether he would accept or refuse. The pure strategy $t$ indicated in Figure 17.5(a) calls for him to refuse every proposal, except that in which he is offered the whole dollar.

The pure strategy pair $(s,t)$ is a Nash equilibrium. To verify this, it is necessary to confirm that $s$ is a best reply to $t$ and $t$ is simultaneously a best reply to $s$. This isn't

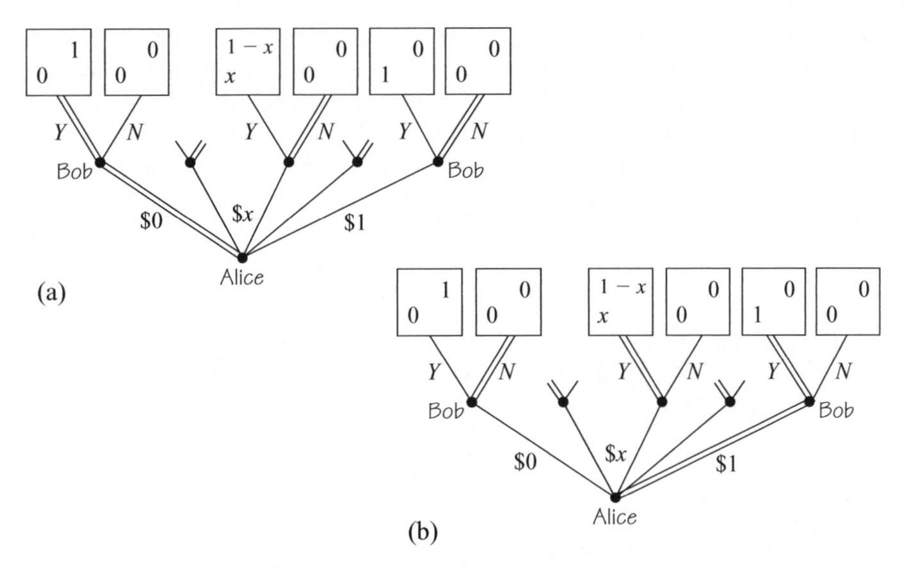


Figure 17.5 The Ultimatum Game.

hard for Bob. He can't get more than the whole dollar, and this is what he gets when he plays $t$ in reply to $s$. Alice's situation is less rosy. She gets nothing by replying to $t$ with $s$. But neither would she get anything if she used any other pure strategy because any other proposal will be refused. Therefore $s$ is a best reply to $t$ since it does at least as well as any other reply.

*Subgame-Perfect Equilibrium.* The pair $(s, t)$ isn't subgame perfect. It calls for Bob to plan to play *irrationally* in subgames that aren't reached when $(s, t)$ is used. Bob might threaten Alice that he will play $t$, but she will find his threat incredible (Section 5.5.2). For example, pure strategy $t$ calls for Bob to refuse when offered ten cents. Ten cents isn't much, but it is better than nothing. If Bob is rational, he will therefore accept ten cents if it is offered. One can argue that he might refuse out of spite, or to "teach Alice a lesson," or because he wishes to establish a reputation as a tough cookie. But all of these arguments require attributing motives to Bob other than a love of money (Section 19.2.2).

Figure 17.5(b) illustrates the use of backward induction in finding the unique subgame-perfect equilibrium. This calls for Bob to plan to accept all offers and for Alice to demand the whole dollar. Matters aren't entirely straightforward because the game is infinite since Alice can choose any real number in the interval [0, 1] as her demand. The procedure will therefore be described carefully. Three steps are required:

**Step 1.** Double all branches corresponding to an acceptance by Bob of a demand $x < 1$ by Alice. To accept such a demand is optimal because $1 - x > 0$.

**Step 2.** Double the branch at which Alice demands 1. No demand $x < 1$ can be optimal because a demand $y$ with $x < y < 1$ would be accepted and hence yield a better payoff than the demand $x$.

**Step 3.** Double the branch corresponding to an acceptance by Bob of the demand 1 by Alice. It is true that refusal is also optimal, but refusal doesn't correspond to a subgame-perfect equilibrium. If Bob plans to refuse the demand of 1 by Alice, then Alice would do better to make some demand $x < 1$ because this will necessarily be accepted. However, we have seen that such a demand can't be optimal.

The basic principle will be met again in more complicated models, and so it is worth stating it explicitly:

> *In equilibrium, a proposer always plans to offer the responder an amount that will make the responder indifferent between accepting and refusing. In equilibrium, the responder always plans to accept such an offer or better and to refuse anything worse.*

### 17.5.2 A Two-Stage Bargaining Game

The principle we just met is now used in a model in which Bob gets to make a counterproposal after rejecting Alice's initial proposal. To make the problem interesting, we suppose that the players not only prefer more money to less but also prefer getting the money sooner rather than later. This is usually a reasonable assumption. After all, if it didn't matter *when* you reached an agreement, it wouldn't matter *whether* you reached an agreement.

*Discount Factors.* Alice's and Bob's attitudes to time are modeled with discount factors satisfying $0 < \delta_i < 1$ (Section 11.3.3). Player $i$'s utility for getting $\$x$ at time $t$ is then taken to be $x\delta_i^t$. Discount factors are a simple way of modeling how impatient players are. A player with a discount factor close to zero is very impatient. A player with $\delta_i = 1$ isn't impatient at all. For such a player, getting fifty cents now is no different from being sure of getting fifty cents in ten years time.

Figure 17.6(a) illustrates the game tree for the two-stage bargaining game. Alice makes the first proposal at time 0. If Bob rejects the proposal, he makes a counter-proposal at time $\tau > 0$. If this is refused by Alice, then both players get nothing. We begin by studying the case $\tau = 1$.

*Subgame-Perfect Equilibrium.* The subgames rooted at nodes where Bob makes a counterproposal are just copies of the Ultimatum Game. If equilibrium strategies are used when such a subgame is reached, then Bob gets the whole dollar. He assigns this outcome a utility of $1 \times \delta_2 = \delta_2$ since he gets the dollar at time 1. Alice assigns the same event a utility of $0 \times \delta_1 = 0$.

Backward induction now tells us to replace each of the subgames by a leaf labeled with the payoff pair $(0, \delta_2)$ that results from equilibrium play in the subgame. This reduces the situation to that shown in Figure 17.6(b).

In this reduced game, Alice's equilibrium proposal makes Bob indifferent between accepting and refusing. She therefore demands $1 - \delta_2$, thus leaving $\delta_2$ for Bob, which is what he gets in equilibrium from refusing. In equilibrium, he accepts Alice's demand of $1 - \delta_2$ for the reasons given in the previous section.

If $\delta_2$ is nearly zero, so that Bob is very impatient, Alice gets nearly all of the dollar. If $\delta_2$ is nearly 1, so that Bob is very patient, he gets nearly all of the dollar.

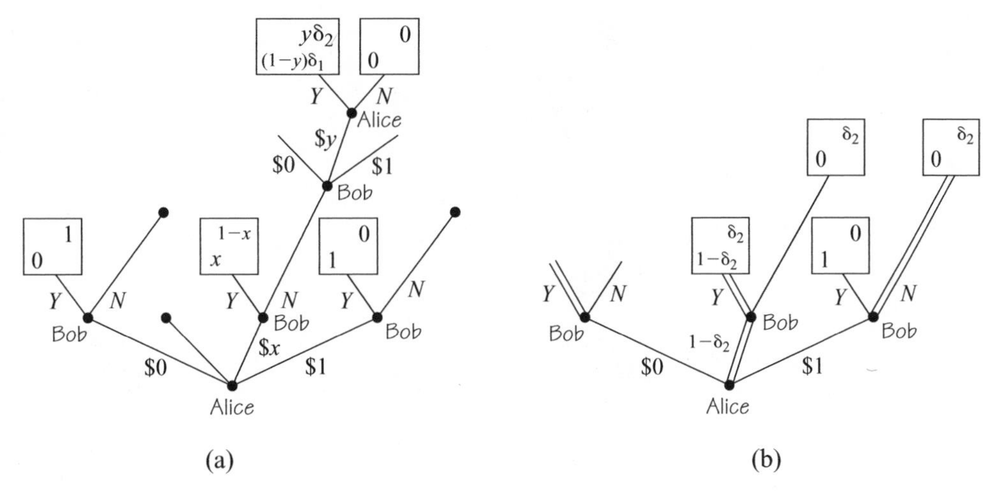


Figure 17.6 A two-stage bargaining game.

The length of the time interval that elapses between one stage of the game and the next is also significant. If this is $\tau$ instead of 1, then $\delta_i$ must be replaced everywhere by $\delta_i^\tau$. Alice will then demand $1-\delta_2^\tau$ in equilibrium, which Bob will accept. Since $\delta_2^\tau \to 1$ as $\tau \to 0$, it follows that Bob gets nearly everything if $\tau$ is sufficiently small. Thus, if Bob could choose how long to wait before making his counterproposal, he would choose to wait as short a time $\tau$ as possible.

### 17.5.3 The Infinite Horizon Game

The simple bargaining games of the preceding subsections were studied in preparation for the model that follows. This is perhaps the most natural of all possible bargaining models. It is therefore a striking vindication of Nash's approach to bargaining that the subgame-perfect equilibrium outcomes can be described using a generalized Nash bargaining solution.

*Preferences.* The basic bargaining problem continues to be Divide-the-Dollar, but we now revert to the general formulation of Section 16.4.2. A deal is a pair $m=(m_1, m_2)$ of money payments in the set $M$ of Figure 16.3. Player $i$'s Von Neumann and Morgenstern utility for the deal $m$ at time $t$ is

$$u_i(m,t) = v_i(m_i)\delta_i^t,$$

where $v_i$ is strictly increasing and concave, so that the players are now risk averse. We also set $v_i(0)=0$ to ensure that a player who is going to get nothing will be indifferent about when the check for zero dollars is delivered. For convenience, we also choose a utility scale with $v_i(1)=1$.

*Rules.* In the bargaining game $G$ to be studied, Alice makes the first proposal at time 0. If Bob rejects the proposal, he makes a counterproposal at time $\tau$. If Alice rejects this proposal, she makes a counter-counterproposal at time $2\tau$. They continue in this

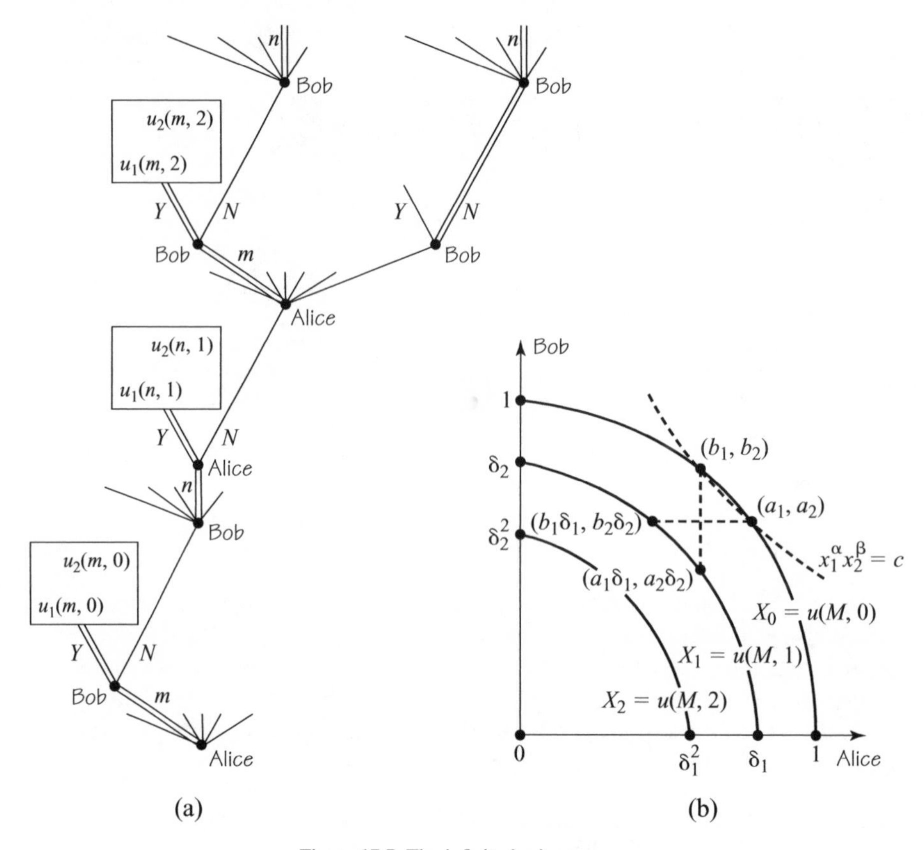


Figure 17.7 The infinite horizon game.

way until a proposal is accepted. However, nothing prevents all proposals being refused, in which case the game will go on forever. Both players attach a utility of zero to this eventuality. As in Section 17.5.2, we take $\tau = 1$ to keep the algebra simple.

The tree for the game $G$ is shown in Figure 17.7(a). Figure 17.7(b) shows the set $X_0 = u(M, 0)$ of feasible utility pairs at time 0. The set $X_1 = u(M, 1)$ of feasible utility pairs at time 1 is smaller. The set $X_2 = u(M, 2)$ is smaller still. The economist Ariel Rubinstein, who first studied this model, likes to think of $X_t$ as a pie that gradually shrinks over time. This shrinkage provides an incentive for the players to reach an early agreement because each proposal that is refused means that there is less pie to be divided.

*Stationary Strategies.* The doubled branches in the tree of Figure 17.7(a) show *stationary* or Markov strategies for each player. Strategies are stationary when they ignore a player's history.

Whatever may have happened in the past, a player using a stationary strategy always plans to play the same in the future. For example, whenever it is Alice's turn to make a proposal, she always proposes the deal *m,* regardless of the history of rejected offers and counteroffers that may have gone before. Similarly, Bob always proposes the deal *n.*

We study the special case in which Bob always plans to accept the deal $m$ (or anything better for him) and to refuse anything worse. Similarly, Alice always plans to accept the deal $n$ (or anything better for her) and to refuse anything worse. Can such pure strategies constitute a subgame-perfect equilibrium? The answer is surprisingly simple.

*A Subgame-Perfect Equilibrium.* Let the vectors $a$ and $b$ be defined by $a = u(m, 0)$ and $b = u(n, 0)$. For example, $a_2 = u_2(m, 0)$ is the utility that Bob gets from accepting the deal $m$ at time 0, and $a_1 = u_1(m, 0)$ is the utility that Alice gets if the deal $m$ is accepted at time 0. We know that an equilibrium offer should make the player who responds indifferent between accepting and refusing. If Alice proposes $m$ at time 0 in game $G$, then Bob will get a utility of $a_2$ from accepting. If he refuses, then he will propose $n$ at time 1, and Alice will accept. He will therefore get a utility of $b_2\delta_2$ from refusing. To make him indifferent between accepting and refusing, the requirement is that

$$a_2 = b_2\delta_2. \tag{17.4}$$

Note that (17.4) implies $a_2\delta_2^t = b_2\delta_2^{t+1}$ for any $t$. Thus Bob is indifferent between accepting and refusing $m$, whenever this may be proposed by Alice.

A similar condition is necessary for Alice. This is most easily formulated by repeating the preceding discussions for the companion game $H$, which is the same as $G$ except that it is Bob who makes the first proposal at time 0. If Bob proposes $n$ at time 0 in game $H$, then Alice will get a utility of $b_1$ from accepting. If she refuses, then she will propose $m$ at time 1, and Bob will accept. She will therefore get a utility of $a_1\delta_1$ from refusing. To make her indifferent between accepting and refusing, the requirement is that

$$b_1 = a_1\delta_1. \tag{17.5}$$

Note again that (17.5) implies $b_1\delta_1^t = a_1\delta_1^{t+1}$ for any $t$. Thus Alice is indifferent between accepting and refusing $n$, whenever this may be proposed by Bob.

The conditions (17.4) and (17.5) are illustrated in Figure 17.7(b). Condition (17.4) just says that the points $(a_1, a_2)$ and $(b_1\delta_1, b_2\delta_2)$ are on the same horizontal line. Condition (17.5) says that the points $(b_1, b_2)$ and $(a_1\delta_1, a_2\delta_2)$ are on the same vertical line.[7]

Equations (17.4) and (17.5) characterize the equilibrium deal in the case when the time interval $\tau$ between successive proposals satisfies $\tau = 1$. However, to work out what the equilibrium deal actually is requires more calculating. Such calculation can be evaded if we turn our attention to the limiting case when $\tau \to 0$. Fortunately, this

[7]Do such $a$ and $b$ exist? Note that $a_1 = v_1(m_1)$, $a_2 = v_2(m_2)$, $b_1 = v_1(n_1)$, and $b_2 = v_2(n_2)$. Since $a$ and $b$ are Pareto efficient, $m_1 + m_2 = 1$ and $n_1 + n_2 = 1$. Thus $b_i = f(a_i)$, where $f(x) = v_2(1 - v_1^{-1}(x))$, so equation (17.4) may be written as $f(a_1) = \delta_2 f(b_1)$. Combining this with (17.5) yields that $f(a_1) = \delta_2 f(a_1\delta_1)$. A condition for existence is therefore that the function $g$: $[0, 1] \to [0, 1]$ defined by $g(x) = f(x) - \delta_2 f(x\delta_1)$ is zero somewhere in $[0, 1]$. Note that $g(0) = 1 - \delta_2 > 0$ and $g(1) = 0 - \delta_2 f(\delta_1) < 0$. But a continuous function that is positive at 0 and negative at 1 must take the value zero somewhere between 0 and 1. Actually, $a$ and $b$ are *uniquely* defined by (17.4) and (17.5) when $v_1$ and $v_2$ are concave since $g$ is then strictly decreasing on $[0, 1]$.

limiting case is the case of greatest interest because, in the real world, nothing constrains a negotiator to keep to a strict timetable, and, given that a player has just refused an offer, the optimal thing to do next is to make a counteroffer as soon as possible.

When $\tau \neq 1$, we must replace $\delta_i$ by $\delta_i^\tau$ in (17.4) and (17.5). It makes life easier if we simultaneously move from the discount factors $\delta_i$ to the corresponding *discount rates* defined by $\delta_i = e^{-\rho_i}$, so that[8]

$$a_2 = b_2 e^{-\rho_2 \tau}, \tag{17.6}$$

$$b_1 = a_1 e^{-\rho_1 \tau}. \tag{17.7}$$

THEOREM 17.1 *Suppose that the stationary subgame-perfect equilibrium specified by* (17.6) *and* (17.7) *leads to the payoff pair* $s(\tau)$. *then*

$$s(\tau) \to s \quad as \quad \tau \to 0,$$

*where* $s$ *is the generalized Nash bargaining solution for the bargaining problem* $(X_0, 0)$ *corresponding to the bargaining powers* $\alpha = 1/\rho_1$ *and* $\beta = 1/\rho_2$.

*Proof.* It follows from (17.6) and (17.7) that

$$\left(\frac{a_2}{b_2}\right)^\beta = \left(\frac{b_1}{a_1}\right)^\alpha = e^{-\tau} \tag{17.8}$$

because $\alpha = 1/\rho_1$ and $\beta = 1/\rho_2$. But (17.8) implies that

$$a_1^\alpha a_2^\beta = b_1^\alpha b_2^\beta,$$

and so the points $a = (a_1, a_2)$ and $b = (b_1, b_2)$ both lie on the same curve $x_1^\alpha x_2^\beta = c$ as illustrated in Figure 17.7(b) for the case $\tau = 1$.

Equation (17.8) tells us that $a_2/b_2 \to 1$ and $b_1/a_1 \to 1$ as $\tau \to 0$. Hence the points $a$ and $b$ converge on the same value $s$.[9] This tells us something interesting about Figure 17.7(b) in the case when $\delta_i$ is replaced everywhere by $\delta_i^\tau$. When $\tau \to 0$, the revised figure reduces to Figure 16.7(a) with $X = X_0$ and $b = d = 0$. Thus $s(\tau) \to s$ as $\tau \to 0$.

### 17.5.4 Uniqueness of Equilibrium

→ 17.5.5

We thought the Nash program was going well in Section 17.3.3 until we found that our model had other equilibria. The same isn't true in Rubinstein's bargaining model. The game has no subgame-perfect equilibria other than the stationary

[8]Discount rates correspond to what economists call the "instantaneous rate of interest." If interest is charged $n$ times a year at a rate of $\rho/n$, then the yearly discount factor is $\delta = (1 + \rho/n)^{-n} \to e^{-\rho}$ as $n \to \infty$.

[9]A prior assumption that $a$ and $b$ converge is unnecessary. The argument shows that all limit points of $a$ are equal to $s$. Thus $a$ can't have different limit points and thus converges. Similarly for $b$.

equilibrium corresponding to the generalized Nash bargaining solution with bargaining powers $\alpha = 1/\rho_1$ and $\beta = 1/\rho_2$.

For a proof of this result, we return to the special case in which an agreement that assigns $x$ of the dollar to Alice and $1-x$ of the dollar to Bob at time $t$ yields a payoff of $x\delta_1^t$ to Alice and $(1-x)\delta_2^t$ to Bob. As in previous sections, the algebra will be simplified by taking $\tau = 1$.

THEOREM 17.2 (Rubinstein) *The infinite-horizon bargaining game G has a unique subgame-perfect equilibrium outcome.*

*Proof* The game $G$ might have many subgame-perfect equilibrium outcomes, from each of which Alice gets a different payoff. Let the largest subgame perfect equilibrium payoff to Alice be $A_1$, and let the smallest be $a_1$.[10] Recall that $H$ denotes the companion game in which Bob makes the first proposal. Let the largest subgame-equilibrium payoff to Bob in $H$ be $B_2$, and let the smallest be $b_2$. The proof consists of showing that $A_1 = a_1$ and $B_2 = b_2$.

**Step 1.** In the game $G$, a subgame-perfect equilibrium can't assign Bob less than $b_2\delta_2$ because he can always refuse whatever Alice proposes at time 0. The game $H$ will then be played starting at time 1. But the smallest subgame-perfect equilibrium outcome for Bob in $H$ is $b_2$, which has to be discounted by a factor $\delta_2$ because of the delay of length 1. If Bob gets at least $b_2\delta_2$ in equilibrium, then Alice can get no more than $1 - b_2\delta_2$ because there is only one dollar to be divided. This justifies the inequality:

$$A_1 \leq 1 - b_2\delta_2. \tag{17.9}$$

**Step 2.** Suppose that $x < 1 - B_2\delta_2$. It will be shown that $x$ isn't a member of the set $S$ of subgame-perfect equilibrium payoffs for Alice in $G$.

Let $x < y < 1 - B_2\delta_2$. Since $1 - y > B_2\delta_2$, a demand by Alice of $y$ at time 0 would necessarily be accepted by Bob in equilibrium. The reason is that, if he refuses, the companion game $H$ will be played at time 1. The largest subgame-perfect equilibrium outcome for Bob in $H$ is $B_2$, which has to be discounted by a factor of $\delta_2$ because of the time delay of length 1. He therefore gets more by accepting $1 - y$ than the largest amount $B_2\delta_2$ he could get by refusing.

It follows that it can't be optimal for Alice to use a strategy that results in her receiving a payoff of $x$ because she can get $y$ at time 0 simply by demanding $y$. Hence $x \notin S$. Since this is true for each $x < 1 - B_2\delta_2$, the smallest element $a_1$ of $S$ must satisfy

$$a_1 \geq 1 - B_2\delta_2. \tag{17.10}$$

**Step 3.** Two further inequalities are obtained by exchanging the roles of $G$ and $H$ in the preceding discussion:

[10] The set $S$ of subgame-perfect equilibrium payoffs to Alice turns out to have a maximum and a minimum. However, it isn't necessary to assume this. The proof works equally well if $A_1$ and $a_1$ are taken to be the supremum and infimum of $S$.

$$B_2 \leq 1 - a_1\delta_1, \tag{17.11}$$

$$b_2 \geq 1 - A_1\delta_1. \tag{17.12}$$

**Step 4.** It follows from (17.12) that $-b_2 \leq -(1 - A_1\delta_1)$. This conclusion may be substituted in (17.9), and so

$$A_1 \leq 1 - b_2\delta_2 \leq 1 - \delta_2(1 - A_1\delta_1) = 1 - \delta_2 + A_1\delta_1\delta_2.$$

We deduce that

$$A_1 \leq \frac{1 - \delta_2}{1 - \delta_1\delta_2}. \tag{17.13}$$

**Step 5.** Similarly, it follows from (17.11) that $-B_2 \geq -(1 - a_1\delta_1)$. This conclusion may be substituted in (17.10), and so

$$a_1 \geq 1 - B_2\delta_2 \geq 1 - \delta_2(1 - a_1\delta_1) = 1 - \delta_2 + a_1\delta_1\delta_2.$$

We deduce that

$$a_1 \geq \frac{1 - \delta_2}{1 - \delta_1\delta_2} \tag{17.14}$$

**Step 6.** Recall that $a_1$ is the minimum and $A_1$ the maximum of the set $S$. Hence $a_1 \leq A_1$. Thus (17.13) and (17.14) and the corresponding inequalities for $B_2$ and $b_2$ imply that

$$a_1 = A_1 = \frac{1 - \delta_2}{1 - \delta_1\delta_2}; \quad b_2 = B_2 = \frac{1 - \delta_1}{1 - \delta_1\delta_2}.$$

This completes the proof of the theorem.

*Equilibrium Strategies?* What subgame-perfect equilibrium strategy pair yields Alice's unique equilibrium payoff of $a_1$? It turns out that the necessary pure strategies are those discussed in Section 17.5.3. Alice proposes the deal $a = (a_1, 1 - a_1)$ at time 0. Bob is indifferent between accepting and refusing (because $1 - a_1 = b_2\delta_2$), but his equilibrium action is to accept. If it were in equilibrium for Bob to refuse, then the game $H$ would be played at time 1. The unique subgame-perfect equilibrium payoff for Alice in $H$ is $1 - b_2$, which needs to be discounted by $\delta_1$ because of the time delay of length 1. But it is easy to confirm that $(1 - b_2)\delta_1 < a_1$, and so a refusal by Bob at time 0 would make it impossible for Alice to get her unique equilibrium payoff.

*Nash Bargaining Solution?* Is the conclusion consistent with the result of Section 17.5.3 that the agreed deal $a$ approximates the generalized Nash bargaining solution with bargaining powers $\alpha = 1/\rho_1$ and $\beta = 1/\rho_2$ when the interval $\tau$ between successive proposals is sufficiently small? If so, the dollar will be split in the ratio $\rho_2$:

$\rho_1$. To verify this, replace $\delta_i$ everywhere by $\delta_i^\tau$, and consider the limiting value of $a_1$ as $\tau = 0$. By L'Hôpital's rule,[11]

$$\lim_{\tau \to 0}\left(\frac{1-\delta_2^\tau}{1-\delta_1^\tau \delta_2^\tau}\right) = \lim_{\tau \to 0}\left(\frac{1-e^{-\tau\rho_2}}{1-e^{-\tau(\rho_1+\rho_2)}}\right) = \lim_{\tau \to 0}\left(\frac{\tau\rho_2}{\tau(\rho_1+\rho_2)}\right) = \frac{\rho_2}{\rho_1+\rho_2}.$$

### 17.5.5 Pros and Cons

→ 17.6

In saying that the bargaining problem is indeterminate (Section 16.6), Von Neumann and Morgenstern\ were simply echoing the orthodoxy of their times. Economists held that bargaining was a branch of psychology to which they had nothing much to contribute. Rubinstein shattered this complacent attitude, but he is the first to deny that his analysis is anywhere near the final answer to the problem of bargaining. So what are the pros and cons of his model?

*Pros*

**1. The Coase theorem.** The unique subgame-perfect equilibrium of the Rubinstein model requires that Alice's opening offer be accepted immediately by Bob. The outcome is therefore Pareto efficient, and so we have a formal proof of the Coase theorem under fairly general conditions.

**2. The Nash program.** The generalized Nash bargaining solution has survived a severe testing by the Nash program. If the interval $\tau$ between successive proposals in the Rubinstein bargaining model is sufficiently small, then a subgame-perfect equilibrium outcome approximates the generalized Nash bargaining solution with bargaining powers $\alpha = 1/\rho_1$ and $\beta = 1/\rho_2$. Why should we pay particular attention to the case when $\tau$ becomes vanishingly small? Because nothing constrains the players to keep to the timetable of our bargaining game in real life, and both players have an incentive to make their counteroffers as soon after refusing an offer as they can (Section 17.5.2).

**3. Applications.** In using the Nash program to test cooperative solution concepts, we also obtain guidance on how to use them in practice. From the Rubinstein bargaining model, we learn that it is the players' attitudes to time that determine their bargaining powers in the generalized Nash bargaining solution. So it is good to be patient when bargaining. In the next section, we will adapt the model to confirm the intuitions about breakdown and deadlock discussed in the previous chapter.

[11]If $f$ and $g$ are continuous at $\xi$, then

$$\lim_{x \to \xi}\frac{f(x)}{g(x)} = \frac{f(\xi)}{g(\xi)},$$

provided that $g(\xi) \neq 0$. What happens if $g(\xi) = 0$? The limit may still be finite, provided that $f(\xi) = 0$. L'Hôspital's rule says that

$$\lim_{x \to \xi}\frac{f(x)}{g(x)} = \lim_{x \to \xi}\frac{f'(x)}{g'(x)},$$

provided that $f$ and $g$ are differentiable close to $\xi$ and the right-hand limit exists.

*Cons*

**1. Information.** People sometimes feel cheated that bargaining in the Rubinstein model is over as soon as it begins. However, why should there be any action when rational bargainers know each other's strengths and weaknesses in advance of the negotiation? It would just waste time—and time is valuable. On the other hand, real people are often irrational. Nor are they often anywhere near as well informed as the Rubinstein model assumes, so there is a lot of ground that the model doesn't cover.

**2. Equilibrium.** Any outcome in the Rubinstein bargaining game can be supported as a Nash equilibrium, and there are problems with using subgame perfection to solve the equilibrium selection problem (Section 14.2.1). We can replace subgame perfection with the idea of a security equilibrium, in which players are only assumed to get no less than their security level in any subgame. Evolutionary arguments can also be marshalled in favor of Rubinstein's result. However, all of these arguments leave room for doubt.

**3. Experiment.** Experiments—including some of my own—show that laboratory subjects don't use backward induction in the Ultimatum Game or in two-stage bargaining games (Section 19.2.2). So why should they in Rubinstein's game? However, for whatever reason, my own experiments show that Rubinstein's theory actually does quite well in the laboratory. The calculus of breakdown and deadlock points developed in the next section is especially good at predicting experimental results.

## 17.6 Going Wrong

→ 17.7

One of the advantages of the theory developed by Nash and Rubinstein is its simplicity, but those of us who try to explain how it works to applied economists haven't always done a very good job. This section tries to redress the balance by pointing out some do's and don'ts.

As the guy who discovered both the connection between Rubinstein's bargaining model and the generalized Nash bargaining solution and also the way that outside options fit into the theory, I find one widespread abuse especially distressing. This is the practice of ignoring the distinction between the deadlock and breakdown points and automatically using the latter as the disagreement point in Nash's original theory. If everything else is modeled right, the Nash bargaining solution will then usually give the wrong prediction. But this error is now "corrected" by using the generalized Nash bargaining solution with whatever bargaining powers best fit the available data.

*Epicycles?* As an example of the kind of error that can result from abusing the generalized Nash bargaining solution in this way, consider the Divide-the-Dollar problem analyzed in Section 16.6.1. If it is right to use the symmetric Nash bargaining solution, we found that the dollar will be split in the ratio $\gamma : \delta$, where $\gamma$ and $\delta$ measure how risk averse the players are. But suppose we failed to notice that it is the players' attitudes toward taking risks that determines how the dollar gets split. We might then make the bad mistake of modeling the players as risk neutral, so that $\gamma = \delta = 1$. We can "reconcile" this error with the data by using the generalized Nash bargaining solution with bargaining powers $\gamma$ and $\delta$. But rescuing a bad modeling judgment in this way is no more respectable than rescuing the theory that heavenly bodies move only in perfect circles by factoring in epicycles when it is found that they don't.

I am not arguing that one should never fit the bargaining powers in a generalized Nash bargaining solution to the data. Nor am I arguing that one should never place the disagreement point in Nash's theory at the breakdown point. The point I am making is simply that one needs to understand the underlying theory well enough to know when one is in one of those rare situations when this is the right thing to do.

*Breakdown or Deadlock?* In Section 16.4.6, we saw that there are often at least two candidates for the disagreement point $\xi$ in Nash's original theory: a deadlock point $d$ and a breakdown point $b$. We argued in favor of recognizing both of these points by expanding Nash's model of a bargaining problem from $(X, \xi)$ to $(X, b, d)$. But how does one convince an applied worker, who is accustomed to ignoring $d$ and taking $\xi = b$?

Such disputes can be settled in principle by looking at sufficiently detailed noncooperative bargaining games. The model we look at here is designed to explore the different implications of *forced* and *unforced* breakdown. An unforced breakdown occurs when a player induces the outcome $b$ by walking off to take up his or her best outside option. In a forced breakdown, the negotiations are interrupted by some outside agency that imposes the outcome $b$, whether the players like it or not. For example, while production is suspended as Alice and the Mad Hatter bargain over the wage she pays him, someone else may steal their market.

We put both of these possibilities into the same Divide-the-Dollar model by introducing some extra moves into the Rubinstein Bargaining Game. After each refusal, a chance move occurs in which the breakdown point $b$ is forced with probability $\lambda\tau$, where $\lambda \geq 0$. Perhaps the philanthropist who donated the dollar becomes impatient with the prolonged negotiation and takes his money back. With probability $1 - \lambda\tau$, the game continues to another new move. At this new move, the player who just refused an offer may now opt out or in.[12]

Opting out induces the breakdown point $b$. If the refuser opts in, the clock advances by $\tau$, and the refuser becomes the proposer in the next period.

THEOREM 17.3 *With the assumptions of Section 17.5.3, a subgame-perfect equilibrium outcome of the modified Rubinstein Bargaining Game converges on a generalized Nash bargaining solution as $\tau \to 0$. The bargaining problem is $(X, b, d)$, where*

$$d_i = \frac{\lambda b_i}{\lambda + \rho_i}. \tag{17.15}$$

*The bargaining powers are $\alpha = 1/(\lambda + \rho_1)$ and $\beta = 1/(\lambda + \rho_2)$.*

The proof just recycles the ideas of the previous section, and so we focus instead on the implications of the result.

[12] It matters that a player must refuse a final offer before opting out, but it is hard to commit yourself not to listen. I recall a light airplane flying over Ann Arbor on a football afternoon towing a banner that said "Marry me, Maisie"!

*Computing the Deadlock Point.* In general, Alice's deadlock payoff $d_1$ is what she gets if there is no agreement, and nobody ever pursues their outside option. For example, if she received an income of $h\tau$ for each period in which the negotiation continues, then her deadlock payoff would be[13]

$$d_1 = h\tau + h\tau\delta_1^\tau + h\tau\delta_1^{2\tau} + \cdots = \frac{h\tau}{1-\delta_1^\tau}, \tag{17.16}$$

which converges to $h/\rho_1$ as $\tau \to 0$, by L'Hôpital's rule (Section 17.5.4). In the modified Rubinstein Game, Alice's deadlock payoff is

$$d_1 = \lambda\tau b_1\delta_1 + \lambda\tau(1-\lambda\tau)b_1\delta_1^{2\tau} + \lambda\tau(1-\lambda\tau)^2 b_1\delta_1^{3\tau} + \cdots = \frac{\lambda\tau b_1}{1-(1-\lambda\tau)\delta_1^\tau}$$

which converges to $\lambda b_1/(\lambda+\rho_1)$ as $\tau \to 0$.

**The case** $\lambda = 0$. This is what we took to to be the standard case in Section 16.4.6. There is no risk that delay will result in the surplus being lost, and we are back to the problem of Section 17.5. The deadlock point corresponds to perpetual disagreement, which means that $d = 0$ here. The bargaining powers are $\alpha = 1/\rho_1$ and $\beta = 1/\rho_2$. However, there is always some small risk that the surplus will be lost, and so we learn that our standard answer to the bargaining problem is only an approximation.

**The case** $\rho_1 = \rho_2 = 0$. The players are infinitely patient, and only the risk of losing the surplus matters. The deadlock point $d$ becomes identical to the breakdown point $b$. The bargaining powers are equal. The practice of using the *symmetric* Nash bargaining solution with a single disagreement point equal to the breakdown point is therefore vindicated. But how often are we entitled to assume that time doesn't matter in a negotiation?

## 17.7 Roundup

The *Nash program* is an invitation to test and refine cooperative solution concepts by studying noncooperative bargaining models. The assumptions we need to write into such models for their equilibrium outcomes to coincide with a cooperative solution concept indicate when and how the concept can be applied in practice. The chief obstacle in applying the Nash program is that we can't handle noncooperative bargaining games with incomplete information. Except for an occasional example, the theory is therefore confined to the case of *complete information,* in which everything of interest to the players is common knowledge when the negotiations begin.

Commitment can be a powerful weapon in bargaining, but it is hard to make credible commitments. Nash's Demand Game models the case in which both players have unlimited commitment power. Each therefore simultaneously makes a take-it-or-leave-it demand. All Nash equilibrium outcomes in a smoothed version of the game approximate the symmetric Nash bargaining solution.

[13]Recall that the formula for the sum of a geometric progression is $1 + x + x^2 + \ldots = (1-x)^{-1}$, provided that $|x| < 1$.

Nash's threat theory deals with the case when a disagreement point isn't given. The players then make irrevocable threats that specify what they will do if agreement isn't reached. This reduces the situation to a strictly competitive threat game that can be solved by the same methods as a two-person zero-sum game. When applied to the problem of collusion in a Cournot duopoly, each player threatens to act as though in a monopoly. These threats determine how much the low-cost player pays the high-cost player to stay out of the market.

The Rubinstein Bargaining Game models the case when only credible threats are heeded. Players exchange offers until agreement is reached. The game has a unique subgame-perfect equilibrium outcome that approximates a generalized Nash bargaining solution as the interval between successive proposals becomes vanishingly small. A player whose discount rate is $\rho_i$ has bargaining power $1/\rho_i$. Patient players therefore get more.

## 17.8 Further Reading

*The Economics of Bargaining*, by Ken Binmore and Partha Dasgupta: Blackwell, New York, 1987. This collection of papers expands on the theory described in this chapter.

*Bargaining Theory without Tears*, by Ken Binmore: *Investigaciones Económicas* 18 (1994), 403–419. This article proves the result of Section 17.6 and offers a generalization to the case of a nonconvex feasible set.

## 17.9 Exercises

1. When bargaining with incomplete information, why might it be rational for one player to open the proceedings by burning some money? Comment on the similarity between strikes and burning money. **phil**
2. The analysis of the smoothed Nash Demand Game given in Section 17.3.2 applies to bargaining problems $(X, b, d)$ in which $b = d$. Explain why a similar argument leads to the same conclusion when $b > d$.
3. Section 9.6.3 considers a demand game played in the Edgeworth box. Adam and Eve simultaneously commit themselves to a price and a quantity. Sketch the region in utility space outside which Adam commits himself not to trade by making such a demand. Indicate the region that corresponds to a best reply by Eve. **econ**
4. Daniel Defoe was a hack who wrote many books other than *Robinson Crusoe.* In *The Compleat English Tradesman,* Defoe explains that Quakers of his time refused to bargain because they thought it dishonest to ask for a deal better than one is willing to take. Why might it be quite profitable to be a Quaker with such a reputation? **phil**
5. The philosopher David Gauthier argues that bargaining should be seen as a two-stage process in which each player makes a claim at the first stage, followed by a concession at the second. He then asserts that rational players will claim their utopian payoffs and then both concede the same percentage of the gain $U_i - d_i$ that they initially claimed. Explain why he thereby unknowingly reinvented the Kalai-Smorodinsky bargaining solution. How does Gauthier's argument measure up to the standards of the Nash program? **phil**

phil

6. Gauthier's presentation of the idea in the previous exercise is preceded by a rejection of Nash's threat theory. He says,

   Maximally effective threat strategies . . . play a purely hypothetical role . . . since Adam and Ann do not actually choose them. But if Adam and Ann would not choose these strategies, they cannot credibly threaten with them. Maximally effective threat strategies prove to be idle.

   Why are maximally effective threats never used? Why do things that don't happen sometimes determine what does happen? Why is the real question whether a threat *would* be used if it *were* to prove ineffective in deterring some unwanted behavior?

math

7. The Nash Demand Game assumes that commitments are irrevocable. To explore a less severe requirement, we look at Divide-the-Dollar with risk-neutral players. Their negotiation game begins with each making a claim. The players can back down from these claims, but doing so is costly. A player who accepts $x < a$ after claiming $a$ $(0 \leq a \leq 1)$ must pay $c(a - x)$, where $c > 0$.

   If Alice claims $a$ and Bob claims $b$, explain why their subsequent bargaining problem is $(X, 0)$, where the feasible set $X$ is shown in Figure 17.8(b). If this problem is solved using the symmetric Nash bargaining solution, show that the players' optimal initial claims are $a = b = \frac{1}{2}$. The players therefore simply claim what they expect to get.

econ

8. Alice is a monopoly seller of an electronic good that she can duplicate at zero cost. Bob values Alice's product at $B$ and Chris at $C$, where $0 < B < C$. It is common knowledge that these are the valuations but that Alice thinks it equally likely that the high-value consumer is Bob or Chris. If she bargains with each separately, use Harsanyi and Selten's version of the symmetric Nash bargaining solution to predict the price she will get from each. An alternative is for Alice to bargain with an agent representing both Bob and Chris for a site license. It is then common knowledge that the agent's valuation is $B + C$. Predict the price in this situation, using the symmetric Nash bargaining solution. Prove that Alice prefers to sell a site license.

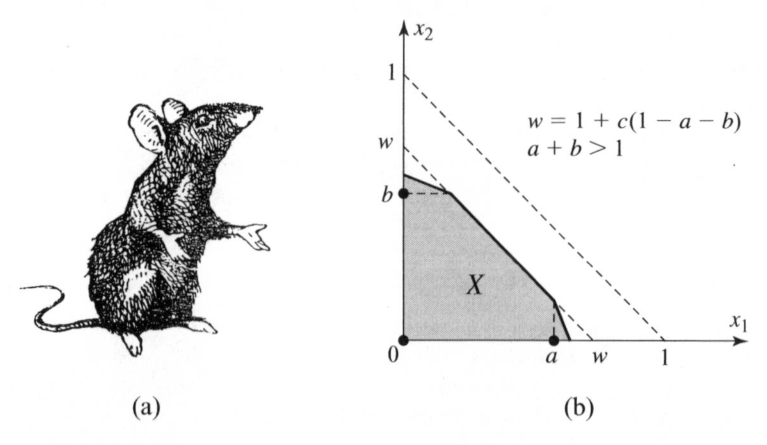


Figure 17.8 Revocable commitments. Figure 17.8(a) shows the posture to adopt when backing down from a commitment.

9. The previous exercise suggests that site licenses are bad for the consumer. But if $2C > 5B$, show that the bargaining game with incomplete information has an equilibrium that Alice prefers, in which she sells only to the high-value consumer. How is a Pareto improvement possible when a site license is sold?

econ

10. Analyze the Chicken game of Figure 6.15(a) using Nash's threat theory.
11. Draw the noncooperative payoff region for the simplified Cournot Game of Section 5.5.1. In Section 16.6.2, we considered both $N = (16, 16)$ and $P = (12, 8)$ as possible disagreement points in a collusive negotiation when bribery is impractical. Why does the former make sense when the players have no commitment power? Show that the latter is only one of a large number of disagreement points consistent with Nash's threat theory but that all of these generate a collusive agreement on the outcome $(20, 15)$.

econ

12. When side payments are allowed, explain why Nash's threat game is necessarily constant sum. Explain why finding the players' optimal threats can be reduced to solving the constant-sum game whose payoff table is obtained by replacing the payoff pair $d$ in each cell of the payoff table of the underlying game $\mathcal{G}$ by the symmetric Nash bargaining solution $s(d)$ of the problem $(Z, d)$.

math

13. Apply the procedure of the previous exercise to the payoff table of Figure 17.9(a), whose noncooperative payoff region was drawn in Exercise 6.9.29. Explain why the optimal threats can be found by solving the threat game of Figure 17.9(b). Confirm that Alice's payoff matrix in this constant-sum game has a saddle point. Determine the optimal threats and the payoffs that the players receive in the resulting agreement.
14. Repeat the previous exercise for the games of Figure 17.10 on the assumption that side payments are allowed. Show that:
    a. Alice should threaten to play her first pure strategy with probability $\frac{1}{6}$ and her second pure strategy with probability $\frac{5}{6}$.
    b. Bob should threaten to play his first pure strategy with probability $\frac{2}{3}$ and his second pure strategy with probability $\frac{1}{3}$.
    c. The result is an agreement in which \$12 is split so that Alice gets $\$8\frac{2}{3}$, and Bob gets $\$3\frac{1}{3}$.
15. Explain why Nash threat games may not be constant sum, but show that they are always strictly competitive. Justify the claim that the players' optimal threats remain the security strategies of the threat game.

math

16. Why is the story that opens Section 17.3 a reversed Ultimatum Game?

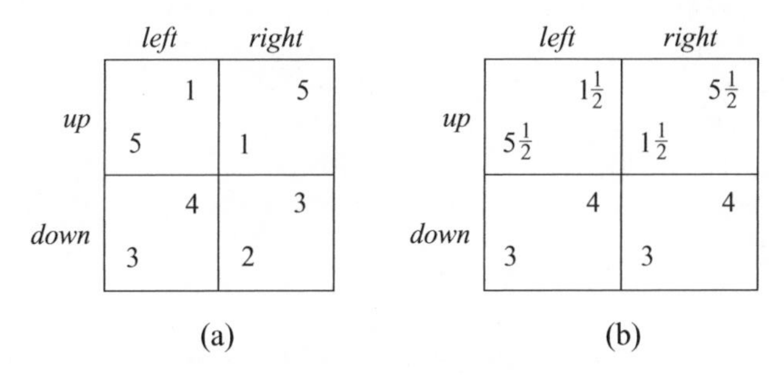


Figure 17.9 Payoff tables for Exercise 17.10.13.

Disagreement game

| | | |
|---|---|---|
| 5, 3 | 12, 0 | 2, 2 |
| 6, 0 | 6, 2 | 9, 1 |

Threat game

| | | |
|---|---|---|
| 7, 5 | 12, 0 | 6, 6 |
| 9, 3 | 8, 4 | 10, 2 |

Figure 17.10 Payoff tables for Exercise 17.10.14.

17. Find all subgame-perfect equilibria of the Ultimatum Game when demands must be made in whole numbers of cents. Show that one of these calls for Alice to demand ninety-nine cents. Deduce that the principle that concludes Section 17.5.1 doesn't apply when demands are discrete.

math

18. Continuing the previous exercise, take the smallest unit of currency to be $\mu > 0$ instead of one cent. In what sense do all the subgame-perfect equilibria of the discrete model converge on the unique subgame-perfect equilibrium of the continuous model as $\mu \to 0$?
19. Find a Nash equilibrium for the Ultimatum Game in which the dollar is split fifty-fifty.
20. Suppose that the two-stage bargaining game of Section 17.5.2 is extended so that it has three stages, with Bob making the first proposal. How much does Bob get in the unique subgame-perfect equilibrium?

fun

21. Suppose that Alice and Bob play the following bargaining game about the division of a dollar donated by a philanthropist. The philanthropist specifies that only the splits 10 : 90, 20 : 80, 50 : 50, and 60 : 40 are to be permitted. Moreover, Alice and Bob are to alternate in vetoing splits that they regard as unacceptable. What split will result if subgameperfect strategies are used and Alice has the first opportunity to veto? What split will result if Bob begins?
22. Show that any division of the dollar corresponds to a Nash equilibrium in the Rubinstein Bargaining Game.
23. Why does the Rubinstein Bargaining Game give the first mover an advantage in Divide-the-Dollar? Show that this advantage disappears in the limit as $\tau \to 0$.

math

24. Suppose the next proposer is always chosen at random in Rubinstein's Bargaining Game. If Alice and Bob are equally likely to be chosen, show that Theorem 17.1 survives. How must the bargaining powers be altered when Alice is chosen with probability $1 - p$ and Bob with probability $p$?

math

25. In Rubinstein's Bargaining Game, the players are restricted to proposing either that Alice should get the whole dollar or that Bob should get the whole dollar.
    a. Find a subgame-perfect equilibrium in which Alice begins by proposing that she get the whole dollar and Bob agrees.
    b. Find a subgame-perfect equilibrium in which Alice begins by proposing that the whole dollar be given to Bob and he agrees.
    c. Use these results to show that there are other subgame-perfect equilibria in which agreement doesn't occur immediately.

math

26. Suppose the players in Rubinstein's Bargaining Game don't discount time at all but pay a fixed cost of $c_i\tau$ for each period of length $\tau$ that passes without an

agreement. Confirm that a subgame-perfect equilibrium analysis generates the following results when $\tau \to 0$:

a. If $0 < c_1 < c_2$, Alice gets the whole dollar.
b. If $c_1 > c_2 > 0$, Bob gets the whole dollar.
c. If $c_1 = c_2 > 0$, any outcome in which both players get at least a zero payoff is possible.

27. If subgame perfection is taken as the criterion of rational behavior, why does the last exercise provide a counterexample to the Coase "theorem"? phil
28. We study the limiting case as $\tau \to 0$ in the Rubinstein Bargaining Game because nothing in real life constrains players to a particular timetable, and it is always best to make your next proposal as soon as possible. Critics argue that this would advantage players with quicker reaction times. Write a model in which Alice has a quicker reaction time than Bob to confirm that the critics are mistaken. Why is the objection a red herring anyway? phil
29. Exercise 17.10.17 shows that there are multiple subgame-perfect equilibria in the Ultimatum Game when proposals must be made in whole numbers of cents. Pushing this kind of argument further, it is possible to show that any division of the dollar corresponds to a subgame-perfect equilibrium in the Rubinstein Bargaining Game when proposals must be made in whole numbers of cents—provided that the interval $\tau$ between successive proposals is sufficiently small. Critics argue that this result torpedoes Rubinstein's theory, but if we are going to have a smallest unit $c > 0$ of currency about which people care, why don't we also have a smallest unit $\tau > 0$ of time about which people care? What is your take on this controversy? phil

# 18
# *Teaming Up*

## 18.1 Coalitions

Power in chimpanzee societies is exercised thorough a constantly shifting pattern of alliances among chimps. We like to pretend otherwise, but matters aren't so different in human societies. Who you know usually counts for more than what you know, even in societies that pride themselves on their egalitarian social contracts.

How should coalitions be modeled? Karl Marx thought that potential coalitions like Capital and Labor could be treated like monolithic players in a mighty game, but such a naive modeling approach overlooks the fact that the cohesion of a coalition depends on the extent to which it succeeds in satisfying the aspirations of its individual members. We need to know what holds coalitions together. How and why does one coalition form rather than another?

Von Neumann and Morgenstern tried to answer these difficult questions in the cooperative half of their *Theory of Games and Economic Behavior.* Their work spawned a large literature in which a bewildering variety of cooperative solution concepts are defended with lists of axioms, all of which seem plausible when considered in the abstract but which can't all be simultaneously true. The Nash program provides a possible way of determining in what context it makes sense to apply which concept, but game theory hasn't gone very far down this road as yet. We therefore say the minimum on this subject, outlining only the three most popular cooperative concepts.

More progress has been made on how coalitions form in markets. Which buyers and sellers will team up in what trading partnerships? On this subject, we find ourselves treading firmer ground.

## 18.2 Coalitional Form

When passing from an extensive form to the corresponding strategic form, we lose some of the structure of the game (Section 5.1). When we leave noncooperative game theory and start considering cooperative solution concepts, we throw away yet more structure since the fact that *binding* preplay agreements are possible in cooperative game theory means that the strategic structure of the underlying game often ceases to be important (Section 16.3). The coalitional form of a game follows this line of reasoning all the way by throwing away all information about a game other than the sets of payoff profiles available to each possible coalition.

*Payoff Regions.* The set of all players in an $n$-player game is denoted by $N = \{1, 2, \ldots, n\}$. A *coalition* is any subset of $N$. For example, the possible coalitions in a two-player game are $\emptyset$, $\{1\}$, $\{2\}$, and $\{1, 2\}$. The set $N$ itself is called the *grand* coalition.

To say that a coalition has formed means that the members of the coalition have signed up to a joint choice of strategies. Where side payments are possible, the agreement will include a deal on how the eventual payoffs the players receive are to be redistributed among the members of the coalition.

Cooperative solution concepts commonly seek to predict which coalitions will form using only the data supplied by a game's coalitional form. The *coalitional form* of a game is a listing of the payoff regions $V(S)$ for each possible coalition $S$. The cooperative payoff region $V(S)$ of a coalition $S$ in an $n$-player game is defined so that an $n$-tuple $x$ belongs to $V(S)$ if and only if the members of $S$ are able to coordinate on a joint strategy that guarantees a payoff of at least $x_i$ to each player $i$ in the coalition $S$.

For example, if Alice is player 1 and Bob is player 2 in a three-player game, then it may be that they can sign binding agreements that guarantee any of the payoff pairs in the set $Y$ of Figure 16.2(b). But $V(\{1, 2\})$ isn't the two-dimensional set $Y$. It is the three-dimensional cylinder parallel to the third player's payoff axis with base $Y$.

Section 16.4.1 made some simplifying assumptions about the cooperative payoff region $X$ of the grand coalition in a two-player game. We make the same assumptions about the cooperative payoff region of any coalition $S$ in an $n$-player game. The set $V(S)$ is therefore convex, closed, and bounded above. We also permit free disposal, although this assumption doesn't matter much.

*Transferable Utility.* Assuming that utility is transferable simplifies things a lot, albeit at a considerable cost in realism (Section 16.3.3).

The set $V(S)$ is then determined by a single number $v(S)$, which is the total utility the coalition can guarantee for itself independently of the behavior of players outside the coalition. This utility can be redistributed among the coalition members however they like, and so $V(S)$ is the set of all $n$-tuples $x$ for which

$$\sum_{i \in S} x_i \leq v(S).$$

Alice and Bob's maximum joint payoff of $35 = 20 + 15$ in Figure 16.2 occurs at $M = (20, 15)$. If side payments are allowed, they can therefore agree on any payoff pair in the set $Z$ of Figure 16.2(c). If this structure is embedded in a three-player game with transferable utility, $V(\{1,2\})$ is therefore the three-dimensional cylinder with base $Z$ that lies parallel to the third player's payoff axis. That is to say, $V(S) = \{(x_1, x_2, x_3) : x_1 + x_2 \leq 35\}$.

The coalitional form of a cooperative game with transferable utility is determined by its *characteristic function*, which simply assigns a coalitional security level or value $v(S)$ to each coalition $S$. We look only at examples in which the characteristic function $v$ is superadditive. This means that $v(S \cup T) \geq v(S) + v(T)$ whenever $v(S \cap T) = \emptyset$.

*Bargaining Set.* Coalitions with only a single player are particularly easy. If $S$ contains only player $i$, then $V(S)$ is the set of all $n$-tuples $x$ for which $x_i \leq v_i$, where $v_i$ is the player $i$'s security level (Section 7.4.1). In cooperative game theory, the proposition that players will never accept less than their security level is called *individual rationality.*

The grand coalition of all the players has a special role because it must usually form if the players are to have access to the Pareto-efficient payoff profiles of a game. A lazy application of the Coase "theorem" suggests that rational players will therefore always form the grand coalition, but we will find important exceptions.

As in Section 16.4.3, we follow Von Neumann and Morgenstern in referring to the set of Pareto-efficient and individually rational payoff profiles as the *bargaining set.*[1]

In the two-player case we defined the bargaining set in terms of the players' breakdown payoffs, which are their best outside options if they fail to make an agreement (Section 16.4.3). In doing the same here, we are are observing that a player's security level is the best that the player can get upon failing to make an agreement with anyone at all. A coalitional form throws away the structure of a game that would allow us to say anything about deadlock payoffs. Anything that hinges on the fact that deadlock payoffs may differ from breakdown payoffs therefore can't be expressed using the coalitional form—although this fact is seldom acknowledged.

## 18.2.1 Do Coalitional Forms Make Sense?

The mention of security levels sounds a cautionary note. Section 7.5.6 makes fun of the absurdly cautious folk who always use the maximin principle as members of the belt-and-suspenders brigade. But coalitions have no more reason to be cautious than individuals. It is therefore necessary to recognize that working with coalitional forms makes sense only when the strategic interaction between coalitions that we pack away in a black box when doing cooperative game theory is of a particularly simple kind.

→ 18.3

*Constant-Sum Games with Transferable Utility.* In this case, Von Neumann and Morgenstern gave a very reasonable justification for throwing away all of the structure of a game except its characteristic function. If a coalition $S$ forms in such a game, then the Coase theorem predicts that the complementary coalition $\sim S$ of all

[1]There is a risk of some confusion. Aumann and Maschler use the same term to describe their refinement of a Von Neumann and Morgenstern stable set. Von Neumann and Morgenstern referred to the payoff profiles in the bargaining set of a constant-sum game as *imputations.*

the other players will form. The two coalitions will then play a two-player, constant-sum game (Section 7.2). Each coalition $S$ will therefore get its value $v(S)$ to divide among its members.

Von Neumann and Morgenstern claimed to be able to reduce all cooperative games to the constant-sum case by introducing a fictional player whose payoff is always equal to the difference between the sum of what the other players get and some constant. But nobody nowadays thinks that this mathematical trick gets us anywhere. Nor do modern game theorists follow Von Neumann and Morgenstern in regarding the concept of transferable utility as a harmless simplifying device (Section 16.3.3).

Critics say that such modern views are heretical deviations from the words of our prophet, but we think it only natural that some ideas should turn out not to work when a science is being developed. Personally, I find it comforting that even the great Von Neumann sometimes got things wrong.

*Trading Games.* In trading games, all the strategic activity is usually over once it has been determined which coalitions will form, so that no question arises of a game being played between $S$ and $\sim S$.

For example, five players have houses to sell for which there are seven potential buyers. After it is decided which house will be sold to which buyer at what price, no other issues need be resolved. With risk-neutral players, the value of a coalition $S$ that contains three sellers is found simply by reassigning their houses efficiently to other members of the coalition. The sum of the new owners' valuations of the houses is then $v(S)$. How this sum of money is divided among the other members of the coalition determines who pays what to whom for which house.

## 18.3 Core

Von Neumann and Morgenstern argued that the outcome of any bargaining problem must lie in its bargaining set (Section 16.4.3). The core is an attempt to generalize this idea.

*Blocking and Domination.* The same reasoning that leads to the Coase theorem says that bargaining won't stop when a payoff profile $y$ is on the table if another payoff profile $x$ (called an *objection*) can be found with $x \succ_S y$. This means that

- $x \in V(S)$
- $x_i > y_i$ for each player $i$ in the coalition $S$

If $y$ were about to be agreed under these circumstances, the members of $S$ would get together to *block* $y$ in favor of an outcome that they all prefer and that they can enforce without the help of anyone outside the coalition.

Coalitional dominance is only loosely related to strategic or Pareto dominance (Sections 5.4.1 and 8.5.1). We say that $x$ dominates $y$ and write

$$x \succ y$$

if and only if $x \succ_S y$ for at least one coalition $S$.

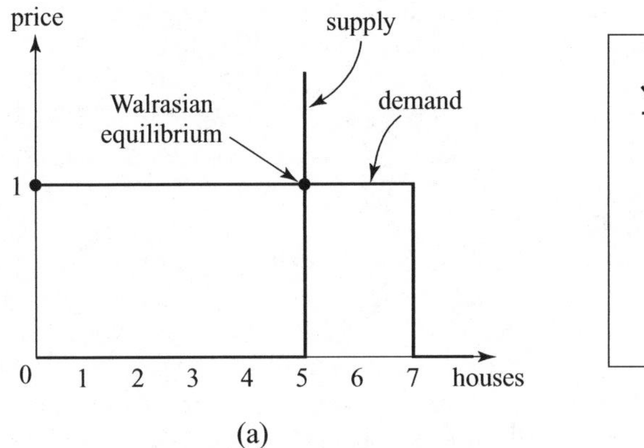


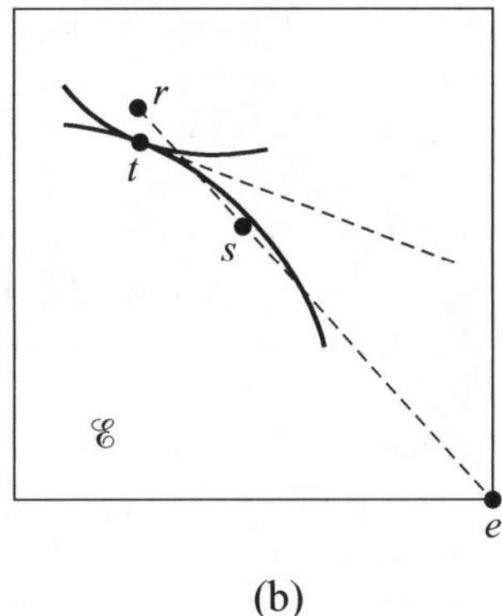


Figure 18.1 Walrasian equilibria and the core.

The *core* of a game is the set of its undominated payoff profiles—those that no coalition can block.

*Buying and Selling Houses.* The core of a two-player game is the same as its bargaining set. In the version of Divide-the-Dollar in which Alice is a seller of a house worth 0 to her and Bob is a buyer to whom the house is worth 1, to say that the outcome will be in the core therefore tells us nothing beyond the fact that the house will be sold to Bob at some price between 0 and 1 (Section 16.4.2). But what if we make a market by replicating Alice five times and Bob seven times?

Figure 18.1(a) shows the supply and demand curves in this market. The Walrasian equilibrium occurs where these curves cross (Section 9.6.1). At this equilibrium, all five houses are sold at price 1, and so the short side of the market appropriates he entire surplus. If a buyer tried to acquire a house for less, he would be undercut by one of the buyers who would otherwise fail to buy a house at all. The Walrasian payoff profile $w$ therefore assigns a payoff of 1 to each seller and 0 to each buyer.

To see what cooperative game theory has to say, model the market as a coalitional form with transferable utility, in which $v(S)$ is the number of houses for which $S$ contains both a buyer and a seller. We show that the core consists of the single point $w$. Any payoff profile $y \neq w$ will be blocked by the coalition of the buyer and seller to whom it assigns the smallest payoffs.

To see this, let Alice and Bob be the seller and buyer whose payoffs $y_A$ and $y_B$ are the smallest. Then $5 \geq 5y_A + 7y_B$ because the total payoff can't exceed the total number of houses for sale. The value of Alice and Bob's coalition is one, and so they will block $y$ if and only if $y_A + y_B < 1$. But if $y_A + y_B \geq 1$, the inequality $5 \geq 5y_A + 7y_B$ implies that $y_A \geq 1$. So $y$ assigns all sellers a payoff of 1. Thus $y = w$.

### 18.3.1 The Core in an Edgeworth Box

→ 18.4

How well does the core work in more complicated markets? The Edgeworth box turns out to be a useful tool in tackling this question (Section 9.4.1). We learn nothing new when Alice and Bob are the only traders. In such a bilateral monopoly, the core corresponds to the whole of the contract curve. However, matters change if we replicate Alice and Bob a very large number of times to simulate a perfectly competitive market (Section 9.6).

Edgeworth showed that such a market must trade at the Walrasian equilibrium because a coalition of traders would otherwise split off and set up their own separate market. In modern terms, an allocation that isn't Walrasian lies outside the core. We offer only a sketch of the proof.

Figure 18.1(b) shows the crucial case of a non-Walrasian allocation on the contract curve that assigns the same bundle $t=(a,f)$ to each Alice and the same bundle $(A-a,F-f)$ to each Bob. Each Alice would rather trade at $r=(b,g)$ than $t$, and each Bob would rather trade at $s=(c,h)$ than $t$.[2]

The point $s$ has been chosen so that positive integers $M$ and $N$ can be found satisfying $(M+N)s=Mr+Ne$, where $e$ is the endowment point. The point $s$ must then lie on the straight-line segment joining $e$ and $r$, which is possible if and only if $t$ isn't Walrasian.

The equation $(M+N)s=Mr+Ne$ ensures that a coalition of $M+N$ copies of Alice and $M$ copies of Bob can redistribute their endowments so that each Alice gets $(b,g$ and each Bob gets $(A-c,F-h)$.[3] Each Alice therefore trades at $r$, and each Bob trades at $s$. It follows that they all prefer the result of the redistribution to trading at $t$. So they block the allocation $t$, which therefore lies outside the core.

### 18.3.2 Condorcet Paradox

Condorcet was a French revolutionary who hoped to create a utopia by mathematical reasoning but was sentenced to the guillotine instead. His paradox of voting was considered in Exercise 4.11.7. We use the same idea here to illustrate a major problem with the core as a cooperative solution concept: It is often empty. It is therefore too strong to demand that the outcome of a cooperative game always lie in the core.

*Odd-Man-Out.* A slightly modified version of the game Odd-Man-Out provides a simple example (Exercise 7.11.36). This is a three-player version of Divide-the-Dollar in which any pair of players can guarantee any outcome they choose, irrespective of the wishes of the odd man out. It may be, for example, that the way the dollar is divided is determined by majority voting.

The bargaining set of the game is $\mathscr{B}=\{x : x_1+x_2+x_3=1 \text{ and } x\geq 0\}$. Figure 18.2(a) represents $\mathscr{B}$ as a triangle whose three vertices each correspond to one of the players getting the whole dollar (Section 6.5.3). The shaded part of the triangle shows the payoff profiles $x$ dominated by the profile $y$. Since we are free to place $y$ anywhere in $\mathscr{B}$, it follows that we can find an objection to any $x$ in the bargaining set. Odd-Man-Out therefore has no undominated payoff profiles, and so its core is empty.

[2]Because Bob prefers $(A-c,F-h)$ to $(A-a,F-f)$.

[3]The vector equation $(M+N)s=Mr+Ne$ reduces to $(M+Nc)=Mb+NA$ and $(M+N)h=Mg$. Rewritten as $(M+N)c+M(A-b)=(M+N)A$ and $(M+N)h+M(F-g)=MF$, these equations say that the coalition as a whole is endowed with just the right number of apples and fig leaves for the redistribution to be possible.

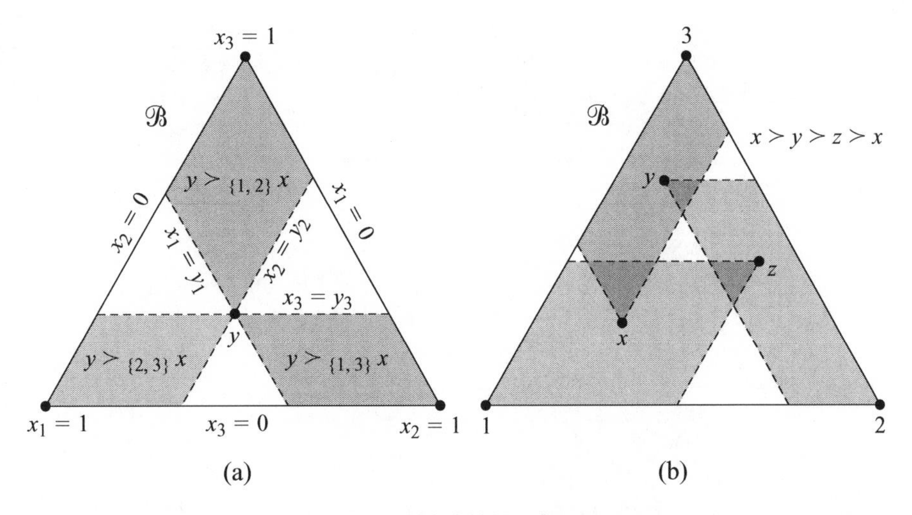


Figure 18.2 Odd-Man-Out.

Figure 18.2(b) shows how to construct a domination cycle

$$x \succ y \succ z \succ x,$$

starting with any payoff profile $x$ in $\mathscr{B}$. The domination relation $\succ$ is therefore intransitive (Section 4.2.2). This example of Condorcet's paradox occurs frequently in cooperative games. When it does, it always creates problems for the stability of any coalitions that may form.

## 18.4 Stable Sets

Von Neumann and Morgenstern didn't bother with the core, presumably because they saw that it can't make sense for every coalition always to plan to block anything it can block. The players in the coalitions $R$, $S$, and $T$ would need to be particularly myopic to plan to block each of $x$, $y$, and $z$ when they form part of a domination cycle of the form: $x \succ_R y \succ_S z \succ_T x$.

Von Neumann and Morgenstern argued that forward-looking players will not cooperate in blocking a profile $x$ unless the objection $y$ they have in mind is itself viable as a possible outcome of the preplay negotiations. But what makes a payoff profile viable? Von Neumann and Morgenstern proposed the following criteria for the set $\mathscr{V}$ of viable payoff profiles:

- An objection inside $\mathscr{V}$ can be found to everything outside $\mathscr{V}$.
- No objection inside $\mathscr{V}$ can be found to anything inside $\mathscr{V}$.

Nowadays, a set $\mathscr{V}$ that satisfies these requirements is said to be *stable*. Any stable set must contain the core of the game.

When a cooperative game has more than one stable set, we are faced with the kind of problem that arises in equilibrium selection (Section 8.5). Just as Americans drive on the right and Japanese drive on the left, so different societies might have different standards of behavior when forming coalitions that are manifested in the choice they make among the available stable sets.

### 18.4.1 The Three-Cake Game

Alice, Bob, and Chris are playing a cooperative game in which only one of three cakes can be eaten. Each coalition of two players controls a different cake. The problem is to decide which cake is eaten and who gets how much to eat. We begin by assuming that utility is transferable.

The coalitional form is specified using a characteristic function $v$ with $v(\emptyset) = v(1) = v(2) = v(3) = 0$. We assume that $0 \leq \alpha \leq \beta \leq \gamma$ and take $v(\{2,3\}) = \alpha$, $v(\{3,1\}) = \beta$, and $v(\{1,2\}) = \gamma$. If the grand coalition were to form, its members would choose to divide the largest of the three cakes, and so $v(\{1,2,3\}) = \gamma$. Odd-Man-Out is the special case when $\alpha = \beta = \gamma = 1$. The bargaining set is $\mathscr{B} = \{x : x_1 + x_2 + x_3 = \gamma \textit{ and } x \geq 0\}$.

*Nonempty Core.* We first discard profiles outside the bargaining set $\mathscr{B}$ shown in Figure 18.3 because these will be blocked either by the grand coalition or by individual players.

The core is nonempty only when $\alpha + \beta \leq \gamma$. The patterns of domination shown in Figure 18.3(a) reveal that the undominated profiles are then those in which the cake of size $\gamma$ is divided so that Chris gets nothing, Alice gets at least $\beta$, and Bob gets at least $\alpha$. The core isn't stable because no profile in the core dominates profiles in which the cake of size $\gamma$ is divided between Alice and Bob in a way that assigns one of the two players only a small amount. Figure 18.3(b) shows one of the many stable sets.

The analysis that leads to such stable sets is orthodox, but does it really make any sense? Chris contributes nothing to his partners if he joins the coalition containing Alice and Bob. So how come he gets in on the action with a positive payoff? The reason is that we have forced the assumption that the grand coalition will necessarily form on the model. But this assumption is safe only if all the players gain something positive from getting together, which isn't true in this case (Section 16.4.3).

The Nash program will be used later in the chapter to explain how and why Chris might be able to squeeze a positive payoff from his strategically weak position (Section 18.6.2). In the interim, we begin the cooperative analysis again on the assumption that the grand coalition will *never* form. The unpalatable assumption that utility is transferable then no longer simplifies the model, and so we throw this away as well.

Figure 18.4(a) shows the three-dimensional diagram we now have to draw when $\alpha + \beta \leq \gamma$. The core remains unchanged in this new situation, but now there are no stable sets at all.[4]

*Empty Core.* The case when $\alpha + \beta > \gamma$ is shown in Figure 18.4(b). Recall that we are now assuming that utility isn't transferable and that the grand coalition won't form.

[4]This isn't surprising since we threw out the grand coalition, but William Lucas showed that an orthodox analysis of largish games can also yield no stable sets.

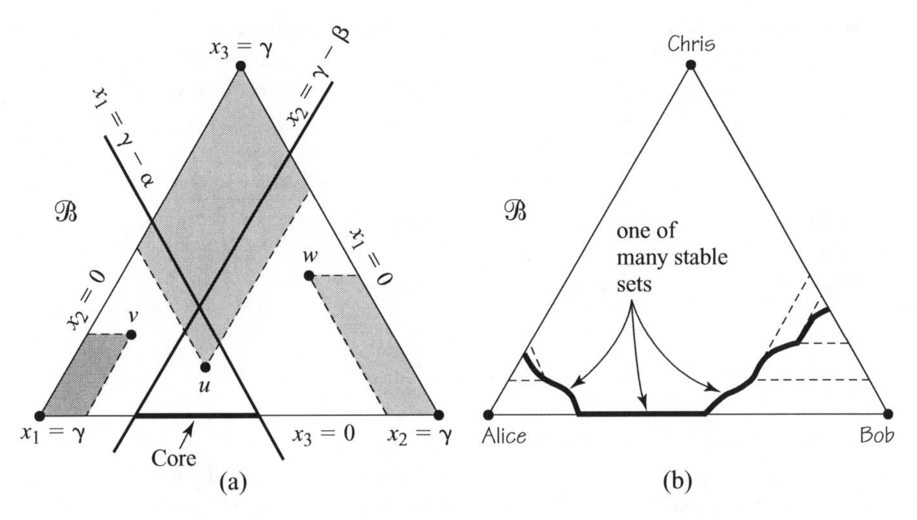


Figure 18.3 The bargaining set in the Three-Cake Game with transferable utility. The shadows cast by $u$, $v$ and $w$ in Figure 18.3 correspond to the profiles blocked by $\{1,2\}$, $\{2,3\}$, and $\{3,1\}$ respectively. We must have that $v_1 \geq \gamma - \alpha$ and $w_2 \geq \gamma - \beta$, but $u$ can be anywhere. The curves that form part of the stable set $\mathscr{V}$ in Figure 18.3(b) mustn't cross the broken lines that radiate from them.

The core is empty, but the triple $\{u, v, w\}$ of payoff profiles shown in Figure 18.4(b) is stable. ($v_1 = w_1 = \frac{1}{2}(\beta + \gamma - \alpha)$, $w_2 = u_2 = \frac{1}{2}(\gamma + \alpha - \beta)$, and $u_3 = v_3 = \frac{1}{2}(\alpha + \beta - \gamma)$.) No profile in the triple is dominated by any profile in the triple. Any profile not in the triple is dominated by a profile in the triple.

Since the game has only one stable set, we have a unique prediction about the way coalitions form in this case. However, the prediction doesn't tell us which player will be the odd man out. Since players get the same payoff with both their potential partners, they don't care who they team up with. They care only that they aren't left out in the cold.

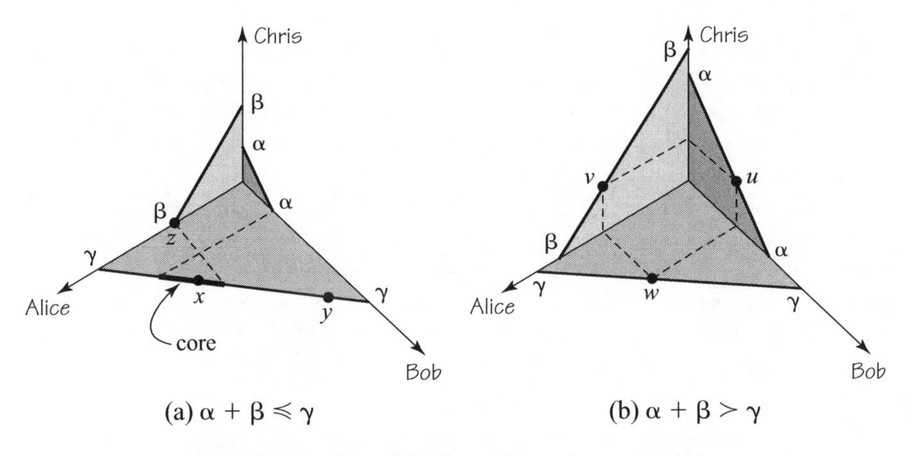


Figure 18.4 The Three-Cake Game without the grand coalition.

Perhaps this is why coalitions in both human and chimpanzee societies can be so fluid. If Alice and Bob form a coalition that excludes Chris, he will offer whoever will listen a little more than their stable share. If Bob takes up Chris's offer and abandons Alice, then Alice will become the odd man out, with an incentive to offer Chris a little more than he is currently getting in order to persuade him to abandon Bob, and so on.

The results can be devastating. For example, the border between England and Wales where I live was fought over for centuries as any two of the Welsh, the king of England, and the Marcher lords[5] shifted alliances to combine against whichever of the three was currently most powerful.

## 18.5 Shapley Value

I was once summoned urgently to London to explain what the French government was talking about when it proposed that the costs of a tunnel under the English Channel be allocated to countries in the European Union using the Shapley value. Economists need to know what the Shapley value is because of its potential application in such cost-sharing exercises, but it isn't easy to defend as a prediction of what will happen on average when a cooperative game with transferable utility is played rationally.

Lloyd Shapley derived his value from three axioms. The first says that "dummy" players who contribute no more to any coalitions they join than they get on their own receive exactly this amount and no more. The second says that interchangeable players get the same payoff. The third says that the Shapley value of a game obtained by adding the characteristic functions of two games is the sum of their Shapley values.

The third and most important of these axioms makes it clear why the Shapley value is a useful cost-sharing device. However, there is no particular reason why the strategic considerations that arise in a game obtained by adding two characteristic functions should be related in any simple way to the strategic considerations in the games from which they are derived.

### 18.5.1 Calculating the Shapley Value

Just as the symmetric Nash bargaining solution picks a unique payoff profile as the solution of a two-player bargaining problem, so the Shapley value picks a unique payoff profile for a game with transferable utility in coalitional form. The easiest way to calculate the Shapley value makes it explicit that it is intended as an *average* of all the possible ways that coalitions might form.

There are $n!$ ways in which the $n$ players of a game can be placed in a particular order. For each such ordering $D$, add the players one by one to a coalition of increasing size that starts with the empty set and finishes with the grand coalition. Along the way, player $i$ will eventually be added to some coalition $S$. Player $i$'s marginal contribution to the value $v(N)$ of the grand coalition when it is built up in this way is then $\Delta_i = v(S \cup \{i\}) - v(S)$.

[5]These powerful magnates were originally intended to guard the English border or marches against raids by the Welsh tribes.

The *Shapley value* of the game is the vector $s$ in which $s_i$ is player $i$'s average marginal contribution. Thus,

$$s_i = \frac{1}{n!} \sum_D \Delta_i(D),$$

where the sum extends over all $n!$ orderings of $N$ and $\Delta_i(D)$ is player $i$'s marginal contribution when the ordering is $D$.

*Two-Player Bargaining.* The simplest case to which the Shapley value can be applied is the two-player game whose characteristic function $v$ is given by $v(\emptyset) = v(1) = v(2) = 0$ and $v(N) = 1$. The possible orderings of the players are $(1,2)$ and $(2,1)$. Player 1's marginal contribution with the first ordering is $v(1) - v(\emptyset) = 0 - 0 = 0$. His contribution with the second ordering is $v(N) - v(1) = 1 - 0 = 1$. Thus

$$s_1 = \tfrac{1}{2}(0+1) = \tfrac{1}{2}.$$

The same is true of player 2, and so $s = (\frac{1}{2}, \frac{1}{2})$.

This outcome is commonly said to coincide with the symmetric Nash bargaining solution, but this is true only when the deadlock point $d$ is the same as the breakdown point $b = (0,0)$ (Section 16.4). Even when utility is transferable, we can therefore only hope that the Shapley value will predict the outcome of rational bargaining when breakdowns are forced rather than optional (Section 17.6).

*Three-Cake Game.* The following table shows the necessary calculation for the Shapley value in the case when utility is transferable and the grand coalition is allowed to form. Notice that Chris contrives to get a positive payoff again, although he contributes nothing to the coalition $\{1,2\}$ that will actually form when $\alpha + \beta < \gamma$.

| Ordering | Player 1 | Player 2 | Player 3 |
|---|---|---|---|
| 123 | $0$ | $\gamma$ | $0$ |
| 132 | $0$ | $\gamma - \beta$ | $\beta$ |
| 213 | $\gamma$ | $0$ | $0$ |
| 231 | $\gamma - \alpha$ | $0$ | $\alpha$ |
| 312 | $\beta$ | $0$ | $0$ |
| 321 | $\gamma - \alpha$ | $\alpha$ | $0$ |
| Shapley value | $\frac{1}{6}(-2\alpha + \beta + 3\gamma)$ | $\frac{1}{6}(\alpha - 2\beta + 3\gamma)$ | $\frac{1}{6}(\alpha + \beta)$ |

## 18.6 Applying the Nash Program

This section briefly explores the extent to which the cooperative solution concepts we have been studying turn out to predict the deals that rational players would reach using various different noncooperative bargaining models.

### 18.6.1 A Bargaining Pit Model

→ 18.6.3

A visit to the Chicago wheat market is a must if you are in the city. Traders mill about in a "bargaining pit" shouting or signaling their offers to anyone who happens to be looking their way.

Onlookers often ask whether all the sound and fury is really necessary, but it isn't just unthinking conservatism that prevents such bargaining pits being replaced by computerized substitutes. Bargaining pits are a kind of chaotic double auction in which both buyers and sellers bid at the same time (Section 9.6.3). However, they differ from computerized double auctions in that all the action is common knowledge among the traders, so that nothing whatever needs to be taken on trust.[6]

In the bargaining pit model for the Three-Cake Game, players rotate in taking turns being active. Active players first decide whether to accept any of the offers they currently have in hand. If they decide to refuse all their current offers, they then shout out an offer of their own that consists of the lowest payoff they will accept to form a coalition with someone. This offer goes to both the other players and remains valid until the player who made the offer becomes active again. As always, we look at what happens in the subgame-perfect equilibrium when the interval between successive proposals becomes vanishingly small.

In the Three-Cake Game, the cooperative theory that dispenses with the grand coalition does rather well in predicting the outcome of the bargaining pit model. When the core is empty, the outcome is one of the triple $u$, $v$, and $w$ that make up the stable set in Figure 18.4(b). Which of these three payoff profiles we observe depends on the order in which the players move, whoever is last to move becoming the odd man out.

The outcome is in the core when this is nonempty. We can even use two-player bargaining theory to predict which point of the core it will be. The result is the same as if Alice and Bob bargained together without taking account of Chris, beyond the fact that he offers an outside option of $\beta$ to Alice and an outside option of $\alpha$ to Bob (Section 16.4.3). Attempts by Alice or Bob to persuade the other to take less than these outside options will always be foiled by Chris's offering more because he knows that he will otherwise be excluded.

*One Seller and Two Buyers.* Alice has only one hat for sale, which Bob and Chris both want to buy. Bob values the hat at \$$V$ and Chris at \$$v$, where $V > v$. How much will Alice will get for her hat?

If the players are risk neutral, we have a version of the Three-Cake Problem in which the cake available to the two buyers is of size zero. If the sale takes place at a market stand with everybody present simultaneously calling out their offers, then the bargaining pit model applies.

We know that the outcome will be in the core shown in Figure 18.5(a). Thus Alice will sell the hat to Bob at a price $p \geq v$.

As is often the case, we can avoid computing subgame-perfect equilibria by getting to the same result using two-player bargaining theory. The bargaining problem

[6]Other than the fact that traders must honor the deals they make in the pit. But traders who welsh on a deal might as well pack up and go home because nobody will ever trade with them again (Section 11.5.1).

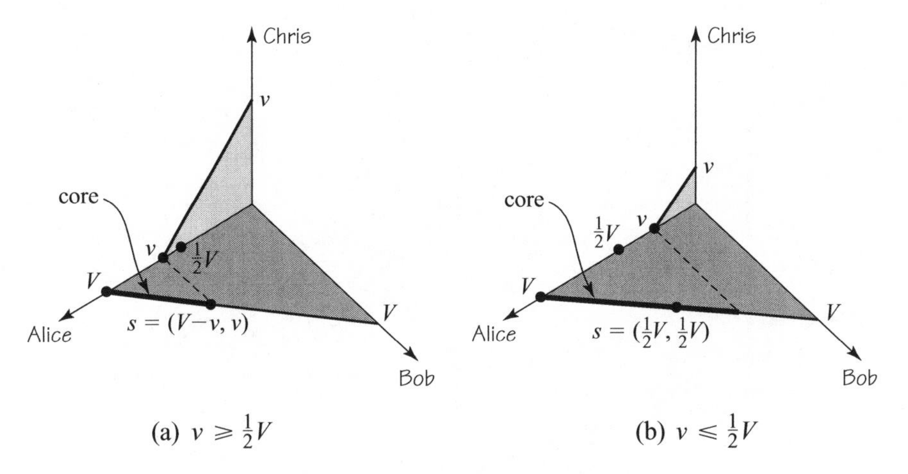


Figure 18.5 One seller and two buyers in a bargaining pit.

faced by Alice and Bob is $(X, b, d)$, where the feasible set $X$ consists of all pairs of payoffs that they can achieve by dividing a cake of size $V$, the breakdown point is $b=(v, 0)$, and the deadlock point is $d=(0, 0)$ (Section 16.4). If the players have equal bargaining powers, the answer depends on whether $v \geq \frac{1}{2}V$ as in Figure 18.5(a) or $v \leq \frac{1}{2}V$ as in Figure 18.5(b).

When $v \geq \frac{1}{2}V$, the Nash bargaining solution is $s=(v, V-v)$, so that Alice sells the hat to Bob at price $p=v$. When $v \leq \frac{1}{2}V$, the Nash bargaining solution is $s=(\frac{1}{2}V, \frac{1}{2}V)$, so that Alice sells the hat to Bob at price $p=\frac{1}{2}V$. Figure 18.6(a) shows that both outcomes correspond to Walrasian equilibria since the supply and demand curves cross at all prices $p$ between $v$ and $V$.

*Monopoly Pricing.* As we saw in Chapter 9, monopolists can sometimes jack up their selling price by restricting supply. Section 9.6.1 considered a simple case in which Dolly irrevocably restricts her production to $W$ ounces of wool, but we now revert to Alice's hat factory. If Alice can convince all her many potential customers that she is never going to have more than $H$ hats to sell, what price will she get for each hat?

We assume that each potential customer wants only one hat and that the maximum amount that each customer is willing to pay for a hat is common knowledge. The customers are represented by a demand curve in Figure 18.6(b). In Chapter 9, two cases were considered. The first is the classic case in which Alice charges the same price $p$ to every customer who buys a hat. The second is the case of a fully discriminating monopolist, who is able to extract all of the surplus lightly shaded in Figure 18.6(b). How much of this surplus will Alice actually be able to extract from her customers if they all gather around her stand in the market trying to get the best bargain they can?

Bargaining theory predicts that Alice will be able to price-discriminate only against her richer customers—those who are willing to pay more for a hat. Any customer whose valuation $V$ of a hat satisfies $p \leq V \leq 2p$ pays the same price $p$. Customers whose valuations $V$ satisfy $V > 2p$ each pay $p=\frac{1}{2}V$.

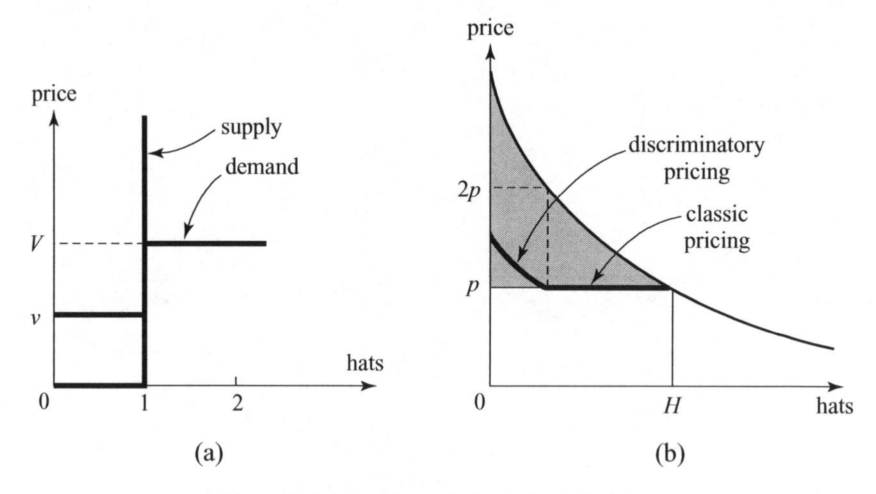


Figure 18.6 Bargaining and discriminatory pricing.

Alice bargains with all her customers at once. When bargaining with Bob, they both know that her alternative to selling a hat to him is to sell it to Chris, who is the potential customer with the highest valuation who will actually fail to get a hat. Figure 18.5 shows that the solution to Alice and Bob's bargaining problem is therefore $p = v$ when $v \geq \frac{1}{2}V$ and $p = \frac{1}{2}V$ when $v \leq \frac{1}{2}V$.

The bold curve in Figure 18.6(b) therefore gives all the prices at which Alice sells her hats. It shows Alice selling at the classical monopoly price to all her poorer customers but managing to partially discriminate among her richer customers. The fraction of the surplus that she thereby manages to appropriate is heavily shaded in Figure 18.6(b). Of course, her life would be much more difficult if she couldn't distinguish between her rich and poor customers.

### 18.6.2 Decentralized Bargaining

The bargaining pit model allows a seller to hold offers from two buyers simultaneously. The resulting informal auction generates a Walrasian selling price. We now continue our attempt to apply the Nash program by looking at a decentralized bargaining model in which Alice negotiates with Bob or Chris one at a time.

Door-to-door salesmen face this problem. So does an employer renegotiating wages with an employee. If the employee withdraws his labor, the employer may threaten to replace him with an outsider, but the threat will be empty if it is optimal for the outsider to behave exactly like the insider he is replacing once he has gotten the job.

*Optional Breakdown.* Consider the game with one seller and two buyers. If Chris were not around at all, Alice and Bob would agree on the symmetric Nash bargaining solution $s = (\frac{1}{2}V, \frac{1}{2}V)$ of their bargaining problem, assuming that their bargaining powers are equal. If Alice sells her wares from door to door or over the telephone, she gains nothing at all from Chris's appearance on the scene. Figure 18.7(a) explains how this simplified version of Diamond's paradox works.

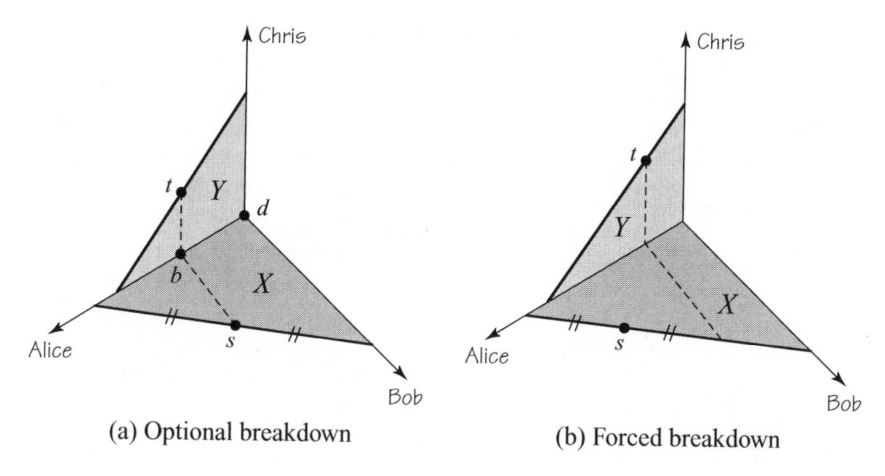


Figure 18.7 Bargaining over the telephone.

With Chris present, Alice can choose to break off her negotiation with Bob temporarily and go next door to talk to Chris. Their bargaining problem is $(Y, b, d)$ in Figure 18.7(a) because Alice can return to Bob and get $s$. The Nash bargaining solution of $(Y, b, d)$ is $t$ because Alice's outside option is active. Chris's presence therefore modifies Alice and Bob's bargaining problem to $(X, b, d)$, but this change doesn't alter the result since the Nash bargaining solution of $(X, b, d)$ remains $s$.

Whether Alice knocks on Bob's or Chris's door, she will therefore sell her hat for $\frac{1}{2}V$, which is the same price she would get if only Bob wanted the hat. Door-to-door salesmen therefore have no monopoly power.

If this cooperative analysis is disputed, it can be defended using a telephone version of the Rubinstein bargaining model, which we outline for the case of the Three-Cake Game. Some player has the initiative at the beginning of each round of bargaining. This player then makes an offer to whoever of the other players he or she chooses. If the offer is accepted, the game ends. If the offer is refused, the player who refuses has the initiative in the next round.

Look at what happens at a subgame-perfect equilibrium of the game with one seller and two buyers when the interval between successive rounds becomes vanishingly small. Alice then sells her hat at the price she would sell it to Bob if Chris were absent altogether. Whether the hat is sold to Bob or Chris depends on who has the initiative at the opening move.

None of the cooperative theories of coalition formation reviewed in this chapter comes anywhere near predicting this result!

*Walrasian Markets.* Generations of economics students have been taught that the core of a large enough market is Walrasian (Section 18.3.1). However, we have seen that only very myopic players will block everything they can in the indiscriminate manner required to justify the core (Section 18.4).

The bargaining pit model with one buyer and two sellers gives a better reason why we should care about Walrasian equilibria, but bargaining pits are highly centralized markets in which all the traders can be in on the action simultaneously if they so choose (Section 18.6.1). What of decentralized markets like the house market? Does the telephone bargaining model say that such markets can't be Walrasian? This would be a rash conclusion to draw from studying such a highly simplified model. More realistic matching-and-bargaining models converge on the Walrasian outcome when the search and bargaining frictions become vanishingly small (Section 9.6.3).

The mechanism that leads such models to the Walrasian outcome remains the informal auctioning that does the same trick in bargaining pit models, but now the auctioning is local rather than global. In each round, unmatched buyers and sellers search for a bargaining partner. The searching process is usually modeled by introducing a Chance move that pairs players up at random. Once they are matched, a buyer and a seller begin to bargain. Their outside options are determined by what they expect to get on returning to the pool of unmatched players if the negotiation breaks down.

If the model allows players to continue searching for a second partner while bargaining with their current partner, we are back with a bargaining pit model after a second or more partners have been found. It then isn't surprising that we end up with a Walrasian conclusion. However, the models featured in the literature usually don't allow simultaneous searching and bargaining. But neither do they mimic the telephone bargaining model by making breakdown optional. Partners who fail to agree are forcibly parted at some point and thrown back into the pool of unmatched players whether they like it or not.

To see why this modeling feature might have a similar effect to allowing local auctioning, we return to the game with one seller and two buyers to examine what happens when the bargaining model combines random matching of infinitely patient players with forced breakdown (Section 17.6). One may imagine that Alice knocks on doors at random if unmatched. When Bob or Chris answers the door, she bargains with him until agreement is reached or their unpredictable wives grow tired and run her off the property.

As always, we consider the outcome of a subgame-perfect equilibrium when the interval between successive proposals becomes vanishingly small. If the players are infinitely patient,[7] the outcome is exactly the same as in the bargaining pit model: Alice's hat is sold to Bob at the larger of the prices $p = \frac{1}{2}V$ and $p = v$ in the bargaining pit model, rather than always being sold to Bob at price $p = \frac{1}{2}V$ as in the telephone bargaining model with optional breakdown.

To verify that this result is consistent with Theorem 17.3, suppose first that Alice and Chris would agree on $t$ in Figure 18.7(b) when matched together. Alice and Bob would then agree on $s$ when they were matched. But then Alice *wouldn't* agree to $t$ because she would always do better to wait patiently until a forced breakdown matched her with Bob. The deadlock point $d$ in Theorem 17.3 therefore needs to be calculated on the assumption that Alice will never do a deal when matched with Chris. However, the existence of Chris still gives Alice an outside option of $v$ that needs to be incorporated in the breakdown point $b$, although Alice will never exercise this option.

[7]In the matching-and-bargaining literature, the effect of this assumption is often achieved by supposing that one of a matched pair is chosen at random to make a take-it-or-leave-it offer to the other, with a refusal resulting in the partnership breaking up so that the two players return to the pool of unmatched players.

It is reassuring that the matching-and-bargaining literature is able to give a rationale for using a Walrasian analysis in decentralized markets when transactions costs are negligible, but my view is that its real potential lies in examining what happens in markets where the transaction costs can't be neglected. My discount rate has certainly always been large when buying and selling houses!

*Implementing the Shapley Value?* The Shapley value in the Three-Cake Game assigns Chris a positive payoff, but none of our noncooperative bargaining models offer him a whisker of a chance of getting anything. A model that generates the Shapley value on average needs to incorporate more than random matching and forced breakdown since we have just seen that Chris can still end up with nothing in such models.

Faruk Gul has shown that this problem disappears when we also reinterpret the payoffs in the profiles that make up a coalition's cake as income *flows* that the members of the coalition can derive by combining resources that they originally own (Section 16.4.7). When Alice is matched with Chris, it may then pay her to buy his resource—not because she plans to permanently enjoy the income flow that she derives from exploiting his and her resources together—but because the fact that she will be enjoying this income flow will improve her bargaining position when she later gets matched with Bob. When she has made a deal with Bob, Chris's resource will be thrown on the scrap heap, but it will have served its purpose in getting a bigger share of the surplus for Alice (Exercise 18.9.23).

### 18.6.3 Lessons

If game theory were a mature science, cooperative solution concepts would come accompanied with answers to all the questions an applied worker might ask. Are coalitions assumed to form once and for all? Or do they form and break apart as the negotiation proceeds? Can a player belong to more than one coalition simultaneously? And so on. Perhaps there will one day be a book that lists all possible noncooperative bargaining models, along with the cooperative solution concepts that describe their equilibrium outcomes. Among many other things, all the informational questions that we have neglected by looking only at games of perfect information would be answered.

However, as things stand, all we can genuinely deduce from our attempt to apply the Nash program to the Three-Cake Game is that cooperative game theory resembles fire in being a good servant but a bad master. Fortunately, the days when economists would simply use one cooperative solution concept or another without feeling the need to justify their choice are now gone. But we still have a long way to go before it will be possible to justify the use of one concept rather than another on genuinely scientific grounds.

## 18.7 Roundup

The coalitional form of a game discards all of the structure of a game except the set of payoff profiles that the members of each possible coalition can secure by acting together. In a game with transferable utility, the coalitional form is specified by a characteristic function $v$ that assigns a single number $v(S)$ to each coalition $S$. One can think of $v(S)$ as the security level of the coalition. In an $n$-player game, the payoff region $V(S)$ of the coalition $S$ is then the set of all $n$-tuples $x$ for which $\sum_{i \in S} x_i \leq v(S)$. Apart from constant-sum games with transferable utility, trading games are the leading case for which it makes sense to restrict attention to what a coalition can secure for certain independently of the behavior of players outside the coalition.

A myopic analysis says that bargaining won't stop when a payoff profile $y$ is on the table if an objection $x$ can be found with $x \succ_S y$. This means that all the players in a coalition $S$ prefer $x$ to $y$ and that they can secure $x$ or better by acting together (so that $x \in V(S)$). It is usual to say that $S$ will block $y$ when $x \succ_S y$, but it is more accurate to say that $S$ *can* block $y$ if it chooses.

If $x \succ_S y$ for some $S$, we write $x \succ y$. The profile $y$ is then said to be dominated. The voting paradox of Condorcet shows that the domination relation can be intransitive, which goes a long way toward explaining why coalition patterns in real life are often very unstable.

The set of all undominated profiles is called the *core* of the game. It coincides with what von Neumann and Morgenstern called the bargaining set of a two-player bargaining problem. It also converges on the Walrasian equilibrium of an exchange economy in which each agent is replicated sufficiently often. However, games often have an empty core. Even when they don't, why would a coalition block $y$ in favor of $x$ when they know that another coalition will block $x$ in favor of $z$?

Von Neumann and Morgenstern addressed this problem with their idea of a stable set $\mathscr{V}$. Not only do we require that an objection inside $\mathscr{V}$ can be found to everything outside $\mathscr{V}$, but we also ask that no objection inside $\mathscr{V}$ can be found to anything inside $\mathscr{V}$. Any stable set must contain the core of the game.

The Shapley value approaches the problem of coalition formation from a completely different angle. Given any ordering of the players, compute the marginal contribution each player makes to the value of the grand coalition of all the players as this is built up one player at a time. The Shapley value assigns each player his or her average marginal contribution taken over all possible orderings.

These and other cooperative solution concepts remain to be properly assessed using the Nash program, but enough is known to be sure that no single concept applies in all situations. For example, studies of the Three-Cake Game make it clear that we can't always assume that the grand coalition will form. But it would be a bad mistake to throw cooperative game theory out the window because it leaves so many questions unanswered. How would applied work be possible at all without cooperative solution concepts that summarize the equilibrium outcomes of noncooperative bargaining models, which are often difficult to analyze directly?

We looked briefly at how three noncooperative bargaining models work in the Three-cake Game, paying special attention to the case of one seller and two buyers. If the seller has a single indivisible object for sale that the two buyers value at $v$ and $V$ respectively, what will the selling price be when $v < V$? If the players have equal bargaining powers, the selling price is the larger of $v$ and $\frac{1}{2}V$ in a bargaining pit model. In a telephone or door-to-door bargaining model with optional breakdown, the selling price is $\frac{1}{2}V$. When the players are infinitely patient and breakdown is forced in the telephone model, the selling price is the same as in the bargaining pit model.

Such results have implications in larger markets. Fortunately for economists, the idea of a Walrasian equilibrium proves to be remarkably robust. However, pricing in a monopoly turns out to hinge on features of the bargaining process that textbooks commonly neglect.

## 18.8 Further Reading

*Axioms of Cooperative Decision Making*, by Hervé Moulin: Cambridge University Press, New York, 1994. An elegant and comprehensive introduction to the axiomatic approach.

*Bargaining and Markets*, by Martin Osborne and Ariel Rubinstein: Academic Press, San Diego, 1990. This book offers another angle on the subject and covers a lot more ground.

## 18.9 Exercises

1. In a two-player game in strategic form, Adam has only one pure strategy. Eve has two pure strategies that yield the respective outcomes $(0, -1)$ and $(1, 0)$. What is the coalitional form of this game? What does it neglect that matters?
2. The characteristic function of Odd-Man-Out is altered so that $v(S) = c$ for each coalition with precisely two players (Section 18.3.2). Draw a diagram like Figure 18.2(a) that represents the new game for the case when $0 \leq c \leq \frac{2}{3}$. Indicate the core in your diagram. Why is the core empty when $c > \frac{2}{3}$?
3. In a seven-player version of Divide-the-Dollar, any majority of the players can divide the dollar as they choose. (Odd-Man-Out of Section 18.3.2 is the three-player case.) Show that the core is empty.
4. In the Dumping Game, each of three players ends up with a bundle $(b, g)$ consisting of $b$ units of a bad commodity and $g$ units of a good commodity. Each player's utility for this bundle is $u(b, g) = 3 + g - b$. They each begin with an endowment $(1, 1)$, which they can alter by irreversibly dumping as much of each commodity as they choose in the backyard of one or more of the other players. Sketch the bargaining set of this game, and show the payoff region $V(\{1, 2\})$ on the assumption that utility is *not* transferable. Use your diagram to show that the core consists of the single point $(3, 3, 3)$. What would the core be if utility were transferable?
5. An archeological expedition discovers a treasure in the Sierra Madre mountains. Each piece of treasure requires two players to carry it away. Explain why the Sierra Madre Game can be modeled as an $n$-player game in coalitional form using the characteristic function

$$v(S) = \begin{cases} \#(S), & \text{if } \#(S) \text{ is even} \\ \#(S) - 1, & \text{if } \#(S) \text{ is odd,} \end{cases}$$

   where $\#(S)$ is the number of players in the coalition $S$. If $n \geq 3$, show that the core is empty when $n$ is odd but consists of the single point $(1, 1, \ldots, 1)$ when $n$ is even.
6. Using Figure 18.2(a), confirm that the set $\{(\frac{1}{2}, \frac{1}{2}, 0), (\frac{1}{2}, 0, \frac{1}{2}), (0, \frac{1}{2}, \frac{1}{2})\}$ is stable in Odd-Man-Out (Section 18.3.2).

7. With the help of Figure 18.8(a), explain why the set $\mathscr{V}$ with $x_3 = c$ is stable in Odd-Man-Out, provided that $0 \leq c \leq \frac{1}{2}$. Von Neumann and Morgenstern called such stable sets discriminatory.

math

8. One can refine away the discriminatory stable sets for Odd-Man-Out discovered in the previous exercise by introducing the idea of a strongly stable set. If $v$ lies in a stable set $\mathscr{V}$, then it can't be dominated by any other profile in $\mathscr{V}$, but it might be dominated by a *heretical* profile $h$ that lies outside $\mathscr{V}$. Say that $h \succ_S v$. Because $h$ lies outside $\mathscr{V}$, it is dominated by at least one profile $w$ inside $\mathscr{V}$. If there is always a player in $S$ who would resist moving from $v$ to any such $w$, the set $\mathscr{V}$ is said to be *strongly* stable. Show that only the nondiscriminatory stable set is strongly stable in Odd-Man-Out (Exercise 18.9.6).

phil

9. Follow Von Neumann and Morgenstern in drawing a picture of a stable set in the game with one seller and two buyers on the assumption that the grand coalition will form (Figure 18.3(b)). Comment on the payoff profiles in the set that assign positive payoffs to all three players.
10. Under what circumstances is the set $\mathscr{V}$ of Figure 18.8(b) stable for the Dumping Game without transferable utility? (Exercise 18.9.4)

phil

11. We argued for replacing the core by the idea of a stable set because the former requires that the players act myopically (Section 18.4). Why are stable sets vulnerable to a more subtle version of the same criticism?

math

12. Prove the following claims:
    a. The empty set can't be stable.
    b. The core is a subset of any stable set.
    c. If the core is stable, then it is the only stable set.
    d. If $\mathscr{V}$ and $\mathscr{W}$ are stable, then $\mathscr{V} \subseteq \mathscr{W} \Rightarrow \mathscr{V} = \mathscr{W}$.

math

13. A game in coalitional form with transferable utility is *simple* if $v(N) = 1$, and $v(S)$ is always 0 or 1 for every coalition $S$. A coalition for which $v(S) = 1$ is a winning coalition. A player who belongs to every winning coalition has a veto.
    a. Why is Odd-Man-Out a simple game in which no player has a veto?
    b. Show that any simple game in which no player has a veto has an empty core.

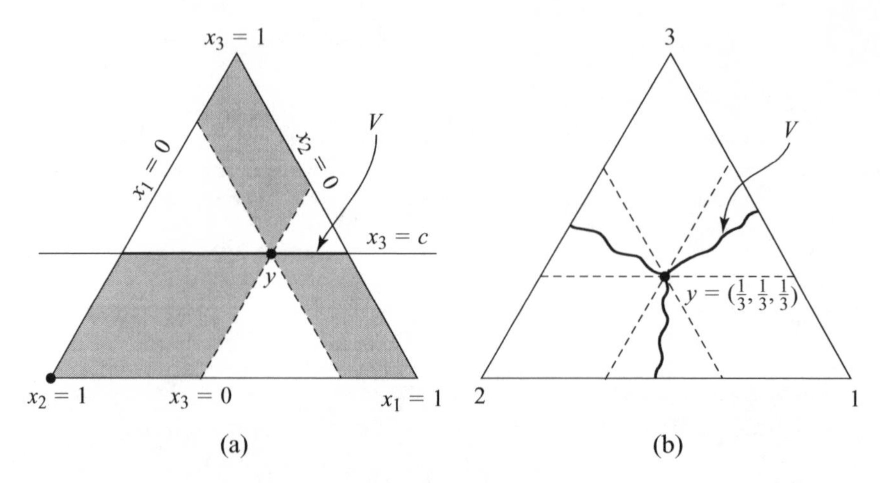


Figure 18.8 Some stable sets.

c. Show that the core in a game in which some players have a veto assigns a zero payoff to all the other players.

14. Let $S$ be a minimal winning coalition in a simple game, so that taking any players away from $S$ converts it into a losing coalition. Show that the set of payoff profiles in the bargaining set that assign zero to each player outside $S$ is stable. **math**
15. A player $i$ is a dummy if $v(S \cup \{i\}) - v(S)$ is always equal to his or her security level for all coalitions $S$ (Section 18.5). Prove that dummies get only their security levels in any stable set. **math**
16. A European parliament has $n$ parties, of which two each have $\frac{1}{3}$ of the seats, and the other $n-2$ parties share the remaining seats equally. Show that the Shapley value assigns a payoff to the larger parties that converges to $\frac{1}{4}$ as $n \to \infty$. If the Shapley value were an appropriate measure of political power, why wouldn't the small parties benefit from forming a single coalition? **math**
17. The punch line of the previous exercise appears in a sharper form as Harsanyi's paradox. The symmetric Nash bargaining solution is used to solve a three-player version of Divide-the-Dollar on the assumption that any disagreement will result in each player getting nothing. Each player then gets a payoff of $\frac{1}{3}$. Players 1 and 2 now form a coalition in which they agree to bargain as a unit with player 3, sharing the spoils equally. Why will they now end up with a payoff of only $\frac{1}{4}$ each? Is this an important insight into something real or an indication of the limitations of the cooperative approach to game theory? **phil**
18. Using the bargaining pit model of Section 18.6.1, find the price at which Alice will sell her hat to Bob or Chris when their respective bargaining powers are $a$, $b$, and $c$. What happens in the case of the telephone bargaining model with optional breakdown? **econ**
19. What happens in Odd-Man-Out when the players use the telephone bargaining model with optional breakdown? Assume that their bargaining powers are all unequal.
20. Suppose that the demand equation in the monopoly pricing example of Section 18.6.1 is $p + h = M$ and Alice's unit cost is $c > 0$. Why is the number $H$ of hats she chooses to take to market the same as in a classical monopoly? (Section 9.5) What would we need to do to the demand curve to get a different result? **econ**
21. The result in the monopoly pricing example of Section 18.6.1 is obtained under the assumption that the bargaining pit model applies. What results do we get by applying the telephone bargaining model with optional or forced breakdown? **econ**
22. Karl Marx would presumably have assumed the least favorable case for the consumer in the previous exercise. He would also have assigned all the bargaining power to the monopolist (Exercise 18.9.21). What are the implications for monopoly pricing? **phil**
23. Section 18.6.2 outlines a matching-and-bargaining model that generates the Shapley value on average. In the case of the Three-Cake Game, confirm that the profile $s$ of expected payoffs for each player at the outset of the game is the same as the Shapley value calculated in Section 18.5.1.

    You can short-circuit a formal calculation of the limiting outcome of the subgame-perfect equilibrium by observing that, once two players have been matched, their bargaining problem is $(X, b, b)$, where $b$ is what they expect if their partnership is broken without an agreement and they are thrown back into the pool of unmatched players, and $X$ is the set of payoff pairs available to them when they take into account the fact that whoever buys the other's resource will then go on to bargain with the third player. Applying the symmetric Nash bargaining solution, you will then be able to show that the outcomes when the original matchings are respectively $\{1,2\}$, $\{2,3\}$, and $\{3,1\}$ are

$$\begin{aligned} u &= (\tfrac{1}{2}\gamma + \tfrac{1}{2}(s_1 - s_2),\ 0,\ \tfrac{1}{2}\gamma - \tfrac{1}{2}(s_1 - s_2)), \\ v &= (\tfrac{1}{4}(\gamma+\alpha) + \tfrac{1}{2}(s_2 - s_3),\ \tfrac{1}{4}(\gamma+\alpha) - \tfrac{1}{2}(s_2 - s_3),\ \tfrac{1}{2}(\gamma - \alpha)), \\ w &= (\tfrac{1}{4}(\gamma - \beta),\ \tfrac{1}{4}(\gamma+\beta) - \tfrac{1}{2}(s_3 - s_1),\ \tfrac{1}{4}(\gamma+\beta) + \tfrac{1}{2}(s_3 - s_1)). \end{aligned}$$

It then remains to solve the equation $s = \frac{1}{3}(u+v+w)$, which is much easier than it looks because $s_1 + s_2 + s_3 = \gamma$.

econ

24. The idea of a Walrasian equilibrium is often defended as the core outcome in an Edgeworth box in which Adam and Eve have been replicated a very large number of times (Section 18.3.1). But if the players negotiate using the telephone bargaining model with optional breakdown, they will reach the same outcome as if each Adam and Eve negotiated independently of the rest of the market. Confirm that using the symmetric Nash bargaining solution in the Edgeworth box can result in an outcome that is different from the Walrasian equilibrium.

# 19 *Just Playing?*

## 19.1 Ethics and Game Theory

What can game theory contribute to ethics? Followers of the philosopher Immanuel Kant say it teaches us nothing at all because ethics is about doing what you don't want to do, and game theory is about getting what you want. But the same guys also think it rational to cooperate in the Prisoners' Dilemma.

However, other traditions in moral philosophy welcome game theory as a potentially useful tool. For example, David Hume argued that no theory of morals can serve any useful purpose unless it can show that the duties it recommends are also in the true interest of each individual. This approach makes it possible to see fairness norms as equilibrium selection devices for the repeated games of real life (Section 11.5). Fairness is then controversially seen as one of a number of possible ways that power might be balanced, rather than as a substitute for the need to balance power at all.

Hume understood that fairness norms can't work unless people are able to *empathize* with each other enough to make it possible for them to compare each other's welfare meaningfully. We can't even express the utilitarian and egalitarian theories of justice proposed by luminaries like Harsanyi and Rawls without presupposing that such interpersonal comparisons of utility are possible.

In welfare economics, interpersonal comparison of utility is taken for granted, but students of microeconomic theory are often simultaneously taught that such comparisons are intrinsically meaningless. This schizophrenic attitude toward interpersonal comparison presumably arose from the observation that von Neumann and Morgenstern's axioms provide no basis whatever for making interpersonal comparisons (Section 4.6.3). But why not follow Harsanyi and add some extra assumptions that do provide a basis?

Everything in this chapter is controversial to some degree. It should therefore be read as a series of sketches of what might one day be a theory, rather than an account of an established literature whose details are sacrosanct.

## 19.2 Do People Play Fair?

What do people think is fair? Do they actually play fair themselves when the chips are down? This section reviews some of the scientific evidence.

### 19.2.1 Experimental Game Theory

Only in recent years have serious attempts been made to see how well game theory does in predicting how real people play games in carefully controlled laboratory experiments. At first, the theory took a beating, and for some time the conventional wisdom was that Nash equilibria have no predictive power at all. However, just as early chemists learned to clean their test tubes, so experimental economists learned that one can expect game theory to work in the laboratory only under certain conditions. The requirements I stress in my own experimental work are these:

- The game is presented to the subjects in a user-friendly way that makes it easy for them to understand what is going on.
- Adequate cash incentives are provided, so that the subjects have good reason to pay attention to the problem with which they are faced.
- Sufficient time is made available for trial-and-error learning. This means that the subjects must be allowed to play the game repeatedly, against new opponents each time.

*Trial and Error.* The importance of the last point can't be stressed too much. It is just silly to think that ordinary people in off the street will play like game theory predicts right away.

For example, Emile Borel was a mathematical genius who anticipated Von Neumann in formulating the minimax theorem but wrongly guessed it to be false. So if Borel had been a subject in an experiment on zero-sum games, he wouldn't have played the maximin strategy right off the bat. Since ordinary people are a lot less clever than Borel, they won't either. If the subjects in an experiment find their way to an equilibrium of the game at all, it has therefore got to be through some process of trial-and-error adjustment.

Whether the subjects get to equilibrium or not depends on the game. In market games or auction games, the convergence is sometimes very fast. In some other games, the convergence is slower but no less sure. In spite of what our critics say, the one-shot Prisoners' Dilemma is one of the latter games. It is true that inexperienced subjects who aren't paid much cooperate about half the time. However, the rate of defection rises remorselessly, both as the subjects gain experience of playing the game and as the rates of pay are increased.

### 19.2.2 The Ultimatum Game

But game theory doesn't always triumph. The Ultimatum Game is the most talked-about exception. Section 17.5.1 shows that the proposer will get nearly everything in a subgame-perfect equilibrium. However, laboratory experiments show that real people play fair. The most likely proposal is for a fifty-fifty split. Proposals for an unfair split like seventy-thirty are refused more than half the time, even though the responder then gets nothing at all.

This is the most replicated result in experimental economics. I have replicated it myself several times. It doesn't go away when the stakes are increased. It holds up even in countries where the dollar payoffs are a substantial fraction of the subjects' annual income.

*Rival Explanations.* How should we react to this experimental data? Some game theorists see no reason for concern. They accept that real people don't use backward induction in the Ultimatum Game, but subgame-perfect equilibria were already under suspicion for theoretical reasons (Section 14.2.1). Moreover, the Ultimatum Game has lots of other Nash equilibria—including one that results in a fifty-fifty split (Exercise 17.10.19). So where's the problem?

But if we throw away backward induction here, why not elsewhere in bargaining theory? What becomes of the calculus of breakdown and deadlock points that other experiments show works rather well in the Rubinstein Game? As for the data being explicable in terms of alternative Nash equilibria, this won't do at all. People who end up with nothing at all after saying *no* to a tenth of their annual income aren't maximizing their monetary payoff!

Behavioral economists address the last point by asking why we insist that the subjects must be maximizing money. Perhaps subjects have "social" preferences that take into account the welfare of other people, as well as their own. Nobody denies that this is true to some extent. Nor do "social" preferences create any problem for game theory (Section 1.4.1). But there is a controversy over the extent to which introducing such exotic utility functions can explain the data. My own view is that the anomalies in the data are too large to be explicable by any realistic perturbation of the money payoffs in a game. If the perturbations were large, they would surely be evident in the many experiments in which the data are predicted rather well by the assumption that subjects maximize expected money.

In any case, my own experiments with two-stage bargaining games show that backward induction still fails, even with utility functions that take account of both a player's own money payoff and that of the opponent.[1] But fitting exotic utility functions to the data from the Ultimatum Game may still be useful. In accordance with the theory of revealed preference, we may thereby summarize the data in a way that allows us to predict what will happen in other games. But as Section 4.2 emphasizes, we then need the players' behavior to be *stable* when the same game or a similar game is played in the future.

However, the evidence shows that the players' behavior *changes* as they gain experience. It changes very slowly in the Ultimatum Game compared to other

[1]Of course, any behavior can be explained by backward induction if any utility function whatsoever can be fitted after the event.

games, but it changes nevertheless. Nor do the utility functions derived from observing the play of one game commonly predict the data of apparently similar games. Just awarding some subjects a meaningless gold star in the Ultimatum Game is enough to change the way the game is played.

*Norms as Habits.* The theory I favor abandons the attempt to rescue rational decision theory by explaining human behavior in one-shot situations as optimization relative to an exotic utility function. I think inexperienced or unmotivated people usually don't behave rationally at all. Nor is a reason hard to find in the case of the Ultimatum Game.

Real people are habituated to responding to ultimata in situations that are quite unlike those they encounter in the laboratory. In real life, the players usually expect to interact again in the future. Even if not, their play is likely to be observed by others with whom they do have an ongoing relationship. In such real-life situations, they are therefore playing some complicated *repeated* version of the Ultimatum Game.

The folk theorem tells us that the equilibria of a repeated game can be totally different from those of the one-shot game. In particular, it can become rational to take account of both *reciprocity* and *reputation.* Players who hope to establish a basis for reciprocal sharing with a prospective long-time associate can't afford to acquire the reputation of being a soft touch.

David Hume would argue that fairness norms evolve in such socialized contexts as equilibrium selection devices that help us coordinate on one of the vast number of available equilibria. Of course, we are seldom conscious that this is what we are doing. We internalize the norms as children and mostly don't even notice that we are playing a game when we use them. We are therefore not equipped to respond readily when an experimenter substitutes a laboratory version of the one-shot Ultimatum Game for the repeated version to which our habituated behavior is adapted. So we initially behave in a way that really makes sense only in a repeated context. It isn't therefore surprising that behavioral economists find that both reciprocity and reputation need to be taken into account when explaining the behavior of inexperienced subjects.

But we aren't robots without the capacity to learn. In the Ultimatum Game, we have to learn the difficult lesson that there is no point in shooting yourself in the foot because you are angry at receiving an unfair offer from someone you are never going to meet again.[2] In the one-shot Prisoners' Dilemma, we have to learn the equally difficult lesson that there isn't any point in trying to establish a reciprocal arrangement with a stranger who will never have the opportunity to return your favor.

But when the rewards are sufficient, we sometimes do eventually adapt to the fact that we are playing a one-shot game, rather than one of the repeated versions to which we are accustomed. Section 19.8 provides a bargaining example, but don't hold your breath waiting for something similar to happen in games like the Ultimatum Game!

[2]Analysis of the testosterone levels in the sputum of responders who refuse unfair offers shows that people do get genuinely angry.

## 19.3 Social Choice Paradoxes

Economists are traditionally impatient with the idea that fairness matters. They argue that there is a trade-off between fairness and efficiency that should be resolved in favor of efficiency. These views are buttressed by a number of paradoxes that seem to say that a rational society must be unfair.

### 19.3.1 Arrow's Paradox

Condorcet's voting paradox says that a society that determines its communal preferences from the individual preferences of its citizens by honest voting over each pair of alternatives will necessarily be collectively irrational since the communal preference will sometimes be intransitive (Section 18.3.2).

Ken Arrow generalized Condorcet's paradox to a whole class of social welfare functions that map the individual preferences of the citizens to a communal preference. As with the Nash bargaining solution, a crucial requirement is that the social welfare function satisfy a version of the Independence of Irrelevant Alternatives (Section 16.6.3).

Arrow's version says that the communal preference between two alternatives $a$ and $b$ should depend only on the individual preferences between $a$ and $b$ (and not on preferences that involve some other alternative $c$). Voting is the archetypal example.

Proposition 19.1 (Arrow's Impossibility Theorem) *With at least three alternatives, a social welfare function that maps any profile of individual rational preferences to a rational communal preference is necessarily dictatorial, provided it is Pareto efficient and satisfies Arrow's Independence of Irrelevant Alternatives.*

The statement of the proposition will be clarified while sketching the proof for the case of two citizens. We focus on three alternatives, represented by the radii in Figure 19.1(a). These are labeled $A$, $B$, and $N$ as in Exercise 4.11.7. The arrows on the outer circle in Figure 19.1(a) show Horace's preferences. Those on the inner circle show Maurice's preferences. The arrows drawn outside the circles show the communal preferences.

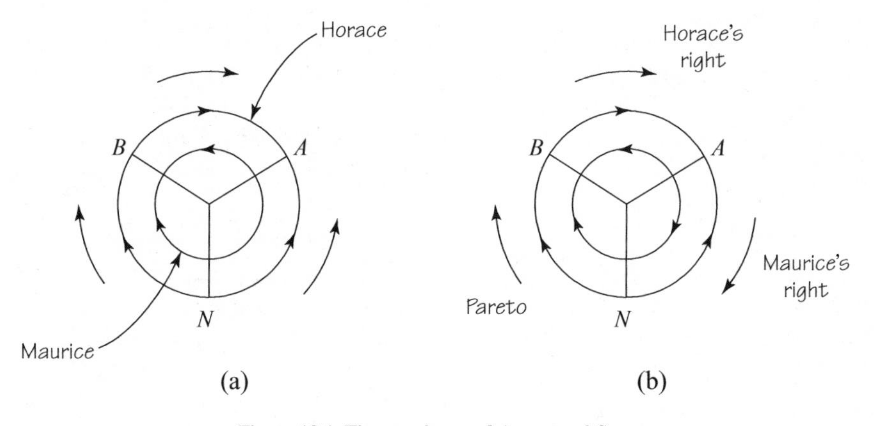


Figure 19.1 The paradoxes of Arrow and Sen.

To draw Figure 19.1(a), we begin by looking at the case when the communal preference favors $B$ over $A$, but Horace and Maurice differ in their preferences over these alternatives. Suppose the communal preference favors Horace in this case. We then pick any other alternative $N$ and show that Horace's individual preference between $A$ and $N$ also determines the communal preference between $A$ and $N$. Since we can play this trick over and over, Horace must then be a dictator over all pairs of alternatives.

To prove that Horace dictates between $A$ and $N$, we use the fact that the social welfare function is defined over the whole domain of individual preference profiles. So we can specify the players' preferences that aren't yet determined however we like. This won't make any difference to the communal preference between $A$ and $B$ because of Arrow's Independence of Irrelevant Alternatives. The individual preferences chosen in Figure 19.1(a) require that the communal preference between $B$ and $N$ favor Horace because of Pareto efficiency. It must therefore favor Horace between $A$ and $N$, whatever Maurice may prefer, because a rational preference must be transitive. Arrow's Independence of Irrelevant Alternatives then tells us that Horace dictates over $A$ and $N$, whatever anybody's preferences between other pairs of alternatives may be.

*Log Rolling.* Popular writers who don't really understand Arrow's theorem sometimes say that it spells doom for a just society. But this isn't true, even in theory. One reason is that the unrestricted domain assumption is rarely satisfied since a broad consensus about fundamentals exists in most societies. A second reason is that Arrow's Independence of Irrelevant Alternatives forbids the kind of "log rolling" that goes on when Horace promises to vote with Maurice on his pet project if Maurice will do the same for him in return. It even forbids reference to a status quo position that allows the bargaining solutions of Chapter 16 to work in seeming defiance of Arrow's theorem. Most importantly, it refuses to admit the kind of interpersonal comparison of utility without which it isn't even possible to say what fairness is.

## 19.4 Welfare Functions

Economists don't believe in telling people what they ought to want. Just as game theorists need to be told the players' utility functions before they can give them sensible advice on how to play a game, so economists who advise governments need politicians to tell them what objective function they should maximize. Such *welfare functions* never satisfy Arrow's harsh requirements, but why would we want them to?

*Bergsonian Welfare Functions.* In a society with $n$ citizens, a Bergsonian welfare function $W : \mathbb{R}^n \to \mathbb{R}$ assigns a real number $W(x)$ to each possible profile $x = (x_1, x_2, \ldots, x_n)$ of individual utilities. Maximizing a Bergsonian utility function over whatever set of possibilities is feasible therefore amounts to aggregating the utilities of the individual citizens in a particular way.

The Nash and the Kalai-Smorodinsky bargaining solutions provide examples. We can implement the symmetric Nash bargaining solution for $x \geq d$ and $n = 2$ by taking our welfare function to be the Nash product:

$$W_N(x) = (x_1 - d_1)(x_2 - d_2).$$

To implement the Kalai-Smorodinsky solution for $x \geq d$ and $n = 2$, the welfare function needs to be more elaborate:

$$W_{KS}(x) = \min\left\{\frac{x_1 - d_1}{U_1 - d_1}, \frac{x_2 - d_2}{U_2 - d_2}\right\}.$$

The Nash bargaining solution is sometimes suggested as a candidate for a fair arbitration scheme, but neither $W_N$ nor $W_{KS}$ should appeal to anyone who cares about social justice. Axiom 16.2 tells us that neither involves any interpersonal comparison of utility units. But how can we decide what is fair without comparing how much people get?

→ 19.5

*Surplus as Welfare.* Many economists deny that utilities can be compared at all, and so comparisons must be made in terms of physical commodities. Money is a popular basis for comparison, with welfare defined as the sum of consumer and producer surplus because of its interpretation as "total money saved" (Section 9.7). It is sometimes said to be the only viable candidate because other welfare functions involve some kind of trade-off between efficiency and equity and hence can't be "socially optimal."

When only one good other than money is being traded, the sum of consumer and producer surplus can be defined in terms of the demand and supply functions as

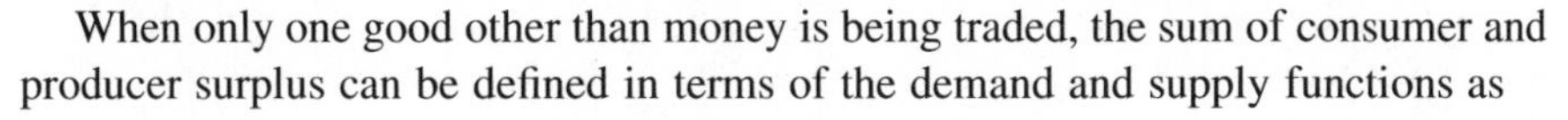

$$W_\$(q) = \int_{q_0}^{q} D(x)dx - \int_{q_0}^{q} S(x)dx,$$

where $q$ is the amount of the good traded and $q_0$ is some convenient benchmark.[3] The derivative of this quantity is zero when $D(q) = S(q)$ and thus demand equals supply at the maximizing value of $q$. With this definition, welfare is therefore largest at a Walrasian equilibrium, and so perfect competition is socially optimal without any need to cheat on the meaning of "socially optimal."

But Figure 16.15 shows that Walrasian equilibria needn't be at all fair.[4] Moreover, we know that there will normally be numerous other efficient outcomes on the contract curve when only Adam and Eve are in the market. In the case when their preferences are quasilinear—which we know is the only time when the interpretation of $W_\$(q)$ as total money saved is genuinely valid—the contract curve is vertical, as in Figure 9.5(a). To get from the Walrasian equilibrium to a fairer but equally

[3]Recall from Section 9.3.2 that $S(x)$ is the producer's marginal cost of producing $x$ (provided that it is profitable to produce at all).

[4]As do the axioms for the Walrasian bargaining solution outlined in Section 16.9.2.

efficient outcome, we therefore only need to transfer some money from one player to the other. A more equitable outcome is thereby achieved without changing the total amount of money saved.

We could restrict ourselves to quasilinear preferences and modify $W_{\$}(q)$ by insisting on a fair side payment, but how many dollars is fair? Since an extra dollar is obviously worth more to a beggar than to a billionaire, how can we possibly say without knowing more about Adam and Eve?

*No Envy.* A Walrasian equilibrium is Pareto efficient.[5] Any Pareto-efficient outcome is a Walrasian equilibrium if the endowment point is suitably chosen.

Given these welfare theorems, why measure welfare in money—or at all? Goods can be transferred until everybody has the same endowment of commodities, and then they can trade. The resulting Walrasian equilibrium even has a pleasant *no envy* property (Exercise 19.11.10). Nobody prefers to swap their final bundle for the bundle assigned to anyone else. The rich will complain about being taxed to subsidize the poor, but surely social justice will be served?

An example shows why this question is naive. An allocation procedure assigns Adam and Eve each a bottle of gin and a bottle of vermouth, for which they have no use except to make martinis. Adam is an unsophisticated soul who shakes together equal measures of gin and vermouth. He therefore ends up with two bottles of martini. Eve is more discerning and tolerates martinis only if made by diluting straight gin with no more than one drop of vermouth. She therefore ends up with only a little more than one bottle of martini.

The allocation is Pareto efficient and satisfies the no-envy criterion, but is it fair? Why are we measuring Adam's and Eve's welfare in terms of gin and vermouth, when what they really care about is martini? Perhaps we should assign $\frac{2}{3}$ of a bottle of gin to Adam and $\frac{4}{3}$ to Eve, so that each will then be able to drink the same number of martinis.

But nothing says that Adam and Eve will *enjoy* an equal number of martinis equally. If it wouldn't make much difference to Adam whether he drank pure vermouth rather than his disgusting martinis, why not give him all but a few drops of vermouth and Eve all but a few drops of gin?

All such discussions finally get around to the same conclusion. What we really need to compare are Adam's and Eve's *utilities.*

## 19.5 Interpersonal Comparison of Utility

In a Nash bargaining problem, the disagreement point can be used as an anchor for a zero point on new utility scales for Adam and Eve. The Nash bargaining solution is therefore able to make use of a comparison of Adam's and Eve's utility *levels* (Section 16.6).

A comparison of utility *units* in a Nash bargaining problem is ruled out by the absence of a meaningful second anchoring point. For a full interpersonal comparison

[5]The conditions necessary for this first welfare theorem to be true are much more stringent than libertarian philosophers like to admit.

of Adam and Eve's Von Neumann and Morgenstern utility functions, $u$ and $v$, we require *two* anchoring points for each player.

If $x_0 \prec_A x_1$, then $x_0$ and $x_1$ will serve as the zero and the unit points for a new utility scale for Adam. If $y_0 \prec_E y_1$, then $y_0$ and $y_1$ will perform the same service for Eve. The Von Neumann and Morgenstern utility functions that correspond to the new scales are:

$$u_{01}(x) = \frac{u(x) - u(x_0)}{u(x_1) - u(x_0)}; \quad v_{01}(y) = \frac{v(y) - v(y_0)}{v(y_1) - v(y_0)}. \tag{19.1}$$

We now require that $V$ utils on the $u_{01}$ scale be counted as worth the same as $U$ utils on the $v_{01}$ scale. With this definition, Adam gains more than Eve when we move from $(a_1, a_2)$ to $(b_1, b_2)$ on the old scales if and only if

$$U\left(\frac{b_1 - a_1}{\eta_1 - \xi_1}\right) > V\left(\frac{b_2 - a_2}{\eta_2 - \xi_2}\right), \tag{19.2}$$

where $\xi$ and $\eta$ are the old payoff pairs at the anchoring points.

This criterion depends on how the anchoring points are chosen but doesn't depend on which particular Von Neumann and Morgenstern utility function we use to represent a player's preferences over lotteries. For example, if our criterion says that 4 utils on Alice's current scale are worth the same as 5 utils on Bob's scale, and we replace $u$ by $2u + 3$, then 8 utils on Alice's new scale are now counted the same as 5 utils on Bob's scale.

*Zero-One Comparison.* Philosophers sometimes gloss over the requirement that the location of the anchoring points be meaningful. When Adam and Eve both agree that $\mathscr{W}$ and $\mathscr{L}$ are the best and worst possible outcomes, they take $x_0 = y_0 = \mathscr{L}$ and $x_1 = y_1 = \mathscr{W}$ in (19.1) and claim that the choice $U = V = 1$ solves the problem of interpersonal comparison. The fact that any normalization whatever creates a viable standard for interpersonal comparison certainly shows that it is silly to say that interpersonal comparison is impossible (Section 4.6.3), but what use is a standard chosen for no good reason? Who would use the zero-one standard if $\mathscr{W}$ and $\mathscr{L}$ mean winning or losing a dollar, when Adam is a billionaire and Eve is a beggar?

→ 19.5.1

We have to dig deeper if we are to come up with a *meaningful* way of making interpersonal comparisons. Harsanyi followed David Hume and Adam Smith in seeking an answer in our capacity for empathizing with others. What follows is a condensed version of my own adaptation of Harsanyi's ideas.

### 19.5.1 Empathy

Eve *empathizes* with Adam when she puts herself in his position to see things from his point of view. Players in a game need to be able to empathize with each other if they are to succeed in coordinating on an equilibrium, but attributing a capacity for empathy to the players says nothing about how they feel about each other's welfare.

If Adam's welfare appears as an argument in Eve's utility function, psychologists say that she *sympathizes* with him. For example, a mother commonly cares more for her baby's welfare than her own. Lovers are sometimes no less unselfish. Many of us even get a warm glow from giving a small fraction of our incomes to relieve the distress of strangers in faraway places. Such sympathetic preferences, however attenuated, need to be distinguished from the empathetic preferences to be discussed next.[6]

To hold an empathetic preference, you need to empathize with what others want, but you may not sympathize with them at all. For example, we seldom sympathize with those we envy, but Eve can't envy Adam without comparing his lot with hers. However, for Eve to envy Adam, it isn't enough for her to imagine having his possessions and her own preferences, as in the no-envy criterion. Even if she is poor and he is rich, she won't envy him if he is suffering from incurable clinical depression. She literally wouldn't swap places with him for a million dollars. When she compares her lot with his, she needs to imagine how it would be to have both his possessions *and his preferences.* Her judgment on whether or not to envy Adam after empathizing with his full situation will be said to reveal an empathetic preference on her part.

*Empathetic Preferences.* We write $x \preceq y$ when talking about the *personal* preferences revealed by Pandora's choice behavior (Section 4.2). We write $(x, A) \preceq (y, E)$ when talking about Pandora's *empathetic* preferences. Such a relation registers that Pandora would rather be Eve in situation $y$ than Adam in situation $x$. Later on, we will take for granted that

$$(x_0, A) \prec (y_1, E),$$

which says that Pandora would strictly prefer to be Eve at her upper anchoring point than Adam at his lower anchoring point.

I believe that players reveal their empathetic preferences whenever they use a fairness criterion to solve an equilibrium selection problem (Section 8.6). This last point is important. Nobody doubts that our sympathy for other human beings can make us more generous or spiteful in some situations. The payoffs in games played in laboratories will therefore be perturbed away from the cash payments the subjects receive at the end of the experiment. Such perturbed payoffs can be modeled by inventing exotic *personal* utility functions. The controversy touched on in Section 19.2.2 concerns the nature of such perturbations. Are they large or small? Are they stable or transient? But all this has nothing to do with empathetic preferences, which are embedded in the *norms* players use to solve the equilibrium selection problem posed by a game.

*Trading-off Utils.* We assume that the set of outcomes to be considered is lott $(\Omega)$—the set of all lotteries over a finite set $\Omega$ of prizes. To keep things simple, we restrict the set $\{A, E\}$ of people with whom Pandora empathizes to Adam and Eve.

Harsanyi makes two simple assumptions about Pandora's empathetic preferences, which we grandiosely describe as axioms.

[6]Tradition doesn't help here. Hume and Adam Smith use the word *sympathy* as we now use the word *empathy.* Arrow, Harsanyi, and other economists who have written on the subject refer to empathetic preferences as extended *sympathy* preferences.

AXIOM 19.1 *Empathetic preference relations are consistent in the sense that they satisfy the Von Neumann and Morgenstern postulates.*

Pandora's empathetic preference relation can therefore be represented by a Von Neumann and Morgenstern utility function (Section 4.5.2):

$$w : \Omega \times \{A, E\} \to \mathbb{R}.$$

We can then define $w_A : \Omega \to \mathbb{R}$ and $w_E : \Omega \to \mathbb{R}$ by

$$w_A(x) = w(x, A) \quad \text{and} \quad w_E(y) = w(y, E).$$

AXIOM 19.2 *Let $u$ and $v$ be Von Neumann and Morgenstern utility functions that represent Adam's and Eve's personal preferences. Let $w$ be a Von Neumann and Morgenstern utility function that represents Pandora's empathetic preferences. Then $w_A$ and $w_E$ represent the same preference relations as $u$ and $v$.*

The second axiom insists that Pandora be fully successful in empathizing with Adam and Eve. My guess is that when attempts to use fairness as a coordinating mechanism in real life go wrong, it is usually because the players fail to achieve Harsanyi's ideal of total empathetic identification.

Harsanyi's favorite example arises when Pandora is wondering to whom she should give a hard-to-get opera ticket that she can't use herself. One consideration that matters is whether Adam or Eve will enjoy the performance more. It may be that Pandora shares my distaste for Wagner, but if Adam prefers Wagner to Mozart, then Axiom 19.2 states that Pandora will choose Wagner over Mozart when making judgments on Adam's behalf.

Because the subject of interpersonal comparison is traditionally controversial, the little piece of bookkeeping that comes next is presented as a formal theorem. It says that Pandora's empathetic preferences determine a rate at which she trades off Adam's personal utils against Eve's. Although the anchoring of the utility functions is arbitrary, this isn't true of Pandora's trade-off rate, which is an intrinsic feature of her empathetic preferences (Section 19.5). As always, the analogy between temperature and utility is useful (Section 4.6.2). The zero and the unit on two temperature scales is arbitrary, but the way one compares degrees on the two scales is not.

THEOREM 19.1 *Anchor Adam and Eve's personal Von Neumann and Morgenstern utility functions at $(x_0, x_1)$ and $(y_0, y_1)$ respectively, so that $u_{01}$ and $v_{01}$ are given in terms of any other Von Neumann and Morgenstern utility functions that represent the same personal preferences by (19.1). Similarly, anchor Pandora's empathetic Von Neumann and Morgenstern utility function so that $w_A(x_0) = 0$ and $w_E(y_1) = 1$. Now take $w_A(x_1) = U$ and $w_E(y_0) = 1 - V$. Then, for all $x$ and $y$ in $\Omega$,*

$$\begin{aligned} w_A(x) &= Uu_{01}(x), \\ w_E(y) &= Vv_{01}(y) + 1 - V. \end{aligned} \tag{19.3}$$

*Proof* Theorem 4.1 says that two Von Neumann and Morgenstern utility functions that represent the same preference relation over lott($\Omega$) are affine transformations of each other. Thus, Axiom 19.2 implies that

$$w_A(x) = \alpha u_{01}(x) + \gamma,$$
$$w_E(y) = \beta v_{01}(y) + \delta,$$

where $\alpha$, $\beta$, $\gamma$, and $\delta$ are constants. The theorem follows on solving the four equations that result from taking $x = x_0$, $x = x_1$, $y = y_0$, and $y = y_1$.

## 19.6 More Bargaining Solutions

The bargaining solutions of Chapter 16 are attempts to predict what will happen when rational players use whatever power they have to try to get the best available deal for themselves. Fairness is irrelevant in such a setting, and so these bargaining solutions are independent of the players' utility units.

We now look at two bargaining solutions from cooperative game theory that compare the players' utility units. Moral philosophers regard the solutions as competing candidates for the welfare function of a just society. The various axiom systems that characterize the two solutions don't seem to help anyone decide whose view should prevail in this philosophical debate, and so they won't be described here. Instead, we bring a version of the Nash program to bear on the choice of a fair bargaining solution (Sections 19.7.2 and 19.7.4).

### 19.6.1 Utilitarian Bargaining Solution

When it is given that $V$ of Adam's utils are worth $U$ of Eve's, the *utilitarian bargaining solution* to the bargaining problem $(X, \xi)$ is the point $h$ at which the utilitarian welfare function

$$W_H(x) = Ux_1 + Vx_2$$

is maximized, subject to the constraint that $x$ is in $X$. Figure 19.2(a) illustrates the idea.

### 19.6.2 Egalitarian Bargaining Solution

The *egalitarian bargaining solution* of the bargaining problem $(X, \xi)$ is the point $r$ in Figure 19.2(b). It is located where the straight line of slope $U/V$ through $\xi$ crosses the boundary of $X$.

Calling the solution egalitarian begs a question or two. To justify this terminology, one needs to choose the anchoring points in (19.1) so that $\xi$ is the status quo of the bargaining problem, and $\eta_1 - \xi_1 = \eta_2 - \xi_2 = 1$. This makes the two sides of (19.2) equal at the egalitarian solution. Less tendentiously, the solution is also called the proportional bargaining solution because the players' gains over the status quo are always in the same proportion. Some bogus authority for the idea can then be derived from Aristotle, who said, "What is just... is what is proportional." Psy-

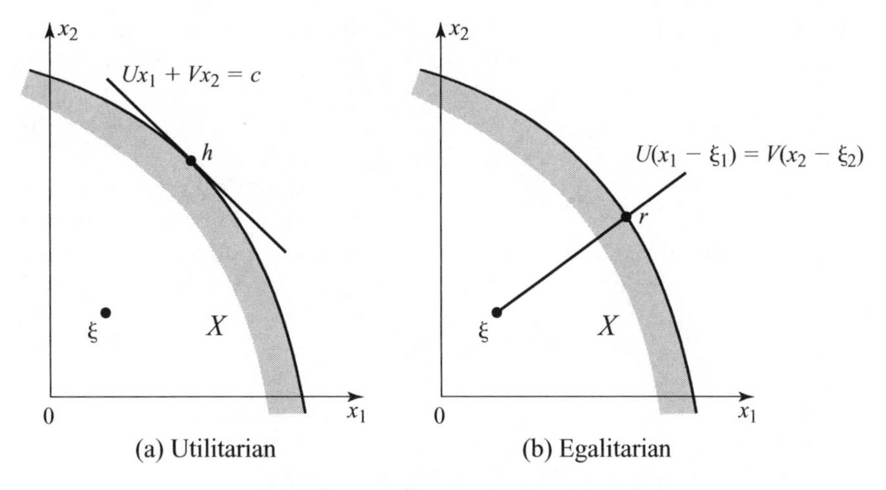


Figure 19.2 Utilitarian and egalitarian bargaining solutions.

chologists who have found that the egalitarian bargaining solution sometimes works quite well in predicting what people treat as fair in laboratories similarly have Aristotle in mind when they refer to their discoveries as "modern" equity theory.

Note that the utilitarian bargaining solution compares only the units on Adam's and Eve's personal utility scales, but the egalitarian solution assumes *full* interpersonal comparison of utility by comparing both zeros and units. Once this assumption has been made into an axiom, it doesn't seem to matter too much what other axioms are proposed. Anything reasonable generates the egalitarian bargaining solution.

## 19.7 Political Philosophy

→ 19.8

Game theorists aren't in the business of telling people what they ought to want or value. This is the territory of moral and political philosophy. Our concept of rationality asks only that people behave *consistently.* As far as game theory is concerned, David Hume may prefer the destruction of the entire universe to scratching his finger—and yet still be rational (Section 1.4.1).

Of course, when we take off our game theory hats, game theorists have moral and political opinions, just like anyone else. Mine aren't hard to guess from the choice of games I think it interesting to study, but we regard it as unprofessional to allow our personal opinions to bias the analysis of a game. Although our critics are sometimes incredulous, we are no more giving voice to a value judgement when we say what follows from what in a game than a mathematician who proves that $2+2=4$. If we sometimes forget ourselves and seem to be claiming a higher function, we should sternly be told to get back in our boxes.

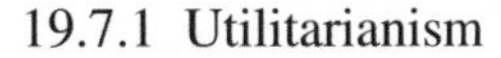

### 19.7.1 Utilitarianism

In accordance with Jeremy Bentham's last will and testament, my college in London keeps his mummified corpse on public display in a glass case. His claim to fame is that he invented *utilitarianism,* which seeks to promote the greatest good for the greatest number.[7] John Stuart Mill is said to have provided proper intellectual foundations for the theory, but one needs to do a lot more than argue that what people really want is happiness! What is happiness? How do we compare how happy two different people are? These are questions he left unanswered.

Harsanyi showed that one can provide firm foundations for utilitarianism by reinterpreting utility in the modern sense of Von Neumann and Morgenstern. If we interpret utilitarianism as a doctrine to be followed by a benevolent government with the power to enforce its rulings, then Theorem 19.1 can be adapted to provide a bowdlerized version of one of his arguments (Exercise 19.11.18).

Imagine that Pandora is a rational philosopher-king ruling Adam and Eve. If her decisions on distributive justice are consistent, then Axiom 19.1 applies. If she takes decisions as though she had half a chance of ending up as either Adam or Eve, her expected utility is then $\frac{1}{2}w(x,A)+\frac{1}{2}w(x,E)$. Unless Pandora is the kind of do-gooder who thinks that she knows what people want better than they know themselves, then Axiom 19.2 also applies. But Theorem 19.1 then says that Pandora will act as though maximizing the utilitarian social welfare function

$$W_H(x) = Ux_1 + Vx_2,$$

provided that all the utility functions are appropriately normalized.[8]

→ 19.7.2

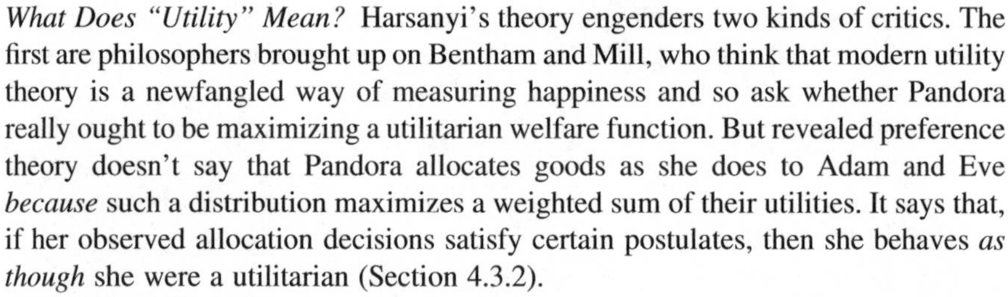

*What Does "Utility" Mean?* Harsanyi's theory engenders two kinds of critics. The first are philosophers brought up on Bentham and Mill, who think that modern utility theory is a newfangled way of measuring happiness and so ask whether Pandora really ought to be maximizing a utilitarian welfare function. But revealed preference theory doesn't say that Pandora allocates goods as she does to Adam and Eve *because* such a distribution maximizes a weighted sum of their utilities. It says that, if her observed allocation decisions satisfy certain postulates, then she behaves *as though* she were a utilitarian (Section 4.3.2).

The other kind of critic argues that Harsanyi's theory isn't utilitarian precisely because he does interpret utility in the modern sense. But I think such critics should face up to having lost the battle for the word *utility*.

### 19.7.2 The Original Position

Why should Adam and Eve surrender power to a philosopher-king whose views on fairness they may not share? Democratic theories of political legitimacy insist instead that governments need a mandate from the people they govern. Pandora should therefore enforce only laws that Adam and Eve make for themselves. But then we need a fair way for laws to get made.

On this subject two great minds had but a single thought. Harsanyi and Rawls both independently proposed that it would be fair if Adam and Eve agree on how to

[7] Only one thing can really be maximized at a time, but creative genius must be allowed some license—especially in the case of the guy who invented the word *maximize*.

[8] Harsanyi always renormalizes again to make $U=V=1$, thereby giving the mistaken impression that the problem of where Pandora gets her empathetic preferences from has somehow been resolved.

split a surplus while temporarily forgetting who is who. Rawls drew a parallel with traditional social contract theory by saying that they would then bargain in the *original position.*

How would the surplus be split in Divide-the-Dollar? (Section 16.4.2.) Here the great minds differ. Harsanyi says the outcome will be utilitarian. Rawls says it won't. Since Harsanyi won a Nobel prize for his work in game theory, it isn't surprising that game theory comes down on his side if we grant him his assumptions.

*Bargaining in the Original Position.* Players I and II are in the original position, having forgotten their identities. If a fair coin lands *heads,* player I will be Adam and player II will be Eve. If it lands *tails,* player I will be Eve and player II will be Adam.

How do players I and II evaluate an agreement in which Adam and Eve get the personal payoff pair $y$ when the coin lands *heads* and $z$ when it lands *tails?* To answer this question, we assume that all the utility functions have been suitably normalized and appeal to Theorem 19.1. If Adam and Eve carry the *same* empathetic preferences with them into the original position,[9] the theorem tells us that players I and II will evaluate the agreement $(y, z)$ as being equivalent to the *empathetic* payoff pair

$$a = \tfrac{1}{2}b + \tfrac{1}{2}c,$$

in which $b = (Uy_1,\ Vy_2 + 1 - V)$ and $c = (Vz_2 + 1 - V,\ Uz_1)$. The constant $1 - V$ is irrelevant when maximizing, and so it wouldn't hurt to throw it away.[10]

The set $X$ in Figure 16.3 shows how Adam and Eve evaluate all possible ways of dividing the dollar using their *personal* preferences. If they were to bargain face to face, their bargaining problem would be $(X, \xi)$, where $\xi$ is the pair of payoffs they assign to the event that the dollar is completely wasted.

The set $B$ in Figure 19.3 consists of all points $b$ that correspond to some $y$ in $X$. The set $C$ consists of all points $c$ that correspond to some point $z$ in $X$. The set $A$ consists of all points of the form $a = \tfrac{1}{2}b + \tfrac{1}{2}c$ and thus shows how players I and II use their empathetic preferences to evaluate all their possible agreements. Their bargaining problem in the original position is therefore $(A, \alpha)$, in which the disagreement point is $\alpha = \tfrac{1}{2}\beta + \tfrac{1}{2}\gamma$, where $\beta$ and $\gamma$ are the points in $B$ and $C$ that correspond to $\xi$.

Since the bargaining problem $(A, \alpha)$ is symmetric, its solution will be the symmetric point $\tilde{a}$ of Figure 19.3 (provided that the bargaining procedure satisfies Axiom 16.4). To achieve the payoff pair $\tilde{a}$, players I and II must agree on the pair $(\tilde{y}, \tilde{z})$ that corresponds to $(\tilde{b}, \tilde{c})$ in Figure 19.3. But $\tilde{b}$ and $\tilde{c}$ are the points in $B$ and $C$ at which $a_1 + a_2$ is maximized, and so $\tilde{y}$ and $\tilde{z}$ are the points in $X$ at which $Ux_1 + Vx_2$ and $Vx_2 + Ux_1$ are maximized.

Whoever player I and player II turn out to be, the dollar is therefore split so that Adam and Eve end up with the payoff pair $h$ in $X$ at which $Ux_1 + Vx_2$ is maximized. In other words, the outcome is utilitarian.

[9]This is a big assumption! But justifying it here would take us too far afield.

[10]Why is $c = (Vz_2 + 1 - V,\ Uz_1)$? When the coin lands *tails,* player I is Eve and player II is Adam. But $z_2$ for Eve counts as $Vz_2 + 1 - V$ for player I. Similarly, $z_1$ for Adam counts as $Uz_1$ for player II.

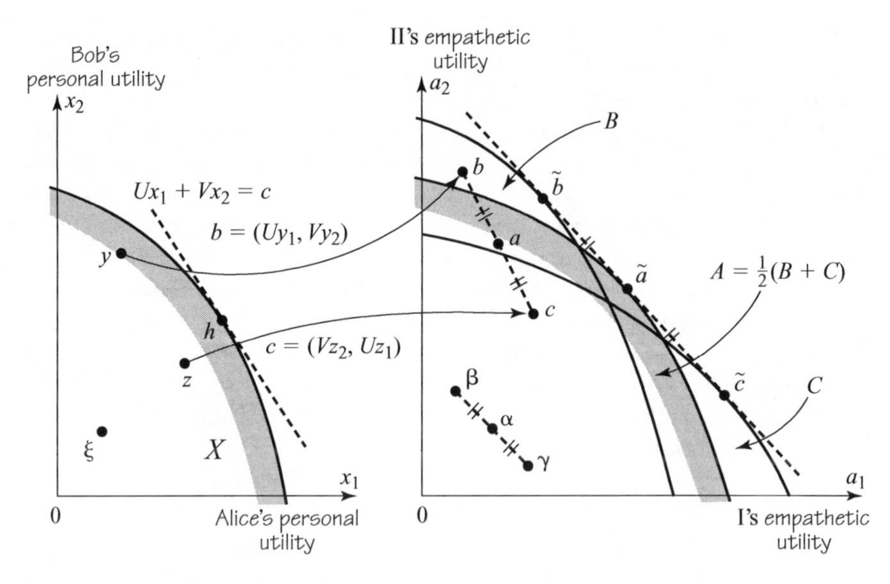


Figure 19.3 Bargaining in the original position. Players I and II face the bargaining problem $(A, \alpha)$, whose solution is $\tilde{a} = \frac{1}{2}\tilde{b} + \frac{1}{2}\tilde{c}$. Because $\tilde{b}$ and $\tilde{c}$ maximize $a_1 + a_2$ on $B$ and $C$ respectively, they both correspond to the utilitarian point $h$ in $X$.

### 19.7.3 Egalitarianism

John Rawls is the most famous modern critic of utilitarianism. After basic human rights have been guaranteed, his own theory of distributive justice calls for maximizing the welfare function

$$W_R(x) = \min\{U(x_1 - \xi_1), V(x_2 - \xi_2)\},$$

where I have replaced Rawls's "index of primary goods" by a commonly accepted standard of interpersonal comparison of utility. Philosophers say that Rawls is an egalitarian because he prioritizes the welfare of the least well off.

Figure 19.2(b) shows that $W_R(x)$ is maximized at the egalitarian bargaining solution $r$, provided that the feasible set in the bargaining problem $(X, \xi)$ is strictly comprehensive.[11] So game theorists also classify Rawls as an egalitarian, although his use of the original position to defend egalitarianism is calculated to break our hearts.

Rawls denies that orthodox decision theory is rational in the original position. Instead of evaluating lotteries in terms of expected utility, he says that players will use the maximin criterion. But such paranoia makes sense only if you think that the aim of the universe is to minimize your payoff (Section 7.5.6). I know that things sometimes seem that way, but logic tells us that this can't simultaneously be true for everybody!

[11]So that its boundary contains no vertical or horizontal straight-line segments.

### 19.7.4 Utilitarianism versus Egalitarianism?

Rawls's analysis of the original position is wrong, but losing a battle doesn't mean the war is lost. Utilitarianism may be hard to fault as an ethos for a paternalistic government with the power to enforce the decisions it thinks right—the scenario envisaged in welfare economics—but Harsanyi asks us to accept it as a system of personal morality. Followers of David Hume interpret this to mean that utilitarianism qualifies as a fair equilibrium selection device. But will Adam and Eve agree?

In the following example, both Adam and Eve need a heart transplant, but only one heart is available. A utilitarian bioethics expert says it doesn't matter who gets the heart because the gain in utility is the same in both cases. The heart is then given to Adam because he is a man. When Eve complains that this is unfair, she is told that she had an equal chance with Adam of being a man when her egg was fertilized in the womb, so what's her gripe?

Who is going to find such an answer acceptable? If some lottery is to decide who gets the heart, let a fair coin be tossed right now! The same goes for the phantom coin tossed in the original position to decide who will be who. Why should Eve accept the original position as a fair procedure if she knows that the hypothetical coin is fated to fall so that Adam is advantaged according to both their empathetic preferences? Even if a real coin is tossed right now, why should the loser accept the outcome? As we saw in Section 6.6.2, an agreement to abide by the fall of a coin can be self-policing only if the outcomes among which it adjudicates are *equilibria.*

*Rawls Redux!* We return to the original position with the proviso that a fairness norm needs to operate by the free consent of all parties. Only equilibria of the underlying game of life are then available. Moreover, the bargaining analysis of Section 19.7.2 must be modified so that agreements in which a toss of the phantom coin might disadvantage one of the players in the original position must be thrown out. Only possibilities with $b = c$ remain, so the feasible set shrinks from $A = \frac{1}{2}(B + C)$ in Figure 19.3 to $A = B \cap C$. We also need $\alpha = \beta = \gamma$ since the players won't consent to use the original position at all if the current status quo is unfair.

The bargaining solution $\tilde{a}$ of $(A, \alpha)$ now lies on the line $a_1 = a_2$. The corresponding pair $r$ of personal payoffs therefore lies on the line $U(x_1 - \xi_1) = V(x_2 - \xi_2)$ and thus is the egalitarian bargaining solution of the problem. In this context, utilitarianism therefore loses out to egalitarianism—even when we load the dice in favor of the former by adopting Harsanyi's basic framework.

## 19.8 Which Fairness Norm?

In real life, there are usually many competing claims about what should count as fair. For example, the two sides in traditional Swedish labor negotiations don't discuss who should get how much but whose fairness norm should prevail. So how do fairness norms get established?

Some colleagues and I explored the issue with an experiment on the bargaining problem $(X, 0)$ of Figure 19.4, with serious money substituting for utility.[12] Subjects

[12] *International Journal of Game Theory* 22 (1993), 381–409.

played a smoothed Nash Demand Game chosen to make any Pareto-efficient outcome an $\varepsilon$-equilibrium with $\varepsilon$ less than one dime (Section 17.3.2). A smoothed Nash Demand Game normally has only one exact Nash equilibrium, but computer technology forced the use of a discrete approximation in which the reaction curves lie on top of each other for a period. The discrete game therefore has the whole band of exact Nash equilibria shown in Figure 19.4.

The experiment began with ten trials in which different groups of subjects knowingly played against robots programmed to converge on one of the outcomes $E$, $N$, $K$, and $U$ that correspond to symmetric versions of the

Egalitarian
Nash
Kalai-Smorodinsky
Utilitarian

bargaining solutions. We hoped that the subjects would then be conditioned to coordinate their demands on the chosen bargaining solution. The conditioning phase was followed by thirty trials in which the subjects played against randomly chosen human opponents from the same group. The results were unambiguous. Subjects started out playing as they had been conditioned, but each group ended up at an exact Nash equilibrium of the game.

In the computerized debriefing that followed their session in the laboratory, subjects showed a strong tendency to assert that the outcome reached by their own group was the "fair" outcome of the game. In fact, the median of the final demands actually made by a group of subjects turned out to be a remarkably sharp predictor of the median of the demands said to be "fair" by members of that group. But different groups found their way to different exact equilibria!

I think the results exemplify David Hume's view of how fairness works. In a situation that doesn't match anything to which they are habituated, people show no sign of having some fairness stereotype built into their utility functions. On the contrary, money works well as a putative motivator in this experiment, as in many

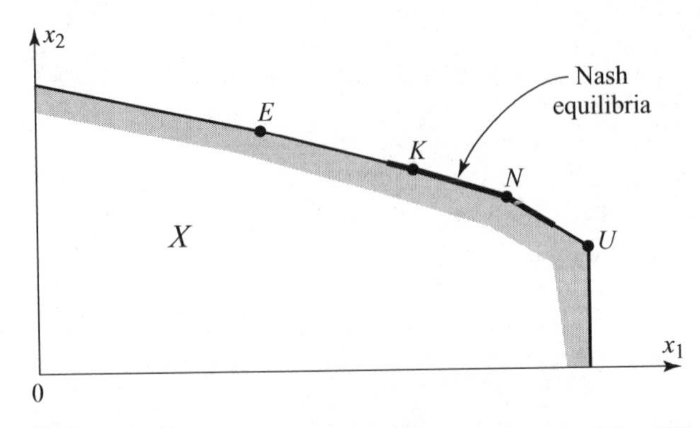


Figure 19.4 What is fair? Groups of subjects were conditioned to play one of four different bargaining solutions to the bargaining problem $(X, 0)$. After thirty repetitions, all groups were playing one of the exact Nash equilibria of the game, irrespective of their initial conditioning.

others. Instead of attributing fair behavior to built-in preferences, we need to think of the subjects in each experimental group as the citizens of a minisociety in which a fairness norm evolved over time as an equilibrium selection device.

It is striking that the different fairness norms that evolved in the experiment selected only *exact* Nash equilibria, even though the egalitarian and utilitarian solutions were available as approximate Nash equilibria, on which some groups were initially conditioned.[13]

Notice that I don't say that this or any other experiment proves that Hume is right, and there are plenty of critics falling over themselves to explain why the right experiments show him to be wrong, but at least nobody in the experimental community says that game theory is irrelevant any more.

## 19.9 Roundup

Critics think of game theory as an instrument of evil that tells selfish people how to exploit their power, but game theorists believe their subject is ethically neutral. Like logic or mathematics, it can be used on either side of any dispute. In ethics, it supports David Hume against Immanuel Kant. It denies that it is rational to cooperate in the one-shot Prisoners' Dilemma but confirms that it is usually rational to cooperate in the repeated games we commonly play. Fairness norms can then be seen as equilibrium selection devices that allow us to coordinate on one of the many efficient equilibria of such games.

Behavioral economists believe that solid experimental results from the Ultimatum Game refute David Hume. The most likely offer in Ultimatum Game experiments is fifty-fifty, and a seventy-thirty offer is more likely to be rejected than not. Such results certainly refute the theory that the subjects are playing any equilibrium of the Ultimatum Game—subgame perfect or otherwise—if we assume that their object is only to make money.

The theory can be rescued to some extent by postulating that the subjects have exotic utility functions with a built-in taste for fairness. This explanation of the data says that we have been using the wrong preferences in the right game. I think this explanation captures only a small part of the truth. The explanation I favor says that the subjects are habituated to playing a different game, so that we have been interpreting the data using approximately the right preferences—but in the wrong game.

Arrow's paradox is said to show that only dictatorships can be collectively rational, but his assumptions exclude Bergsonian welfare functions that aggregate the utilities of the citizens of a society and hence provide a putative objective function for a benign government. Some economists deny that the interpersonal comparison of utility assumed by such welfare functions can be meaningful. Others insist that there is a necessary trade-off between equity and efficiency. Neither view survives critical scrutiny. Nor do traditional attempts to measure welfare only in terms of consumer goods.

[13]The results don't discredit the egalitarian and utilitarian solutions because nobody claims that this is the kind of situation in which we should expect to see them in action. As observed in Section 19.6.2, there is experimental evidence supporting the egalitarian norm in more favorable contexts.

John Harsanyi's theory of empathetic preferences provides a way of making sense of interpersonal comparison that is entirely compatible with orthodox economics. Pandora expresses an empathetic preference when she says that she would prefer to be Adam wearing a fig leaf than Eve eating an apple.

Harsanyi requires that Pandora's empathetic preferences be consistent in the sense of Von Neumann and Morgenstern and that her attempt at empathetic identification with Adam and Eve be entirely successful. That is to say, Pandora's *personal* preferences may be for eating an apple rather than wearing a fig leaf—but if Adam is modest, then Pandora's *empathetic* preferences will accept that, if she were Adam, then she would be modest, too. With these assumptions, Harsanyi argues that Pandora will always regard $V$ of Adam's personal utils as having the same worth as $U$ of Eve's.

By making Pandora a benign philosopher-king with the power to enforce her edicts, we can generate a defense of utilitarianism that makes sense for welfare economics. But we must remember that Pandora's empathetic utils aren't the units of happiness postulated by Bentham and Mill. She doesn't make distributive decisions *because* they maximize a utilitarian welfare function. On the contrary, we assign a utilitarian welfare function to her behavior in order to make sense of the distributive decisions we observe her making.

The device of the original position provides a possible means of building a theory of social justice on democratic principles of political legitimacy. The idea is that Adam and Eve will split a surplus fairly if they bargain without knowing who is who. Game theory confirms Harsanyi's claim that the result will be utilitarian if the players are somehow committed to honor the procedure.

A bowdlerized version of John-Rawls's approach claims that bargaining in the original position will lead to the use of the egalitarian welfare function

$$W_R(x) = \min\{U(x_1 - \xi_1),\ V(x_2 - \xi_2)\},$$

If the feasible set in the bargaining problem $(X, \xi)$ is strictly comprehensive, the outcome is then the same as the egalitarian bargaining solution. This is the Pareto-efficient point of $X$ that lies on the straight line of slope $U/V$ that passes through $\xi$.

The assumption that Adam and Eve will refuse to honor the original position if it involves anything unequal according to their empathetic preferences yields a trivial defense of Rawls's position. Egalitarianism therefore remains a candidate as a fair equilibrium selection device. Modern equity theory is a small branch of psychology that offers some empirical support for this claim.

## 19.10 Further Reading

*Handbook of Experimental Economics*, edited by John Kagel and Al Roth: Princeton University Press, Princeton, NJ, 1995. John Ledyard's survey of the experimental literature explodes the myth that real people seldom free ride. The level of contribution to a public good declines steadily with experience and the amounts involved. The subjects who never learn to free ride make up only a small fraction of the total population.

*Game Theory and the Social Contract*. Vol. 2: *Just Playing*, by Ken Binmore: MIT Press, Cambridge, MA, 1998. This book explains how game theory can be used to flesh out David Hume's insights into how morality works.

*Rational Behavior and Bargaining Equilibrium in Games and Social Situations*, by John Harsanyi: Cambridge University Press, New York 1977. This neglected book is packed with creative insights.

*A Theory of Justice*, by John Rawls: Clarendon Press, Oxford, UK, 1972. This book is widely regarded as last century's most important contribution to moral philosophy.

## 19.11 Exercises

1. A widely circulated story concerns two young economists who ignored the evidence from the Ultimatum Game when bargaining with the driver of an unmetered taxicab in peacetime Jerusalem. Instead of agreeing on a price at the outset of the journey, the driver insisted that he would give them a fair price at their destination. The economists agreed, thinking their bargaining position would then be stronger. But when they refused to pay what he thought was fair, he drove them back to where they came from and turned them out into the street. What mistakes did the young economists make?
2. Why is it particularly important in welfare economics that exotic utility functions attributed to people are stable? phil
3. Condorcet's paradox of voting works only if there is a sufficient lack of consensus among the voters (Section 19.3.1). Show that no intransitivity can arise in the communal preference revealed by majority voting if all of the voters have single-peaked preferences over three possible alternatives that represent increasing amounts of public expenditure. This means that no voter's utility function can have a strict minimum at the middle expenditure (since this would imply that the utility function would have twin peaks at the extreme expenditures).
4. A utilitarian welfare function fails to satisfy the requirements of Arrow's Impossibility Theorem because it requires the interpersonal comparison of utility (Section 19.7.1). Explain why the conditions of the theorem rule out such interpersonal comparison. phil
5. Why does the Nash bargaining solution fail to satisfy the assumptions of Arrow's Impossibility Theorem?
6. Sen's paradox says that a rational social welfare function that gives each citizen the right to decide between at least two social alternatives can't be Pareto efficient. Sketch a proof of Sen's paradox with the help of Figure 19.1(b), on the assumption that a citizen who has a right to decide between $a$ and $b$ is a dictator over these alternatives.
7. Sen's definition of a right in the previous exercise has been criticized on the grounds that his model doesn't allow citizens to exercise their rights independently. An alternative model has been proposed in which citizens have a right to use any of their pure strategies in whatever civic game is being played. Imagine that all such civic games can be found by restricting the strategies each citizen can play to subsets of their space of pure strategies in some fixed strategic-form game. If the social outcome is determined as an equilibrium of the current civic game, show that the revealed communal preference will be intransitive in general. Sen's paradox therefore just gets worse in this setting. phil
8. Show that the welfare function $W_{KS}$ of Section 19.4 is maximized at the Kalai-Smorodinsky bargaining solution. Show that the welfare function $W_R$ of phil

Section 19.7.3 is maximized at the egalitarian bargaining solution when the feasible set $X$ is strictly comprehensive.

econ

9. Explain why the argument of Section 19.4 against the claim that only surplus makes sense as a measure of welfare works equally well when there are many Adams and Eves.

econ

10. Draw an Edgeworth Box in which Adam and Eve have identical bundles at the endowment point $E$. Show a Walrasian equilibrium $W$. Explain why the point $V$ in which Adam's bundle at $W$ is swapped with Eve's lies on the straight line through $E$ and $W$. Why does it follow that $W$ satisfies the no-envy condition?

econ

11. One way of finding an allocation that satisfies the no-envy criterion is described in the previous exercise. Is it the only such allocation?

phil

12. A story told by the philosopher Nozick features an old-time basketball star named Wilt Chamberlain. A leveling government is assumed to redistribute wealth between Wilt and his fans until all have the same bank balance. Wilt then sells tickets for an exhibition of his skills slightly below the price that each fan is actually willing to pay. He thereby becomes rich, and so the distribution of wealth ceases to be egalitarian. But everybody is better off than before. Relate this story to the no-envy criterion. Does it really imply that there is a necessary trade-off between equity and efficiency?
13. The classic cut-and-choose method of fair division is of ancient vintage.[14] Adam divides a cake into two pieces, and Eve chooses whichever piece she prefers. In this exercise, Adam and Eve have different preferences over nuts and raisins, which are unevenly distributed in the cake.[15]
    a. If Adam cuts the cake so that he is indifferent between the two parts, show that the final outcome satisfies the no-envy criterion.
    b. If Adam prefers nuts and Eve prefers raisins, how can Adam do better for himself if he knows Eve's preferences? Does the outcome still satisfy the no-envy criterion?
14. Section 19.4 discusses measuring the welfare of martini drinkers in terms of the gin and vermouth they consume. Why is it Pareto efficient to divide the available bottles of gin and vermouth equally between Adam and Eve? Why does this allocation satisfy the no-envy criterion?

phil

15. In the example of the previous exercise, one school of philosophical thought argues that Eve should be penalized for having "champagne tastes" that are expensive to satisfy. Can this argument survive the consideration that Eve might need martinis for urgent medical purposes?

phil

16. Longinus records the following conversation between Alexander the Great and his general Parmenio after the Persians proposed a peace treaty following their first major defeat:

*Parmenio:* If I were Alexander, I would accept this treaty.
*Alexander:* If I were Parmenio, so would I!
Who understood the point of Axiom 19.2 better?

[14]In one of Aesop's fables, a donkey, a fox, and a lion have to divide the spoils of a hunt. The donkey splits the kill into three equal piles and invites the lion to make the first choice. The lion responds by eating the donkey. The fox then offers the lion a choice between a very large pile and a very small pile.

[15]If the cake could be deconstructed altogether, the problem can be solved by assigning Adam all the nuts and Eve all the raisins. But all the normal courtesies of cake cutting are to be observed.

17. We simplified in the text by considering only situations in which Pandora had to contemplate lotteries that might result in her becoming either Adam in situation $x$ or Eve in situation $y$. We now abandon the requirement that lotteries are necessarily involved and simply interpret the pair $(x, y)$ to mean that Adam is in situation $x$ and Eve is in situation $y$. Assuming appropriate versions of Axioms 19.1 and 19.2, show that Pandora's empathetic utility function $w : \Omega \times \Omega \to \mathbb{R}$ satisfies

math

$$w(x, y) = Au(x)v(y) + Bu(x) + Cv(y) + D$$

for some constants $A$, $B$, $C$, and $D$.

18. Show that if Pandora's empathetic preferences are independent of Adam and Eve's utility *levels* (Section 19.5), then $A = 0$ in the previous exercise. Use this result to obtain a more satisfactory version of the defense of utilitarianism given in Section 19.7.1. That is to say, replace the requirement that Pandora makes decisions as though she is equally likely to end up as Adam or Eve by the less restrictive requirement that she doesn't regard the location of the status quo as relevant.

phil

19. Section 11.3.3 evaluates an income stream $m_1$, $m_2, \ldots$ using the discounted sum

math

$$u(m_1) + \delta u(m_2) + \delta^2 u(m_3) + \cdots .$$

The method of the previous exercise can be used to justify such utility functions by interpreting Adam and Eve as Pandora's own future selves. To obtain the discount factor $\delta$, it is necessary to append a stationarity assumption that says that Pandora's future selves always treat the same future in the same way. Formulate this requirement as an axiom.

20. Exercise 16.12.28 shows that neither the Nash nor the Kalai-Smorodinsky bargaining solution is monotonic. Show that the egalitarian bargaining solution, but not the utilitarian solution, is monotonic.
21. By choosing the weights $U$ and $V$ suitably, we can make any two of the Nash, egalitarian, and utilitarian solutions equal for a particular bargaining problem $(X, \xi)$. Using Section 16.6.2 or otherwise, show that all three are then equal.
22. If a fairness norm is used in Divide-the-Dollar, will Alice want her weight $U$ to be large or small compared with Bob's weight $V$? Show that utilitarianism and egalitarianism yield different answers.
23. Adam and Eve are bargaining without outside options. Player $i$ discounts time at rate $\rho_i > 0$ and must also pay at a rate $c_i$ while deadlocked. Section 17.5.3 then argues for using the generalized Nash bargaining solution with bargaining powers $1/\rho_i$ of the bargaining problem $(X, \xi)$, in which $\xi_i = -c_i/\rho_i$ (Equation 17.16). Show that this Nash bargaining solution converges on the utilitarian bargaining solution with weights $U = r_1/c_1$ and $V = r_2/c_2$, when $\rho_1 \to 0$ and $\rho_2 \to 0$ so that $\rho_1/\rho_2 = r_1/r_2$.

math

24. Zero-one utilitarianism is a version of utilitarianism that makes interpersonal comparisons of utility by calibrating everybody's personal utility scales so that their worst possible outcome gets a utility of zero and their best gets an outcome of one (Section 19.5.1). Show that majority voting implements zero-one

phil

utilitarianism in the case when there are only two candidates. Is this good for zero-one utilitarianism or bad for voting?

phil

25. Explain why the Kalai-Smorodinsky bargaining solution can be seen as a version of the egalitarian bargaining solution in which interpersonal comparisons of utility units are made by taking the utopian point as a second anchoring point when using a zero-one methodology (Section 19.5). Comment on the claim that we should regard the Kalai-Smorodinsky solution as Rawlsian rather than the egalitarian solution.

phil

26. Surgeons can sometimes use a healthy kidney to save the life of someone suffering from kidney failure. But what if the kidney must come from a live person who is unwilling to surrender it? How might Harsanyi and Rawls differ on this subject?

phil

27. Give an example to show that the utilitarian solution $h$ to a bargaining problem $(X, \xi)$ needn't satisfy $h \geq \xi$. Why does this property create an enforcement problem?
28. Section 19.7.3 mentions how Rawls uses the maximin criterion in the original position when defending his version of egalitarianism. This exercise pursues the point.A lottery assigns the empathetic payoff pairs $b$ and $c$ to players I and II with equal probabilities. Explain why expected utility theory identifies this lottery with the point $a$ midway between $b$ and $c$, but the maximin criterion makes it equivalent to the southwest corner $a$ of a rectangle with $b$ and $c$ at the other corners. Why must the set $A = \frac{1}{2}B + \frac{1}{2}C$ in Section 19.7.2 therefore be replaced by $A = B \cap C$ when the maximin criterion replaces expected utility maximization? Is this enough to extract the egalitarian bargaining solution from the original position?
29. Explain how the Rawlsian welfare function $W_R$ of Section 19.7.3 can generate very unequal outcomes if applied in cases when the feasible set is neither convex nor comprehensive.

# 20 *Taking Charge*

## 20.1 Mechanism Design

In one of her adventures, Alice starts out as a pawn in a chess game but eventually makes it to the eighth rank, where she becomes a queen. We now similarly take charge by abandoning the viewpoint of the player for that of the game master—the guy who makes and enforces the rules. In the grandest applications, the game master represents the government, and the players in the game are all the citizens in the country. The aim is then to design games that will be played in a way that maximizes the government's welfare function.

When talking about design, the rules of a game are called a *mechanism.* In some areas, the term is particularly apt since one can reasonably think of mechanism design as a branch of economic engineering. Its application to the design of auctions has been especially successful. This doesn't mean that every exotic theorem always works in practice, but some designs are so reliable that they can be tested in the laboratory with every expectation that the theory will predict well. Such tried-and-tested designs have raised billions of dollars in extra revenue for governments wise enough to take proper advice when selling valuable public assets (Chapter 21).

## 20.2 Principals and Agents

A simple example of mechanism design appears as a "solution" to the problem of the Tragedy of the Commons in Section 1.10.2. Too many goats are overgrazing the available pasture. A social planner has the power to control the number of goats but doesn't know much about goat herding. She therefore confiscates all the milk produced and redistributes it equally among the goat-herding families. They then become players in a game designed by the planner. At a Nash equilibrium of this game, the number of goats each family chooses to keep makes the total number of goats socially optimal.

In the language of economics, the social planner in the story is a *principal,* and the goat-herding families are *agents*. In such a principal-agent problem, the principal has a utility function she wants to maximize, but the decisions on the ground are made by the agents, whose preferences and beliefs are unlikely to coincide with hers. It is bad news that the agents would prefer not to do what she tells them, but it could be good news that the agents' beliefs differ from hers because this may mean that they are better informed on how to achieve her aims than she is herself.

The principal deals with her problem by inventing a game for the agents to play that provides them with incentives to work toward her aims rather than their own. Her problem often reduces to deciding what incentives will persuade the agents to reveal their private information. For example, in the Tragedy of the Commons, the social planner needs to know how many goats grazing on the common will maximize total milk production, but this fact is known only to the goat herders.

### 20.2.1 The Judgment of Solomon

→ 20.3

The Bible provides an early example of a principal-agent problem. When confronted by two women disputing the motherhood of a baby, King Solomon famously proposed that the baby be sliced in two, so that each claimant could have half. The false mother agreed to the judgment, but the true mother's "bowels yearned upon her son" so that she begged for the baby to go to her rival rather than being hacked in two (1 Kings: 3: 26). Solomon then knew the true mother and awarded her the baby.

Actually, the biblical story doesn't support Solomon's proverbial claim to wisdom particularly well. His plan would have failed if the false mother had been more strategically minded. Thinking up a better plan will provide a preview of how game theorists structure problems in mechanism design.

Solomon is the principal. The two agents are the plaintiff and the defendant. These are the two roles in the script for the Bayesian game that Solomon must construct. The two possible actors are Trudy and Fanny. Trudy is the true mother, and Fanny is the false mother. A chance move casts either Trudy in the role of the plaintiff and Fanny in the role of the defendant or Trudy in the role of the defendant and Fanny in the role of the plaintiff. Solomon's aim is to maximize the probability of awarding the baby to the true mother, but he doesn't know what type each agent is.

To keep things simple, we assume it is common knowledge that Trudy would pay all she has in the world for her baby, but Fanny will pay only some lesser amount. The precise probability with which Trudy is the plaintiff and Fanny is the defendant turns out not to matter.

The following mechanism achieves the first-best outcome of awarding the baby to the true mother for certain. Figure 20.1(a) shows the script. The plaintiff moves first by saying whether or not she claims to be the mother. If she denies being the mother, the baby is given to the defendant. If she claims to be the mother, the defendant must say whether or not she claims to be the mother. If she denies being the mother, the baby is given to the plaintiff. If both women claim to be the mother, the baby is given to the defendant, and both women are fined.

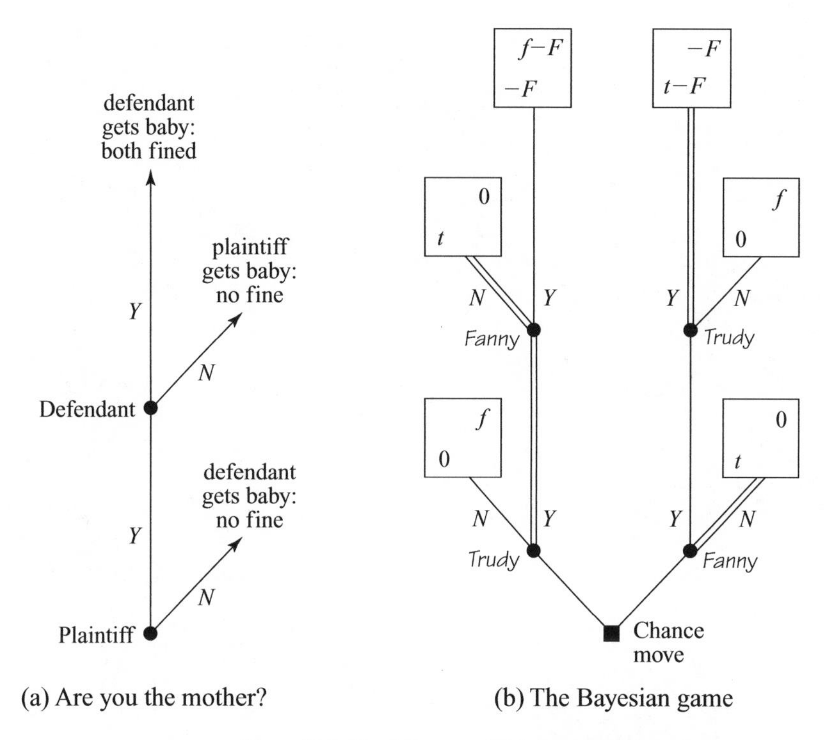


Figure 20.1 The Judgment of Solomon.

In Figure 20.1(b), $f$ is Fanny's valuation of the baby, and $t$ is Trudy's valuation. Solomon has used his famed wisdom to set the fine $F$ so that $f < F < t$. The doubled lines show the result of applying backward induction. When the actors use this subgame-perfect equilibrium, Trudy *always* gets the baby, and no fine is paid.

## 20.3 Commitment and Contracting

This section explains when one can use mechanism design to solve a principal-agent problem. Figure 20.2 indicates the procedure for a simple case.

### 20.3.1 Moral Hazard

The game of Figure 20.2 opens with a casting move that determines the types of the actors. One doesn't need to assume either that the actors then learn only their own types or that the principal learns nothing at all, but these are the most common assumptions. The principal then chooses a mechanism that specifies the rules of a game that the agents are to play.

If the principal isn't a government, she will normally need the agents' signatures on a contract guaranteeing that they will abide by her rules. If the agents had the power to commit themselves to the terms of any contract whatsoever that the principal might choose to write, her problem would become relatively straightforward,

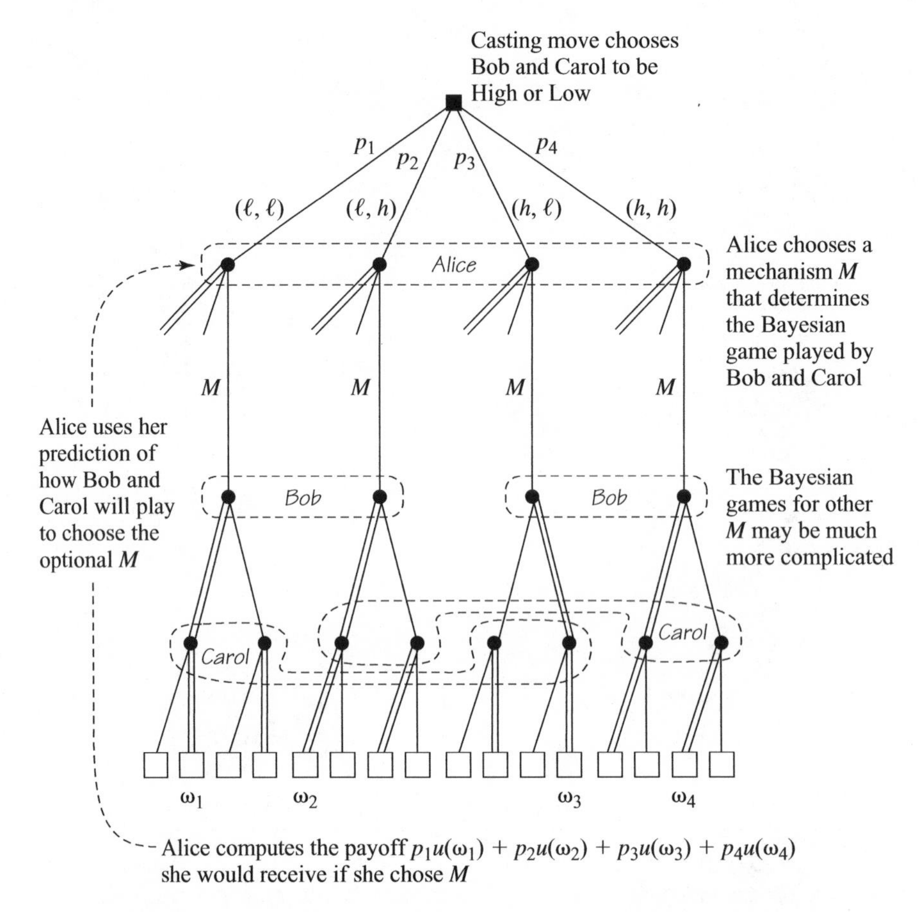


Figure 20.2 Choosing an optimal mechanism. In the case shown, a casting move first determines which actors will play the roles of Bob and Carol. The role of Bob may be played by either Mr High or Mr Low, and the role of Carol by either Ms High or Ms Low. Alice learns nothing about the outcome of the casting move. The actors playing Bob and Carol learn their own type but not the type of the actor playing the other agent.

but agents almost never have such commitment power (Section 5.5.2). The only effective contracts are therefore those that can be enforced because they are written in terms of events that can be *verified* in a court of law.

For example, an employer might wish that her workers would exert more effort but is unable to monitor them closely. The workers may promise to work harder if they are paid more, but experience suggests that the employer would be unwise to risk relying on their moral scruples to enforce such an agreement. Economists therefore say that the employer faces a *moral hazard* problem.

### 20.3.2 Adverse Selection

One usually speaks of moral hazard in *hidden-action* problems, in which effort or some other variable under the control of the agents can't be verified. The principal then needs to link the incentives she offers to some variable that can be verified—such as output. However, the same issues also arise in *hidden-type* problems.

A benign government wishes to relieve the distress of the poor by distributing welfare checks. But who are the poor? The government could circulate a questionnaire asking all citizens whether they are rich or poor, but it would be foolish to treat the results as gospel. We all know that some people will lie if they get a welfare check from saying *yes* and a tax bill from saying *no*. Nor is it easy for honest folk to sustain their moral scruples once the system has been brought into disrepute by the dishonesty of others.

A principal who recognizes that there is no point in simply asking agents to tell her their type will understand the need to offer incentives to induce them to reveal this private information. Problems of *adverse selection* arise when the incentives fail to result in the different types of agent being adequately distinguished. For example, if an insurance company isn't careful in designing its policies, only high-risk types will choose to buy them. Its clientele will then be self-selected in a manner that is adverse for the insurance company's profits.

### 20.3.3 Predicting the Agents' Play

Alice makes her choice of mechanism in Figure 20.2 without knowing the types of the agents, Bob and Carol, who will play the Bayesian game her choice of mechanism determines. She predicts the expected utility she will derive from their play of each game that she might choose by computing a Bayes-Nash equilibrium of the game.

A theoretical problem is that games commonly have multiple equilibria (Section 8.5). Alice can approach this problem in various ways. She might restrict her attention to mechanisms with only one equilibrium. If she is ambitious, she might look only at mechanisms in which Bob and Carol each have a strategy that strongly dominates its rivals. But imposing such restrictions will usually result in Alice's losing some utility overall. Her alternative is to tolerate mechanisms with multiple equilibria in the hope that the agents will accept her as a coordinating device when she nominates the particular equilibrium she would like them to play. If Bob has no reason to doubt that Carol will follow Alice's advice on how to play, it is then optimal for him to follow her advice as well.

*Rogues and Knaves?* Real-life principals sometimes fail to see the need to compute an equilibrium at all because they underestimate the ease with which human beings can adapt their behavior to new regimes. Rather than compute an equilibrium of the new game created by their choice of mechanism, they prefer to assume that people will continue to use the strategies they were using in whatever game was being played before. As the Red Queen is explaining to Alice in Figure 20.3(a), this short-term attitude is a major factor in the widespread failure of most attempts to apply social planning on a large scale.

Sometimes principals oppose using equilibria to predict human behavior on "moral" grounds. When David Hume is quoted to the effect that officers of the state should always be regarded as rogues and knaves, they respond by saying that building such assumptions into the constitution ensures that people will indeed behave like rogues and knaves. Fortunately the authors of the U.S. Constitution were less

Figure 20.3 Who pays? In Figure 20.3(a), the Red Queen is explaining to Alice that she needs to use Nash equilibria to predict how Bob and Carol will respond to her design of a mechanism. In Figure 20.3(b), Bob is explaining to Alice and Carol that his unfortunate personal circumstances require that Carol should shoulder the burden of funding Alice's public project.

naive. They understood that rogues and knaves will necessarily gravitate to positions of authority unless active measures are taken to discourage such adverse selection.

A story from real life will illustrate both points. The new chair of the controlling body of a university health plan argued in favor of abolishing its copay arrangements. These require you to pay the first hundred dollars or so of any claim you make. Their purpose is to discourage frivolous use of the service. To make up the lost revenue, the chair proposed that the premiums be increased by enough to cover the copay bills from the previous year. When the economist on the committee objected that the premiums would need to be increased by more than this, a vote was taken on whether anyone else thought that "people would visit their doctor when they didn't need to." Only the economist voted *yes* to this loaded question, but there wasn't enough money to pay the bills in the following year. The lesson is that reformers are likely to do more harm than good if they aren't realistic about how people will respond to the changes they make.

### 20.3.4 Commitment by the Principal

After the principal has found her expected utility for the equilibrium outcome of each Bayesian game that she might set the agents to play, she returns to the information set in Figure 20.2, at which she chooses a mechanism and selects whichever mechanism maximizes her total expected utility.

It is important for everybody to believe that Alice is *committed* to the mechanism she finally chooses (Section 5.5.2). For example, at the end of the process, the agents may have revealed private information that makes the principal wish that she had chosen a different mechanism, but it is vital that she not be free to change her mind.

If she were able to change her mind, Bob and Carol wouldn't believe that they are playing the game that Alice says they are playing, and all her calculations would be worthless.

The town council of Hamelin is the leading example of a principal that failed to honor the incentives it proposed. After the Pied Piper cleared Hamelin of its plague of rats, the council refused to pay the agreed-upon fee. Even if the Piper hadn't been someone whom it was obviously unwise to antagonize, the town council would have been stupid to lose their reputation for honest dealing. Who would act as their agent tomorrow after seeing them cheat their agent today? Unfortunately, modern governments often make the same mistake and so put the methods of mechanism design beyond their reach. Life is easier on this front for smaller principals, who can realistically add their signature to an enforceable contract with their agents.

## 20.4 Revelation Principle

As in the Judgment of Solomon, optimal mechanism design often reduces to finding a way to persuade the agents to tell the truth about their types.

In the goat-herding mechanism of Section 1.10.2, the principal found out the type of the goat herders *indirectly*. Instead of simply asking the agents to tell her how many goats grazing the common would maximize milk production, she deduced this fact from the number of goats each family chose to keep after the rules of her mechanism were imposed. Solomon's mechanism is an example of a *direct* mechanism. A strategy for an agent simply consists of stating which of the possible types the agent claims to be.

The following theorem is called the *revelation principle*. Although it is absurdly easy to prove, it is hard to conceive how one would go about designing optimal mechanisms in its absence.

Theorem 20.1 (Gibbard) *Any outcome $x$ that can result from the play of a Bayes-Nash equilibrium in a game $G$ derived from some mechanism $M$ can also be achieved through the play of a* truth-telling *Bayes-Nash equilibrium in a game $H$ derived from a* direct *mechanism $D$.*

*Proof* To implement $x$ with a direct mechanism, the principal need only announce that the mechanism $M$ is to be used but that she proposes to save the agents the trouble of figuring out what strategies to use in the game $G$ that the choice of $M$ forces them to play. They need only tell her their types, and, for each type of actor, she commits herself to using the strategy that such an actor would play in the Bayes-Nash equilibrium of $G$ that yields the outcome $x$.

No actor then has an incentive to do other than tell the truth about his or her type, provided all other actors also plan to tell the truth. Nothing can be gained from misleading the principal. The last thing you want her to do is to play a strategy on your behalf that you wouldn't have chosen yourself if given the opportunity.

The revelation principle doesn't provide a magical way of getting to a first-best outcome (Section 1.10.3). The revelation principle says only that, *if* something can be done at all, then it can somehow be done by asking people to reveal their true types. In the Judgment of Solomon, the first-best outcome happened to lie in Solomon's feasible set, but the limitations on what a principal can know or do in a given situation usually force a second-best outcome on the designer of an optimal mechanism.

## 20.5 Providing a Public Good

Political philosophers sometimes argue that we consent to be taxed in order to enjoy the public goods that a government can then provide. As an exercise in the use of the revelation principle, we look at two versions of a simple model of the principal-agent problem implicit in this story.

### 20.5.1 A Street Lamp Problem

Bob and Carol are risk-neutral neighbors who would benefit from a street lamp outside their houses. It is worth $b$ to Bob and $c$ to Carol, but it costs 1 (thousand dollars) to put one up. Do Bob and Carol want a lamp enough to pay this much? If so, how should they share the cost?

Alice plans to help Bob and Carol out by designing a mechanism that solves these problems. She might begin by asking them how much they would each like a lamp, but if the players realize that their answers will influence their contributions to providing the lamp, all Alice is likely to hear is the kind of sob story with which Bob is regaling Alice and Carol in Figure 20.3(b). However, if both Bob and Carol try to persuade Alice that they don't want the lamp enough to pay more than a few dollars for it, Alice won't know whether it is a good idea for her to provide the lamp at all (Section 1.4). She needs to provide Bob and Carol with appropriate *incentives* if they are to reveal the information she needs to know in order to engineer an outcome she likes better.

*The Principal's Motive.* It isn't a game theorist's business to tell Alice what she should be trying to achieve when designing a mechanism for Bob and Carol to play (Section 19.7). Alice will have personal or ethical reasons for choosing one welfare function rather than another, and game theorists offering advice on mechanism design must simply take this welfare function as one of the axioms that determines their problem (Section 19.4).

In our first version of the problem, we assume that Alice aims to maximize her own expected dollar profit, subject to the constraint that the lamp is provided if and only if the joint benefit $b+c$ to Bob and Carol is no less than its cost of 1.

*The Principal's Constraints.* If Alice knew Bob's and Carol's valuations, her task would be simple. She would provide the lamp if and only if $b+c \geq 1$ and charge Bob $b$ and Carol $c$ for her services. This first-best outcome is unattainable because Alice is constrained by her lack of knowledge of Bob's and Carol's valuations, but she will seldom know nothing at all.

We assume it is common knowledge that the agents' valuations are independently chosen to be $l$ or $h$ with equal probability, where $0 < l < h < 1$. To make the problem interesting, we also assume that $2l < 1 \leq h + l$, so that Alice needs to find a mechanism that results in the lamp always being provided, except when Bob and Carol both have the low valuation of $l$.

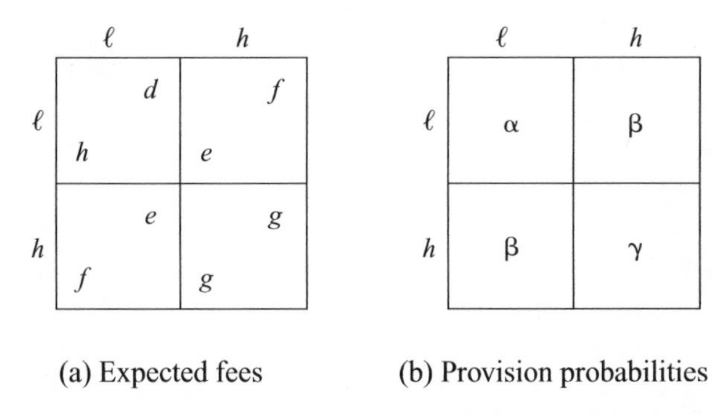


(a) Expected fees (b) Provision probabilities

Figure 20.4 Symmetric design parameters in the street lamp problem.

The revelation principle helps Alice a lot in finding an optimal mechanism. Instead of searching the impossibly large space of all possible mechanisms, she need only consider the truth-telling equilibria of direct mechanisms. In such a mechanism, the actors cast in the roles of Bob and Carol simply say whether their valuations are $l$ or $h$.

Figure 20.4 shows the parameters that Alice controls, on the assumption that Bob and Carol are treated symmetrically. In the case we are currently considering, Alice must begin by setting $\alpha = 0$ and $\beta = \gamma = 1$. She must next choose the table of fees that Bob and Carol expect to pay under each possible contingency so that neither actor will have an incentive to lie. Unless both actors have low valuations, she will then always provide the lamp. From all mechanisms with these properties, Alice wants to choose whichever maximizes her expected profit.

*Incentive Compatibility.* If the actors always tell the truth, an actor with a low valuation will have to pay Alice $L = \frac{1}{2}(d+e)$ on average. An actor with a high valuation will have to pay $H = \frac{1}{2}(f+g)$ on average. But is it in their interest to incur these costs?

If all the actors believe that the other actors will tell the truth, it is optimal for them to tell the truth themselves if the following *incentive constraints* are satisfied:

$$\tfrac{1}{2}l - L \geq l - H, \tag{20.1}$$
$$h - H \geq \tfrac{1}{2}h - L. \tag{20.2}$$

Figure 20.5(a) shows the region $I$ to which the pair $(L, H)$ is restricted by the inequality $\frac{1}{2}l \leq H - L \leq \frac{1}{2}h$, to which the two constraints reduce.

Where do the incentive constraints come from? The left-hand side of (20.1) is what Mr. $l$ will get from telling the truth. He expects to pay $L$, but he will be matched with Ms. $h$ only half the time, so his benefit is only $\frac{1}{2}l$. The right-hand side is what Mr. $l$ will get from pretending that he is Mr. $h$. He will then expect to pay $H$, but now the lamp will always be provided, and so his benefit is $l$. Equation (20.2) similarly says that Mr. $h$ will get as least as much from telling the truth as pretending to be Mr. $l$. (Since everything is symmetric, we don't need another two inequalities for Ms. $l$ and Ms. $h$.)

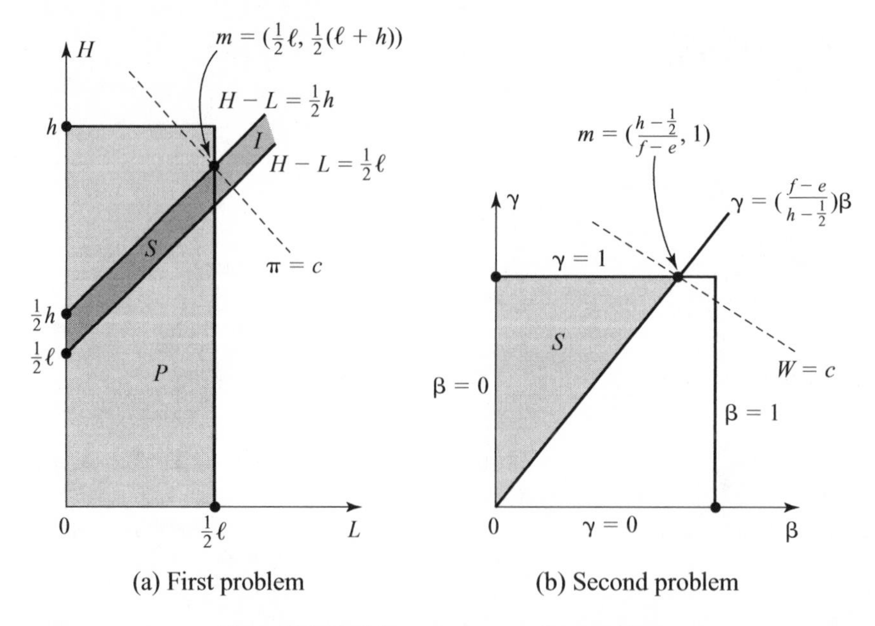


Figure 20.5 Feasible sets in the street lamp problem.

*Participation.* Government by consent is a myth to the extent that nobody offers us a choice about paying our taxes. Real governments punish citizens who try to opt out. But nothing compels Bob and Carol to sign on to Alice's mechanism in our street lamp problem. In order to get them to state their types, a direct mechanism must offer them an expected payoff that is no less than their outside option of zero. We therefore need to impose the following *participation constraints:*

$$\tfrac{1}{2}l - L \geq 0, \qquad (20.3)$$

$$h - H \geq 0. \qquad (20.4)$$

Figure 20.5(a) shows the region $P$ to which the pair $(L, H)$ is restricted by these constraints. Alice must get $(L, H)$ into the feasible set $S = I \cap P$ in order to satisfy both the incentive and the participation constraints at once.

*Optimal Profit.* Alice's expected fee revenue is

$$R = \tfrac{1}{4}\{2d + 2(e+f) + 2g\} = L + H.$$

Her expected cost of providing the street lamp is $C = \frac{3}{4}$. Maximizing her expected profit $\pi = R - C$ is therefore the same as maximizing her revenue.

Since the linear function $R$ must be maximized subject to the four linear inequalities that determine the feasible set $S$, we have a simple linear programming problem to solve (Section 7.6). Figure 20.5(a) shows that the maximum occurs at the point

$$m = (L, H) = (\tfrac{1}{2}l, \tfrac{1}{2}(l+h)). \qquad (20.5)$$

The active constraints are the high actor's incentive constraint and the low actor's participation constraint. This makes sense because a high actor is more likely to want to pretend to be a low actor, and a low actor is more likely not to want to be involved at all. Guessing which constraints will be active in advance can often eliminate a lot of tiresome algebra.

Figure 20.6(a) shows one possible scheme of fees compatible with the optimizing condition (20.5).

Alice's optimal expected profit turns out to be

$$\pi = l + \tfrac{1}{2}h - \tfrac{3}{4},$$

which needn't be positive (for example, if $l = \frac{1}{3}$ and $h = \frac{2}{3}$). Alice will then be offering a subsidy to Bob and Carol to make the mechanism work.

*Alternative Equilibria?* Truth telling is a Bayes-Nash equilibrium in the Bayesian game that Alice creates by imposing the fee scheme of Figure 20.6(a). But how will Bob and Carol solve their equilibrium selection problem in this game if there are alternative equilibria? Figure 20.7(a) allows us to make a start on this problem. It shows all the gory details of the Bayesian game that Bob and Carol have to play (Exercise 20.9.2).

→ 20.5.2

With all the detail we have added, it isn't obvious any more that truth telling is an equilibrium in the Bayesian game. To confirm that it is, begin by noting that lying is a weakly dominated strategy for a low type. For example, most of Mr. $l$'s potential payoffs are zeros, but he sometimes gets a negative payoff of $-\Delta$ if he chooses the lying strategy of announcing $h$.

Deleting the weakly dominant strategy for both low types, we are led to the game played between the high types shown in Figure 20.7(b). For example, the payoff to Mr. $h$—when both high types choose the truthful strategy of announcing $h$—is $\frac{1}{2}\Delta$ because the probability he will be playing Ms. $l$ is one-half. In this reduced game, lying is a weakly dominated strategy for both high types. We therefore obtain the truth-telling equilibrium after two rounds of deleting weakly dominant strategies.

There are no other symmetric equilibria, but it isn't hard to find asymmetric equilibria. For example, the fact that $(l, h^*)$ is a Nash equilibrium in the reduced game of Figure 20.7(b) shows that it is a Bayes-Nash equilibrium in the whole game if Mr. $h$ lies about his type, provided all the other actors are truthful. Alice doesn't

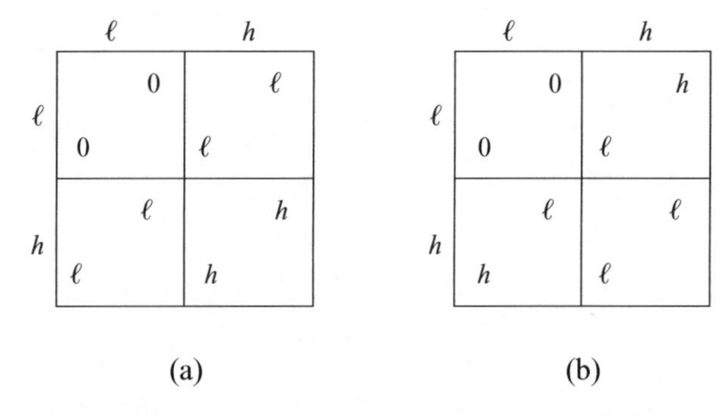


Figure 20.6 Optimal fee schemes in the street lamp problem.

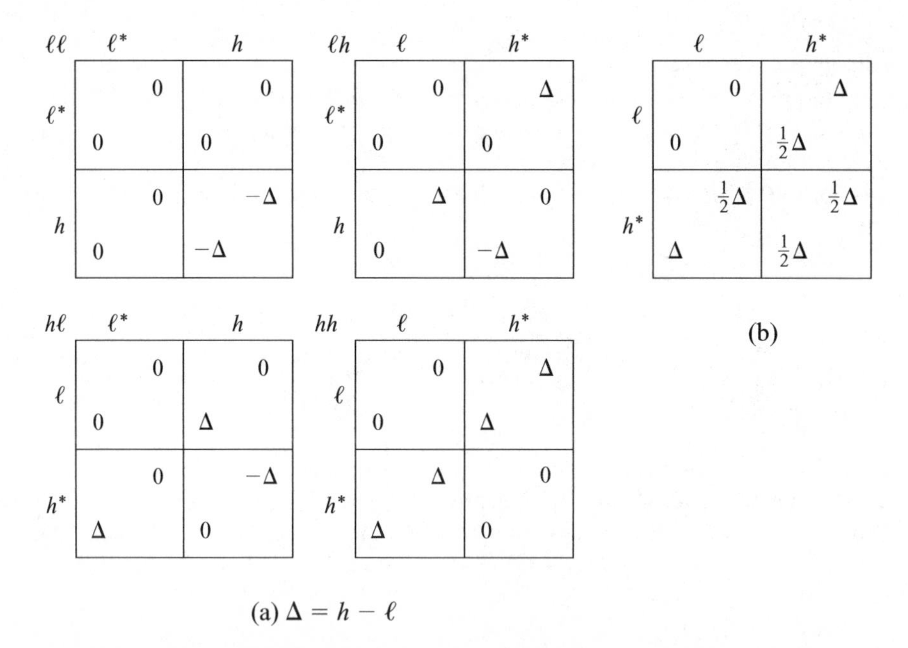


Figure 20.7 A Bayesian game for the fee plan of Figure 20.6(a): The four tables in Figure 20.7(a) correspond to the four possible outcomes of the casting move: *ll*, *lh*, *hl*, *hh*. The starred actions are the actors' truthful claims in each case. Figure 20.7(b) shows the game played between the high types that results when the low types play their weakly dominant strategy.

like this alternative equilibrium because she fails in her primary aim of always providing the street lamp when this is socially optimal.

What can Alice do about the equilibrium selection problem that we have uncovered? Alice may feel that telling the truth is so focal for Bob and Carol that she needn't worry about it. If Alice is less sanguine, she may emphasize that truth telling is focal by issuing a brochure that advertises the advantages of telling the truth when playing her game. The fact that alternative equilibria are eliminated by the successive deletion of weakly dominated strategies might figure in the argument that she uses to press this claim.

But will Bob and Carol be convinced? With Alice's fee scheme, the alternative equilibrium is a Pareto improvement on the truth-telling equilibrium for the actors.[1] If they can, they will therefore be tempted to collude in advance on a deal in which Carol tells the truth but Bob always claims to be low.

*Dominant Strategies?* Alice would really like there to be no alternative to the truth-telling equilibrium when designing a mechanism. Ideally, she would like for lying always to be a strictly dominated strategy, so that no actor would even have to know anything about the casting move or the plans of the other actors when deciding that telling the truth is optimal. How far toward this ideal can she get in the street lamp problem?

[1]Ms. $h$ improves her payoff from $\frac{1}{2}\Delta$ to $\Delta$, while the other actors all stay the same. Alice's expected profit is reduced from $l+\frac{1}{2}h-\frac{3}{4}$ to $l-\frac{1}{2}$.

Alice can nearly obtain the truth-telling equilibrium after one round of deleting *weakly* dominated strategies by using the alternative fee plan of Figure 20.6(b) (Exercise 20.9.2).[2] However, there are now symmetric, lying equilibria that represent a Pareto improvement for Bob and Carol on the truth-telling equilibrium (Exercise 20.9.3).

Such alternative equilibria arise sufficiently often in the mechanism design problems that the usual practice of ignoring their existence is a minor scandal in the literature.

### 20.5.2 Balancing the Budget

In a second version of the street lamp problem, we change Alice's objective function so that she becomes more benign. She now gives up trying to make a profit and insists instead only on balancing her budget under all possible contingencies. Assuming that the fees are payable only when the lamp is provided, this means that $d = g = \frac{1}{2}$ and $e + f = 1$ in Figure 20.4(a).

Subject to always balancing her budget, Alice tries to maximize the utilitarian welfare function given by

$$W = \tfrac{1}{2}\{\alpha(l - \tfrac{1}{2}) + \beta(l + h - 1) + \gamma(h - \tfrac{1}{2})\}, \qquad (20.6)$$

which we can interpret as the expected net benefit to Bob and Carol when they are treated symmetrically.

Alice would like to set $\alpha = 0$ and $\beta = \gamma = 1$ in Figure 20.4(b), but we know from our analysis of the preceding version of the street lamp problem that she won't then always be able to balance her budget when $l + \frac{1}{2}h < \frac{3}{4}$. We therefore add this limitation on $h$ and $l$ to the requirements that $0 < l < \frac{1}{2} < h < 1$ and $l + h \geq 1$. An optimal mechanism will now still presumably make $\alpha = 0$ and $\gamma = 1$, but we will have to live with $\beta < 1$ to keep the high types honest.

*Solving the Second Street Lamp Problem.* We first gather together all the constraints. The first column in the following list consists of the incentive constraints. The second column contains the participation constraints. The third column prevents Alice from designing a mechanism that is physically impossible:[3]

→ 20.6

$$\begin{array}{lll} pl - L \geq ql - H & pl - L \geq 0 & 0 \leq \alpha \leq 1 \\ qh - H \geq ph - L & qh - H \geq 0 & 0 \leq \beta \leq 1 \\ & & 0 \leq \gamma \leq 1 \end{array}$$

[2]But not quite, since a high type is never made *strictly* better off by telling the truth (Exercise 20.9.4).

[3]In the previous version of the street lamp problem, we were lucky to be able to get away without considering such constraints, but Alice obviously can't be allowed to incorporate perpetual motion machines or other physically impossible devices into her mechanism. It is hard not to notice that probabilities must lie between 0 and 1, but physical constraints aren't always so easy to identify. Sometimes you just have to hope that you have included all the constraints that matter. To confirm that you have, always write down a specific mechanism that implements the supposedly optimal outcome to verify that it really is feasible.

In this list, $p = \frac{1}{2}(\alpha+\beta)$ and $q = \frac{1}{2}(\beta+\gamma)$ are the probabilities that the low and high types respectively attach to the event that the lamp will be provided when they all tell the truth. The quantities $L = \frac{1}{4}\alpha + \frac{1}{2}\beta e$ and $H = \frac{1}{2}\beta f + \frac{1}{4}\gamma$ are the fees that the low and high types respectively expect to pay when they all tell the truth.

The participation constraints in the second column reduce to

$$\beta(e - \tfrac{1}{2}) \leq (\alpha+\beta)(l - \tfrac{1}{2}), \qquad (20.7)$$

$$\beta(f - \tfrac{1}{2}) \leq (\beta+\gamma)(h - \tfrac{1}{2}). \qquad (20.8)$$

Since $l < \frac{1}{2}$, (20.7) tells us that $e \leq l$ (unless $\beta = 0$).

The incentive constraints in the first column reduce to

$$(l - \tfrac{1}{2})(\gamma - \alpha) \leq \beta(f - e) \leq (h - \tfrac{1}{2})(\gamma - \alpha). \qquad (20.9)$$

In the optimum mechanism, $f - e = 1 - 2e \geq 1 - 2l > 0$. The left inequality is therefore automatically satisfied, and so we forget about it from now on.

We can now show that $\alpha = 0$ in the optimum mechanism (Exercise 20.8.5). As $\alpha$ increases from 0, the objective function W gets smaller for fixed values of the other parameters. At the same time, the set of feasible values for the vector of the other parameters shrinks because this feasible set is determined by $\alpha$ through (20.7) and the right inequality of (20.9). If we optimize for the other parameters keeping $\alpha = 0$, we therefore optimize overall.

After writing $\alpha = 0$ in (20.7), the inequality ceases to have any bite, and so we can throw it away. The same goes for (20.8) because this is implied by the right inequality of (20.9) when $\alpha = 0$. Our set of constraints is then

$$\beta(f - e) \leq (h - \tfrac{1}{2})\gamma,$$

$$0 \leq \beta \leq 1,$$

$$0 \leq \gamma \leq 1.$$

Figure 20.5(b) shows the feasible set $S$ determined by these linear inequalities for fixed values of $e$ and $f$. Note that $f - e > h - \frac{1}{2}$ because $l + \frac{1}{2}h < \frac{3}{4}$. It follows that $W = \frac{1}{2}\beta(l + h - 1) + \frac{1}{2}\gamma(h - \frac{1}{2})$ is maximized at

$$m = (\beta, \gamma) = \left( \frac{h - \frac{1}{2}}{f - e}, 1 \right).$$

It remains to choose $e$ and $f$ optimally. With the constraints $e + f = 1$ and $e \leq l$, we make $W$ largest by taking $e = l$ and $f = 1 - l$. Figure 20.8 summarizes the optimal design parameters. The expected net benefit to Bob and Coral is then the positive number

$$W = \tfrac{1}{4}(h - l)\left( \frac{2h - 1}{1 - 2l} \right).$$

It is worth noting that the optimal direct mechanism requires that Alice randomize her decision on whether to supply the public good when one agent claims to

| | $\ell$ | $h$ |
|---|---|---|
| $\ell$ | $\frac{1}{2}$, $\frac{1}{2}$ | $\ell$, $1-\ell$ |
| $h$ | $1-\ell$, $\ell$ | $\frac{1}{2}$, $\frac{1}{2}$ |

(a) Expected fees

| | $\ell$ | $h$ |
|---|---|---|
| $\ell$ | 0 | $\frac{1}{2}\left(\frac{2h-1}{1-2\ell}\right)$ |
| $h$ | $\frac{1}{2}\left(\frac{2h-1}{1-2\ell}\right)$ | 1 |

(b) Provision probabilities

Figure 20.8 Optimal parameters in the second street lamp problem.

be low and the other high, but if you tried to explain why to a government official, you would quickly be shown the door (Section 6.3). On the other hand, it isn't so distant from what happens in practice. We vote politicians into power and pay their tax demands, but people seldom think that the probability of their delivering on their election promises is high.

### 20.5.3 Clarke-Groves Mechanism

→ 20.6

The traditional method for extracting the truth about their types from reluctant agents when providing a public good is called the Clarke-Groves mechanism. It makes sense to use it in the crude form described here only when neither a balanced budget nor participation constraints are a concern.

Alice is the principal in a version of the street lamp problem that has $n$ risk-neutral home owners. She decides to provide the lamp if and only if

$$b_1 + b_2 + \cdots + b_n \geq C,$$

where $b_i$ is the benefit in dollars that the $i$th home owner attaches to the provision of a lamp, and $C$ is the cost of providing it. She gets the truth from the agents by using the following procedure:

**Step 1:** Knowing that they may lie, require each agent to state the dollar value $\beta_i \geq 0$ that he or she attaches to the provision of a lamp. The announced total benefit to the community excluding agent $j$ is then $B_j = \sum_{i \neq j} \beta_i$.

**Step 2:** Provide the lamp if and only if $B = \beta_1 + \beta_2 + \cdots + \beta_n \geq C$.

**Step 3:** Agent $j$ is pivotal (in favor of providing the lamp) when $B_j < C$ but $B \geq C$ (Section 13.2.4). Make each such agent pay $C - B_j$, which can be interpreted as the net cost that his pivotal action imposes on the community.

In the Bayesian game that this mechanism creates, a casting move fills the role of the $j$th home owner with an actor, Mr. $b_j$, who puts a dollar value of $b_j \geq 0$ on the provision of a street lamp. If you are Mr. $b_j$, it turns out to be always optimal for you to tell the truth about your valuation—whatever anyone may or may not know about the valuations the casting move has assigned to the other agents.

We show that this is true in each of three cases that cover all the possibilities. In the first case, you can't make yourself pivotal (because $B_j \geq C$). In the second case, you can make yourself pivotal, but you don't want to (because $B_j < C$ and $b_j - (C - B_j) < 0$). In the third case, you can make yourself pivotal and want to (because $B_j < C$ and $b_j - (C - B_j) \geq 0$).

**Case 1:** You might as well tell the truth because the lamp will be provided without your paying anything whatever you do.

**Case 2:** You might as well tell the truth because you get the same payoff of 0 however you avoid becoming pivotal. Telling the truth prevents your becoming pivotal in this case because $B = \beta_j + B_j < C$ when $\beta_j = b_j$.

**Case 3:** You might as well tell the truth because you get the same payoff of $b_j - (C - B_j)$ however you make yourself pivotal. Telling the truth makes you pivotal in this case because $B = \beta_j + B_j \geq C$ when $\beta_j = b_j$.

Whatever happens, you therefore always maximize your payoff by telling the truth. Telling the truth is a weakly dominant strategy for the game.[4]

## 20.6 Implementation Theory

Why can't people cooperate as well as ants or bees? As Hobbes explained long ago, the reason is that the workers in an insect colony all have the same preferences, but each human is different. We therefore can't hope to achieve the first-best solutions that evolution has found for the social insects. Our principal-agent problems usually have only second-best solutions (Section 20.5.2).

Utopians don't like our belonging to a second-best species any more than they like defection being rational in the Prisoners' Dilemma. Usually nothing can be done to please them because they are asking for a circle to be squared, but implementation theory studies the exceptional cases. It asks when mechanisms can be constructed so that the equilibrium outcomes of the games they induce always satisfy a predetermined welfare criterion.[5]

As with coalition formation, the newly hatched subject of implementation theory has a long way to go before all the wrinkles are ironed out, but this fact isn't as widely acknowledged as it should be.

### 20.6.1 Social Decision Rules

In Arrow's paradox, the principal maps the agents' profile of preferences onto a social preference (Section 19.3.1). A Bergsonian principal similarly aggregates the

[4]We haven't quite proved this. We also need to show that telling the truth is sometimes strictly better than any other strategy (Section 5.4.1). But you don't want to announce $\beta_j < b_j$ in those cases when $\beta_j < C - B_j \leq b_j$. Nor do you want to announce $\beta_j > b_j$ in those cases when $\beta_j \geq C - B_j > b_j$.

[5]Note the causal reversal as compared with the Nash program. In principle, the latter begins by modeling the noncooperative games that people are already playing. Cooperative solution concepts are then conceived as being constructed in an attempt to predict the equilibrium outcomes of these games (Section 17.2).

agents' utility functions to obtain a social utility or welfare function (Section 19.4). Such principals end up with social criteria that can be applied to all the possible feasible sets that society might face. Implementation theory isn't so ambitious. It restricts its attention to implementing social decision rules.

A *social decision rule* maps profiles of preferences onto subsets of the available alternatives. For example, if the set of alternatives consists of all trades in an Edgeworth Box, one social decision rule maps Adam's and Eve's preferences (as represented by their indifference curves) onto the contract curve (Figure 9.3(d)). Another maps their preferences onto the set of all Walrasian equilibria. If one wants a social decision rule that maps onto the Nash bargaining solution, it is necessary to expand the set of alternatives to include all lotteries over possible trades, so that we can talk about von Neumann and Morgenstern utility functions (Section 4.5.2).

Note that a social decision rule pays no attention to some aspects of the agents' types (Section 15.3.1). Their beliefs are ignored altogether. If the set of alternatives is small, the same may also be true of some aspects of the agents' preferences. For example, if preferences over lotteries are excluded, a social decision rule can't take into account how risk averse an agent may be. One therefore has to be careful not to use a particular social decision rule when the features of the agents' types it ignores are socially significant.

### 20.6.2 Implementing a Social Decision Rule

In mainstream mechanism design, a mechanism is said to be optimal if it is the best the principal can do *ex ante*—which is how economists express the fact that her choice is made before the outcome of the casting move is known. It is because her choice is made ex ante that the outcome is usually only second best. If she could optimize *ex post*—after the outcome of the casting move is known—her choice would be first best.[6] In implementation theory, we look for situations when a mechanism chosen ex ante turns out to be optimal ex post. We then have to contemplate a situation in which the mechanism determines the rules of a whole collection of games—one for each way in which the casting move can assign types to the agents. If the set of equilibrium outcomes of each of these games always coincides with the set of alternatives chosen by a social decision rule, then the mechanism is said to *implement* the social decision rule.

The Judgment of Solomon is a good example. Solomon's social decision rule is utilitarian, his aim being to maximize the total expected payoff to the two women claiming the baby. As is common in implementation problems, although the agents' types are hidden from the principal, they are common knowledge among the agents themselves. Solomon's choice of mechanism determines two possible games. In one game, the plaintiff is Trudy and the defendant is Fanny. In the other game, the plaintiff is Fanny and the defendant is Trudy. Each game has only one subgame-perfect equilibrium. In each case, the subgame-perfect equilibrium outcome coincides with the unique alternative chosen by the principal's social decision rule. The mechanism is therefore said to implement the social decision rule in subgame-perfect equilibria.

[6]Agents' choices are made neither ex ante nor ex post but in the *interim* because they learn their own types before moving.

*What Counts as an Equilibrium?* Usually the equilibria used in implementation problems are Nash, but mixed strategies are mostly disallowed because lotteries over the alternatives aren't being considered. Sometimes some refinement of Nash is used instead, as in the Judgment of Solomon. However, the Gibbard-Satterthwaite theorem severely restricts the opportunities for implementing in strongly dominated strategies. It says that this is possible if and only if the social decision rule is dictatorial.[7]

→ 20.6.3

*Weak Implementation.* Why must the set of equilibrium outcomes *coincide* with the set of alternatives chosen by the social decision rule? Who cares if some of the alternatives among which society is indifferent might never get played? With *weak* implementation, we ask only that each equilibrium outcome always be socially optimal—which is still a lot more than in straight mechanism design, where we are delighted if *some* equilibrium outcome turns out to be first-best.

The orthodox answer is that weak implementation is included in the case of strong implementation because one can always replace a social decision rule $f$ that can only be weakly implemented by a social decision rule $g$ that can be strongly implemented and always selects a subset of the alternatives chosen by $f$. This is true—but until we know everything there is to know about strong implementation, it seems perverse to neglect the more interesting case of weak implementation altogether.

### 20.6.3 Maskin Monotonicity

A *monotonic* social decision rule has the property that an alternative $a$ in its outcome set remains there when the preference profile changes so that everybody ranks $a$ at least as high as before.

For example, if Adam's preferences over the set $S = \{a, b, c, d\}$ change from $a \sim_1 b \prec_1 c \prec_1 d$ to $a \prec_2 c \prec_2 b \sim_2 d$, then $b$ is ranked at least as high in his second set of preferences as in the first. If a monotonic social decision rule includes $b$ in its outcome set when Adam and Eve's preference profile is $(\preceq_1, \preceq_1)$, then $b$ must therefore also be in the outcome set when their preference profile becomes $(\preceq_2, \preceq_1)$.

THEOREM 20.2 (Maskin) *Any Nash implementable social decision rule is monotonic. If the number of agents is at least three, then any monotonic social decision rule that gives nobody a veto is Nash implementable.*

*Proof* A Nash equilibrium remains a Nash equilibrium when Adam's preferences change so that his alternatives to playing the equilibrium strategy become worse. It follows that an implementable social decision is necessarily monotonic.

In the clever mechanism whose construction supposedly proves the sufficiency part of the theorem, each agent announces a preference profile, an alternative, and an integer. The rules then specify the following procedure:

[7]As in Arrow's paradox, the proof requires that *all* preference profiles over at least three alternatives be admissible. Moreover, each alternative must be chosen by the social decision rule for some profile of preferences.

**Step 1:** If all agents make the same announcement, implement the alternative they all announce.

**Step 2:** If all agents but one make the same announcement, look at the alternative $a$ proposed by the majority and the alternative $b$ proposed by the deviant. Implement $a$ unless $b$ is no better for the deviant than $a$ according to the preferences the majority assigns to the deviant, in which case implement $b$.

**Step 3:** If neither of the preceding steps applies, implement the alternative proposed by the agent who announces the highest integer (breaking ties in some predetermined way).

*Criticism.* Rather than give the details of the proof (Exercise 20.9.18), we continue by casting doubt on the relevance of such theorems.

The necessity part of the theorem is of doubtful practical relevance because almost any social decision rule—monotonic or otherwise—is *approximately* Nash implementable. The sufficiency part is even more dubious because it assumes that the agents can play an *infinite* game in a rational way.

We have seen lots of infinite games in previous chapters, so why start complaining now? The difference is that the earlier infinite games were all approximations to games that can actually be played in a rational way by real people. Our only reason for looking at infinite approximations to such games was to simplify the mathematics. But games in which the winner is whoever chooses the largest integer or whoever shouts the loudest aren't approximations to a genuinely playable game. If the players explore any chain of best replies, they will inevitably be led off to infinity. However, the time has to arrive when it is physically impossible to go any farther. We could of course alter the integer game to take account of the physical limitations of the players, but then it wouldn't work in implementation proofs any more (Exercise 20.9.18).

In brief, just as there is no place in a zoology book for griffins and unicorns, so there is no place in a game theory book for the kind of integer games used in the proofs of orthodox implementation theorems.

## 20.7 Roundup

In a principal-agent problem, the principal tries to persuade one or more agents to assist her in achieving her objectives. The agents may know more or be able to do more than the principal, but they are unlikely to share her goals. The principal therefore designs a mechanism that specifies the rules of a game for the agents to play. It is important that both the principal and the agents be effectively committed to honoring the rules. In selecting an optimal mechanism, the principal considers the Bayes-Nash equilibria of all the games she might have the agents play and then chooses the rules that generate the equilibrium outcome that maximizes her expected utility.

In moral hazard problems, the principal can't observe the agents' actions. In adverse selection problems, she can't observe the agents' types. Our examples were all of problems with hidden types.

The revelation principle allows mechanism design problems to be solved without undertaking the impossible task of considering all possible mechanisms. It says that we need consider only the truth-telling equilibria of *direct* mechanisms, in which the agents are simply asked to state their types. However, nothing says that these truth-telling equilibria aren't accompanied by lying equilibria that the agents may end up playing instead.

In looking for the optimal direct mechanism, the principal must usually take account of three kinds of constraints:

incentive constraints
participation constraints
physical constraints

The incentive constraints say that no agent has an incentive to lie about his type. The participation constraints say that no agent can do better by not signing on to the mechanism. The physical constraints say that the principal isn't allowed to do anything physically impossible.

In mechanism design problems, we usually end up with second-best outcomes. Implementation theory studies the special case when first-best outcomes can be achieved. Each mechanism creates the rules for a collection of games—one for each way the casting move can assign types to agents. We then ask that the set of equilibrium outcomes of each of these games coincide with the set of alternatives assigned by a social decision rule. If the equilibria are Nash, we then say that the social decision rule is Nash implementable.

The Judgment of Solomon is an example in which a utilitarian social decision rule turns out to be implementable in subgame-perfect equilibria. It would be ideal if a social decision rule could be implemented in strongly dominant strategies, but the Gibbard-Satterthwaite theorem says that this is usually impossible.

Some theorems in orthodox implementation theory are flawed because they ignore the possibility of approximate implementation or else involve "integer games" that can't really be played. But the intrinsic importance of the subject guarantees that they will eventually be replaced by more practical results.

## 20.8 Further Reading

"A Crash Course in Implementation Theory," by Matthew Jackson: *Social Choice and Welfare* 18 (2001), 655–708. A careful assessment of the successes and failures of current implementation theory.

*Game Theory: Analysis of Conflict*, by Roger Myerson: Harvard University Press, Cambridge, MA, 1991. One of the prime movers in mechanism design reveals his secrets.

## 20.9 Exercises

1. In the first street-lamp problem of Section 20.5.1, Alice maximizes her expected profit subject to the constraint that the lamp is provided whenever the benefit $b+c$ to Bob and Carol is no less than the cost 1 of providing the lamp. It is claimed that the optimal design results in an expected revenue of $l+\frac{1}{2}h$ to Alice.

What is wrong with the following argument, which claims that Alice can increase her revenue to $h$? Set an entry fee marginally less than $h$. The low types will then choose not to participate. Anyone who participates must therefore be a high type. So supply the lamp whenever someone participates, and pick up the entry fees. Your expected revenue is then $\frac{1}{4}h+\frac{1}{4}h+\frac{1}{4}2h = h > l+\frac{1}{2}h$.

2. When considering how to implement an optimal mechanism in the first street lamp problem of Section 20.5.1, Alice considers the general fee plan of Figure 20.4(a) with $d=0$. Explain why the resulting benefits and costs can be summarized as in Figure 20.9.

   If it is always optimal for actors to participate and tell the truth about their types, show that $e \leq l$, $f+g \leq 2h$, and $e+l \leq f+g \leq e+h$. If Alice is also to achieve her maximal expected payoff of $l+\frac{1}{2}h-\frac{3}{4}$ subject to providing the lamp when this is socially beneficial, show that $e=l$ and $f+g=l+h$, as in both of the fee plans of Figure 20.6. If we also require that it be a *weakly dominant* strategy for actors to participate and tell the truth about their types, show that $e=g=l$ and $f=h$ as in the fee plan of Figure 20.6(b).

3. In the first street lamp problem of Section 20.5.1, with the fee plan of Figure 20.6(b), there are alternative symmetric equilibria to that in which all actors tell the truth. Why is it an equilibrium if all types claim to be low or if all types claim to be high? To what extent does Alice achieve her aims in each case? How might Bob and Carol profit by colluding on the equilibrium in which all actors claim to have a high valuation? What difficulties would Bob and Carol face in negotiating an incentive-compatible agreement in advance of playing Alice's game?

4. In the first street lamp problem of Section 20.5.1, with the fee plan of Figure 20.6(b), explain why it isn't a weakly dominant strategy for a high type to tell

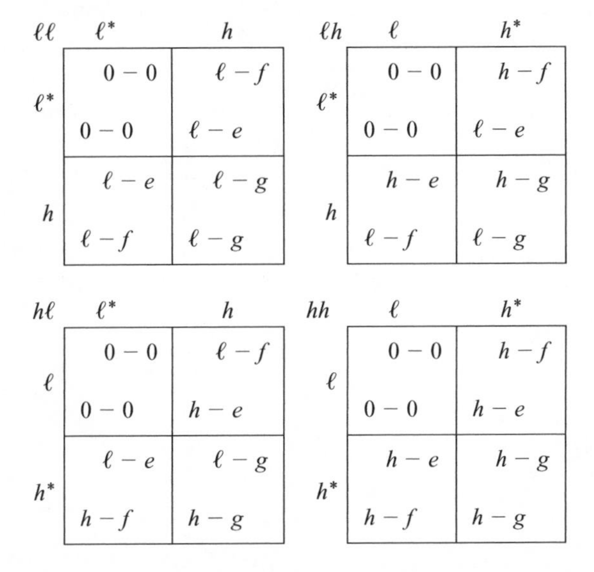


Figure 20.9 Summarizing benefits and costs. The four tables correspond to the four possible outcomes of the casting move. The stars show the truthful claims for actors chosen to play.

the truth—even though it is always optimal for him to play this way, whatever the plans of the other actors may be.

math

5. Give a more detailed account of why $\alpha = 0$ in the second street-lamp problem of Section 20.5.2.

econ

6. By fiddling with the Clarke-Groves mechanism of Section 20.5.3, one might hope to ensure that the public good is supplied if and only if the net benefit is nonnegative, without violating the participation or balanced-budget constraints. Why is this sometimes impossible?

econ

7. Suppose that the agents playing the Clarke-Groves mechanism all decide to announce that their benefit is $C/(n-1)$. The lamp will then always be provided, and no agent will ever have to pay anything. Why is this an equilibrium? Why do all the agents like this lying equilibrium at least as much as the truth-telling equilibrium, whatever their type may be? How does the principal feel about the prospect of the lying equilibrium replacing the truth-telling equilibrium?

econ

8. Section 16.4.3 quotes a result of Myerson and Satterthwaite in which they use mechanism design to show that the outcome of rational bargaining will often be very inefficient—even when the bargaining process is chosen to maximize expected surplus. This exercise looks at a simple version of the result.

   A buyer of a house is equally likely to value it at $b$ or $B$, where $b < B$. A seller is equally likely to value it at $s$ or $S$, where $s < S$. Their valuations are independent. If this is all common knowledge, find a mechanism that maximizes the expected surplus created by the sale when $s < b < S < B$. Symmetrize by taking $s = -B$ and $b = -S$, and restrict attention to the symmetric mechanisms characterized as in Figure 20.10. Show that the first-best outcome can be achieved only when $S \leq \frac{1}{2}B$. If $S > \frac{1}{2}B$, show that the expected surplus is maximized when $\alpha = \frac{1}{2}B/S$.

phil

9. Discuss the relevance of the preceding exercise to the Coase "theorem" (Section 16.4.3). When a buyer and seller bargain, how will the inefficiency that is inevitable when $S > \frac{1}{2}B$ be manifested in their behavior when they employ the kind of indirect bargaining mechanism that is actually used in practice?

math

10. What happens in Exercise 20.9.8 with $\beta$ instead of 1 in Figure 20.10(b)?

econ

11. In Exercise 15.10.21(b), a benevolent government steps in to ensure that the public good is *always* provided. However, it still insists that the cost of pro-

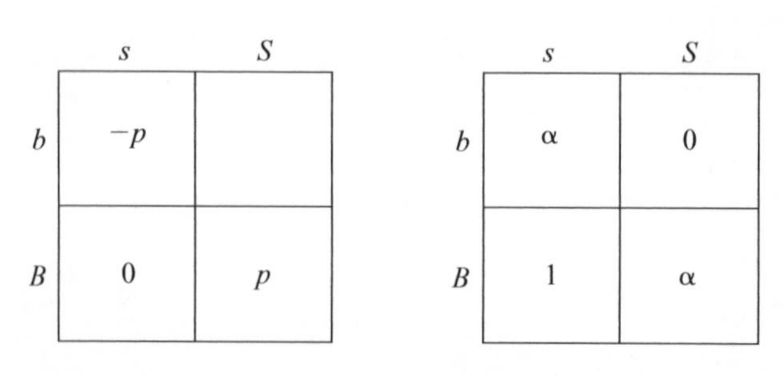


Figure 20.10 Symmetric design parameters in a bargaining problem.

viding the good be born by the players. This cost can't be shared between the players, so the government's task is to decide who should pay. It would prefer to assign the cost to a player whose cost is low, but only the players themselves know their true costs. An economist is therefore employed to design a direct mechanism that induces both players to report their true costs of providing the public good. Why will the government be disappointed with the plan the economist designs?

12. Alice has one unit of a good to divide among Bob, Carol and David. It is common knowledge that Bob and Carol want as much of the good as they can get, but David may be a High type or a Low type. The High type wants as much of the good as he can get, up to $\frac{1}{2}$ of what is available, after which he is made worse off if given more. The Low type has similar preferences, but they peak at $\frac{1}{3}$. Alice doesn't know David's type, but it is common knowledge between the agents. David would like to conceal his type from Alice because Alice's social decision rule picks $(\frac{1}{4}, \frac{1}{4}, \frac{1}{2})$ when he is Mr. Low, and $(\frac{1}{3}, \frac{1}{3}, \frac{1}{3})$ when he is Mr. High.

    Sjöström's mechanism implements Alice's social decision rule by simultaneously asking each agent to state David's type, with the threat that lying about a neighbor's type will result in a bad payoff. The incentives are shown in Figure 20.11, where Bob chooses a row, Carol chooses a column, and David chooses a payoff table. Show that the implementation requires one round of eliminating weakly dominated strategies by Mr. High and Mr. Low, followed by a further round in which Bob and Carol eliminate strongly dominated strategies.

13. Let $(b, c, d)$ represent the mixed-strategy profile for the game of Figure 20.11 in which Bob, Carol, and David say *High* with probabilities $b$, $c$, and $d$. Whatever David's type, show that $(0, 0, d)$ is a Nash equilibrium when $d \leq \frac{3}{7}$. Show that $(1, 1, d)$ is a Nash equilibrium when $d \geq \frac{1}{2}$. How safe are we in discarding the Nash equilibria that don't survive the iterated deletion of dominated strategies in the previous exercise?

14. Show that Solomon's social decision rule isn't Maskin monotonic (Sections 20.2.1 and 20.6.3) and is therefore not Nash implementable.



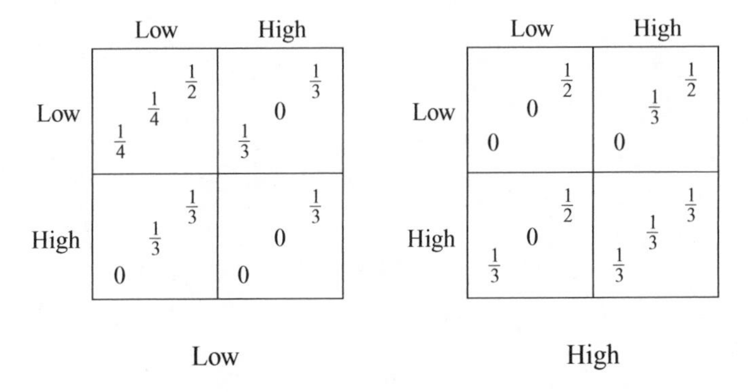


Figure 20.11 An example illustrating Sjöström's mechanism.

15. Horace, Boris, and Maurice have the Condorcet preferences of Exercise 4.11.7. For a fair social decision rule in a voting context, the standard proposals that aggregate such preferences all agree that no alternative can then be singled out from the others, and so the social outcome should be the whole set $\{A, B, N\}$ of alternatives. Change Maurice's preferences to $N \prec_4 B \prec_4 A$. Why do the standard proposals all agree that the social outcome should now be $\{A\}$? Why does it follow that none of the standard proposals are Maskin monotonic? (Look at the ranking of $N$ in each player's preferences both before and after the change.)
16. It is common knowledge between Adam and Eve that their terms of trade are described by an Edgeworth box under the conditions of Chapter 9. Pandora would like to implement the set of Walrasian outcomes as the Nash equilibria of a suitable game, but she doesn't know Adam's and Eve's utility functions. Show that the trading game described at the beginning of Section 9.6.3 has Nash equilibrium outcomes that aren't Walrasian and hence won't suffice—unless Pandora is willing to make do with implementation in Pareto-efficient, Nash equilibria.
17. Alice would like to implement the regular Nash bargaining solution in a game of Divide-the-Dollar between Bob and Carol. The players' Von Neumann and Morgenstern utility functions are common knowledge between them but are unknown to Alice. Show that Alice's social decision rule is Maskin monotonic. Show that the same isn't true if her aim is to implement the Kalai-Smorodinsky bargaining solution.

math

18. Write a formal proof of Theorem 20.2. (No veto power means that if all but one player like $a$ at least as much as any other alternative, then $a$ is one of the alternatives chosen by the social decision rule.) Why wouldn't the argument work if the players were restricted to stating integers smaller than the number of electrons in the universe?

phil

19. Rubinstein's bargaining model is an infinite game (Section 17.5.3). Why don't game theorists direct the same criticism at this infinite game as at the integer games used in implementation theory? (Section 20.6.3)

| | | idle | | busy | |
|---|---|---|---|---|---|
| | | \$0 | \$10 | \$0 | \$10 |
| idle | \$0 | $\frac{8}{12}$ | $\frac{1}{12}$ | $\frac{3}{12}$ | $\frac{5}{12}$ |
| idle | \$10 | $\frac{1}{12}$ | $\frac{2}{12}$ | $\frac{1}{12}$ | $\frac{3}{12}$ |
| busy | \$0 | $\frac{3}{12}$ | $\frac{1}{12}$ | $\frac{2}{12}$ | $\frac{1}{12}$ |
| busy | \$10 | $\frac{5}{12}$ | $\frac{3}{12}$ | $\frac{1}{12}$ | $\frac{8}{12}$ |

Figure 20.12 The table for Exercise 20.9.20.

math

20. This exercise illustrates the problems raised by moral hazard. A risk-neutral manager is responsible for two workers who can operate at two effort levels: idle ($E=0$) or busy ($E=8$). Their effort levels can't be directly monitored. The manager therefore constructs an incentive plan based on a worker's output. Each worker can either output a satisfactory item worth \$10 or a reject item worth \$0. The manager pays the worker \$$X$ in the first instance and \$$Y$ in the second. When a worker's wage is $W$ and his effort level is $E$, his utility is $U(W,E) = 10\sqrt{W} - E$. Each worker has a reservation utility of 10.

    Figure 20.12 shows how the workers' effort levels are related to their outputs. The entries are probabilities that reflect factors in the production process outside anyone's control. Find the principal's optimal values for \$$X$ and \$$Y$, given that her aim is to induce both workers to remain in her employ operating at the busy effort level. Compare the principal's expected payoff from the use of the optimal plan with the first-best payoff that she could obtain if she were able to monitor the workers' effort levels directly.

# 21 *Going, Going . . .*

## 21.1 Telecom Auctions

The theory of auctions is the branch of game theory in which the most progress has been made. It is also an area in which game theory has been used with spectacular success to solve applied problems. The amount of money raised in telecom auctions designed by game theorists is astronomical. The telecom auction that Paul Klemperer and I designed for the British government raised a total of $35 billion, all by itself.

The fat cats of the telecom industry squealed like stuck pigs at being made to pay a substantial proportion of their own valuations of the worth of a license to use certain airwave bands for cellular telephones. They had gotten accustomed to the licenses being handed out for peanuts to whoever kept the appropriate government officials happy. But why should taxpayers part with a valuable public asset for less than its market price?

After the collapse of the NASDAQ index in 2001 and the consequent bursting of the hi-tech bubble, the squealing got even louder as telecom executives tried to blame their own failure to assess the market properly on the Machiavellian maneuvering of the designers of the auctions that revealed how much the executives thought the licenses were worth. As the leader of a team that designed many telecom auctions in Belgium, Britain, Denmark, Greece, Israel, and Hong Kong, I came in for a lot of personal abuse. In *Newsweek* magazine, I was described as a "ruthless, poker-playing economist" who favors auctioning off beds in public hospitals! But I think all the brouhaha only serves to highlight how effective game theorists can be when allowed to apply the discipline of mechanism design on a large scale.

*Put Your Money Where Your Mouth Is.* Allocating telecom licenses is an example of a principal-agent problem with hidden types. Whether a government is trying to assign the available licenses efficiently or simply trying to raise money, its problem is that the people most likely to know the information it needs to maximize its welfare function are the companies that are candidates for a license themselves.

Once upon a time, governments used to organize what became known as 'beauty contests' when allocating valuable public assets to private companies. Each company would submit a mighty document explaining why it should get the asset rather than one of its rivals. A committee of officials would then decide whose document they liked the best.

But why would anyone tell the truth in such a beauty contest? As we saw in the previous chapter, the agents need to be offered appropriate incentives before they will part with the information the principal needs. The way auctions persuade candidates to tell the truth is by making them put their money where their mouths are.

However, an auction can't make a candidate pay more for a license than the candidate thinks it is worth—as the telecom fat cats shamelessly claim—because there is also a participation constraint. Nobody has to bid anything at all if they don't want to—and sometimes they don't, as in the telecom auction I designed for the Belgian government.[1]

*Auction Design.* If auctions are to be used, why consult game theorists? Why not go to the real auction experts? This is how the U.S. government reasoned when it hired Sotheby's to sell six satellite transponders. The result was a mess because it isn't a good idea to auction six identical objects in the same way that you would sell six old masters. The design of a big money auction needs to be tailored to the specific problem at hand—as the Dutch government discovered when it copied the basic format for its big telecom auction from the design that Klemperer and I developed for the British government, without apparently recognizing that it was facing a very different design problem.

When the British government thought it had the same problem as the Dutch,[2] we proposed a sealed-bid design that worked rather well when used later in Denmark. But the ascending-price auction that was eventually used so successfully in Britain merely exacerbated the Dutch problem when used in the Netherlands.

Just as you can't simply take an auction design off the shelf, so the theorems in the game theory literature never quite fit an applied problem one is asked to solve. Sometimes nobody has proved any relevant theorems at all. It is therefore much more important to understand the basic principles of auction design than to learn lots of theorems. This is just as well for this book since there would need to be a large jump in the level of the mathematics used if this chapter were to attempt anything approaching a comprehensive survey of the available literature.

[1]The three incumbent operators bid the minimum for a license. The fourth license remained unsold.

[2]When the number of available licenses is equal to the number of incumbent operators, entry to the auction by other bidders becomes problematic because they recognize that their probability of winning a license is small, and so they are reluctant to spend money on preparing a business case.

## 21.2 Types of Auctions

People with objects to sell usually want to sell them for the highest price they can get. Sometimes there is an established market that takes the problem of price setting out of their hands. Sometimes a seller has no choice but to bargain with the prospective buyers. But when the conditions are right, the situation reduces to a principal-agent problem in which the seller is the principal and the buyers are the agents (Section 20.3). We then say that the mechanism chosen by the principal is an *auction,* although no auctioneer with a block and gavel need necessarily be involved.

In discussing the auctions that are commonly considered, we shall keep things simple by always assuming that everybody is risk neutral.

*Take-It-or-Leave-It Auctions.* Retail outlets use this type of auction. A price is written on the object, and prospective buyers can take it or leave it. But how credible is a store manager who tells you she is committed to her mechanism when you try to bargain down the price of an expensive item? (Section 17.3)

*First-Price, Sealed-Bid Auctions.* This is the standard format for government tenders. Each potential buyer privately writes his bid on a piece of paper and seals it in an envelope. The seller commits herself to selling the object to whoever makes the highest bid at the price he bid. (The seller needs some means of breaking ties. Our assumption will always be that she chooses the winner at random from those who make the highest bid.)

*English Auctions.* Sotheby's uses this kind of auction to sell old masters. The same format was presumably used in ancient Babylon when wives were reportedly sold to the highest bidder by their husbands. But the price probably got a lot higher when the Pretorian Guard auctioned off the Roman Empire to Didius Julianus in AD 193.

In a so-called English auction, an auctioneer invites oral bids. The bidding continues until nobody is willing to bid any more. The auctioneer traditionally cries out, "Going, going, gone!" If nobody interrupts with a new bid, he brings down his gavel, and the object is sold to the buyer who made the last bid.

*Dutch Auctions.* The auctioneer begins by announcing a high price. This is then lowered gradually until a buyer calls a halt. The first buyer to do so then acquires the object at the price outstanding when he or she intervened.

Dutch auctions are quick and so are used to sell perishable goods like fish or cut flowers. In Amsterdam's flower auction, a seller may fly cut flowers in from Zimbabwe, and the buyer may ship them out to sell them in Chicago all in a single day. However, slow-motion Dutch auctions are sometimes operated by used furniture stores that reduce the price of unsold items by 10% each month.

*All-Pay Auctions.* Instructors in game theory courses are fond of auctioning a dollar according to the following rules. The bidding is as in an English auction, with the highest bidder getting the dollar, but *everyone* pays their highest bid, *including* those who don't win the dollar. Watching the expression on students' faces when the bidding reaches one dollar and the losers realize that it is now worth their while to bid *more* than one dollar can be quite entertaining!

Bribing corrupt politicians or judges is rather like an all-pay, sealed-bid auction. Everyone pays, but only one bribe can be successful. If there is honor among thieves, it will be largest bribe that carries the day.

*Vickrey Auctions.* William Vickrey is the hero of auction theory. He was advocating the use of specially designed auctions for the sale of major public assets long before the idea became popular. He was belatedly awarded the Nobel prize but died a few days later. The Hong Kong government chose to use a Vickrey auction for its big sale of telecom licenses in 2002.[3]

In a Vickrey auction, the object is sold to the highest bidder but at the highest price bid by a *loser*. This will be the *second*-highest price unless there is a tie for first place, in which case the winner is chosen at random from the highest bidders.

We always assume that the bids are submitted using the sealed-bid mechanism and thus identify a Vickrey auction with a second-price, sealed-bid auction.

*The Journalists' Fallacy.* At first sight it seems crazy for a seller to choose a Vickrey auction. Why should she settle for the second-highest price? Why not use a first-price, sealed-bid auction and sell the object to the highest bidder at the price he bid?

Journalists are particularly fond of this question. What they fail to see when criticizing the use of a Vickrey auction is that the buyers will adapt their behavior to the choice of game the seller makes them play (Section 20.3.3).

How high the buyers will bid depends on what type of auction is used. Once this point is grasped, it becomes obvious that bidders will bid higher in a second-price auction than in a first-price auction because you have to pay your own bid in a first-price auction, but you only have to pay some lower bid in a second-price auction.

But how much higher will the buyers bid in a Vickrey auction? To answer this question in a satisfactory way, we need to review a little integration theory.

## 21.3 Continuous Random Variables

→ 21.4

A fair pointer is spun. You win $\sqrt{\omega/10}$ dollars when it stops spinning, where $\omega$ is the clockwise angle measured in degrees from where the pointer starts to where it finishes. Since $0 \leq \omega < 360$, your winnings will be between \$0 and \$6. What is the probability you win no more than \$3?

The discrete random variables we met in Section 3.4.1 won't suffice for this problem because the sample space $\Omega = [0, 360)$ isn't finite. However, we can still define a random variable $X : \Omega \to \mathbb{R}$ by writing $X(\omega) = \sqrt{\omega/10}$.

The probability distribution of this continuous random variable is specified by a function $P : \mathbb{R} \to [0, 1]$ defined by

$$P(x) = \text{prob}\{\omega : X(\omega) \leq x\}.$$

We want $P(3)$, but $P(x)$ will be calculated for all values of $x$.

[3]The plan was that each of the four licenses would be sold to the four highest bidders at the price bid by the highest loser, but only four buyers chose to bid!

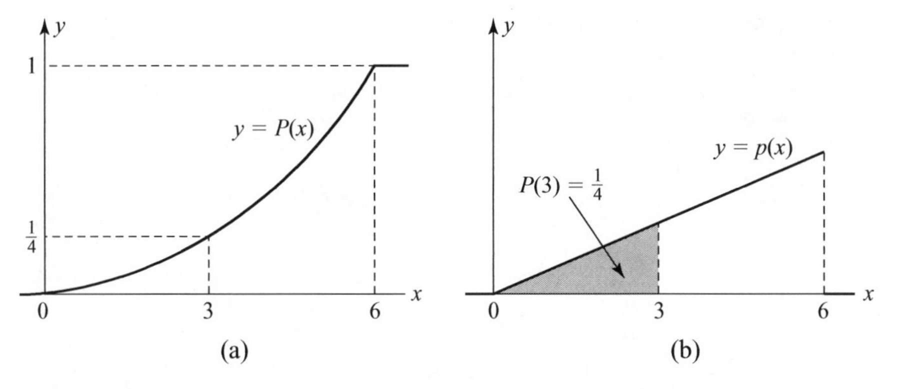


Figure 21.1 Probability distribution and density functions. The probability that $a \leq X \leq b$ is the area under the graph of the probability density function $p$ between $x = a$ and $x = b$. Alternatively, it is $P(b) - P(a)$, where $P$ is the probability distribution function.

Notice first that $P(x) = 0$ when $x < 0$ because it is impossible that you will win less than \$0. Similarly, $P(x) = 1$ when $x > 6$. When $0 \leq x \leq 6$, $P(x)$ can be calculated using the fact that $X(\omega) \leq x$ if and only if $\omega \leq 10x^2$.

The value of $P(x)$ is therefore the probability that $\omega$ lies in the interval $[0, 10x^2]$. Since each angle is equally likely with a fair pointer, this probability must be proportional to the length of the interval. Hence

$$P(x) = 10x^2/360 = \left(\frac{x}{6}\right)^2 \quad (0 \leq x \leq 6).$$

In particular, the probability $P(3)$ that you win \$3 or less is $(3/6)^2 = \frac{1}{4}$.

Figure 21.1(a) shows the graph of the probability distribution function $P : \mathbb{R} \to [0, 1]$. Sometimes a random variable also has a probability *density* function. When this is true, the probability density function is just the derivative of the probability distribution function, wherever the derivative is defined.

For example, the probability density function $p : \mathbb{R} \to \mathbb{R}_+$ for the random variable $X$ we have been considering is defined by

$$p(x) = P'(x) = \frac{2x}{36} = \frac{x}{18},$$

when $0 < x < 6$. When $x < 0$ or $x > 6$, $p(x) = P'(x) = 0$. When $x = 0$ or $x = 6$, it doesn't matter how $p(x)$ is defined.

Probability density functions are useful because they allow probabilities to be expressed as integrals.[4] For example, $\text{prob}(0 < X \leq 3)$ is equal to the shaded area in Figure 21.1(b).

[4]Discrete random variables don't have probability density functions. A graph of the probability distribution function of a discrete random variable looks like a flight of steps. Such a step function can be differentiated except where it jumps. It therefore has a zero derivative "almost everywhere." But this zero derivative is useless as a candidate for a probability density function $p$ because we won't recover the probability distribution function $P$ by integrating the zero function.

In general, the probability that $X$ lies between $a$ and $b$ is equal to the area under the graph of the probability density function between $a$ and $b$. To see this, note that $\text{prob}(a < X \leq b) = P(b) - P(a)$. But, since integrating a derivative takes you back to where you started,

$$\int_a^b p(x)\,dx = \int_a^b P'(x)\,dx = [_a^b P(x)] = P(b) - P(a).$$

In particular,

$$\text{prob}\,(0 < X \leq 3) = \int_0^3 \tfrac{1}{18}x\,dx = \tfrac{1}{4}.$$

### 21.3.1 Uniform Distribution

The random variable $\omega$ in the spinner problem is *uniformly distributed* over the interval $[0, 360]$ because its probability density function is constant.

People usually say that a uniformly distributed random variable is equally likely to take any value in its range, but it is more accurate to say that the probability it will take a value in any interval lying within its range is proportional to the length of the interval.

For example, the probability that a random variable $X$ that is uniformly distributed on the interval $[a, b]$ will take a value less than $c$ is

$$\text{prob}\,(X < c) = \frac{1}{b-a}\int_a^c dx = \frac{c-a}{b-a} \qquad (a \leq c \leq b).$$

### 21.3.2 Fundamental Theorem of Calculus

The fundamental result of calculus is that integration is the opposite of differentiation. This fact was just now used in demonstrating how to use probability density functions. If $p$ is continuous on $[a, b]$ and $Q'(x) = p(x)$ for each $x$ in $(a, b)$, then the fundamental theorem tells us that

$$\int_a^b p(x)\,dx = \int_a^b Q'(x)\,dx = [_a^b Q(x)] = Q(b) - Q(a).$$

The function $Q$ can be anything whose derivative is $p$. Such a $Q$ is called a *primitive* or an *indefinite integral* for $p$. Primitives are never unique. If $Q$ is a primitive, then so is $Q + c$, where $c$ is any constant.

The simplest example of a primitive for $p$ is the function $P$ defined by

$$P(x) = \int_a^x p(y)\,dy.$$

To verify that $P$ is a primitive, one need remember only that the fundamental theorem tells us that differentiation is the opposite of integration, and so

$$P'(x) = \frac{d}{dx}\int_a^x p(y)\,dy = p(x). \tag{21.1}$$

It may seem superfluous to state something so obvious as (21.1), but the appalling notation commonly used for primitives often leads to confusion.

The bad notation consists of writing $Q(x) = \int p(x)\,dx$ in specifying a primitive $Q$. This notation invites beginners to imagine that when they use the fact that $Q'(x) = p(x)$, they have somehow contrived to carry out the absurd operation of differentiating with respect to a variable of integration.[5] However, once this misunderstanding is cleared up, nothing could be simpler than differentiating an indefinite integral as in (21.1). One simply evaluates the integrand at the upper limit of integration.

### 21.3.3 Integrating by Parts

Let $u$ and $v$ be functions that are continuous on $[a, b]$ and differentiable on $(a, b)$. Let $U$ and $V$ be primitives for the two functions. The formula for differentiating a product says that $(UV)' = U'V + UV' = uV + UV'$. From the fundamental theorem, it follows that

$$\int_a^b (uV + UV')\,dx = \int_a^b (UV)'\,dx = [{}_a^b UV]$$

$$\int_a^b uV\,dx = [{}_a^b UV] - \int_a^b UV'\,dy.$$

This is the formula for integrating by parts. It is useful whenever a product has to be integrated. You must decide which of the terms of the product to be integrated to call $u$ and which to call $V$. Usually you will want to make $V$ the more complicated term since it may become simpler when differentiated. You also get a choice about which primitive $U$ for $u$ to use. Usually it is best to choose a primitive that vanishes at one of the limits of integration.

### 21.3.4 Expectation

Section 3.5 explains how expected values are calculated by multiplying each possible value of the random variable by its probability and then summing. The equivalent definition for a continuous random variable $X$ with a probability density function $p$ is

$$\mathcal{E}X = \int xp(x)\,dx,$$

[5] Think about differentiating $\int_0^3 y^2\,dy = 9$ with respect to $y$.

where the range of integration extends over all values taken by $X$. For example, your dollar expectation in the spinner problem is \$4 because

$$\mathcal{E}X = \int_0^6 xp(x)\,dx = \int_0^6 \tfrac{1}{18}x^2\,dx = \tfrac{1}{18}\Big[_0^6\,\tfrac{1}{3}x^3\Big] = 4.$$

### 21.3.5 Independence

To find the probability that two independent events will both occur, we *multiply* their probabilities (Section 3.2.1). The same goes for the probability density functions of two independent random variables.

Thus, if $X$ and $Y$ are two independent random variables with probability density functions $f$ and $g$, then the expected value of $\phi(X, Y)$ is

$$\mathcal{E}\phi(X, Y) = \iint \phi(x, y) f(x) g(y)\,dxdy,$$

where the double integral extends over all values taken by $X$ and $Y$. For example, if $X$ and $Y$ are independent and uniformly distributed on $[0, 1]$, then

$$\mathcal{E}XY = \int_0^1\int_0^1 xy\,dxdy = \left\{\int_0^1 x\,dx\right\}\left\{\int_0^1 y\,dy\right\} = \tfrac{1}{2}\times\tfrac{1}{2} = \tfrac{1}{4}.$$

## 21.4 Shading Your Bid

Alice's house is worthless to her if she can't sell it. Bob and Carol are the potential buyers. It is common knowledge that their valuations are independent and equally likely to be anything between 0 and 1 (million dollars).

This is the archetypal example of a single-unit auction in which the buyers have independent private values. We are studying the symmetric case on the assumption that everybody is risk neutral.

*English Auction.* One of the advantages of an English auction is that nobody has to think very hard about the optimal way to bid. If the auction is run with a price clock that raises the price continuously from zero until only one buyer is still willing to bid, then it is a weakly dominant strategy for both Bob and Carol to plan to keep bidding until the price reaches their valuation. Players who foolishly bid more than their valuations will take a loss if they win. Players who stop bidding before their valuations are reached pass up a positive probability of making a gain.

The house will then be sold to the bidder with the higher valuation. But Alice won't get paid the winner's valuation of the house because the auction is over when the *loser* stops bidding. The house is therefore sold to the winner at the loser's valuation. With many bidders, it would be sold to the winner at the valuation of the *second*-highest bidder.

When Bob's and Carol's valuations are $v$ and $w$, Bob expects to win with probability $P(v) = \text{prob}(w < v) = v$ and to make an expected payment of

$$F(v) = \int_0^v w\,dw = \tfrac{1}{2}v^2.$$

His expected gain is therefore $G(v) = vP(v) - F(v) = \frac{1}{2}v^2$. Since the problem is symmetric, the same is true of Carol. Because each of the two buyers expects to pay $\frac{1}{2}v^2$ when they value the house at $v$, Alice's expected revenue from the sale is

$$R = 2\int_0^1 \tfrac{1}{2}v^2\,dv = \tfrac{1}{3}.$$

*Vickrey Auction.* Journalists don't complain when the winner in an English auction pays only the highest bid made by the loser, so why should they complain when the same thing happens in a Vickrey auction? In fact, a Vickrey auction is simply the direct mechanism to which Alice would be led by applying the revelation principle to an English auction (Section 20.4).

It is a weakly dominating strategy for both buyers to seal their *true* valuations of the house into the envelopes they submit to Alice in a Vickrey auction. You can never benefit from bidding *below* your true valuation because, whatever bids the other buyers may have sealed in their envelopes, bidding below your own valuation can only lessen your probability of winning the auction without altering the amount you will pay if you win. Equally, you can never benefit from bidding *above* your true valuation because, if this is necessary in order to win, it must be because some other player has submitted a bid that is at least equal to your true valuation. In this case, you would have to pay at least your true valuation if you won, and perhaps you would have to pay more.

The lesson is that English and Vickrey auctions are essentially the same in this setting. Both auctions elicit the truth from the buyers in the sense that they don't shade their bids down from their valuations at all. But the same isn't true of the auction types we consider next.

*Dutch Auction.* If Alice sells her house using a Dutch auction, Bob would be foolish to stop the auction while the price remains above his valuation, but what should he do if the price falls until it reaches his valuation? If he stops the auction at that point and claims the house, his profit will be zero. It would therefore be better for him to hold on for a while, hoping that Carol won't beat him to the punch. The difficult question is to determine by how much he should shade his bid down from his valuation.

This question is best answered by asking the same question of a first-price, sealed-bid auction. Just as a Vickrey auction is equivalent to an English auction in this setting, so a first-price, sealed-bid auction is equivalent to a Dutch auction. The reason is that buyers in a Dutch auction might as well write down the price at which they plan to stop the auction before anything else happens. If they seal these prices in envelopes and hand them to the auctioneer running a first-price, sealed-bid auction, the result will be exactly the same as if they used them as stopping prices in a Dutch auction.

*First-Price, Sealed-Bid Auction.* By how much should Bob and Carol shade their bids in a first-price, sealed-bid auction? We look for a symmetric Bayes-Nash equilibrium in which the buyers bid $B(v)$ when their valuation for Alice's house is $v$. We assume that $B$ is strictly increasing and differentiable.

If Carol always bids $B(w)$ when her valuation is $w$, then Bob's expected gain when his valuation is $v$ and he bids $\beta$ is

$$(v - \beta)\operatorname{prob}(\beta > B(w)) = (v - \beta)\operatorname{prob}(C(\beta) > w),$$

where $C = B^{-1}$ is the inverse function to $B$. Since $w$ is uniformly distributed on the interval $[0, 1]$, $\operatorname{prob}(w < C(\beta)) = C(\beta)$. To find his optimal bid, Bob therefore differentiates $(v - \beta)C(\beta)$ and sets the result equal to zero:

$$-C(\beta) + (v - \beta)C'(\beta) = 0. \qquad (21.2)$$

In equilibrium, the maximizing value of $\beta$ occurs when $\beta = B(v)$. Write $\beta = B(v)$ in (21.2). Since $b = B(v)$ if and only if $v = C(b)$, we obtain

$$\begin{aligned} -v + (v - b)\frac{dv}{db} &= 0, \\ v\frac{db}{dv} + b &= v, \\ \frac{d}{dv}\{vb\} &= v. \end{aligned}$$

Integrating this differential equation with the boundary condition $b = 0$ when $v = 0$ yields

$$vB(v) = vb = \int_0^v u\,du = \tfrac{1}{2}v^2,$$

from which it follows that $B(v) = \frac{1}{2}v$.

We have therefore found an equilibrium in which buyers always shade their valuations by 50%—which is a lot more than people usually guess! What does everybody expect to get in this equilibrium?

When Bob's and Carol's valuations are $v$ and $w$, Bob expects to win with probability $P(v) = \operatorname{prob}(w < v) = v$. So his expected payment is

$$F(v) = \tfrac{1}{2}v \times v = \tfrac{1}{2}v^2.$$

His expected gain is therefore $G(v) = vP(v) - F(v) = \frac{1}{2}v^2$. Since the problem is symmetric, the same is true of Carol.

Just as in an English or a Vickrey auction, each buyer expects to pay $\frac{1}{2}v^2$ when they value the house at $v$. Alice's expected revenue of $R = \frac{1}{3}$ in a Dutch or first-price, sealed-bid auction is therefore exactly the same as in an English or a Vickrey auction. The journalists who argue that first-price auctions will obviously make more money than second-price auctions therefore get it altogether wrong. In this case, the buyers shade their bids down in a first-price auction just enough to make the seller *indifferent* between using a first-price or a second-price auction.

*Does It Matter Which Auction?* The revenue equivalence among all the auctions analyzed so far is a striking phenomenon. As a consequence, people who know a little more than the economic correspondents of newspapers sometimes quote such revenue equivalence results when wrongly arguing that it *never* matters what kind of auction the seller uses. But the result goes away if Bob and Carol are budget constrained or risk averse. Alice then gets more in a Dutch auction than an English auction. Nor does it hold if Bob's and Carol's valuations cease to be independent. Alice will then expect more in an English auction than a Dutch auction.

### 21.4.1 Revenue Equivalence

Although not all auctions generate the same expected revenue for the seller in all situations, a revenue equivalence theorem always holds for standard auctions in standard private-value scenarios.

In our definition of a standard auction, Alice has a single object for sale. Bob and Carol make dollar bids, and the object is awarded to whoever makes the higher bid, with ties broken at random. The amount the winner pays is always a nondecreasing, continuous function of the winning bid.

In our definition of a standard private-value scenario, Bob and Carol are risk neutral. Their valuations are chosen independently from the same interval using the same probability density function. The density function can be anything, provided it is positive on the interval, so that there are no gaps in the spectrum of valuations that need to be considered.

The number of buyers is unimportant, but some of the other conditions that don't look as though they matter are needed to ensure that the bid $B(v)$ that a buyer with valuation $v$ makes in a symmetric equilibrium is *strictly increasing*—a fact that we just assumed when analyzing a first-price, sealed-bid auction (Exercise 21.11.15). This is important since the secret of the revenue equivalence theorem is that the probability with which a buyer wins the auction doesn't depend on the type of standard auction that Alice chooses. To see this, observe that

$$\text{prob}(B(w) < B(v)) = \text{prob}(w < v) \qquad (21.3)$$

for valuations that lead Bob and Carol to enter the auction.

THEOREM 21.1 (Revenue Equivalence). *In a standard private-value scenario, all standard auctions with the same participation constraint have the same expected selling price at a symmetric equilibrium, providing everything is continuously differentiable.*

*Proof* In the auctions we have analyzed up to now, we paid close attention to the incentives of the buyers but ignored their participation constraints (Section 20.5). However, sellers commonly set reserve prices or charge an entry fee that can make it costly for a buyer to take part in the auction. In our symmetric setting, such an entry cost is captured by a single number $\underline{v}$, which is the lowest valuation at which buyers can enter the auction without expecting to take a loss. The theorem is therefore telling us something about a class of auctions that all have the same $\underline{v}$.

**Step 1.** As with a first-price, sealed-bid auction, we first find a differential equation that the bidding function $B$ must satisfy in a symmetric equilibrium.

If Carol bids according to $B$, then Bob's expected gain from bidding $\beta$ when his valuation is $v \geq \underline{v}$ is $g(\beta) = vp(\beta) - f(\beta)$, where $p(\beta)$ is his probability of winning the auction and $f(\beta)$ is the overall amount he expects to pay. So the choice of $\beta$ that maximizes his expected gain satisfies $vp'(\beta) - f'(\beta) = 0$. In equilibrium, this equation must be satisfied with $\beta = B(v)$, and so

$$vp'(B(v)) - f'(B(v)) = 0.$$

Multiplying through by $B'(v)$, we find that

$$vP'(v) - F'(v) = 0, \tag{21.4}$$

where $P(v) = p(B(v))$ is the probability that Bob wins when his valuation is $v \geq \underline{v}$ and $F(v) = f(B(v))$ is what he expects to pay Alice.

**Step 2.** To find $F(v)$, we use the fundamental theorem of calculus to solve the primitive differential equation (21.4). For $v \geq \underline{v}$,

$$F(v) - F(\underline{v}) = \int_{\underline{v}}^{v} F'(u)\,du = \int_{\underline{v}}^{v} uP'(u)\,du. \tag{21.5}$$

**Step 3.** We need a boundary condition to determine $F(\underline{v})$. This is derived from observing that a buyer whose valuation is exactly $\underline{v}$ will be just indifferent between entering the auction and staying out. The expected payoff of such a marginal buyer must therefore be zero, and so $G(\underline{v}) = \underline{v}P(\underline{v}) - F(\underline{v}) = 0$.

**Step 4.** We now write $F(\underline{v}) = \underline{v}P(\underline{v})$ in (21.5) and then integrate by parts to get a neater expression. For $v \geq \underline{v}$,

$$F(v) = vP(v) - \int_{\underline{v}}^{v} P(u)\,du. \tag{21.6}$$

**Step 5.** The final step is the heart of the proof. The right-hand side of (21.6) depends only on $\underline{v}$ and the way the valuations are distributed because $P(v) = \text{prob}(w < v)$ by (21.3). The type of standard auction that Alice chooses therefore makes no difference to what she expects to get paid.

COROLLARY 21.1 *With the assumptions of the revenue equivalence theorem, Alice's expected revenue is equal to the expected value of the smaller of Bob's and Carol's valuations.*

*Proof* Since this is true in a Vickrey auction, it is also true for all standard auctions by the revenue equivalence theorem.

### 21.4.2 Reserve Prices

The *reserve* or *reservation price* in an auction is the minimum bid that the principal is willing to recognize. Setting a reserve price in real auctions is often a major headache. Sometimes principals have to be argued out of setting a reserve price *below* their own valuation of the object they have for sale! More often, they want to allow themselves the freedom to set a high reserve price and then run a second auction if their object isn't sold at the first auction. But who is going to believe a principal in the future if she is known to have lied about her minimum selling price in the past? (Section 20.3.4)

In the following example, Alice continues to value her house at 0, and Bob's and Carol's valuations are still independently chosen from a uniform distribution on [0, 1]. What is Alice's expected revenue from a symmetric equilibrium in a standard auction if she sets a reserve price of $r$, with $0 \leq r \leq 1$?

We can get the answer by writing $r = \underline{v}$ in (21.6). More elegantly, the Revenue Equivalence Theorem tells us that the answer is the same as in an English auction. Recycling the analysis of Section 21.4, Bob now expects to pay $F(v) = \frac{1}{2}(v^2 + r^2)$ when $v \geq r$. Thus Alice's expected revenue is

$$R = 2\int_r^1 \tfrac{1}{2}(v^2 + r^2)\,dv = \tfrac{1}{3}\{1 + 3r^2 - 4r^3\},$$

which is maximized when $r = \frac{1}{2}$. Alice then expects $R = \frac{5}{12}$, which is more than the revenue $R = \frac{1}{3}$ she expects when $r = 0$.

If Alice is choosing from among the standard auctions with the expectation that a symmetric equilibrium will be played, she usually does best to set a positive reserve price (Exercise 21.11.19). The revenue she loses when both Bob and Carol have valuations below the reserve price is outweighed by the extra revenue she derives on average by making Bob and Carol pay more when their valuations exceed the reserve price. However, in real life, the information that Alice needs to set the reserve price according to the principles outlined in this section is almost never available.

### 21.4.3 Nonstandard Equilibria and Entry

A Vickrey auction is a version of the Clarke-Groves mechanism (Section 20.5.3). The incentive for the bidders to reveal their true valuations is that the winning bidder pays the social cost of making the other bidder a loser.

→ 21.5

This identification reminds us of two things. In a Vickrey auction, it is a *weakly dominant* strategy to bid your true valuation. This is true no matter how valuations are assigned to the buyers or what anybody knows about this chance move. It is similarly always a weakly dominant strategy to plan to bid up to your true valuation in an English auction—provided that your valuation is never going to be altered by the bids other buyers may make (Section 21.6).

The second thing to recall about the Clarke-Groves mechanism is that it has lying equilibria as well as truth-telling equilibria (Exercise 20.9.7). Auction theorists sometimes brush such equilibria aside on the grounds that they involve the use of weakly dominated strategies and are therefore not trembling-hand perfect (Section 14.2.3). However, these neglected equilibria can be crucial in big-money telecom auctions.

*Entry in Telecom Auctions.* If Bob is an incumbent operator and Carol is a potential invader, it will be common knowledge that Bob has a higher valuation for a new license than Carol. So why should Carol bother to bid at all in an English or a Vickrey auction since she is doomed to lose? If Carol plans to bid zero and Bob plans to bid up to his valuation, Alice is looking at an equilibrium in which Bob gets the license for nothing. It is true that Carol also gets nothing, but she comes out smelling like roses in practice because she avoids paying a million dollars or so to prepare a business case to put before her financial backers (Section 6.1.1). Sometimes her decision to stay out is sweetened by commercial concessions made to her by the incumbent on the understanding that she won't compete for a license. One can't call these concessions bribes because they aren't necessarily illegal, but the effect is to coordinate Bob's and Carol's behavior on an equilibrium that would normally be regarded as collusive.

Klemperer and I therefore believe that sealed-bid auctions have an important role when the number of incumbents equals the number of licenses for sale. Even when it is common knowledge that Bob's valuation is more than the most that Carol's valuation can be, Carol will still pay a cost to enter a first-price, sealed-bid auction if her valuation is sufficiently high (Exercise 21.11.20).

## 21.5 Designing Optimal Auctions

Auctions are the jewel in the crown of mechanism design. One never has all the information one needs to design an auction according to the recipes that appear in textbooks, but without the guiding principles that these recipes exemplify, one wouldn't know where to begin when faced with a practical problem.

### 21.5.1 The Principal's Motives

Three reasons for using auctions to allocate assets are usually given:

- They are quick.
- They are hard to corrupt.
- They elicit information about the buyers' valuations.

Economists usually concentrate on the third of these reasons, which works because the buyers are forced to put their money where their mouths are.

If Alice aims to maximize her expected revenue, her problem is simple if she knows the buyers' valuations in advance. She can simply use a take-it-or-leave-it auction in which the reserve price is equal to the *higher* of the two valuations (or one penny less) and hence achieve the first-best outcome. But her ignorance will normally force her to accept a second-best outcome. For example, in the case of the standard private-value scenario, it turns out that the best she can do is to use a standard auction, in which case her expected revenue is equal to the expected value of the *lower* of the two valuations (Exercises 21.11.23–21.11.26). However, since we have explored this ground a great deal already, we will study an example where the optimal design is more of a surprise.

Journalists always assume that governments are interested only in raising revenue when they sell a public asset, but another reason why they care about finding out the buyers' valuations is so that they can assign the public asset efficiently. Efficient allocation requires assigning the asset to whichever buyer has the highest valuation, on the assumption that this is the buyer with the best business plan. Although journalists are skeptical, greed seems to have been the primary motivation of a seller in only one big telecom auction so far. In the other auctions, maximizing revenue played second fiddle to promoting competition in the telecom industry by maximizing the number of viable operators and assigning the licenses among these operators efficiently.[6]

Although economists focus on information revelation in auctions, the fact that auctions are quick and hard to corrupt is often much more important to the principal, particularly when previous allocations have been the object of open scandal or their implementation has been delayed for years by legal hassle. It is easy to underestimate the importance of avoiding even the appearance of corruption. In my experience, governments are therefore often eager to promote entry to an auction *for its own sake* so that the public can see that there is open competition for the asset, rather than some back-room deal.

### 21.5.2 An Exercise in Mechanism Design

The following problem in optimal auction design will provide an excuse for reviewing the general principles of mechanism design.

As in Section 21.4, Alice has an upscale house to sell that is worth nothing to her if it can't be sold. The prospective buyers are Bob and Carol. They both know their own valuations of the house, but this information is unknown to anybody else. Everybody is risk neutral.

The set $\mathscr{M}$ of actors who may be cast in Bob's role consists of just two individuals, Mr. High and Mr. Low. Similarly, the set $\mathscr{F}$ of actors who may be cast in Carol's role consists of just Ms. High and Ms. Low. High actors value the house at \$4 million. Low actors value the house at \$3 million. It is common knowledge that the casting move selects male and female actors independently and that the probability that the Low actor is chosen in each case is $p$.

Alice begins by *committing* herself to a mechanism. Her choice of mechanism constitutes a *script* with roles for Bob and Carol. Harsanyi's theory converts this script into a Bayesian game of imperfect information. If the agents are rational, Alice will be able to predict how the game she has invented for them will be played. In particular, she will be able to predict her own expected revenue. In spite of what has been said about the complexity of the motives of real principals, Alice is assumed to choose the mechanism that maximizes this expected revenue. She will then have designed a mechanism for the problem that is optimal, given her aims.

Alice's first-best outcome would be to sell the house for \$4 million when one of the agents is a high type and to sell for \$3 million when both agents are low types. Since both agents are low types with probability $p^2$, Alice's expected payoff with

[6]Of course, even a greedy government can afford to be relaxed about not giving priority to revenue maximization when the only way to assign licenses efficiently requires letting the bidding increase until the inefficient operators cease to compete!

this first-best outcome is $3p^2 + 4(1-p^2) = 4 - p^2$. However, Alice's ignorance of the agents' true valuations means that she won't be able to achieve this first-best result. It is instructive to see how close she can come by using some of the simple auctions discussed earlier.

*Take It or Leave It.* If Alice decides simply to post a take-it-or-leave-it price, she would be foolish to consider any prices except 3 or 4 (million dollars). If she sets the price at 3 (or one penny less), the house will be sold at that price regardless of who the agents may be, and her expected payoff will therefore be 3. This is a second-best result because $3 < 4 - p^2$ except when $p = 1$.

If she sets the price at 4 (or one penny less), the house will be sold at that price unless both agents are low types. If both are low types, the house won't be sold at all. With this arrangement, her expected payoff is therefore $4(1-p^2)$. This is also second-best because $4(1-p^2) < 4 - p^2$ except when $p = 0$.

If Alice were confined to take-it-or-leave-it auctions, she would choose to post a price of 3 when $3 > 4\,(1-p^2)$, which happens if and only if $p > \frac{1}{2}$. If $p < \frac{1}{2}$, she would choose to post a price of 4.

*Vickrey or English.* In the truth-telling equilibrium, the highest price bid by a loser is 3 unless both agents are high types. In the latter case, the highest losing bid is 4. (Recall that ties are broken at random.) Alice's expected payoff is then $4(1-p)^2 + 3(1-(1-p)^2) = 3 + (1-p)^2$. This is a second-best result because $3 + (1-p)^2 < 4 - p$ unless $p = 0$ or $p = 1$.

However, a Vickrey or an English auction is better than posting a take-it-or-leave-it price of 3 unless $p = 1$. It is better than posting a take-it-or-leave-it price of 4 when $3 + (1-p)^2 > 4(1-p)^2$. This occurs when $\frac{2}{5} < p \leq 1$.

*First-Price, Sealed-Bid or Dutch.* Here we get into deep waters because agents with a high valuation will necessarily randomize their bids. Although the revenue equivalence theorem of Section 21.4.1 doesn't apply directly, we still obtain that the expected revenue in a Bayes-Nash equilibrium is $3 + (1-p)^2$, just as in a Vickrey auction (Exercises 21.11.21 and 21.11.22).

*Modified Vickrey.* The Vickrey auction is looking good when $\frac{2}{5} < p \leq 1$, but we can improve on it by restricting the agents to bids of 3 or 4 and making the winner pay the *average* of the winning and losing bids.

It remains an equilibrium for all actors to plan to bid their true valuations. To see this, consider Mr. Low first. If he bids 3, he gets nothing if he loses and nothing if he wins (because he has to pay his true valuation). If he bids 4, he gets nothing if he loses and at most $3 - \frac{1}{2}(3+4) = -\frac{1}{2}$ if he wins. Thus he optimizes by bidding 3, and so does Ms. Low.

Now consider Mr. High. If he bids 4, he gets nothing when he wins and nothing when he loses on those occasions when his opponent is Ms. High. When his opponent is Ms. Low, he will win and gain $4 - \frac{1}{2}(3+4) = \frac{1}{2}$. Thus his expected payoff from bidding 4 is $\frac{1}{2}p$. If he bids 3, he gets nothing when his opponent is Ms. High. When his opponent is Ms. Low, he will win with probability $\frac{1}{2}$. Thus his expected payoff from bidding 3 is $\frac{1}{2}(4-3)p = \frac{1}{2}p$. It follows that Mr. High has no incentive to switch from bidding 4 to bidding 3 because he is indifferent between the two bids. The same goes for Ms. High.

What does Alice get? Her expected payoff is $4(1-p)^2 + \frac{1}{2}(3+4)p(1-p) + \frac{1}{2}(3+4)(1-p)p + 3p^2 = 4 - p$. This is still second-best because $4-p<4-p^2$ except when $p=0$ or $p=1$. But it is better than the regular Vickrey auction, unless $p=0$ or $p=1$. It is also better than posting a take-it-or-leave-it price of 4 provided $4-p>4(1-p^2)$. This occurs when $\frac{1}{4}<p\leq 1$.

*Summary.* Of the auctions considered, posting a take-it-or-leave-it price of 4 does best when $0\leq p\leq\frac{1}{4}$, and the modified Vickrey auction does best when $\frac{1}{4}\leq p\leq 1$. In fact, using these plans is *optimal* for Alice. To see why, we need to consider her mechanism design problem in the abstract.

### 21.5.3 The Optimal Design

The auctioning plans we have been considering are just a few of a bewildering variety of possibilities open to Alice. She might set entry fees to be paid by all bidders. She might even seed the auction room with shills primed to push the bidding up if things seem slow. But the revelation principle tells us that, in considering what outcomes can be achieved, all the possibilities that don't arise from a truth-telling equilibrium in a *direct* mechanism can be ignored (Theorem 20.1). To keep things simple, attention will be restricted to the case of symmetric equilibria of mechanisms that treat Bob and Carol symmetrically.

*Characterizing a Direct Mechanism.* Recall that players in a direct mechanism are simply asked to state their type (Section 20.4).

Provided that the other actors tell the truth, an actor who announces *High* wins the auction with some probability and makes some payment. Let the probability with which he or she wins the auction be $h$. How much the actor pays will depend on who wins the auction and possibly on other things as well. The parameter $H$ is therefore taken to be the *expected value* of the amount the actor will pay. An actor who announces *Low* in the same circumstances wins the auction with probability $l$ and expects to pay $L$.

*Objective Function.* Alice doesn't know the buyers' types. She therefore expects each agent to pay her

$$F = (1-p)H + pL. \tag{21.7}$$

Her goal is to maximize the quantity $F$ by choosing $h$, $l$, $H$, and $L$ suitably.

*Incentive Constraints.* A High actor who announces *High* will gets an overall expected payoff of $4h-H$ if cast in the role of a buyer. A High actor who announces *Low* gets $4l-L$. For truth telling to be optimal for a High actor, it is therefore necessary that $4h-H\geq 4l-L$. If we also write down the condition that makes truth telling optimal for a Low actor, we obtain the incentive constraints (Section 20.5.1):

$$4h - H \geq 4l - L\,, \tag{21.8}$$

$$3l - L \geq 3h - H\,. \tag{21.9}$$

A simple consequence is that $h \geq l$ and $H \geq L$. Thus a High actor, wins more often than a Low actor but expects to pay more.

*Participation Constraints.* For a High actor to be willing to play, we need that $4h - H \geq 0$. If we also write down the condition for a Low actor to be willing to play, we obtain the participation constraints (Section 20.5.1):

$$4h - H \geq 0\,, \tag{21.10}$$

$$3l - L \geq 0\,. \tag{21.11}$$

*Physical Constraints.* In a symmetric auction, the probability $(1-p)h + pl$ that Bob wins can't exceed $\frac{1}{2}$. Nor can a High actor do better than win all the time against a Low opponent and half the time against a High opponent, so that $h \leq p + \frac{1}{2}(1-p) = \frac{1}{2}(p+1)$. Similarly, $l \leq (1-p) + \frac{1}{2}p = 1 - \frac{1}{2}p$. The physical inequalities constraining $h$ and $l$ are therefore

$$(1-p)h + pl \leq \tfrac{1}{2}\,, \tag{21.12}$$

$$h \leq \tfrac{1}{2}(p+1)\,, \tag{21.13}$$

$$l \leq 1 - \tfrac{1}{2}p\,. \tag{21.14}$$

Figure 21.2(a) shows the set $S$ of pairs $(h, l)$ that satisfy the physical constraints.[7]

*Linear Programming.* Perhaps more constraints on Alice's choice of $h$, $l$, $H$, and $L$ are necessary, but let us try to solve her linear programming problem with the constraints listed so far (Section 7.6).

Alice's aim is to maximize the linear objective function $(1-p)\,H + pL$ of (21.7), subject to the linear inequalities (21.8), (21.9), (21.10), (21.11), (21.12), (21.13), and (21.14).

*Active Constraints?* It isn't hard to guess that (21.8) and (21.11) must be active in our problem. The intuition is that High actors have the greater incentive to lie and Low actors have the greater incentive not to participate. The intuition can be confirmed by examining Figure 21.2(b), which shows the set $T$ of all feasible pairs $(H, L)$ for a pair $(h, l)$. Observe that, whenever $h \geq l$, the expression $F = (1-p)H + pL$ is maximized at the point $(H, L)$, satisfying

$$H - L = 4(h - l)\,, \tag{21.15}$$

$$L = 3l\,. \tag{21.16}$$

[7]It may be helpful to note that $\frac{1}{2}(p+1) \leq \frac{1}{2}(1-p)^{-1}$ and $1 - \frac{1}{2}p \leq \frac{1}{2}p^{-1}$.

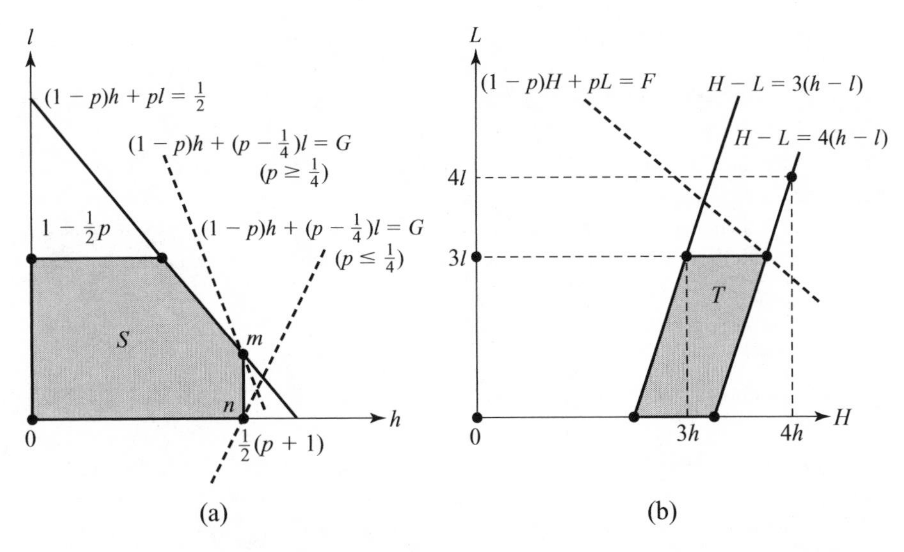


Figure 21.2 Designing an optimal auction. Figure 21.2(a) shows the set $S$ of all pairs $(h,\ell)$ that satisfy the physical constraints (21.12), (21.13), and (21.14). For each possible $(h,\ell)$ that might be chosen, Figure 21.2(b) shows the set $T$ of all pairs $(H,L)$ that satisfy the incentive and participation constraints (21.8), (21.9), (21.10), and (21.11).

This observation simplifies the problem immensely. Substitute $H=4h-l$ and $L=3l$ into (21.7). We then have to maximize

$$G=(1-p)h+(p-\tfrac{1}{4})l\,,$$

subject to the constraints (21.12), (21.13), and (21.14). The location of the maximum depends on whether $p\geq\frac{1}{4}$ or $p\leq\frac{1}{4}$. Figure 21.2(a) shows that the maximum is achieved at $m$ in the former case and at $n$ in the latter case.[8]

*The Case $p\geq\frac{1}{4}$.* Since $m=(\frac{1}{2}(p+1)\,,\,\frac{1}{2}p)$, the optimal values for $h$ and $l$ when $p\geq\frac{1}{4}$ are $\tilde{h}=\frac{1}{2}(p+1)$ and $\tilde{l}=\frac{1}{2}p$. The corresponding values for $\tilde{H}$ and $\tilde{L}$ are $\tilde{H}=4\tilde{h}-\tilde{l}=\frac{3}{2}p+2$ and $\tilde{L}=3\tilde{l}=\frac{3}{2}p$. Alice then gets an expected payoff of $2\tilde{F}=2(1-p)\tilde{H}+2p\tilde{L}=4-p$.

*The Case $p\leq\frac{1}{4}$.* Since $n=(\frac{1}{2}(p+1),0)$, the optimal values for $h$ and $l$ are $\tilde{h}=\frac{1}{2}(p+1)$ and $\tilde{l}=0$ when $p\leq\frac{1}{4}$. The corresponding values for $\tilde{H}$ and $\tilde{L}$ are $\tilde{H}=4\tilde{h}-\tilde{l}=2(p+1)$ and $\tilde{L}=3\tilde{l}=0$. Alice then gets an expected payoff of $2\tilde{F}=2(1-p)\tilde{H}+2p\tilde{L}=4(1-p^2)$.

*What Is Optimal?* In Section 21.5.2, we learned that Alice can get an expected payoff of $4(1-p^2)$ by posting a price of 4 in a take-it-or-leave-it auction. Now we know that this outcome is optimal when $0\leq p\leq\frac{1}{4}$.

[8]If $p>\frac{1}{4}$, the line $G=(1-p)h+(p-\frac{1}{4})l$ slopes down more steeply than $(1-p)h+pl=\frac{1}{2}$. If $p<\frac{1}{4}$, the line $G=(1-p)h+(p-\frac{1}{4})l$ slopes upward. If $p=\frac{1}{4}$, any point on the line segment joining $m$ and $n$ is optimal.

Notice that we don't need to worry any more that we might have overlooked some relevant constraint because we can point to an auction that actually achieves the maximum $4(1-p^2)$ of the optimization problem we set ourselves. It follows that it is impossible to make the maximum less than $4(1-p^2)$ by adding an extra constraint to the optimization problem. Any further constraints that might be proposed are therefore irrelevant.

We also learned in Section 21.5.2 that Alice can get an expected payoff of $4-p$ using a modified Vickrey auction. Now we know that this outcome is optimal when $\frac{1}{4} \leq p \leq 1$.

## 21.6 Common-Value Auctions

So far, we have considered only auctions in which the buyers' values are private. Actors with private values learn their valuation once and for all before the auction begins, and nothing they might learn during the auction will lead them to change their valuations.

*Common values* sit at the other end of the valuation spectrum. In a common-value auction, it is common knowledge that the value of the object being sold is the same for all the prospective buyers.

For example, when licenses to drill for oil in undersea tracts are auctioned, the amount of oil in a tract is the same for everybody. The interest in such common-value auctions is that different buyers will have different beliefs about what the common value is.

### 21.6.1 Winner's Curse

In oil-tract auctions, the buyers' estimates of how much oil is likely to be in a tract will depend on their geological surveys. Such surveys aren't only expensive but also notoriously unreliable. Some prospective buyers will therefore receive optimistic surveys, and others will receive pessimistic surveys. So who will win the auction?

If Bob treats his survey's estimate of the value of the tract as a private value, then he will win whenever his survey is the most optimistic. But when Bob realizes that his winning the auction implies that all the other surveys are more pessimistic than his, then he will curse his bad luck at winning! If he had known at the outset that all the other surveys were more pessimistic than his, he wouldn't have bid so high.

As with the all-pay auction, game theory instructors like to fool their students by trapping them with a common-value auction. A glass jar filled with coins and crumpled bills of various denominations is auctioned off to the highest bidder, who usually falls prey to the winner's curse and thus takes a substantial loss.

But not only students get duped. As we have seen, the winner's curse is a real phenomenon in oil-tract auctions. To avoid it, you need to *condition* the probability you attach to each possible value of the object for sale on the hypothetical event that you win the auction.

*The Wallet Game.* Bulow and Klemperer's Wallet Game is a simple example that demonstrates how to avoid the winners' curse.

Alice confiscates Bob's and Carol's wallets and uses an English auction to sell their combined contents back to whoever bids higher. Bob's wallet contains $b$ dollars, and Carol's wallet contains $c$ dollars, but they know only how much was in their own wallet. We look for a symmetric equilibrium in which an actor whose wallet contained $x$ plans to bid up to $B(x)$, where $B(x)$ is a strictly increasing function of $x$.

If Carol plays according to the equilibrium and quits at price $p$, then Bob will know that the value for the two wallets is $b+B^{-1}(p)$. He should therefore plan to stay in the auction as long as $p<b+B^{-1}(p)$. He should plan to quit when $p=b+B^{-1}(p)$. Writing $p=B(b)$ in this equation, we obtain

$$B(b) = b + B^{-1}(B(p)) = 2b.$$

In equilibrium, Bob and Carol should therefore each bid twice the value of their own wallet. The winner is therefore the moneybags with the larger bankroll—even if it is only a tiny bit larger.

### 21.6.2 Toeholds

I used a version of Bulow and Klemperer's Toehold Game when arguing on behalf of a client that Rupert Murdoch shouldn't be allowed to take over Manchester United soccer team because of the advantage such a toehold in the soccer cartel would give him in their periodic auctions of exclusive television rights to England's top soccer games. The players in the following simplified version are therefore called Rupert and Sophia.

In a common-value auction with a zero reserve price, Rupert is assumed to have an advantage because a small percentage of his bid is returned to him if he wins. In an attempt to nullify his advantage, we assume that Sophia is better informed than Rupert. It is common knowledge that she knows the precise value $v$ of the object being sold, whereas Rupert believes it is uniformly distributed on [0,1]. However, Rupert comes out decisively on top in an English auction if Sophia plays her weakly dominating strategy of bidding up to her valuation since Rupert now enjoys what one might call the winner's blessing.

Rupert should always overbid Sophia by a small amount because, if winning at the current price is profitable for her, it is even more profitable for him. He therefore does best to make sure of always winning by planning to bid all the way up to the maximum.

So what is the point of Sophia bidding at all—especially if there is some cost of entry?[9] But if Sophia stays out of the auction, the seller makes nothing at all. Everybody therefore loses except for Rupert, who can now congratulate himself on his foresight in having invested in gaining a toehold in the seller's revenues. The lesson again is that apparently insignificant entry costs can have large strategic implications.

[9]Even when billions of dollars are involved, losers in what seems like a hopeless battle sometimes claim to have bid way beyond their own valuation in order to make the winner pay more. But we don't have to believe what they say when they claim to have taken such an incredible risk! In any case, if Rupert knew that Sophia was motivated by malice, he would factor this consideration into his bidding strategy.

## 21.7 Multiunit Auctions

→ 21.8

In the big telecom auctions used to motivate this chapter, several licenses were sold simultaneously, but no two telecom licenses are ever totally interchangeable. This section focuses on multiunit auctions in which all the items offered for sale are precisely the same. The leading examples are treasury auctions, which governments use to borrow many billions of dollars every year by selling bonds. Another example is the recent series of auctions in which the British government sold off a substantial fraction of its gold reserves.

In describing the various types of multiunit auction that are commonly considered, our seller will continue to be Alice, and the buyers will be Bob and Carol. The units offered for sale will be called bonds.

### 21.7.1 Sealed-Bid Auctions

As with single-unit auctions, we can classify multiunit auctions by whether their format is sealed bid or open. In a sealed-bid format, Alice begins by publicly committing herself to a supply curve. Bob and Carol are then each asked to submit a demand curve in a sealed envelope. How many units Bob and Carol get is then always determined by finding the allocation that equates total demand and supply. Different auction types differ in the amount that Bob and Carol are required to pay for their allocation.

I think that governments would be wise to think harder about the supply curve they use in treasury auctions, but the relevant officials are too timid to try anything even mildly innovative. And who can blame them? As we will see, the one time that the U.S. Treasury took what must have seemed the best possible advice from economists, it was sold a lot of hogwash. In any case, it is standard for the supply curve to take the "L" shape that results from announcing that a fixed number $S$ of bonds will be sold at the best prices that can be obtained above a stated reserve. We simplify further by always assuming that the reserve price is zero.

*Geometric Representation.* A bundle $(x, p)$ in Figure 21.3 corresponds to Bob's getting $x$ bonds at a price of $p$ dollars per bond. He therefore buys $x$ bonds for $px$ dollars. We assume that his preferences over such bundles are determined by a quasilinear utility function $u(x) - px$, in which $u$ is strictly increasing, differentiable, and concave. (Similar assumptions also apply to Carol.)

Bob's true demand curve—not the phony demand curve he announces to Alice—then has equation $p = u'(x)$, as shown in Figure 21.3(a) (Section 9.3.2). Some of Bob's indifference curves are also shown (Section 9.3). Notice that the indifference curves are horizontal only where they cross the demand curve and that Bob always wants to move *below* an indifference curve.

Since only $S$ bonds are available for sale, if Bob gets $x$ bonds, then Carol gets $S - x$ bonds when the market clears. Plotting Carol's true demand curve on the same diagram as Bob's, we therefore obtain Figure 21.3(b). Supply is equal to total demand at the point $(X, P)$ where these two demand curves cross because we then have $X + (S - X) = S$. Thus, $(X, P)$ corresponds to the Walrasian equilibrium in this setup.

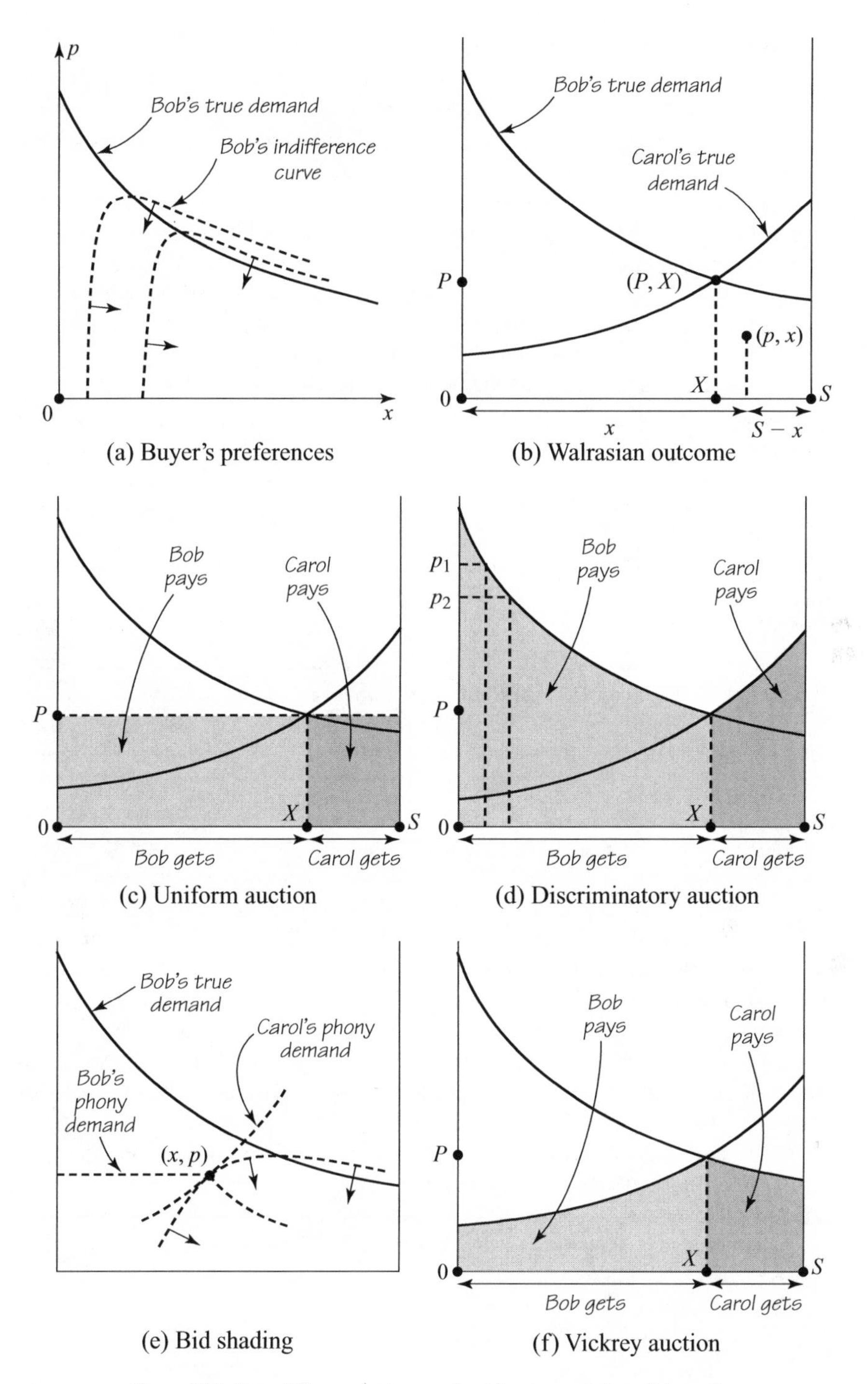


Figure 21.3 How different auctions work with true revelation of demand.

*Uniform Auction.* A uniform-price auction is modeled on a perfectly competitive market. Figure 21.3(c) shows what would happen if Bob and Carol were to ignore the advantages of shading their bids and submit their true demand curves. The buyers pay the clearing price $P$ for each bond they are allocated. The seller's revenue is then represented by the shaded rectangle.

*Discriminatory Auction.* In a discriminatory-price or pay-your-bid auction, Bob and Carol are made to pay the most that their stated demand curve says that they are willing to pay for each unit they are allocated. Figure 21.3(d) shows what would happen in a discriminatory auction if the buyers were to submit their true demand curves. Bob pays $P_1$ for the first bond, $P_2$ for the second bond, and so on. He pays $P$ only for the last bond he buys. If the number $S$ of bonds is very large, the seller's revenue is therefore approximately equal to the shaded area in Figure 21.3(d) (Section 9.5.2).

*Bid Shading.* Since the seller's revenue in the uniform auction of Figure 21.3(c) is less than his revenue in the discriminatory auction of Figure 21.3(d), it seems as though the seller should prefer the discriminatory auction. But this is the journalists' fallacy (Section 21.4).

Bob and Carol won't bid their true demand curves. They will shade their bids down for strategic reasons. This is immediately evident in the case of a discriminatory auction. If Bob knows that Carol will announce a phony demand curve of the type shown in Figure 21.3(e), then his best reply is to find the bundle $(x, p)$ he likes best on her phony demand curve and to announce a phony demand curve of his own that passes through $(x, p)$ and is horizontal to the left.

*Multiunit Vickrey Auction.* In the finance literature, a uniform auction is called a *second-price auction.* This misleading terminology commemorates a memorable fiasco in which the Nobel Prize winners Milton Friedman and Merton Miller mounted a successful campaign to persuade the U.S. Treasury to switch from using the traditional discriminatory format to a uniform format in some bond auctions.

A discriminatory multiunit auction corresponds to a first-price, sealed-bid, single-unit auction, and so Friedman and Miller thought that a uniform auction would correspond to a second-price, sealed-bid, single-unit auction. They therefore advocated using a uniform auction on the grounds that the buyers would then be induced to bid their true demands.[10]

A multiunit Vickrey auction employs the Clarke-Groves mechanism (Section 20.5.3). Each buyer therefore pays the amount under the *other* buyer's demand curve, as indicated in Figure 21.3(f). With private values, it is therefore a weakly dominant strategy to bid your true demand curve—but the same isn't true of a uniform auction!

### 21.7.2 Open Auctions

As with single-unit auctions, each conventional, sealed-bid, multiunit auction has an open counterpart.

[10] *Wall Street Journal,* 28 August 1991; *New York Times,* 15 September 1991.

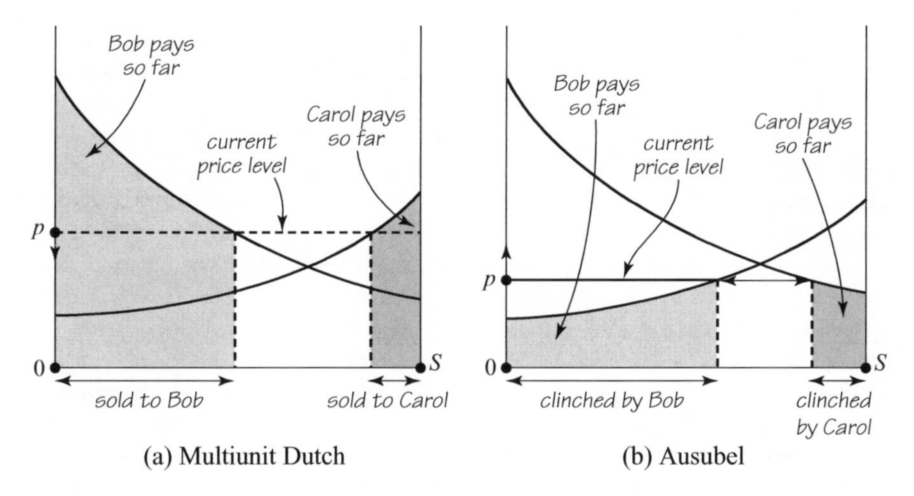


Figure 21.4 Two open multiunit auctions.

*Multiunit English.* The price starts low and is gradually raised. At each price, the buyers signal how many units they are willing to buy at that price. The auction stops when the total demand reduces to the available supply, and the buyers are then allocated their current demand at the final price.[11] Its sealed-bid counterpart is a uniform auction.

*Multiunit Dutch.* The price starts high and is gradually lowered. When a buyer signals, he or she is awarded a unit at the current price, and the auction continues. Figure 21.4(a) indicates why its sealed-bid counterpart is a discriminatory auction.

*Ausubel Auction.* I have tried to sell Ausubel's clever idea to several clients without success, but it will be widely used one day. The price starts low and is gradually raised. As the price rises, the buyers reduce their demands. Eventually, the demand from one buyer gets too small to meet the supply by itself. The *other* buyer is then said to have *clinched* the number of bonds necessary to make up the difference. Whenever a buyer clinches a new bond, he or she pays the current price for that bond. Figure 21.4(b) illustrates why the sealed-bid counterpart of an Ausubel auction is a multiunit Vickrey auction.

### 21.7.3 Strategic Behavior in Multiunit Auctions

The general properties of multiunit auctions aren't very well understood at present. The analysis offered here is essentially an exercise in Nash implementation theory (Section 20.6.2).

We assume that the buyers' preferences are common knowledge between Bob and Carol but are unknown to Alice. Her aim is to implement a Walrasian outcome. Which multiunit auction should she choose?

[11]Such auctions differ from the double auctions mentioned in Section 9.6.3 because only the buyers bid.

*Uniform?* Figure 21.5(a) shows that it can be an equilibrium for both buyers to announce their true demand curves in a uniform auction only if these are flat where they cross. Whatever demand curve Carol announces, Bob's best reply is to choose a phony demand curve that cuts Carol's demand curve at whatever point $(x, p)$ the first buyer likes best. At this point, one of Bob's indifference curves touches Carol's announced demand curve. If $(x, p)$ lies on Bob's true demand curve, then Carol must have announced a demand curve that is flat at $(x, p)$.

Figure 21.5(b) shows that the phony demand curves in a uniform auction can't cross in equilibrium at a point $(x, p)$ above Bob's true demand curve. If they did, the drawn indifference curve indicates that Bob would then prefer to announce a phony demand curve that crosses Carol's phony demand curve farther to the left. In equilibrium, the buyers' phony demand curves can therefore cross only at a point in the shaded region of Figure 21.5(b). It follows that Alice can't anticipate selling her bonds for more than the true clearing price in a uniform auction.

Figure 21.5(c) shows why each point $(x, p)$ in the shaded region of Figure 21.5(b) corresponds to a Nash equilibrium in which Bob is allocated $x$ bonds and the selling price is $p$. The phony demand curves in the illustrated equilibrium have a kink at $(x, p)$. Any price between 0 and the true Walrasian price $P$ can be sustained as an equilibrium in this way.

*Discriminatory?* Figure 21.5(d) shows two possible phony demand curves for Carol in a discriminatory auction. As we saw in Figure 21.3(e), each must be horizontal to the right of $(x, p)$ in equilibrium. The indifference curves drawn are Bob's. They show that an equilibrium can't occur when the phony demand curves cross at a point $(x, p)$ either above or below Bob's true demand curve. In the first case, Bob would prefer to announce a phony demand curve that crosses Carol's phony demand curve farther to the left. In the second case, he would prefer that the curves cross farther to the right.

The only possibility for a Nash equilibrium in a discriminatory auction is therefore that the phony demand curves cross at a point $(x, p)$ lying on both Bob's and Carol's true demand curves, so that the Nash equilibrium implements a Walrasian outcome $(X, P)$. Figure 21.5(e) shows that to sustain such a Nash equilibrium, each buyer's phony demand curve must lie above the indifference curve of the other buyer that passes through the point $(X, P)$.

*Vickrey?* In a multiunit Vickrey auction, it is a weakly dominant strategy for all buyers to announce their true demand curves, thereby implementing a Walrasian outcome. But here, as in other cases we have studied, there are also lying equilibria on which the buyers may collude to the seller's disadvantage. Figure 21.5(f) shows a lying equilibrium in which Alice's revenue is nearly zero.

→ 21.8

*The Friedman Fiasco.* It is ironic that the traditional discriminatory auction condemned by Milton Friedman and Merton Miller should prove best at implementing a Walrasian outcome. The uniform format Friedman favors would seem particularly vulnerable to collusive manipulation by the buyers (which needn't be in the least illegal).

The usual response is that treasury auctions are common-value rather than private-value events because of the existence of a secondary market in which bonds

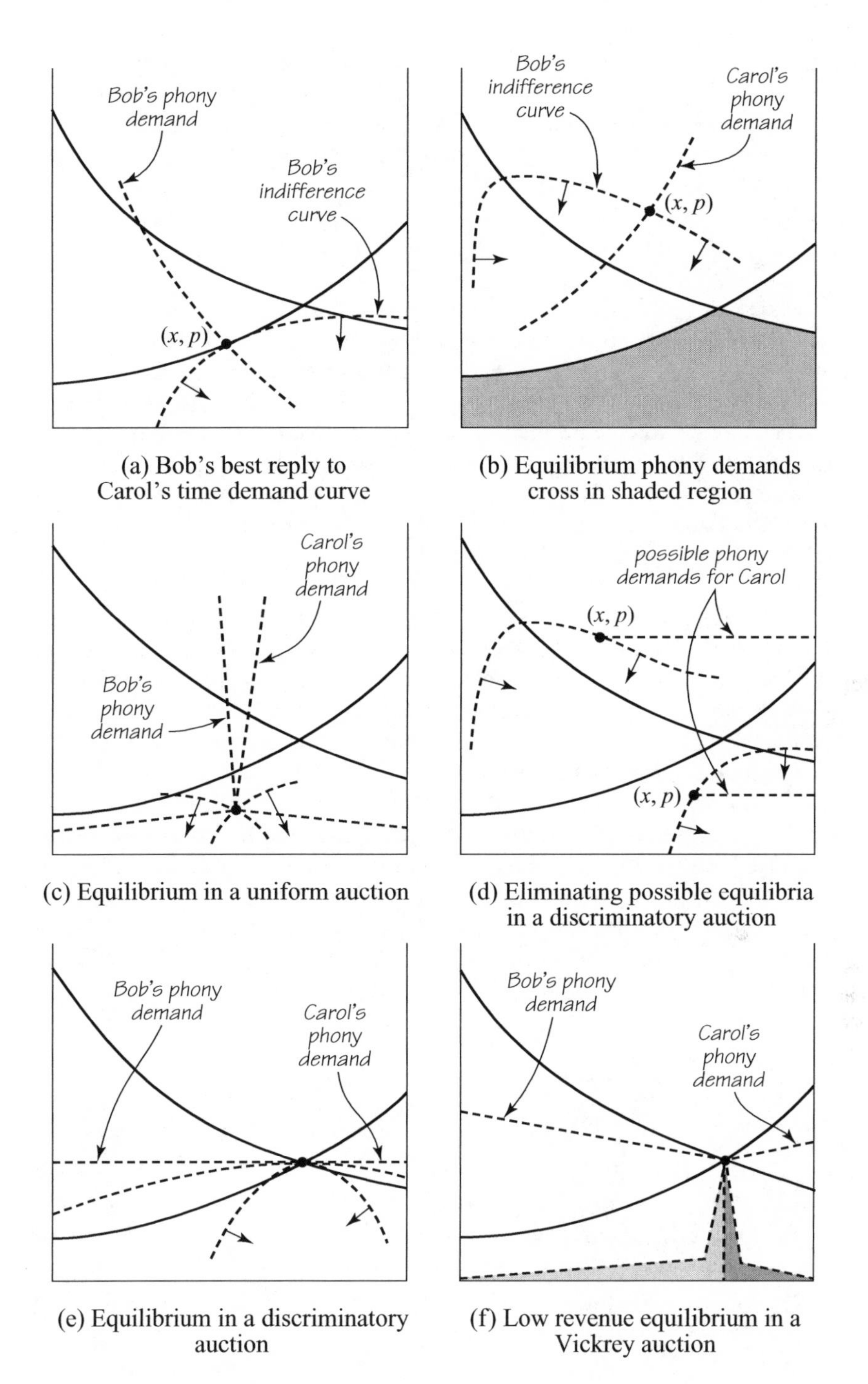


Figure 21.5 Nash equilibria in discriminatory and other auctions.

are freely traded. But resale is another area in which economists of the old school have been too quick to jump to conclusions. For example, it most certainly isn't true that the possibility of resale guarantees an efficient outcome independently of the type of auction used. If efficiency is your goal use an efficient auction in the first place!

## 21.8 The Chopstick Auction

→ 21.9

The Chopstick Auction has been chosen to end this book to illustrate why game theorists enjoy their subject. Who would have guessed at such a bizarre solution? But the problem itself arose in an entirely mundane context.

In a telecom auction, the bidders needed to buy several of the small frequency packages offered for sale in order to be able to operate a viable business. What should buyers do if the packages are sold simultaneously in independent, sealed-bid auctions: concentrate their money in a few of the auctions or spread it out across them all? Bob Rosenthal didn't know the answer when asked by his client, and so he constructed the following model to investigate the question.

*Selling Chopsticks.* As in Figure 21.6(a), Alice has three chopsticks for sale. The potential buyers are Bob and Carol. They are risk neutral, and so we measure their utility in dollars. Figure 21.6(b) shows the utility function that they share. One chopstick is no better than no chopsticks. Bob and Carol value both outcomes as worth nothing. Three chopsticks are no better than two chopsticks. Bob and Carol value both outcomes as worth one dollar.

Alice decides to sell the three chopsticks simultaneously, using three independent, first-price, sealed-bid auctions. What strategy should Bob and Carol use in the game that Alice thereby creates for them to play? Should they concentrate all their money on two auctions or spread it out across all three?

A pure strategy for Bob in the Chopstick Auction is a triple $(x, y, z)$ that lists how much he is to bid in each of the three auctions. We assume that $x \geq 0, y \geq 0, z \geq 0$,

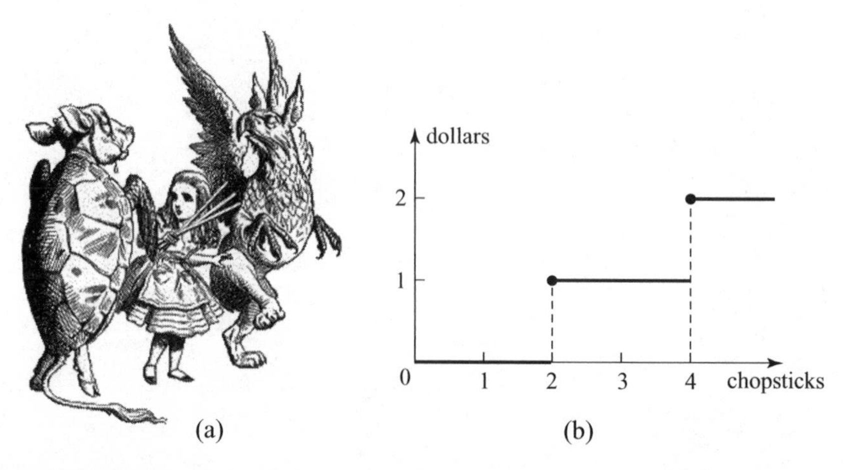


Figure 21.6 Selling chopsticks. Alice offers three chopsticks for sale. Bob and Carol each want exactly two chopsticks.

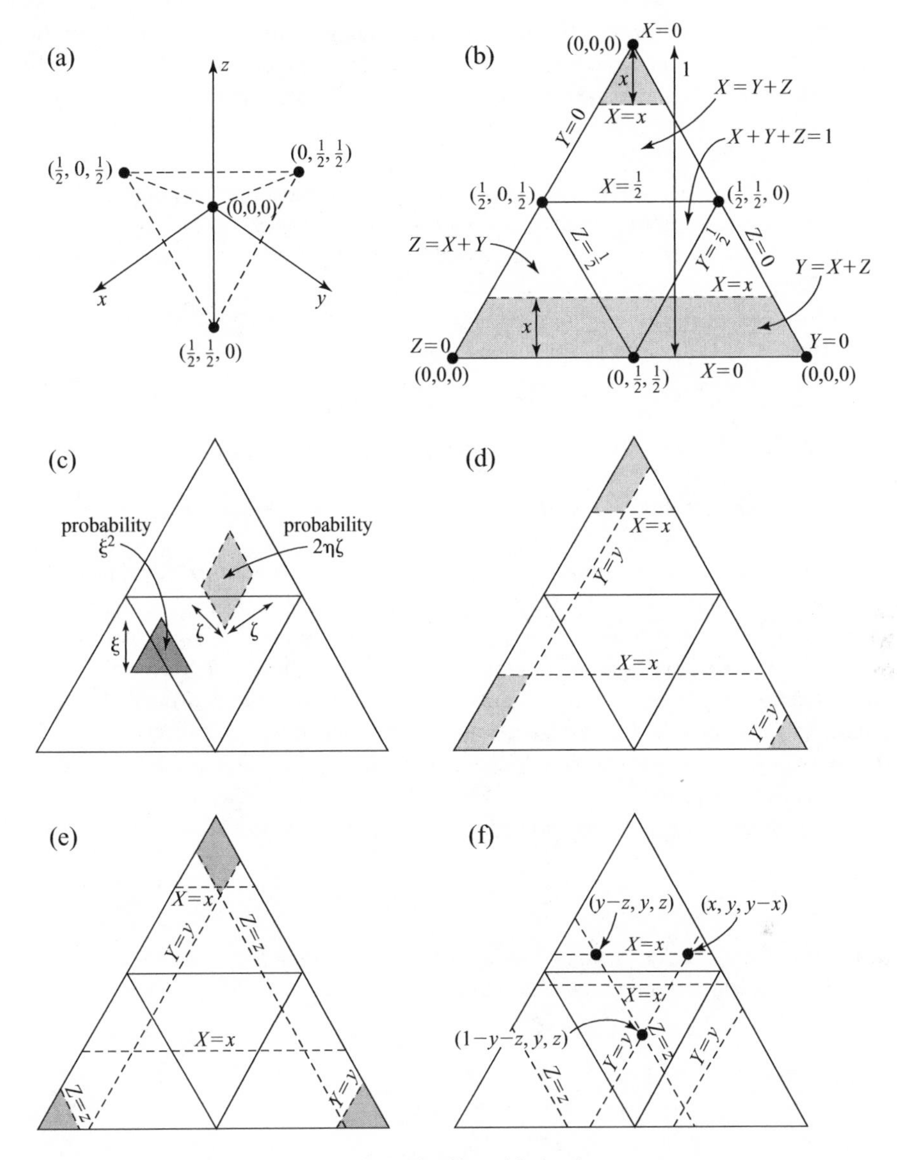


Figure 21.7 The Chopstick Auction.

and $x+y+z \leq 1$. Because the players can always undercut their rival in the two auctions in which the rival plans to bid least, there is no hope of finding a Nash equilibrium in pure strategies.

*A Mixed Equilibrium.* The first step in finding a mixed equilibrium is to construct the regular tetrahedron $T$ of Figure 21.7(a), whose vertices are the pure strategies $s_0 = (0,0,0)$, $s_1 = (0, \frac{1}{2}, \frac{1}{2})$, $s_2 = (\frac{1}{2}, 0, \frac{1}{2})$, and $s_3 = (\frac{1}{2}, \frac{1}{2}, 0)$. The second step is to introduce a uniform probability distribution $\mu$ over the surface of the tetrahedron $T$. If Bob and Carol each independently choose their bids according to $\mu$, then we are looking at a Nash equilibrium of the Chopstick Auction. Since the support of $\mu$ is the *surface* of $T$, rather than its whole *volume*, Bob and Carol neither put all their money on two auctions nor spread it evenly among all three. In the mixed equilibrium, they somehow contrive to do both at once!

*Proof* We need to confirm that any of Bob's pure strategies $(x,y,z)$ that lie on the surface of $T$ is a best reply to Carol's use of the mixed strategy $\mu$. In fact, whenever $(x,y,z)$ lies inside or on the surface of $T$, Bob's expected payoff is zero. When $(x,y,z)$ lies beyond $T$, his expected payoff is negative.

If Bob uses the pure strategy $(x,y,z)$, his expected payoff when Carol plays according to $\mu$ is the probability that he wins at least two chopsticks minus his expected payment:

$$\pi(x,y) = \{p_{12} + p_{23} + p_{31} - 2p_{123}\} - \{xp_1 + yp_2 + zp_3\}, \tag{21.17}$$

where $p_1$ is the probability he wins the first auction, $p_{12}$ is the probability he wins the first and second auctions, and so on.

To work out the probabilities in (21.17), unfold the tetrahedron $T$ as explained in Figure 6.13(d). This yields the two-dimensional representation $S$ of its surface shown in Figure 21.7(b). Although we have unfolded $T$, we continue to use the coordinate system of $\mathbb{R}^3$. For example, the two line segments in Figure 21.7(b) marked $X = x$ show where the plane $X = x$ in $\mathbb{R}^3$ cuts the surface of the tetrahedron $T$. The distances and areas marked are proportional only to the real distances and areas on $S$, but the constants of proportionality cancel out from the relevant equations, and so we forget about them.

The probability $p_1$ is the shaded area in Figure 21.7(b), and so

$$p_1 = x^2 + \{1 - (1-x)^2\} = 2x.$$

It is easiest to work out such areas using the facts about the areas of parallelograms and triangles illustrated in Figure 21.7(c). Using the Figure 21.7(d), which has been drawn on the assumption that $x \geq y$,

$$p_{12} = y^2 + 2xy + x^2 - (x-y)^2 = 4xy.$$

Since the answer is symmetric in $x$ and $y$, the result also holds when $y \geq x$.

The probability $p_{123}$ is more troublesome. A number of assumptions are built into Figure 21.7(e). The first is that $x \geq y \geq z$, from which it follows that $z + x \geq y$ and $x + y \geq z$. The second assumption is that $y + z \geq x$. The third assumption is that $x + y + z \leq 1$. (Figure 21.7(f) shows how things change if $x + y + z > 1$.) With all these assumptions

$$\begin{aligned} p_{123} &= z^2 + y^2 - (y-z)^2 + x^2 - (x-y)^2 - (x-z)^2 \\ &= 2xy + 2yz + 2zx - x^2 - y^2 - z^2. \end{aligned} \tag{21.18}$$

Since the result is symmetric in $x, y$, and $z$, we can abandon the assumption that $x \geq y \geq z$, provided that we maintain the assumptions

$$x + y \geq z, \quad y + z \geq x, \quad z + x \geq y, \quad x + y + z \leq 1,$$

which together imply that (21.18) holds when $(x, y, z)$ is inside or on the surface of the tetrahedron $T$.

Substituting our formulas for all the probabilities into (21.17), we find that Bob's expected payoff is

$$\pi(x, y, z) = 0$$

when $(x, y, z)$ is inside or on the surface of $T$. This completes the proof since it is easy to see that $p_{123}$ gets bigger than our formula outside the tetrahedron, and so $\pi\,(x, y, z)$ is then negative.

## 21.9 Roundup

Auction design is one of game theory's great successes. A major lesson from big-money telecom auctions is that it is dangerous to take a design off the shelf. Each new situation requires a design tailored to its special circumstances.

The traditional formats for a single-unit auction are either sealed bid or open. The object for sale always goes to the highest bidder. In a first-price, sealed-bid auction, winners pay their own bid. In a second-price, sealed-bid auction, winners pay the bid of the highest loser. The open formats are called English and Dutch. In a Dutch auction, the price falls until a bidder calls a halt. Dutch auctions are equivalent to first-price, sealed-bid auctions. In an English auction, the price rises until only one bidder remains. In the private-value case, English auctions are equivalent to second-price, sealed-bid auctions.

Second-price, sealed-bid auctions are called Vickrey auctions to honor the pioneer of auction theory. They are essentially a version of a Clarke-Groves mechanism, and accordingly it is a weakly dominant strategy for buyers to bid their true valuations. However, a seller would be unwise to ignore the existence of lying equilibria on which the buyers may collude—especially in cases in which some buyers have only a small probability of winning in the truth-telling equilibrium but face significant entry costs.

The revenue equivalence theorem says that symmetric equilibria in all standard auction formats with the same participation condition generate the same expected revenue for the seller in a standard private-value scenario. This expected revenue, which is equal to the expected value of the second-highest valuation, turns out to be maximal under the same conditions. The participation condition is normally determined by the reserve price chosen by the seller. It is almost never optimal to choose a zero reserve price.

Auction design is a special case of mechanism design. One writes down incentive, participation, and physical constraints and follows where they lead.

Buyers with private values learn their valuation once and for all before the auction begins, and nothing they might learn during the auction will lead them to change their valuation. In a common-value auction, it is common knowledge that the value of the object being sold is the same for all the buyers, but different buyers will have different beliefs about what the common value is.

The winner's curse arises when a winner in a common-value auction fails to consider the implications that winning the auction has for the information about the value of the object for sale that the losers must have received. To avoid the winner's curse, you need to condition the probability you attach to each possible value of the object for sale on the hypothetical event that you win the auction.

The toehold model applies when a buyer is advantaged through having a stake in the seller's revenues. Such a buyer can afford to bid more aggressively, which multiplies the impact of the winner's curse on the other buyers, who are therefore forced to shade their bids even more. In extreme cases, the disadvantaged buyers may not bid at all.

Auctions of treasury bonds are multiunit auctions in which each unit offered for sale is identical. In sealed-bid versions, the buyers submit demand curves. Each buyer is then allocated the number of bonds that equates total demand and supply. In uniform auctions, each buyer pays the clearing price. In discriminatory auctions, each buyer pays the area under his or her demand curve that corresponds to the number of bonds each buyer is allocated. In a multiunit Vickrey auction, buyers pay the amount under the aggregate demand curve of the other buyers. Each type of sealed-bid auction has an open counterpart, which are respectively, the multiunit English, the multiunit Dutch, and the Ausubel auction. As the price rises in the latter, buyers clinch units at the current price as soon as the total demand from the other buyers reduces to a level that would result in the unit's otherwise being unsold.

Milton Friedman mistakenly thought that a uniform auction is the generalization of a single-unit Vickrey auction to the multiunit case. As a consequence the finance literature misleadingly refers to a uniform auction as a second-price, multiunit auction. The moral is a good one for the end of this book. If you are going to let someone else do your thinking for you, make sure it's someone like Von Neumann.

## 21.10 Further Reading

*Auctions and Auctioneering*, by R. Cassady: University of California Press, Berkeley, 1967. Lots of good stories.

*Auctions: Theory and Practice*, by Paul Klemperer: Princeton University Press, Princeton, NJ, 2004. A lively introduction from an original angle.

*Auction Theory*, by Vijay Krishna: Academic Press, San Diego, 2002. A careful account of existing theory with no hype.

"The Biggest Auction Ever: The Sale of British 3G Licences," by Ken Binmore and Paul Klemperer, *Economic Journal* 112 (2002), C74–C96. Who would have thought making so much money could be so dull?

*Putting Auction Theory to Work*, by Paul Milgrom: Cambridge University Press, New York, 2004. The pioneer in auctioning radio-spectrum licenses reveals his secrets.

## 21.11 Exercises

1. If Bob's and Carol's valuations are independently drawn from a uniform distribution on $[0, 1]$, show that the probability of making a sale in a take-it-or-leave-it auction by posting a price $p$ with $0 \le p \le 1$ is $1 - p^2$. Deduce that Alice maximizes her expected revenue by posting the price $p = 1/\sqrt{3}$.

econ

2. Explain why a take-it-or leave-it auction is analogous to a classic monopoly if we identify the probability of making a sale in the former with the amount sold in the latter. Why is a Bertrand duopoly a kind of Dutch auction? (Section 10.4)
3. Find a weakly dominant strategy for Bob in an English auction when he can't bid more than the amount of cash he has in his wallet.
4. Why might the remaining buyers in an English auction revise their valuations of the object for sale after one of their rivals quits? Why does this consideration destroy the strategic equivalence of English and Vickrey auctions when there are three buyers whose valuations are interrelated?
5. A painting is to be sold at a Vickrey auction. It is common knowledge that all the dollar valuations of the painting by potential buyers are different positive integers, but the valuations themselves are unknown. It is a weakly dominating strategy if all buyers bid their true valuations. Explain why it is also a Nash equilibrium if all the buyers bid one dollar less than their true valuations. Why is this lying equilibrium a Pareto improvement on the truth-telling equilibrium for the buyers?
6. Calculate the probability that a random variable that is uniformly distributed on the interval $[3,5]$ takes a value in the interval $[2,4]$.

review

7. Compute

review

(a) $\frac{d}{dx}\int_0^x (1+y^{10})^{-20}\,dy$ (b) $\frac{d}{dx}\int_{-23}^x (1+y^{10})^{-20}\,dy$

(c) $\frac{d}{dx}\int_x^{67} (1+y^{10})^{-20}\,dy$ (d) $\frac{d}{dx}\int_0^{x^2} (1+y^{10})^{-20}\,dy$

8. Let $F : [0,1] \to \mathbb{R}$ be continuous on $[0,1]$ and differentiable on $(0,1)$. Assume that $F(0)=0$. Integrate by parts to show that

review

$$\int_0^1 F(v)dv = -\int_0^1 (v-1)F'(v)dv.$$

9. Why must a probability distribution function $P : \mathbb{R} \to [0,1]$ be increasing? Why does it follow that a continuous[12] probability density function $p : \mathbb{R} \to \mathbb{R}_+$ must be nonnegative? If $P(a)=P(b)$ and $a<b$, why must it be true that $p(x)=0$ for $a<x<b$?

review

10. If the random variable $X$ is uniformly distributed on the interval $[a,b]$, confirm that $\mathscr{E}X = \frac{1}{2}(a+b)$.

review

11. Under the conditions of Section 21.4, two buyers in a first-price, sealed-bid auction shade their bids by 50%. If there were $n$ bidders, show that $B(v) = (n-1)v/n$.
12. Analyze an all-pay, sealed-bid auction under the conditions of Section 21.4. (You will be led to the differential equation $db/dv = v$, which needs to be solved with the boundary condition $b=0$ when $v=0$.) Confirm that the seller's

[12]Without this proviso it would be strictly necessary to qualify the statements that follow with the phrase "almost everywhere."

expected revenue is $R = \frac{1}{3}$, as with the other auctions analyzed in Section 21.4.

13. If there are $n$ buyers in an all-pay, sealed-bid auction, show that they each bid $(n-1)v^n/n$ under the conditions of Section 21.4.
14. Verify that first-price and second-price sealed-bid auctions are standard in the sense of Section 21.4.1. Why isn't the same true of a take-it-or-leave-it auction? What about an all-pay auction?

math

15. The revenue equivalence theorem for standard auctions in a standard private-value scenario hinges on the bid function $B(v)$ being strictly increasing in $v$ (Section 21.4.1). Prove this result in three steps:
    a. By considering optimal play at valuations close to a possible jump, show that $B$ must be continuous.
    b. If $B(u) \leq B(v)$ for $v < u < w$, explain why $B(u) = B(v)$ for $v < u < w$.
    c. if all buyers whose valuations lie in some interval make the same bid, explain why a deviant buyer from the interval who bids a little more will gain as a result.
16. Use equation (21.5) to show that, in a symmetric equilibrium of a standard auction in a standard private-value scenario with $\underline{v} = 0$, Bob expects to pay $P(v)\,\mathcal{E}\,\{w \mid w < v\}$ when his valuation is $v$.

phil

17. Why is it important that a seller be perceived as being committed to her reserve price? Reserve prices are sometimes kept secret from the buyers until after the auction. Why does such secrecy complicate the commitment issue?
18. If Bob's and Carol's valuations are independently drawn from a uniform distribution on $[0, 1]$, confirm that a standard auction with an optimal reserve price generates a larger expected revenue for Alice than an optimal take-it-or-leave-it auction (Exercise 21.11.1).

math

19. If Alice chooses a reserve price of $r$, explain why her expected revenue from the play of a symmetric equilibrium in a standard auction in a standard private-value scenario is

$$R = 2\int_r^{\bar{v}} \left\{ vP(v) - \int_r^v P(u)\,du \right\} P'(v)\,dv\,,$$

where $P(v)$ is the probability that a buyer with valuation $v$ has the higher of the two valuations and $\bar{v}$ is the largest possible valuation. Deduce that Alice's expected revenue is maximized when $1 - P(r) + P'(r) = 0$.[13]

20. It isn't hard to analyze a first-price, sealed-bid auction when it is common knowledge that the buyers' valuations are independently drawn from the *same* distribution, but little is known about the case when their valuations are drawn

[13]When differentiating $R$ with respect to $r$, write the first $r$ in the integral as $a$ and the second as $b$. You can then use the fact that

$$\frac{\partial I}{\partial r} = \frac{\partial I}{\partial a}\frac{da}{dr} + \frac{\partial I}{\partial b}\frac{db}{dr}.$$

Don't be afraid to differentiate under the integral sign, which goes wrong only in pathological cases. And don't forget the fundamental theorem of calculus.

from different distributions (Section 21.4). Vickrey studied one of only two cases that have been analyzed adequately up to now. Bob's valuation for Alice's house is 1. It is common knowledge that Carol's valuation is uniformly distributed on $[0, 1]$. You aren't asked to replicate Vickrey's analysis but to explain why we should expect that Bob will choose a mixed strategy in which his bid is chosen according to a probability density function whose support lies inside $[0, 1]$. Why might it then make sense for Carol to pay a fee to enter the auction if her valuation is sufficiently high?

21. Alice uses a first-price, sealed-bid auction in the private-value scenario of Section 21.5.2 when $p = \frac{1}{2}$. What is her expected revenue $R$?
   a. If Bob and Carol submit their true valuations, show that $R = \$3\frac{3}{4}m$. If Alice sensibly sets a reserve price of 3 (or fractionally less), why will the Low buyers bid their true valuations? Why won't the High buyers bid their true valuations?
   b. If Bob has a high valuation, his cheapskate strategy is to bid fractionally more than 3. He will then pick up the house cheaply when Carol is Low, but he will lose half the time if Carol bids $B > 3$ when she is High. Bob can then win all the time by switching to the strategy of bidding fractionally more than $B$. Show that the outbidding strategy is better for Bob than the cheapskate strategy when $B < 3\frac{1}{2}$ but worse when $B > 3\frac{1}{2}$.
   c. Confirm that the cheapskate strategy strongly dominates all bids of more than $3\frac{1}{2}$ but that no bid of less than $3\frac{1}{2}$ can be in equilibrium because Carol will just bid a fraction more.
   d. Use the random tie-breaking rule to eliminate the possibility that bidding exactly $3\frac{1}{2}$ can be in equilibrium.
   e. Having ruled out all possible pure equilibria, confirm that there is a mixed equilibrium in which High buyers choose their bids from between 3 and $3\frac{1}{2}$ so as to make the probability of bidding less than $B$ precisely $(B-3)/(4-B)$.
   f. Confirm that Alice's expected revenue is $\$3\frac{1}{4}$ million when this mixed equilibrium is used.

22. If Alice sells her house using a first-price, sealed-bid auction under the conditions of Section 21.5.2, Bob and Carol will use a mixed equilibrium. Reason as in the previous exercise to confirm that Alice's expected revenue is then $3 + (1-p)^2$.

23. The next five exercises show that a standard auction maximizes Alice's expected revenue when Bob's and Carol's valuations are independently drawn from a uniform distribution. As in Section 21.4.1, we consider only the symmetric case. We also restrict our attention to the case in which the object is always sold if some buyer's valuation $v \geq \underline{v}$ and is never sold otherwise, where $\underline{v}$ is the valuation at which a buyer is indifferent between entering the auction and staying out. To simplify the algebra, we make $\underline{v} = 0$ and make the maximum possible valuation $\overline{v} = 1$.

   In a symmetric direct mechanism, a buyer who announces a valuation of $v$ expects to pay $F(v)$ and to win with probability $P(v)$. Explain why the incentive constraint $vP(v) - F(v) \geq vP(w) - F(w)$ must be satisfied for all admissible $v$ and $w$. Deduce that $vP(v) - F(v)$, $P(v)$, and $F(v)$ are all increasing. Explain why

math

$$vP'(v) = F'(v) \tag{21.19}$$

is a necessary condition for optimality when $P$ and $F$ are differentiable.

math

24. The participation constraint in the previous exercise is $vP(v) - F(v) \geq 0$ for all admissible $v$. Explain why $\underline{v}P(\underline{v}) - F(\underline{v}) = F(0) = 0$, and deduce that the participation constraints are necessarily satisfied when the same is true of the incentive constraints.

math

25. The next step in continuing the optimal design problem of the preceding exercises is to consider the physical constraints that govern the values that the probabilities $P(v)$ can assume. Write

$$Q(v) = \int_0^v P(w)\,dw\,.$$

Suppose Alice is told that Bob has a valuation between 0 and $v$ but is otherwise uninformed. Why does she assign a probability of $v^{-1}Q(v)$ to the event that he will win the auction? Explain why $Q(1) \leq \frac{1}{2}$ says it is impossible that both agents can win in a symmetric mechanism. Explain why $v^{-1}Q(v) \geq \frac{1}{2}v$ says it is impossible that both agents will lose, given that both valuations are between 0 and $v$.

math

26. This exercise continues the optimal design problem of the preceding exercises by observing that the principal expects to collect

$$R = 2\int_0^1 F(v)\,dv\,.$$

Integrate by parts as in Exercise 21.11.8, and then use (21.9). Integrate what results *twice* by parts to obtain that Alice's expected revenue is

$$R = 2Q(1) - 4\int_0^1 Q(v)\,dv\,.$$

Insert the inequalities of the previous exercise into the preceding equation, and deduce that $R \leq \frac{1}{3}$. Why does this imply that a standard auction optimizes Alice's expected revenue? (Section 21.4)

27. With three players in the Wallet Game of Section 21.6.1, show that there is a symmetric equilibrium in which the first player to quit has valuation $a$, where $p = 3a$ is the quitting price. Show that the second player to quit has valuation $b$, where $q = a + 2b$ is the quitting price.
28. Show that the Wallet Game of Section 21.6.1 has an infinite number of asymmetric equilibria in which Bob's and Carol's bidding functions take the form $B(b) = \beta b$ and $C(c) = \gamma c$, where $\beta\gamma = \beta + \gamma$.
29. What is Rupert's expected gain in the Toehold Game of Section 21.6.2 after Sophia bids up to $x$?
30. Alice has three items for sale. Bob values the first item at \$5 and each extra item at \$1. Carol values each item at \$2. What will happen in an Ausubel auction if Bob and Carol bid their true demands?

31. With the assumptions of Section 21.7.3, find a Nash equilibrium in a uniform auction that generates the Walrasian outcome. In what sense is this equilibrium less risky for the buyers than a collusive equilibrium with a low price?
32. Indian Book is a game invented by Ralph Miles for a Caltech classroom experiment on Dutch books. The professor plays the part of a bookie. Each student is separately asked to announce a probability $p$ for some outside event (Section 13.3.3). Students whose $p$ is above the median are required to bet that the event will occur. They win $1-p$ dollars if it does and lose $p$ dollars if it doesn't. Students whose $p$ is below the median are required to bet that the event won't occur: They win $p$ dollars if it doesn't and lose $1-p$ dollars if it does.

fun

   a. The class contains only Alice and Bob. If Alice announces $p=a$ and Bob announces $p=b$, show that the bookie makes $|a-b|$ however the event turns out. What must Alice and Bob do to escape having a joint Dutch book made against them?
   b. The outside event is the fall of a fair coin. If the coin lands *heads*, Alice and Bob each receive independent signals, $s$ and $t$, which are distributed on the interval $[0,1]$ according to the probability density function $h(x)=2x$. If the coin lands *tails*, the density function is $t(x)=2(1-x)$. Use Bayes's rule to show that $\text{prob}(H\,|\,x)=x$ and $\text{prob}(T\,|\,x)=1-x$.
   c. If Alice knows that Bob will announce $b$, show that her optimal reply is to bid marginally more or less than $b$, thereby making a payoff of $|b-s|$. How much does the bookie get?
   d. Follow the steps below to find a symmetric Bayes-Nash equilibrium of the game played by Alice and Bob, in which a player who receives the signal $s$ announces the probability $p=f(s)$. Assume that $f$ is strictly increasing and differentiable. (The method is the same as that used in Section 21.4 for a first-price, sealed-bid auction.)

math

**Step 1.** If Bob plays according to the equilibrium, show that Alice's expected payoff from choosing $p$ when she receives the signal $s$ is

$$s(1-p)\,(2F(p)^2-1)+(1-s)p(2(1-F)p))^2-1),$$

where $F$ is the inverse function to $f$ (so that $s=F(p)\Leftrightarrow p=f(s)$).

**Step 2.** Differentiate Alice's expected payoff with respect to $p$, and set the derivative equal to zero. For an equilibrium, the resulting equation holds when $p=f(s)$ or $s=F(p)$.

**Step 3.** You now have a differential equation in $p$ and $s$. Exploit its symmetry by writing $p=\frac{1}{2}+y$ and $s=\frac{1}{2}+x$. You will then be able to reduce the equation to the linear form

$$x\frac{dy}{dx}+2y=\frac{4x}{1+4x^2}\,.$$

**Step 4.** Solve the differential equation, and hence show that Alice's equilibrium announcement is $p=f(s)$, where

$$p - \tfrac{1}{2} = \frac{(s - \frac{1}{2}) - \frac{1}{2}\arctan 2(s - \frac{1}{2})}{(s - \frac{1}{2})^2}.$$

e. Why does the preceding analysis imply that neither Alice nor Bob will announce their true subjective probability for the outside event?
f. Why is it optimal for Alice and Bob to tolerate a Dutch book being made against them jointly?
g. Suppose that Alice and Bob both receive each of the signals $s$ and $t$. If this fact is common knowledge between them, why is it now an equilibrium for them both to announce

$$p = \frac{st}{st + (1-s)(1-t)},$$

which is then their true subjective probability for the event?

phil

h. What does all this imply about common priors? (Section 13.7.2)

# Index